Gallium Arsenide and Related Compounds 1984

Gallium Arsenide and Related Compounds 1984

Invited and contributed papers from the Eleventh International Symposium on Gallium Arsenide and Related Compounds held in Biarritz, France, on 26–28 September 1984

Edited by B de Cremoux

Conference Series Number 74

Adam Hilger Ltd, Bristol and Boston

CODEN IPHSAC 74 1–695 (1985)

British Library Cataloguing in Publication Data

International Symposium on Gallium Arsenide and Related Compounds (11th: 1984: Biarritz, France)
    Gallium Arsenide and Related Compounds 1984: invited and contributed papers from the Eleventh International Symposium on Gallium Arsenide and Related Compounds held in Biarritz, France, on 26–28 September 1984.—(Conference series/Institute of Physics; ISSN 0305-2346; no. 74)
    1. Gallium arsenide semiconductors
    I. Title   II. Cremoux, B. de   III. Series
    537.6'22   TK7871.15.G3
ISBN 0-85498-165-9

Organising Committee
    J Magarshack (Chairman), J C Carballès, J P Duchemin, P Fourest, F Gérin, L Hollan, H Martinot, J P Noblanc, D Pavlidis, R Rieux

Permanent Advisory Commitee
    J V Lorenzo (Chairman), L Eastman, W Heywang, C Hilsum, N Holonyak, J Magarshack, T Sugano, H Thim, M Uenohara, H Yanai

Technical Programme Committee
    J P Duchemin (Chairman), G A Acket, E Constant, B de Cremoux, G Gibbons, H Grimmeis, J Hallais, S Hersee, H Launois, S Laval, N T Linh, A Mircea, G Nuzillat, G Rey, M Rocchi, K Zschauer, M Voos, C E Wood

Honorary Editor
    B de Cremoux

Supporting organisations
    Centre National d'Etudes des Télécommunications, Centre National de la Recherche Scientifique, Compagnie Générale d'Electricité, Direction des Recherches et Etudes Techniques, European Office of Aerospace Research and Development, European Research Office of the United States Army, Laboratoire d'Etudes Physiques, Office of Naval Research, RTC La Radiotechnique Compelec, Thomson-CSF

Published on behalf of The Institute of Physics by Adam Hilger Ltd
Techno House, Redcliffe Way, Bristol BS1 6NX, England
PO Box 230, Accord, MA 02018, USA

Printed in Great Britain by J W Arrowsmith Ltd, Bristol

Preface

The 11th International Symposium on Gallium Arsenide and Related Compounds was held in Biarritz, France on 26–28 September 1984.

The choice of Biarritz as the conference town was made because it is the location of a new experimental multiservice fibre optic system, illustrating the use of GaAs based devices in communications. Biarritz is also a beautiful city on the shore of the Atlantic Ocean.

This symposium is increasingly successful as demonstrated by the number of submitted papers (250 regular plus 12 late news) and the 420 attendees from several countries all around the world. The technical programme committee thus had the difficult task of making a selection among papers with a generally high level of quality and decided to arrange poster sessions for the first time in this series. 113 papers were accepted for oral presentation, including 9 late news, and 25 were accepted as posters depending on their suitability for each type of presentation. To illustrate the capabilities of the Biarritz fibre network, each poster paper author was allowed to record a 5 minute 'video clip' which was afterwards made available on the connected TV sets.

It is worth noting that, besides the now classical fields of interest in III–V compounds, an increasing amount of attention is being paid to two-dimensional structures, properties and applications: 5 papers in this area had been selected for the previous symposium in Oiso, Japan, while 13 have been presented in Biarritz.

The symposium proceedings give the papers accepted for oral or poster presentation arranged in 9 chapters. The first comprises two introductory invited papers aimed at showing the industrial impact of GaAs based devices and describe the Biarritz optical fibre system and the commercial propects of GaAs IC's, respectively. Chapters 2 to 8 are respectively devoted to bulk growth and implantation, epitaxial growth, characterization, two-dimensional structures, optoelectronic devices, discrete microwave devices and IC's. Finally chapter 9 gathers together the late news papers.

B de Cremoux

GaAs Symposium Award and Heinrich Welker Gold Medal

The Gallium Arsenide Symposium Award was established in 1976. Candidates for the award are selected by the GaAs Symposium Award Committee from those who have distinguished themselves in the area of III–V compound semiconductors. The award consists of $1000 and a plaque citing the recipient's contribution to the field. In addition, the Heinrich Welker gold medal, sponsored by Siemens, is also presented to the award recipient.

The first Award and medal were presented to Nick Holonyak of the University of Illinois for his work leading to the development of the first practical light-emitting diode and his continuing research in III–V compound semiconductors. Cyril Hilsum of the Royal Signals and Radar Establishment (now at GEC Research Laboratories) received the second Award and medal in 1978 for his contributions in the field of transferred-electron devices and his leadership in III–V compound semiconductor research. The 1980 Award and medal were presented to Hisayoshi Yanai, University of Tokyo, for his contributions to progress in transferred-electron logic devices and the advancement of GaAs MESFETs. In 1981 the GaAs Symposium Award and Heinrich Welker gold medal were presented to Gerald L Pearson, Stanford University, for his research and teaching in the field of compound semiconductor physics and new device technology. In 1982 the Award and medal were presented to Herbert Kroemer, Professor of Electrical Engineering at the University of California at Santa Barbara (UCSB) for his contributions to hot-electron effects, the Gunn oscillator, and III–V heterojunction devices including the heterojunction laser.

The 1984 GaAs Symposium Award and Heinrich Welker gold medal were made to Dr Izuo Hayashi for his many contributions to the development and understanding of room temperature GaAlAs/GaAs double heterojunction lasers.

Dr Izuo Hayashi graduated from the Department of Physics, University of Tokyo and MIT in 1964, and then joined Bell Telephone Laboratories. There he first worked on semiconductor high energy particle detectors, while taking a growing interest in the basic understanding of semiconductor lasers. He invented the GaAlAs/GaAs double heterojunction laser, and was the first in the world to demonstrate continuous laser oscillation at room temperature. In 1971, he returned to Japan and joined the Central Research Laboratories, NEC Corporation, to lead semiconductor laser research and development. His great contributions in understanding laser degradation and the means of improving laser life are well known. In 1981, when the Optoelectronics Joint Research Laboratory was established as a Japanese national project, he joined as chief scientist. Under his leadership, the laboratory has become world famous.

Dr Hayashi is a Fellow of the IEEE and a recipient of the Ichimura Prize and Achievement Award of the Institute of Electronics and Communication Engineers of Japan for his breakthrough on long life laser diodes.

Contents

Chapter 3: Epitaxial growth

Chapter 4: Characterisation

Chapter 5: Two-dimensional structures

Chapter 6: Optoelectronic devices

Chapter 7: Discrete microwave devices

xviii   *Contents*

Inst. Phys. Conf. Ser. No. 74: Chapter 1
Paper presented at Int. Symp. GaAs and Related Compounds, Biarritz, 1984

1

From the multiservice fiber optic system in Biarritz to the development plan for videocommunication

Francois GERIN. Delegue aux Videocommunications
DGT Ministère des PTT (FRANCE)

Abstract: The Biarritz fiber optic system is, at the same time, the
first step in the field of local videocommunication networks and a long
term futuristic performance in terms of services, either distributed or
switched. It prepares for the short term introduction of optronics in
local telecommunication networks, while it provides a realistic test for
new services, ranging from CATV to videophony, and from video on request
to picture banks.

As the French Minister of PTT, Mr Mexandeau, has just said to you, there
is consistency between the multiservice fiber optic system in Biarritz
and the development plan for videocommunication that was decided two
years ago on his proposal. Firstly, the services provided by the
Biarritz network anticipate those to be offered in future
videocommunication networks. Secondly, the building of this system has
launched the French telecommunications industry into the age of
optronics.

The first original item of the Biarritz network is the <u>switched services</u>
proposed to subscribers.

The switched services at Biarritz provide for the establishment of
bidirectional audio and audio plus picture connections. The main
applications are:

   - Videophony, in which two people can not only talk to one another,
as in telephony, but also see one another, if they wish. Communication
is visual only if both parties agree. When one of the parties does not
want to be seen, the call is voice-only, preventing any unwanted
intrusion into subscribers' private lives.

Picture quality is good enough for one correspondent to read a document
placed by the other in front of his or her videophone camera. For
greater flexibility in this respect, the videophone cameras have
adjustable positions.

The videophone cameras are monochrome, because lighting in private homes
and apartments tends to be inadequate for good color filming. It is,
however, possible to use color cameras in professional premises with
suitably designed lighting. In addition, subscribers can connect their
own color video cameras or VCRs to the videophones and use them in
transmitting over the picture channel. All the videophones at Biarritz

have color tubes and are thus capable of receiving and displaying both monochrome and color pictures. In addition they can send to the TV set the image and sound that they receive for a better display.

   - Telephony. For calls to and from locations inside or outside the Biarritz optical fiber network, including intercity and international calls, subscribers use either the videophones (in the voice-only mode) or ordinary telephones predating the new network, as they prefer.

   - Videotex. Using their videophones, subscribers have access to a range of information and transactional services similar to those developed for the French videotex system TELETEL (1), plus Biarritz network directory information.

   - Supervideotex, following the installation of an automated videobank in the Biarritz network at the end of 1984. Supervideotex and the use of the videobank are explained later.

The preceding list of services is not exhaustive. The Biarritz network is a testbed and has been built for easy integration of additional services. In the final resort, the services offered and their exact scope will depend partly on the use made of the network and on the opinions and wishes of the participating subscribers.

The second original item, in terms of services, is the simultaneous distribution of TV and HiFi sound channels to the subscribers on the same fiber optic transmission system.

The network is engineered for interactive distribution of 15 (2) moving picture channels and 12 stereo sound channels. The moving picture channels are used for signals delivered to subscribers' television sets. Most of the channels are dedicated to TV broadcasts, including the three French public service channels, 2 channels from nearby Spain, a French-language channel from Belgium, one channel from Switzerland and BBC 1 from the UK transmitted through a microwave link, and two satellite-transmitted channels (Sky Channel and TV 5). However, users will also have access to a large choice of films, documentaries, etc, in a local program bank.

This program bank has been installed since May 1984, initially with manual operation. It will now be progressively automated until the middle of 1985. Subscribers will use the videotex terminals of their videophones to interact with the program bank for "à la carte" and "on request" program selection. The corresponding videotex host computer will have a capacity of 80 simultaneous calls.

One of the 15 moving picture channels provides a split-screen "mosaic" display of all on-cable programs for pushbutton selection by users through the handheld control unit. The same channel will be employed for announcing future programs and pay-per-view events, for explaining to subscribers how to make optimal use of their terminals and of the network, and so on.

The sound channels, now giving six French and Spanish public service channels, will also carry "on request" programs within the next few months for testing this new way of using sound material.

(1)   500 000 terminals at the end of 1984
(2)   with a possible extension to 30

In practice, in the subscriber's premises, the standardised sockets of
the local bus allow for easy changes of location of the different
terminals, either videophone, TV1, TV2 (1) or HiFi (1) interfaces.
These last interfaces generally deliver baseband video and/or audio
signals to TV Peritel plugs or audio amplifier plugs, providing
excellent quality.

The third original item is the first French operation of <u>automated banks
of pictures</u>, which combine the switched and distributed services for a
program bank, and videodisk players with videotex machines for
videobanks.

A <u>program bank</u> contains films and other moving picture programs. It is
connected to several cable channels and subscribers have access to the
stored programs on a pay-per-view basis. This access can be "à la
carte" or "on request". The "à la carte" mode will be used during heavy
viewing hours, when for obvious reasons of limited channel availability
each subscriber cannot have totally individualized access to the bank.
Instead, advance schedules will offer a list of possible programs on
each channel for defined time slots in each day. The decision as to
which of the programs is actually transmitted over the channel may be
determined by subscriber voting. Several arrangements are possible.
For example, each family can be given a number of "voting points" for
each month, then freely consuming these points in influencing
programming as a function of purely personal criteria. In other words,
the content of these local programming channels will reflect the
collective preferences and desires of a community.

On the other hand, during light viewing hours, a subscriber will be able
to communicate with the program bank for immediate connection to a
channel of any one out of a catalog of several hundred stored videograms
(feature films, documentaries, educational films, etc.) ("on
request"). No voting is involved and this is thus the ultimate in
personal control of viewing. (In Biarritz it may also be possible
through the switched part of the network.)

Dialog between the subscriber and the program bank will be by Teletel
videotex over the cable network. The computerized program banks being
developed in France will feature sophisticated hardware and software for
automatic management of videogram access, connection to cable channels,
voting procedures, subscriber billing, and so on (even automatic loading
of videogram readers).

<u>Videobanks</u> are different. Also known as audiovisual databases, their
use is scheduled as part of a vastly enhanced videotex service, referred
to as <u>supervideotex</u>. In basic videotex, as now being implemented
throughout France, users have switched access to nationwide databases –
supplying alphanumeric information and fairly simply graphics.
Videobanks on the other hand will transmit photographic still images,
moving picture sequences, and sound sequences stored on videodisks.
Until such time as there is a nationwide switched broadband network, the
videobanks will be installed locally within each fiberoptic cable
network. On the other side, they will be linked to a local videotex
host computer or through the Transpac packet-switched data network to
hosts on the premises of videotex service providers. Users will
interact with these local or remote hosts for access to conventional

(1) an option

videotex services: teleshopping, news reports, weather, seat
reservations, local cinema and theater schedules, video games,
telebanking, and so on. When appropriate, the videotex host computer
will also command transmission to the user's terminal of single pictures
or sequences stored at the local videobank, this transmission taking
place over the fiberoptic cables for display at the user's television
set or videophone terminal.

Applications will include: animated teleshopping catalogs (e.g. a short
moving picture sequence showing a model wearing a dress or how easy it
is to use a new lawn mower); non-animated catalogs but with a spoken
commentary for each still image; promotional sequences for new products
and services, combining moving pictures and sound; user and maintenance
guides explaining step-by-step how to use a product or perform simple
repairs and servicing operations; filmed illustrations of hotels and
holiday locations proposed by travel organisations; and so on through a
long list.

The videobanks developed by French engineers are built around a
microcomputer and can each have 1 to 16 videodisk readers, with one or
several output channels. Apart from the described applications in cable
networks, mainly for home reception, they have considerable potential
for stand-alone or networked business use.

Thus we have seen that the Biarritz system is the field test of a
complete range of new services, from CATV to videophony, parallel to
video on request and picture banks. These services, with the exception
of the large scale use of videophony, are proposed for the future
videocommunications networks that were ordered from French industry at
the end of 1983 and which will begin operation at the end of 1985. Most
of them will use the optic fiber star architecture, which gives the best
potential of technical evolution and the largest range of services.

Generally, by a combination of distributed channels, with possible
access control, and videotex systems, banks of pictures may be operated
on all these networks, for enlarging the classical attraction of CATV,
and thus the revenues of the future commercial operators, as well as the
satisfaction of the subscribers.

The tests that have or shall be performed in Biarritz will obviously be
important to this extension, either for technical aspects, or cultural
and commercial points of view. (Recall that the Biarritz subscribers,
who are volunteers, are paying for the different services that they can
get from the broadband telecommunications network.)

At last, as you are all eminent specialists in the field of Gallium
Arsenide and related compounds, I want to say that the lasers and LED in
operation in Biarritz use this technology on a very large scale, the
greatest in France, and probably in the world for one network (1). But
it is a simple step, when you consider the future use of LED in French

(1) more than 1500 of each category in service next year!

videocommunications networks: by mid-1985, potentially more than
300 000 shall have been ordered from industry, with a better efficiency
and reduced costs in comparison to the starting point of the Biarritz
project. It will be, in fact it is now, the industrialisation of these
components and their mass production. This is probably the beginning of
their use, opening up large changes in the telecommunication network,
especially in terms of cost/efficiency ratio and reliability, also
improved by the industrialisation of wave length multiplexing and the
development of digital transmission systems.

All these R and D aspects, and the industrial forces, including the
field of audiovisual content or program production, are consistent
elements of the French development plan for videocommunications. It
would never have been possible without the competence of you and your
colleagues, working on GaAs and related compounds!

Inst. Phys. Conf. Ser. No. 74: Chapter 1
Paper presented at Int. Symp. GaAs and Related Compounds, Biarritz, 1984

7

The commercial prospects for gallium arsenide ICs

A.E. Pyne and P.C. Newman,
Mackintosh International Ltd.,
Napier Road,
Luton,
Beds.
UK

Abstract. The commercial prospects for Gallium Arsenide ICs are
compared with the current and future situation for Silicon ICs. The
pressure for changing from Si to GaAs comes from the demand of users
for higher speeds. GaAs offers such an improvement, but with a cost
penalty. Devices and ICs made from GaAs should obtain a small, but
significant, share of the available market, amounting to $800M in 1990
and $1,500M in 1992 (about $1\frac{1}{2}$% of sales for Si devices).

1. Introduction

In order to put the commercial prospects for GaAs ICs into perspective,
it is advisable to start with a few words about the markets for Silicon
ICs; these devices have been in existence for about 25 years, so that the
markets for them have been both observed and forecast for similar lengths
of time. Discrete devices in Si, of course, were available even earlier;
Figure 1 shows the total usage of Si wafers (both for ICs and discretes)
from 1982 to 1986, in which period it rises from 1 to 2 billion square
inches, an average growth rate of about 25% per annum. Some parts of the
business, such as MOS ICs fabricated on epitaxial layers, are growing at
twice this average rate. At the same time, the diameter of the largest
Si wafer used in processing is also growing rapidly, from 4" to 5" last
year to 6" now and to 8" by 1990. This is just one factor in helping the
production of ICs to keep ahead of the demand, which is doubling every
four years.

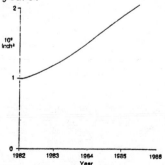

FIGURE 1. Silicon Usage, 1982 to 1986.

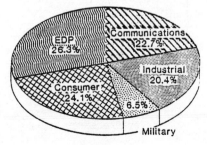

$12.08 Bn.

FIGURE 2. The World Merchant IC
Market by Application sector, 1983.

The produced output of ICs can be defined as captive (entirely for in-house use) or merchant (sold to any user, including those in-house); the split of the market for sales of merchant ICs is given in Figure 2. In 1983, data processing was the major consumer of ICs, with consumer equipment a close second. The smallest identified sector was that of Military (including Aerospace), with only 6½% of the market. These applications sectors are not likely to change drastically in the next year or two, although locations of production of electronic equipment may do so.

Why should the users of ICs wish to change from Silicon, which is well-established, to Gallium Arsenide, which is certainly not yet in the same state? One reason is the pressure from data processing - itself the largest sector of the market. Figure 3 shows the increasing power of computers (measured in MIPS) from before 1970 to nearly 2000. The line goes inexorably upwards. For operational speeds of 3,000 MIPS (say, 10 Gbit/sec in the logic circuits), GaAs is needed. The operations demanding this turn of speed include real-time, continuous speech recognition and the processing of radar returns, visual images and meteorological data. On the analog side, operation at 10GHz, or higher, is demanded by telephone repeaters, instrumentation and direct broadcasting from satellites (DBS).

However, operation at these high speeds brings its own problems. Some of these are the ability to achieve high packing density and to dissipate the heat generated, when it is achieved. Delays due to external wiring will dominate at the smaller scales of integration (see Figure 4) which are all that can be achieved with GaAs at present. This raises the question of whether ICs in GaAs can really be five times faster than those in Si; the answer seems to be 'Yes' for analog circuits, but 'No' for digital, where the factor is nearer to 2.

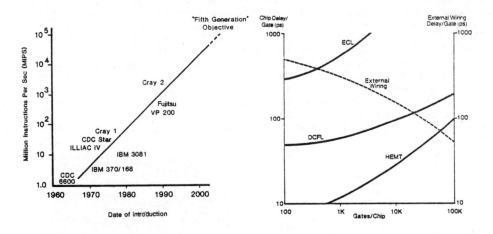

FIGURE 3.   The Increasing Demand   FIGURE 4.   Internal and External
            for Processor Speed   Wiring Delays

However, GaAs has some other instrinsic advantages over Si, which will
provide it with certain niches in the market. These are, the ability to
work at higher temperatures, a better resistance to radiation and the
efficient emission (and detection) of infra-red radiation. At the same
time, those who are working on the development of ICs in GaAs can call on
the improvements in processing (particularly in lithography) that are set
up to satisfy the demands of Si. However, it must always be borne in mind
that some physical limits are set to the performance of any semiconductor
device, by such considerations as the limiting velocity of electrons, the
transit time for a signal from one side of a chip to the other and the
random fluctuations in the total number of impurities in very small
volumes.

2. GaAs Devices

The first attempts to make amplifying devices in GaAs in the late 1950's
were based on bipolar devices, but the quality of the material was not
good enough for this. At the same time, it was known that the native
oxide formed on GaAs was quite unsuitable for the fabrication of MOSTs,
while the deposition of insulators of quite a different composition was
insufficiently advanced. So, by 1963, attention had turned to the Field-
Effect Transistor, using first of all a diffused p-n junction as the gate,
but moving quite rapidly on to metal-GaAs Schottky diodes.

Parallel to the work on three-terminal amplifying devices, two two-
terminal devices were under development. One of these was an optical
device, which started as the humble diode emitter of incoherent
radiation (the LED), but was in 1962 transformed into the injection laser.
The second was a diode with negative resistance, arising from the transferred electron effect in GaAs. The theory of this effect in GaAs was
proposed in 1962 and the effect was observed in 1963. Due to this
negative resistance, such a diode can be used in an amplifier, but the
existence of only two terminals makes it extremely difficult to handle in
a circuit. Such 2-terminal devices will not be considered in the rest
of this paper.

Discrete devices, for use at microwave frequencies, represent over 90% of
all sales of GaAs devices at present. Analog ICs are now being developed
in GaAs, to take over where the discrete devices leave off. Typical
examples are mixers for microwave receivers and amplifiers for instruments.
Somewhere between analog and digital ICs come pre-scalers, for operation
in frequency-synthesis radios (transmitting or receiving) at microwave
frequencies.

For fully digital circuits, five different types of circuit have been
evolved so far, and these are shown in Figure 5. The first three types of
circuit shown require two power rails, of differing and opposing voltages;
this complicates the construction of systems using these circuits. The
buffered FET logic (BFL) arrangement has, at least until recently, been
the favourite type of circuit for all digital operations, such as simple
gates, memory, adders and multipliers. However, it will need the successful development of a low-power single-rail circuit, such as Direct
Coupled FET Logic (DCFL), to ensure that the markets for GaAs ICs develop
along the lines forecast in this paper.

Moreover, designing ICs in GaAs (whatever
type of circuit arrangement is used) is
different from (and more difficult than)
designing the same function in Si.
Particular attention has to be paid
to keeping leads short, to avoiding
reflections and to maintaining impedances
at near-constant values throughout the
circuit. On the other hand, GaAs helps
in the design process, by being a lower
loss substrate than Si at these
frequencies.

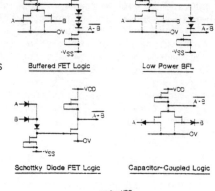

FIGURE 5.   The Main GaAs
Circuit Approaches

3. The Markets for GaAs Devices

GaAs is, and always will be, a more expensive material than Si. This is
reflected in the present and forecast prices for raw wafers (Table 1) and
the costs for processing a wafer of GaAs, when compared with similar
costs for Si (Table 2). The processed wafer costs are based on what is
generally considered to be a quite large throughput for a Pilot Plant now.
With chip sizes not far different for a 1K SRAM in GaAs, NMOS and ECL
the Si devices gain a great cost advantage from the number of chips that
can be started on a 6" wafer, compared with the equivalent number on a 3"
wafer, which is the best that can be managed with GaAs in 1984 (see Table 3).

	1984	1987	1990
SILICON	0.78	0.69	0.61
GaAs (LEC)	35	25	15

TABLE 1. Prices of Wafers, $/inch^2

PROCESS	CHIP SIZE MM^2	POSSIBLE
GaAs	1.8 x 1.8	1300/3"Wafer
NMOS	1.8 x 1.8	5200/6"Wafer
ECL	2.0 x 2.0	4400/6"Wafer

TABLE 2. Cost for 250 wafers per week

PROCESS	NMOS	ECL	GaAs	
DIAMETER	4	4	3	INCHES
CAPITAL	4.72	5.64	4.04	$M
RUNNING	3.17	3.89	5.64	$M/Year
PER WAFER	293	360	588	$
WAFER YIELD	90	90	80	%

TABLE 3. Device per Wafer, 1K SRAM

Figure 6 shows the overall, worldwide merchant markets for GaAs devices, both Analog and Digital. It is not until the early 1990s that the market for digital devices will surpass that for analog. There is, at present, captive production of GaAs devices, that at least doubles the sizes given in Figure 6, to reach the value of the total market. These figures are based on the most probable scenario for high-speed ICs, rather than on unduly optimistic or pessimistic ones - at least as far as the division between GaAs, NMOS and ECL goes.

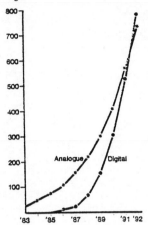

FIGURE 6. Worldwide Merchant GaAs Device Markets

Figure 7 shows the split of the digital GaAs IC market by 1992, both in terms of geographical location and of application sector. The largest sector is EDP, but this varies from 48% in Europe, through 59% in the USA, to 72% in Japan. By way of contrast, the Military sector, which accounts for 18% in Europe and 16% in the USA, drops to 1% in Japan. The use of digital GaAs ICs in Communications ranges from 18-26%, while their use in Professional equipment is level and fairly constant with location at 6-8% of the total GaAs digital market.

Figure 8 shows a less well-researched set of values for the analog GaAs IC market in 1992. While the order of market consumption remains USA-Japan-Europe, there are two significant differences from Figure 10. The use of these circuits in EDP has disappeared, while Consumer applications (such as DBS) now constitute an important sector. However, in all regions, communications dominates as the major applications sector for analog GaAs ICs.

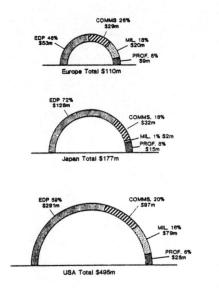

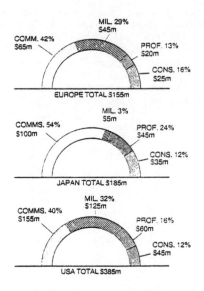

Figure 7. Digital GaAs IC
Market 1992

Figure 8. Analog GaAs IC
Market 1992

Finally, to put the markets for GaAs devices into perspective, Table 4
shows rounded figures for the worldwide merchant markets for all Si
devices, for semi-custom ICs in Si and for GaAs devices. While semi-
custom ICs will grow to about 20% of all Si devices by 1990, GaAs
devices will still be around 1% - although this still amounts to $800M
for merchant sales, with a possible increase arising from greater
penetration of the presently captive market.

$ BILLION

	1984	1987	1990
SILICON DEVICES	25	41	68
SEMI-CUSTOM ICs	1	4	15
GaAs DEVICES	0.05	0.20	0.80
(Merchant)			

TABLE 4. Worldwide Markets

Inst. Phys. Conf. Ser. No. 74: Chapter 2
Paper presented at Int. Symp. GaAs and Related Compounds, Biarritz, 1984

13

The preparation of active layers for GaAs ICs: is there any problem left?

Gérard-Marie MARTIN, Marc DUSEAUX, Jose MALUENDA

Laboratoires d'Electronique et de Physique Appliquée
3, avenue Descartes, 94450 LIMEIL-BREVANNES, France

Abstract : The technology of preparation of active layers for GaAs IC's, using direct implantation into semi-insulating substrates, has progressively matured in the last five years. This progress has been achieved in different steps, related to an improvement in the quality of starting substrates, the engineering of wafers, and the control of the implantation-annealing process. After recalling the main physical effects which have been seen to provide the necessary improvements, we will focus our attention on the remaining problems, which are thought to limit the performance of circuits. In particular, the problem of homogeneity of the active layer conductivity will be reviewed as well as their possible solutions.

Introduction

A good control of the incorporation of impurities in GaAs bulk Bridgman (Bg) or Liquid Encapsulated Czochralski (LEC) ingots has now been achieved, which means that semi-insulating materials are commonly obtained with a pretty good yield. The main effort, during the last years, has been devoted to decreasing the background concentration of donor and acceptor impurities, in such a way that the role of residual stoichiometric defects has become more and more important in the electrical properties of "modern" materials. In parallel, IC's substrate requirements have become more and more severe, in terms of homogeneity of active layers. Results obtained in these two fields will be first reviewed. Then data coming from detailed investigations of defect distributions and of possible sources of inhomogeneities will be presented, together with some elements of interpretation. Finally, present and future solutions to overcome the present difficulties will be discussed.

1. Compensation mechanisms

Till the end of the 70's, most of the Bg, but also LEC materials, presented a residual n type level, mainly due to Si, which has to be compensated by deep Cr impurities (Lindquist 1977, Zucca 1977, Martin 1980a). Many efforts have been pursued by suppliers to improve the purity of starting materials with various success, as shown for instance by Farges (1982). As a result, Se, Te, Cr, Fe, Ag, Hg, Mn, Mg are in the 10^{13} - 10^{14} cm^{-3} range, Zn in the 10^{14} - 10^{15} range typically (Farges 1982, Thomas 1984). Sulfur still seems to lie in the 10^{14} - 10^{16} range, depending on As suppliers for instance. Growth of Bg crystals with low Si content (a few 10^{15}cm^{-3}) has been shown to be achievable when monitoring the oxygen overpressure during growth

(Ainslie 1962, Martin 1982, Kaminska 1982). But still lower values of Si down to a few 10^{14} cm^{-3} can be obtained in LEC pulling. This may be achieved either by using a quartz crucible with a "wet" B_2O_3 encapsulant or using a PBN crucible whatever the B_2O_3 encapsulant (Akai 1981, Oliver 1981, Thomas 1984). Boron incorporation is inversely proportional to the humidity of B_2O_3, and its concentration may reach a few 10^{17} cm^{-3} (Thomas 1984). It may be dangerous only when on As site, since it may then be responsible for the rather shallow double acceptor at 78 and 203 meV (Elliot 1984, Bishop 1984). Due to the improved control of donor incorporation, the residual concentration of shallow levels is now commonly p type with $N_A - N_D \simeq$ a few 10^{15} cm^{-3}, the main acceptor impurity being carbon in most of the cases (Holmes). The compensation is then due to the deep donor level EL2 located near mid-gap (Martin 1980a) and proved to be due to a defect and not to oxygen (see below). Later on, we will essentially discuss the results obtained with that type of non Cr doped material.

After direct implantation of donors, in-depth redistribution of implanted impurities and defects occurs during the annealing at 850°C necessary to cure the implantation defects and to get the electrical activation of implanted species. Si seems to be among the slow diffusing donors (see for instance Kasahara 1982) and is thus a very good candidate for achieving the n and n$^+$ regions necessary in the active layers of FET's and IC's. On the other hand, the EL2 concentration vanishes in the surface region after annealing, leaving residual acceptors uncompensated (Makram-Ebeid 1982). As a matter of fact, a correlation has been recently quantitatively established between the C concentration initially present in substrates and the drop σV_T of threshold voltage of FET's measured on active layers prepared from different substrates in exactly the same implantation-annealing (I.A.) procedure : a variation of 185 mV in V_T is noticed for a 1 x 10^{16} cm^{-3} change in C concentration (Chen 1984). This is of course a cause of difficulties in getting reproducible results if both C and EL2 concentrations vary from one ingot to another.

Many studies have been devoted to the analysis of the free carrier profiles $n(x)$ obtained after direct ion-implantation in substrates. As reported above, in-diffusion and out-diffusion of implanted or residual impurities and defects pursued during the last 5 to 6 years have been shown to satisfactorily explain the main features of the in-depth problem. Lateral variation of properties of the active layer has only been considered recently. There are two reasons for that : i) a rather good control of $n(x)$ is now achieved in most laboratories whether the cap or capless annealing is used after implantation, in such a way that the study of smaller but still dangerous effects can now be started, ii) this study requires technological processes which are time consuming and require heavier equipment. The importance of these lateral fluctuations of the active layers doping level $n(x)$ will be reviewed in the next paragraph.

2. Short range inhomogeneities : FET's properties and dislocations density fluctuation

Besides the problem of reproducibility, the one of homogeneity of the active layer on a given wafer is crucial. This problem has deserved special attention since very complex and dense circuits, like multipliers or memories, have been designed in several laboratories which were based on normally-off FET's : for lowest achievable consumption. For that application, severe applications concerning the maximum allowed fluctuation of micro-FET's (typically 5 x 5 μm^2 active area) properties are required. It was then first reported by Nanishi (1983) that very large variation of micro-FET's properties (current I_{DSS} or V_T) can be obtained on standard LEC substrates. This team at NTT laboratories noticed short range ($\approx$ 10 to 100 μm) variations which they attributed to the presence of dislocations. Later on, they published a curve showing a correlation between values of V_T of quasi normally-off FET's measured in a 1 x 24 mm^2 area and the distance of the corresponding FET's from a dislocation (Miyazawa 1983). More recently, another group at Hughes Laboratories published contradictory results : they did not observe any correlation, except maybe a decrease of the average value of V_T with the dislocation density in the concerned area. They have suggested that the reported correlation might exist, not with isolated dislocations, but rather with the local density of dislocation which might not have been constant in the long and narrow stripe investigated previously (Winston 1984). According to that hypothesis, the variation of V_T should then be related to $1/(DD)^3$, with DD the dislocation density, rather than to the value of d. Takebe (1984) reported an overall quantitative analysis which supported this remark. In the meanwhile, the NTT Japanese teams have obtained more detailed observations and refined their diagnostic. Matsuoka et al (1984) have shown that FET's properties are not so much modified by some isolated dislocations present with a random distribution, but strongly changed if clusters of dislocations are observed near-by. Thus, the density of dislocation by itself is not a very relevant parameter when averaged over too large areas. Ishii et al (1984) seem to have obtained the most conclusive data by probing V_T and σV_T, its standard deviation, over a 2" wafer and assessing its dislocation distribution. Their conclusions reconcile the previously reported data : V_T is smaller when DD increases, but σV_T is clearly bigger in regions where dislocations are clustered and arranged in a network pattern of typically 400 μm in diagonal cell size. This is mainly the case in the central region of the wafer. One can notice that, in standard LEC materials, V_T and σV_T do not show the same variation all over a wafer.

An analysis of the variation of V_T has been made using C-V measurements of $n(x)$ profile of free carriers from fat FET's. Of course, some averaging then occurs over a few 10^3 μm^2 in that case. It was concluded (Takebe 1984) that the thickness of the active layer does not vary too much, but that the observed V_T variations correspond to n_{peak} variations of the implanted layer of the order of 10^{16} cm^{-3}. This is actually the order of magnitude for the residual concentration of level (carbon and EL2) in starting material, and it is thus thinkable that this is also the order of magnitude of their fluctuations.

These observations have initiated a very large number of studies dealing with dislocations distribution (extended defects) and with EL2 (a point defect). Of course, it would have been very useful to also assess the distribution of impurities, and in particular C, on a short range but available techinques, like SIMS, are not very sensitive for light elements, like Si and C. We only have an idea of rather long range inhomogeneities for Si (Kamejima 1982) and for C (Chen 1984). On the other hand, well established techniques were available for the assessment of dislocations and EL2. Since both are defects, many laboratories have thought that annealing studies should provide some changes or improvements in the properties of wafers, or at least should increase our understanding of the related phenomena. It has turned out that both occur.

3. Elements of understanding

3.1 Parameters controlling the as-grown EL2 concentration

A long range variation of EL2 was reported by Martin et al (1980 b). Brozel et al (1983) were the first to observe large inhomogeneities on a short range in optical absorption of wafers, that Skolnick (1984) undoubtedly related to variation of EL2. Two sorts of pattern were evidenced : a kind of network superimposed on a cross-like structure in 2" slices. Indeed careful analysis of dislocations also revealed dislocation lineages arranged in a cross pattern superimposed on the dislocation network mentioned above. The similarity of dislocation and EL2 pictures is striking and later on, has been observed by other workers (Brozel 1984, Leigh 1984). It has been difficult to find a point to point correspondence between both distributions, because in standard material, dislocations can only be studied in thin slices ($\simeq$ 100 μm) while EL2 related absorption measurement requires thicknesses larger than 1 or 2 mm. Nevertheless, all studies conclude that the EL2 concentration [EL2] is low inside the dislocation network cells, is large at the cell walls or at the lineage walls, while it seems to present a minimum in some cases, at a certain distance from the lineage wall for instance (Leigh 1984). In view of the small extension of the volume of the cells, this last effect might exist, but might not be visible since optical beam averages the concentration over a few cells.

EL2 was also measured as a function of the stoichiometry of the starting LEC melt (Holmes 1982) ; [EL2] drops down to zero when the Ga/(Ga + As) ratio decreases behind 0.50 to $\sim$ 0.46.

Holmes and co-workers (Holmes 1984) have proposed that EL2 can be generated by two different mechanisms : a thermodynamical equilibrium of defects near growth temperature T_G related to the melt stoichiometry, leading to a homogeneous concentration in a cross-section, plus a post-generation related to dislocation climb, as was first proposed by Weber et al (1982). This second process would be more efficient in regions where dislocations density is larger. This explains why i) macroscale EL2 concentration and DD have similarly relative variations (Martin 1980 b) on a given wafer, ii) stoichiometry effects are detected for ingots to another or even in a same ingot (Holmes 1982), since the stoichiometry of the melt practically never remains the same as a function of the solidified fraction. Recent data from Duseaux et al (1984), which show that EL2 is 2 x 3 times lower in dislocation free ingots, also support this assumption. On the other hand, this hypothesis focuses the attention on the post-growth cooling of ingots, i.e. on the thermal history of the material.

As a matter of fact, this aspect has been the object of special attention and results of recent annealing studies shed a new light on the material preparation science.

3.2 Annealing equipment

Surprising results of annealing were obtained by Woodall as soon as 1966 who remarked reversible variations of electrical properties of undoped bulk high resistivity Bg materials following annealing at different temperatures. The first significant data on LEC undoped GaAs was presented by Rumsby et al in 1983. They noticed an improvement of the homogeneity of the resistivity and Hall mobility μ_H of as-grown wafers after an annealing at 950°C for 5 hours. The very large improvement of μ_H from 2500 to

4500 cm^2/Vxs in these undoped wafers (Rumsby 1984) is, in our opinion, a clear indication of the homogenisation of the resistivity of the material on a short range, as discussed by Martin et al (1980 a). Still more recently, two teams at Rockwell and LEP have specially investigated the behaviour of EL2 during 650 to 950°C annealings. According to Holmes et al (1984), EL2 in standard LEC materials increases on average by 20 to 30 % after a short 3 hour period at 950°C and then decreases to a steady state value (after a 240 hour long period) which was observed to be identical for two materials grown in two different conditions of stoichiometry. The long range average value of EL2 becomes more and more homogeneous in that process (Holmes 1984). Smoothing of EL2 absorption at a short range has even been clearly observed by Leigh (1984), after a similar procedure applied on standard LEC material. Both long range and short range variations have been assessed by Martin S et al (1984) who proved that the same absolute concentration of EL2, close to 1.2×10^{16} cm^{-3} can be achieved after long annealing at 900°C in either undoped Bg and standard LEC (all of them being dislocated materials) and In doped dislocation free material. The concentration of EL2 averaged over a wafer in these as-grown materials was quite different, i.e. 1.0×10^{16}, 1.1×10^{16} and 4×10^{15} cm^{-3} respectively. The In : GaAs LEC dislocation free material is the most homogeneous after this treatment.

More work is still necessary to get an overall picture of the main resulting effects, but one can now suggest that these are three different effects leading to the EL2 distribution in as-grown ingots :
  i) a thermodynamical equilibrium at a given temperature which depends on boundary conditions, like the melt stoichiometry at $T = T_G$. Nevertheless, for $T \ll T_G$, this last condition does not seem to prevail, neither does the stoichiometry at the surface. This means that this is a bulk effect of minimization of the lattice energy by the creation of intrinsic defects and should be fully homogeneous. This opens a thrilling field of future investigations.
  ii) a generation by dislocation climb, leading to long range fluctuations created in the temperature range of dislocation propagation.
  iii) some gettering around dislocation clusters, once the dislocation pattern is frozen in, leading to short range accumulation at the cluster walls.

3.3 Nature of the EL2 defect

It is now agreed that EL2 is actually a defect. A review of all the research concerning this level has been published elsewhere (Martin G.M. and Makram-Ebeid 1984 a). The question has turned to the assignment of EL2 to the quadruplet (Q) signal detected in electron paramagnetic resonance (EPR) and first attributed to As$_{Ga}$ by Lagowski(1982). Many studies have been devoted to this question, and there is not enough place here to mention them all. Many similarities were noticed (Schneider 1982) but, till now, there were two major discrepancies in related observations which prevented a definite assignment. These two problems have been clarified and solved. The first problem consisted in a direct correlation between EL2 and the EPR Q signal in the same bulk material. No correlation was never reported and sometimes attributed to the insufficient limit of detection of EPR measurement. Elliott et al (1984) have further proved that there does not exist any correlation between the EL2 concentration and the Q signal (corresponding to the particular electronic stage As$_{Ga}^{4+}$ of the antisite) detected in absorption and known to be due to neutral EL2, called EL2^0. But, there is a clear correlation between the Q signal and the carbon concentration in same samples. Since carbon is compensated by EL2, this is equivalent to a correlation between As_{Ga}^{4+} and the ionized part of EL2, called $EL2^i$. The reported lack of correlation is only due to the fact that EPR and absorption are sensitive to two different charge states of the same level.

The second problem lies in the totally different behaviour of the Q signal and the EL2 DLTS (Deep Level Transient Spectroscopy) signal after irradiation, such as neutron irradiation, as shown by Martin et al (1983). These authors (Makram-Ebeid 1984, Martin 1984 b), and Weber (1984 a) in parallel have proposed an hypothesis which reconcile the apparently inconsistent observation. The EL2 defect might be surrounded by other defects which strongly modify the electron reemission from the isolated EL2 defect level and thus its DLTS manifestation which has not anymore the "pure" EL2 signature. But these defects would be located sufficiently apart, i.e. not in the nearest neighbours shell, in such a way that the Q signal is hardly modified. This hypothesis has been given an experimental verification by the recent observations of Goltzene et al (1984 a, b) who proved that, in all neutron irradiated materials and sometimes in as-grown material, the Q As_{Ga} signal is detected together with an S (singlet) signal, both defects having exactly the mutual influence expected. Furthermore, their annealing behaviour is also as predicted for both of them, the Q signal behaving like the EL2 isolated defect (Martin, 1984 b).

This discards the main doubts about the EL2-EPR Q defects assignment, as discussed at the last 13th Conference on Defects in Semiconductors (Weber 1984 b). The increase of the Q EPR As_{Ga} signal under annealing up to 900°C (Kaufmann 1984), quite similar to EL2, still supports the assignment.

4. Investigated solutions for improving material homogeneity

From all the reported observations it is clear that the present difficulties mainly come from both extended and point defects. This is because the main effort has been devoted so far to the reduction and control of impurity incorporation, in such a way that their concentration is not larger than that of point defect levels anymore. Thus, the guide lines for ongoing actions focus on dislocation density reduction and control of EL2 defect via adjustment of growth conditions and post-growth annealing.

4.1 Bridgman material

Ishii et al (1984) have not found any variation of V_T of micro FET 's as a function of distance d from dislocation, and the fluctuation σV_T is about twice smaller than for LEC substrates. Matsuoka et al (1984) gave more quantitative results over wafers showing that σV_T is actually smaller in Bg wafers when measured either over a diameter of 2" wafer or inside a 400 x 400 μm^2 array. These authors have used a pattern of 2 micro FET's (1 N-OFF, 1 N-ON), distance   15 μm from one another, repeated every 320 μm. Correlation of two ON-OFF FET properties is noticed for both 15 and 320 μm distances in Bg wafers, but is not maintained over the 320 μm distance in LEC substrates. This better homogeneity of Bg wafers is consistent with EL2 absorption data which only show, in that case, a vague network of larger cell size (Martin S. 1984). This can be understood by the about 10 times lower dislocation density (near 1-4 x 10^3 cm^{-3}) in Bg ingots and the growth conditions which lead to a long annealing of the solidified fraction of the ingot and to a perfect and constant stoichiometry of the melt as well. But it will be difficult to increase much further the diameter of Bg ingots, and 3 inch now seems an upper limit.

4.2 LEC materials

Attempts to reduce the dislocation density by reduction of the temperature gradient has been made in a high pressure puller (Matsumoto 1984), but the lowest DD value is still around 10^4 cm^{-3}. Similar growth conditions, in low pressure puller, still lead to dislocations and EL2 distributions very close to standard LEC ones (Leigh 1984). One might guess that there will be very small $\sigma\, V_T$ improvement with this kind of material, if any.

On the other hand, very distinct improvement of $\sigma\, V_T$ by a factor of 1.8 has already been measured on standard LEC material annealed at 800°C for more than 12 hours (Ishii 1984). These authors measured that the typical W shape DD variation did not significantly vary during that process. Even if no very detailed analysis of the annealing effect or the cellular dislocation structure is available yet (some data can be found in Leigh 1984), one may guess that this result is correlated with EL2 homogenisation. Whether impurities, previously gathered around dislocation, diffuse and their concentration becomes more homogeneous is questionable, since the dislocation induced stress should anyway stay in place.

Magnetic field equipped, 2" and 3" high pressure pullers have recently become available (Terashima 1984, Osaka 1984) and the results show that their hazardous fluctuations of temperature at the growth interface can be totally suppressed when either the vertical or horizontal magnetic field reaches about 1000 Oersteds. This means that a better control of the growth interface is then achievable. Cornier et al (1984) have shown that precipitates observed in standard LEC materials are insulars (GaAs polycrystal grain) and are probably one of the dislocation sources. It is likely that smaller temperature fluctuations should minimize the probability of their formation and, as such, the concentration of further generated dislocation. Nevertheless no large decrease of DD has been reported, so far.

Doping with isovalent impurities to strengthen GaAs material has been studied by Mil'vidskii (1978). Jacob et al (1983) proved for the first time that low dislocation GaAs could actually be pulled this way and that In doping is the most efficient element for that purpose. Then Duseaux (1984) showed that In doping can lead to at least 2", dislocation free (i.e. DD 10^2 cm^{-2}) wafers. The validity of that approach has been verified by other laboratories during last year (Kimura 1984, Hobgood 1983, Nakanishi 1984). According to Winston (1984), who reports on In : GaAs wafers on which the DD was zero in certain areas and $\simeq 10^3$ cm^{-2} in others, there is actually an improvement of $\sigma\, V_T$, but these authors noticed that this material was grown at a slower growth rate and that the improvement might also be related to the corresponding longer annealing of the ingot. Nevertheless, this team (Kimura 1984) published a value of $\sigma\, V_T$ = 10 mV for 25 micro-FET's located in a dislocation free area. This has to be compared to $\sigma\, V_T$ $\simeq$ 100 meV and 400 mV for the so-called SAINT and the usual aligned technologies respectively (Matsuoka 1984). There are still long range fluctuations on this particular wafer (which was not dislocation free over the whole area), but these fluctuations are not hazardous since the same shape of V_T variation can be noticed on several parallel rows of FET's distance 300 µm from each other. This could be due to the presence of In doping striations generated by periodic fluctuation of In concentration related to growth temperature variation.

As a summary, one could imagine a combination of all these ideas to come out with an ingot, grown under magnetic field conditions to smooth out striations and get a better stability of growth interface, doped with In to suppress the dislocations and further annealed to stabilize EL2 at a high and constant level. This would combine all the improvements, noticed so far, and should lead to high grade substrate suitable for IC's.

References

Ainslie N G, Blum S E, Woods J F, 1962 J. Appl. Phys. 33 2391
Akai S, Fujita K, Sasaki M, Tada K, 1982 Inst. Conf. Ser. 63 512
Bishop S G, Shanabrook B V, Moore W J, 1984 t. b. p. in J. Appl. Phys.
Brozel M R, Grant I, Ware R M, Stirland D J, 1983, Appl. Phys. Lett. 42 610
Brozel M R, Grant I, Ware R M, Stirland D J, Skolnick M S, 1984 J. Appl.
    Phys. to be published in September
Chen R T, Holmes D E, Asbeck P M, 1984 to be published in Appl. Phys. Lett.
Cornier J P, Duseaux M, Chevalier J P, 1984, to be published in Appl.
    Phys. Lett. and this volume
Duseaux M, Martin S, 1984 3rd Conf. on III-V Semi-insulating Materials
    (Nantwich : Shiva) to be published
Elliott K R, 1984 a J. Appl. Phys. 55 3856
Elliot K, Chen RT, Greenbaum SG, Wagner RJ, 1984 b Appl. Phys. Lett. 44 907
Farges J P, Jacob G, Schemali C, Martin G M, Mircea-Roussel A and Hallais J
    P, 1982, 2nd Conf. on S.I. III-V Materials (Nantwich: Shiva) pp 45-54
Goltzene A, Meyer B, Schwab C, 1984 a, Proc. of 3rd Conf. on III-V
    Semi-insulating Materials (Nantwich : Shiva) to be published
Goltzene A, Meyer B, Schwab C, 1984 b, Proc. of 13th Conf. on Defects in
    Semi-conductors Inst. Phys. Conf. Ser. 74 35
Hobgood H M, Barrett, Ta C B, Eldridge G W, Thomas R N, paper A-2 1983
    Electron. Mat. Conf. (Burlington USA)
Holmes D E, Chen R Y, Elliott K R, Kirkpatrick C G, Yu P W, 1982 IEEE
    Trans. ED 29 1045
Holmes D E, Kuwamoto H, Kirkpatrick C G, Chen R T, 1984, 3rd Conf. on III-V
    S.I. Materials (Nantwich : Shiva) to be published
Ishii Y, Miyazawa S, Ishida S, 1984 IEEE Trans. on ED 31 800
Jacob G, Duseaux M, Farges J P, Van den Boom M M, Roksnoer P J, 1983 J. of
    Cryst. Growth 61 417
Kamejima T, Shimura F, Matsumoto Y, Watanabe H, Matsui J, 1982 Japan. J.
    Appl. Phys. 21 L 721
Kaminska M, Lagowski J, Parsey J, Wada K, Gatos H C, 1982 Inst. Phys.
    Conf. Ser. 63 197
Kasahara J and Watanabe N, 1982 Proc. of the 2nd Conf. on Semi-insulating
    III-V materials (Nantwich : Shiva) pp 238-48
Kaufmann U, Windscheif J, Baeumler M, Schneider J, Kohl F, 1984 Proc. of
    3rd Conf. on III-V Semi-insulating Materials (Nantwich : Shiva) to be
    published
Kimura H, Afable C B, Olsen H M, Hunter A T, Miller K T, Winston H V, 1984
    Proc. of Conf. on Solid State Devices and Materials, Kobe Japan, to be
    published
Lagowski J, Gatos H C, Parsey J N, Woda K, Kaminska N, Walukiewicz W, 1982
    Appl. Phys. Lett. 40 342
Leigh P A, Hall I P, Elliott C R, Wakefield B, Lyons M H, 1984 to be
    published in J. Phys. C
Lindquist P F, 1977 J. Appl. Phys. 48 1262
Makram-Ebeid S, Gautard D, Devillard P, Martin G M, 1982 Appl. Phys. Lett.
    40 161

Makram-Ebeid S, Langlade P, Martin G M, 1984, Proc. of 3rd Conf. on III-V
    Semi-insulating Materials (Nantwich : Shiva) to be published
Matsumoto K, Morishita H, Sasaki M, Nishine S, Yokogawa M, Sekinobu M, Tada
    K, Akai S, 1984, Proc. of 3rd Conf. on III-V Semi-insulating Materials
    (Nantwich : Shiva) to be published
Matsuoka Y, Ohwada K, Hirayama M, 1984 IEEE Trans. on ED 31 1062
Mil'vidskii M G, Bochkarev E P, 1978, J. Cryst. Growth 44 61
Miyazawa S, Ishii Y, Ishida S, Nanishi Y, 1983 Appl. Phys. Lett. 43 853
Martin G M, Farges J P, Jacob G, Hallais J P, Poiblaud G, 1980 a J. Appl.
    Phys. 51 2840
Martin G M, Jacob G, Goltzene A, Schwab C, Poiblaud G, 1980 b, Inst. Phys.
    Conf. Ser. 59 281
Martin G M, Jacob G, Hallais J P, Grainger F, Roberts J A, Clegg B, Blood
    P. Poiblaud G, 1982 J. Phys. C : Solid State Phys. 15 1841
Martin G M and Makram-Ebeid S, 1983, Physica 116 B 371
Martin G M and Makram-Ebeid S, 1984 a, in "Deep Defects in Semiconductors"
    (New-York : Gordon and Breach) to be published
Martin G M, Esteve E, Makram-Ebeid S, 1984 b, to be published in J. Appl.
    Phys. (October)
Martin S and Duseaux M, 1984, Int. Conf. on GaAs and Related Compounds
    Inst. Phys. Conf. Ser. 74 53
Nakamishi H, Kohda H, Yamada K, Hoshikawa K, 1984 5th Conf. on Solid State
    Devices and Materials Kobe Japan
Nanishi Y, Ishida S, Miyazawa S, 1984 Japan. J. Appl. Phys. 22 L 54
Oliver J R, Fairman R D, Chen R T, Yu P W, 1981 Electron. Lett. 17 840
Osaka J and Hoshikawa K, 1984 Proc. of 3rd Conf. on III-V Semi-insulating
    Materials (Nantwich : Shiva) to be published
Rumsby D, Ware R M, Smith B, Tyjber M, Brozel M R, Foulkes E J, 1983 GaAs
    IC Symposium Tech. Digest p 34
Rumsby D, Grant I, Brozel M R, Foulkes E J, Ware R M, 1984, 3rd Conf. on
    III-V Materials (Nantwich : Shiva)
Schneider J, 1982 Proc. of 2nd Conf. on III-V Semi-insulating Materials
    (Nantwich : Shiva) 144-154
Skolnick M S, Brozel M R, Reed L J, Grant I, Stirland D J and Ware R M,
    1984 J. Electron. Mater. 13 107
Takebe T, Shimazu M, Kawasaki A, Kotani T, Nakai R, Kikuchi K, Murai S,
    Tada K, Akai S, Susuki T, 1984, 3rd Conf. on S.I. III-V Materials
    (Nantwich : Shiva) to be published
Terashima K, Katsumata T, Orito F, Fukuda T, 1984 Japan. J. Appl. Phys. 23
    L 302
Thomas R N, Hobgood H M, Eldridge G W, Barrett D L, Braggins T T, Ta L B,
    Wang S K, 1984 Semiconductors and semi-metals pp 1-86
Weber E R, Ennen H, Kaufmann U, Windschief J, Schneider J, 1982 Appl.
    Phys. 53 6140
Weber E R, 1984 a, Proc. of 3rd Conf. on III-V Semi-insulating Materials
    (Nantwich : Shiva) to be published
Weber E R, 1984 b, Concluding remarks of 13th Conf. on Defects in
    Semiconductors Inst. Phys. Conf. Ser. 74
Winston H V, Hunter A T, Olsen H M, Bryan R P, Lee R E, 1984 a, Appl.
    Phys. Lett. (August)
Winston H V, Hunter A T, Olsen H M, Bryan R P, Lee R E, 1984 b, Proc. of
    3rd Conf. on III-V Semi-insulating Materials (Nantwich : Shiva) to be
    published
Woodhall J M, Woods J F, 1966, Solid State Commun. 4 33
Zucca R, 1977, J. Appl. Phys. 48 1987

Inst. Phys. Conf. Ser. No. 74: Chapter 2
Paper presented at Int. Symp. GaAs and Related Compounds, Biarritz, 1984

23

Growth and properties of constant composition bulk GaAsP ternary alloys

Y. Takeda, Sg. Fujita, and A. Sasaki

Department of Electrical Engineering, Kyoto University
Kyoto 606, Japan

Abstract. Bulk GaAsP ternary alloys of constant composition were
grown in a closed quartz ampoule by the iodine transport method
using a vertical furnace. GaAsP alloy powder was used as the source
material. The growth temperature was 850°C and the temperature
gradient for the transport was 1°C/cm. It was found by   x-ray
diffraction that the grown ternary crystals have the same
composition as that of the source powder. The constant and uniform
composition was confirmed by   Auger Electron Spectroscopy with
the simultaneous sputter etching and also by observing the colour
uniformity of the transmitted light.

1. Introduction

III-V ternary and quaternary semiconductors have been grown on binary
compound substrates by various epitaxial growth techniques (LPE, VPE,
OMVPE, MBE)(e.g., Pearsall 1982). Very close lattice-matching of the
epitaxial alloy layers with the substrate is required to avoid the
serious degradation of the crystal quality due to lattice-mismatch
(Takeda and Sasaki 1978, Olsen and Ettenberg 1978). Because of this
severe requirement, alloy crystals only on the constant
lattice-parameter line can be utilized for most of the device
applications. It is needless to say that the combination of the alloys
and binary compounds on the constant lattice-parameter line has enabled
fabrication of many devices in the optoelectronics(DH lasers, LED's,
Heterojunction APD's) and electronics(HEMT's, HPT's, HJBT's).

However, if it is possible to choose arbitrary lattice parameters, the
whole area, except for miscibility gap(Onabe 1982), of alloy
semiconductors can be utilized and new device structures may become
possible. As an example, the (x,y) compositional dependence of
energy-gap values in $Ga_xIn_{1-x}As_yP_{1-y}$ is shown in Fig. 1 (Sasaki et al.
1980, Glisson et al. 1978, Casey and Panish 1978). Only the
compositions on the line for InP lattice-parameter and that for GaAs
lattice-parameter are meaningful. In the figure, if the arbitrary
lattice-parameter between InAs and GaAs is available, i.e., the bulk
ternary InGaAs of arbitrary composition can be used, most of the area,
except for a triangle area at the GaP corner, is covered. GaAsP
ternary alloy covers the remaining GaP corner area. With GaAsP
substrate, the highest direct energy-gap among III-V semiconductors
which can form the double heterostructure is obtained in GaInP/AlInP.

Ternary GaInP and InAsP are other possible alloys to cover the whole area of the (x,y) compositional plane. Thus, it is a natural extrapolation to use ternary alloy substrates in order to be free from the limit imposed by the discrete lattice-parameters available and to cover the wider area of the alloy compositional planes for making new device structures. To do this, constant composition ternary bulk crystals are required. Growth of GaAsP was taken as the first step toward this idea.

In this paper, we describe the growth procedure of GaAsP by the iodine transport method and some properties of single crystalline bulk GaAsP with constant composition.

2. Growth Procedure

2.1 Preparation of source alloy powder

GaAsP alloy powder was used as the source material to ensure composition uniformity of the source. The alloy powder was formed by solid-state interdiffusion between well-mixed GaP and GaAs powder in a vacuum sealed ampoule at 850°C for 48 h. The temperature and the interval were sufficient to form homogeneous alloy powder of GaAsP. This was confirmed by the x-ray diffraction as shown in Fig. 2. Diffraction peaks of the binary compounds at 45.34°(GaAs) and at 47.13°(GaP) from (220) reflections were not detected. From the peak position of the alloy powder in Fig. 2 the composition was estimated to be $GaAs_{0.4}P_{0.6}$, assuming Vegard's law. The size of the binary compound powder was not larger than 100 μm. Before the ampoule was sealed, the

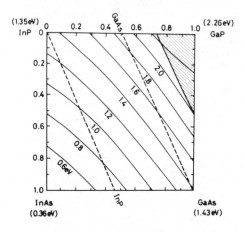

Fig. 1 Compositional dependence of energy-gap values in $Ga_xIn_{1-x}As_yP_{1-y}$. Compositions on the lines of InP and GaAs lattice-parameters are used so far. If the ternary InGaAs and GaAsP substrates are available, the whole area of the quaternary alloy can be covered.

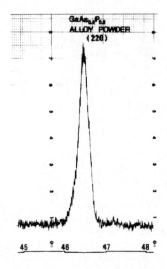

Fig. 2 X-ray diffraction peaks from $GaAs_{0.4}P_{0.6}$ alloy powder used for the source material in the iodine transport growth.

mixed powder was purified in a Pd-diffused flowing hydrogen at 800°C
for several hours. By the alloying the colour of the powder changed
from gold of the mixed powder to red of the alloy powder.

2.2 Growth

The quartz ampoule was opened in a nitrogen atmosphere to avoid the
oxygen adsorption to the alloy powder in it. A seed crystal(GaAs) was
set at the neck of the ampoule and fixed by the tip of a quartz rod.
The schematic drawing of the ampoule for the growth is shown in Fig. 3
together with a part of the temperature profile of the furnace. The
shape of the ampoule is the same as that used for the growth of ZnS and
ZnSe by iodine transport(Fujita 1979, Kaldis 1974). After the ampoule
was pumped down to $\sim 10^{-6}$ Torr, the iodine which was kept in a
liquid-nitrogen trap to reduce its vapor pressure was transported into
the ampoule. The sealed ampoule was set in a vertical furnace. The
volume of the ampoule was 25 cm^3, the weight of the source powder was 5
g, and the amount of iodine was 125 mg. The temperature at the source
was kept at 850°C. The temperature gradient for the transport growth
was 1°C/cm. The growth was continued for $2 \sim 3$ weeks.

3. Results and Discussion

The transport rate of $GaAs_{0.4}P_{0.6}$ and $GaAs_{0.6}P_{0.4}$ was approximately 5
mg/h. In the GaAs growth the transport rate was 3.9 mg/h. In the
growth of GaAs and InAs binary compounds, larger single crystals($\sim 5 \times 6 \times 7$
mm^3) were easily obtained. However, for the ternary alloys the crystal
size was not easy to control. The ampoule and the grown crystals of
GaAs and $GaAs_{0.4}P_{0.6}$ are shown in Fig. 4. Round plates inside the
ampoule are the remaining source materials sintered during the growth
process. The largest single crystalline alloy obtained so far was
approximately $3 \times 4 \times 5$ mm^3. Although a seed crystal was used, the
lattice-mismatch between seed GaAs and grown GaAsP was relieved at the

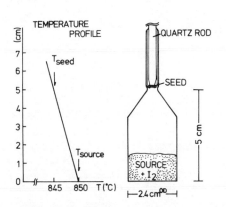

Fig. 3 Schematic illustrations
of the growth ampoule and the
temperature profile of the
vertical furnace.

Fig. 4 Growth ampoules and
grown crystals of GaAs and
$GaAs_{0.4}P_{0.6}$ after two weeks
growth. Round plates are
the sintered source powder.

neck portion of the ampoule. Polycrystals grew on the seed and below
the neck portion a few larger crystals grew preferentially. Since the
grown single crystals did not touch the inside surface of the ampoule,
the stress between the grown crystals and the quartz tube was avoided.
The crystal surfaces were the (111) and (100) low index planes, which
were found by x-ray diffraction.

(400) x-ray reflection peaks from $GaAs_{0.6}P_{0.4}$ and $GaAs_{0.4}P_{0.6}$ are shown
in Fig. 5. Cu $K\alpha_1$ and $K\alpha_2$ peaks are clearly resolved, indicating good
uniformity of the composition in both alloy crystals within the
penetration depth of the x-ray($\sim$20 μm). (400) reflection from
VPE GaAsP grown on GaAs substrate is shown in Fig. 6 for comparison.
This epitaxial layer is a 25 μm-thick constant composition layer of
arsenic composition 0.6 on a 28 μm-thick composition grading layer.
For x-ray diffraction measurement the substrate and the composition
grading layer were removed by a selective etching with the solution of
$H_2SO_4:H_2O_2:H_2O$. With the x-ray diffraction system, which had the angle
resolution of 0.06°, no difference in the half width of the diffraction
pattern was observed between bulk GaAsP and VPE GaAsP. Auger Electron
Spectroscopy with simultaneous Ar-sputter etching was also employed for
both bulk and VPE GaAsP to confirm the composition and the composition
uniformity observed by x-ray diffraction.

The composition of the source alloy powder and that of the grown
crystal were the same. This was found by x-ray diffraction of both the
source alloy powder(Fig. 2) and the grown alloy crystal($GaAs_{0.4}P_{0.6}$ in
Fig. 5). This is the greatest advantage of the present growth method
for obtaining constant composition bulk alloys. When alloy

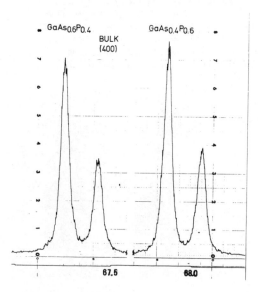

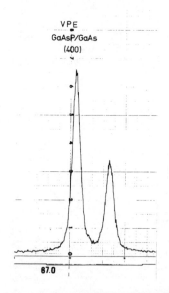

Fig. 5 X-ray diffraction
peaks from bulk $GaAs_{0.6}P_{0.4}$
and $GaAs_{0.4}P_{0.6}$. Cu $K\alpha_1$ and
$K\alpha_2$ are clearly resolved.

Fig. 6 X-ray diffraction from
VPE-grown GaAsP. Substrate
GaAs and composition grading
layer were removed for the
measurement.

semiconductors are grown from solutions, it suffers compositional change along the growth direction because of the difference between the liquidus and solidus. This compositional change accompanies a change in lattice parameter and a degradation of crystal quality(e.g., Bachmann et al. 1979). If the growth is limited by diffusion in the iodine transport method, there should be a compositional variation along the growth direction. We considered the growth was limited by convection in the vertical arrangement of the source and the seed in the present experiment. In the growth of ZnS and ZnSSe the transport mechanism was concluded to be the convective flow from the ampoule diameter dependence of the growth rate(Fujita et al. 1979).

In Fig. 7 (a) uniformity (of the colour in the actual sample) of the transmitted light through bulk $GaAs_{0.4}P_{0.6}$ with thickness of 3 mm is demonstrated. The colour of the transmitted light represents the energy gap and thus the composition of the alloy. Constant colour from the neck portion to the tip of the grown alloys was observed. Although the composition uniformity in the constant composition layer of VPE GaAsP is good, dark lines and textures are observed at this magnification(x200) as shown in Fig. 7 (b). Most of the lines correspond to the cross-hatch pattern of the epitaxial layer which was caused, even with the composition grading layer, by the lattice-mismatch.

Although the lattice-mismatch between seed GaAs and grown GaAsP is removed during the growth, the mismatch causes the growth of polycrystals at the beginning. Use of single crystal GaAsP which has the natural grown surface may be desirable to avoid this effect and to grow larger single crystals of the alloy. All the grown alloys showed n-type conductivity with a high electron concentration of $\sim 2 \times 10^{18}$ cm^{-3}. This is considered to be due to the iodine incorporated into the grown crystals. Elimination of donor iodine will be very difficult by this growth method. However, as the substrate for epitaxial growth this will not be a serious problem.

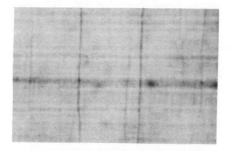

(a) BULK GaAsP   (b) VPE GaAsP ON GaP

Fig. 7 Uniformity of the transmitted light (uniformity of the colour in the actual sample) through 3-mm thick $GaAs_{0.4}P_{0.6}$ is demonstrated in (a). Black line at bottom right was used for focusing. VPE-grown GaAsP shows dark lines and textures due to lattice-mismatch (b). (x200 for both.)

4. Summary

Constant composition GaAsP ternary alloy crystals were grown in a closed quartz ampoule by the iodine transport method in a vertical furnace. The composition of the source alloy and the grown alloy crystal was found to be the same. This is the greatest advantage of this growth method to obtain constant composition alloy crystals. Composition uniformity was confirmed by x-ray diffraction, Auger Electron Spectroscopy, and optical transmission.

Acknowledgment

The work was supported in part by Iwatani Naoji Foundation and in part by Kudo Science Foundation.

References

Casey H C and Panish M B 1978 Heterostructure Lasers Part B (New York: Academic Press) pp 32-48

Fujita S, Mimoto H, Takebe H and Noguchi T 1979 J. Cryst. Growth 47 326

Glisson T H, Hauser J R, Littlejohn M A and Williams C K 1978 J. Electron. Mat. 7 1

Kaldis E 1974 Crystal Growth-Theory and Techniques Vol. 1 ed Goodman C H L (New York and London: Plenum Press) pp 49-191

Olsen G H and Ettenberg M 1978 Crystal Growth-Theory and Techniques Vol. 2 ed Goodman C H L (New York and London: Plenum Press) pp 1-56

Onabe K 1982 Jpn. J. Appl. Phys. 21 L323

Pearsall T P (ed) 1982 GaInAsP Alloy Semiconductors (Chichester: John Wiley & Sons) pp 11-106

Sasaki A, Nishiuma M and Takeda Y 1980 Jpn. J. Appl. Phys. 19 1695

Takeda Y and Sasaki A 1978 J. Cryst. Growth 46 257

Inst. Phys. Conf. Ser. No. 74: Chapter 2
Paper presented at Int. Symp. GaAs and Related Compounds, Biarritz, 1984

29

Effects of whole ingot annealing on photoluminescence and electrical properties in 2-inch/3-inch undoped/lightly Cr doped LEC-grown GaAs

M. Yokogawa, S. Nishine, K. Matsumoto, H. Morishita, K. Fujita and S. Akai*

Sumitomo Electric Industries, Ltd., 1-1-1, Koya-Kita, Itami-shi 664, Japan

*Sumitomo Electric Industries, Ltd., 1-1-3, Shimaya, Konohana-Ku, Osaka-shi 554, Japan

Abstract. The effects of whole ingot annealing (WIA) on PL and electrical properties were studied in 2-inch/3-inch undoped/lightly Cr doped LEC-GaAs. PL measurements were performed on the major 1.49eV band using a newly developed computer-controlled system for two-dimensional mapping of PL intensities in a complete wafer. It has been found that WIA considerably increases PL intensities, and drastically reduces both the radial and axial variations. Moreover, the mobilities of 2-inch and 3-inch annealed undoped ingots were found to be consistently larger than 4,000 and 5,000 cm^2/v.s, respectively, throughout the ingot. It has also been found that the improvement of substrate quality has a good influence on device characteristics.

1. Introduction

Liquid-Encapsulated-Czochralski (LEC) grown, semi-insulating GaAs wafers are widely used as the substrate in the fabrication of GaAs ICs together with the direct ion implantation technique. Usually LEC growth of GaAs is carried out in a severe thermal environment, which results in a high dislocation density with the characteristic W-shaped distribution pattern across the wafer. Thus, there have been many investigations of the crystal inhomogeneities in LEC-grown GaAs substrates, and it has been found that the radial variations of photoluminescence (PL) intensities (Tajima 1982; Kitahara et al. 1982), electrical resistivities and mobilities closely correlate to those of the dislocation densities. In addition, it has been found that the properties of the substrate significantly affect the characteristics of FETs fabricated on it by direct ion implantation (Nanishi et al. 1982a; Nanishi et al. 1982b). Recently it has been reported that whole ingot annealing (WIA) improves the radial uniformity of electrical resistivity and mobility in undoped LEC semi-insulating GaAs with no significant change in its dislocation density (Rumsby et al. 1983). In this paper we will describe the effectiveness of WIA in improving not only the material quality but also the device characteristics. The cause of the large increase in Hall mobility will also be discussed.

2. Experimental procedure

The effects of WIA on the PL and electrical properties were investigated using twenty-one ingots. The as-grown ingots were cut perpendicular to growth axes and divided into three parts labeled S, M and T from the seed end. The middle part (M) was subjected to heat treatment under various conditions for each ingot. Annealing temperature was in the range of 700 to 900°C and annealing time was in the range 6 to 40 hours. The wafers used in this experiment were sliced from part S, M and T, and were lapped and chemically etched to obtain a mirror-like surface.

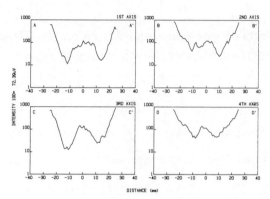

3. Photoluminescence studies

A newly developed computer-controlled system has been used for mapping PL intensities on complete wafers up to 3 inches in diameter. The PL measurement was performed at 4.2K with the sample immersed in liquid helium in a metal Dewar with a 100mm∅ quartz window. The sample was excited by a 5145A e-mission from an Ar laser. The PL light was focused onto the entrance slit of a 3/4m SPEX-1702 monochromator equipped with a 600-groove mm⁻¹ grating blazed at 1.2μm. The output of a photomultiplier (HTV-7102) was detected with a lock-in amplifier (PAR-5205). Further details of this system have been described elsewhere (Yokogawa et al. 1984a). PL measurements were performed on the major 1.49eV band attributable to the carbon

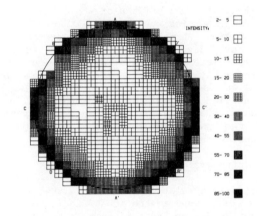

Fig.1. One- and two-dimensional distribution maps of 1.49eV PL intensities at 4.2K in a (100) wafer obtained from part S of an as-grown undoped LEC ingot. A-A', B-B', C-C', and D-D', correspond to ⟨011⟩, ⟨001⟩, ⟨01̄1̄⟩ and ⟨010⟩ axes, respectively. One-dimensional distribution was mapped at 1 mm pitch, and two-dimensional map at 2.5 mm pitch.

impurity (Ashen et al. 1975). Figure 1 shows one- and two-dimensional distribution maps of PL intensities in a (100) wafer obtained from part S of an as-grown 2-inch undoped LEC ingot. Measurements of one-dimensional distributions were carried out along four directions: the ⟨011⟩, ⟨01̄1̄⟩, ⟨001⟩ and ⟨010⟩ axes. It is clearly seen that the radial variation has a characteristic W-shaped distribution and four-fold symmetry, which are both in good agreement with the distribution pattern of dislocation densities.

Figure 2 shows the distribution maps in the wafer obtained from part M, which was annealed at 800°C for 24 hours. It can be seen that the characteristic W-shaped radial variation has been almost extinguished and a rather uniform distribution has been obtained. It can be also seen that the annealing causes a large increase in PL intensity. It should, at this

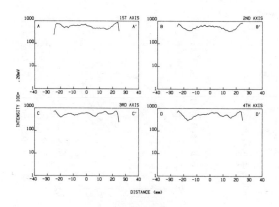

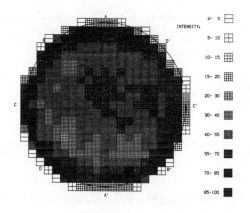

Fig.2. One- and two-dimensional distribution maps of 1.49eV PL intensities at 4.2K in a (100) wafer obtained from annealed part M.

point, be noted that these changes occurred not only in the periphery of the ingot but also deep in the ingot. The axial variation of the PL intensities was measured and it has been found that this increase is greater near the tail end than near the seed end of part M,resulting in a more uniform axial distribution of PL intensities. Similar drastic changes in the 1.49eV PL properties were observed in the remaining ingots. Further details of PL studies on WIA have been described elsewhere(Yokogawa et al. 1984b).

4. Electrical properties

The mobility,which is one of the criteria of the crystal quality,was measured at room temperature for a number of ingots,particularly undoped ones, using Hall-effect measurements with Van Der Pauw geometry. Average mobilities of 2-inch and 3-inch as-grown undoped ingots were in the range of 1,000 to 4,000 cm^2/v·s and in the range of 3,000 to 5,500 cm^2/v·s,respectively, and the radial variation showed the W-shaped distribution,especially in the seed end,as shown in Fig.3. The mobilities were increased considerably by WIA. The obtained values of 2-inch and 3-inch annealed ingots were consistently larger than 4,000 cm^2/v·s and 5,000 cm^2/v·s,respectively, throughout the ingot with no significant degradation of electrical resistivity. Figure 4 shows the result obtained in a 2-inch undoped ingot annealed at 800°C for 24 hours. The characteristic W-shaped distribution of mobility across the wafer was almost extinguished and more uniform distribution was obtained. The s.d. of mobility variation across wafers obtained in the annealed ingots became less than 10%,which is far superior to those of the as-grown undoped ingots. The quality of the substrate has a great effect on the device properties. To demonstrate the feasibility of WIA in terms of device characteristics,the electrical properties of an active layer formed by direct ion implantation technique were investigated. The pinch-off voltage V_p was defined as the applied voltage intersecting the capacitance of 1pF on the C-V profile. It can be seen in Fig.5 that the difference between the pinch-off voltages at the seed end and the tail end was much reduced, even though all ingots used were lightly Cr doped.

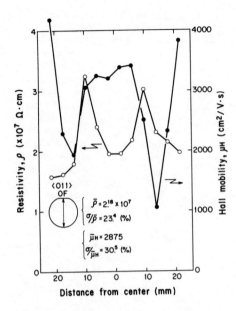

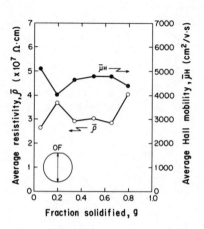

Fig.4. Axial variation of average Hall mobility in a 2-inch undoped ingot annealed at 800°C for 24 hours.

Fig.3. Radial variations of resistivity and Hall mobility in a 2-inch as-grown undoped ingot.

Figure 6 shows that Hall mobilities in the active layer formed in the wafer obtained from the annealed ingot are superior to those obtained from the as-grown ingot,especially at liquid-nitrogen temperature. Further investigations are now in progress.

5. Discussion

In the mixed conduction model the apparent Hall mobility u_H is given by:

$$\mu_H = \mu_n \frac{n + p/b^2}{n + p/b} \qquad (1)$$

where

μ_n is the true electron mobility
$b = \mu_n/\mu_p$ is the electron-to-hole mobility ratio
n is the conduction band electron concentration
p is the valence band hole concentration

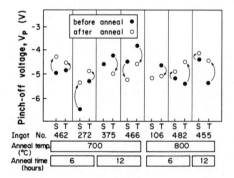

Fig.5. Variations of pinch-off voltage V_p between the seed end and tail end, obtained in the as-grown and annealed ingots.

The true electron mobility μ_n and the electron-to-hole mobility ratio b have been calculated by Walukiewicz et al.(1982) as a function of temperature and ionized impurity concentration N^i. n and p depend on the Fermi level E_F,which can be obtained

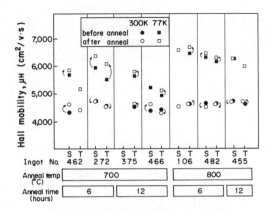

Fig.6. Hall mobilities in the active layers formed in the wafer obtained from the as-grown and annealed ingots.

by solving the equation of charge neutrality(Martin):

$$N_{SA} + Cr \; f(Cr) = N_{SD}^+ + [EL2] \; (1 - f(EL2)) \quad (2)$$

where N_{SA} and N_{SD} are the concentrations of shallow acceptors and donors,and both are fully ionized in semi-insulating GaAs. In the case of undoped semi-insulating GaAs in which the deep donor EL2 acts as a main compensating center for shallow impurity,E_F is given as a solution of the following equation:

$$(N_{SA} - N_{SD})/[EL2] = 1 - f(EL2) \quad (3)$$

When $(N_{SA} - N_{SD})/[EL2]$ decreases,μ_H increases as a result of the Fermi level shift towards the conduction band. It is thought from equations (1) and (3) that an increase of μ_H observed in WIA arises from the increase of μ_n or the reduction of $(N_{SA} - N_{SD})/[EL2]$. However,the former explanation is hardly acceptable because it requires a significant reduction of the ionized impurity concentration. For example,according to the calculations of Walukiewicz et al.(1982), an increase of μ_n from 2,000 to 5000 corresponds to a reduction of N^i from $10^{17}cm^{-3}$ to $10^{16}cm^{-3}$. Thus,it is implied that an increase of μ_H results from a decrease of $(N_{SA} - N_{SD})/[EL2]$. There are two possibilities to explain the decrease of $(N_{SA} - N_{SD})/[EL2]$. One is a reduction of $N_{SA} - N_{SD}$ and the other is an increase of $[EL2]$. Since the increase of μ_H in WIA occurs not only in the periphery of the ingot but also deep in the ingot, the former case is not acceptable. Hence, the latter case is the most appropriate explanation. In fact this agrees well with the experimental results of infra-red absorption measurement(Rumsby et al. 1983). They have observed that the deep donor EL2 concentration shows the W-shaped distribution across wafer in an as-grown ingot,and is increased and uniformly distributed across a wafer by WIA. Therefore,it can be deduced that an increase of $[EL2]$ is responsible for the increase of μ_H in WIA. Moreover,the W-shaped distribution of μ_H across wafers in an as-grown ingot can be explained by this radial variation of $[EL2]$.

6. Impact of WIA

The development of a GaAs IC has made rapid progress in the past few years, especially in the field of memory and logic devices. Several GaAs ICs have been successfully fabricated with LSI-level complexity. However,the emergence of a commercial market for GaAs ICs is still far away,mainly due to the poor device yield. To increase and stabilize the device yield,it is extremely important to obtain ingots that are homogeneous not only radially but also along the growth axis,and that are easily reproducible. It is likely that WIA will become one of the key technologies employed to obtain satisfactory homogeneity and reproducibility.

In summary,we have studied the effects of whole ingot annealing on PL and electrical properties in 2-inch/3-inch undoped/lightly Cr doped LEC semi-insulating GaAs. It has been found that the quality of substrates can be greatly improved and that this has a good influence on the device characteristics. It has been deduced that the large increase in the apparent Hall mobility in the annealed ingot is due to the increase of the deep donor EL2 concentration.

References

Tajima M,   1982 Jpn.J.Appl.Phys.21,L227
Kitahara K, Nakai K and Shibatomi S, 1982 J.Electrochem.Soc.129,880
Nanishi Y, Ishida S, Honda T, Yamazaki H and Miyazawa S, 1982a Jpn.J.
    Appl.Phys.21,L335
Nanishi Y, Ishida S and Miyazawa S, 1982b Jpn.J.Appl.Phys.22,L54 .
Rumsby D, Ware R M, Smith B, Tyjberd M, Brozel M R and Foulkes E J,
    1983 Technical Digest of 1983 GaAs Symposium, 34
Yokogawa M, Nishine S, Matsumoto K, Akai S and Okada H, 1984a
    Jpn.J.Phys.23,663
Ashen D J, Dean P J, Hurle D T J, Mullin J B and White A M 1975
    J.Phys.Chem.Solids 36,1041
Yokogawa M, Nishine S, Sasaki M, Matsumoto K, Fujita K and Akai S,
    1984b Jpn.J.Appl.Phys.23,L339
Walukiewicz W, Pawalowicz L, Lagowski J and Gatos H C, 1982
    Semi-insulating III-V Materials Evian 1982 ed. S. Makram-Edeid
    and B. Tuck (Shiva publishing) 121
Martin G M, 1980 Semi-insulating III-V Materials Nottingham 1980
    ed. G. J. Rees (Shiva publishing)13

Inst. Phys. Conf. Ser. No. 74: Chapter 2
Paper presented at Int. Symp. GaAs and Related Compounds, Biarritz, 1984

35

Modelling of fast neutron induced defects in GaAs

A. Goltzene, B. Meyer, C. Schwab

Laboratoire de Spectroscopie et d'Optique du Corps Solide
Unité Associée au C.N.R.S. n° 232 - Université Louis Pasteur
5, rue de l'Université - 67000 STRASBOURG France

S.G. Greenbaum*, R.J. Wagner

Naval Research Laboratory, Washington, D.C. 20375, U.S.A.

Abstract. EPR on fast neutron irradiated GaAs allow us to ascribe the main features of the experimental spectra namely a quadruplet superimposed to a singlet, to As_{Ga}^{4+} and V_{Ga}^{2-} centers which are interacting.

These centers are directly correlated to electrical and optical signals linked to the EL2 and U bands. Even more the small variations in the values of the parameters are fully consistent with the picture of associated defects where the As_{Ga}^{4+} defect has either an intrinsic or an extrinsic partner, located beyond the first shell.

1. Introduction

In undoped semi-insulating GaAs, grown-in intrinsic defects are likely to be the major electronic traps. Fast neutron irradiation allows to generate them at the high concentrations requested for an eventual identification by conventional electron paramagnetic resonance (EPR).

A detailed analysis of the EPR spectra of neutron irradiated material (Goltzene et al., 1983) shows that they consist in two main features, namely a quadruplet and a singlet respectively ascribed to As_{Ga}^{4+} and V_{Ga}^{2-} centers (Goltzene et al. 1984b), which are associated in a complex (Goltzene et al., 1984 a,b). Further experiments have permitted to separate these two signals by shifting to higher microwave frequencies, by annealing and by selective saturation (Goltzene et al., 1984 c,d).

On the other hand, deep level spectroscopy experiments (DLTS) have evidenced two strong bands which could be linked to the neutron irradiation, the electron trap U and the hole trap L (Martin and Makram-Ebeid, 1983). These broad bands appear also after boron implantation (Martin et al., 1982). It has been pointed out that the annealing kinetics of U suggest that it disappears in favour of the deep donor level EL2 (Martin et al., 1984 a). As EL2 remains the main level in any melt-grown bulk GaAs, it

* National Research Council Postdoctorate Associate.
  Present address : Department of Physics and Astronomy,
  Hunter College (CUNY), NEW YORK, N.Y. 10021, U.S.A.

has occasionally been suspected to be a native defect particularly since the report of its non assignment to oxygen (Huber et al., 1979). Its attribution to an isolated As_{Ga}^{4+} (Lagowski et al., 1982) raises several problems despite the similarity of quenching behaviour in DLTS (Vincent and Bois, 1978) and in EPR (Weber et al., 1982) measurements.

Instead of entering in the very details of the controversies raised by the novel structures in the EL2 DLTS spectra (Taniguchi and Ikoma, 1982 ; Lagowski et al., 1984) and photosensitivity of the EPR spectra (Weber and Schneider, 1983) we shall revisit the identification problem of fast neutron induced defects in GaAs and discuss its modelling within the framework set by the general properties displayed by the defect related to the EL2 level.

2. Correlation between EPR, optical absorption and transport data

At 4 K, the EPR quadruplet and singlet are proportional to the neutron fluence and the ratio of their intensities is about one (Goltzene et al., 1983). It corresponds to an introduction rate of some 10 paramagnetic defects/cm^3 per incident neutron/cm^{-2} for each signal (Goltzene et al., 1984 d). This yield is similar to the introduction rate of the traps corresponding to the U band in DLTS experiments (Martin and Makram-Ebeid, 1983 ; Magno et al., 1984). At 5 K, the optical absorption coefficient at 1 μ ($\sim 1,2$ eV), which is essentially due to the presence of EL2 (Martin, 1981), is also proportional to the neutron fluence (Coates and Mitchell, 1975 ; Weber, 1984).

However, recent experiments have shown that the photoquenchable fraction of the EPR signal (Goltzene et al., 1984) is not proportional to the neutron fluence.

At this time, we cannot conclude that either the quadruplet or the singlet is correlated with U or the absorption at 1 μ since the same argument would hold for both.

As a function of recording temperature, the EPR quadruplet and singlet follow the paramagnetic CURIE law, expected for a system with a constant number of spins, only for T < 100 K. For T > 100 K, there is an apparent loss of spins with increasing temperature (Goltzene et al., 1983). Furthermore, below 100 K, there is evidence of a lineshape variation which has been ascribed to a transition from a static to a dynamic JAHN-TELLER regime (Goltzene et al., 1984b).

It is interesting to compare these critical temperatures with similar ones observed in transport and optical measurements. The recovery temperature for the metastability of EL2 is around 120 K (Bois and Vincent, 1977 ; Martin and Makram-Ebeid, 1984 b). An unexpected temperature dependence of the spectral distribution of the optical cross-section for the population of the metastable state has been found around 50 K (Omling et al., 1984). Surprisingly the thermal dependence of the photoluminescence excitation spectra of the EL2 emission band presents a threshold at about 25 K (Shanabrook et al. 1983).

Finally, both the scatters of the activation energies in the transition between the two states of EL2 and of the recovery temperature have been explained by the existence of a family of traps of similar nature

(Taniguchi and Ikoma, 1984 a ; Taniguchi et al., 1984 b).

3. Correlation between annealing behaviour

Isochronal annealing experiments of irradiated GaAs show a parallel decrease of the EPR quadruplet (Goltzene et al., 1984 c) and of the DLTS U
and L bands (Martin and Makram-Ebeid, 1983) with a main decay between 400°
and 600° C. Contrary to earlier EPR data (Wörner et al., 1982), a similar
residual signal for both the quadruplet and the U band (Martin et al.,
1984 a) is observed for T > 600°C.

The corresponding absorption spectra (Coates and Mitchell, 1975) reveal
a band at 0.27 eV remaining visible up to 350°C and a band at 1.2 eV,
whose annealing temperature is in the 450°-600°C range. These results are
in close analogy with the respective annealing temperatures observed for
the singlet and quadruplet (Goltzene et al., 1984 c).

To our knowledge, the possibility of a trap annealing out in the 200-
400°C range has not yet been reported.

Martin et al. (1984 a) have found the EL2 peak increasing between 400°
and 500°C, whereas EL6, after an initial increase sharply vanishes between 450° and 500°C. Slight increases are also found for the singlet and
the quadruplet before their annealing (Goltzene et al., 1984 c). However,
below 350°C, the concentration· of the former corresponding centers remains always lower by one order of magnitude in comparison with U or L.

Finally, the successive anneals alter the lineshapes of the signals both
in EPR and DLTS experiments, even in the temperature range where they are
not decreasing.

4. Modelling

The EPR quadruplet contains the signature of an s-electron (A,state) on an
As ion ; in principle, either a substitutional antisite As_{Ga}^{4+} or an interstitial As could yield such a signature. The final attribution is actually
made by a proper scaling with the P_{Ga} center in GaP, where the signature
is complete (Goswami et al., 1981) or by reference to theoretical calculations (Meyer et al., 1984).

The singlet has an even more reduced signature. However we may ascribe it
to p-holes using the following approaches. This signal is almost the same
as one observed in amorphous As, where it has been ascribed to a p-orbital
extending over less than three As atoms (Bishop et al., 1976). In a GaAs
lattice, this configuration corresponds to a hole trapped at a Ga vacancy,
i.e. V_{Ga}^{2-}. Furthermore the g-value of the singlet may be extrapolated from
the values of the cation vacancies in ZnS and ZnSe (Goltzene et al.
1984 b) ; i.e. V_{Zn}^- or in GaP (Kennedy et al., 1978), i.e. V_{Ga}^O.

As a result, we are left with two electronic states, i.e. V_{Ga}^O versus V_{Ga}^{2-}.
Since the first state does not allow a JAHN-TELLER effect due to strong
exchange interaction in its high spin configuration (Kennedy et al., 1983)
we keep the second one. EPR experiments under uniaxial stress would un-
undoubtedly help to strengthen this point.

Both the As_{Ga}^{4+} and V_{Ga}^{2-} models assume local tetrahedral symmetry, which
means no other defect on the first shell of neighbouring atoms. However,

these representations do not exclude interaction with centers beyond this first shell, since most of the parameters would only be slightly affected.

To sum up, we respectively ascribe the quadruplet and the singlet to an s-like electron and a p-electron, both being localized on As nuclei, the latter near a vacancy.

Evidence that these two elemental defects are interacting is brought by their strong correlation versus recording temperature and the thermal variation of the hyperfine constant (Goltzene et al., 1984 b), which requires a paramagnetic neighbour undergoing a JAHN-TELLER distortion. Further support of this interaction is also given by comparison of the microwave saturation of neutron and electron generated defects and spin hamiltonian parameter variation with annealing temperature.(Goltzene et al., 1984 c).

In brief, radiation induced defects are essentially As_{Ga}^{4+} and V_{Ga}^{2-} centers interacting in complexes of $As_{Ga}-V_{Ga}$ type, where the relevant parameter is likely their distance.

At this point, we may now revisit the identification problem of the main DLTS signals in neutron irradiated GaAs.

The correlative variation between U and the quadruplet as a function of fluence allows its attribution to anionic antisite-like defects. Since U and EL2 are linked together, this conclusion holds obviously for EL2.

The discrepancies are easily understood in terms of the associated defect model, where one partner is always a As_{Ga}^{4+} center and the second one a V_{Ga}^{2-} (or another acceptor) at varying distances. For example in neutron irradiated material, the dominant defects are intrinsic since the overall defect content can amount 2.10^{18} cm^{-3} which is well above any contaminant concentration. In as-grown material, there is no definite argument against an extrinsic partner. To the contrary, the similarity of the EL2 and EL2-0 induced by oxygen (Lagowski et al., 1984) is a strong hint for this to happen. A second indication is the difference of the annealing behaviour of EL2 according to the growth method (Taniguchi et al., 1984 b). However, tracking these centers by conventional EPR might be particularly difficult even for the major impurities like B, C or Si, although one can predict g-values equal or lower than for V_{Ga}^{2-}.

Referring to the theoretical calculations [see reviews by Baraff (1984), Pötz and Ferry (1984), Van der Rest and Pecheur (1984)], there is no contradiction with the former models since for both As_{Ga}^{4+} and V_{Ga}^{2-}, deep levels are found within the bandgap. The antisite should act as a donor with a A_1 ground state, whereas the vacancy is an acceptor with a T_2 ground state. However, one may remark that the calculations assume isolated defects in an otherwise perfect GaAs lattice, certainly an unrealistic picture at least for heavily neutron irradiated or ion-implanted material.

Acknowledgement

This work has been supported by the Centre National d'Etudes des Télécommunications.

References

Baraff G A 1984 Proceedings 3rd Conf. on Semiinsulating III-V Materials, Warm Springs, Oregon (USA) April 24-26, 1984. In press.
Bishop S G, Strom U and Taylor P C 1976 Solid State Comm. 18 573
Bois D and Vincent G 1977 J. Phys. Lett. 38 351
Coates R and Mitchell E W J 1975 Advances in Physics 24 593
Goltzene A, Meyer B and Schwab C 1983 Rev. Phys. Appliquée 18 703
Goltzene A, Meyer B and Schwab C 1984a Phys. Stat. Solidi b123 K125
Goltzene A, Meyer B and Schwab C 1984b Proceedings 3rd Conf. on Semiinsulating III-V Materials, Warms Springs, Oregon (USA) April 24-26, 1984. In press.
Goltzene A, Meyer B and Schwab C 1984c Proceedings 13th Conf. on Defects in Semiconductors, Coronado, California (USA) August 12-17, 1984. In press.
Goltzene A, Meyer B, Schwab C, Greenbaum S G and Wagner R J 1984d. To appear in J. Appl. Phys.
Goltzene A, Meyer B, Schwab C 1984e. Unpublished.
Goswami N K, Newman R C and Whitehouse J E 1981 Solid State Comm. 40 473
Huber A M, Linh N T, Valladon M, Debrun J L, Martin G M, Mitonneau A and Mircea A 1979 J Appl. Phys. 50 4022
Kennedy T A and Wilsey N D 1978 Phys. Rev. Lett. 41 977
Kennedy T A, Wilsey N D, Krebs J J and Stauss G H 1983 Phys. Rev. Lett. 50 1281
Lagowski J, Gatos H C, Parsey J M, Wada K, Kaminska M and Walukiewicz W 1982 Appl. Phys. Lett. 40 342
Lagowski J, Lin D G, Ayoyama T and Gatos H C 1984 Appl. Phys. Lett. 44 336
Magno R, Spencer M, Giessner J G and Weber E R 1984 Proceedings 13th Conf. on Defects in Semiconductors, Coronado, California (USA) August 12-17, 1984. In press.
Martin G M, Mitonneau A and Mircea A 1977 Electron. Lett. 13 191
Martin G M 1981 Appl. Phys. Lett. 39 747
Martin G M, Secordel P and Venger C 1982 J. Appl. Phys. 53 8706
Martin G M and Makram-Ebeid S 1983 Physica 116B 371
Martin G M, Esteve E, Langlade P and Makram-Ebeid S 1984a. To appear in J. Appl. Phys.
Martin G M and Makram-Ebeid S 1984b in Deep Defects in Semiconductors Ed. S. Pantelides, Gordon and Breach, New York. In press.
Meyer B K, Spaeth J M and Scheffler M 1984 Phys. Rev. Lett. 52 851
Omling P, Samuelson L and Grimmeiss H G 1984 Phys. Rev. B29 4534
Pötz W and Ferry D K 1984 Phys. Rev. B29 5687
Samuelson L, Omling P, Weber E R and Grimmeiss H G 1984 Proceedings 3rd Conf. on Semiinsulating III-V Materials, Warm Springs, Oregon (USA) April 24-26, 1984. In press.
Shanabrook B V, Klein P B, Swiggard E M and Bishop S G 1983 J. Appl. Phys. 54 336
Taniguchi M and Ikoma T 1982 Proceedings 2nd Conf. on Semiinsulating III-V Materials, Evian (F) April 19-21. Ed . Makram-Ebeid S and Tuck B, Shiva Publishing Ltd, Nantwich (UK) p. 283
Taniguchi M and Ikoma T 1984a Appl. Phys. Lett. 45 69
Taniguchi M, Mochizuki Y and Ikoma T 1984b Proceedings 3rd Conf. on Semiinsulating III-V Materials, Warm Springs, Oregon (USA) April 24-26, 1984. In press.
Van der Rest J and Pecheur P 1984 J. Phys. C Solid State Phys. 17 85
Vincent G and Bois D 1978 Solid State Comm. 27 431

Weber E R, Ennen H, Kaufmann U, Windscheif I, Schneider J and Wosinski T
  1982 J. Appl. Phys. 53 6140
Weber E R and Schneider J 1983 Physica 116b 398
Weber E R 1984 Proceedings of the 3rd Conf. on Semiinsulating III-V Mate-
  rials, Warm Springs, Oregon (USA) April 24-26, 1984. In press.
Wörner R, Kaufmann U and Schneider J 1982 Appl. Phys. Lett. 40 141

Inst. Phys. Conf. Ser. No. 74: Chapter 2
Paper presented at Int. Symp. GaAs and Related Compounds, Biarritz, 1984

41

Oxygen in GaAs; direct and indirect effects

H. C. Gatos, M. Skowronski, L. Pawlowicz and J. Lagowski

Massachusetts Institute of Technology, Cambridge, Massachusetts 02139

Abstract. Oxygen has profound effects on the key electronic properties and point defects of GaAs crystals. Thus, when added in
the growth system, it decreases the free electron concentration
and enhances the concentration of deep donors in the resulting
crystals. Both of these effects are highly beneficial for achieving
semi-insulating material, and they have been utilized for that purpose. They have been attributed to the tendency of oxygen to get-
ter silicon impurities during crystal growth. Only recently, we
have found that oxygen in GaAs introduces also a midgap level, ELO,
with essentially the same activation energy as EL2 but with four
times greater electron capture cross section. In the present report
we reassess the electrical and optical properties of the midgap
levels in GaAs crystals grown by the horizontal Bridgman (HB) and
the Czochralski-LEC techniques. Emphasis is placed on the identification of the specific effects of ELO.

1. Introduction

The importance of oxygen in the growth of GaAs crystal has been recognized
since the early 60's, primarily as a means of minimizing silicon contamination originating in the dissolution of the silica crucible by the GaAs
melt (Cochran and Foster 1962, Woods and Ainslie 1963). Addition of gallium or arsenic oxides to growth melt or enrichment of the growth ambient
with oxygen were utilized and remain practical solutions to Si contamination (Akai et al 1981, Kaminska et al 1981, Martin et al 1982). In 1978
the pyrolytic BN crucibles were introduced (Swiggard et al 1978) in the
liquid encapsulated Czochralski (LEC) technique which reduced contamination of Si below $10^{15} cm^{-3}$. However, even in LEC oxygen still remains a
common contaminant, since boric oxide used as the encapsulant of the GaAs
melt serves also as a source of oxygen.

In this paper we discuss the role of oxygen in the growth of GaAs from the
melt. We summarize the current views on the indirect effects of oxygen
on the crystal properties which relate to the Si gettering action, and we
present new results on oxygen-related midgap level ELO.

2. Indirect Effects of Oxygen

The concentration of oxygen dissolved in the GaAs melt is expected to be
very low due to the high volatility of gallium oxide. Thus, even if Ga_2O_3
is intentionally added to the GaAs charge material, most of the oxygen
will be transferred to a vapor phase according to the reaction:

$$Ga_2O_3 + 4Ga_{(in\ melt)} = 3Ga_2O_{(vapor)} \qquad (1)$$

The partial vapor pressure of Ga_2O in the growth system is critical in the suppression of the dissolution of the quartz boat or the purification of the melt (Cochran and Foster 1962, Woods and Ainslie 1963):

$$SiO_2 + 4Ga_{(in\ melt)} \rightleftarrows Si_{(in\ melt)} + 2Ga_2O_{(vapor)} \qquad (2)$$

When the reaction proceeds from right to left, the melt is being purified (i.e., silicon gettering). In a closed ampoule, utilized in HB growth, the silicon concentration in the melt, C_{Si}^{melt}, typically varies from $10^{18}cm^{-3}$ to $< 10^{15}cm^{-3}$, depending upon the partial pressure of added Ga_2O (Kaminska et al 1981). The corresponding range of Si concentration in the grown crystals ($C_{Si}^{GaAs} = K_{eff}C_{Si}^{melt}$ where the effective segregation coefficient $K_{eff} \simeq 0.5$, Kaminska et al 1981) becomes $5 \times 10^{17}cm^{-3}$ to $<5 \times 10^{14}cm^{-3}$. It should be noted that the removal of Si from the melt is by no means restricted to the HB technique. The addition of Ga_2O_3 to LEC growth systems will also lead to purification of GaAs crystals in accord with reactions 1 and 2.

Si in GaAs is an amphoteric dopant which can occupy either Ga or As sites and thus act as a donor or an acceptor, respectively. In melt-grown GaAs crystals, Si acts predominantly as a donor. Thus, the addition of oxygen to the growth system has an important electronic consequence, namely it decreases the free electron concentration and increases the free carrier mobility. Other effects of oxygen listed in Table I are far more complex, although the suppression of silicon donors governed by reaction (2) is probably directly or indirectly involved. Effects 3 to 5 have been recognized only recently, and they are the subject of current studies.

Table I. Indirect effects of oxygen on GaAs properties.

Effect	Probable Cause	Reference
1. Decrease in electron concentration & increase in mobility	Decrease of Si donor concentration brought about by reaction (2)	Cochran & Foster 1962; Kaminska et al 1981; Akai et al 1981; Martin et al 1982
2. Elimination of deep acceptor luminescence band (1.18 eV)	Decrease of the concentration of Si complexes with native defects	Kaminska et al 1981
3. Increase of dislocation density	Enhancement of dislocation loop formation due to Fermi energy shift brought about by elimination of Si donors	Lagowski et al 1984 b
4. Appearance of deep level luminescence band (0.63 eV)	Oxygen-related midgap levels	Yu and Walters 1982
5. Enhancement of midgap traps	Enhancement of EL2 concentration; oxygen-related midgap level ELO	Kaminska et al 1981; Lagowski et al 1984 a; present work

The oxygen effect on the EL2 and the dislocation concentrations can be generally classified as "Fermi Energy Effects". They are due to the Fermi energy change brought about by the decrease of the Si (donor) concentration, which in turn affects the charge state of the native defects, their subsequent interactions during cooling of the crystal, and thus the final defect structure in as-grown GaAs crystals (see Lagowski and Gatos 1984).

3. Oxygen-Related Midgap Level ELO

The complex electrical and optical properties of "undoped" semi-insulating GaAs have been generally interpreted in terms of "the one deep level model" with only one midgap donor referred to as EL2. However, recent photoluminescence (Yu and Walters 1982) and photocapacitance studies (Taniguchi and Ikoma 1982) have implied the presence of midgap levels other than EL2, and especially that of the oxygen-related level ELO in melt-grown GaAs.

The primary difficulty in studying ELO stems from the striking similarity between its parameters and those of EL2. Accordingly, when both ELO and EL2 are simultaneously present, their unique distinction requires reference to standard material containing only one midgap level. In our study the crystals grown without oxygen doping and containing only EL2 were used as reference. DLTS spectra of such crystals are shown in Fig. 1a, while Fig. 1b presents the corresponding DLTS spectra of oxygen-doped GaAs. In spectra 1 to 4 filling pulse duration decreases from 200 ns (which saturates the deep levels, spectrum 1) to about 10 ns (spectrum 4). In Fig. 1a the decrease of the filling pulse duration causes a decrease in the peak height; however, there is no change in peak position nor in its half-width; this result is consistent with involvement of only one deep level.

In oxygen-doped GaAs shortening of the filling pulse causes a shift of the DLTS peak to lower temperatures and a decrease of the peak of half-width. This behavior proves that in oxygen-doped GaAs there are two levels, EL2 and ELO present together. ELO exhibits a higher value of electron capture cross section, and its relative contribution to the overall DLTS signal is enhanced by short duration filling pulses, because it is preferentially filled by such pulses.

Using this type of DLTS measurement as well as direct analysis of capacitance transients, we have studied over 1,000 samples from about 100 GaAs crystals grown in our laboratory using HB and LEC techniques. We have found (see Fig. 2) that for both crystal growth methods EL2 and ELO exhibit exactly the same emission rates. It is also seen from Fig. 2 that both levels have very similar activation energies of electron emission rate; however, ELO has about four times larger electron capture cross section than EL2.

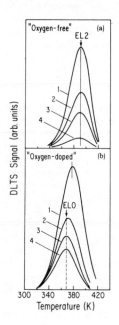

Fig. 1. DLTS spectra of midgap levels in (a) oxygen-free and in (b) heavily oxygen-doped GaAs. Spectra 1-4 correspond to decreasing filling pulse duration.

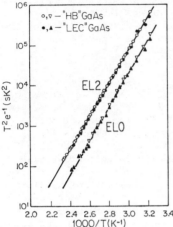

Fig. 2. Thermal activation plot of emission rate for EL2 and ELO

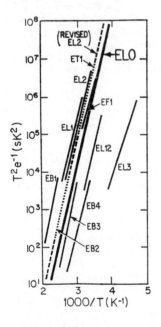

Fig. 3. Emission rate for deep electron traps in GaAs (see text)

In Fig. 3 we compare literature data on deep levels (compiled by Martin et al 1977) with results of our study on EL2 and ELO. Our results on EL2 are in excellent agreement with early results obtained at Bell Laboratory ("EB2") and at the University of Tokyo ("ET1"). They are, however, slightly different than those designated as EL2 (Laboratoires d'Electronique et de Physique Appliquée) and taken so far as "standard". We consider our results (which span about seven orders of magnitude, i.e., a much greater range than those of any previous study) as more accurate than the results of Martin et al 1977. We propose that they be adopted as standard revised parameters of EL2 and also for ELO.

4. Optical Properties of Midgap Levels

Optical absorption in GaAs has been utilized as a convenient method for the determination of EL2 concentration and also as a fundamental tool for the study of the origin of EL2 (Martin 1981). It has not been realized, however, that in the GaAs crystals studied both levels EL2 and ELO rather than just EL2 might have been present.

We have carried out an extensive study of optical absorption of oxygen-doped and oxygen-free crystals which has shown that the optical properties of EL2 and ELO are very similar. As shown in Fig. 4, both crystals exhibit a characteristic optical absorption band which extends from about 0.8 to 1.3 eV. The absorption and the photoionization cross section spectra are different due to the contribution from intracenter transitions (dotted areas) to the absorption but not to the photoionization. A high resolution study of these intracenter transitions has shown that the band begins with a no-phonon line followed by transverse acoustic phonon replicas (Kaminska et al 1983). As shown in Fig. 5, the no-phonon line in oxygen-doped crystals is composed of two symmetrical gaussian lines separated by about 1.5 meV. This result clearly indicates the presence of two centers which could not be resolved in the overall absorption spectra.

From a practical point of view it is of significance to emphasize the ambiguity involved in the routine determination of the EL2 concentration from optical absorption measurements (Martin 1981). The fact that more than one deep level is involved leads to a

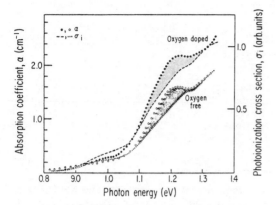

Fig. 4. Absorption and photoionization cross section spectra for oxygen-doped and oxygen-free GaAs.

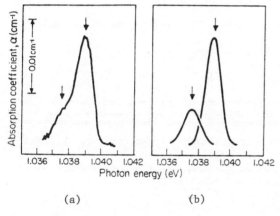

(a)   (b)

Fig. 5. No-phonon line of intracenter midgap level transitions in oxygen-doped GaAs; (a) - experimental, (b) - deconvoluted into two gaussian lines.

considerable spread in the experimental data of the absorption coefficient vs. the total concentration of midgap levels. Still, as seen in Fig. 6, the absorption coefficient at $\lambda \approx 1.09$ μm exhibits an overall tendency to increase with increasing total concentration of midgap levels, and thus it can be used only for a rough estimation of such concentrations.

Midgap levels other than EL2, but with optical properties very similar to those of EL2 (levels referred to as "EL2 family"), have recently been reported to exhibit a photocapacitance quenching effect (Taniguchi and Ikoma 1984), which has been treated in the past as the "fingerprint" of EL2. In our study we have also found that ELO (as well as EL2) exhibits photocapacitance quenching.

5. Comments on the Possible Origin of ELO in the Light of Results on Isoelectronically Doped GaAs Crystals

On the basis of the results discussed above, one could conclude that the oxygen is directly involved in the formation of ELO. Such a conclusion, however, cannot be fully justified in view of the results of our most recent study with GaAs crystals doped heavily with In and with oxygen. In these crystals (dislocation free) no ELO could be detected, and the EL2 concentration was reduced to the low 10^{15}cm^{-3} range. Since these crystals are expected to contain reduced concentrations of native defects, it is conceivable that ELO is a complex involving native defects, in which case the role of oxygen must be considered indirect.

It should be pointed out, however, that theoretical calculations (Jaros 1980) have indicated a midgap level of oxygen impurity located on the arsenic site. Recent theoretical studies of oxygen levels in GaAs$_{1-x}$P$_x$ indicated that an energy level of O_{As} in GaAs (i.e., x = 0) would be located at about 0.79 eV below the conduction band edge (Wolford et al 1983). This energy is very similar to the ELO activation energy of 0.825 eV. SIMS analysis of our crystals has indicated an oxygen concentration of about 1 to 2 x 10^{16}cm^{-3}, which is comparable to the concentration of midgap levels.

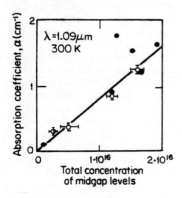

Fig. 6. Optical absorption of midgap levels: open circles after Martin 1981; closed circles - present results.

Further studies with GaAs doped with oxygen isotope ^{18}O rather than ^{16}O, utilizing the sensitivity of SIMS determination of ^{18}O concentrations, will possibly resolve the questions regarding the origin of ELO.

Acknowledgement

The authors are grateful to the National Aeronautics and Space Administration and to the Office of Naval Research for financial support.

References

1. Akai S, Fujita K, Sasaki M and Tada K 1981 Inst. Phys. Conf. Ser. 63 13.
2. Cochran C N and Foster L M 1962 J. Electrochem. Soc. 109 149.
3. Jaros M 1980 Adv. Phys. 29, 409.
4. Kaminska M, Lagowski J, Parsey J, Wada K and Gatos H C 1981 Inst. Phys. Conf. Ser. 63 197.
5. Kaminska M, Skowronski M, Godlewski M and Kuszko W 1983 presented at 4th Lund Conf. on Deep Level Impurities in Semiconductors, Eger, Hungary.
6. Lagowski J, Lin D G, Aoyama T and Gatos H C 1984a Appl. Phys. Lett. 44 336, 1984b Appl. Phys. Lett. in press.
7. Lagowski J and Gatos H C 1984 Proc. 13th Int. Conf. on Defects in Semiconductors, Coronado, California, in press.
8. Martin G M 1981 Appl. Phys. Lett. 39 747.
9. Martin G M, Jacob G, Hallais J P, Grainger F, Roberts J A, Clegg B, Blodd P and Poiblaud G, 1982 J. Phys. C. 15 1841.
10. Martin G M, Mittonneau A and Mircea A 1977 Electron Lett. 13 191.
11. Swiggard E M, Lee S H and Von Batchelder F W 1978 Inst. Phys. Conf. Ser. 45 125.
12. Taniguchi M and Ikoma T 1982 Inst. Phys. Conf. Ser. 65 65.
13. Taniguchi M and Ikoma T 1983 J. Appl. Phys. 54 6448.
14. Taniguchi M and Ikoma T 1984 Appl. Phys. Lett. 45 69.
15. Wolford D J, Modesti S and Streetman B G 1983 Inst. Phys. Conf. Ser. 65 501.
16. Woods J F and Ainslie N G 1963 J. Appl. Phys. 34 1469.
17. Yu P W and Walters D C 1982 Appl. Phys. Lett. 41 863.

Inst. Phys. Conf. Ser. No. 74: Chapter 2
Paper presented at Int. Symp. GaAs and Related Compounds, Biarritz, 1984

47

Microscopic distribution of deep and shallow levels around dislocations in undoped SI-GaAs

T. Kikuta, T. Katsumata, T. Obokata, and K. Ishida

Optoelectronics Joint Research Laboratory
1333 Kamikodanaka, Nakahara-ku, Kawasaki 211, JAPAN

Abstract Microscopic distributions of deep and shallow level concentrations around dislocations in undoped SI-GaAs have been studied
by low temperature photoluminescence (PL), two dimensional infrared
absorption imaging and resistivity mapping. In the cell boundaries
of dislocation networks, the concentration of EL2 is high, while
that of the killer center for near band emissions is low. The
analyses of shallow PL bands suggest the increase in the shallow
donor and acceptor concentrations around dislocations. The
resistivity in the cell boundary is more than one order as low as
that in the interior of the cell. This is explained by the increase
in the EL2 concentration.

1. Introduction

Undoped semi-insulating (SI) GaAs is a promising substrate for GaAs IC
fabrication using the direct ion implantation technique. Attainment of
microscopic as well as macroscopic uniformity in the substrate is
required for the GaAs ICs.

Microscopic distributions around dislocations of undoped SI-GaAs have
been examined using low temperature photoluminescence (PL) (Heinke and
Queisser 1974, Kitahara et al. 1982), two dimensional near infrared (IR)
imaging (Skolnick et al. 1984) and resistivity mapping (Matsumura et al.
1984) methods. Results of these studies demonstrated the presence of
microscopic inhomogeneity associated with dislocations. However,
variations of the deep and shallow level concentrations around
dislocations have not been reported. In the present paper, we describe
the microscopic variations of the deep and shallow level concentrations
by low temperature PL and near IR absorption, and discuss the correlation
between these variations and resistivity decrease observed around
dislocations.

2. Experimental

The wafers used in this study were taken from the body of undoped SI-GaAs
grown from PBN crucibles by the direct synthesis LEC technique; the 3 – 4
mm thick wafer was examined by near IR absorption and the adjacent wafers
were studied by PL, optical microscopy/chemical etching and resistivity
mapping. Photoluminescence measurements were carried out at 15 K. The

samples were mounted in a cryostat with a closed-cycle refrigerator, which was set on a micrometer stage. A Kr laser operating at a wavelength of 647.1 nm with a maximum intensity of ∼120 mW was used as an excitation source. The beam diameter was ∼100 μm. The excitation intensity was varied about two orders of magnitude by means of neutral density filters. The shallow and deep emission spectra were analyzed by 1 m and 20 cm monochromators, respectively. Etch pit patterns of dislocation networks were revealed by photoetching with diluted AB etchant and molten KOH. Two dimensional infrared absorption images at 1 μm were observed using a silicon visicon TV camera. The video signal from the camera was stored in the memory and the quality was improved using the integration and enhancement techniques. Two dimensional resistivity mapping was performed by the three-electrode guard method (Amey and Hamburger 1949) using automated test equipment. The diameter of the electrode was 45 μm and the interval of the center of each electrode was 70 μm.

3. Results and Discussion

Figure 1 shows a photograph of the etch pit pattern revealed by the photoetching and the PL intensity variations of shallow (1.49 eV) and deep (0.80, 0.65 eV) emission bands along the broken line. In the cell boundary region, where the dislocation density is high, the intensities of the 0.65 eV and 1.49 eV bands increase, while that of the 0.80 eV band

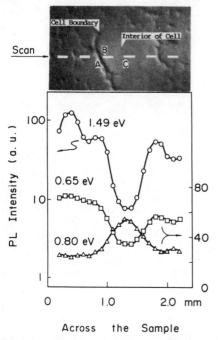

Fig. 1   PL intensity variations of 1.49 eV, 0.80 eV and 0.65eV emission bands and etch pit pattern. PL scan is performed along the broken line.

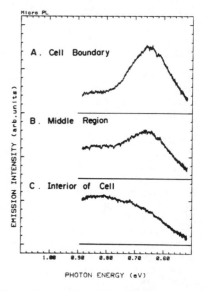

Fig. 2   Deep PL spectra at the position A, B and C indicated by arrows in Fig. 1.

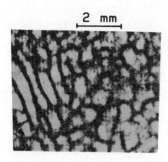

Fig. 3   Infrared absorption image at 1 μm band. The dark regions, where abosrption coefficient is large, correspond to the cell boundaries.

decreases. The opposite tendency is observed in the dislocation free region of the interior of the cell. The spectra of the deep emission bands, observed at the position A, B and C indicated by the arrows in Fig. 1, are shown in Fig. 2. The distinct change of the dominant emission peak is observed.

The infrared absorption image is shown in Fig. 3. A comparison with the etch pit pattern showed that the dark regions, where the absorption coefficient is large, correspond to the cell boundaries. Since the absorption coefficient at 1 μm band is proportional to the EL2 concentration N_{EL2} (Martin 1981), the above result shows that the EL2 concentration is high in the cell boundary, while low in the interior of the cell.

The 0.65 eV band has been reported to be associated with the EL2 occupied by electrons (Mircea-Roussel and Makram-Ebeid 1981). Therefore, the increase in the 0.65 eV band intensity in the cell boundary is consistent with high EL2 concentration observed by near IR absorption. This tendency has been also observed for the macroscopic variation along the wafer diameter (Tajima 1982, Kikuta et al. 1983, Holmes et al. 1982). Furthermore, we have confirmed that the photoquenching effect by YAG laser irradiation was consistently observed in both the 0.65 eV PL band (Leyral et al. 1982) and 1 μm absorption band (Martin 1981).

As shown in Fig. 1, the intensity of the 1.49 eV band, which is reported to be related to carbon (Ozeki 1974), varies inversely to that of the 0.80 eV band. This correlation has also been observed macroscopically (Tajima 1982) and the intensity of the 1.49 eV band in the 0.65 eV dominant crystals is about 20 times as strong as that in the 0.80 eV dominant crystals (Kikuta et al. 1983). However, we have found that the carbon concentration obtained by FT-IR is usually higher in the 0.80 eV dominant crystals than in the 0.65 eV dominant crystals. This result clearly shows that the intensity of the 1.49 eV band is not proportional to the carbon concentration, and that the level related to the 0.80 eV band acts as a killer center for the 1.49 eV emission band. Therefore, these results suggest that the concentration of the killer center is high in the interior of the cell, while low in the cell boundary.

Three emission bands at 1.514 eV, 1.494 eV and 1.491 eV are usually observed in undoped SI-GaAs. These are the neutral-donor-to-valence-band (D^{o}, V), conduction-band-to-neutral-acceptor (e, A^{o}), and neutral-donor-to-neutral-acceptor (D^{o}, A^{o}) transitions, respectively. In Fig. 4(a), the intensity variations of these bands are shown as a function of excitation intensity. The emission intensities in the cell boundary are about ten times as strong as those in the interior of the cell.

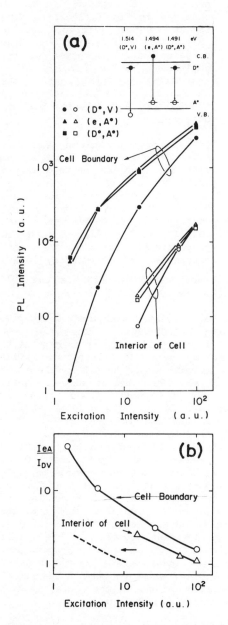

The emission intensities of (D^0, V) and (e, A^0) are approximated by (Nam et al. 1977)

$$I_{eA} = C_1 \cdot n \cdot n_A ,$$
$$I_{DV} = C_2 \cdot p \cdot n_D ,$$

where n_A, n_D, n and p are the concentrations of the neutral acceptors, neutral donors, electrons in the conduction band, and holes in the valence band, respectively; C_1 and C_2 are constant. In the high excitation limit, it is expected that $n_A \to N_A$, $n_D \to N_D$ and $p \to n$, and thus, N_A/N_D can be obtained by measuring I_{eA}/I_{DV} as a function of the excitation intensity (Yu 1978). Figure 4(b) shows the I_{eA}/I_{DV} of the data in Fig. 4(a). In the high excitation region, both curves exhibit a similar slope and the value of I_{eA}/I_{DV} in the cell boundary is slightly larger than that in the interior of the cell. However, since the concentration of the killer center is high in the interior of the cell, the effective excitation intensity in the interior of the cell is expected to be lower than that in the cell boundary, as shown by the broken line, and I_{eA}/I_{DV} becomes even smaller in the interior of the cell. This tendency seems to keep in the high excitation limit. Therefore, the above results suggest that N_A/N_D in the cell boundary is larger than that in the interior of the cell. However, it is difficult to estimate the value of N_A/N_D quantitatively (Kamiya and Wagner 1977), since the concentration of the killer center is inhomogeneously distributed in SI-GaAs crystals.

Recently, Kuramoto et al. (1983) have shown that the residual shallow donor Si and S concentrations measured by SIMS closely correlate to the etch pit density along the wafer diameter. This result suggests the shallow donor concentration N_D increases

Fig. 4(a) Intensity variation of (D^0, V), (e, A^0) (D^0, A^0) transitions in the cell boundary and the interior of the cell, (b) intensity ratio of (e, A^0) to (D^0, V) as a function of excitation intensity.

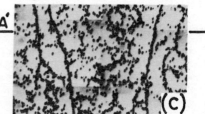

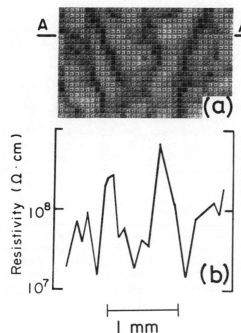

Fig. 5(a) Two dimensional resistivity distribution. The
resistivity is low in the dark
regions. (b) One dimensional line
scan taken along the line A-A'.
(c) Etch pit pattern revealed by
molten KOH.

in the cell boundary. Therefore,
the increase in N_A/N_D and N_D
shows that the shallow acceptor
concentration N_A and N_A-N_D also
increase in the cell boundary.

Figure 5(a) shows the two dimensional resistivity distribution.
The resistivity is low in the
dark regions. The one dimensional line scan taken along the
line A-A' is shown in Fig. 5(b).
The etch pit pattern revealed by
molten KOH correlates to the
resistivity distribution, as
shown in Fig. 5(c). The
resistivity in the cell boundary
is about one orde as small as
that in the interior of the cell.
Figure 6 shows the calculated
data of resistivity as a function
of the EL2 concentration N_{EL2}
when N_A-N_D equals 5×10^{14} cm^{-3} and
2×10^{15} cm^{-3} (Johnson et al. 1983,
Martin et al. 1980, Walukiewicz
et al. 1983). Considering the
observed increase in N_{EL2} and N_A-N_D
in the cell boundary and the
calculated results in Fig. 6, we
conclude that the EL2 concentration in the cell boundary should
be more than ten times as high as
that in the interior of the cell
to account for the decrease in
resistivity.

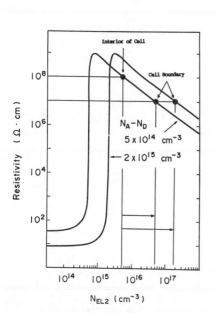

Fig. 6 Calculated data of resistivity as a function of the EL2
concentration N_{EL2} when N_A-N_D is
5×10^{14} cm^{-3} and 2×10^{15} cm^{-3}.

4. Summary

Microscopic distributions of deep and shallow level concentrations around dislocations were studied by low temperature PL, two dimensional infrared absorption imaging and resistivity mapping. In the cell boundary, where dislocation density is high, the concentration of EL2 is high, while that of the killer center for near band emissions is low. The opposite tendency is observed in the dislocation free regions of the interior of the cell The resistivity in the cell boundary is more than one order as low as that in the interior of the cell. The analyses of the intensity ratio (e, A^O)/(D^O, V) of the shallow PL bands indicate that the concentration ratio of the shallow acceptors to donors is large in the cell boundary. To explain the resistivity decrease, the EL2 concentration in the cell boundary should be more than ten times as high as that in the interior of the cell.

Acknowledgement

The authors wish to thank T. Iizuka and M. Hirano for useful discussions. The present research effort is part of a major research and development project on optical measurement and control systems, conducted under a program set up by the Ministry of International Trade and Industry's Agency of Industrial Science and Technology.

References

Amey W G and Hamburger F Jr 1949 Proc. Am. Soc. Testing Mater. 49 1079
Heinke W and Queisser H J 1974 Phys. Rev. Lett. 33 1982
Homes D E, Elliott K R, Chen R T and Kirkpatrick C G 1982 Semi-
    Insulating III-V Materials, Evian (Shiva Publishing Limited England)
    p.19
Johnson E J, Kafalas J A and Davies R W 1983 J. Appl. Phys. 54 204
Kamiya T and Wagner E 1977 J. Appl. Phys. 48 1928
Kikuta T, Terashima K and Ishida K 1983 Jpn. J. Appl. Phys. 22 L409
Kitahara K, Nakai K and Shibatomi S 1982 J. Electrochem. Soc. 129, 881
Kuramoto K, Nakajima M, Kikuta T, Orito F, Emori H and Ishida K 1983
    Proc. of Fourth International Conference on Secondary Ion Mass Spectro-
    metry, Osaka (Springer Verlag) p.288-290
Leyral P, Vincent G, Nouaillat A and Guillot G 1982 Solid State Commun.
    42 67
Martin G M, Farges J P, Jacob G, Hallais J P and Porblaud G 1980
    J. Appl. Phys. 51 2840
Martin G M 1981 Appl. Phys. Lett. 39 747
Tajima M 1982 Jpn. J. Appl. Phys. 21 L227
Matsumura T, Obokata T and Fukuda T 1984 to be published in J. Appl.
Phys.
Mircea-Roussel A and Makram-Ebeid S 1981 Appl. Phys. Lett. 38 1007
Nam S B, Langer D W, Kingston D L and Luciano M J 1977 Appl. Phys.
    Lett. 31 652
Ozeki M, Nakai K and Ryuzan O 1974 Jpn. J. Appl. Phys. 13 1121
Skolnick M S, Brozel M R, Reed L J and Grant I 1984 J. Electronic
    Mater. 13 107
Walukiewicz W, Lagowski J and Gatos H C 1983 Appl. Phys. Lett. 43 192
Yu P W 1978 Solid State Commun. 27 1421

Distribution of EL2 concentration in dislocated and dislocation-free GaAs substrates before and after annealing

Sylvie MARTIN, Marc DUSEAUX, Marko ERMAN

Laboratoires d'Electronique et de Physique Appliquée
3, avenue Descartes, 94450 LIMEIL-BREVANNES, France

Abstract :A comparative study of the distribution of the dominant deep trap EL2 in different crystals before and after annealing is presented. The dislocated as-grown Bridgman crystal is more homogeneous than the LEC one, but they both present the same average value ($11.10^{15}cm^{-3}$) of EL2 concentration. The dislocation-free In doped LEC crystal presents a EL2 concentration two or three times lower. Annealing for 60 hours at 900°C leads to an identical fully homogeneous EL2 concentration around $12.10^{15}cm^{-3}$ for all these crystals. This result is not consistent with the hypothesis already advanced of diffusion of the EL2 defect during annealing and strongly supports the relationship between the EL2 defect density and the thermodynamical equilibrium of the lattice. It is shown that GaAs substrates can be obtained free of dislocations by indium doping and homogeneous from the defect's point of view by further post-growth annealing at 900°C. These materials should meet the severe requirements of homogeneity for IC fabrication.

1. Introduction

Recently, it has been found that the characteristics of FET's elaborated on dislocated LEC GaAs wafers were inhomogeneous at the macroscale (Nanishi 1982 a, b) and at the microscale (Miyazawa 1983 a) as well.

For example, the normalized threshold voltage of FET's follows a W-shape across a diameter of the slice (macroscale variation). This W-shape is well known to crystal growers since many physical parameters of the ingot follow this shape : the dislocation density, the EL2 concentration (Martin G.M. 1980), or 1.49 eV photoluminescence intensity (Yokogawa 1984). These variations can themselves induce other variations such as that of the resistivity (Bonnet 1982).

So, the inhomogeneities of FET's characteristics can obviously be correlated with at least one of these fluctuations. The problem is to determine with which one. The situation is more complicated if one reasonably assumes that the physical fluctuations are themselves correlated together.

However, some experiments bring some light to this question. Actually, if the crystal is annealed in the 800-900°C range for several hours, the inhomogeneities of FET's and of EL2 distribution are smoothed (Niyazawa 1983, Rumsby 1982). On the other hand, the dislocation distribution is not modified. This fact supports the correlation between FET's properties and EL2 (resistivity) concentration. Moreover, it has been shown (Duseaux 1984) that the EL2 concentration is very low ($4-5 \times 10^{15}cm^{-3}$) in dislocation free

In doped crystals. It was suggested that this low concentration level corresponds to the background thermodynamical equilibrium of EL2. Larger values noticed in standard materials would then correspond to an excess of EL2 generated by dislocations.

All these facts suggest that it is interesting to study the concentrations and distributions of EL2 depending on the growth conditions, and the dislocation density before and after annealing. We present here the results we obtained by infrared absorption measurements and infrared imaging on dislocated Bridgman, dislocated and In doped dislocation free LEC GaAs crystals.

2. Experimental techniques

The distribution of EL2 has been deduced from whole slices infrared imaging using a silicon vidicon television set-up (Skolnick 1984). The images on the monitor were directly photographed. Furthermore, quantitative assessment of the same slices has been obtained from the 1.1 µm absorption coefficient determination (Martin 1980) in a Varian 2300 system. The measurements were made every 5 mm. The positions of the Fermi level given by Hall effect measurements indicated that in all cases, EL2 is practically 100 % occupied. 4 mm thick slices coming from four different ingots have been studied before and after annealing at 900°C for 60 hours. The growth conditions and physical properties of the as-grown slice are summarized in Table I.

Slice	Crystal growth conditions	Electrical properties	Dislocation density (cm^{-2})	Position in the ingot
1	LEC undoped	SI n tendency	10^4-10^5	(100) cross section
2	LEC undoped	SI n tendency	10^4-10^5	(110) longitudinal section
3	LEC In doped ($9.10^{19}at/cm^3$)	SI n tendency	< 500	(100) cross section
4	Bridgman Si doped	$N_D-N_A = 10^{16}cm^{-3}$	$< 10^4$	(110) section

Table I : Growth conditions and physical properties of ingots which have been studied.

3. Results

The contour plots of the EL2 distribution on as-grown and annealed slices are presented in fig. 1. The a, c, e, g maps (left part of each slice) concern the n° 1, 2, 3, 4 as grown slices respectively. The b, d, f, h maps (right part of each slice) concern the same slices after annealing. The two dimensional infrared images obtained on the as-grown n° 1, 2, 3, 4 slices are presented in fig. 2a, b, c, d respectively. The fig. 1 gives quantitative values, but averaged over a surface of 5 mm², and it is thus useful to look at   figure 2 which gives a more detailed picture of the wafer uniformity.

Our results on the LEC dislocated as-grown material are in agreement with those of Holmes (1984) and Skolnick (1984). We also obtain the four-fold symmetrical pattern of EL2 concentration, its W-shaped radial variation across a (100) slice reported by Holmes (fig. 1a, 1c), and the straight lines of high infrared absorption lying in [110] directions reported by Skolnick (fig. 2a). Nevertheless, our reported EL2 concentrations are lower.

In the dislocation-free LEC indium doped crystal, the EL2 concentration is two or three times lower (Duseaux 1984) than in dislocated material and seems to be more homogeneous (fig. 1c and 2c). Nevertheless, since a fluctuation of 10 % in EL2 concentration corresponds only to a variation of about 4.10^{14} cm^{-3} in the dislocation-free slice (against 10^{15} cm^{-3} in dislocated slice), our assessment is less accurate in that case. A high infrared absorption zone lying on an [110] direction is seen near the periphery: it corresponds to the location where some dislocations appear.

Those results support the hypothesis of the formation of EL2 in a two step process : a low EL2 concentration (certainly not very different of the concentration obtained in the dislocation-free indium doped crystal) is determined by a thermodynamical equilibrium and corresponds to a background of EL2 which is further enhanced by the presence of the dislocations (Duseaux 1984, Holmes 1984).

The figure 1g leads to the conclusion that the EL2 concentration in the dislocated Bridgman slice is, at the macroscopic scale, fully homogeneous. Moreover, the value of this concentration is equal to the average one present in the dislocated LEC crystal. Nevertheless,   fig. 2d shows that fluctuations at short distances are still present.

The results presented in fig. 1b, d, f, h prove that annealing leads to an identical and fully homogeneous EL2 concentration around 12.10^{15} cm^{-3} for all the slices. That means that the growth conditions have little, if any, effect on the EL2 concentration obtained after annealing. This observation has never been reported so far, and may turn out to have important practical applications.

4. Discussion and conclusion

So far, the studies on the EL2 distribution before and after annealing have been made on LEC semi-insulating dislocated material. Our results presented in fig. 1a and 1b are in agreement with those published before. Since, in that case, the average EL2 concentration is quite the same before and after annealing, the conclusions were that the annealing induces a diffusion of EL2 which then becomes homogeneously distributed (Rumsby 1983). Our results obtained on the dislocated-free LEC crystal which presents after annealing a mean enhancement of 8.10^{15} cm^{-3} in the EL2 concentration are not consistent with this interpretation. Moreover the fact that the slices, quite different before annealing, become equivalent from the point of view of the EL2 concentration after annealing, brings a new light on the mechanisms of EL2 formation. Since the annealing procedure has no effect on the dislocations (no significant difference in their distribution on the LEC dislocated crystal, no creation in the LEC dislocation-free crystal) it is clear that EL2 is not only related to the dislocations. Furthermore, the fact that the slices stayed 60 hours at 900°C allows us to say that at this temperature a thermodynamical equilibrium was achieved and that the concentration of EL2 observed is typical of this equilibrium and completely independent on the presence of the dislocations.

On the other hand, it is well known that gettering effects take place around dislocations (Kamegima 1982, Kirkby 1975). Some defects can be created during dislocation motion (Holmes 1983, 1984, Weber 1982) at high temperature and impurities or defects can pin the dislocations in order to minimize the strain energy. Since EL2 is enhanced around the dislocations in the as-grown dislocated crystal, EL2 is certainly one of these defects. Our results show that annealing erases the segregation of EL2 around the dislocations, but it is not sure that the annealing homogenizes the distribution of the other impurities. Then we cannot yet make sure that the annealing on the dislocated slices suppresses all the inhomogeneities which could have detrimental effects on the integrated circuit performance.

In that sense, the best material is clearly a materia which is free of dislocations and homogeneous from the point defect point of view. We have shown that this can actually be achievable by indium doping and further post-growth annealing at 900°C.

References

Bonnet M, Visentin N, Gouteraux B, Lent B, Duchemin J P, 1982 Proc. of the GaAs IC symposium, New Orleans.

Duseaux M. and Martin S. 1984 Proc. of the S.I. III-V Materials Conference Oregon.

Holmes D E, Chen R T, Elliot K R and Kirkpatrick C G, 1983 Appl. Phys. Lett. 43 pp 305-7

Holmes D E and Chen R T 1984 J. Appl. Phys. 55 pp 3588-94

Kamegima T, Shimura F, Matsumoto Y, Watanabe H and Mitsui J, 1982 J. Appl. Phys. 21 p L721

Kirkby P A, 1975 IEEE J. Quant. Elect. 11 pp 562-8.

Martin G M, Jacob G, Gotzene A and Schwab C   1980   Proc. 11th Int. Conf. on Radiation Effects in Semiconductor, Oiso, Tokyo.

Miyazawa S, Ishii Y, Ishida S and Nanishi Y 1983a Appl. Phys. Lett. 43 pp 853-5

Miyazawa S, Honda T, Ishii Y and Ishida S   1983b   Appl. Phys. Lett. 44 pp 410-2

Nanishi Y, Ishida S, Honda T, Yamazaki H and Miyazawa S 1982a Jpn J. Appl. phys. 21 pp L335-7

Nanishi Y, Ishida S and Miyazawa S 1982b Jpn J. Appl. Phys. 22 pp L54-6

Rumsby D, Ware R N, Smith B, Tyjberg M, Brozel N R and Foulkes E J 1983 Technical Digest of 1983 GaAs IC Symposium, Phoenix (Arizona) pp 34-6

Skolnick M S, Brozel N R, Reed L J, Grant I, Stirland D J and Ware R N 1984 J. Elec. Mat. 13 pp 107-125

Weber E R, Ennen H, Kaufmann V, Windscheif J, Schneider J and Wosinski T, 1982, J. Appl. Phys. 53 pp 6140-3

Yokogawa M, Nishine S, Sasaki M, Matsumoko K, Fujita K and Akai S   1984 Jpn J. Appl. Phys. 23 pp L339-41

Figure captions

FIGURE 1 : EL2 maps
    a) : Dislocated LEC (100) slice before annealing
    b) : after annealing
    c) : dislocated LEC longitudinal slice before annealing
    d) : after annealing
    e) : Dislocation-free LEC (100) slice before annealing
    f) : after annealing
    g) : Dislocated Bridgman (110) slice before annealing
    h) : after annealing
FIGURE 2 : Infrared images
    a) : dislocated LEC (100) slice
    b) : dislocated LEC longitudinal slice
    c) : dislocation-free LEC (100) slice
    d) : dislocated Bridgman (110) slice

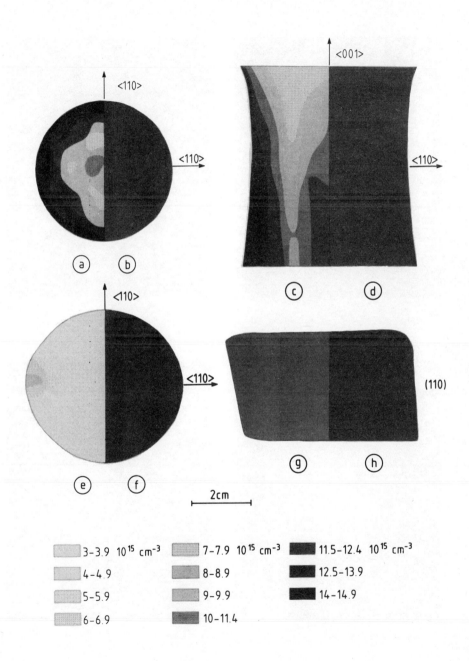

3-3.9 10^{15} cm^{-3} 7-7.9 10^{15} cm^{-3} 11.5-12.4 10^{15} cm^{-3}

4-4.9   8-8.9   12.5-13.9

5-5.9   9-9.9   14-14.9

6-6.9   10-11.4

figure 1

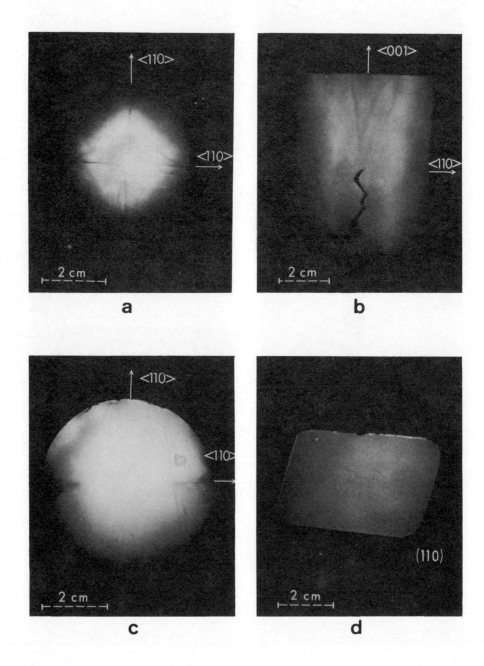

figure 2

Inst. Phys. Conf. Ser. No. 74: Chapter 2
Paper presented at Int. Symp. GaAs and Related Compounds, Biarritz, 1984

59

Three-dimensional imaging of the distribution of 1 μm absorption (EL2) in undoped, semi-insulating, LEC GaAs

M R Brozel*, E J Foulkes*+ and D J Stirland**

*Department of Electrical and Electronic Engineering, Trent Polytechnic, Burton Street, Nottingham, NG1 4BU, U.K.

present address
+Cambridge Instruments Ltd., Rustat Road, Cambridge, CB1 3HQ, U.K.

**Plessey Research(Caswell) Ltd., Allen Clark Research Centre, Caswell, Towcester, Northants, NN12 8EQ, U.K.

Abstract.   A novel technique employing infra-red absorption imaging has been devised for the examination of dislocation arrangements in semi-insulating GaAs.   By use of a rotational stage specimens can be viewed continuously over angular spreads of ∿ 100°, enabling three-dimensional images at varied angles of observation to be produced.   Complex lineage structures (sheets and streamers) and cellular networks have been studied.   Possible dislocation interactions responsible for the complex defects are considered, with reference to processes linking the occurrence of high localised [EL2] with high dislocation densities.

1. Introduction

Previously we have shown (Brozel et al 1983) that the concentration of the deep donor level EL2,[EL2], as determined by the variation of optical absorption at 1 μm (Martin 1977 and 1981) is very non-uniform across {001} semi-insulating, LEC substrates.   Fluctuations in absorption were equated with fluctuations in dislocation density: specifically well-defined regions of bunched dislocations (lineage structures) were unambiguously correlated with features of high absorption (Stirland et al 1983).   Initially, information about the various dislocation groupings such as cellular networks (Clark and Stirland 1981) and the lineage structures which we have named streamers (Stirland et al 1983) and sheets (Skolnick et al 1984) was obtained by optical Nomarski contrast micro-scopy of {001} and/or {110} A/B etched surfaces.   This method provides direct information on dislocation arrangements within surface regions (typically ∿ 10-15 μm depth).   By use of contiguous {001}wafers and opposite faces of the same wafers or by examination of etched {110} sections, volumetric arrangements could be indirectly determined. Confirmation of these arrangements resulted from optical absorption observations.   However, since in this case specimens were 3-5 mm thick, direct correlation between strongly absorbing features and etched features was only possible by their favourable configurations (Brozel et al 1984).   Essentially, the lineage structures lie along <110> directions in {001} surfaces and extend down the <001> ingot growth axis for many millimetres so that thin {001} sections from any position within this depth show the same relative configuration of lineage structures. Recent etching observations (Stirland et al 1984) have shown that when <110> streamers merge into <110> sheets the transition is not apparent in the cellular networks surrounding them. However, in 1 μm infra-red absorption images the transition is clearly marked, because the sheets (but not the streamers) are flanked by regions of very low absorption. These have been named "Extended Denuded Zones" (EDZs).

The present work arose from a study of EDZs and lineage structures by infra-red absorption, when it was noted that additional, three-dimensional, information could be obtained from examinations of thick specimens.

2. Experimental details

The basic infra-red method has been described previously (Brozel et al 1983; Skolnick et al 1984). In order to obtain three-dimensional (3-D) information specimens were mounted in a holder, which could be rotated about a vertical axis, while being viewed by a vidicon/CCTV monitor system. Transmission infra-red images were either recorded directly using a 35 mm camera, from the monitor screen, as static frames at different measured tilt angles, or were recorded on a video recorder whilst the sample was rotating.

Substrates from several manufacturers have been examined. The majority consisted of {001} orientation slices of 2" diameter and 3 mm thickness cut from close to the seed end of the ingots; some 5 mm or 7 mm thickness slices have also been examined, and one 3" diameter specimen of 5 mm thickness. Each specimen was prepared in a similar way prior to exam- ination: both faces were polished using a chemico-mechanical bromine- methanol method.

3. Three-dimensional stereoscopic imaging

There are some unusual aspects of the 3-D method we have employed which result from the particular material under study. A stereoscopic image, which provides 3-D information from a pair of 2-D images ("stereo-pairs") is usually obtained by generating two images of the object from different viewpoints ($\pm\theta^0$ from the specimen normal). This requires two imaging systems. A single imaging system can be used if the images can be recorded sequentially either by moving the imager through $2\theta^0$ or by tilting the specimen through $2\theta^0$, the method used here.

The recorded images are simultaneously presented to the eyes of the observer and this may require a special viewing system ("stereo-viewer"). Assuming an eye separation of $\sim$ 6 cm and typical viewing distance of 50 cm then $\theta \sim 3.4^0$, so that the tilt angle is $\sim 7^0$. However, for an object within a dielectric medium, such as GaAs, with refractive index n $\sim$ 4, it is necessary to use a larger tilt angle $\pm\Phi^0$ (= $\sin^{-1}$ (4 $\sin\theta$)) to obtain comparable stereoscopic images. For $\theta = 3.4^0$, $\Phi = 12.5^0$, so that the tilt angle is now 25^0 .

The use of a rotating specimen stage enables images to be formed over a wide range of tilt angles ($\pm$ 50^0). This mean that any pair of images, separated by $\sim$ 2 Φ, can be used to form a stereo-pair. Thus a stereo- pair can be **formed** by combining the image at $-2\Phi^0$ with that at 0^0 , or that at 0^0 with that at $+2\Phi^0$. The resulting stereo images are then equivalent to viewing the specimen along axes inclined at $-\Phi^0$ and $+\Phi^0$

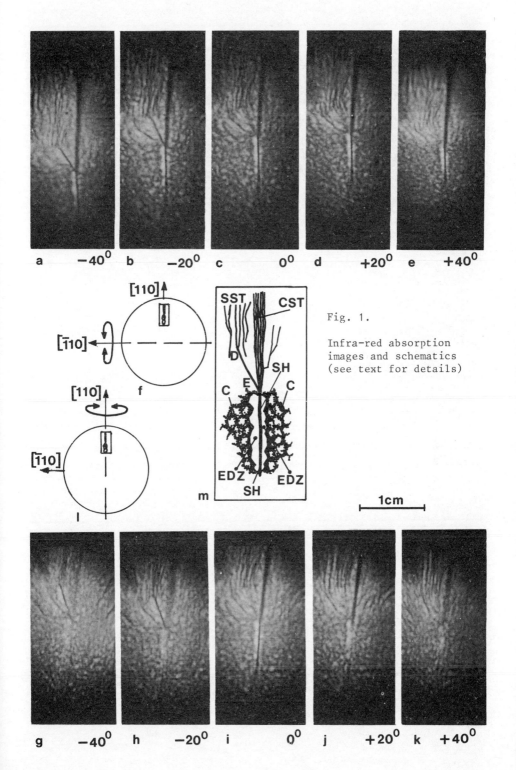

Fig. 1.

Infra-red absorption
images and schematics
(see text for details)

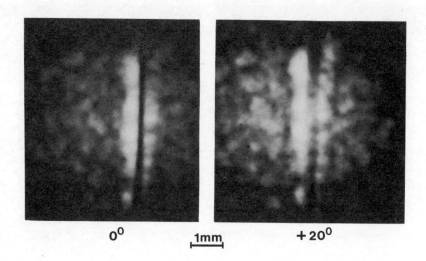

$$0^0 \qquad \underset{\text{1mm}}{\rule{2cm}{0.4pt}} \qquad +20^0$$

Fig.2;   Stereo-pair of central region of Fig. 1(i) and Fig.1(j)

respectively to the specimen normal.

4.   Experimental results

Fig.1 shows infra-red micrographs of the same area of a 5 mm thick, 3"
diameter specimen.   The schematics of Figs. 1(f) and 1(ℓ) indicate the
position of the area on the specimen and the axes of rotation used to
obtain the five different angular positions shown in Figs. 1(a) - 1(e)
and Figs. 1(g) - 1(k) respectively.   (In practice, the specimen rather
than the rotation axis was turned through 90^0 between the two sets of
observations.   The one set of micrographs have therefore been displayed
with a 90^0 rotation about <001> to facilitate comparison with the other
set.)   The schematic of Fig. 1(m) indicates the principal defects within
the selected area: C = cellular networks; EDZ = extended denuded zones;
CST = clustered streamers: SST=separated streamers (including DE): SH =
sheet.   These defects are typical of those encountered in every specimen
we have studied.

For reasons of space it is not possible to display the micrographs of
Fig.1 as stereo-pairs at correct separation.   However a higher magnification stereo-pair of the central region of Figs. 1(c) and 1(j) is shown at
correct separation in Fig.2 for a 130 mm focal length binocular viewer and
eye separation of 65 mm.   Most of the observations described can be
obtained by study of individual micrographs, but the interpretation is
easier if the 3-D images can also be seen.

Angular rotations from $+40^0$ to -40^0 about [$\bar{1}10$], illustrated in Figs.1(a)-
1(e), show no appreciable change in the contrast and dimensions of SH, CST
or the EDZs which flank SH.   There are differences in the cellular
structure C and some of the SST, including DE.   Similar angular rotations
about [110],illustrated in Figs. 1(g) - 1(k), show considerable changes in
all the defect structures, however.   These observations can be interpreted
as follows.   The cellular structure (C) evident in the regions next to

the EDZs in Fig.2, and in the micrographs of Fig.1, can be resolved into individual cells distributed through the specimen thickness in a 3-D image. We have shown previously (Stirland et al 1984) that typical (etched) cellular network structures are 200-300 μm in extent, so that a 5 mm thick sample will contain stacks of ∿ 15-25 cell thicknesses, if the cells are equi-axed rather than columnar. The present observations indicate that the cells are predominantly equi-axed.

The EDZs flanking the highly absorbing sheet SH are regions of low absorption bounded by cellular networks on one side and SH on the other. Figs.1(a)-1(e) show that rotation about [$\bar{1}$10] does not alter either the relative contrast of the EDZs and SH, or their widths (EDZ ∿ 200-300μm; SH ∿ 50 μm). Thus both of these defects must extend exactly down [00$\bar{1}$]. This deduction was confirmed by examination of Figs. 1(g) - 1(k).

The EDZ region is seen to smear and spread slightly for tilts ∿ 20°. The SH region whilst well defined at zero tilt is seen to disappear rapidly at tilt angles greater than 5°. The clustered streamers CST do not show such a sharp reduction in absorption contrast for a 20° rotation. In fact the CST contrast at 20° tilt is stronger than at 0°. This supports the view that the streamers are loosely clustered dislocation arrays lying off exact <110> and <001>, in contrast with the precisely oriented sheet arrays. This is also evident for the separated streamers SST. Although the majority of streamers lie approximately along <110> a few deviate considerably from these directions. Figs.1(a) and 1(b), 1(g) and 1(h) show an example (DE on Fig. 1(m)) of a streamer considerably inclined to <110>, which is not visible for other tilt angles. The implication is that the narrow dislocation array, comprising this streamer, is also considerably inclined away from <001>.

5. Discussion

The 3-D infra-red examinations have confirmed and extended earlier observations which indicated that there are significant non-uniformities in the distributions of dislocations, or 1 μm absorbing regions,throughout {001} GaAs specimens. A cellular network structure appears to be the most general configuration adopted by dislocations: previous examination of {110} sections in addition to {001} sections had suggested that the cellular arrangement was equi-axed and not columnar. The present results confirm this. There has been no satisfactory explanation either for the occurrence of the sheet and streamer lineage structures, or any description of their compositions. Their existence seems to have been first noted by Chen and Holmes (1983). Their proposed explanation of polygonised dislocations forming walls defined by {110} planes normal to {111} slip planes (intersecting {001} along <110>) was rejected by Stirland et al (1983) on the experimental evidence that the lineage structures lay along the {110} planes which are normal to {001}. The present results confirm that the latter is the correct configuration of the sheet structures. Recent work by Brown et al (1984) has indicated that lineage features similar to the sheets are low angle boundaries (0.01°) with a strong tilt component. We propose tentatively that these lineage structures could be formed by reactions involving glide dislocations on intersecting {111} planes producing sessile edge dislocations. For example, reactions such as

$$\tfrac{1}{2}a[011] + \tfrac{1}{2}a[\bar{1}0\bar{1}] \rightarrow \tfrac{1}{2}a[\bar{1}10] \qquad (1)$$

and
$$\tfrac{1}{2}a[\bar{1}01] + \tfrac{1}{2}a[01\bar{1}] \rightarrow \tfrac{1}{2}a[\bar{1}10] \qquad (2)$$

represent 60° dislocations gliding on ($1\bar{1}1$) and ($\bar{1}11$), with axes [110]

intersecting to give edge dislocations with b = $\frac{1}{2}$a[$\bar{1}$10] and axes [110].
Because the glide plane of the resultant dislocations is (001), they are
expected to be sessile. Similar reactions have been proposed by Abrahams
and Ekstrom (1960) to account for observations of cleavage in GaAs. It is
clear that experimental evidence in the form of Burgers vector analyses
of dislocations comprising the sheet structures, by electron microscopy
or X-ray topography methods, is necessary to test this hypothesis.

The present results have confirmed our previous brief observations
(Stirland et al 1984) on the nature of the extended denuded zones (EDZs).
They represent regions of low 1 μm infra-red absorption, hence low EL2
and low dislocation density. They are always observed in association with
sheet structure but their formation mechanism remains unclear.

Finally it is important to note that none of these observations can
establish whether dislocation-associated EL2 centres detected by 1 μm
absorption result from 'gettering' of EL2 by dislocations (Holmes and
Chen 1984) or whether EL2 is generated by dislocation climb (Weber et al
1982) or glide (Figielski 1982) processes.

Acknowledgements

The authors would like to acknowledge the help and encouragement of
R W Burns and technical assistance from D G Hart and I Grant. This work
has been carried out with the partial support of Procurement Executive,
Ministry of Defence, sponsored by DCVD.

References

Abrahams M S and Ekstrom L 1960 Acta Metall. 8, 654
Brown G T, Skolnick M S, Jones G R, Tanner B $\bar{K}$ and Barnett S J 1984 to be
    to be published,presented at Semi-Insulating III-V Materials Conf.
    Kah-nee-Ta, Oregon 1984.
Brozel M R, Grant I, Ware R M and Stirland D J 1983 Appl.Phys.Lett.42,610.
Brozel M R, Grant I, Ware R M, Stirland D J and Skolnick M S 1984
    J. Appl. Phys. 56, 1109.
Chen R T and Holmes D E 1983 J. Crystal Growth 61, 111.
Clark S and Stirland D J 1981 Micros.Semicond.Mater.Conf. Oxford 1981
    Inst.Phys.Conf. Ser No 60, 339.
Figielski T, 1982, Appl. Phys. A, 29 , 199.
Holmes D E and Chen R T 1984 J.Appl.Phys. 55, 3588.
Martin G M, Mitonneau A and Mircea A, 1977, Elect.Lett.13, 191
Martin G M, 1981, Appl. Phys Lett. 39, 747
Skolnick M S, Brozel M R, Reed L J, Grant I, Stirland D J and Ware R M,
    1984 J. Elec. Mats. 13, 107.
Stirland D J, Augustus P D, Brozel M R and Foulkes E J, 1984 to be
    published, presented at Semi-Insulating III-V Materials Conf.
    Kah-nee-Ta, Oregon, 1984.
Stirland D J, Grant I, Brozel M R and Ware R M 1983 Micros. Semicond.
    Mater. Conf. Oxford, 1983, Inst.Phys.Conf. Ser.No.67, 285.
Weber E R, Emmen H, Kaufmann V, Windschief J, Schneider J and Wosinski T
    1982,J. Appl. Phys. 53, 6140.

A new model for the origin of mid-gap electron traps (EL2 family) in liquid encapsulated Czochralski GaAs

T. Ikoma, M. Taniguchi and Y. Mochizuki
Institute of Industrial Science, University of Tokyo
7-22-1, Roppongi, Minatoku, Tokyo 106, Japan

Abstract. Peculiar properties of mid-gap levels in GaAs (EL2 family) are summarized, and to explain these properties a new model for EL2 family, arsenic aggregates, is presented and discussed.

1. Introduction

Mid-gap electron traps often called EL2 (Martin et al 1977) play a dominant role in compensating shallow acceptors in undoped semi-insulating GaAs, which is used as a substrate for high speed integrated circuits. It is important to elucidate the origin of EL2 , which is not **clearly** identified. In early years, it was believed that EL2 was oxygen-related. This model was once refuted but recently it has been shown that high concentration doping of oxygen during Bridgman growth induces a mid-gap level (Lagowski et al 1984).

After the oxygen-related model was refuted, the As/Ga antisite model was proposed (Wagner et al 1980). Very recently, the detailed experiments made by Meyer et al (1984) showed that the isolated antisite defect was not EL2. They also showed that there existed a **variety of** As/Ga antisite defects. Goltzene et al (1984) also reported that the antisite defect was associated with gallium vacancy.

The difficulty to identify the origin of EL2 lies in the fact that the data reported by various authors are different and contradictory. The present authors have shown that the mid-gap traps detected in various GaAs crystals are not the same level (Taniguchi and Ikoma 1983a). They have pointed out that there exist more than two mid-gap levels which have very similar properties. These levels were named "EL2 family" (Taniguchi and Ikoma 1983b).

In this paper we will show that properties of the mid-gap levels vary among various crystals and with heat treatment and deposited metals, and deduce specific features for EL2 family. Then, a new model for the origin of EL2 family will be presented and discussed.

2. Creation and Annihilation of EL2

2.1 Creation

EL2 exists in bulk (both liquid encapsulated Czochralski (LEC) and horizontal Bridgman (HB)) and vapor phase epitaxy (VPE) GaAs but neither in liquid phase epitaxy (LPE) nor molecular beam epitaxy (MBE) GaAs. Non-existence of EL2 in LPE GaAs is explained by the fact that a crystal is grown under

Ga-rich condition, since EL2 is related with excess arsenic. However, there is no reasonable explanation why an MBE crystal does not contain EL2 even if it is grown under arsenic over-pressure.

The density of EL2 increases with increase of excess arsenic both in LEC (Holmes et al 1982) and VPE growth (Samuelson 1981). However, we have observed that increase of 0.35 eV trap (EL6) was greater than that of EL2 in a Czochralski GaAs when applying arsenic pressure. EL6 is a very common electron trap existing in both HB and LEC materials and sometimes the density is higher than that of EL2. Therefore, when the ESR signal is analyzed in terms of existing traps in LEC GaAs, EL6 can not be ignored.

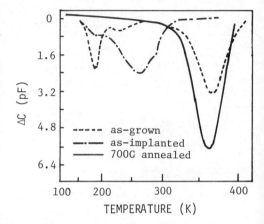

Fig. 1 DLTS spectra for O-implanted and annealed HB wafer

EL2 is created by ion implantation and successive annealing as shown in Fig.1. In an as-implanted sample, the DLTS exhibited a U-shape spectrum, the peak of which was around 260 K. After annealing at 400 C for 15 min. this U-shape disappeared quickly and a large peak appeared at around 360 K after 700 C annealing. The position of the peak was almost the same as that of EL2 in HB GaAs but slightly lower. Such annealing behavior was also observed for O, Ga, As, Si, and N implantation (Taniguchi and Ikoma 1983a). All these levels show the photoquenching effect which is regarded as a finger print of EL2 (Taniguchi and Ikoma 1984). They are thermally unstable and annihilated by longer time or higher temperature annealing.

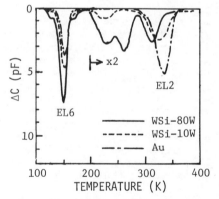

Fig. 2 DLTS spectra for WSi and Au Schottky contacts

Electron beam (EB) irradiation does not create EL2. The 0.75 eV level which was created by EB irradiation and appeared in DLTS at the position similar to EL2 did not show photoquenching. Neutron irradiation which is known to increase the antisite ESR signal (Weber et al 1982) does not create EL2 in as-irradiated samples (Magno et al 1984).

2.2 Annihilation

The density of EL2 decreases after annealing in hydrogen atmosphere. Its dependence on annealing temperature is different in HB and LEC crystals. Hydrogen plasma enhances decrease of EL2 (Lagowski et al 1982). In LEC

crystals annealing induces change
of the electron emission rate
(Taniguchi and Ikoma 1983b).

Ion implantation also annihilates
EL2 as shown in Fig.1. An as-
implanted sample contains no EL2.
Recently we have found that EL2 was
annihilated or changed its property
with depositing WSi (Makimoto et al
1984). Figure 2 shows the DLTS
spectra measured in WSi/GaAs
Schottky contacts which were made
at different sputtering input
powers. For reference, the DLTS
spectrum for a Au/GaAs Schottky
contact made on the same wafer is
included. For the sample fabricated
by sputtering, EL2 decreased and
the peak temperature shifted to
lower temperatures. The shift of
peak temperature is greater for
higher sputtering power.

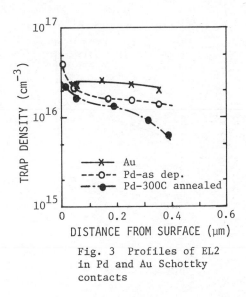

Fig. 3 Profiles of EL2
in Pd and Au Schottky
contacts

Change of EL2 was also observed for Pd evaporation. The density profiles
of EL2 in Pd/GaAs and Au/GaAs Schottky contacts are shown in Fig.3. It is
clear that Pd evaporation changed the profile of EL2 density. The
effect was also observed for Al/GaAs Schottky contacts (Yahata et al 1984).

Furthermore, we have found (Makimoto et al 1984) that EL2 near the interface has level broadening, which is extended as deep as 50 nm from the
metallurgical interface. EL2 reacts with metals selectively at very low
temperature.

3. Variety of EL2 Properties

3.1 DLTS

When we examine the DLTS data closely, we find that some mid-gap levels are
actually different. Ikoma and Takikawa (1977) found a step-like increase of
the density of the mid-gap level (#10+#11) in a VPE layer as shown in
Fig.4. This step-like increase was found to be due to appearance of
another level (#11), which has the same emission rate but different photoionization threshold energy. Mircea et al (1977) also reported existence
of two levels. Ozeki et al found in VPE layers two levels which had
different activation energies depending on growth temperature. The similar
observation was reported for MOVPE layers (Watanabe et al 1981).

Oxygen-implantation-induced level has a photoionization threshold energy
different from that of EL2 in HB GaAs (Taniguchi and Ikoma 1982). In an HB
wafer annealed at 400C for 15min two levels simultaneously exist; one
existed in the original HB wafer before implantation and the other a newly
generated level. Such simultaneous existence of two mid-gap levels was
also found in as-grown LEC and oxygen-doped HB GaAs (Lagowski et al 1984).

In LEC GaAs we have observed two groups of mid-gap levels; one is stable
and has the same electron emission rate as that of EL2 in HB GaAs and the

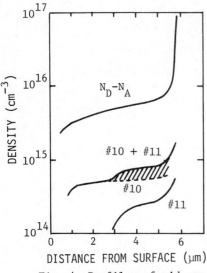

Fig. 4 Profiles of mid-gap
traps showing the existence
of two levels

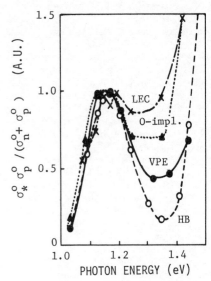

Fig. 5 Photoquenching
spectra for four mid-gap
levels (EL2 Family)

other changes its emission rate and density with low temperature annealing
(Taniguchi and Ikoma 1983b). Furthermore, we have found that mid-gap levels
in front and tail sections of an LEC ingot are very different; the difference of the DLTS peak temperatures was about 25 K. This indicates that the
mid-gap level in a front section was annealed and the emission rate and
density were reduced. Properties of a mid-gap level in LEC GaAs different
from those in VPE and HB GaAs were also reported by Li et al (1984).

3.2   Photoquenching

Photoquenching effect is a very peculiar characteristic and hence regarded
as a finger print for EL2. We have found that all the mid-gap levels
detected in our samples showed photoquenching except the one created by EB
irradiation. All these mid-gap levels, though they have different activation energies and capture cross sections, are considered to have similar
atomic structures and be classified in the same group: EL2 family.

It is important, however, to note that the time constants of transition
from the normal to metastable states are different for these mid-gap
levels and also the photoquenching spectra have variation as shown in
Fig.5, which shows normalized photoquenching spectra for four levels in
various crystals. We have found photoquenching efficiency increases for
photon energy higher than 1.4 eV, which was not reported (Vincent and Bois
1978).

It should be noted that the capacitance transients due to both quenching
and recovery are non-exponential. This non-exponential transient is most
remarkable for the oxygen-implantation-induced level. These observations
suggest that there exist multiple metastable states and EL2 has various
electron-lattice couplings.

3.3 Photoluminescence

The typical photoluminescence bands in semi-insulating GaAs are 0.64, 0.68 and 0.78 eV. The correlation of these emission bands with EL2 has been studied but not certainly identified. We have observed no correlation of the 0.64 and 0.68 eV bands with EL2 density. According to Tajima (1984), the photoquenching of these bands depends on samples and he sometimes observed partial photoquenching, suggesting variation of EL2 levels. Recently photoquenching of the 0.78 eV band was reported (Yu 1984). However, this emission band is not always observed in samples which contain EL2.

4. Model and Discussion

We summarize the properties of EL2 family as follows.
1) EL2 is related with excess arsenic.
2) EL2 appears at the intermediate stage of annealing, during which clusters of native defects created by heavy particle bombardment dissociate and migrate to form more stable forms.
3) EL2 has a large lattice distortion as revealed by the configuration coordinate picture and also has a metastable state which is considered to be an atomic structural change due to charge states. This implies that atomic bonds surrounding the defect might be "soft" and that the defect would not be a simple point defect like an isolated antisite.
4) EL2 has variation in thermal and optical properties, which suggest the defect might be a sizable defect consisting of clusters or complexes of point defects.
5) EL2 changes its form by annealing and reaction with Schottky metals at very low temperature. This suggests that constituents of the defect would be fast diffusers.
6) EL2 appears as broadened energy states at the interface between metal and GaAs. This also indicates that the defect would have a complex form.

There have been proposed various models for EL2. The present authors have proposed that EL2 family should be As aggregates, which perturb a periodical crystal potential in GaAs network. Sizes of aggregates might vary. The smallest one might be a five-atom cluster which could be an As/Ga antisite. Large ones might have a size of the order of 10 nm or more. These defects could be regarded as a complex of an As/Ga antisite with interstitial arsenic atoms or amorphous arsenic atoms or arsenic precipitation. Involvement of interstitial arsenic atoms is conceived from the item (5) cited above. Interstitial atoms migrate faster than vacancies and react with metals. In bulk GaAs, existence of excess arsenic was suggested.

In amorphous phase arsenic aggregates, lattice distortion is considered to be large, and electron-lattice coupling can vary according to possible variety of atomic structures. Furthermore, configurational change due to charged states (transition from normal to metastable states) can take place. Actually, in amorphous arsenic, quenching of photoluminescence was observed by illumination of 1.1 eV light (Bishop et al 1976).

Change of EL2 by deposition of certain metals is explained as follows. When arsenic reacts with a Schottky metal to form an intermetallic compound, arsenic aggregates dissociate near the interface and are extracted to the interface, resulting in reduction of EL2. Reappearance of EL2 after annealing might be ascribed to diffusion of excess arsenic atoms from the bulk region into the vicinity of the interface.

The idea that atomic aggregates perturb the periodical crystal potential to generate localized states is somewhat similar to generation of surface states. The similarity of EL2 level with surface states in GaAs can be pointed out. The As aggregate model also explains the non-existence of EL2 in MBE wafers. In MBE mono-layer growth takes place and therefore there is little possibility for aggregates to be formed.

5. Conclusion

A new physical model for EL2 family, As aggregates, has been proposed to explain very peculiar properties of EL2 which are not well explained with existing models. To understand how As aggregates create localized states near the mid-gap and why they show very peculiar characteristics such as photoquenching effect, further study is needed.

References

Bishop S G, Storm V and Taylor P C 1976 Solid State Commun. 18 573
Goltzene A, Meyer B and Scwab C 1984 13th Int. Conf. on Defects in Semi-
   conductors, Coronado, USA.
Holmes D E, Chen R T, Elliott K R and Kirkpatrick C G 1982 Appl. Phys.
   Lett. 40 46
Ikoma and Takikawa 1977 unpublished work
Lagowski J, Lin D G, Aoyama T and Gatos H C 1984 Appl. Phys. Lett. 44
   336
Lagowski J, Kaminska M, Parsey J M Jr, Gatos H C and Lichtensteiger M,
   1982 Appl. Phys. Lett. 41 1078.
Li G P, Wu Y, and Wang K L 1984 Int. Conf. on Defects in Semiconductors,
   Coronado, USA
Magno R, Giessner J G and Spencer M 1984 Int. Conf. on Defects in Semicon-
   ductors, Coronado, USA
Makimoto T, Taniguchi M, Ogiwara K, Ikoma T and Okumura T 1984 Int. Conf.
   on Solid State Devices and Materials, Kobe, Japan, D-3-2
Martin G M, Mitonneau R C and Mircea A 1977 Electron. Lett. 13 191
Meyer B K, Hofmann D M, Lohse F and Spaeth J M 1984 13th Int. Conf. on
   Defects in Semiconductors, Coronado, USA
Mircea A, Mitonneau A, Hallis J and Jaros M 1977 Phys. Rev. B16 3665
Ozeki M, Komeno J, Shibatomi A and Ohkawa S 1979 J. Appl. Phys. 50 4808
Samuelson L, Omling P, Titze H and Grimmeiss H 1981 J. Cryst. Growth 55
   164
Tajima M 1984 private communication, to be published in Jap. J. Appl.
   Phys. Sept. 1984
Taniguchi M and Ikoma T 1982 Semi-Insulating III-V Materials
   (Cheshire:Shiva) pp283-289
Taniguchi M and Ikoma T 1983a Inst. Phys. Conf. Ser.#65 65
Taniguchi M and Ikoma T 1983b J. Appl. Phys. 54 6448
Taniguchi M and Ikoma T 1984 Appl. Phys. Lett. 45 69
Vincent G and Bois P 1978 Solid State Commun. 27 731
Wagner R J, Krebs J J, Stauss G H and White A M 1980 Solid State Commun.
   36 15
Watanabe M O, Tanaka A, Nakanishi T and Zohta Y 1981 Jap. J. Appl. Phys.
   20 L429
Weber E R, Ennen H, Kaufmann U, Windscheif J, Schneider J and Wosinski T
   1982 J. Appl. Phys. 53 6140
Yahata A and Nakajima M 1984 Jap. J. Appl. Phys. 23 313
Yu P W 1984 Int. Conf. on Phys. of Semiconductors, San Francisco, USA

Inst. Phys. Conf. Ser. No. 74: Chapter 2
Paper presented at Int. Symp. GaAs and Related Compounds, Biarritz, 1984

71

Lateral spread of ion implants in GaAs

R. G. Wilson

Hughes Research Laboratories, Malibu, CA 90265, USA

Abstract. We have measured the lateral range straggle ΔR_ℓ for ^{1}H, ^{2}H, [9]Be, [11]B, [28]Si and [80]Se ions implanted into (100) GaAs at 200 keV and compare them with the corresponding values of ΔR_p. The values of ΔR_ℓ were determined by varying the implantation angle, measuring the implanted depth distributions using secondary ion mass spectrometry, and deconvolving the values of ΔR_p and ΔR_ℓ. We have also measured values of projected range R_p and straggle ΔR_p for several energies of the same ions in GaAs. By combining these data, we can predict experimental lateral spreads as a function of implantation ion energy for these ions used to fabricate FETs or waveguides in GaAs.

1. Introduction

The current thrust of integrated circuit technology toward submicrometer dimensions to achieve high speed and high density, including narrow gate GaAs FETs, where ion implantation is used in high resolution fabrication steps, requires consideration of lateral or transverse spread in implanted depth profiles. Lateral spreading may place an ultimate limit on the narrowness of FET gates that can be achieved practically and reproducibly using either large area masked or maskless direct-write focused beam ion implantation technologies.

Ions scatter laterally during the stopping process, producing a lateral distribution that can be characterized by a deviation or straggle much as the standard deviation or straggle ΔR_p about the projected range R_p in the longitudinal direction. Thus an impurity, doping, or displacement damage distribution is created beneath the edge of a mask or surface layer or radially outward from a tiny ion beam with a lateral straggle ΔR_ℓ. This edge might be of a gate, source or drain, isolation, channel wave-guides or a resist or metal implant mask. This effect should be considered when devices and circuits with small dimensions are designed. The magnitude of lateral spread that results from a 7 or 8° tilt ($\sim$1/8 R_p) or from electrostatic beam scanning ($\leq$ 1/12 R_p) used during implantation is small compared with this effect.

Theoretical treatments of lateral spreading during ion stopping have been set forth by Winterbon (1972, 1976) and by Furukawa et al. (1973). Winterbon (1972, p. 223) has indicated that the effect can be studied best by oblique angle implantation. Winterbon (1976) also shows that the ratio of lateral-to-longitudinal straggle $\Delta R_\ell / \Delta R_p$ should depend on ion energy and on the difference between the masses of the ions and the target, being larger for larger mass differences. Oblique angle

implantation and RBS analysis of the resulting depth distributions of
damage (disorder) in Si have been used by Furukawa and Matsumura (1973),
by Matsumura and Furukawa (1976), and by Grant et al. (1976). Pan and
Fang (1974) used electrical measurements of FET gate structures in Si.
Okabayashi and Shinoda (1973) and Matsumura and Stephens (1977) have
demonstrated that the combination of oblique angle implantation and
differential C-V profiling provides data that allow the determination of
lateral spreading of electrical profiles in Si and GaAs, respectively,
within the accuracy of the differential C-V technique. They indicate
agreement with calculations of Furukawa (1973). Akasaka et al. (1972)
used the angle lap and stain technique to delineate junctions. Sakurai
et al. (1979) used an SEM to study stained regions of lateral spread and
Tsuchiya (1980) used electrical measurements of Si MOSFETs. Miyauchi
employed stain and etch patterns in GaAs. Lateral spread under the
edges of masks is specifically discussed in the work of Matsumura and
Stephens (1977), Sakurai et al. (1979), and Tsuchiya (1980). Shimizu
et al. (1977) discuss lateral spreading but report no measurements. Only
two papers discuss GaAs, namely, Matsumura and Stephens (1977) (protons)
and Miyauchi (1983) (Be and Si).

2. Experimental Techniques

We implanted ^{1}H, ^{2}H, ^{9}Be, ^{11}B, ^{28}Si, and ^{80}Se ions into (100)LEC bulk
GaAs at room temperature and at tilt angles of 7, 45, 60, and 72.5 degrees
to the normal at energies of 200 or 250 keV, and of fluences of 2×10^{13}
cm^{-2} for all ions except ^{1}H and ^{2}H, for which the fluence was 4×10^{15}cm^{-2}.
The as-implanted depth distributions were measured using secondary ion
mass spectrometry (SIMS) using cesium ion bombardment for ^{1}H, ^{2}H, ^{28}Si,
and ^{80}Se, and oxygen ion bombardment for ^{9}Be and ^{11}B. The first and
second moments of these depth distributions (the mode μ, or projected
range R_p, and the standard deviation or range straggle ΔR_p of a modified
Gaussian) were determined by Pearson IV fitting. The values of lateral
straggle ΔR_ℓ were determined using the technique of Furukawa and
Matsumura (1973) by deconvolving ΔR_p and ΔR_ℓ from the value of R_θ for the
implantation tilt angles θ, as also done by Matsumura and Furukawa (1976),
Grant et al. (1976), Matsumura et al. (1976), and Matsumura and Stephens
(1977). The equation is $\Delta R_\ell = (\Delta R_\theta^2 - \Delta R_p^2 \cos^2\theta)^{1/2}(\sin\theta)^{-1}$.

Ion energies are accurate to within about $\pm 2\%$, ion fluences, to within
about $\pm 10\%$, and implant angles, to within about ± 1 degree. The depth
scales in the SIMS profiles were determined from measurements of the
crater depth by surface profilometry and are accurate to within about
$\pm 7\%$. The atom density scales of the SIMS profiles were determined by
equating the integral of the 7 degree profiles to the implant fluence,
and for the larger angles, were obtained from the SIMS sensitivity
factors that resulted from the 7° profiles (i.e., normalized through the
matrix ion intensities).

3. Experimental Results

The depth distributions of the six ions implanted into GaAs at several
angles to the normal and measured by SIMS profiling are shown in Figs. 1
through 5. The proton and deuteron profiles are combined in Fig. 1, and
show that at 250 keV, the range of protons slightly exceeds the range of
deuterons. This relationship is reversed for ion energies below 200 keV.
The implant fluence was increased above 10^{15} cm^{-2} for the hydrogen iso-
topes because of the anticipated greater background level in SIMS

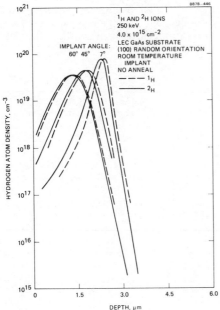

Fig. 1. Depth distributions for ^{1}H and ^{2}H implanted into GaAs

Fig. 2. Depth distributions for ^{9}Be implanted into GaAs

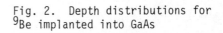

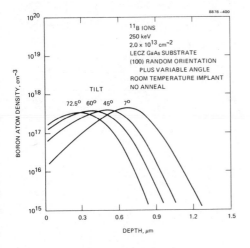

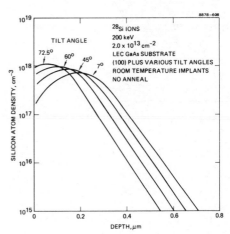

Fig. 3. Depth distributions for ^{11}B implanted into GaAs

Fig. 4. Depth distributions for ^{28}Si implanted into GaAs

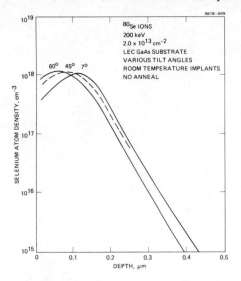

Fig. 5. Depth distributions for ^{80}Se implanted into GaAs

measurements. The data of Fig. 1 show that we might have been able to reduce the fluence for ^{2}H but not for ^{1}H. The profiles for H and Be show the effect of a larger lateral straggle ΔR_ℓ compared with the longitudinal straggle R_p by the increase in profile width with increasing tilt angle.

The profiles of Figs. 1-5 show that the depth distributions are not Gaussian. The projected ranges and range straggles or first and second moments of the distributions can be determined only by a fitting procedure, such as for a modified Gaussian, because of the presence of third and fourth moments. The values of R_p do not occur at the peaks of the distributions and ΔR_p is not exactly the half-width at $e^{-1/2}N_{max}$. We chose to determine the values of $\mu(R_p)$ and $\sigma(\Delta R_p)$ by Pearson IV fitting, which also determines values of the third moment γ_1 (skewness) and fourth moment β_2 (kurtosis); other appropriate fitting routines could be used. Pearson IV fitting was performed for each profile at each implant angle. The values of R_p and ΔR_p were obtained from the fit for the 7° implants. The values of ΔR_ℓ were obtained for each remaining angle θ and the values of ΔR_ℓ were calculated by using the equation given in Section 2. The resulting values R_p, ΔR_p, and ΔR_ℓ are given in Table I, together with the ions and their atomic numbers and energies. The ratios of lateral straggle-to-longitudinal straggle $\Delta R_\ell / \Delta R_p$ are given in another column of Table I.

Table I. Range parameters and lateral spread data for implants in GaAs.

Ion	Z_1	Energy keV	R_p μm	ΔR_p μm	ΔR_ℓ μm	$\dfrac{\Delta R_\ell}{\Delta R_p}$	$\dfrac{\Delta R_\ell}{R_p}$	$\dfrac{3\Delta R_\ell}{R_p}$
^{1}H	1	250	2.30	0.256	0.50	1.9	0.22	0.7
^{2}H	1	250	2.17	0.287	0.50	1.7	0.23	0.7
^{9}Be	4	200	0.63	0.271	0.37	1.4	0.59	1.7
^{11}B	5	250	0.60	0.216	0.27	1.3	0.45	1.4
^{28}Si	14	200	0.15	0.174	0.19	1.1	1.27	3.8
^{80}Se	34	200	0.082	0.102	0.11	1.1	2.1	4.1

Matsumura et al. (1977) have calculated values of range and straggle for protons (^{1}H), which they compare with their experimental values determined indirectly by measuring the C-V profiles of carrier removal in GaAs caused by the proton bombardment. We compare our directly measured values with their calculated values for 250 keV protons in Table II. The agreement is felt to be quite good for a system as difficult to calculate as the light ion ^{1}H in the heavy target GaAs. The best agreement is for ΔR_p (3% difference) and the worst is for ΔR_ℓ (25% difference), the values for R_p, $\Delta R_\ell/\Delta R_p$, and $\Delta R_\ell/R_p$ all being within 15%. We have performed these same measurements and determined the same parameters for 250 keV ^{1}H and ^{2}H in Si. The ratio of $\Delta R_\ell/\Delta R_p$ for 250 keV ^{1}H in Si is 1.7, closer to their calculated value for GaAs.

Table II. Experimental and calculated range and straggle parameters for 250 keV protons in GaAs.

	R_p(μm)	ΔR_p(μm)	ΔR_ℓ(μm)	$\Delta R_\ell/\Delta R_p$	$\Delta R_\ell/R_p$
Experimental, this work	2.30	0.256	0.50	1.9	0.22
Calculated, Matsumura et al. (1977)	2.00	0.248	0.386	1.6	0.19

The data given in Table I show that the value of lateral straggle ΔR_ℓ is only about 10% greater than the longitudinal straggle ΔR_p for ions of mass equal to or greater than Si. This ratio increases for increasing mass difference approaching the value of about 1.8 for hydrogen, in agreement with prediction of Winterbon (1972). The ratio of ΔR_ℓ to the projected ion range R_p increases strongly, beginning at ∿0.2 for hydrogen, and reaching 2 for heavy ions like Se.

Experimental values of ΔR_ℓ measured by Matsumura et al. (1977) for 400 keV protons are 0.54 and 0.48 μm. Our value for 250 keV protons is 0.50 μm. Because they show relatively little dependence of ΔR_p and ΔR_ℓ on ion energy in the range from 200 to 400 keV, we feel that the agreement seen here is good. We have also determined values of the projected range R_p for several ion energies in the range from 40 to 600 keV for the ions for which lateral spread measurements were made. Curves of R_p versus ion energy obtained from these data are given in Fig. 6. The curve for ^{28}Si is taken from the calculations of Brice (1975).

4. Discussion

We can compare our results for the direct measurement of the atom depth distributions in both the longitudinal and lateral directions for protons implanted in GaAs with the results of Matsumura and Stephens (1977) who measure the corresponding distributions of carrier removal in GaAs, and with the results for Be and Si implanted in GaAs reported by Miyauchi et al. (1983), who used stain etching patterns. No one has reported results for deuterons, boron, or selenium. Boron, along with oxygen, is used for implanted isolations in GaAs. Matsumura and Stephens (1977) report measured values of 0.54 and 0.48 μm for ΔR_ℓ for 400 keV protons, assuming a Gaussian depth distribution, and also 3.3 μm for R_p. Our results for protons in GaAs at 200 keV given in Table I yield a value of 0.50 μm for ΔR_ℓ, agreeing favorably with the results of Matsumura and Stephens (1977), assuming little dependence on energy between 250 and 400 keV. Our curve of R_p versus ion energy (Fig. 6) gives 3.6 μm for 400 keV protons in GaAs.

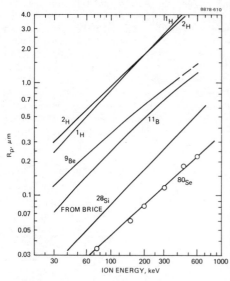

8878-610

Fig. 6. Ion ranges R_p in GaAs

The value of R_p given by Matsumura and Stephens (1977), inferred from carrier removal depth distributions is 3.3 μm. The agreement is good, and even better if the peak of the damage depth distribution should lie just shallower than the R_p of the atom depth distribution.

Miyauchi et al. (1983) report full width unannealed and annealed lateral spreads as a function of implant fluence for 160 keV Be and Si in GaAs, using focused ion beams. We can compare our results with theirs at our fluence of $2 \times 10^{13} cm^{-2}$ for Be and Si. Our values of lateral straggle are approximately independent of implant fluence. The number of ΔR_ℓ's that are in appropriate correspondence with their lateral spread dimensions measured by their stain etch technique is an issue to be dealt with. They discuss implanted doping densities of 10^{14}, 10^{15}, and $10^{16} cm^{-3}$ for different implant fluences. At $2 \times 10^{13} cm^{-2}$ fluence in their Fig. 2, the values of lateral spread are about 1.3 μm for Be and 0.5 μm for Si. These data are for annealed implants, but should be the same for unannealed samples according to their discussion and in agreement with our similar measurements for both Si and Be. If we assume that their lateral spread measurements are delineated at 4% of the peak density, then full width corresponds to $5\Delta R_\ell$. Our measurements for 200 keV Be and Si yield $5\Delta R_\ell$ values of 1.85 μm for Be and 0.95 μm for this case, somewhat higher than their values. If we assume that their lateral spread delineation occurs at 14% of the peak density value, then the corresponding full width is $4\Delta R_\ell$, and our values are 1.5 μm for Be and 0.76 μm for Si, closer, but still 0.2 μm high.

5. References

Akasaka Y, Horie K, and Kawazu 1972 Appl. Phys.Lett. 21, 128
Brice D K 1975 Ion Implantation Range and Energy Deposition Distributions
   IFI/Plenum New York Vol. 1
Furukawa S and Matsumura H 1973 Appl. Phys. Lett. 22, 97
Grant W A, Williams J S, and Dodds D 1976 Radiat. Effects 29,189
Matsumura H and Furukawa S 1976 J. Appl. Phys. 47, 1748
Matsumura H, Nagatomo M, Furukawa S, and Stephens K G 1977 Radiat.
   Effects 33, 121
Matsumura H and Stephens K G 1977 J. Appl. Phys. 48, 2779
Miyauchi E, Arimoto H, Bamba Y, Takamori A, Hashimoto H,and Utsumi T
   1983 Jap. J. Appl. Phys. 22, L423
Okabayashi M and Shinoda D 1973 J. Appl. Phys. 44, 4220
Pan E and Fang F F 1974 J. Appl. Phys. 45, 2801
Sakurai T, Kawata H, Sato T, Hisatsugu T, Hashimoto H, and Furuya T 1979
   J. Appl. Phys. 50, 1287
Shimizu R, Kang S T, Ogata H, Kanayama K, Ogata Y, Akasaka Y, and Horie K
   1977 J. Appl. Phys. 48, 1745
Tsuchiya T 1980 J. Appl. Phys. 51, 5773
Winterbon K B 1972 Radiat. Effects 13, 215
Winterbon K B 1976 Radiat. Effects 30, 85

Inst. Phys. Conf. Ser. No. 74: Chapter 2
Paper presented at Int. Symp. GaAs and Related Compounds, Biarritz, 1984

77

Optical furnace annealing of Be implanted GaAs

N J Barrett, D C Bartle, R Nicholls and J D Grange

GEC Research Laboratories, Hirst Research Centre, Wembley UK

Abstract. High hole concentrations with abrupt interfaces have been
produced by Be implantation and optical furnace annealing (OFA) of GaAs.
Doses ranged between 1×10^{14} cm^{-2} and 1×10^{15} cm^{-2} and energies from
25 keV to 360 keV. Samples were OFA using 18 kW of incoherent
irradiation from tungsten halogen lamps and compared with conventional
furnace annealing. To prevent the decomposition of the GaAs, three
methods have been used; annealing in an arsine ambient, face to face
proximity using silicon substrates and plasma deposited Si_3N_4
encapsulation.

Introduction

Be ion implantation has generated a great deal of interest for fabricating
p-type layers in GaAs devices (Kräutle et al 1982, K V Vaidyanathan et al
1983 and Y Umemoto et al 1984) and GaAlAs devices (Narozny and Beneking
1984 and Shtrikman and Febete 1984).   Post ion implantation annealing has
generally been carried out using a conventional furnace anneal between 700
and 900°C. However, recently rapid annealing methods of the order of
seconds rather than minutes have been used to electrically activate Be
implants for GaAs pin diodes (Tabatabaie-Alavi et al 1983) and
heterojunction bipolar transistors (Asbeck et al 1983).

As Be is the lightest p-type dopant in GaAs, the lattice disorder produced
by implantation is significantly less than heavier ions. Consequently the
electrical activation of implanted Be is obtained with anneal temperatures
as low as 500°C (Hunsperger et al 1972, Anderson and Dunlap 1979 and Sugata
et al 1983). However, to regain crystallinity an anneal at 900°C is
necessary (Chatterjee et al 1975 and McLevige et al 1977) which results in
the loss of Be by out-diffusion at the surface and a broadening of the Be
atomic distribution (McLevige et al 1978). A solution to this problem is
rapid thermal annealing (Banerjee et al 1983). For certain devices it is
necessary to implant both p- and n-type dopants. Si implants into GaAs for
n-type layers ideally require post ion implantation anneals of 850°C or
above (Grange and Wickenden 1983). Therefore, most of the Be implants have
been annealed under similar conditions as Si implanted GaAs.

Experimental

$^9Be^+$ ions generated from a BeO ceramic source (Sugata et al 1983) were
implanted nominally at room temperature, 7° to the normal into

semi-insulating undoped LEC (100) GaAs. Unimplanted substrate samples maintained a sheet resistivity of greater than 10^7 Ω/square after an 850°C/20 minute anneal in an arsine ambient (Grange and Wickenden 1983). These SI substrates had also been qualified for Si implant activation.

Ion doses of 1×10^{14} to 1×10^{15} cm^{-2} were implanted with energies in the range from 25 to 360 keV. Post implantation annealing was undertaken using 20 minute anneal cycles with a plasma deposited Si_3N_4 encapsulant (Bartle et al 1984) or capless in an arsine ambient. Rapid anneal cycles were carried out in an optical furnace with a Si_3N_4 encapsulant or with a silicon proximity encapsulant positioned on the GaAs. Differential Hall and step measurements (Stewart et al 1984) were used to determine electrical hole concentrations and Hall mobility profiles.

The optical furnace consists of twelve $1\frac{1}{2}$ kW linear tungsten halogen lamps, positioned above and underneath the wafer chamber. GaAs slices were loaded onto a carbon disc supported by a quartz sledge (Figure 1). A thermocouple was secured to the $\frac{1}{2}$ mm thick carbon disc to measure the sample temperature. A quartz cylinder 356 mm long with a 200 mm inside diameter encloses the wafer chamber and all the electrical and gas connections. The Eurotherm controller has the ability to select three temperature/time regimes with adjustable temperature rise rates upto 200°C/s with a maximum temperature of about 1300°C.

The anneal times quoted are the dwell times at temperature, for example 850°C/30 s. Temperature rise rates of about 150°C/s were used from a background temperature of 600°C which was maintained for a few seconds. Optical furnace anneals at 950 and 1000°C took approximately 3 s from 600°C to reach temperature at which point, the power was turned off. The dwell time is zero in these cases and therefore the anneals have the notation 950°C/0 s and 1000°C/0 s.

Results

Standard furnace annealing results

Figure 2 is a 750°C anneal for 20 minutes in a furnace using a Si_3N_4 encapsulant. Be was implanted with an energy of 40 keV using three doses; 1×10^{14}, 5×10^{14} and 1×10^{15} Be$^+$ cm^{-2}. The electrical activation is similar to an 800°C anneal, with hole concentrations between 5 and 6 x 10^{18} cm^{-3} and mobilities greater than 100 cm^{2}V^{-1}s^{-1}. However, once the anneal temperature reaches 850°C and above, the electrical hole concentration decreases using either a Si_3N_4 encapsulant or capless in an arsine ambient. The anneal in Figure 3 is at 880°C with the same implant energy and doses as in Figure 2. There is little difference between anneals at 850°C and 880°C with an equilibrium hole concentration of 1 to 1.4×10^{18} cm^{-3}. The sheet hole concentration begins to saturate at 5×10^{14} cm^{-2}.

Fig 1: The optical furnace with the wafer holder extended for loading

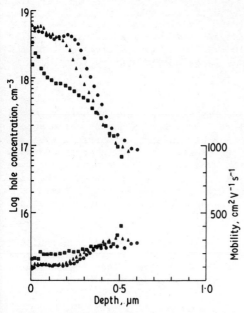

Fig 2: 750°C/20 m anneal for 40 keV Be using a Si_3N_4 encapsulant with doses, ■ $1x10^{14}$ cm^{-2},' ▲ $5x10^{14}$ cm^{-2} and ● $1x10^{15}$ cm^{-2}.

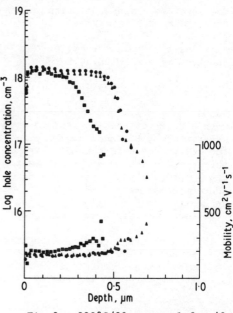

Fig 3: 880°C/20 m anneal for 40 keV Be in an arsine ambient with doses, ■ $1x10^{14}$ cm^{-2}, ▲ $5x10^{14}$ cm^{-2} and ● $1x10^{15}$ cm^{-2}.

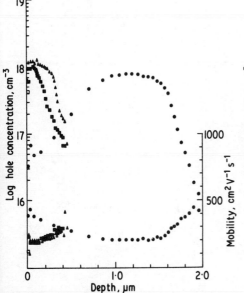

Fig 4: 880°C/20 m anneal with $1x10^{14}$ Be cm^{-2} in an arsine ambient with energies, ■ 25 keV, ▲ 40 keV and ● 360 keV.

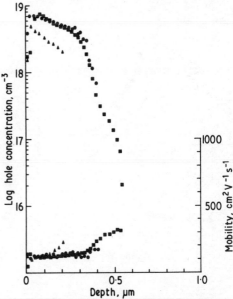

Fig 5: OFA for 40 keV Be with doses of $5x10^{14}$ cm^{-2} using a Si_3N_4 cap with temperature/times at temperature of ▲ 850°C/60 s, ■ 900°C/5 s and ● 950°C/0 s.

The sheet hole concentration increases with increasing energy. Figure 4 shows hole profiles for an 880°C/20 minute anneal in an arsine ambient for a dose of 1×10^{14} cm^{-2}. These profiles gave electrical activities of 19,35 and 74% for implant energies of 25,40 and 360 keV respectively. For the 360 keV implant, doubly charged Be^{++} was used, accelerated by 180 kV.

Optical furnace annealing results

For a dose of 5×10^{14} cm^{-2} at 40 keV, very similar results were obtained with a Si$_3$N$_4$ encapsulant for an anneal at 900°C for 5s and a rapid anneal straight to 950°C and down (Figure 5). Based on other electrical profiles the slight knee in the back edge of the profile may be interpreted as a second diffusion front. Increasing the dose to 1×10^{15} cm^{-2} gave the same sharp profile as with the 5×10^{14} cm^{-2} dose but increased the hole concentration to 1×10^{19} cm^{-3}. The 850°C/60 s anneal (Figure 5) has a hole concentration of 5×10^{18} cm^{-3} which is significantly higher than the electrical profile shown in Figure 3.

Almost identical results are obtained if the Si$_3$N$_4$ encapsulant is replaced by silicon on the GaAs surface to act as a proximity encapsulant (Figure 6). The anneal conditions 900°C/5 s or 950°C/0 s gave identical results. A slight improvement is made by a rapid anneal to 1000°C.

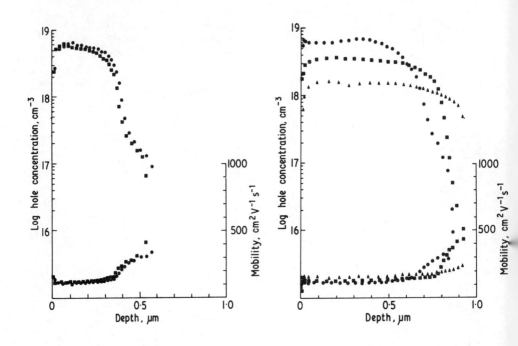

Fig 6 : OFA for 40 keV Be with doses of 5×10^{14} cm^{-2} using a silicon proximity encapsulant with temperature/times at temperature of ■ 950°C/0 s 900°C/5 s and ● 1000°C/0 s.

Fig 7 : OFA for 75 keV Be with doses of 5×10^{14} cm^{-2} compared with a standard anneal using a Si$_3$N$_4$ encapsulant with temperature/times at temperature of ● 1000°C/0 s, ■ 900°C/25 s and ▲ 850°C/20 minutes.

Increasing the implant energy from 40 to 75 keV increases the maximum activation obtained for a 5×10^{14} cm^{-2} implant from 50 to 70%. Figure 7 emphasises the disadvantage of a long anneal time, such as a standard furnace anneal for 20 minutes. Even a dwell time of 25 s as compared with 5 s at 900°C (Figures 5 and 6) is detrimental to the anneal.

Discussion

The highest hole concentration (6×10^{18} cm^{-3}) for 20 minute furnace anneals was achieved at 750°C, the lowest temperature studied. There was very little difference to the electrical profiles by annealing with a Si_3N_4 encapsulant or in an arsine ambient. For furnace anneals the percentage electrical activation decreases with increasing temperature. However, if the dwell time is reduced to 60 s or less (Figure 5) the high anneal temperatures are an advantage rather than a disadvantage.

The encapsulants Si_3N_4 or Si proximity made very little difference to the electrical profiles using an optical furnace anneal. The electrical activation of Be is very insensitive to temperature (for an optimum dwell time) and only a marginal improvement in the electrical hole concentrations is obtained by increasing the temperature from 900°C to 1000°C for example. Be is electrically activated in less than 5 s at 900°C and at 950°C it is only necessary to go straight up to temperature and down again with a zero dwell time. Consequently if the dwell time is longer than necessary, then the results indicate that Be is lost from the surface, in agreement with McLevige et al (1978). This may be one reason why electrical activation increases with implant energy (Figure 4) due to deeper Be profiles and less loss from the surface.

Conclusions

The three methods used to prevent the GaAs surface from degrading have produced very similar electrical profiles, indicating that the activation or diffusion of Be is not effected by surface strain generated at the Si_3N_4/GaAs interface for the atomic concentrations used. The results also suggest that there is Be out diffusion from the GaAs surface as well as through the Si_3N_4.

The hole concentrations obtained by optical furnace annealing at 850°C or greater have been approximately six times higher than conventional furnace anneals. This is principally due to the short anneal times limiting the diffusion of Be.

Acknowledgement

Part of this work has been carried out with the support of the Procurement Executive, Ministry of Defence, sponsored by DCVD.

Dr Colin Wood is thanked for his continued support and technical interest in this optical furnace programme.

References

Asbeck P M, Miller D L, Babcock E J and Kirkpatrick C G, 1983, IEEE Elect Dev Lett, EDL-4 81
Bartle D, Andrews D C, Grange J D, Harris P, Trigg A D and Wickenden D, 1984, Vacuum, <u>34</u> 315

Banerjee S K, DeJale R Y, Soda K J and Streetman B G, 1983, IEEE Elect Dev, ED-30 1755

Chatterjee P K, Vaidyanathan K V, McLevige W V and Streetman B G, 1975, Appl Phys Lett, 27 567

Grange J D and Wickenden D K, 1983, Sol State Elec, 25 313

Hunsperger R G, Wilson R G and Jamba D M 1972, J Appl Phys, 43 1318

Kräutle H, Narozny P and Beneking H, 1982, IEEE Elec Dev Lett, EDL-3 315

Anderson C L and Dunlap H L, 1979, Appl Phys Lett, 35 178

McLevige W V, Helix M J, Vaidyanathan K V and Streetman B G, 1977, J Appl Phys, 48 3342

McLevige W V, Vaidyanathan K V, Streetman B G, Comas J and Plew L 1978, Sol. State Comm, 25 1003

Narozny P and Beneking H, 1984, Elec Lett, 20 442

Shtrikman H and Febete D, 1984, J Appl Phys, 56 1298

Stewart C, Medland J D and Wickenden D K, 1984, Semi-insulating III-V Materials Conference, Oregon

Sugata S, Tsukada N, Nakajima M, Kuramoto K and Mita Y, 1983, Jpn J Appl Phys, 22 L470

Tabatabaie-Alavi K, Masum Choudbury A N M, Kanbe H, Fonstad C G and Gelpey J C, 1983, App Phys Lett, 43 647

Umemoto Y, Takahashi S, Matsunaga N and Nakamura M, 1984, Elec Letts, 20 98

Vaidyanathan K V, Jullens R A, Anderson C L and Dunlap H L, 1983, Sol State Elec, 26 717

Inst. Phys. Conf. Ser. No. 74: Chapter 2
Paper presented at Int. Symp. GaAs and Related Compounds, Biarritz, 1984

83

Rapid thermal annealing of Zn^{64} and Mg^{24} implants in GaAs

S. Tiwari, J. C. DeLuca and V. R. Deline

IBM Thomas J Watson Research Center PO Box 218, Yorktown Heights, NY 10598

Abstract. Zn^{64} and Mg^{24} implants and their annealing were studied under conditions of rapid thermal annealing for 200 keV energy implantation using SIMS and differential Hall measurement. Both species exhibit very little or no diffusion in the temperature range studied (800 °C to 1110 °C) for fluences up to 10^{14} cm^{-2}, with near ideal activations in the optimal temperature range and Hall mobilities close to 100 cm^{2}v^{-1}.s^{-1}. At a fluence of 1 x 10^{15} cm^{-2} or more, both species show anomalous diffusion. At high concentrations (>1 x 10^{19} cm^{-2}) they show a high concentration dependent diffusion with significant accumulation of dopant at the surface. This is postulated to be due to either point defects generated during implantation and those formed at the surface, or due to defects generated when recrystallization occurs during annealing. At the tail end, zinc shows a diffusion that is temperature and dose dependent, and is postulated to be due to an interaction with point defects. Magnesium, however, shows a rapid initial diffusion even at the lowest temperature (800 °C) followed by a slow diffusion. We consider this to be due to the interstitial distribution of magnesium at the tail end which diffuses rapidly until it becomes substitutional. Magnesium activates at temperatures as low as 770 °C and shows a broad temperature range for constant activation.

1. Introduction

This study is motivated by the need for understanding acceptor redistribution and incorporation in GaAs for applications such as the heterojunction bipolar and the heterojunction p-channel field effect transistors. Rapid thermal annealing (RTA, less than 8 sec. rise time) for time durations of 3 to 6 seconds was used to minimize dopant redistribution. Beryllium and zinc have been the preferred species, to date, for acceptors. Beryllium being a small atom exhibits channelling during implantation. It also, possibly, would diffuse faster than any other acceptor species because of its significantly large interstitial nature of movement. Zinc has been the subject of contradictory evidence [Kendall]. In numerous reports - it appears to show concentration dependent diffusion and significant pile up at the surface even at low concentrations. For the purpose of this investigation, magnesium was chosen for its preferred channelling properties, and relatively deep implants and zinc was chosen for a reassessment of the evidence. Relatively shallow and deep implantations were achieved by using a constant implantation energy of 200 keV for the two species. The samples were annealed for various peak temperatures and were characterized using secondary ion mass spectroscopy (SIMS) and differential Hall measurements.

2. Experiment and Discussion

Samples of undoped semi-insulating GaAs were implanted with fluences varying from 1 x 10^{13} cm^{-2} to 1 x 10^{16} cm^{-2} at 7 degrees angle and 200 keV energy and rapid thermal annealed at peak temperatures for 3-6 seconds. The set up was designed to have a pure ambient through the use of cryo-pumped high vacuum (less than 1 x 10^{-6} torr) prior to introduction of dry and deoxidised argon to a pressure of 700 torr. Two banks of infrared quartz lamps were used on both sides of the wafer. The samples are placed with implanted side on a silicon wafer.

At a fluence of 1 x 10^{13} cm^{-2} for zinc, the measured SIMS depth profiles show no measurable diffusion up to an annealing temperature of 1060 $\overset{\circ}{}$C. Activations (defined as percentage of holes measured by Hall measurements divided by implant fluence) of 80 percent or more were measured at temperatures above 950 $\overset{\circ}{}$C. However, mobilities are consistently lower than those in bulk GaAs at similar concentrations [Sze] indicating residual damage and incomplete crystal recovery. These temperatures are higher than those used in conventional low temperature annealing of GaAs. As in silicon, during RTA, a higher temperature is needed for recovery because of the small times employed. Fig. 1 and Fig. 2 show the observed dopant profiles at various annealing temperatures for a fluence of 1 x 10^{14} cm^{-2} and 1 x 10^{15} cm^{-2}. At the fluence of 1 x 10^{14}cm^{-2}, only the highest temperatures (950 $\overset{\circ}{}$C and more) indicate any diffusion of zinc and this enhanced diffusion itself appears to be characterized by a single temperature dependent diffusion coefficient. No diffusion is observed at the lower fluence of 1 x 10^{13} cm^{-2} at similar temperatures suggesting a dose dependent diffusion related to increased lattice defects via simple or elaborate complexes. The simplest complex for such an enhanced diffusion effect is zinc in **interstitial** site coupled with a vacancy. This complex has been postulated as the source of anomalous diffusion effects in superlattices [Laidig]. Activations in these samples is 50 percent or more at temperatures above 900 $\overset{\circ}{}$C and reaches above 90 percent at temperatures of 1000 $\overset{\circ}{}$C or more. This indicates a predominantly substitutional nature for zinc after annealing. Since heavy doping effects are not well understood in p-GaAs, it is not clear if the observed activations are ideal or not.

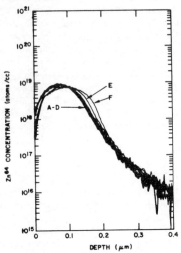 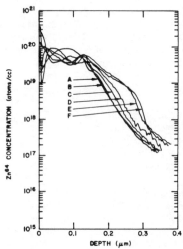

Fig. 1 Distribution of Zn (fluence 1 x 10^{14} cm^{-2}) measured by SIMS after annealing. The peak temperatures are A - 760 $\overset{\circ}{}$C, B - 818 C, C - 860 $\overset{\circ}{}$C, D - 908 $\overset{\circ}{}$C, E - 950 $\overset{\circ}{}$C and F - 1040 $\overset{\circ}{}$C.

Fig. 2 Distribution of Zn (fluence 1 x 10^{15} cm^{-2}) measured by SIMS after annealing. The peak temperatures are A - 760 $\overset{\circ}{}$C, B - 818 C, C - 860 $\overset{\circ}{}$C, D - 908 $\overset{\circ}{}$C, E - 950 $\overset{\circ}{}$C and F - 1060 $\overset{\circ}{}$C.

The depth profiles of zinc at a fluence of 1 x 10^{15} cm^{-2} (Fig. 2), however, show a very different behavior with multiple inflection points. These observed profiles are similar in nature but with much smaller diffusion distances than earlier data [Kular]. Significant diffusion occurs both at maximum concentrations and at the tail end. The most pronounced effect is redistribution at the peak which accumulates at the surface and moves rapidly as a function of temperature into the bulk, resulting in the knee. A larger fraction of this zinc is, however, accumulates at the surface. As the temperature of annealing increases the samples show a broad plateau at 4 - 5 x 10^{19} cm^{-3}. Because of the higher fluence, the material in this case has either a significantly higher concentration of lattice defects or the surface is amorphized. The latter is a strong possibility because GaAs is expected to have a lower amorphization threshold than silicon because of higher entropy change per entering ion [Sadana]. In the case of phosphorus implants in silicon at this fluence and energy, amorphization is known to occur. The concentration, dose and implant energy dependent diffusion near the surface can be caused by enhanced diffusion through complexes such as the vacancy-interstitial zinc complex. These point defects already exist due to implantation, are created at the surface (e.g. a vacancy) or are formed during the regrowth of crystalline layer from the amorphous layer. At the tail end, with lower temperatures, slower diffusion is observed which is seen to increase until it slows down again at 1050 °C. This is consistent with the model used to explain the behavior at tail end of Fig. 1. A higher concentration of point defects and a higher concentration of dopant atoms resulting from the redistribution at the peak is available in this case compared to that of Fig. 1. This results in enhanced concentration dependent diffusion.

The enhanced diffusion due to the complex is temperature dependent and proceeds rapidly until a temperature is reached where the point defects disappear due to recombinations. This results in slowing down of diffusion with temperatures greater than 1030 °C. This is consistent with the results of Fig. 3 that show a rise in mobility at these temperatures. In interpreting Fig. 3, careful consideration has to be given to degeneracy effects because although the density of states in valence band is relatively high (7×10^{18} cm^{-3}), the acceptor degeneracy factor is 4, and the ionization energy is actually higher than the thermal energy even at room temperature (.028 eV for Mg and .031 eV for Zn). In addition, heavy doping effects in GaAs are not well understood. The differential Hall measurements in Fig. 3 indicate near complete activation, including at the surface indicating that most zinc is in substitutional sites.

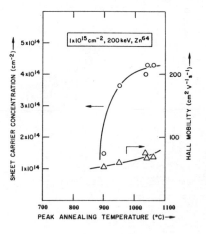

Fig. 3 Hole properties upon RTA of Zn implants at 1×10^{15} cm^{-2} fluence and 200 keV energy.

The behavior of magnesium is documented in Fig. 4 and 5 for atomic distribution using SIMS for similar fluences as those used in Fig. 1 and Fig. 2. The corresponding hole properties are reported in Fig. 6 and 7. It can be seen from Fig. 4 that magnesium shows no noticeable diffusion upto 1060 °C. The activation as shown in Fig. 6 is greater than 70 percent for optimal conditions. The mobility of the material is close to bulk values, and is indicative of less damage than in case of zinc. This is to be expected because the lighter Mg atom causes less damage to the lattice than the heavier Zn atom. At 1×10^{14} cm^{-2} fluence, diffusion is detected only at 1060 °C - a temperature which is significantly above the temperatures where the highest activation was obtained. The mobility of the material shows a consistently lower mobility as the fluence is increased in the range 1×10^{13} cm^{-2} to 1×10^{16} cm^{-2}. This is consistent with higher lattice damage at higher fluence and higher impurity scattering. The hole concentration is indicative of near complete occupancy of substitutional Ga sites by magnesium. The high activations appear at considerably lower temperatures than that in zinc (800 °C compared to 920 °C).

At 1×10^{15} cm^{-2} fluence, considerable anomalous diffusion is observed - similar in part to the case of zinc. Even at the lowest temperature of annealing (770 °C), a rapid initial movement of magnesium atoms occurs at the tail - a characteristic distinctly different from zinc anneal. As the annealing temperature is raised further the diffusion of the tail is much less and is characterized by a simple diffusion coefficient.

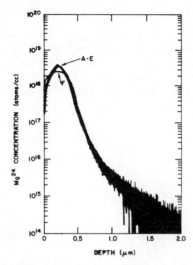

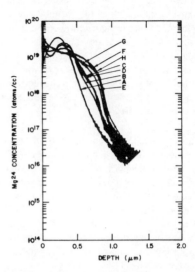

Fig. 4 Distribution of Mg (fluence 1 x 10^{14} cm^{-2}) measured by SIMS after annealing. The peak temperatures are A - 770 °C, B - 815 °C, C - 850 °C, D - 920 °C, E - 980 °C and F - 1040 °C.

Fig. 5 Distribution of Mg (fluence 1 x 10^{15} cm^{-2}) measured by SIMS after annealing. The peak temperatures are A - as implanted, B - 770 °C, C - 815 °C, D - 920 °C, E - 980 °C and F - 1040 °C, G - 1065 °C.

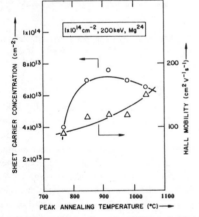

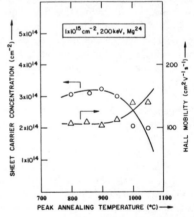

Fig. 6 Hole properties upon RTA of Mg implants at 1 x 10^{14} cm^{-2} fluence and 200 keV energy.

Fig. 7 Hole properties upon RTA of Mg implants at 1 x 10^{15} cm^{-2} fluence and 200 keV energy.

This case is very similar to the case of boron diffusion at the tail in silicon during rapid thermal annealing [Hodgson]. Being of a small atomic radius both the species can have large diffusion coefficients corresponding to interstitial movement. Implantation leaves point defects deeper than the dopant implantation. As the temperature is raised during the rapid annealing, point defect agglomeration is prevented by the rapid temperature rise and interstitial atoms diffuse rapidly until they recombine with a vacancy and become substitutional. This process occurs even at 770 °C because of the rapid interstitial movement and a larger point defect density due to the high fluence and possible amorphous to crystalline transition. The accumulation of magnesium at the surface is indicative of a very similar mechanism as in the case of zinc, and is considered to be due to similar causes. The activation of the implants (Fig. 7) shows a plateau of activation at greater than 30 percent for temperatures above 780 °C. The mobilities show a degradation compared to the lower fluence samples and also show the need for higher temperatures for better damage annneal.

Fig. 8 shows results of differential Hall measurements for hole concentration with depth with the SIMS distribution of magnesium as reference at 815 °C anneal. The measured hole distribution which may have an error of nearly 20 percent, closely follows the magnesium distribution. If degenerate statistics were employed, it indicates near complete activation at the rapid thermal annealing temperature of 815 °C.

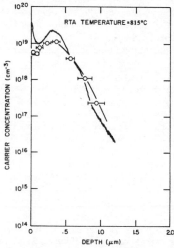

The residual damage in the material can have considerable effect on device performance because it acts as a generation and recombination center. The behavior of the effect of these traps as generation centers in devices can be gauged by observing the behavior of current voltage characteristics in p^+/n diodes since this damage will extend into the underlying n material.

Fig. 8 Hole distribution as a function of depth for a Mg, 1×10^{15} cm^{-2} fluence, 200 keV energy implant annealed at 815 °C. SIMS profile is drawn for reference.

GaAs doped to 1×10^{17}cm^{-3} with silicon was implanted with magnesium at fluences of 1×10^{15} and 1×10^{16} cm^{-2}. Samples were annealed at 850 °C, a temperature at the center of activation plateau during annealing of 1×10^{15} cm^{-2} implant. This temperature was, however, found to give only about 5 percent activation for 1×10^{16} cm^{-2} implantation. The damage and amorphization during the extremely high fluence need higher temperatures for complete activation. The diodes were measured for their temperature dependent current voltage characteristics.

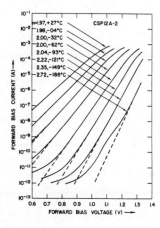

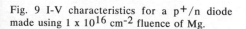

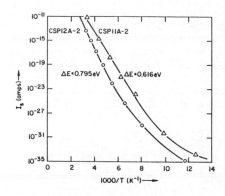

Fig. 9 I-V characteristics for a p^+/n diode made using 1×10^{16} cm^{-2} fluence of Mg.

Fig. 10 Arrhenius plots for current for p^+/n diodes made using 1×10^{15} cm^{-2} and 1×10^{16} cm^{-2} Mg implantations.

Fig. 9 and 10 show example of diode characteristics and Arrhenius plot for diodes. For the case of 1 x 10^{16} cm^{-2} fluence (Fig. 9) the forward current characteristics have an ideality factor close to 2 indicated by the dotted line over a wide temperature range. This indicates that generation current may be the dominant current mechanism in this structure. For the plots of Fig. 10 the temperature prefactor term has not been included because, a-priori, its temperature dependence is not known. The ideality factor for diodes with 1 x 10^{15} cm^{-2} fluence over the same temperature range is found to vary between 1.16 and 1.38. This shows that the diodes although not ideal, are not dominated by traps. The latter would have resulted in an ideality factor of 2. The activation in the Arrhenius plot (Fig. 10) of the saturation current indicates an activation energy of .616 eV which is considerably less than half of the band-gap supporting this conclusion. The 1 x 10^{16} cm^{-2} fluence diodes, however, show an activation energy of .795 eV which is closer to half the band gap for these temperatures showing that saturation current is proportional to the intrinsic carrier concentration. The ideality factor and the activation energy make us conclude that in this case the diode behavior is dominated by generation recombination centers. Also, while the 815 °C temperature is sufficient for annealing damage of the 1 x 10^{15} cm^{-2} fluence, it is not sufficient for 1 x 10^{16} cm^{-2}. The lower activation of 1 x 10^{16} cm^{-2}, and its lower mobility both indicate the incomplete annealing of damage.

3. Conclusions

In this paper we have studied the nature of rapid thermal annealing for magnesium and zinc implants over a variety of fluences to study their diffusion properties, activation properties and the effect of residual damage on active device properties. Both species exhibit a threshold in fluence beyond which a larger anomalous diffusion behavior occurs. At high fluences, both species show an anomalous accumulation at the surface relatable to either lattice point defects produced during implantation and from the surface or due to amorphized - single crystal material transition during annealing. Unlike zinc, magnesium at high fluence shows a fast initial diffusion at the tail even at very low annealing temperatures, which is followed by a slower diffusion at higher annealing temperatures. This may be related to the interstitial nature of atoms at the tail after implantation. Magnesium activates with high efficiency at all fluences upto 1 x 10^{15} cm^{-2}, shows a broad temperature range of high activation, and shows higher mobility than annealed zinc implants. In all cases considered, RTA is found to have lower diffusion and better activations than those reported in the literature for conventional anneals.

Acknowledgements

We wish to thank Marc Albert and Alyce Ginzberg who helped in this study, Conrad Lanza for making available his measurement facilities and Ralph Fiorio for providing the implantations.

References

Hodgson R. T., Deline V. R., Mader S., and Gelpey J. C., 1984, Appl. Phys. Lett., V44, p.589.
Kendall D. L., 1968, in Semiconductors and Semimetals, ed. Willardson R. K., and Beer A. C., Academic Press, New York, V4, Ch. 3.
Kular S. S., Sealy B. J., Ono Y., and Stephens K. G., 1984, Sol. State Electron., V27, p. 83.
Laidig W. D., Holonyak N., Camras M. D., Hess K., Coleman J. J., Dapkus P. D., and Bardeen J., 1981, Appl. Phys. Lett., V38, p.776.
Sadana D. K., Choksi H., Washburn J., Byrne P. F., and Cheung N. W., 1984, Appl. Phys. Lett., V44, p.301.
Sze S. M., 1981, Physics of Semiconductor Devices, John Wiley and Sons, p.29.

Inst. Phys. Conf. Ser. No. 74: Chapter 2
Paper presented at Int. Symp. GaAs and Related Compounds, Biarritz, 1984

89

New tools to investigate large diameter LEC semi-insulating undoped GaAs

B. GOUTERAUX, N. VISENTIN, B. LENT, M. BONNET

THOMSON-C.S.F., D.H.M., Domaine de Corbeville, Boîte Postale n° 10
91401 ORSAY (France)

ABSTRACT :

   Two original experiments have been set-up to characterize semi-
   insulating GaAs crystals grown in a pyrolitic boron nitride crucible
   using the LEC technique. The EL2 optical absorption coefficient and
   the Hall characteristics distribution are measured automatically over
   a two-inch wafer surface with 1.5 x 1.5 mm² resolution. From the Hall
   measurements, the EL2 population rate at 296 K can be deduced and
   then the EL2 concentration distribution. The evolution of the EL2
   concentration pattern from seed-end to tail-end is analysed. Also the
   influence of the B_2O_3 encapsulant humidity on the average EL2
   concentration is discussed.

INTRODUCTION :

Large diameter semi-insulating undoped GaAs crystals are grown at high
pressure using the liquid encapsulated Czochralski technique, for
integrated circuit applications. The initial deviation from stoïchiometry
in the melt is minimized by controlling the in-situ compounding of GaAs
in a pyrolitic boron nitride (PBN) crucible.

The electrical properties of GaAs are explained by the schematic model
(Martin 1980) : two shallow levels E_A and E_D are compensated by a donor -
type deep level EL2 which pins the FERMI level E_F near the mid-gap. The
concentrations of these levels are N_A, N_D, N_{EL2}, respectively.

It is of prime importance to know the distribution of this electrically
active centre, through a crystal, in order to determine its influence
on the inhomogeneity of the I.C. performance (Miyazawa 1983).

With this aim we have set-up two original experiments which automatically
scan a two-inch wafer surface with 1.5 X 1.5 mm² resolution, measuring
the Hall characteristics (Bonnet 1982) and the optical absorption
coefficient of more than 800 elementary samples.

THEORY :

1) The occupation rate of the EL2 centre is given by :

$$f_{EL2}(T) = \left[1 + \frac{N_c(T)}{n(T)} \exp\left(-\frac{E'_n(T)}{kT}\right)\right]^{-1} \tag{1}$$

$$\begin{cases} E'_n(T) = 0.759 - 2.37 \times 10^{-4} \times T \text{ eV (Martin 1980)} \\ n(T) = N_c(T) \exp\left(-\frac{E_c - E_F}{kT}\right) \end{cases}$$

For n-type semi-insulating GaAs, the neutrality equation :

$$N_A - N_D = N_{EL2}(1 - f_{EL2}) \tag{2}$$

gives :

$$n(T) = \left(\frac{N_{EL2}}{N_A - N_D} - 1\right) N_c(T) \exp\left(-\frac{E'_n(T)}{kT}\right)$$

The slope of the curve $\frac{1}{n(T)} T^{3/2} \left(\frac{1}{T}\right)$ is 0.759 eV, then if the activation energy E_A of the Hall constant is close to 0.759 eV, we have :

$$n_H(T) = \frac{1}{q R_H(T)}$$

which means that the compensation is only due to the electrons and that the EL2 centre is responsible for the compensation.

Therefore if we measure E_A between 350 K and 450 K it is possible to compute R_H (296 K) and then f_{EL2} (296 K) for each elementary sample .

2) The optical absorption coefficient is given by the expression :

$$\alpha = A N_{EL2} \sigma^\circ_n f_{EL2} \left[1 + \frac{\sigma^\circ_P}{\sigma^\circ_n}\left(1 - \frac{1}{f_{EL2}}\right)\right] \tag{3}$$

where σ°_n (σ°_p) is the electron (hole) photoionisation cross-section for the EL2 centre. At 296 K, $\frac{\sigma^\circ_P}{\sigma^\circ_n}$ (E = 1.13 eV) < 1 and for the samples we have studied f_{EL2} (296 K) > 0.75, then the last factor of (3) is very close to 1. To assess the transition responsible for the absorption we also verified that the spectra α (E) are similar to the curves σ°_n (E) (Bois 1980). The $A\sigma^\circ_n$ factor is determined by the calibration of the system : we used the curves N_{EL2} (α) given by Martin (1980). So :

$$N_{EL2} = \frac{1}{A\sigma^\circ_n} \frac{\alpha}{f_{EL2}} \tag{4}$$

and from the α and f_{EL2} cartographs we can easily obtain a N_{EL2} cartograph.

EXPERIMENTAL

We are interested in determining the electrical characteristics and the
position of the FERMI level in the band gap.

The Hall measurements, made on an elementary square-shaped sample using
the Van Der Pauw method, and the displacements of the wafer are controlled
by a mini-computer (Bonnet 1982).

Another automatic experiment is used to measure the optical absorption
coefficient of a 5 mm - thick wafer polished on both faces. The incident
beam is square - shaped and has an energy of 1.13 eV. The measurements
and the displacements are also controlled by a mini - computer.

All these data are stored in a memory unit and subsequently analysed by
the computer and a cartograph is plotted.

RESULTS

The influence of the mass fraction solidified, of the stoïchiometry, and
of the humidity of the boric oxide on the distribution of α, n_H and N_{EL2}
across a wafer has been analysed.

1) Near the seed-end of the ingot, the systematic four-fold symmetry of
the distribution of α controls the N_{EL2} distribution pattern. Both n_H and
α have the same distribution across most of the studied wafers : four
minima located at the middle of the < 110 > radii and a maximum at the
centre and at the periphery.

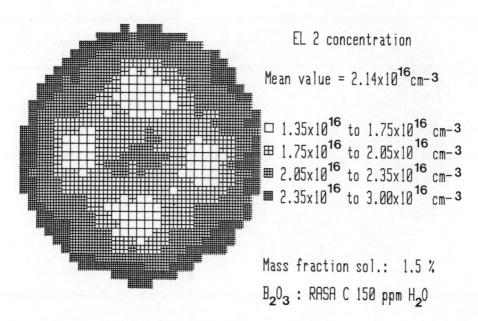

EL 2 concentration

Mean value = 2.14×10^{16} cm-3

□ 1.35×10^{16} to 1.75×10^{16} cm-3
⊞ 1.75×10^{16} to 2.05×10^{16} cm-3
▦ 2.05×10^{16} to 2.35×10^{16} cm-3
▩ 2.35×10^{16} to 3.00×10^{16} cm-3

Mass fraction sol.: 1.5 %

B_2O_3 : RASA C 150 ppm H_2O

Figure 1 : Distribution of the concentration of the
EL2 center across a seed-end wafer.

Therefore the total dispersion of N_{EL2} is less than that of α (see equations (1) and (4)). In some wafers the resistivity ρ_H and n_H do not follow the four-fold symmetry : then the W - shaped pattern of ρ_H and n_H along the < 110 > direction is not present systematically.

When 50% of the crystal has been pulled, the quadratic symmetry evolves towards an axial symmetry. Then, when the mass fraction solidified increases, the maximum located at the center of the wafer spreads out and progressively shifts towards the edge.

Near the tail-end of the crystal any noticeable pattern vanishes.

The average EL2 concentration, $\overline{N_{EL2}}$ across a wafer increases by up to 30% from the seed-end to the tail-end of the same ingot ; the As atom fraction in the melt was 50.0% and did not vary more than 0.2% along the growth axis.

2) This quantity was computed using the mass balance method. An As loss of 11g during the growth, starting from a 3 kg melt, has been determined experimentally ; the remaining calculated loss was assumed to be due to As evaporation during the synthesis.

$\overline{N_{EL2}}$ increases rapidly up to a threshold value of the As atom fraction in the melt equal to 49% and then saturates in the 1.0×10^{16} - 3×10^{16} cm^{-3} range.

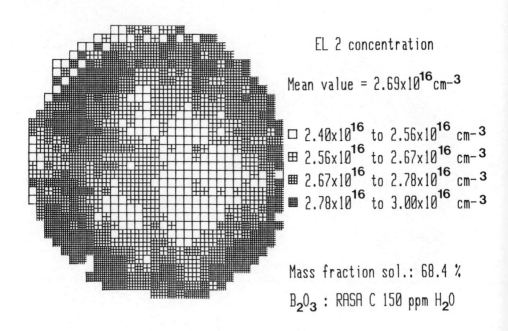

EL 2 concentration

Mean value = 2.69×10^{16} cm^{-3}

□ 2.40×10^{16} to 2.56×10^{16} cm^{-3}
⊞ 2.56×10^{16} to 2.67×10^{16} cm^{-3}
⊞ 2.67×10^{16} to 2.78×10^{16} cm^{-3}
■ 2.78×10^{16} to 3.00×10^{16} cm^{-3}

Mass fraction sol.: 68.4 %

B_2O_3 : RASA C 150 ppm H_2O

Figure 2 : Distribution of N_{EL2} across a tail-end wafer coming from the same ingot as Fig. 1.

3) The role of the humidity of the encapsulant has been studied on wafers coming from a stoïchiometric melt (As atom fraction in the melt : $50.00 \pm 0.1\%$). Two different specifications of B_2O_3 from RASA have been used : $\overline{E}$ - 150 ppm H_2O and C - 300 ppm H_2O. When comparing wafers corresponding to the same mass fraction solidified, it is found that N_{EL2} (n_H) decreases (increases) with the water content of the encapsulant.

DISCUSSION

Considering the As atom fraction in melt, two values are of importance : 47.5% (Holmes 1982) and 49.0%.

1) Up to 47.5% the material is p-type and the compensation is governed by the acceptor with energy 78 meV (Elliott 1982). The spectra α (E) and σ°_n (E) are no longer similar and no absorption seems to be due to EL2.

2) Between 47.5% and 49%, the material is semi-insulating and n-type, the activation energy of the Hall constant is close to 0.76 eV and $\overline{N_{EL2}}$ is very low (in the range of 1.0×10^{16} to 1.2×10^{16} cm^{-3}).

High precision D.L.T.S. measurements have shown the existence of a family of traps with a thermal activation energy centred around 0.820 eV (Taniguchi 1983, Lagowski 1984). If these traps have an activation energy of the thermal capture cross section close to 0.060 eV then they can account for the dispersion of the activation of the Hall constant. Provided that these traps have superimposed optical absorption bands, then we have measured the total concentration of these traps, and not solely that of EL2.

3) Between 49.0% and 51.0%, EL2 is systematically present with concentrations superior to 1.6×10^{16} cm^{-3}. The most important parameter seems, then, to be the humidity of the boric oxide. For very dry B_2O_3 (150 ppm H_2O), $\overline{N_{EL2}} > 2 \times 10^{16}$ cm^{-3} and $\overline{n_H}$ is low, which means that the concentration of acceptors is high. In this case $f_{EL2} = 75\%$ and the optical absorption measurements are not sufficient to deduce the total EL2 concentration.

The main acceptor has been determined to be carbon (Holmes 1982) and to have a concentration dependent on the humidity of the B_2O_3 (Hunter 1984). The presence of carbon can be explained by a competition between the hygroscopic nature of the B_2O_3 and the formation of CO - CO_2 during the synthesis.

It is tempting to correlate the increase of the carbon and of the EL2 concentrations. Assuming that the EL2 centre is related to the Ga_{As} antisite defect (Lagowski 1982), a possible mechanism would be an increase of this defect due to the preferential incorporation of carbon on an As site (Theis 1983), and in consequence a decrease of Ga vacancies in the crystal.

Near the seed-end of the ingot, the distribution of EL2 and of the free carriers is ruled by the crystallographic symmetry. The minima observed may reflect the facetting of the solid-liquid interface along (111) planes (Mullin 1980) due to anisotropic growth rates and to possible convection cells.

As the melt volume is reducing the stirring becomes different, so the convection pattern is modified progressively, leading to an axial symmetry.

CONCLUSION

These two experiments have allowed us to determine more precisely the distribution of the EL2 centre across two-inch wafers of undoped GaAs and thus to point out the importance of the water content of the boric oxide.

The existence of EL2 is governed by the stoichiometry of the melt, its concentration by the humidity of the B_2O_3 and the mass fraction solidified ; its distribution across a wafer is only dependent on this latter parameter and is probably an image of the solid-liquid interface.

REFERENCES

1   Bois D and Chantre A ; Rev. Phys. Appl. 15 631 (1980)

2   Bonnet M, Visentin N, Gouteraux B, Lent B and Duchemin JP ;
    Proceedings of the GaAs IC Symposium New Orleans November 1982

3   Elliott K R, Holmes D E, Chen R T and Kirkpatrick C G ;
    Appl. Phys. Lett. 40 898 (1982)

4   Holmes D E, Chen R T, Elliott K R and Kirkpatrick C G ;
    Appl. Phys. Lett. 40 46 (1982)

5   Hunter A T, Kimura H, Baukus J P, Winston H V and Marsh O ;
    J Appl. Phys. Lett. 44 74 (1984)

6   Lagowski J, Gatos H C, Parsey J M, Wada K, Kaminska M and
    Walukiewicz W ; Appl. Phys. Lett. 40 342 (1982)

7   Lagowski J, Lin D G, Aoyama T and Gatos H   ;
    Appl. Phys. Lett. 44 336 (1984)

8   Martin G M, Farges J P, Jacob G and Hallais J P ;
    J. Appl. Phys. 51 2840 (1980)

9   Miyazawa S, Ishii Y, Ishida S and Nanishi Y ;
    Appl. Phys. Lett. 43 853 (1983)

10 Mullin J B and Hulme K F ; J. Phys. Chem. Solids 17 1 (1960)

11 Taniguchi M and Ikoma T ;
    J. Appl. Phys. 54 6448 (1983)

12 Theis W M, Bajaj K K, Litton C W and Spitzer W G ;
    Physica 117 B & 118 B 116 (1963)

Inst. Phys. Conf. Ser. No. 74: Chapter 2
Paper presented at Int. Symp. GaAs and Related Compounds, Biarritz, 1984

Evidence for polycrystalline insulars in dislocated undoped GaAs single crystals; comparison with near dislocation free crystals

Jean-Pierre CORNIER*, Marc DUSEAUX* and Jean-Pierre CHEVALIER**

* Laboratoires d'Electronique et de Physique Appliquée
  3, avenue Descartes, 94450 LIMEIL-BREVANNES, France
** C.E.C.M. - C.N.R.S. 15 rue G. Urbain, 94400 VITRY, FRANCE

Abstract : The formation of dislocations in Czochralski GaAs single crystals is a problem which is still unsolved. We present the results of an extensive transmission electron microscopy study of as grown undoped GaAs single crystals, both dislocated and virtually dislocation-free. These have been produced from either deliberately Ga or As rich melts, or doped with In, as well as in "standard" conditions. In dislocated crystals, we have observed a low density ($\sim 10^8 - 10^9$ cm^{-3}) of small (1000 Å) particles; through careful electron diffraction experiments, and using X-ray microanalysis, we have deduced that these particles consist of polycrystalline GaAs insulars. We discuss how such insular crystals can be formed during growth, and how they can act as sources for dislocations which would then be propagated through thermal stresses.

1. Introduction

For a number of electronic devices, notably lasers, the existence of dislocations is detrimental to the device performance. For integrated circuits, dislocations are thought to affect homogeneity. This is the motivation in attempts to produce dislocation free single crystals. However, dislocation densities in usual semi-insulating GaAs Liquid Encapsulated Czochralski (LEC) crystals is typically in the $10^4 - 10^5$ cm^{-2} range. Indium doping reduces this to less than 5.10^2 cm^{-2}, which can be considered as being virtually dislocation free (Jacob et al, 1983 ; Duseaux and Martin, 1984). On the whole, it can be said that the mechanisms of dislocation nucleation during crystal growth are not well understood, and that thermal stresses, existing in the crystal during LEC growth, which are often invoked for the production of such dislocations (Jordan et al, 1980 ; Duseaux, 1983), cannot be the sole mechanisms (Duseaux and Jacob, 1982). This suggests that some kind of as-grown crystal defect (e.g. clusters of vacancies, antisites, Ga or As (Cullis et al, 1980) precipitates, etc.) could act as sources for dislocations.

The guiding lines in the study presented here are to examine a large number of specimens corresponding to different regions of different single crystals (see Table 1) and look for as-grown defects ; the observations are then tentatively correlated to the observed dislocation densities. The observations are made by electron microscopy, although this technique is not well suited to the observation of low ($< 10^{10}$ cm^{-3}) densities of defects. To overcome the problems involved in sampling, it was necessary to observe considerable numbers of thin foils : in this case over 200 foils, which enables a rough statistical study to be made.

2. Experimental procedure

A full description and discussion of the LEC method of crystal growth can be found elsewhere (Jacob et al, 1982) and will not be presented here. Tables 1 and 2 summarise the different single crystals studied, together with the number of thin foils examined in each case. The (100) single crystal ingots are sliced perpendicularly to the growth direction and 2.3 mm discs are cut from each slice, using an ultrasonic drill. Three slices from each ingot are examined (top, middle and bottom) whenever possible. In the case of ingots grown from either Ga or As rich melts, it is possible to calculate approximate values (Holmes et al 1982) for the Ga:As ratio (see Table 1). Thin foils have been prepared using chemical thinning (Adachi, 1981) in a suitably modified twin-jet thinning machine. The foils have been observed at 200 kV in a JEOL 200 CX transmission electron microscope. Analysis was carried out using an energy dispersion X-ray detector in a vacuum generator HB 501 Scanning Transmission Electron Microscope (STEM) at 100 kV.

TABLE 1

Crystal	Slice	Dislocation density	Dopant	As : Ga ratio	Comment
A	middle	$10^4 - 10^5$ cm^{-2}	undoped	50 : 50	"standard" crystal
B B B	top middle bottom	$10^4 - 10^5$ cm^{-2} " "	undoped " "	50.4 : 49.6 50.56 : 49.44 50.8 : 49.2	As rich melt
C C	top bottom	$10^4 - 10^5$ cm^{-2} "	undoped "	48.1 : 51.9 46.15 : 53.85	Ga rich melt
D	middle close to top	0	undoped	50 : 50	∅ 20 mm crystal pulled with the necking procedure
E F	cut.lengthwise "	500 "	Indium "	-	In doping ~ 8 x 10^{19} cm^{-3}

We have estimated the average volume observed in the microscope by assuming that it corresponds to a torus with a wedge section. The wedge angle and maximum thickness can be obtained using two-beam thickness fringes (Hirsch et al, 1977). The value obtained for the average volume observed per thin foil is :

$$V = (1 \pm 0.5) \times 10^{-9} cm^3.$$

3. Results

In the dislocated single crystals, we have observed particles of about 1000 Å in size and with a very low number density. We present these results in two sections below : first, a description of the statistics of particle observation as a function of the nature of the single crystal and then a brief summary of the analysis of what we have considered to be a typical particle.

3.1 Statistics of particle observation

The results can be obviously separated into two categories : specimens in which particles have been observed and specimens for which none have been observed. The results are summarised in Table 2. Number densities of particles range from about 10^8 to 10^9 cm^{-3} for the following crystals : A ("standard GaAs"), B (grown from an As rich melt) and C (grown from a Ga rich melt). In all three cases, particle density is less by a factor of about 10 for specimens from the bottom slice with respect to the top slice. On the other hand, no particles have been observed for the In doped crystals (E,F) (which is virtually dislocation-free) nor for the rather special dislocation free small diameter crystal (D).

TABLE 2

Crystal	Slice	Nbre thin foils observed	Total nbre of particles observed	Particle density
A	middle	5	5	10^9 cm^{-3}
B	top	22	19	10^9 cm^{-3}
B	middle	34	25	7×10^8 cm^{-3}
B	bottom	46	3	7×10^7 cm^{-3}
C	top	29	28	10^9 cm^{-3}
C	bottom	14	4	3×10^8 cm^{-3}
D	middle close to top	5	0	$< 2 \times 10^6$ cm^{-3}*
E	slice cut	70	0	$< 2 \times 10^5$ cm^{-3}*
F	Lengthwise			

* These values are calculated using the Poisson theorem (e.g. Papoulis, 1965) for a probability of 99 % of not finding particles in the specimens observed.

3.2 Analysis of a typical particle

All the particles have very similar morphology and sizes, suggesting that the complete analysis of what we consider to be a typical particle can be generalized to the others. This analysis has been published elsewhere (Cornier et al, 1984) and the main findings are summarised below.
- Particles are of size $\sim$ 1000 Å.
- About 4 out of 5 particles are associated with a dislocation.
- The particle consists of several grains.
- There is no matrix strain around the particle.
- The cavity occupied by the particle (negative crystal, c.f. Hartman, 1973) is bounded by $\{111\}$ faces (cf fig. 1).
- Diffraction patterns from the particle show extra spots which can be indexed as GaAs.
- X-ray microanalysis shows no detectable changes in composition and no traces of impurities (with atomic number greater than Z = 11).

We deduce from these findings that the particles consist of polycrystalline GaAs insulars.

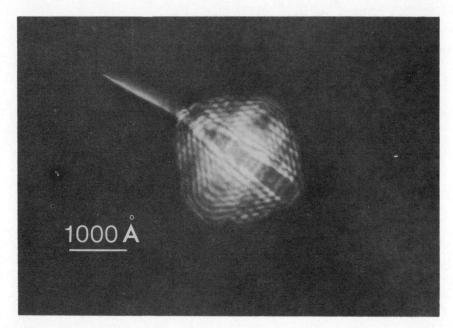

FIGURE 1 : (220) weak-beam image showing thickness fringes and an impinging dislocation. The fringes correspond to a thickness gradient in the single crystal matrix, thus revealing the morphology of the cavity occupied by the particle.

4. Discussion

It is obviously tempting to correlate the observation of insulars with the presence of dislocations. Figure 2, which shows a dislocation (perhaps) bowing out from an insular can also encourage this correlation. That is, given that the insular and the matrix together produce a grain boundary, dislocations could be readily nucleated at this preferential site and then be propagated by the thermal stress. Since the nature of the grain boundary will depend on the relative orientation of insular and matrix and our observations suggest that this can be near random, we can expect a spectrum of resolved stresses for the propagation of dislocations form these insulars.

A closer examination of the results poses further problems. The apparent slight decrease in insular density (about a factor 10) on going from top to bottom is contrary to the variation of dislocation density, which tends to increase towards the bottom of the crystal. This can be tentatively explained by the multiplication of propagating dislocation. Furthermore, the "W" dislocation density profile can be accounted for by the propagation of dislocations due to the thermal stresses, which have also been shown to have a "W" distribution in magnitude (Jordan et al, 1980 ; Duseaux, 1983).

This decrease in insular density, from top to bottom, corresponds to an increase in the deviation from stoichiometry of the melt especially in the case of Ga or As rich melts ; this will increase the degree of undercooling and it could be expected that this might increase the density of insulars. The reverse is observed and this will be discussed below.

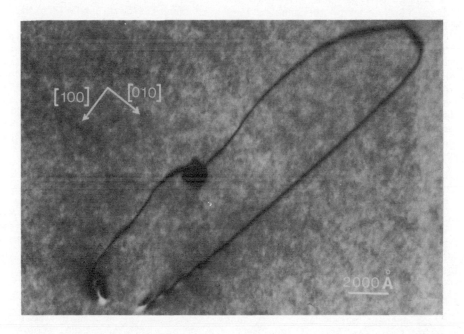

FIGURE 2 : (220) bright field image showing a bowed dislocation and an insular grain.

Examination of the electron micrographs suggests a plausible mechanism for the creation and inclusion of the insular grains in the single crystal matrix : the shape of the cavity (negative crystal bounded by $\{111\}$ facettes) and the absence of strain around the insular indicate that it is likely that it is formed at the growth interface. This (100) growth interface almost certainly consists of a corrugated, or rather facetted surface, growth occurring on $\{111\}$ facettes. Fluctuations in the composition of the melt will necessarily lead to a lowering of the solidification temperature. It is thus possible that a GaAs droplet may be trapped in one of the facetted dimples. This droplet would subsequently crystallise, nucleating on the $\{111\}$ facettes. To avoid strain due to volume change on crystallisation of the droplet, this has to occur before it is completely surrounded by the matrix. The shape of the negative crystal would then be a truncated pyramid with its vertex pointing towards the seed.

It is interesting to examine this simple mechanism in the light of the dislocation free undoped small diameter crystal. Usually GaAs is directly synthesized in the crystal puller from Ga and As in the appropriate proportions and allowing for a slight loss of As. In this case, GaAs was prepared previously and a stoichiometric charge was introduced in the crystal puller. Furthermore, this particular puller (MSR2) produces a strong temperature gradient at the growth interface. It is possible that the combination of these two factors prevents the formation of a substantial number of insulars and hence of dislocations. The stoichiometric charge would reduce the risks of strong composition fluctuations, and in any case, the strong temperature gradient would reduce the effects of melt undercooling fluctuations.

Indeed, if the key of the problem lies with composition fluctuations in the melt, leading to fluctuations in undercooling, then for the proposed mechanism to be consistent with the observed density of insular grains requires anomalies in the liquidus in both the Ga-As and the GaAs-InAs equilibrium phase diagrams. In both cases the slope of the liquidus would have to be very steep for small ($<$ 0.1 %) variations in composition around the ideal stoichiometric value, and would also have to flatten off for larger variations. Fluctuations in undercooling would then be reduced for In doping and also for relatively large Ga or As deviations in the melt. In the latter cases lower densities of insular grains would be in competition with the tendency to produce macroscopic precipitates of Ga or As (of size several µm).

4. <u>Conclusions</u>

In dislocated GaAs single crystals, we have observed a low density of particles (ranging from 10^8 to 10^9 cm^{-3}) of size 1000 Å. These have been clearly identified as GaAs polycrystalline insular grains. In dislocation free crystals, we have not observed any such insulars, implying that their density is substantially lower in such cases. We have proposed that the insulars are incorporated in the single crystal matrix as late-crystallising droplets, induced by fluctuations in melt composition leading to fluctuations in undercooling.

It appears very likely that these insular grains act as dislocations sources and that the dislocations are propagated under the effect of thermal stresses.

References

Adachi S, Kawaguchi H and Iwane G, 1981 J. Mat. Science <u>16</u> 2449
Cornier J P, Duseaux M and Chevalier J P, 1984, to be published in Appl. Phys. Lett.
Cullis A G, Augustus P D and Stirland D J, 1980 J. Appl. Phys. <u>51</u> 2556
Duseaux M, 1982 Thèse Paris 6
Duseaux M and Jacob G, 1982 Appl. Phys. Lett. <u>40</u> 790
Duseaux M, Schiller C, Cornier J P, Chevalier J P and Hallais J P, 1983 Journal de Physique C4 397
Duseaux M, 1983 J. Cryst. Growth <u>61</u> 576
Duseaux M and Martin S, 1984 to be published in "Proc. 3rd Int. Conf. on Semi-insulating III-V Materials, Kah-Nee-Ta (Oregon).
Hartman P, 1973 Ch. 14 of "Crystal Growth : an Introduction" Edit. P. Hartman, Publ. North Holland/American Elsevier, Amsterdam (1973)
Hirsch P B, Howie A, Nicholson R B, Pashley D W and Whelan M J, 1977 "Electron Microscopy of Thin Crystals", 2nd Edition Pub. Robert E. Krieger, Huntingdon New-Yorl (1977).
Holmes D E, Chen R T, Elliott K R, Kirkpatrick C G, 1982 Appl. Phys. Lett. <u>40</u>, 46
Papoulis A, 1965 "Probability, Random variables and Stochastic processes" Publ. Mc Graw-Hill Kogakusha, Tokyo.

* The work at C.N.R.S. on microscopy analysis has been supported by a contract with the "Ministère de la Recherche et de la Technologie".

Inst. Phys. Conf. Ser. No. 74: Chapter 2
Paper presented at Int. Symp. GaAs and Related Compounds, Biarritz, 1984

101

Lattice defects investigation of semi-insulating LEC GaAs crystals by X-ray diffractometry including usage of synchrotron radiation

J. Matsui, T. Kitano, T. Kamejima, and T. Ishikawa[*]

Fundamental Research Laboratories, NEC Corporation, Miyazaki 4-1-1,
Miyamae-ku, Kawasaki, 213 Japan.
[*]National Research Laboratory for High Energy Physics, Oho-machi,
Tsukuba-gun, Ibaraki, 305 Japan.

Abstract   It is described in this paper that lattice orientation in
undoped semi-insulating GaAs varies to a large extent at lineages and to
a small extent at cellular structure walls.   In In-doped GaAs, however,
a striated pattern due to lattice spacing variation is clearly observed.

1. Introduction

Crystalline defects such as dislocations in semi-insulating LEC-grown GaAs
crystals have recently been of much interest due to their possible relation
to FET characteristics variations (Nanishi et al 1983).   It is well known
that the dislocations rearrange into equilibrium configurations in such a
way that they form cellular structure walls (Clark and Stirland 1981) and
lineages, the latter lying most likely along $<110>$ directions in a single
(001) wafer.

Although highly In-doped LEC-grown GaAs crystals (Jacob et al 1983) have
recently been grown for obtaining a bulk with a reduced dislocation density,
there seems little knowledge especially about lattice distortion arising
from nonexistence of the dislocations.

In the present investigation, lattice distortion relating to those defects
are examined in detail by X-ray diffractometry, i.e., transmission and
double- or triple-crystal asymmetric reflection topographies including usage
of synchroton radiation with the requirement of an extremely intense and
parallel x-ray beam, so-called plane wave topography.

2. Experimental

$<001>$ -oriented GaAs crystals used in the present paper are undoped and also
highly ($\sim 5 \times 10^{19}$ cm^{-3}) In-doped LEC-grown with a two-inch diameter.
Several (001) full-size wafers were cut out from various portions of each
ingot with a thickness larger than 3 mm for the reflection topography and of
~ 0.3 mm for the usual transmission topography.

Prior to X-ray measurements, both surfaces of cut wafers were mechano-
chemically polished with special care taken with warpage due to surface
damage induced by the polishing.   MoKα_1 and CuKα_1 X-ray radiations from
rotating anodes were used for the transmission and the (+, -) double-
crystal reflection topographies, respectively.   For a precise measurement
of the reflection intensity distribution, we adopted a triple-crystal

arrangement in the (+, +, -) setting with   radiation ($\lambda \sim 0.7$ angstrom) from a 2.5 GeV synchrotron of National Research Laboratory at Tsukuba. Both for the double- and triple- crystal setting, asymmetric reflections were employed to obtain highly parallel incoming X-ray beams and also to reduce image distortion on a film.

In the reflection case, a series of topographs was taken successively by rotating the specimen wafer with an angular separation of about $3 \sim 5$ sec. of arc around the vertical axis normal to the X-ray beam and parallel to the wafer surface.   After that, the wafer was rotated by 180° around the axis normal to the (001) surface plane, and then, another series of topographs was taken in the same way as before.   By careful inspection of the topographs taken at various angles before and after 180° rotation, one can clarify whether contrast change in reflection intensity is due to lattice orientation variation i.e., lattice rotation or due to lattice spacing variation (Kikuta et al 1966).   To avoid a mistake in understanding   a sign of lattice rotation (concave or convex) in a wafer,   special attention was paid for the seed-end surface to be faced to the incoming X-ray.   Because vertical angular divergence of the incoming X-ray is not restricted in the present case, obtained reflection topographs do not give any information about lattice rotation around a horizontal axis.

3. Results and Discussions

3.1 Lineages and cellular structure walls in GaAs

Figure 1 shows an enlarged transmission topograph of near the < 110 > edge of un- doped wafer, where lineages as designated by an arrow A and walls of cellular structure by an arrow B coexist.   As seen in Fig. 1, transmission topography indicates similarity in nature between the lineage and the cellular structure wall.   From the observation of transmission topographs taken with various diffraction planes, it can be said that   both types of defect are basically bundles of dislocations with different Burgers vectors, since their images are not governed by the extinction rule of dislocation image in any diffraction cases.   It is likely that they are formed essentially by rearrangement of dislocations already generated during crystal growth.

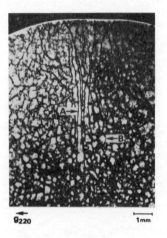

The cellular structures are mostly polygonal in shape in the area other than near <110> radii, while those near <110> radii are rectangular with relatively long sides along the <110> radius.   It is clear, as seen in Fig. 1, that the lineage is a connection of those long segments of the cells.

Fig. 1 X-ray trans- mission topograph of an undoped GaAs wafer, showing lineages indicated by an arrow A and cell walls by an arrow B.

A plausible mechanism ruling in the lineage formation is as follows.   A growing crystal will undergo a tensile stress in the edge and a compressive one in the center of bulk, which cause generation of slip dislocations at

the periphery.   In a (001) wafer, for example, a number of slip dislocations coming from the $\langle 100 \rangle$ edges in the [110] direction would become less
mobile at around half [1$\bar{1}$0] radii and easily rearrange themselves forming
dislocation bundles lying along [1$\bar{1}$0] direction (as A in Fig. 1).   On the
other hand, a shear stress necessary for the dislocations to arise from the
[1$\bar{1}$0] edge may be minimized and few dislocations come into the [1$\bar{1}$0]
direction, resulting a relatively long distance between the [1$\bar{1}$0]-aligned
bundles.   The dislocation rearrangement mentioned above may be performed
not only by slip but also by climb, though no clear evidence for the latter
mechanism has been reported yet.

3.2 Lattice rotation in undoped wafers

It is safely said that, when dislocations of a similar character, at least,

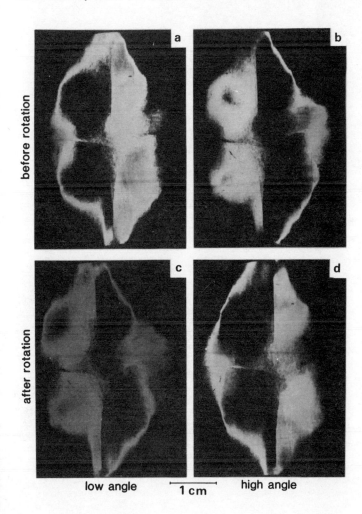

Fig. 2 X-ray
double-crystal
reflection topo-
graphs of an un-
doped GaAs wafer
cut out from a
middle part of
an ingot, showing
contrast change
due to lattice
rotation.
(a) and (b) are
taken at a lower
and a higher
angle, respec-
tively, than the
Bragg peak.
(c) and (d) are
taken in the same
way after 180°
rotation around
the normal to the
(001) surface.

of the same sign, line up above the other, a small variation of lattice orientation, i.e., lattice rotation occurs as in the case of small-angle grain boundary.   Figure 2 (a)-(d) show double-crystal reflection topographs of a wafer cut out from a middle part of an undoped ingot.   (a) and (b) are taken at both sides of the Bragg peak in a rocking curve before 180° rotation, of which angular separation is about 20 sec. of arc, and (c) and (d) are taken in the same way after 180° rotation.   It was confirmed by the analytical method described above, that changes in contrast between those topographs are attributed to lattice rotation around the lineages existing almost along the $\langle 1\bar{1}0 \rangle$ vertical axis.   From those figures, a somewhat four-fold pattern of lattice rotation like a four-leafed clover is also recognized.

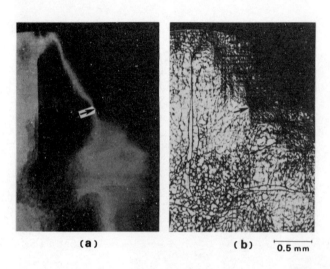

(a)   **(b)**   0.5 mm

Fig. 3 (a) X-ray double-crystal reflection and (b) transmission topographs of a quadrant of the same specimen as in Fig. 2.

Figure 3 (a) and (b) represent a reflection and a transmission topograph of a quadrant of the same specimen as in Fig. 2.   It is established by comparing the features in Fig. 3 (a) and (b), that the lattice rotates quite suddenly at the vertical lineages.   Furthermore, one can observe that the lattice rotation also occurs at an area where the dislocations pile up, as indicated by an arrow in Fig. 3 (a) and (b).

Although a small change in contrast between cells due to a slight lattice rotation is recognized in Fig. 3 (a), a triple-crystal reflection topograph by use of synchrotron radiation, as reproluced in Fig. 4, shows lattice features in the cell structures in more detail. Besides a contrast change due to a slight lattice rotation between the cells, some striated patterns are seen inside the

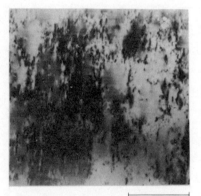

0.5 mm

Fig. 4 X-ray triple-crystal reflection topograph of an undoped GaAs wafer by use of synchrotron radiation, showing striations in the cells.

cells.   However, no further information on its origin has   been obtained
yet.

It is evident from the above results that the lattice orientations are
strongly influenced by dislocation distribution which has a four-fold
symmetry as theoretically given by Jordan et al (1980).   It is stated from
our present experiments that the (001) lattice plane orientation varies to a
large amount (10 ~ 20 sec. of arc) at the lineages and to a small amount
(less than several sec. of arc) at the cell walls.

Figure 5 shows a pair of double-crystal reflection topographs taken for a
wafer cut out from the seed end.   In addition to the usual lattice rotations at the lineages, some circular patterns are seen around the periphery.
After etching by molten KOH, it was found that those are due to a severe
occurrence of slip lines from the wafer edge.   Although no clear

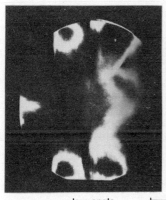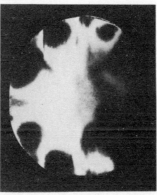

low angle ├── 1 cm ──┤ high angle

Fig. 5 A pair of
X-ray double-
crystal reflection
topographs of an
un-doped GaAs wafer
cut out from the
seed end. They are
taken at angles
lower (left) and
higher (right) than
the Bragg peak,
showing lattice
rotations due to
slip occurrence at
the edges.

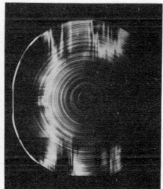

low angle ├── 1 cm ──┤ high angle

Fig. 6 A pair of
X-ray double-
crystal reflection
topographs of an
In-doped GaAs wafer
cut out from the
seed end. They are
taken at angles
lower (left) and
higher (right) than
the Bragg peak,
showing lattice
spacing variations.

explanations for the slip lines to arise more frequently from the edges
other than < 100 > can be given at the present time, that may be responsible
to a complicated stress distribution at the seed end.

3.3 Lattice spacing variations in In-doped GaAs

In contrast to undoped wafers, lattice rotations in highly In-doped ones
are thought to be minor, since those crystals are intended to grow with a
reduced dislocation density or free from dislocations. As a matter of
fact, a usual dislocation density in highly In-doped crystals is lower than
10^3 cm^{-2}.   Figure 6 shows a pair of double-crystal reflection topographs
of a seed-end wafer cut out from an In-doped ingot.   It should be noted
that the lattice rotations due to slippage at the wafer edges, being similar
to Fig. 5, are observed even in this In-doped wafer. We understand that
these slippages will be simply minimized by decreasing temperature gradients
during crystal growth.

In the remainder however, appears a characteristic concentric pattern
(striation pattern).   This was successfully correlated with a variation of
lattice spacing, not to lattice rotation as in undoped wafers.   In the case
of Fig. 6, the spacing variation between two adjacent regions with reversal
contrast was roughly estimated to be about 0.0006 angstrom.   From a simple
calculation based on Vegard's law, this value corresponds to about 0.15 % of
In-content variation, which seems not to be negligibly small.   Accordingly,
such local In-content fluctuation in   bulk is required to be suppressed
by controlling liquid convection and hence, shape of solid-liquid interface
during crystal growth.

4. Conclusion

By using X-ray diffractometry, mainly plane wave topography, it was
confirmed that lattice rotates in undoped LEC-grown GaAs wafers to a
relatively large extent at lineages and to a small extent at cellular
structure walls, both basically being dislocation bundles.   In highly
In-doped wafers, however, severely striated patterns were observed instead
of the lattice rotation, which are identified to be due to a lattice
spacing variation probably arising from local In-content fluctuations.

Acknowledgement

The authors would like to express their thank to Drs. D. Shinoda and
H. Watanabe for their encouragement.

References

Clark S and Stirland D J 1981   Microscopy of Semiconducting Materials
   (London:The Institute of Physics) pp 339-344
Jordan A S, Caruso R and Von Neida A R 1980 Bell System Tech. J. 59 573
Kikuta S, Kohra K and Sugita Y 1966 Jpn. J. Appl. Phys. 5 1047
Nanishi Y, Ishida S and Miyazawa S 1983 Jpn. J. Appl. Phys. 22 L54
Jacob G, Duseaux M, Farges J P, Van Den Boom M M B and Rocksnoer P J 1983
   J Crystal Growth 61 417

Study of α-cristobalite micro-inclusions in III−V compound semiconductor crystals

Sun Guiru, Liu Ansheng

General Research Institute for Non-Ferrous Metals, Beijing
2 Xin Jie Kou Wai Daijie Beijing, China

Abstract. The morphologies of Si-O micro-inclusions in single crystals, such as HB Si-, Cr-doped and undoped GaAs, LEC Cr-doped GaAs,
Sn-doped InP and Te-doped GaP, have been studied with a high voltage
electron microscope (HVEM). The electron diffraction patterns indexed
by Niggli cell method have shown that they are tetragonal α-cristo-
balites (lattice constant a=4.97 A, c=6.92 A). From the HVEM observation, electron and X-ray diffraction analysis, secondary electron
images and thermodynamical calculation, we suggested that there are
two origins whence the α-cristobalites in crystal have come. It was
also found that the α-cristobalite micro-inclusion is a nucleus for
forming As or Cr_2As, CrAs precipitates.

1. Introduction

In recent years, with the rapid development of compound semiconductor
devices and integrated circuits, uniform and perfect, doped or undoped
bulk single crystals with better quality are needed, so that more and
more authors are concerned about the precipitates and micro-inclusions
in III-V compound crystals. It has been reported (Kressel 1969, Thompson
1973, Tainman et al 1980) that precipitates in crystals would have an
influence on the performance of devices. Lin et al (1978) and Shi et al
(1977) indicated that silicon, oxygen and other compound precipitates or
precipitate regions might cause the degradation of devices or low mobility at room temperature. Kressel (1969), Miki et al (1973) and Zhu et
al (1981) reported that there were SiO_2 precipitates in heavily Si-doped
GaAs crystals. Udagawa et al (1982) shown that anomalously large amounts
of SiO_2 existed in Cr-doped GaAs crystal, because doped Cr acted as a
nucleation agent for α-cristobalite formation on the inner surface of
the quartz container. In this paper we study the morphologies of the
Si-O micro-inclusion, its structure and distribution in III-V compound
semiconductor crystals. Discussions about their origins are presented.
To help understand the origin of Si-O micro-inclusions, we have adopted
R. Kiessling's (1978) definition, that is an exogenous inclusion is a
inclusion which results from mechanical incorporation of slags, refractory or other materials with which the melt comes in contact; an indigenous inclusion is a inclusion which occurs as a result of reactions
taking place in the melting or condensing process, or by addition or by
change of solubility in the melt during cooling and freezing of the melt.
In other words, the latter, according to C.E. Sims's (1958) definition,
is a precipitate.

2. Experiment

The list of the samples examined in our experiments, together with their main inclusions, is given by Table 1. For the ingots 81-SD-7, 81-SD-8,

Table 1. Samples examined in experiments

Crystal	Number	Dopant	Method	Material	Container	Inclusion
GaAs	81-SD-7	Si	HB	1	3	α -
GaAs	81-SD-8	Si	HB	1	3	α -
GaAs	80-SD-28	Si	HB	2	3	α -
GaAs	81-SD-17	no	HB	1	3	α -
GaAs	82-CD-5	no	HB	1	3	α -
GaAs	80-CD-48	Cr	HB	1	3	α-, Cr_2As, CrAs
GaAs	BO-78	Cr	HB	1	3	α-, Cr_2As, CrAs
GaAs	A-5	Cr	HB	1	4	Cr_2As, CrAs
GaAs	A-9	Cr	HB	1	4	Cr_2As, CrAs
GaAs	79-CD-140	Cr	HB	2	3	α-, Cr_2As, CrAs
GaAs	BO-73-4	Cr	LEC	1	5	α-, Cr_2As, CrAs
GaAs	VG-81-17	Cr	LEC	2	5	α-, Cr_2As, CrAs
InP	80-ty	Sn	LEC	1	5	α -
GaP	ty-2	Te	LEC	1	5	α -

Note: 1- Synthesis, 2- Regrown, 3- Quartz boat, 4- Specially treated
      boat, 5- Quartz crucible, α-: α-cristobalite

80-SD-28 and 81-SD-17, 4 or 5 wafers were cut from each ingot at different positions along the ingot (Fig. 1(a)), only 1 or 2 wafers were cut from other ingot. Every wafer was lapped and thinned to 200 μm in thickness with abrasive, then polished with another abrasive cream (both without SiC and diamond) on both sides. After cleaving the thinned wafer into small specimens, their positions in the wafer are marked in Fig. 1(b), chemical thinning was carried out, then the specimen was mounted on a copper support grid for transmission electron microscopy observation. We used the selected area electron diffraction technique with the high voltage electron microscope (JEM-1000 HVEM) operating at 1M voltage and a low beam current. During observation we took bright field image photos of the second phase and electron diffraction patterns with different zones and dark field images using specific diffraction spots. The diffraction patterns were indexed by Niggli cell method. For analysis of diffraction patterns the Niggli cell data tables were plotted by computer according to the crystal lattice constant on the ASTM cards.

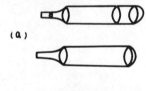

(a)

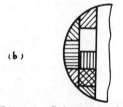

(b)

Fig. 1 Schematic diagram
of (a) a wafer positioned
in a ingot, (b) a specimen
for HVEM positioned in a
wafer

3. Results and Discussion

The morphologies of Si-O micro-inclusions in Si-, Cr-doped and undoped GaAs crystals grown by HB have been studied. It was found that there were some different shapes of Si-O micro-inclusions in the crystals (Fig. 2). The electron diffraction analysis indicated that the inclusions were tetragonal α-cristobalite (lattice constant a=4.97 A, c=6.92 A). It was

Fig. 2

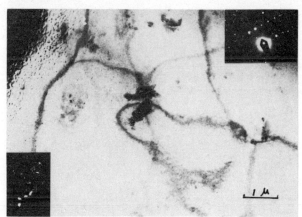

Normal topograph of the α-cristobalite
in Si-doped GaAs.
right: [125] diffraction pattern
left : dark field image by (12$\bar{1}$)reflection

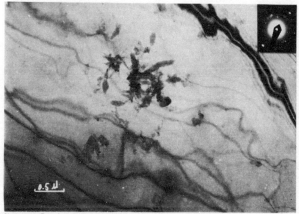

Normal topograph of the α-cristobalite
in GaAs grown with re-grown material.
right: α- diffraction pattern

Fig. 2

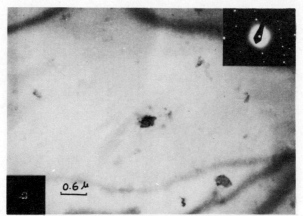

α-cristobalite in Cr-doped GaAs grown
by HB.
right: α- [031] diffraction pattern
left : dark field image by (1̄11)reflection

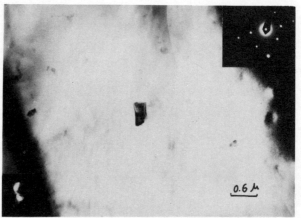

α-cristobalite in undoped GaAs grown
by HB.
right: α- [521] diffraction pattern
left : dark field image by (1̄31)reflection

observed that there were more ⍺-cristobalites in both HB and LEC crystals
grown from a regrown material than from a directly synthsized one; only
for HB crystals, were the number of ⍺-cristobalites in the wafer at the
tail position of an ingot more than at the head position; in the same wafer
the ⍺-cristobalites in the region contacting the quartz boat were more than
in the centre region. The experiment indicated that there were also some
⍺-cristobalites in Cr-doped GaAs, Sn-doped InP and Te-doped GaP crystals
grown in a quartz crucible by LEC (Fig. 3). To find the responsibility of
quartz for the formation of ⍺-cristobalites we studied the crystals grown
in a specially treated boat where the melt was prevented from directly
contacting the quartz, no ⍺-cristobalite micro-inclusions have been observed.
These facts showed that the presence of the ⍺-cristobalites in the crystals
did not directly depend on dopants, crystal growth method and crystal grown,
but rather on the history of melt contact with a quartz container. To study
the origin of ⍺-cristobalites, white powders from the inner surface of a
used quartz boat have been analysed with JEM-1000 HVEM and TUR-M62 X-ray
diffraction instrument. It was confirmed the ⍺-cristobalites are formed on
the inner surface of the boat. From the secondary electron image (SEI) of
the section of a used quartz boat (Fig. 4), we suggested that the conjunction between cristobalite and quartz may break under mechanical and/or
chemical actions, so that cristobalite particles fell down in a melt during
the

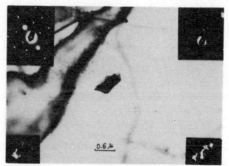

Fig. 3 ⍺-cristobalite and CrAs in Cr-
doped GaAs grown by LEC
right: CrAs (orth) [$\bar{2}21$] , by ($01\bar{2}$)
left : ⍺-cristobalite [542], by ($0\bar{1}2$)

Fig. 4 SEI of the section of a used boat

growth process. Cristobalite with a melting point of 1710°C was still in the solid state in the melt at about 1250°C. This is one of the sources of the α-cristobalites (called exogenous inclusions). The other α-cristobalites (called indigenous inclusions) are from the reactions of a melt with the quartz container. We used data given by Cochran (1962), Kubaschewski (1979) and Hultagren (1973) and calculated the standard Gibbs energies of the following reactions

$$4Ga_{(1)}\text{(or GaAs)} + SiO_{2(s)} = Ga_2O_{(g)} + Si_{\text{(in GaAs)}} \tag{1}$$

$$\Delta F_1^\circ = 163190 - 65.71\ T \qquad\qquad \text{(cal/mole)}$$

$$Si_{\text{(in GaAs)}} + SiO_{2(s)} = 2SiO_{(g)} \tag{2}$$

$$\Delta F_2^\circ = 169600 + 12.9\ T\,l_g T - 124.3\ T \qquad \text{(cal/mole)}$$

$$2Ga_{(1)}\text{(or GaAs)} + SiO_2(s) = Ga_2O_{(g)} + SiO_{(g)} \tag{3}$$

$$\Delta F_3^\circ = 166400 + 6.\ 45\ T\,l_g T - 95.01\ T \qquad \text{(cal/mole)}$$

$$3Ga_2O_{(g)} + 2As_{2(g)} = Ga_2O_{3(s)} + 4GaAs_{(s)} \tag{4}$$

$$\Delta F_4^\circ = -116600 + 364.44\ T \qquad\qquad \text{(cal/mole)}$$

$$3Ga_2O(g) + As_{4(g)} = Ga_2O_{3(s)} + 4GaAs_{(s)} \tag{5}$$

$$\Delta F_5^\circ = -55500 + 334.16\ T \qquad\qquad \text{(cal/mole)}$$

Using the standard Gibbs energies and the formula of Gibbs energy in a chemical reaction $\Delta F = \Delta F^\circ + RT\,l_n J$ for reaction (4) and (5), we obtained

$$\Delta F_{(4)} = \Delta F^\circ_{(4)} + RT\,l_n \frac{a^4_{GaAs} \cdot a_{Ga_2O_3}}{P^2_{As_2} \cdot P^3_{Ga_2O}}$$

$$\Delta F_{(5)} = \Delta F^\circ_{(5)} + RT\,l_n \frac{a^4_{GaAs} \cdot a_{Ga_2O_3}}{P_{As_4} \cdot P^3_{Ga_2O}}$$

We calculated $\Delta F_{(4)}$ and $\Delta F_{(5)}$, both of the values were negative (Sun 1984). From thermodynamical calculation Ga_2O_3 could be formed under the conditions of crystal growth. In fact, we have found the α-Ga_2O_3 precipitate (hexagonal, lattice constant a=4.98 A, c=13.43 A) in the GaAs crystals (Fig. 5). A three-heating-zone furnace was used to suppress the silicon contamination, but the reaction (4) and (5) have occurred introducing more Si and O impurities into the melt. For LEC InP similar reactions also took place in the synthesis process. During crystal growth, Si in the melt may take the position of Ga (or In) or As (or P) in the crystal. These silicon atoms then would be combined with oxygen atoms in the interstitial position to form β-cristobalite (cubic, a=7.05 A, see Fig. 6) which transforms to α-cristobalite at 270 – 180°C.

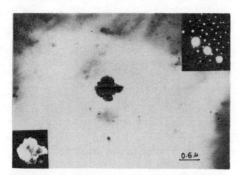

Fig. 5 α-Ga$_2$O$_3$ micro-precipitate
in undoped GaAs
right: [33$\bar{1}$] diffraction pattern
left : dark field image by ($\bar{1}$1$\bar{6}$)

Fig. 6 Structure of
β-cristobalite

The α-cristobalite in the crystal plays the role of heteronuclei for
arsenic or a dopant precipitation as well as influencing the purity and
uniformity of the crystal. α-cristobalite accompanying As micro-precipi-
tates (Fig. 7), also Cr$_2$As, CrAs (Fig. 3), have been observed with HVEM.

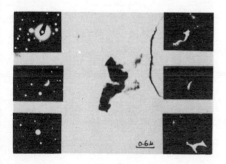

Fig. 7 As micro-precipitate, α-
cristobalite and C micro-inclusion
in undoped GaAs
up: As left [$\bar{1}$4$\bar{1}$] diffraction
       right by (1$\bar{1}$3)
middle: α-cristobalite
       left [124]
       right by (0$\bar{2}$1)
down C left [245]
       right by (2$\bar{1}$0)

Therefore, development of a container material should receive more
attention for improving the uniformity and perfection of crystals.

4. Conclusion

1. It has been found that tetragonal α-cristobalite (lattice constant
a=4.97 A, c=6.92 A) micro-inclusions exist in III-V compound crystals
grown in a quartz container by HB or LEC.
2. The α-cristobalites formed on the inner surface of a quartz container
are the origin of the exogenous micro-inclusions. The reactions of a
melt with a quartz container are the main origin of indigenous micro-
inclusions.
3. The α-cristobalite micro-inclusions in the crystals mainly depend on
the history of a melt directly contacting a quartz container, not on
dopants and crystal growth method.
4. The distribution of α-cristobalite in the crystals shows there are
more α-cristobalites in crystals grown from regrown materials than in
crystals grown from a directly synthesized one. For the crystals grown
by HB, there are more α-cristobalites in the wafer at the tail position
of an ingot than at the head position; in a wafer the α-cristobalites in

a region contacting the quartz boat are more than in the centre region.

Acknowledgments

The authors wish to thank Tan Lifang for samples prepared from a specially treated boat and SIMS data. The authors are also very grateful to the crystal growth groups of our institute who provided all wafers for the experiments.

References

Cochran C N and Foster L M 1962 J. Electrochem. Soc. 109 149
Hultagren P et al 1973 Selected Values of the Thermodynamic Properties of the Elements (American Society for Metals)
Kressel H 1969 Appl. Phys. 40 3589
Kiessling R 1978 Non-metallic Inclusions in Steel (The Metals Society London) Part III-1
Kubaschewski O and Alcook C B 1979 Metallurgical Thermochemistry 5th (Pergamon)
Lin Lanying et al 1978 Rare Metals (in chinese) No1 2
Miki H et al 1973 Applied Physics (Jap) 42 432
Shi Huiying et al 1977 presented at Conf. on Gallium Arsenide and Related Compounds China
Sims C E 1959 Trans. AIME. 215 367
Sun Guiru 1984 (to be published)
Thompson A G 1973 J. Electron Materials 2 47
Tairman R D and Ouver J R 1980 SI III-V Materials Conf. (Nottingham) 83
Udagawa T and Nakanish T 1982 Gallium Arsenide and Related Compounds 1981 (Inst. Phys. Conf. Ser. 63) 19
Zhu Jian et al 1981 presented at Conf on Gallium Arsenide and Related Compounds China

Inst. Phys. Conf. Ser. No. 74: Chapter 2
Paper presented at Int. Symp. GaAs and Related Compounds, Biarritz, 1984

115

Brillouin scattering investigation of the surface acoustic wave propagation in implanted GaAs

J. SAPRIEL, J.C. MICHEL, P. HENOC and Y.I. NISSIM
Laboratoire de Bagneux C.N.E.T. - 196 rue de Paris - 92220 Bagneux - FRANCE

Abstract. Brillouin scattering from Rayleigh surface phonons have been performed on GaAs substrates implanted with Te^+ and P^+. In the case of Te, the implanted layer was amorphized over a thickness of 3000Å and the dispersion of the Rayleigh surface waves have been measured. We also report the first determination of the RSW velocity in a-GaAs.

1. Introduction

Rayleigh surface waves (RSW) are acoustic waves which propagate parallel to the free surface of an infinite half space. The amplitude of their displacement undergoes an exponential decay with depth below the free boundary and completely vanishes within a distance of the order of one acoustic wavelength. (G.W. Farnell 1973).

Surface and localized acoustic vibrations in layered media have been investigated through the propagation of earthquakes by Lord Rayleigh. More recently the development of RSW technology led to a resurgence of interest in these excitations. Instead of investigating the RSW produced by an electromechanical transducer, we preferred to probe the RSW which are thermally excited (phonons) at room temperature through their coupling with an incident laser beam. The scattering of light from acoustic surface phonons can be performed only with the help of a high contrast and high sensitivity Brillouin spectrometer, i.e. a sophisticated experimental set-up is needed. But at least two advantages can be obtained. Firstly Brillouin scattering (BS) is a versatile and non destructive tool for measuring the acoustic properties and secondly the frequency of the involved RSW are high enough ($\sim$ 10 GHz) so that small volumes of matter can be probed. Actually BS is a unique tool for acoustic investigation of microstructures like superlattice crystals grown on a substrate, whose total thickness is of the order of $1\mu m$ (J. Sapriel et al 1983).

2. Propagation of surface acoustic waves in implanted GaAs :

GaAs is a cubic crystal which is anisotropic with respect to the propagation of acoustic waves. The anisotropy parameter $\eta=2C_{44}/(C_{11}-C_{12})$ is equal to 1.86 ; [100] and [110] are two high symmetry directions in the plane (001) which correspond to RSW polarized in the sagittal plane (i.e. the plane perpendicular to the free surface and containing the propagation direction). The velocities v_R are given by

$$C_{11}(v_R^2-A)\ (v_R^2+B-C) = C_{44}(v_R^2-C)\quad \text{for } [100]$$

$$C_{11}(v_R^2-A)\ (v_R^2+B-D) = C_{44}(v_R^2-D)\quad \text{for } [110]$$

here $A = C_{44}/\rho$ $\qquad B = C_{12}^2/\rho C_{11}$
$\qquad C = C_{11}/\rho$ $\qquad D = B(1+\eta)/2 + C_{12}(1-\eta)/2\rho$
One obtains the velocity of the RSW in an isotropic material by taking $C_{44}=(C_{11}-C_{12})/2$.

In a homogeneous crystal the velocity in a given direction is non dispersive, in that v_R is independent of the frequency. If we consider now a thin layer on a substrate, of a thickness e of the order of the acoustic wavelength or less, the acoustic wave becomes dispersive. For a given couple thin layer-substrate, the velocity of the RSW is a function of the product Ke, K being the wavevector of the RSW (G.W. Farnell and E.L. Adler). One can consider an implanted material as a system consisting of a layer where a large number of defects are created by the incident ions on the top of a substrate. This perturbed layer has acoustic properties different from those of the substrate, which is interesting to investigate. At our knowledge nothing has been published on the acoustic propagation in implanted GaAs.

3. Sample preparation and characterization

Our experiments have been performed on selected implanted GaAs (001) substrates. The experimental conditions of the implantation have been chosen in order to obtain a damaged layer extending from the surface to a depth of several thousands Angstroms. The first sample had been implanted with Te ions at low temperature and low current at 250KeV with a dose of $1 \times 10^{15}/cm^2$. The second sample was bombarded with two P ion beams to create a more uniform distribution of defects, one at 200KeV with 2.10^{14}ions/cm^2 and the other at 460KeV with a dose $5.10^{14}/cm^2$.

After thinning the samples from the substrate side with a methanol bromide solution, the implanted layer is observed by electronic microscopy and diffraction. In the case of the first sample an amorphous layer has been evidenced at the surface, whose depth is larger than 2.500Å, which is the upper limit of the observable thickness, given by the mean inelastic free pass of the incident electrons (100KeV). One can see in Fig.1 the annular diffraction pattern caracteristic of the amorphous state.

No amorphous structure has been detected from our investigations of the layer implanted with P (Fig.2). The GaAs crystal was nevertheless strongly damaged and Fig.3 which corresponds to a dark field [220] shows a high crystalline defects density (dislocations and microprecipitates) of the order of 10^{12}cm^{-2}.

Brillouin scattering from RSW in implanted GaAs

The propagation of the RSW induces ripples (undulations) of the free surface and variation of the refractive index in the zone just below the surface where the RSW is localized (photoelastic effect).Both the ripples and the refractive index variations act as a periodic grating and can scatter the light (J. Sapriel 1979). In the case of semiconductors Mishra and Bray have shown that the dominant mechanism was the scattering by the ripples.

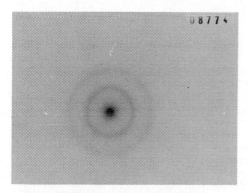

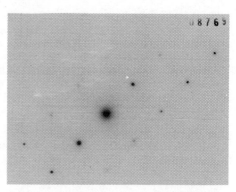

Fig.1:Diffused ring of the GaAs
layer amorphized by high dose
Te ions,obtained on a micro-
diffraction pattern

Fig.2:Microdiffraction pattern of
the GaAs layer implanted with P^+
corresponding to the (001) plane
of the reciprocal lattice.

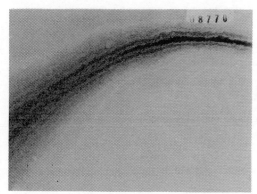

Fig.3:Dark field (220) of the P
implanted GaAs substrate of Fig.2.
The crystalline defects are clearly displayed (Magnification
X 10,000)

Contrary to bulk acoustic waves whose Brillouin lines are severely broadened (J.R. Sandercock 1972) by the optical absorption, the coupling of
light with RSW gives rise to sharp Brillouin lines and the velocity v_R of
RSW can be determined with a good accuracy ($\sim$ 0.5 %) from the Brillouin
line frequency shift v_B.

The wavevector $K=2\Pi v_B/v_R$ of the phonon involved in the scattering process
must fulfil the wavevector conservation requirement (J.R. Sandercock
1978)

$$K = k(\sin\theta_i + \sin\theta_s) \qquad (1)$$

Here $k=2\Pi/\lambda$ refers to the wavevector of the light in the vacuum, θ_i and
θ_s are the angle of incidence and the angle of scattering respectively.
Thus the acoustic wavelength $\lambda_a=2\Pi/K$, which is approximately equal to the
penetration depth of the RSW are determined by the scattering geometry
and varies between 2500 Å and 5500 Å in our experimental conditions.

The experimental set-up is a high contrast and high resolution Brillouin
spectrometer which consists of a four-pass plane Fabry-Perot (PFP) and a

confocal Fabry-Perot (CFP) in tandem. The total contrast of the system is of the order of 10^{10} and the maximum resolving power is equal to 10^8. The PFP acts like a filter whose maximum transmission is adjusted and maintained at the frequency of the Brillouin line under investigation (Vacher et al). Besides the Rayleigh line due to light scattered at the frequency of the incident laser is strongly attenuated. The CFP is scanned over an interference order and the frequency ν_B of the Brillouin line is measured from the frequency shift $\Delta\nu$ (fig.4).

Two directions of the (001) surface were investigated by means of our Brillouin spectrometer : the $[100]$ and $[110]$ ones. The measured values of v_R in GaAs are 2690 and 2817 m/s along $[100]$ and $[110]$ respectively. Sample 1 was particularly investigated. The results of the velocity variations are given in Fig.5 as a function of Ke, the value of e being taken equal to 3 000 Å. One can see that the velocities along $[100]$ and $[110]$ tend to the same limit v_R=2270 m/s ± 30 m/s, a value which just corresponds to the velocity of the Rayleigh waves in amorphous GaAs. The relative variation $\Delta v_R/v_R$ between a-GaAs and c-GaAs is thus of the order of 20 %.

Preliminary measurements on sample 2 in the same range of acoustic wavelength (2500 Å to 5500 Å) show less variations (~ 10 %) and an anisotropy between $[100]$ and $[110]$ characteristic of the crystalline state.

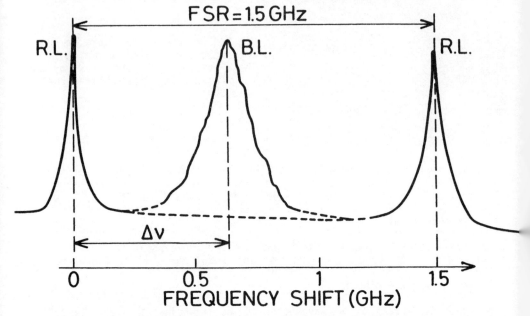

Fig.4 : Typical Brillouin spectrum obtained by scanning the confocal Fabry-Perot, B.L.→Brillouin line, R.L.→Rayleigh line (which is at the frequency of the laser) FSR→free spectral range. ν_B=$\Delta\nu$+n FSR. ν_B is the Brillouin line frequency, n is an integer. The acoustic velocity is given by $v=\lambda_a\nu_B$, λ_a being the acoustic wavelength.

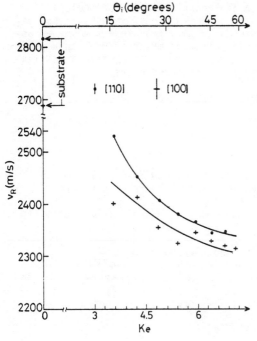

Fig.5:Variations of the velocity v_R of the RSW in the amorphized GaAs layer of thickness e=3000 Å at the surface of a crystalline GaAs substrate as a function of Ke (K being the wavevector of the phonon involved in the scattering process). K varies with the scattering geometry;θ_i and θ_s are the angle of incidence and scattering respectively ; $\theta_s - \theta_i$ is kept fixed at 24.6°.

DISCUSSION

In this lecture are presented the first results on the acoustic properties of implanted GaAs through the study of the RSW propagation. The
knowledge of its acoustical properties might be of practical interest in
integrated piezo-optoelectronics where several functions are gathered on
the same substrate. Obviously the acoustic properties of GaAs can be
tailored over a certain range by varying the implantation conditions. For
high dose implantation one obtains an amorphous structure and a velocity
variation of the RSW of the order of 20 % is observed with respect to the
crystalline state. Surface wave guides, lenses, reflective arrays and
cavities working at high frequency could be advantageously realized using
implantation technique (P. Hartemann). The piezoelectric properties
(electromechanical coupling coefficient, attenuation, temperature stability) of GaAs have been investigated by J. Henaff et al for different
cuts and direction of propagation. Although the piezoelectric effect in
GaAs is of sufficient strength for practical acoustoelectrical devices,
conventional interdigital transducers are limited to an operation band
width of less than 1 %, though ZnO overlays lead to bandwidths wide
enough (T.W. Grudkowski). Recently it has been demonstrated theoretically
and experimentally that edge-bonded transducers increase the bandwidths by
a factor of 5 over ZnO (D.E. Oates and R.A. Becker).

Besides the potential application of implantation to RSW technology (frequency filters, oscillators, delay-lines, convolutors, sensors, decoders, etc...) the study of the defects induced by this technique and the acoustical probing of the transition crystal-glass by means of Brillouin scattering are, in our opinion, promising directions for further investigations, specially in GaAs and related III-V semiconductors.

It is a pleasure to thank N. Duhamel, P. Krauz and D. Descouts for kindly performing the P^+ implantation.

REFERENCES

Farnell G.W. 1973 Physical acoustics (Academic, New York) vol.6,p.109
Farnell G.W. and E.L. Adler 1972 Physical Acoustics Principles and
    Methods (Academic, New York vol.9 p.35)
Grudkowski J.W. 1975 Microwave Laboratory Report 2440, Stanford
    University
Hartemann P. 1982 Ferroelectrics 42, 197
Henaff J., Feldmann M. and Kirov N.A. 1982 Ferroelectrics 42,161
Oates D.E. and Becker R.A. 1980 Ultrasonics Symposium Proceedings 367
Sapriel J., Michel J.C., Tolédano J.C., Vacher R., Kervarec J. and
    Regreny A. 1983 Phys. Rev. B 28, 2007
Sapriel J. 1979 Acousto-optics (Wiley, New York)
Sandercock J.R. 1972 Phys. Rev. Lett. 28, 237
Sandercock J.R. 1978 Solid State Comm. 26, 547

Growth and characterization of heteroepitaxial GaAs on mixed alkaline-earth fluorides

C. FONTAINE, S. SISKOS, A. MUNOZ-YAGUE
Laboratoire d'Automatique et d'Analyse des Systèmes du C.N.R.S.
7, avenue du Colonel Roche - 31077 TOULOUSE CEDEX - France
R. CARLES, J.B. RENUCCI
Laboratoire de Physique des Solides, associé au C.N.R.S.
U.P.S. - 118, route de Narbonne - 31400 TOULOUSE - France

Abstract : Lattice matched GaAs/$(Ca,Sr)F_2$/GaAs structures grown by
molecular beam epitaxy are reported. The results presented show better
material and interface quality for (100) orientation than for (110)
orientation. The photoluminescence efficiency of the GaAs layers is
shown to be increased by the fabrication of a periodic GaAs - $(Ca,Sr)F_2$
structure between the substrate and the top GaAs layer. Raman spectra
show the absence of any disorder induced effects and high structural
quality of these layers.

1. Introduction

A great interest exists in the growth of good quality semiconductor films
on insulators, and alkaline earth fluorides could well be good candidates
for building these structures, which present potential applications for
the realization of three dimensional integrated circuits and integrated
optoelectronic devices. Fluorides are indeed known to be good insulators
with breakdown fields and resistivities of about 10^6 V/cm and $10^{13}\Omega$ cm
respectively. On the other hand, it has already been shown that it is
possible to grow epitaxial fluoride films on semiconductors.

The cubic fluorite structure of the fluorides studied, CaF_2, SrF_2, BaF_2,
is closely related to those of the semiconductors Si, Ge, GaAs, InP, GaP,
which facilitate epitaxial relationships between these different materials.
In addition, for some semiconductors, it is possible to minimize the lattice
mismatch by choosing purposely the fluoride to be grown and, even, to obtain
lattice match by using a mixed fluoride. For example, $(Ba,Sr)F_2$/InP (Tu et
al,1984) and $(Ca,Sr)F_2$/GaAs (Siskos et al, 1983) lattice matched structures
have already been realized successfully.

The growth technique used is molecular beam epitaxy (MBE), which appears
to be well adapted to such growth. Indeed, fluorides have low vapor pressures and they sublime as molecular units because of their high free energies of molecular dissociation. The use of the MBE technique renders also
possible the realization of semiconductor/fluoride/semiconductor structures
in good experimental conditions. As a matter of fact, the different materials can be successively grown in the same run in an ultrahigh vacuum
system. Growth of Si/CaF_2/Si (Asano and Ishiwara, 1983), $Ge/(Ca,Sr)F_2$/Si
(Ishiwara and Asano, 1983), InP/CaF_2/InP, $InP/(Ba,Sr)F_2$/InP (Tu et al,1983)
has been reported. The growth of GaAs/$(Ca,Sr)F_2$/GaAs lattice matched structures has been also reported by the authors (Siskos et al, 1984) showing
promising electrical characteristics of the GaAs layers. In this paper,
we focus our attention namely on the influence of the crystallographic
orientation and that of the intermediate layer between the substrate and
the GaAs layer under study.

2. Experimental

The choice of a mixed lattice matched fluoride was done in order to improve
the quality of the interfaces and of the uppermost semiconductor layer.
CaF_2 and SrF_2, whose lattice parameters make possible lattice match to GaAs,
were preferentially used because a stable solid solution of their mixture
can be achieved for the entire range of composition. In addition their
thermodynamic properties are very close, and they have similar vapor pressures in the temperature range we are concerned with.

For the work reported here, GaAs substrates were prepared with standard MBE
procedures. During growth, substrate temperature was fixed at 550C. The
fluoride growth rate was in the range of 1 µm/h ; for the uppermost GaAs
layer, a growth rate of 1 µm/h was used and the As to Ga flux ratio was
kept at 10.

Under these experimental conditions, $(Ca,Sr)F_2$ and GaAs films were successively grown with thicknesses in the range of 0.1 - 0.3 µm and 0.2 - 5 µm
respectively (see Table I).

Sample N°	GaAs Substrate Orientation	$(Ca,Sr)F_2$ layer thickness (µm)	Undoped GaAs layer thickness (µm)	Doped GaAs layer thickness (µm)
A 56 A	(100)	0.25	-	0.5
A 56 B	(110)	0.25	-	0.5
B 54	(100)	0.1	-	4.5
B 55	(100)	0.1	3	1.5
B 57	(100)	Alternate layers	3	1.5

Table I : Characteristics of the GaAs/$(Ca,Sr)F_2$/GaAs heterostructures
studied.

The layers were investigated using secondary ion mass spectrometry (SIMS)₀
(Cameca IMS 3000), photoluminescence spectroscopy (Argon laser, λ = 5145 A,
f = 1 m monochromator), and Raman spectroscopy (Argon laser, λ = 4880 Å,
Coderg T800 triple monochromator).

3. Results

As far as $(Ca,Sr)F_2$ growth is concerned, Raman spectroscopy confirmed that
a solid solution of the mixed fluoride was obtained (Siskos et al, 1984b).
SIMS profiles of the different elements in seven successively grown layers
were recorded and the composition was found to be uniform and reproducible.
Smooth and mirror like surfaces were routinely obtained for both surface
orientations studied. The fluoride layers were found to be monocrystalline
by transmission electron diffraction. $(Ca,Sr)F_2$ layers appear thus to present necessary qualities to be used as starting material for subsequent
GaAs growth.

The surface of the GaAs layers was observed by Nomarski microscopy to be
lightly rough, whatever surface orientation of the substrates. Transmission
electron diffraction patterns of these layers indicate that they are monocrystalline and oriented as the substrate.

GaAs/(Ca,Sr)F$_2$/GaAs structures were simultaneously grown on two GaAs substrates with (100) and (110) orientations. Figures 1a and 1b correspond
respectively to the Raman spectra obtained. On Figure 1a, two phonon peaks
are observed ; one is associated with the allowed longitudinal optical phonons (LO) and the other is due to the forbidden transverse optical mode (TO);
the origin of this forbidden peak is discussed below. Taking into account
the relative intensity of these peaks and their width (4 cm^{-1}), the layer
crystalline quality appears to be good.

As far as (110) orientation is concerned, two peaks of the Raman spectrum
shown on Figure 1b correspond to GaAs, one (TO) which is allowed for this
orientation, and the other (LO) now forbidden ; in addition, two supplementary forbidden peaks are present, which are related to the presence of As-As
bonds (Zitter, 1971) and denote poor quality of the GaAs.

These results were confirmed using SIMS. Figure 2a and 2b show the profiles
obtained for the heterostructures already studied by Raman spectroscopy.
For the (100) orientation, the composition of the different elements appears
uniform inside each layer. In addition, interdiffusion near the interfaces
is not relevant. Conversely, in the case of the (110) orientation (profiles
shown on Figure 2b), a strong accumulation of Sr and Ca at the interfaces
is observed. The diffusion of As into the (Ca,Sr)F$_2$ layer was also noticeable. Then, the (100) orientation appears to be more suitable for the realization of good quality GaAs/(Ca,Sr)F$_2$/GaAs heterostructures.

Now, we focus our attention on three Si-doped layers fabricated on (100)
substrates in order to obtain information concerning the influence of the
intermediate layer on the characteristics of the GaAs grown on top of the
heterostructure (samples B 54, B 55 and B 57 of Table I).

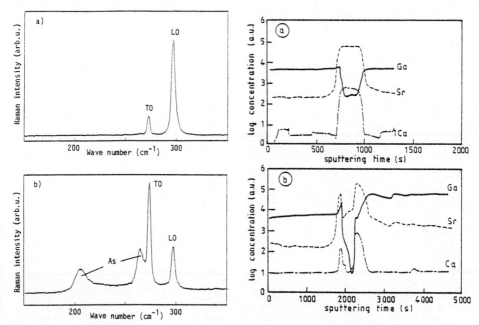

Fig. 1 - Raman spectra of the uppermost GaAs layers of a) sample A 56 A ;
b) sample A 56 B

Fig. 2 - SIMS profiles of two simultaneously grown heterostructures :
a) sample A 56 A ; b) sample A 56 B

In the case of sample B 57, between
the substrate and the GaAs layer under
study, there is a series of alternate
layers of GaAs (500 Å) and $(Ca,Sr)F_2$
(300 Å) ; five of such periods were
successively grown. This is the first
alternate structure of this kind to
be reported, which was made to improve the elastic accommodation of the
stress due to the differences in the
linear expansion coefficients of GaAs
and $(Ca,Sr)F_2$. Figure 3 shows a composition profile obtained using SIMS.
The actual profile is much more
abrupt, the relative smoothness of
the profile shown being mainly related to experimental limitations of
the apparatus.

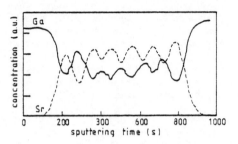

Fig. 3 - SIMS profile of the
alternate layers of GaAs and
$(Ca,Sr)F_2$ in sample B 57.

Figure 4 shows the photoluminescence (PL) spectra obtained at 10K for the
samples under study. The main emissions observed are :
. excitonic transitions with full width at half maximum of about 10 meV :
1.515 eV (donor bound exciton), 1.503 eV (growth induced defect-related
exciton, observed by Kunzel and Ploog (1980))
. carbon acceptor (1.493 eV), not systematically observed
. donor-acceptor transitions (1.48 eV) related to $Si_{Ga}-C_{As}$ and/or $Si_{Ga}-Si_{As}$
. the transition at 1.466 eV, observed by Briones and Collins (1982) and
attributed to carbon related defects. We have often observed the simultaneous presence in other samples of the two emissions at 1.503 and 1.466 eV
as it has also been reported by Mendez et al (1983) in Si-doped MBE GaAs
. the emissions at 1.44 and 1.40 eV, which are often observed on Ge and
Si-doped GaAs. They involve complex defects associating an acceptor
(Ge_{As}, Si_{As}) and point defects (see for example Bafleur et al, 1983).

In order to obtain some trends regarding the PL behaviour, we consider the
intensity ratio between the near band gap emission (1.515 eV) and deep
levels associated with defects or impurity-defect complexes.

We notice that for sample B 55 the emissions between 1.44 and 1.40 have
roughly the same intensity than the near band gap emission, whereas for
sample B 54 those low energy emissions are considerably smaller (by a factor of about 4) than the 1.515 eV peak.

If we now consider sample B 57 we observe a real improvement that can be
ascribed to a better mechanical matching between substrate and GaAs layer
brought by the GaAs - $(Ca,Sr)F_2$ multilayer structure. Indeed, the intensity
of the low energy emissions is still lower when compared to the high energy
emission ; in addition, this spectrum is dominated by excitonic recombination, which denotes good PL characteristics, as well as a low impurity
compensation level.

Another aspect of the efficiency of the alternate layer structure is given
on Figure 5, which compares the 300K photoluminescence spectra of samples
B 54 and B 57 ; it can be observed that the PL intensity of layer B 57 is
about 12 times higher than that of sample B 54, denoting a much better
radiative/non radiative ratio of recombination transitions.

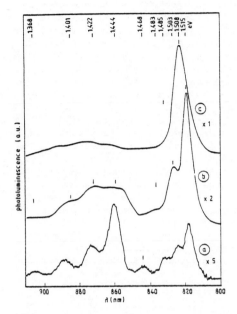

Fig. 4 - Photoluminescence spectra (10K) of a) sample B 55 ; b) sample B 54 ; c) sample B 57.

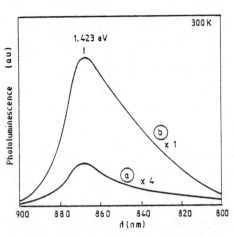

Fig. 5 - Photoluminescence spectra (300K) of a) sample B 54 ; b) sample B 57.

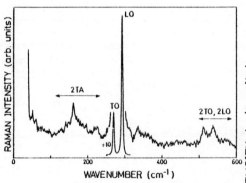

Fig. 6 - Raman spectrum recorded in the backscattering geometry from the (100) face of the B 57 sample, at room temperature with λ_i=4880 A laser excitation.

Fig. 7 - First order optical modes. The resolution was 3 cm^{-1}.

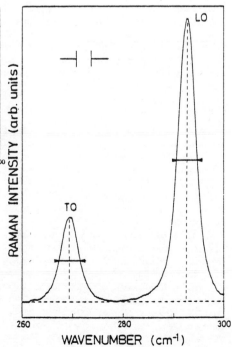

These samples were also investigated by Raman spectroscopy in the true back-scattering configuration using a small 45° tilted mirror kept in front of the scattering (100) face of the samples, in order to avoid any misorientation of the incident probe inside the crystal. The results were rather the same for all the samples and so we comment only on the B 57 spectrum which is reported on Figures 6 and 7. Any disorder-induced effect is absent : i) for instance, the disorder activated transverse and longitudinal acoustical scattering (D.A.T.A. and D.A.L.A.) in the low frequency range of the spectrum (around 80 and 190 cm^{-1}) does not show up ; ii) as a matter of fact, the structure and the high efficiency of either the acoustical (130-230 cm^{-1}) or optical (500-580 cm^{-1}) second order bands reveal the non-softening of the vibrational modes ; iii) finally, no shift, asymmetry or broadening of the first optical modes is detected (see Figure 7). In the limit of the experimental sensitivity these results manifest the structural high quality of the layers. Moreover, according to the usual first order selection rule, only the LO scattering is allowed. Its Γ_{15} symmetry is well confirmed experimentally, within few percent, by polarization measurements in four configurations (respectively crossed and parallel to the <010> and <011> axes). The unexpected TO scattering which shows up with no well-defined symmetry cannot be interpreted as disorder or misorientation induced effects (Biellmann et al, 1983). Its activation lies probably in doping effects (Olego and Cardona, 1981).

4. Conclusion

Lattice matched GaAs/$(Ca,Sr)F_2$/GaAs structures fabricated by molecular beam epitaxy have been reported. The experimental results presented were obtained using secondary ion mass spectrometry, photoluminescence and Raman spectroscopies, to study the characteristics of the uppermost GaAs layer. The comparison between (100) and (110) oriented structures has shown superior material and interface quality for the (100) orientation. An investigation of the intermediate layer between the substrate and the GaAs layer has shown that a drastic improvement of the GaAs photoluminescence efficiency is brought by the use of a periodic structure comprising a series of alternate layers of GaAs and $(Ca,Sr)F_2$ with thicknesses of 300-500 Å. The Raman spectra show the absence of any disorder induced effects, and high structural quality of the GaAs layers grown on top of these structures.

References

Asano T and Ishiwara H , 1983, Appl. Phys. Lett. 42 (6), p. 517
Bafleur M Munoz-Yague A , Castano JL and Piqueras J, 1983, J. Appl. Phys. 54 (5), p. 26
Biellmann J ,Prevot B and Schwab C ,1983,J. Phys. C, Sol. St. Phys.16,p.1135
Briones F and Collins DM , 1982, J. Electr. Mat. 11 (4), p. 847
Ishiwara H and Asano T 1983,Jap. J. Appl. Phys. 22, Sup. 22-1, p. 201
Kunzel H and Ploog K, 1980, Appl. Phys. Lett. 37 (4), p. 416
Mendez EE, Heiblum M ,Fischer R ,Klem J ,Thorne RE and Morkoc H , 1983,
    J. Appl. Phys. 54 (7), p. 4204
Olego D and Cardona M ,1981, Phys. Rev. B24, p. 7217
Siskos S ,Fontaine C and Munoz-Yague A ,1983,Insulating Films on Semiconductors Ed. JF Verveij and DR Wolters (Amsterdam - NH), p. 71
Siskos S ,Fontaine C and Munoz-Yague A ,1984a Appl. Phys. Lett.44(12)p.1146
Siskos S ,Fontaine C and Munoz-Yague A ,1984b J. Appl. Phys. 56(6)p.1642
Tu CW ,Forrest SR and Johston WD , 1984, J. Vac. Sci. Technol. B2(2),p.212
Zitter RN , 1971 Physics of Semimetal and Narrow Gap Semiconductors, ed. DL Carter and RT. Bate (Pergamon Press), p. 285.

Inst. Phys. Conf. Ser. No. 74: Chapter 3
Paper presented at Int. Symp. GaAs and Related Compounds, Biarritz, 1984

127

Lattice location of Si in MBE-grown highly Si-doped GaAs layers

Tadashi Narusawa, Yoko Uchida, Keisuke L. I. Kobayashi, Tsuneaki Ohta, Masato Nakajima and Hisao Nakashima

Optoelectronics Joint Research Laboratory
1333 Kamikodanaka, Nakahara, Kawasaki, 211 JAPAN

Abstract. Displacement-sensitive channeling PIXE measurements have been carried out about MBE-grown highly Si-doped GaAs layers. Results indicate that, most Si atoms are incorporated into substitutional sites, both Ga and As sites with about equal probabilities, however, some of them occupy slightly displaced lattice sites. The displacement is ~ 0.1 Å and the fraction of displaced Si atoms is $\sim 20\%$ in the case of a doping level of 8×10^{19} cm^{-3}, which tends to increase with the doping level. These provide direct evidence that the electrical inactivity of doped Si is associated with such a structural distortion.

1. Introduction

Silicon is an excellent n-type dopant in molecular beam epitaxial growth of GaAs crystals. Particularly, a high sticking coefficient and no surface accumulation during growth justify Si when compared with other materials such as Sn (Cho 1975). Doped Si atoms are generally supposed to occupy the Ga sites in the crystal and donate one free electron each. However, in the heavy doped region over $\sim 5 \times 10^{18}$ cm^{-3}, the carrier density saturates or even decreases with increasing doping level. Where are the excess Si atoms? This problem has been attributed to Si precipitates (Drummond et al. 1982) or autocompensation (Neave et al. 1983), i.e. the Si atoms occupy the As sites also and act as acceptors. Such amphoteric nature of Si in GaAs crystals has been known, and on the basis of infrared absorption spectra which showed various localized vibration modes, Spitzer and Allred (1968) have suggested the existence of Si_{Ga}-Si_{As} pairs in melt grown Si doped GaAs crystals.

The present paper addresses the issue where the doped Si atoms are located in MBE-grown GaAs layers. We have applied the channeling PIXE (Particle Induced X-ray Emission) technique, which is highly sensitive to atomic displacements as small as ~ 0.05 Å (Feldman et al. 1982). Our results indicate that the Si atoms in the high-doped region occupy both the Ga and As sites with about equal probabilities, and further, such high level doping induces a significant lattice distortion which is apparently related to Si_{Ga}-Si_{As} pairing in the GaAs crystal.

2. Experimental

The samples were prepared by MBE on Cr-doped semi-insulating GaAs(100) substrates. The substrate temperature, the As$_4$/Ga flux ratio and the growth rate were kept constant at ~ 670 °C, ~ 6, and ~ 0.6 μm/h, respectively. The doping level of Si was varied by controlling the Si effusion cell temperature. The thickness of Si doped layer ranged from 0.3 to ~ 3 μm. Figure 1 shows the free carrier concentration at room

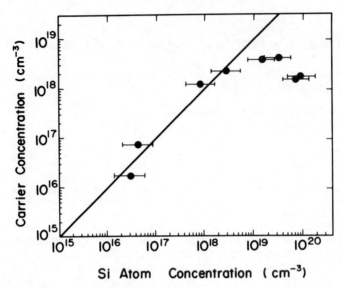

Fig. 1   Free carrier concentration as a function of Si atom concentration in MBE grown GaAs layers.   Hall and SIMS measurements.

temperature vs. the Si atom concentration for various samples. Data plots were obtained by Hall and SIMS measurements. In agreement with previous reports, the activated fraction of doped Si is nearly unity up to a doping level of $\sim 2 \times 10^{18}$ cm^{-3}, nevertheless it saturates and decreases thereafter.

Channeling PIXE measurements were carried out for heavily Si doped samples

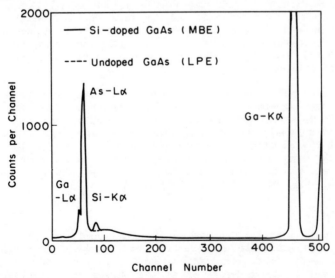

Fig. 2   Typical PIXE spectra from Si doped and undoped GaAs samples.

using 4 MeV-He^{++} ion beam and Si(Li) x-ray detector with an energy resolution of ~150 eV. In order to suppress the effect of surface contamination, which is known to often interfere with analyses of trace Si (Price et al. 1976), all measurements in the present study were done in UHV conditions below 10^{-10} Torr. Figure 2 shows typical PIXE spectra from Si doped and undoped samples. The Si-Kα line is clearly seen for the doped sample but only a flat background for the undoped sample. In these measurements we used a polyimide foil of 25 μm in thickness as an x-ray filter located just in front of the Si(Li) detector. This cuts off the Bremsstrahlung radiation originated from secondary electrons inside the crystal, and greatly enhances the signal-to-background ratio in the low energy spectral region.

Angular scans of He ion backscattering and x-ray yields were made around the ⟨100⟩, ⟨110⟩, and ⟨111⟩ axial channeling directions. As is familiar in the standard procedure of channeling analysis, the interstitial fraction of impurity atoms in a crystal, S, is expressed in terms of the channeling minimum yield, χ_{min}, as

$$S = \frac{\chi_{min}^{i} - \chi_{min}^{h}}{1 - \chi_{min}^{h}}$$

where the superscripts i and h represent the impurity and the host, respectively. Also, the transverse atomic displacement, r_x, of the impurity atoms on the average from the host lattice determines the half width of the channeling dip, $\psi_{1/2}$, as

$$\frac{\psi_{1/2}^{i}}{\psi_{1/2}^{h}} = \left\{ \frac{\ln[(Ca/r_x)^2 + 1]}{\ln[(Ca/\rho)^2 + 1]} \right\}^{1/2}$$

where $C^2 \sim 3$ is a constant, a is the screening distance, and ρ is the

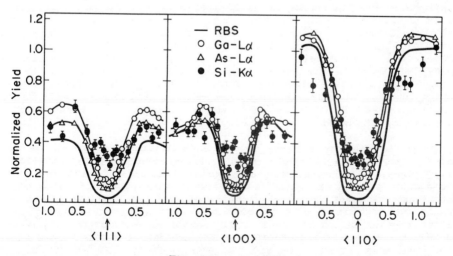

Fig. 3 Angular scans of He ion backscattering (RBS) and characteristic x-ray emission yields. The doping level of Si is 8×10^{19} cm^{-3}. The error bars show the statistical uncertainty.

transverse rms thermal vibration amplitude of the host atoms. We used these formulae in the analysis of our angular scan data.

3. Results and discussion

The experimental results of angular scans are shown in Fig. 3 for the sample doped with Si by 8×10^{19} cm^{-3}. Scanning was made in the $\{110\}$ plane for the $\langle 111 \rangle$ and $\langle 100 \rangle$ axes, and in the vicinity of the $\{100\}$ plane for the $\langle 110 \rangle$ axis. The PIXE results show shallower and narrower profiles in general as compared with the backscattering profiles. This is mostly because of the integral nature of PIXE measurements when compared with RBS which has a depth resolution of ~ 200 Å whereas PIXE has a very poor depth resolution. Actually, PIXE and RBS measurements in our case are estimated to yield averaged information from the surface layers of ~ 1000 Å and ~ 1 μm in thickness, respectively. Since the channeled ion beam travelling through the surface layer gradually loses its directional uniformity, our results can be understood qualitatively.

In any case, we notice that the minimum yield for RBS is quite low and we can hardly find any difference from that of GaAs substrates or undoped samples. This is evidence that the MBE-grown GaAs layer maintains fairly good crystalline quality even when heavily doped with Si. On the other hand, the Si-Kα profiles certainly show the channeling dips as the Ga and As profiles do. However, the value of χ_{min}^{Si} is much higher than that of χ_{min}^{Ga} or χ_{min}^{As} and is almost constant at $\sim 30\%$ independent of the channeling direction. This indicates that the Si atoms do not occupy such specific interstitial sites as the tetrahedral sites but occupy slightly displaced lattice sites. If we apply the analysis formula to the data in Fig. 3, we can conclude that $\sim 20\%$ of doped Si atoms occupy the displaced sites and the displacements of those atoms from the normal lattice sites are ~ 0.1 Å on the average. The latter conclusion is derived from the slightly shallower Si-Kα profile in the $\langle 110 \rangle$ angular scan as compared with the Ga or As profile. However, in other angular scans our data are not good enough to discuss the dip width. Taking all together, we can merely set the upper limits of displaced atoms fraction and displacements as $\sim 20\%$ and ~ 0.1 Å, respectively. There are infinite possibilities to fulfill these requirements, but there is no doubt about the existence of such considerable amount of displaced atoms. These tend to increase with the doping level of Si. For instance, the value of χ_{min}^{Si} reaches $\sim 45\%$ for the sample doped with Si of 5×10^{20} cm^{-3}, which indicates in turn that $\sim 37\%$ of Si atoms are displaced.

In order to discriminate the Ga and As substitutional sites, we made careful measurements of angular scan profiles about the $\langle 110 \rangle$ axis in the $\{110\}$ plane. As Bhattacharya and Pronco (1982,1983) have demonstrated, the angular scan in this condition allows discrimination of two kinds of substitutional sites from the asymmetry of profiles. Figure 4 shows the result for the sample with Si of 5×10^{20} cm^{-3} within the top epitaxial layer of ~ 0.3 μm in thickness. This sample was made intentionally thin to keep the directional uniformity of the incident 4 MeV-He^{++} beam within the interested layer. In Fig. 4, the As-SP profile corresponds to backscattering from surface layer As atoms, which we could barely separate from the bulk contribution and the Ga scattering signal. As expected, the As-SP profile shows the asymmetry being most evident in the yields at ~ 0.9 deg. apart from the [101] axis. In contrast, the Si-K profile shows a fairly symmetric shape, thus indicating that the Si atoms do not prefer

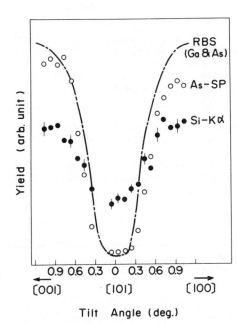

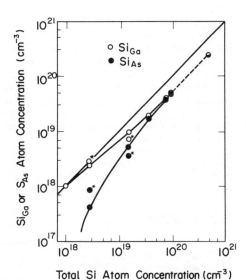

Fig. 4   Angular scan of RBS and PIXE yields about the [101] axis in the {110} plane. The doping level of Si is 5×10^{20} cm^{-3}.

Fig. 5 Si_{Ga} and Si_{As} atom concentrations as a function of total Si atom concentration. The plots marked with * indicate donor and acceptor concentrations calculated from the mobility and carrier concentration data.

the Ga or As sites in this heavily doped sample, but occupy both sites with about equal probabilities.

On the basis of these experimental results, we can explain the electrical characteristics of Si-doped GaAs crystals as follows. When the doping level is increased, the Si atoms increasingly tend to occupy the As sites and compensate the free carrier concentration. Assuming this simple autocompensation mechanism, we can draw the relationship between the Si_{Ga} or Si_{As} atom and the total Si atom concentrations as shown in Fig. 5. The curves are nicely smooth and indicate that the Si atoms start to occupy the As sites at $\sim 10^{18}$ cm^{-3}, and then the fraction of Si_{As} increases gradually to ~ 0.5 at $\sim 10^{20}$ cm^{-3}. Application of the results of Hall measurements to the theoretical relationship (Rode and Knight 1971) between the free carrier concentration and the mobility gives the donor and acceptor concentrations as also shown in Fig. 5, which are in reasonable agreement with the curves. Displacements of Si atoms are doubtlessly related to this increase in the number of Si_{As} atoms. We believe the interaction between the Si_{As}^- acceptors and Si_{Ga}^+ donors acts as a driving force of displacements. Particularly, the Si_{As}-Si_{Ga} nearest neighbors, which must increase in number with the doping level, will occupy displaced sites since the Si-Si bonding is much stronger than that of Si-Ga or Si-As.

4. Summary

We have applied the channeling PIXE technique to atomic structure analysis of MBE-grown highly Si-doped GaAs layers and shown that;
1) Specific interstitial sites do not exist for Si in GaAs, almost all Si atoms occupy the substitutional or slightly displaced sites.
2) The Si atoms occupy both Ga and As sites with about equal probabilities in the high dope region over $\sim 10^{20}$ cm^{-3}.
3) The fraction of displaced Si atoms amounts to several tens of a percent for highly doped samples, and the displacements range ~ 0.1 Å in a rough estimation.
These provide direct and new evidence about the atomic structure, which is useful for further understanding of the electrical characteristics of MBE grown GaAs layers.

Acknowledgement

The present research effort is part of a major research and development project on optical measurement and control systems, conducted under a program set up by the Ministry of International Trade and Industry's Agency of Industrial Science and Technology.

References

Bhattacharya R S and Pronco P P 1982 Appl. Phys. Lett. **40** 890
Bhattacharya R S and Pronco P P 1983 Appl. Phys. Lett. **42** 880
Cho A Y 1975 J. Appl. Phys. **46** 1733
Drummond T J, Lyons W G, Fischer R, Thorne R E, Morkoc T J, Hopkins C G and Evans C A Jr. 1982 J. Vac. Sci. Technol. **21** 957
Feldman L C, Mayer J W and Picraux S T 1982 Material Analysis by Ion Channeling (New York: Academic) Chap. 1-5
Neave J H, Dobson P J, Harris J J, Dawson P and Joyce B A 1983 Appl. Phys. **A32** 195
Price P B, Cooke B E, Ewan G T and Whitton J L 1976 Ion Beam Surface Layer Analysis, ed. O. Meyer et al (New York: Plenum) Vol.2 pp727-734
Rode D L and Knight S 1971 Phys. Rev. **B3** 2534
Spitzer W G and Allred W 1968 J. Appl. Phys. **39** 4999

Inst. Phys. Conf. Ser. No. 74: Chapter 3
Paper presented at Int. Symp. GaAs and Related Compounds, Biarritz, 1984

MBE growth and characterization of $Ga_{0.47}In_{0.53}As/Al_{0.48}In_{0.52}As$ multiquantum well

L. GOLDSTEIN, A.M. JEAN-LOUIS, J.Y. MARZIN, M. ALLOVON
Laboratoire de Bagneux C.N.E.T., 196 rue de Paris, 92220 Bagneux, FRANCE

C. ALIBERT, S. GAILLARD
Equipe de Microélectronique de Montpellier, Unité associée au CNRS n°392
USTL - 34060 Montpellier, FRANCE

Abstract. $Al_{0.48}In_{0.52}As/Ga_{0.47}In_{0.53}As$ multiquantum well lattice matched on InP (100) have been grown by MBE. A 0.5µm buffer layer of $Ga_{0.47}In_{0.53}As$ was first deposited on the substrate followed by (i) a 1000A $Al_{0.48}In_{0.52}As$ layer, (ii) the MQW structure and (iii) a InAlAs cladding layer. Photoluminescence and electroreflectance studies performed at room and low temperature reveal the high optical quality of these samples. We observed the allowed MQW associated transitions and thus discuss the condition band discontinuity.

1. Introduction

The ternary compound $Ga_{0.47}In_{0.53}As$ lattice matched on InP substrate has been widely used for fabricating photodetectors and light sources in the wavelength range of 1.67µm. Using quantum size effects, the energy can be shifted either in the 1.3µm or 1.55µm range (Welch D, 1983). Furthermore quantum well structures exhibit some advantages over normal double heterostructure i.e. : a lower threshold current, an increase of To (Tsang T, 1981). Consequently $Ga_{1-x}In_xAs$ quantum wells are of great interest for optical communication.

$Al_{0.48}In_{0.52}As$ lattice-matched on InP has a band gap of 1.45eV at room temperature. The band gap difference between AlInAs and GaInAs is 0.71eV which leads to a large confinement in the quantum well. In the first part we present MBE growth of the MQW system. In the second part photoluminescence results are reported followed by electroreflectance measurements at low temperature. Finally band discontinuity between $Al_{0.48}In_{0.52}As$ and $Ga_{0.47}In_{0.53}As$ is discussed.

2. MBE growth of $Al_{0.48}In_{0.53}As/Ga_{0.47}In_{0.53}As$ quantum well

On a thermally cleaned (100) InP substrate, we grow a buffer layer of $Ga_{0.47}In_{0.53}As$ of 0.5µm at 520°C. This substrate temperature is kept constant during the growth time. It was found that the use of this $Ga_{0.47}In_{0.53}As$ buffer layer has a smoothing effect and improves the subsequent growth. This is quite similar to the GaAs/$Ga_xAl_{1-x}As$ system where a thin GaAs layer improves the quality of the quantum well (Miller R, 1982) by impurity trapping or a smoothing effect. In contrast, thick $Ga_xAl_{1-x}As$ tends to broaden photoluminescence spectra due to a rough interface. We expect that the same phenomena occur with AlInAs. High electron energy diffraction (RHEED) patterns obtained during growth of GaInAs show well defined and continuous streaks. When a thick $Al_{0.48}In_{0.52}As$ film is deposited the streaks tend to be discontinuous. This is an indication for a roughening effect. For AlInAs film thicknesses lower than 2000Å, GaInAs deposition will restore the RHEED pattern. AlInAs layers thinner than 2000Å are seen to be mirror like by optical microscopy. When the thickness is increased up to 5000Å, there are many more defects and the surface becomes rough. In order to obtain a sharp interface, the first confinement layer was 1000Å thick.

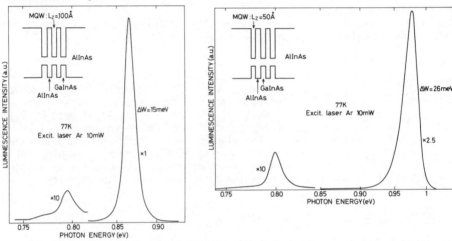

Fig.1:Photoluminescence of MQW GaInAs/AlInAs at 77 K.

3. Photoluminescence

We present optical data obtained on two samples consisting of 5 periods of $In_{0.53}Ga_{0.47}As$ (thickness 50Å (a) or 100Å (b)) and $Al_{0.48}In_{0.52}As$ (thickness 65Å) sandwiched between two 1000Å thick $Al_{0.48}In_{0.52}As$ layers. Theses structures are grown on an InP (100) substrate followed by a 0.5μ $Ga_{0.47}In_{0.53}As$ buffer layer.

Fig.1 shows the 77 K photoluminescence spectra of these two samples. In both spectra, the 0.8 eV lines are due to the InGaAs buffer layer, while the highest energy lines arise from the superlattices. The energy shifts corresponding to the quantum effect are 75 meV and 180 meV for the 100 and 50Å wells respectively. Their widths (Δw = 15 and 26 meV) are relatively small compared to the published results and show the good optical quality of these samples.

The small thin thickness wells exhibit a wider photoluminescence line (Fig.1.b), and this fact can be attributed to thickness inhomogeneities. However, the overall intensities are comparable, and this tends to indicate that, if the levels are more sensitive to these inhomogeneities for narrower wells, the radiative efficiency is not affected. The energies of the observed transitions are in agreement with the results obtained by electroreflectance (ER), which will now be presented.

4. Electro-reflectance

The Schottky-barrier ER technique is well suited to the spectroscopy of semiconductors (ASPNES D, 1972). At low fields, the third-derivative nature of low-field ER spectra allows critical-points structures closely spaced in energy to be resolved (ASPNES D, 1973). ER has been used recently on GaAs/Ga_xAl_{1-x}As heterostructure and Q.W. in order to observe the main optical transitions (ALIBERT C, 1983). We present here an E.R. study of the previously described samples. The ER was measured after deposition of a Cu_2S film used as a biasing electrode. The experimental system has been described elsewhere (ALIBERT C, 1972).

The 35K ER spectra of samples a and b, in the energy range from 0.7 eV to 2 eV are shown in fig.3. Numerous structures can be seen which arise from optical transitions between valence and conduction levels. Between 0.7 eV and 1.4 eV the low field ER spectra of (a) sample shows excitonic transitions associated to the quantum levels : $HH_1 \rightarrow E_1$ (0.872 eV), $LH_1 \rightarrow E_1$ (0.901 eV), $HH_2 \rightarrow E_2$ (1.07 eV) and $HH_3 \rightarrow E_3$ (1.32 eV). Between 1.4 and 1.5 eV the structures cannot be clearly explained.

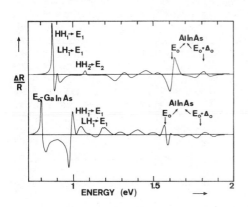

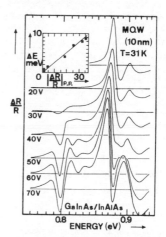

Fig.2: ER of MQW GaInAs/AlInAs   Fig.3:ER for different applied voltage

Near 1.6 and 1.8 eV, the Al_xIn_{1-x}As E_0 ($\Gamma_{8V} \rightarrow \Gamma_{6C}$) and $E_0 + \Delta_0$ ($\Gamma_{7V} \rightarrow \Gamma_{6C}$) transitions are well resolved and situated with the three-point method (ASPNES D, 1973) at 1.61±0.01 eV and 1.81±0.01 eV respectively. This Al_xIn_{1-x}As E_0 gap value can be used to obtain the aluminium composition. The first structure which appears in the ER spectra (b), comes from the E_0 structure of the Ga_xIn_{1-x}As buffer. The 10 meV energy width of this structure corroborates the Ga_xIn_{1-x}As quality. Excitonic transitions associated to the QW levels are seen between 1 and 1.4 eV : $HH_1 \rightarrow E_1$ (0.993 eV), $LH_1 \rightarrow E_1$ (1.045 eV). The Al_xIn_xAs E_0 and $E_0 + \Delta_0$ transitions for this sample are respectively 1.57±0.01 eV and 1.78 eV ± 0.01 eV. The shift in energy on the E_0 transition between the a and b sample corresponds roughly to a variation of 1.5 % in Al composition. Nevertheless the high resolution of the E_0 AlInAs structure observed in the two samples involves in each sample an Al inhomogeneity under 0.3 %.

Fig.4 shows 35 K ER spectra of sample a taken with different applied voltages. For low voltage, i.e. low electric field, the modulated longitudinal electric field does not reach the $Ga_xIn_{1-x}As$ buffer layer so no structure appears near the E_0 (0.8 eV) $Ga_xIn_{1-x}As$ band gap. As the applied voltage is increasing one can observe two effects. (i) The shift to lower energy as function of increasing voltage or longitudinal electric field is clearly observed for the $HH_1 \rightarrow E_1$ and $LH_1 \rightarrow E_1$ excitonic transitions. (ii) Near 0.8 eV an increasing sharp structure appears. This structure arises from the E_0 GaInAs transition. In insert of fig.4 we have shown that the shift to lower energy of the QW transitions is linearly dependent of the amplitude of the ER E_0-spectra associated to the buffer layer. As the ER spectra amplitude is known to be proportional to the electric field squared, we can conclude that the theoretically predicted (BASTARD G, 1983) quasi-quadratic variation of the QW levels as function of the electric-field is verified.

5 - DISCUSSION

Using a three band envelope function model (Bastard G 1981), we have calculated the expected transition energies in these superlattices. We have used for $In_{0.53}Ga_{0.47}As$ the following effective masses : me=0.041mo, mhh=0.465mo and mlh=0.05mo (Alavi K 1980) and an energy gap of 0.813 eV, and for $In_xAl_{1-x}As$ interpolated effective masses and the energy gaps measured by ER. The highest energy transitions related to the superlattice are difficult to locate precisely on the ER spectra and are also very sensitive to the well thicknesses which are only known with a 10 % relative precision. Moreover, coupling between successive wells through the narrow $In_xAl_{1-x}As$ barriers results in the existence of rather wide minibands for these levels, which can be situated as for as 5 or 6 hundred meV from the InGaAs gap, in an energy range where the calculation is less valid.

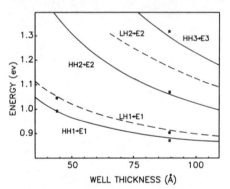

WELL THICKNESS (Å)

Fig.4:Caculated energy positions (not including excitonic effects) for the low energy limit of the different transitions as a function of the well thickness (65 Å barriers). The experimental values deduced from ER are indicated. The value of conduction band discontinuity we used is 75 % of the energy gap difference.

For all these reasons, it is, in our case, more reliable to estimate the band discontinuities from the positions of the lowest levels. In fact, the energy difference between the first heavy and light hole related transitions is, at the same time, less sensitive to the well thickness (in the range 40-50Å) than their absolute positions and rapidly varying with the band discontinuity. Taking into account the experimental results for sample a, and the possible difference between the corresponding exciton binding energies, the conduction band discontinuity can be estimated as lying between 70 % and 85 % of the energy gap difference.

Fig.4 shows the calculated first transitions as a function of the well thickness. They are in satisfying agreement with the experimental data if we assume that the well thicknesses are ~ 45 and 90 Å for sample a and b respectively. In sample b, the third heavy hole transition is also correctly accounted for. This range of values which has to be refined by further experiments is consistent with the previously reported data (Welch D 1984).

6 - CONCLUSION

In summary, we have grown $In_{0.53}Ga_{0.47}As-In_{0.52}Al_{0.58}As$ multiquantum well of high optical quality, as shown by photoluminescence data. We have used ER which appears to be a powerful technique allowing us to determine the energy gaps of the constituents as well as all the superlattice related allowed transitions. The behaviour of these structures under an electric field applied along the growth direction has also been investigated by ER. The calculation of the levels in our superlattice enables us to give an estimate of the band discontinuity in this system.

REFERENCES

- C. ALIBERT, FAN JIA HUA, M. ERMAN, P. FRIJLINK, P. JARRY and J.B. THECTEN, Revue Phys. Appl.18, 709, 1983
- C. ALIBERT, G. BORDURE, A. LAUGIER, J. CHEVALLIER, Phys. Rev.B6, 1301 (1972)
- K. ALAVI, R.L. AGGRAWAL and J.H. GROVES, Phys. Rev.B21, 1311 (1980)
- G. BASTARD, E.G. MENDEZ, L.L. CHANGAND, L. ESAKI, Phys. Rev.B28, 3241 (1983)
- G. BASTARD, Phys. Rev.B24, 5693 (1981)
- DE ASPNES, Phys. Rev. Lett. 28, 913 (1972)
- DE ASPNES, Surf. Sci. 37, 440 (1973)
- R.C. MILLER, W.T. TSANG and MUNTEANY, Appl. Phys. Lett. 41, 375 (1982)
- D.F. WELCH, G.W. WICKS and L.F. EASTMAN, Appl. Phys. Lett.43, 762 (1983)
- D.F. WELCH, G.W. WICKS and L.F. EASTMAN, J. Appl. Phys. 55, 3176 (1984)

Inst. Phys. Conf. Ser. No. 74: Chapter 3
Paper presented at Int. Symp. GaAs and Related Compounds, Biarritz, 1984

139

Luminescence of Ga$_{.47}$In$_{.53}$As/Al$_{.48}$In$_{.52}$As quantum well heterostructures grown by molecular beam epitaxy

W. Stolz, K. Fujiwara*, L. Tapfer, H. Oppolzer**, and K. Ploog

Max-Planck-Institut für Festkörperforschung, Heisenbergstr. 1,
D-7000 Stuttgart 80, Federal Republic of Germany

Abstract We report detailed luminescence (4.2 K < T < 300 K), X-ray diffraction, and TEM measurements of MBE grown (GaIn)As/(AlIn)As quantum well heterostructures (QWH) lattice-matched to InP. The character of the QWH-luminescence is determined by the properties and the thickness of the (GaIn)As well material. At low temperatures excitonic recombination is enhanced with decreasing well width. The carrier confinement in the quantum well leads to a high energy shift of the band to band lumines-cence transition of 10 to 365 meV (L_z = 20 to 1.5 mm) with respect to the (GaIn)As band gap. The assignment of the various luminescence transi-tions in (GaIn)As bulk-type layers and in QWH has been achieved by care-ful study of the temperature dependence of the specific recombination mechanism.

1. Introduction

In the last few years quantum well heterostructures (QWH) have attained considerable interest both in fundamental research for the study of quasi-two-dimensional systems and in device development owing to e.g. the possi-bility to select the emission wavelength of a QWH by varying the layer thickness of the lower band gap material. The wavelength range between 1.3 and 1.65 μm is important for optical communication systems because of the com-bination of minimum dispersion and attenuation in silica fibres. The feasi-bility of QWH emitting mainly in the 1.5 to 1.65 μm range was demonstrated in the (GaIn)As/(AlIn)As-system grown by molecular beam epitaxy (MBE), (Temkin et al. 1983, Alavi et al. 1983, Welch et al. 1983) and in (GaIn)As/InP-system grown by MBE (Tsang, 1984) by metal organic chemical vapor deposition (MOCVD)(Razeghi et al. 1983) and by vapor phase epitaxy (VPE)(Kodama et al. 1983, DiGiuseppe et al. 1983). The large band gap difference between (AlIn)As and (GaIn)As lattice-matched to InP (1.47 vs. 0.75 eV at 300 K) allows the variation of the emission wavelength of the respective QWH between λ=850 nm and λ≈1.65 μm at room temperature. We have investigated the light emission based on confined particle transitions of (GaIn)As/(AlIn)As QWH in the temperature range of 4.2 to 300 K. These de-tailed results will be compared to the luminescence properties of bulk-type (GaIn)As and (AlIn)As layers.

* on leave from: Central Research Lab. Mitsubishi Electric Corporation,
   Amagasaki, Hyogo 661, Japan

** permanent address: Siemens AG, Research Lab., Otto-Hahn-Ring 6,
   D-8000 München 83, Federal Republic of Germany

2. Experimental

MBE growth was performed in a three-chamber UHV system for sample exchange
and storage which is equipped with an azimuthally rotating substrate holder. The (100)-InP substrates were etched in $H_2SO_4 : H_2O_2:H_2O$ (3:1:1) and
in 0.3% Br_2-CH_3OH solution according to a procedure described by Nishitani
and Kotani (1979). The quality of the InP substrate has a significant influence on the luminescence intensity of the QWH. Evaluation of different
substrates from various sources (InP:Fe and InP:Sn) mounted side by side on
the Mo heater block leads to a scattering of the reproducibility of the
luminescence intensity by a factor of 10. The ratio of Al to In and Ga to In
in the molecular beams was calibrated by double crystal X-ray diffraction
of (GaIn)As and (AlIn)As test layers grown with precise temperature setting
and monitoring of the beam flux from the effusion cells. We were able to
improve the flux stability of Al, Ga and In considerably by using a computer controlled temperature regulation which kept the range of actual
temperature fluctuation to below 0.02% during the whole growth run. A lattice mismatch between the epilayers and the InP substrate of less than
1.10^{-3} at 295 K was achieved. The rotary velocity of the substrate holder
was carefully synchronized to the shutter movement, i.e. an integer number
(≥ 2) of rotations per growth time of one quantum well, in order to keep
lateral composition variations especially of the quantum well width as low
as possible. In different growth runs we have grown series of (GaIn)As/
(AlIn)As multi and single QWH lattice-matched to InP with well widths L_Z
systematically varied from 20 down to 1.5 nm. All samples were grown with
a growth rate of 1 μm/h at substrate temperatures of 550 °C $<$ T_{sub} $<$ 570 °C
using As_4 molecular beams. The non-intentionally doped bulk-type layers of
2 μm thickness exhibit the following electrical properties:

N_D-N_A = $5 \cdot 10^{16}$ cm^{-3}/μ = 5000 cm^2/Vsec for (GaIn)As and
N_D-N_A = $5 \cdot 10^{15}$ cm^{-3}/μ = 750 cm^2/Vsec for (AlIn)As at room temperature.

The layer sequence and thickness of the QWH are schematically shown in
Fig. 1a.
For the photoluminescence (PL) investigations the samples were mounted in
a variable temperature cryostat and excited with an unfocused laser beam
of a cw Ar^+ laser (λ=514.5 nm). We used an excitation intensity of 5 mW
(125 mW/cm^2) for QWH samples and 125 mW (3 W/cm^2) for bulk-type (GaIn)As
and (AlIn)As epilayers. The excitation conditions were kept constant over
the entire temperature range. The luminescence signal was spectrally analysed with a 1 m monochromator (SPEX 1704) and recorded by a liquid nitrogen cooled Ge-detector (Northcoast Scientific Corp.) using standard lock-
in amplification. The spectra were corrected for the spectral response of
the measurement system if necessary.

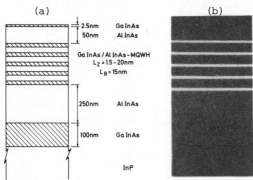

(a)   (b)

Fig. 1a and b:
(a) Schematic arrangement of
layer sequence in (GaIn)As/
(AlIn)As multi QWH with constant
barrier width L_B = 15 nm;

(b) TEM micrograph of thinned
cross-section through part of
layer structure ((200)-darkfield
image)

100 nm

3. Results and Discussion

The excellent quality of the layer sequence of MBE grown (GaIn)As/(AlIn)As QWH is revealed in the TEM micrograph in Fig. 1b for a sample with L_z = 2.5 nm. The interfaces are abrupt within the resolution of the TEM micrograph of 0.5 nm. The well width which was previously determined by extrapolation of bulk-type epilayer growth rates with an error of 10% coincides to the L_z measured by TEM within the errors of the two methods. The most important results obtained from PL investigations of 13 samples are summarized in Table 1. In more detail we will present the results by means of Fig. 2 to 5.
In Fig. 2 we show low temperature PL spectra of 2 μm thick (AlIn)As and (GaIn)As epilayers and of a QWH with L_z = 2.5 nm. The PL spectrum from (AlIn)As consists of a near-band-gap line at 1.538 eV with 16 meV FWHM and a broad band in the range of 1.1 to 0.7 eV due to deep level recombination. (GaIn)As is characterized by a luminescence line at 790 meV with a FWHM of 20 meV. This line shifts slightly to higher energies and becomes narrower with increasing excitation intensity. We ascribe this line to a donor to acceptor pair recombination at very low temperatures and to a band to acceptor transition at higher temperatures in accordance with the results of Marzin et al (1983) obtained on LPE grown (GaIn)As. The QWH luminescence is dominated by a very intense line which shifts systematically to higher energies with decreasing well width L_z. With an excitation density of 125 mW/cm^2 only weak additional PL from deep level recombination in (AlIn)As and from the (GaIn)As buffer layer can be detected. If the excitation density is increased to 3 W/cm^2 however, additional peaks on the high energy side of the spectrum appear which may be attributed to higher subband transitions and to the near-band-gap luminescence of (AlIn)As at 1.538 eV. The luminescence intensity of a QWH is at least one order of magnitude higher than that of bulk-type (GaIn)As layers. A comparison between multi and single QWH grown on high quality substrate material shows no differences in PL intensity (less than a factor of two) and FWHM.

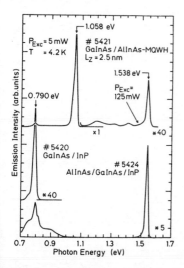

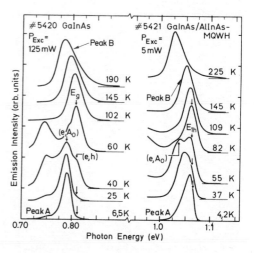

Fig. 2 Low temperature PL spectra obtained from bulk-type (GaIn)As and (AlIn)As layers and from a multi QWH with L_z = 2.5 nm.

Fig. 3 Variation of PL spectra obtained from a bulk-type (GaIn)As layer and from a multi QWH with L_z = 2.5 nm in the temperature range of 4.2 to 225 K.

The temperature dependence of the luminescence observed in (GaIn)As and in a
QWH with L_z = 2.5 nm is compared in Fig. 3. We ascribe the high-temperature
line (T>80 K, Peak B) to a band to band transition for the bulk-type (GaIn)As
layer and to an E_{1h}-subband transition for the QWH, respectively. In both
cases the energy shift of this line with increasing temperature is determined
by the temperature dependence of the (GaIn)As band gap   (see also Fig. 4).
The arrows in Fig. 3 indicate the position of the band gap respectively the
E_{1h}- subband transition taking the results of Penna et al (1984) of the
temperature dependence of the (GaIn)As band gap into account. The (e,A°)-
peak of the (GaIn)As shifts only slightly to lower energies with increasing
temperature but decreases rapidly in intensity and vanishes at 60 K. At 40 K
the band to band recombination appears as a high-energy shoulder. The PL of
the QWH with L_z = 2.5 nm at low temperatures is composed of two lines, one
line near the E_{1h}-subband transition, which we ascribe to an excitonic
recombination, and the (e,A°) line. This composed line is broadened by energy
variations due to well width fluctuations. With increasing temperature the
PL intensity decreases and the peak broadens to lower energies, which results
from a more rapid decrease of the intensity of the excitonic than of the
(e,A°) recombination. At higher temperature, the (e,A°) peak vanishes again
and the dominant peak up to room temperature is due to the E_{1h}-subband transition.

In Fig. 4 we show the temperature dependence of the peak energies of the low
temperature line which to our interpretation is a superposition of the excitonic and the (e,A°) line (peak A,T<50K,
full signs) and the high temperature
E_{1h}-subband transition (peak B, open signs)
for QWH samples with different well widths.
Fig. 4 clearly reveals the high-energy
shift of the E_{1h}-subband transition with
decreasing well width L_z. The temperature dependence of the high temperature
luminescence is determined by the band
gap dependence of the (GaIn)As well
material for all QWH samples. The peak of
the low temperature line is shifting
towards the subband transition energy with
decreasing well width. This result is consistent with an increasing amount of excitonic relative to the (e,A°) recombination
in quantum wells with small L_z. For a pure
(e,A°) transition one would expect an opposite shift to lower energies, because of
an increase of the acceptor binding energy
with decreasing well width according to
calculations by Bastard (1981).

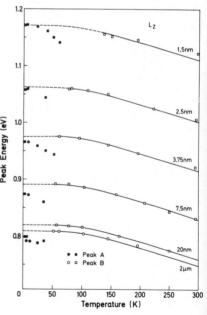

Fig. 4 Temperature dependence
of PL peak energy of two distinct transitions in a multi
QWH having different well width
L_z and in a bulk-type (GaIn)As
layer.

In Fig. 5 we compare the E_{1h}-subband transition as a function of the well width L_z
obtained in this work with data from
Temkin et al (1983), Alavi et al (1983),
and Welch et al (1983). The slightly
higher energies reported by Welch et al
(1983) can be explained by the higher Al-content in their (AlIn)As barrier
layers, which leads to a total band gap discontinuity approximately 100 meV
higher than for (AlIn)As closely lattice matched to InP, and thus to a
higher energy shift of the quantum well luminescence.

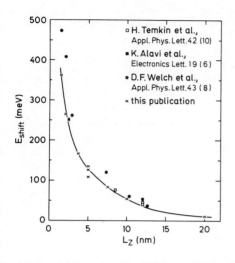

Fig. 5 Shift of emission energy in
(GaIn)As/(AlIn)As QWH as a function
of well width L_z. For comparison,
experimental data from recent publications are also included.

Optical communication based on silica fibres requires light emitting devices
operating at $\lambda = 1.3$ μm and $\lambda = 1.55$ μm. QWH composed of (GaIn)As and
(AlIn)As lattice-matched to InP with well widths $L_z = 3$ nm ($E_{1h} = 950$meV)
and $L_z = 10$ nm ($E_{1h} = 800$ meV) emit at these wavelengths at room temperature.
The feasibility to grow the structures reproducibly by MBE makes this material system interesting for the application in those devices.

4. Summary

(GaIn)As/(AlIn)As QWH lattice-matched to InP with systematically varied
well widths L_z were grown by MBE. TEM micrographs revealed the high quality
of QWH achievable with the MBE-growth technique. The character of the QWH-
luminescence is determined by the properties and the thickness of the
(GaIn)As well material. At low temperatures the excitonic recombination is
enhanced with decreasing L_z. The carrier confinement in the quantum well
leads to a high-energy shift of the band to band luminescence transition of
10 to 365 meV ($L_z = 20$ to 1.5 nm) with respect to the (GaIn)As band gap. The
assignment of the various luminescence transitions in (GaIn)As bulk-type
layers and in QWH has been achieved by study of the temperature dependence
of the specific recombination mechanism. The high efficiency of quantum
well luminescence does not degrade at increased sample temperature, and
there are no differences detectable in linewidth and intensity between multi
and single QWH which indicates the high quality of our (GaIn)As/(AlIn)As
single QWH. The high luminescence intensity and the ability to obtain QWH
emitting at $\lambda = 1.3$ μm and at $\lambda = 1.55$ μm at room temperature make the
(GaIn)As/(AlIn)As-system very promising for the application in optical
fibre communication.

Table 1   Summary of luminescence peak energy shift and of halfwidth at 4.2 K
(FWHM) observed in 13 samples of different well width.

No.	L_z (nm) nom. ±10%	L_z (nm) TEM ±0.5	E_{shift} (meV) $E_{1h}-E_g$(GaInAs)	FWHM (meV) 4.2 K P_{laser} = 5 mW
5420 (GaIn)As				20
5424 (AlIn)As				16
5417 MQWH	20		10	21
5396 MQWH	10		54	28
5423 MQWH	7.5	8.5	83	26
5416 MQWH	5		110	24
5422 MQWH	5		126	27
5429 SQWH	5		135	31
5427 MQWH	3.8		167	34
5421 MQWH	2.5	3.0	253	24
5428 SQWH	2.5		250	32
5426 MQWH	2.1		265	37
5425 MQWH	1.5		365	48

5. Acknowledgements

The authors are indebted to E.O. Göbel for critical reading of the manuscript, to A. Fischer and J. Knecht for expert help with the sample preparation, to W. Heinz and W. Krause for technical assistance with the luminescence measurements, and to S. Schild for expert sample preparation for
the TEM.
Part of this work was sponsored by the "Stiftung Volkswagenwerk".

6. References

Alavi K., Pearsall T.P., Forrest S.R., Cho A.Y., 1983,
    Electronic Letters 19(6), 227-229
Bastard G., 1981, Phys. Rev. B 24(8), 4714-4722
DiGiuseppe M.A., Temkin H., Peticolas L., Bonner W.A., 1983,
    Appl. Phys. Lett. 43(10), 906-908
Kodama K., Ozeki M., Komeno J., 1983, J. Vac. Sci. Techn. B 1(3), 696-699
Marzin J.Y., Benchimol J.L., Sermage B., Etienne B., Voss M., 1983,
    Solid State Comm. 45(2), 79-82
Nishitani Y., Kotani T., 1979, J. Electrochem. Soc. 126(12), 2269-2271
Penna A.F.S., Shah J., DiGiovanni A.E., Dentai A.G. 1984,
    Solid State Comm. 51(4), 217-220
Razeghi M., Hirtz J.P., Ziemelis V.O., Delalande C., Etienne B., Voos M.,
    1983, Appl. Phys. Lett. 43(6), 585-587
Temkin H., Alavi K., Wagner W.R., Pearsall T.P., Cho A.Y., 1983,
    Appl. Phys. Lett., 42(10), 845-847
Tsang W.T., 1984, Appl. Phys. Lett. 44(3), 288-290
Welch D.F., Wicks G.W., Eastman L.F., 1983, Appl. Phys. Lett. 43(8),
    762-764

Inst. Phys. Conf. Ser. No. 74: Chapter 3
Paper presented at Int. Symp. GaAs and Related Compounds, Biarritz, 1984

145

Optical and electrical characterisation of GaAs$_y$–Sb$_{1-y}$ (0 < y < 1) grown by molecular beam epitaxy

T D McLean, T M Kerr, D I Westwood, J D Grange and I J Murgatroyd[*]

GEC Research Laboratories, Hirst Research Centre, Wembley, UK
[*] Dept of Metallugry and Science of Materials, University of Oxford, UK

Abstract. GaSb and GaAs$_y$Sb$_{1-y}$ have been grown by molecular beam epitaxy. Undoped GaSb hetero-epitaxial layers are p-type. with 300K hole concentrations of 5 x 10^{15} cm^{-3} to 5 x 10^{16} cm^{-3} and narrow 4K PL spectral features. GaAs$_y$Sb$_{1-y}$ has been grown throughout the entire composition range. Silicon doping of GaAs$_y$Sb$_{1-y}$ gives n-type conduction only for y > 0.8 while beryllium was found to be a controllable p-type dopant over the entire composition range.

Introduction

Relatively little is known of the epitaxial growth of antimony containing compounds. However, increasing interest in the type II superlattice system consisting of GaSb-AlSb-InAs (Chang 1982, Esaki 1981), the potential requirement for long wavelength light sources for fluoride based low loss optical fibres (Tsang 1984), and the suitability of GaAs$_y$Sb$_{1-y}$ and Ga$_x$Al$_{1-x}$Sb for optoelectronic devices (Cho 1977, Ohmori 1984, Temkin 1984), has highlighted the need for a greater understanding of the growth of antimonide alloys.

GaSb grown from the melt has typically 10^{17} native acceptor centres, presumably due to gallium antisite defects (Effer 1964). GaAs$_y$Sb$_{1-y}$ grown by near equilibrium techniques such as liquid phase epitaxy appears to have a miscibility gap for 0.2 < y < 0.8 at 600°C (Cherng 1984, Pessetto 1983), and spinodal decomposition of GaAs$_y$Sb$_{1-y}$ grown by MOCVD (generally regarded as a kinetically controlled technique) has recently been demonstrated (Stringfellow 1984). The substrate temperature (T_s) and flux ratio (JSb$_4$/As$_4$) dependence of the incorporation efficiency of group V elements makes control of alloy composition in molecular beam epitaxy (MBE) non-trivial (Wood 1982). Data relating the incorporation coefficients of group V atoms as a function of absolute and relative group V fluxes and substrate temperature is available only for y > 0.8 (Chang 1977, Waho 1977, Klem 1983). This report describes the controlled MBE growth and doping of GaSb and its alloys with GaAs across the entire composition range, and presents electrical and optical characterisation of the material.

Experimental

Epitaxial layers were grown in a Vacuum Generators V80H system. Background pressures were typically 10^{-9} torr during growth of GaSb and 10^{-7} torr during growth of GaAs$_y$Sb$_{1-y}$, the latter being a result of the high As$_4$ background.

GaSb was grown homo-epitaxially onto (100) p-type Bridgman grown substrates and onto (100) undoped (semi-insulating) LEC GaAs substrates. GaSb substrates were prepared prior to growth by degreasing with trichloroethylene, acetone and isopropanol followed by free etching in 0.1% Br_2-MeOH solution. In the absence of semi-insulating GaSb substrates, electrical assessment of GaSb and $GaAs_ySb_{1-y}$ was made on material grown onto GaAs substrates. These were prepared in the normal manner (Wood 1978).

Mobility and carrier concentration profiling was performed by the successive hall measurement and stripping technique of Stewart (1984), which employs an $HCl : H_2O_2 : H_2O :$ Sodium Potassium Tartrate etch. The composition of $GaAs_ySb_{1-y}$ epitaxial layers was calculated from measurements of the layer lattice parameter using single crystal diffractometry. XPS analyses of freshly prepared and as-grown surfaces were performed using the XPS equipment situated in the preparation chamber of the growth system.

Results and discussion

Growth and doping of GaSb

XPS analysis of GaSb substrates after Br_2-MeOH etching reveals a thin surface layer of gallium and antimony oxides. This is easily removed by heating in UHV at 580°C in an antimony flux. This treatment produces a surface free of oxygen and carbon impurities to below the detection limits of the XPS analyser (Figure 1).

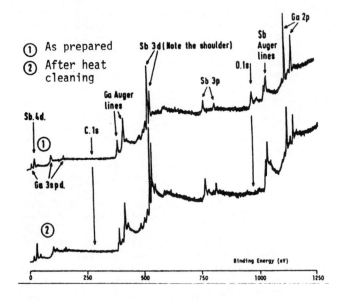

Fig 1: XPS spectra from a GaSb substrate

GaSb films were grown at T_s ~550°C under an $Sb_4 : Ga$ flux ratio of 3:1; determined by measurement of beam equivalent pressures at the substrate position. The 15 kV reflection electron diffraction (RED) pattern normally observed under these conditions is the Sb-stabilised (3x1) reconstruction reported previously by Wood (1983). On cooling in an Sb_4 flux after growth this reverts to a (5x1) reconstructed pattern, which we attribute to an antimony rich surface (Figure 2).

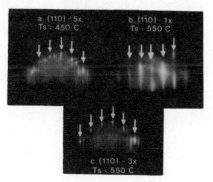

Fig 2: RED patterns showing Sb-stabilised surface atom reconstructions on GaSb (100) surfaces along the <110> azimuths

Nominally undoped layers of GaSb have room temperature free hole concentrations in the range 8×10^{15} cm^{-3} to 5×10^{16} cm^{-3} and mobilities of 1000 to 1400 cm^{2}v^{-1}s^{-1}. These results compare favourably with the high residual acceptor levels found in melt grown GaSb (Effer 1964, Baxter 1965).

4K photoluminescence (PL) spectra of heteroepitaxial GaSb exhibit major near band-gap emissions around ≈ 0.796 eV with a FWHM of 5 meV, and a shallow acceptor related emission with $E_a \simeq 30$ meV. Transmission electron microscopy (TEM) analysis of these layers reveal large numbers of misfit dislocations and stacking fault arrays (Figure 3a) with a density of 10^8–10^{10} cm^{-2}. This is the defect density that would be expected in order to relieve the mismatch related strain between GaSb ($a_0 = 6.082$Å) and GaAs ($a_0 = 5.654$Å). There are however no more defects in homoepitaxial layers than in the starting substrates (10^4 cm^{-2}) (Figure 3b).

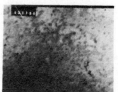

g220

(a) Heteroepitaxial GaSb   (b) Homoepitaxial GaSb

   on GaAs

defect density 10^8 cm^{-2} 10^4 cm^{-2}

Fig 3: Plan view electron micrographs of GaSb

Beryllium has been used to give controllable p-type doping in the range 10^{16} cm^{-3} to 10^{18} cm^{-3} (Figure 5a). Comparison with beryllium doped GaAs grown under the same flux conditions suggests an electrical incorporation coefficient of unity for beryllium in GaSb, whereas doping GaSb with silicon produces p-type but closely autocompensated material.

Growth and doping of GaAs$_y$Sb$_{1-y}$

GaAs$_y$Sb$_{1-y}$ alloys were grown onto (100) oriented GaAs substrates throughout the entire composition range. Since Sb is expected to have a much higher incorporation coefficient than As, reproducible composition control should be facilitated by maintaining the antimony to gallium flux ratio $(J_{Sb}/J_{Ga})<1$ whilst using arsenic as a flood source. Any group V sublattice sites not occupied by Sb will therefore be available for arsenic atoms. At T_s below 550° this is the case. Figure 4 shows the composition dependence on flux ratio (J_{Sb}/J_{Ga}) at T_s 550°C and 600°C.

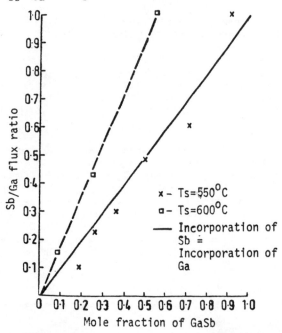

Fig 4: Sb/Ga flux ratio versus composition in GaAs$_y$Sb$_{1-y}$

As Ts is raised to 600°C arsenic incorporation becomes more efficient as a result of the increased re-evaporation of Sb$_4$ from the growing surface. There may also be some contribution from the increased efficiency of As$_4$ dissociation relative to that for Sb$_4$. RED patterns observed during the growth of GaAs$_y$Sb$_{1-y}$ are complex and dependent on layer composition. Compositions near GaAs exhibit the (2x4) reconstruction normally observed for As stabilised GaAs. As the Sb$_4$ mole fraction is raised an increasing asymmetry is observed in the RED pattern, eventually tending to a (3x1) structure for compositions near GaSb. XPS analysis of freshly grown layers shows that the top few (5-10) monolayers of GaAs$_y$Sb$_{1-y}$ cooled after growth in a flux of As$_4$ ($P_{Be}=10^{-6}$ torr) are enriched with respect to As.

2K PL spectra of GaAs$_y$Sb$_{1-y}$ for y = 0.75 show a broad near band gap emission at 1.035 eV with a FWHM of 23.6 meV. A shallow acceptor with Ea ≃ 25 meV is also observed. At higher Sb concentrations (0.3<y<0.7) no PL could be observed. Castano (1983) has suggested that failure to observe PL in GaAs$_y$Sb$_{1-y}$ for y<0.7 is due to the high, mismatch induced, stacking fault density. However, the observation of PL from heteroepitaxial GaSb reported here suggests that this explanation is unlikely to be correct. The possibility that spinodal decomposition is responsible is currently under investigation.

Nominally undoped GaAs$_y$Sb$_{1-y}$ is p-type over the entire composition range with free hole concentrations ~5 x 10^{15} cm^{-3} to 5 x 10^{16} cm^{-3}. Free hole densities for these alloys represent significant improvements on those observed from any other growth technique (Cherng 1984 , Chang 1977). The room temperature mobility of this material is however lower than expected at 50 cm^{2}V^{-1}s^{-1}, possibly as a result of mismatch strain related effects.

Beryllium was found to be a reproducible p-type dopant for all compositions of GaAs$_y$Sb$_{1-y}$,while silicon is a well behaved n-type dopant only for y > 0.8 (Figure 5b). Tin has been shown to change from acceptor to donor behaviour at similar compositions of GaAs$_y$Sb$_{1-y}$ (Chang 1977). At higher antimony concentrations silicon becomes more amphoteric, yielding closely autocompensated p-type epitaxial layers.

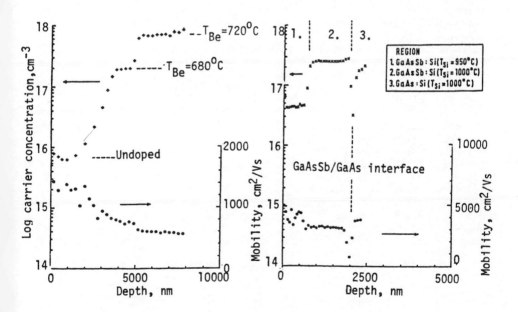

Fig 5a: Graph of log carrier concentration vs depth for a two level GaSb:Be doping staircase

Fig 5b: Graph of log carrier concentration vs depth from a GaAs$_{1-y}$Sb$_y$ layer with y = 0.2

Conclusions

Homoepitaxial GaSb grown during this study is of high crystalline quality whereas heteroepitaxial material shows stacking faults and dislocation densities of 10^{10} cm^{-2}. The composition of GaAs$_y$Sb$_{1-y}$ is readily controlled at low substrate temperatures by using effective trivial rates of Sb/Ga less than unity. Reliable p-type doping has been obtained in GaAs$_y$Sb$_{1-y}$ using beryllium while silicon can only be used as a donor for $y > 0.8$.

Acknowledgements

The authors wish to thank R Nichols and C P Stewart for Hall measurements, B R Brown for X-ray measurements on GaAs$_y$Sb$_{1-y}$ layers and J D Medland and Dr L Canham for low temperature PL measurements. We are grateful to Dr R Booker for useful discussions on TEM. C E C Wood is thanked for his advice and enthusiastic encouragement during this work. This project is partially funded by the Department of Industry. One of us (IJM) acknowledges the support of the SERC and Hirst Research Centre through a CASE studentship.

References

Baxter R D, Bate R T, Reid F J, 1965, J. Phys. Chem. Solids 26, 41
Castano J L, Piqueras J, 1983, J. Appl. Phys. 54, 3422
Chang C A, Ludeke R, Chang L L, Esaki L, 1977, Appl. Phys. Lett. 31, 759
Chang C A, Takaoka H, Chang L L, Esaki L, 1982, Appl. Phys. Lett. 40, 983
Cherng M J, Stringfellow G B, Cohen R M, 1984, Appl. Phys. Lett. 44, 677
Stringfellow,G B,1984, Electronic Materials Conference, Santa Barbara, USA
Cho A Y, Casey H C, Foy P W, 1977, Appl. Phys. Lett. 30, 397
Effer D, Etter P J, 1964, J. Phys. Chem. Solids, 25, 451
Esaki L, Chang L L, Mendez E E, 1981, Japan. J. Appl. Phys., 20, L529
Klem J, Fischer R, Drummond T J, Morkoc H, Cho A Y, 1983, Elect. Lett, 19 453
Ohmori Y, Tarucha S, Horikoshi Y, Okamoto H, 1984, Japan. J. Appl. Phys. Lett. 23, L94
Pessetto J R, Stringfellow G B, 1983, J. Cryst. Growth, 62, 1
Stewart C P, 1984, Semi-insulating III-V Materials Conference Oregon (to be published)
Temkin H, Tsang W T, 1984, J. Appl. Phys. 55, 1413
Tsang W T,1984, 4th international conference on MBE San Francisco
Waho T, Ogawa S, Maruyama S, 1977, J. Appl. Phys. 16, 1875
Wood C E C,Joyce B, 1978, J. Appl. Phys. 49, 4854
Wood C E C, 1982, in GaInAsP alloy semiconductors , Ed Pearsall TP, Wiley-interscience. And references therein.
Wood C E C, Singer K, Ohashi T, Dawson L R, Noreika A J, 1983, J. App. Phys. 54, 2732

Inst. Phys. Conf. Ser. No. 74: Chapter 3
Paper presented at Int. Symp. GaAs and Related Compounds, Biarritz, 1984

151

Growth and optical assessment of low loss GaAs waveguides fabricated by localized vapor phase epitaxy

N. Vodjdani, M. Erman, J.B. Theeten, M. Ketata
Laboratoires d'Electronique et de Physique Appliquée
3, avenue Descartes - 94450 Limeil-Brévannes (France)

1. Introduction

An important step towards the fabrication of III-V semiconductor integrated optic circuits is the interconnection between optoelectronics devices (emitters, detectors) and optical guides (passive or active). The reliability of such devices can be improved if these optical components are integrated on a same substrate including their driving electronic circuits. As a consequence, the optical waveguide has to be compatible with the technology of optoelectronic and electronic devices. For high packing densities one has to deal with the length of the waveguides which can be important (several mm) since most of the active optical devices are based on the electro-optical effect needing long interaction distances. Therefore, waveguiding structures including bends are compulsory, with severe requirements on the overall optical losses.

Most of the waveguiding structures in III-V semiconductors are obtained by etching techniques [1,2,3,4]. These treatments lead to lateral roughness, which is responsible for the high losses currently obtained (4db/cm) and makes the fabrication of low attenuation waveguides very critical. We present an alternative where very low attenuation (1db/cm) waveguides are fabricated by selective epitaxy through an SiO_2 mask, using the chloride vapor phase epitaxy [5]. Using this technique, one can grow either embedded waveguides (by combining in situ etching and epitaxy), or "external" waveguides grown over the substrate plane. Both types of structure are bounded by exact crystallographic planes, the main difference being in the way the optical confinement is obtained. Generally speaking, the optical confinement is stronger in the external waveguides. In contrast, some further technological processes may be more easily done on embedded waveguides (planar technology).

The high anisotropic growth of the vapor phase epitaxy (VPE) gives rise to various prismatic forms depending on the substrate orientation, stripe azimuth and growth conditions. In section 2, a simulation of the influence of the stripe orientation, width and growth conditions on the waveguide cross-section is presented, using a Wulff type modelling of the growth morphology. The fabrication of the straight waveguides and their optical characterization are reported in section 3. Zig-zag optical waveguiding structures have been realized by integrating several etched mirrors with a straight line embedded waveguide, allowing 90 deg. bends with reflection coefficient at each mirror of about 50 %. Results are discussed in section 4.

2. Morphological analysis of selective epitaxy
2.1. Sample preparation

The semi insulating or n^+ doped GaAs substrates within 0.5 ° from (100) orientation are chemically cleaned and etched. A film of 0.1 μm of SiO_2 is then deposited. Stripes of 1 to 8μm width and 1 to 10mm long are opened

in the dielectric film using optical lithography. These stripes are oriented along [110], [1̄10] or [010] exact directions. Prior to growth, the GaAs inside the opened windows is etched either by in situ HCl etching or by chemical etching. For external waveguides, a slight chemical etching is sufficient, while for embedded guides the depth of the etching and its shape determine the inner profile of the waveguide.

2.2. Anisotropic growth in the VPE process

A conventional horizontal VPE (Ga,$AsCl_3$ and H_2) reactor is used. In this system, the growth rate is highly orientation dependent since a surface controlled mechanism is involved. This anisotropy is very sensitive to the growth temperature and the GaCl partial pressure [6]. Growth is also selective i.e. no deposition occurs on the dielectric overlayer.

Fig. 1 is a log log plot of the growth rate for the (100), (110), (111)As and (111)Ga planes vs $AsCl_3$ molar fraction in our experimental conditions [7]. The different intersections of these curves roughly delimit the experimental domains corresponding to different shapes of the overgrowth, the initial stripes being oriented along the zone axis of some of these planes. Fig. 2 sketches the waveguides cross-sections for a strip oriented along [11̄0] direction corresponding to the experimental domains 1, 2 and 3 of Fig. 1.

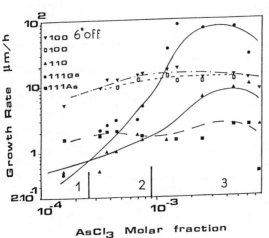

2.3. Simulation of localized growth

In localized structures the edges of the mask define a limit beyond which anisotropic growth takes place. The initial shape of the growth front(converging or diverging) will determine the type of planes delimiting the structure : for a converging growth front,

Fig. 1 - Deposition rate for (100) 6° off, (100), (110), (111)As, (111)Ga orientations vs $AsCl_3$, molar fraction.

the shape is limited by high growth rate orientations and in the case of a planar or diverging growth front, the final profile is bounded by the lowest growth rate orientations (which are the low index planes). We will consider here the latter case. Fig. 2a (corresponding to domain 3 of Fig.1) is bounded by (111)As and (110) facets. In this experimental domain the

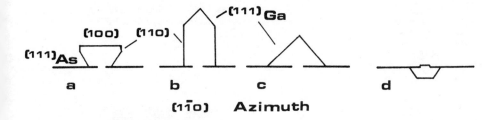

Fig. 2 - Cleaved cross-sections of waveguides oriented along [1̄10] azimuth for various AsCl₃ molar fraction corresponding to :
a/ region 3 of Fig. 1   -   b/ region 2 of Fig. 1
c/ region 1 of Fig. 1   -   d/ embedded waveguide

(111)Ga orientations do   not appear because of their very fast growth rate. For experimental conditions corresponding to domain 1 the (111)Ga plane becomes a limiting orientation (Fig. 2c).

The final growth morphology can be simulated by using a Wulff type modelling [7,8]. In order to predict cross-section of a structure grown on a stripe of width 'W' and oriented along a 'Z' axis, one needs a polar diagram of the growth rate. This polar diagram represents the growth rate versus crystallographic orientation of all planes in zone with the 'Z' direction, and is determined experimentally for each growth condition. The right and the left halves of the polar diagram have to be considered on the corresponding edges of the exposed area. The cross-section is then obtained by considering the normal to the rate vectors for each orientation and plotting only the first normals encountered from a point of the exposed stripe. On the basis of a limited number of experimental data it is possible to predict with this type of modelling [7] :

- the shape of structures oriented along different azimuths.
- the effect of the substrate misorientation.
- the influence of the stripe width on the shape and on the size of structures for a given time of growth.
- the different growth stages of a structure initiated in a stripe of width 'W'.

The computed structure of Fig. 3 shows the morphological evolution of the

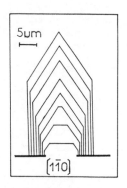

different growth steps for a [1T0] oriented stripe. The (100) higher growth rate plane is progressively overwhelmed by (111)Ga planes which under these conditions correspond to a lower deposition rate orientation.

Fig. 3 - Simulation of the different growth stages for a [1Ī0] oriented wave-guide.

3. Straight waveguides
3.1. External waveguides

Examples of external waveguides are shown in Fig. 2 (a,b,c), while Fig. 4 gives the S.E.M. micrograph of the cleaved cross-section of waveguides oriented along [110] azimuth. The guides are optically isolated from the substrate by a first n$^+$ doped layer which can be the substrate itself. The strong index difference between GaAs and air provides a high lateral confinement. The non cleaved end of the waveguide is bounded by facets oriented towards the substrate. These terminal facets have been used as prisms for injecting light through the substrate backside [5] . In this configuration, light can be coupled to seve-ral guides of different lengths under the same injection conditions. This type of injection alleviates the Fabry Perot type of interference which usually occurs when the two extremities of the guide are cleaved. Attenuations near 1 dB/cm have been obtained.

Fig.4 - S.E.M. micrograph (x1250) of cleaved waveguides oriented along [1Ī0] azimuth.

Furthermore by butt coupling, the overall attenuation (including losses due to the reflection of the incoming beam at the air/GaAs interface, and losses due to mode mismatch between the laser beam and the eigenmodes of the waveguide) have been measured. Depending on the structure, total atte-nuation from 5 to 8 dB for 8 mm long waveguides have been found.

3.2. Embedded waveguides

The size and geometrical shape of the embedded guides is a function of the chemical etching prior to growth. In all cases an in situ HCl etch is per-formed. This provides a groove limited by crystalline facets (Fig. 2d). Depending on the substrate doping level, this groove is filled by a first n$^+$ layer then a n$^-$ layer or only an n$^-$ layer. Under appropriate growth conditions, the top of the structure can be located at the substrate sur-face. Fig. 5 is a S.E.M. micrograph of the cleaved cross-section of embedded guides.

Fig. 5 - S.E.M. micrograph (x640) of embedded wave-guides oriented along [110] direction.

The overall output intensity of embedded guides is comparable to those obtained for the external waveguides, indicating that low attenuations can be achieved also in the case of embedded waveguides. Both type of waveguides maintain well the polarization of TE and TM modes.

4. Integration of mirrors with straight waveguides

In order to change the direction of the waveguide, and thus increase the packing density, bend waveguides with radius of curvature from 100 to 500 μm have been tried by several authors. Up to now, results obtained are not yet satisfactory, since important losses (several db/radian) are currently measured. An other solution would be the use of etched mirrors [9], the change in the direction of the propagation of the light being obtained by total reflection on the GaAs/air interfaces. We present here integration of two mirrors on the same waveguide (embedded waveguides grown by localized epitaxy). A schematic representation of the structure is presented in Fig. 6. Waveguides are grown along [110] and [1$\bar{1}$0] azimuths,

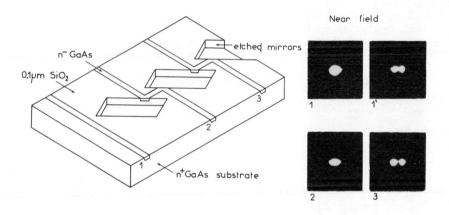

Fig. 6 - Integration of two mirrors with straight waveguides and their corresponding near field output intensity.

while the mirror is positionned at 45° of the waveguide axis. A reaction limited etchant (H_2SO_4 - H_2O_2 - H_2O) provides etched walls perpendicular to the substrate in the [010] direction. Fig. 7 shows a S.E.M. micrograph of an etched mirror integrated to the embedded guide. In order to obtain high reflection coefficient, and due to the waveguide eigenmode penetration into the cladding layer (n^+ GaAs), the etched mirrors have to be larger than the waveguide dimensions.

Fig. 7 - S.E.M. micrograph (x1250) of an etched mirror integrated to an embedded guide.

The structure we have used, allows for direct comparison of the intensity as well as the near field for both straight waveguide and waveguide with mirrors. The latter case represents an overall absolute change in the propagation direction of 180°. In Fig. 6 four near field intensity distributions are shown corresponding to straight waveguide (1 and 1'), and to waveguides with two mirrors (2 and 3). The near fields 1 and 1' are obtained on the same straight waveguide, changing only the injection conditions (laser beam position). Two modes can propagate in the as-grown linear waveguide. In contrast, the near fields 2 and 3 are obtained on two different "zig-zag" waveguides. For the guide 2 only the fundamental mode can be observed, while for the waveguide 3 only the second mode can be observed. The difference can be explained by a slightly different mirror position. It comes out that :

- the positioning of the mirror is critical
- depending on their positioning mirrors can act as mode filters.

This illustrates that integration of more than one mirror is a difficult task. In our case reflection coefficient of 50 % per mirror have been achieved. The residual loss can be attributed to the roughness of the chemically etched facets. Results should be enhanced by improving the etching process.

We wish to thank A. Goutelle and S. Delvallée for their technical assistance in fabricating and assessing the guides. This work was supported by DESTI/RE 84 B0285.

References

1 S. Somekh, E. Garmire, A. Yariv, H. Garvin and R. Hunsperger, Appl. Opt. 13, 327 (1974).

2 F.A. Blum, D.W. Shaw and W.C. Holton, Appl. Phys. Lett. 25, 116 (1974).

3 F.J. Leonberger, C.O. Bozler, R.W. McClelland and I. Melngallis, Appl. Phys. Lett. 38, 313 (1981).

4 L.M. Johnson, Z.L. Liau and S.H. Groves, Appl. Phys. Lett. 44 (3) 278 (1984).

5 M. Erman, N. Vodjdani, 7th Topical Meeting on Integrated and Guided-vave Optics, Kissimmee, Florida (April 1984).

6 J.P. Chané, L. Hollan and C. Schiller, J. Cryst. Growth 13/14, 325 (1972).

7 N. Vodjdani, M. Erman, J.B. Theeten : to be published.

8 D.W. Shaw, J. Cryst. Growth 47 (1979) 509-517.

9 T.M. Benson, Journal of Lightwave technology, Vol. LT-2, n° 1, February 1984.

Inst. Phys. Conf. Ser. No. 74: Chapter 3
Paper presented at Int. Symp. GaAs and Related Compounds, Biarritz, 1984

157

Hydride VPE growth for high-sensitive InGaAs/InGaAsP/InP heterostructure APDs and 1.4–1.5 μm InGaAs/InP MQW LDs

Y. Kushiro, Y. Noda, Y. Matsushima and S. Akiba

KDD Research and Development Laboratories, Meguro-ku, Tokyo 153, Japan

Abstract   Hydride VPE growth of InP-based alloys for optoelectronic devices is reported. The growth system consists of a double-chamber reactor, an isolated waiting chamber and a magnetic sample-loader. The reduction of impurities and defects in the growth layer has been realized with this apparatus. The VPE-grown InGaAs/InGaAsP /InP heterostructure APD has shown the receiver sensitivity higher than that of a Ge APD by 2–7 dB. InGaAs/InP MQW-lasers were also prepared and room-temperature pulsed operation has been achieved.

1. Introduction

InP-based alloys are currently being exploited as light sources and detectors in optical-fiber communication systems. Much of the growths of the optoelectronic devices have been performed by liquid phase epitaxy (LPE). A few studies have been reported on the hydride vapor phase epitaxy (VPE) of InGaAsP/InP heterostructure devices using a double-chamber reactor (Olsen and Zamerowski 1979 and Mizutani et al. 1980). Recently, the authors reported a high receiver sensitivity of the VPE-grown InGaAs/InP-heterostructure avalanche photodiode (HAPD) (Matsushima et al. 1984). This paper gives a detailed description on the VPE-growth technique for the fabrication of the InGaAs/InGaAsP/InP HAPD. The advantage of the present technique is also demonstrated by the growth of InGaAs/InP multiquantum-well (MQW) laser diodes (LDs).

2. Growth Apparatus

The hydride VPE system used is shown schematically in Fig. 1 and consists of : (1) a double-chamber reactor, (2) a waiting chamber isolated from the reactor by a gate valve, (3) and a magnetic sample-loader. The upper growth chamber is used for the InP growth and the lower for InGaAs and InGaAsP. The metal sources are kept at 830 °C, while the temperatures of the growth and pre-heat zones are kept at 700 °C. Arsine and/or phosphine enters the reactor tube from the lower-temperature growth-zone. HCl is also introduced through the AsH_3/PH_3 lines to etch off wall deposits. Electronic-grade 10 % AsH_3, PH_3, HCl and 10 ppm H_2S distilled in H_2 were supplied from gas cylinders. The waiting chamber is isolated from the reactor tube by closing a gate valve, during the period for exchanging sample or etching-off wall deposits. The sample-transfer in the reactor is carried out by a magnetic sample-loader. Two permanent magnet pieces are attached inside a loader-ring which surrounds the waiting-chamber tube. As counterparts, two iron pieces are buried in a teflon-slider which is

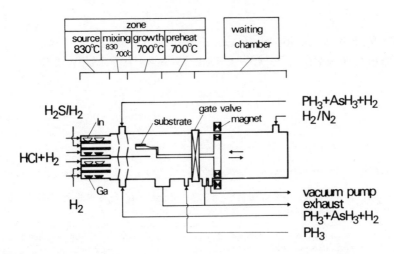

Fig. 1 Schematic diagram of double-chamber VPE system.

held in chamber tube and connected to a substrate-holder rod.   By
transferring or rotating the outer ring, the substrate-holder can
follow the movement. Switching of the sample between two growth-
chambers is completed within 1.5 s.   This sample-transfer apparatus
removes usual O-ring seal for a pushing rod and eliminates air leak and
grease contamination.

3. InP and InGaAs Growth

The single-layer growth of InP and InGaAs was performed on an Fe-doped
semi-insulating (100)-InP substrate in order to evaluate the growth
layers for HAPD.   Table I shows the electrical properties of InP and
InGaAs layers grown by using the present growth system.   The
compensation ratio N_A/N_D is estimated to be as low as 0.2 for InP and
0.5 for InGaAs from the relation between the 77 K mobilities and
carrier concentrations. These low carrier concentrations with low
compensation have been achieved by improving the sample loader from the
O-ring seal method to the present magnetic one.

Table I Electrical properties of InP(B) and InGaAs(T).

Sample	$n_{300}(cm^{-3})$	$\mu_{300}(cm^2/Vs)$	$n_{77}(cm^{-3})$	$\mu_{77}(cm^2/Vs)$
73-B	7.6×10^{15}	3400	5.5×10^{15}	30000
113-B	3.7×10^{15}	3700	2.5×10^{15}	39000
116-B	4.7×10^{15}	3600	3.0×10^{15}	33000
118-B	4.6×10^{15}	3850	3.4×10^{15}	36000
36-T	5.4×10^{15}	9400	4.4×10^{15}	28000
41-T	4.4×10^{15}	9500	3.7×10^{15}	34000
42-T	4.2×10^{15}	9600	3.3×10^{15}	37000

4. Growth of InGaAs/InGaAsP/InP APD Structure

Heterostructure APDs separate a light absorbing layer from a carrier multiplication layer (Nishida et al. 1979). The present VPE-grown wafer structure, shown in Fig. 2, consists of (i) an undoped InGaAs light absorption layer, (ii) an undoped InGaAsP buffer layer, (iii) an S-doped InP carrier multiplication layer, and (iv) an undoped InP layer for the self-guard-ring effect. The four layers were grown on an S-doped (100)-InP substrate. The introduction of the quaternary buffer layer (Matsushima et al. 1981) is essential to reduce a hole pile-up at the valence band discontinuity between InGaAs and InP (Forrest et al. 1982). The sulfer doping in InP was achieved

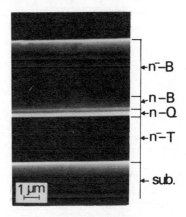

Fig. 2 SEM Photograph cleaved face of VPE-grown HAPD structure.

reproducibly by using H$_2$S. The VPE growth conditions, thickness and carrier concentration in each layer are summarized in Table II. The growth procedure for the heterostructure is illustrated in Fig. 3. At the early stage, the substrate position during etch-off of wall deposits was in one growth chamber or in a pre-heat zone where only AsH$_3$ and/or PH$_3$ was flowing. It was found, however, the photo-luminescence intensity of the samples thus grown had a long response time, as shown in Fig. 4(a). The PL intensity at the peak wavelength of 1.67 μm was measured immediately after exciting with a 15 mW Ar laser. The slow time-response of PL intensity is attributed to hole-trapping by deep defects (Pankove 1971). Any carrier multiplica-tion in the diodes fabricated from those wafers could not be observed by hole injection. On the other hand, as shown in Fig. 4(b) any slow time-response of PL intensities was not observed for the wafers grown with the procedure shown in Fig. 4, and normal carrier multiplication was obtained in APDs fabricated from these wafers. Therefore, it is considered that the deep defects were formed by the contamination of HCl which diffused into a sample during the wall-etch process.

Table II Growth conditions, thicknesses(d) and carrier concentrations (N$_d$) for each layer in APD heterostructure.

	In-HCl (10%) (sccm)	Ga-HCl (10%) (sccm)	AsH$_3$ (10%) (sccm)	PH$_3$ (10%) (sccm)	H$_2$S (10ppm) (sccm)	H$_2$ (sccm)	d (μm)	N$_d$ (10^{16}cm^{-3})
n InP	70			70		1500	3.5	0.6
n InP	70			70	3.3	1500	1	2
n InGaAsP	62.2	2.8	15	50.5		1000	0.5	1
n InGaAs	57.4	7.3	65			1000	3	0.5

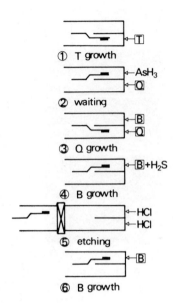

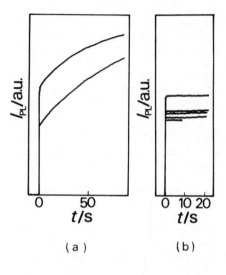

Fig. 4 Time-response of 1.67 μm PL intensity for (a) waiting in growth chamber and (b) in waiting chamber during wall etch.

Fig. 3 Growth procedure for InGaAs/ InGaAsP/InP HAPD structure.

5. Planar-Type HAPD

Planar-type InGaAs/InP HAPDs with an InGaAsP buffer layer were fabricated from the heterostructure wafer described above. A schematic cross-section of the HAPD is shown in Fig. 5. The guard-ring was made by Cd-diffusion at 430 °C which formed a linearly graded junction. The abrupt p^+-n junction for the light-sensitive area was made by a conventional Zn diffusion to the n^--InP/n-InP interface. A surface of the diode was coated and passivated by a plasma CVD SiN film. Ohmic contacts to the p^+- and n^+-InP were made by evaporated AuCr and AuSn. Characteristics of the HAPD were reported elsewhere (Matsushima et al. 1984) and they are summarized in Table III. The sensitivity obtained was better than that of a p^+-n Ge APD by 2 dB at 1.52 μm and 7 dB at 1.59 μm.

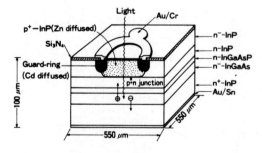

Fig. 5 Schematic cross-section of planar-type InGaAs/InGaAsP/InP HAPD.

Table III Characteristics of InGaAs/InGaAsP/InP HAPD.

Breakdown Voltage V_B	65–70 V
Dark Current at $0.9V_B$	50–80 nA
Optimum Gain	10
η_e at 1.55 μm	82 %
C (with package)	1.5 pF
Receiver Sensitivity	−43 dBm
280Mbit/s, return to zero	
λ=1.55 μm	
Bit error rate=10^{-9}	

6. InGaAs/InP MQW Growth

InGaAs/InP well/barrier MQW structures are of great interest and the preparation of the laser diode by MBE has been reported by Tsang(1984). In this chapter, hydride VPE growth of InGaAs/InP MQW with thickness down to 70 Å is demonstrated. The growth conditions for InGaAs/InP MQW structures were similar to those shown in Table II, and the growth rate was 25-30 Å/sec. The typical structure consists of : (i) a 5 μm thick InP buffer layer, (ii) ten InGaAs wells and nine InP barrier layers with the same thickness, and (iii) a 1.3 μm thick InP top layer. All the grown layers were unintentionally doped. The well thickness dependence of the PL peak energy is shown in Fig. 6. The solid line in the figure corresponds to the transition energy from the n=1 electron-to-heavy hole state, which was calculated by using the envelope-function approximation (Bastard 1981). It is clear that the observed increase of the transition energy with the decrease of thickness is due to quantum size effect and not due to the formation of InGaAsP alloys.

7. MQW Laser Diodes

For current injection laser fabrication, a p-n junction was formed by Zn diffusion in the MQW structure wafers. It was observed, however, that Zn diffusion through the MQW layers resulted in a degradation of PL peak intensity. This indicates that diffusion-induced disordering (Laidig et al. 1981) also occurs in InGaAs/InP MQW. Room temperature lasing under pulsed operation has been observed from the planar stripe diodes with a remote p-n junction in which Zn-diffusion front is stopped in the InP cap layer. The peak wavelength of a dominant lasing mode was 1.431 μm for a 70 Å thick well, and 1.500 μm for a 120 Å thick well. The lasing threshold was about 1.25 A for the 1.50 μm laser with the lasing area of a 18 μm x 485 μm stripe, and 2.0 A for the 1.431μm laser with a 18 μm x 890 μm stripe. Fig. 7 shows lasing spectra. Each longitudinal mode is well-resolved and the wavelength

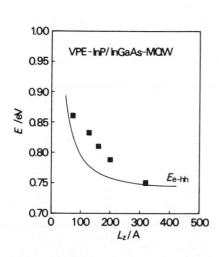

Fig. 6 PL peak energies of InGaAs/
InP MQW structure as a function
of well thickness.

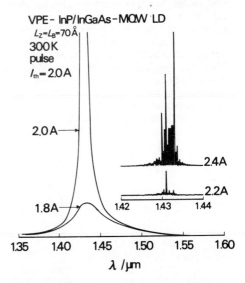

Fig. 7 Lasing spectra of InGaAs/InP
MQW laser with 70 A thick wells.

difference between two adjacent modes corresponds to that of Fabry-Perot modes. The relatively high threshold current density above 10 kA/cm^2 can be attributed to incomplete current confinement due to the preliminary laser structure.

8. Conclusions

Hydride VPE of InP-based alloys was studied for optoelectronic device fabrications. The use of a magnetic sample-loader eliminated the contamination problems and reduced background impurity concentrations in the growth layer. The deep traps observed in InGaAs/InGaAsP/InP heterostructure grown in the early study were eliminated by improving wall-etching process with the use of an isolated waiting chamber. VPE-grown planar-type InGaAs/InGaAsP/InP heterostructure APDs have been made and the high receiver sensitivity improved by 2 to 7 dB over a Ge-APD has been obtained at 1.5 μm wavelength range. Room temperature pulsed operation of InGaAs/InP MQW LDs was also demonstrated.

Acknowledgement

The authors would like to thank N. Seki for the HAPD sensitivity measurements. They wish to thank Dr. T. Yamamoto, Y. Iwamoto, A. Okada, T. Tanaka, Dr. K. Sakai and Dr. K. Utaka for their helpful discussion. They are also grateful to Drs. H. Kaji, K. Nosaka, K. Amano and C. Ota for their continual encouragement.

References

Bastard G 1981 Phys. Rev. B34 5693
Forrest S R, Kim O K and Smith R G 1982 Appl. Phys. Lett. 41 95
Ladig W D, Holonyak N Jr and Camras M D 1981 Appl. Phys. Lett. 38 776
Matsushima Y, Sakai K and Noda Y 1981 IEEE Electron Device Lett. EDL-2 179
Matsushima Y, Noda Y, Kushiro Y, Seki N and Akiba S 1984 Electron. Lett. 20 236
Mizutani T, Yoshida M, Usui A, Watanabe H, Yuasa T and Hayashi I 1980 Japan. J. Appl. Phys. Lett. 19 L113
Nishida K, Taguchi K and Matsumoto Y 1979 Appl. Phys. Lett. 35 251
Olsen G W and Zamerowski T J 1979 Progress in Crystal Growth and Characterization (London:Pergamon) vol II pp 309-375
Pankove J I 1971 Optical Process in Semiconductors (New Jersey: Princeton) pp 370-375
Tsang W T 1984 Appl. Phs. Lett. 44 288

Inst. Phys. Conf. Ser. No. 74: Chapter 3
Paper presented at Int. Symp. GaAs and Related Compounds, Biarritz, 1984

Liquid phase epitaxy and characterization of thin (<100Å) InGaAsP layers

R.M. CAPELLA, J.L. BENCHIMOL, J.F. BRESSE and M. QUILLEC

Laboratoire de Bagneux - C.N.E.T.
196 rue de Paris - 92220 Bagneux - FRANCE

Introduction

Reduction of the thickness of a semi-conductor layer leads to specific and remarkable properties such as carrier energy quantization or two-dimensional transport properties. Thin layer structures have thus received considerable interest in the last few years, when low growth rate techniques(M.B.E., M.O.C.V.D.), which are particularly suited to their elaboration, were developed. However, such expensive and complex techniques are not yet well controlled for the growth of InGaAsP alloys on InP substrates.

This material is almost exclusively grown by liquid phase epitaxy (LPE), which is not considered as a convenient techniques for growth of ultra thin layers (<1000Å). Some work was however done in this field a few years ago.

Multiple quantum well structures in $In_{0.88}Ga_{0.12}As_{0.26}P_{0.74}$ ($\lambda=1.1\mu m$) were grown by Rezek et al (1977) using a specially designed cylindrical boat ; thicknesses as low as 140 Å were obtained. Bauser et al (1977) also developed a LPE apparatus for multiple thin layers utilizing centrifugal forces. These studies, to our knowledge, have not been continued.

In this paper, we present the LPE growth of $In_{0.73}Ga_{0.27}As_{0.62}P_{0.38}$ ($\lambda=1.3\mu m$) layers thinner than 100Å, using a conventional boat. The grown structure was specially designed for photoluminescence characterization of the quantum layer. Layer thicknesses and chemical transition width of interfaces could be accurately measured using Auger microscopy on a chemical bevel.

Liquid phase epitaxy of thin layers

Supersaturation ΔT and growth time t are the main parameters that control epilayer thickness e. Growth rate in LPE is about one or two order of magnitude larger than in MBE, so that a reduction of t to very small values is necessary in order to obtain ultra thin layers.

Using a cylindrical boat, Rezek et al (1977) could get down to t=18 ms and grow at 635°C, from a 10°C supersaturated solution, a stack of 140Å thick InGaAsP ($\lambda=1.1\mu m$) layers. In the present work, we used a conventional "linear" graphite boat. The boat slide was moved by a

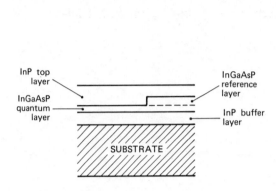

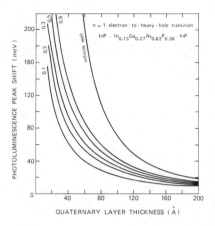

Fig.1:LPE grown structure including both a thin and a thick (>1500 Å) InGaAsP layer. The thick layer, grown from the same melt as the thin layer, served as a real "bulk" material reference.

Fig.2:Calculated energy shift versus quaternary layer thickness, for different energy barrier ratios $\Delta E_C/\Delta E_G$ in the InP/InGaAsP heterojunction. ($\Delta E_G = \Delta E_C + \Delta E_V$)

Microcontrole translation table, which allowed a precise control of growth time. 70Å thick InGaAsP ($\lambda=1.3\mu$m) layers were grown at 740°C, with $t\approx50$ ms and $\Delta T=6$°C.

However, when t becomes smaller than the duration of the melt perturbation introduced by the substrate displacement, a transient growth occurs, which is no longer diffusion-limited. The transition time between the two regimes was found to be about 100 ms in InGaAsP (Rezek et al, 1981).

Rezek et al (1977) and Brunemeier et al (1983) have also shown that the alloy composition obtained from a transient growth is not the same as that obtained from a diffusion-limited growth. The transient growth leads to an InGaAsP alloy with a larger lattice constant and a smaller energy gap. The differences are as large as respectively $+16.10^{-4}$ (relative mismatch) and 20meV. Thus, a comparison of physical properties between thin and thick layers becomes problematic when growth times are smaller than 100 ms.

In order to get the same composition in the two layers, both growths must be performed in a diffusion-limited regime. In that case, the relation between layer thickness and growth time is $e=K.\Delta T.\sqrt{t}$ if a ΔT step-cooling is used (K=constant). Obviously, the growth rate can be reduced by decreasing the supersaturation ΔT. Using a growth temperature of 650°C and $\Delta T \approx 2$°C, InGaAsP epilayers thinner than 100Å could be grown in two seconds. The experimental e(t) plot was found to obey a $\sqrt{t}$ law, as expected from a diffusion-limited growth described by the above expression, with K=0.132μm°C^{-1}s$^{-\frac{1}{2}}$.

The grown structure was made of an InP buffer layer, a thin and a thick (>1500Å) In$_{0.73}$Ga$_{0.27}$As$_{0.62}$P$_{0.38}$ layer grown in the same run on the same wafer (fig.1), and an InP confining layer.

Photoluminescence

In an InGaAsP/InP double heterojunction with a quaternary layer thickness small enough to cause carrier quantization, the first allowed transition between the n=1 energy levels of electrons and heavy holes, corresponds to a higher energy than that obtained on the "bulk" material. This transition is usually the only one observed by photoluminescence in an isolated quantum well. If we compare the photoluminescence peak energies E_{Thin} and E_{Thick} of the thin and thick layers respectively, then the shift $\Delta E = E_{Thin} - E_{Thick}$ will be characteristic of a quantum effect.

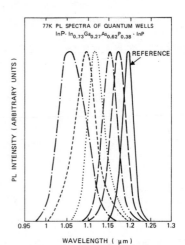

Fig.3:Dependence of photoluminescence peak shift and broadening on quaternary layer thickness.

This shift can be calculated using the standard theory of finite potential wells. The electron and hole barrier heights are determined by the discontinuities ΔEc and ΔEv of the conduction and valence band respectively at the InGaAsP/InP heterojunction. A ratio $\Delta Ec/\Delta Ev$ = .66 was found experimentally in the $In_{0.87}Ga_{0.13}As_{0.29}P_{0.71}$ ($\lambda=1.15\mu m$) alloy by Chin et al (1979). This is the only published value in the quaternary material. Fig.2 shows the expected shift versus layer thickness for different energy barrier heights in the heterojunction $InP/In_{0.73}Ga_{0.27}As_{0.62}P_{0.38}/InP$. The uncertainty on $\Delta Ec/\Delta Ev$ does not allow a precise determination of layer thickness from photoluminescence peak shift.

Energy shifts as high as 140 meV were observed at 77K in our structures ; this is to our knowledge the largest quantum shift reported in the InGaAsP/InP system. (A value of $\Delta E=287meV$ was reported by Razeghi et al (1984) in the InGaAs/InP system).

On a same sample, E_{Thin} was found to vary continuously along the direction of the temperature gradient in the furnace, as shown on fig.3. This variation in E_{Thin} was attributed to a thickness gradient in the thin layer. Indeed, reduction of the growth rate was obtained by decreasing supersaturation ΔT to very low values ; the disadvantage of such a procedure is that a slight temperature gradient yields a large relative error on ΔT and consequently in e. This difficulty makes it necessary to control the furnace temperature profile to within 0.1°C at least.

It also appears on fig.3 that the full width at half maximum increases with decreasing thicknesses, from 37 to 87 meV. This phenomenon was also observed by other authors (Welch, 1983) and attributed to a local variation of layer thickness δe. An estimation of the peak broadening δE can be made as follows : if E_{Thin} is proportional to e^{-2} as in an infinite potential well, then $\delta E=2.E_{thin}.\delta e/e$. Table 1 shows the calculated broadening δE for a thickness variation of $\delta e=10\text{Å}$. These estimations are consistent with our experimental results.

Table 1

Experimental energy shift	Corresponding layer thickness for $\Delta E_c/\Delta E_G=.6$	Broadening calculated	experimental
14.6 meV	195 Å	1.5 meV	9 meV
137 meV	35 Å	78 meV	57 meV

Auger electron spectroscopy

Thin layer thicknesses and interfaces can be directly measured by Auger electron spectroscopy. Ion sputtering is routinely used for such analysis, but resolution of in-depth profiles is limited by ion knock on mixing, sputter redeposition and sputter induced roughness. There are also specific problems in the InGaAsP/InP system, such as cone formation (Williams, 1980). The best resolution reported with this technique is 90Å (Cook, 1980).

The sputtering problem can be avoided by scanning the electron beam on a chemically angle lapped sample. The bevel was made by linear displacement of a bromine-methanol solution in front of the sample (Mellet, 1982). The bevel angle depends on the bromine concentration and liquid displacement rate ; angles of 1/1000 rad were used. This bevel technique requires very good surface morphology.

Each Auger line was scanned with a very narrow electron beam (0.2-1μm) over a length of 50μm along the bevel, which corresponds to 2048 pixels, with a count time of 100 ms per point. Prior to each measurement, the sample surface was cleaned with a scanned ion gun.

The chemical etch used to make the bevel might introduce steps at interfaces. However, binary and quaternary materials were found to have about the same etching rate ; reproducible Auger profiles were also observed when using different electron beam sizes, indicating that no significant step was present at the InGaAsP/InP interfaces.

The best Auger profiles were obtained with the phosphorus element, as shown on fig.4 for a 100Å thick InGaAsP layer. The chemical interface width was equal to about 20Å on both sides : this is the best value reported for LPE. The interface width symmetry indicates that no dissolution occurred during the growth of the InP top layer, unlike other reported observations (Feng, 1979) ; it also shows that there are no measurable steps on the bevel. Layer thicknesses as small as 70Å could be also measured.

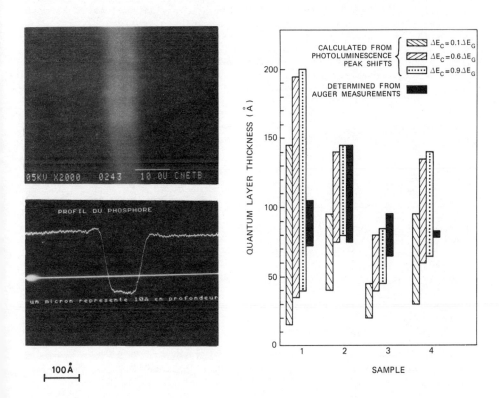

Fig.4:Scanning electron micrograph
and phosphorus Auger line profile
of the quaternary layer, visualized
on a chemical bevel.

Fig.5:Determination of quantum
layer thickness : correlation
between Auger and photoluminescence measurements for several
samples.

A thickness gradient was also found on samples, as observed by
photoluminescence. Fig.5 shows the correlation between Auger and photoluminescence measurements. For each sample are reported the thicknesses
measured by Auger profiles and those calculated from photoluminescence
energy shifts, for various $\Delta Ec/\Delta E_G$ values. Although both measurements
were not exactly made at the same position, a fairly good agreement is
observed between the two techniques.

Conclusion

We have shown that liquid phase epitaxy could be used to elaborate InP/InGaAsP double-heterostructures with quaternary layers as thin as
70Å and interface widths as small as 20Å. Such structures were grown in a
conventional LPE boat. Using a thick layer, grown on the same substrate
from the same melt, as a "bulk" material reference, energy shifts as high
as 140meV were observed by photoluminescence. This quantum size effect
was fairly well correlated to Auger thickness measurements. The resolution of Auger profiles was improved by using chemically angle lapped
samples. All these results show that LPE is still a promising technique for
growth of quantum InGaAsP layers.

Acknowledgements   The authors wish to thank Mrs A.M. Jean-Louis for her
assistance in photoluminescence measurements.

REFERENCES

Bauser E., Schmidt L., Loechner K.S. and Raabe E. (1977) Jap. J. Appl.
   Phys. 16 457
Brunemeier P.E., Roth T.J., Holonyak N. Jr and Stillman J.E. (1983) Appl.
   Phys. Lett. 43 373
Chin R., Holonyak N. Jr, Kirchoeffer S.W., Kolbas R.M. and Rezek E.A.
   (1979) Appl. Phys. Lett. 34 862
Cook L.W., Feng M., Tashima M.M., Blattner R.J. and Stillman G.E. (1980)
   Appl. Phys. Lett. 37 173
Feng M., Cook L.W., Tashima M.M., Stillman G.E. and Blattner R.J. (1979)
   Appl. Phys. Lett. 34 697
Mellet R. (1982) private communication
Razeghi M., Nagle J. and Weisbuch C. (1984) International Symposium on
   Gallium Arsenide and Related Compounds, Biarritz
Rezek E.A., Holonyak N. Jr, Vojak B.A., Stillman G.E., Rossi J.A.,
   Keune J.L. and Fairing J.D. (1977) Appl. Phys. Lett. 31 288
Rezek E.A., Vojak B.A., Chin R., Holonyak N. Jr and Samman E.A. (1981) J.
   Electron. Mater. 10 255
Welch D.F., Wicks G.W. and Eastman L.F. (1983) Appl. Phys. Lett. 43 762
Williams R.S., Nelson R.J. and Schlier A.R. (1980) Appl. Phys. Lett. 36
   827.

Inst. Phys. Conf. Ser. No. 74: Chapter 3
Paper presented at Int. Symp. GaAs and Related Compounds, Biarritz, 1984

169

Non miscibility in III-V alloys

H. Launois[*] and M. Quillec

Centre National d'Etudes des Télécommunications, 196 rue de Paris,
92220 Bagneux, France

Abstract. Interest has been shown recently on III-V ternary and quaternary alloy immiscibility and instability. The case of InGaAsP has been
studied experimentally. The usual models predicted a miscibility gap
(MG) which does occur, even at temperatures commonly used in liquid
phase epitaxy (LPE). Strains were found to play an essential role in
epitaxy of such alloys. New experiments using zinc enhanced auto-dif-
fusion show that epilayers are stable. Growth dynamics however lead to
compositional fluctuations, which is explained in terms of interfacial
spinodal decomposition. Such fluctuations cancel as expected at high
growth temperature and a consequence is a large improvement of electron
hall mobilities.

1. Introduction

It is well known that in III-V ternary and quaternary alloys, as in any
solid mixture, there is a critical temperature T_C under which decomposition
occurs at thermodynamical equilibrium. Only when T_C lies in the usual growth
temperature range, however, would one refer to the existence of a miscibility gap (MG). Long range order is well defined in these alloys ; they
verify Vegard law and show narrow band edges (luminescence and absorption).
However, EXAFS measurements (Bellessa et al, 1982, Mikkelsen et al, 1982)
of the covalent bonds have shown that large size effects occur, leading to
strong local departure from a virtual crystal description. The result is
always a repulsive pair interaction, as is expressed in the DLP model
through the positive interaction parameter Ω (Stringfellow, 1972); as a
consequence, MG exist in many systems. MG were observed some years ago in
the systems involving both As and Sb on the element V sublattice, i.e.
GaAsSb (Gratton et al, 1973), InGaAsSb (Nakajima et al, 1977) and AlGaAsSb
(Nahory et al, 1978). The III-V solid mixtures models (Ilegems et al, 1974,
Stringfellow, 1974) which were available since the early 70's indicate the
existence of MG in many ternary and quaternary compounds. Due perhaps to
contradictions with liquid phase epitaxy experiments and probably also to
a lack of confidence in these simple models, it took about one decade before this was explicitly written. De Cremoux et al (1981) noticed that the
DLP model (Stringfellow, 1974) predicted a MG in the important InGaAsP
system. Experimental work by Quillec et al (1982) proved together the existence of a large intrinsic MG in that system and its reduction by substrate induced strains as liquid phase epitaxy is performed. Springthorpe
(1983) confirmed experimentally the extension of the MG. Several calculations were published extending the MG prediction to other III-V alloys

[*] Present adress : Laboratoire de Microstructures et de Microélectronique
C.N.R.S., 196, rue de Paris 92220 Bagneux.

(Stringfellow, 1982a and 1983a, Onabe 1982a, 1982b, 1983). The strains induced by the substrate, as first suggested by Nahory et al (1978) play an essential role in epitaxial growth of the alloys near or under the critical
temperature. Near T_c, they modify the solid composition ("latching" effect)
as to minimize (relatively) the total Gibbs free energy (Stringfellow 1972,
Quillec et al 1983a). Under T_c, the strains allow epitaxial growth even in
near equilibrium conditions (liquid phase epitaxy) (Quillec et al 1982).
However, as long as $T_c > 300K$, the solid should remain metastable at room
temperature. Due to its importance for optoelectronic devices, the
$In_xGa_{1-x}As_yP_{1-y}$ alloy has been extensively grown, mostly from the liquid
phase, on InP and more recently on GaAs substrates. When grown by LPE
in substrate stabilization conditions, this alloy shows very interesting
microscopic features (Henoc et al, 1982), quite similar to patterns attributed to spinodal decomposition in unstable alloys (see for instance
Hilliard, 1970). Extensive studies were performed on this surprising phenomenon (since excellent devices have already been produced using that
material).The question of the equilibrium state of the alloy will be discussed in the last section, following two paragraphs : on thermodynamical
aspects and microscopic effects.

2. Thermodynamical aspects

A clear review of the MG predictions in the DLP description of the alloy
can be found in Stringfellow (1983c). The strictly regular description
gives similar results (Onabe, 1982 and 1983). Growth inside a MG is possible by kinetically controlled epitaxial growth, like MBE and MOCVD.
Highly metastable alloys were grown at low temperatures by these non-
equilibrium techniques, like $GaAs_ySb_{1-y}$ (y $\cong$ 0.5) (Waho 1977, Stringfellow,
1983b), although in the case of MOCVD, near equilibrium conditions (and MG)
seem to be reached in certain conditions (i.e. when the III/V pressure
ratio is $\ll$ 1) as discussed by Stringfellow (1983c). In the near equilibrium growth process involved in LPE, it was also possible to successfully
grow "a priori" unstable alloys.

Epitaxy of "a priori" unstable alloys

The $In_xGa_{1-x}As_yP_{1-y}$ compound has been grown in many laboratories in the
whole InP lattice-matched range (y $\cong$ 2.2(1-x)),at temperatures 620 to
650°C. Apparently no one suspected any instability problem until de Cremoux
et al (1981) pointed out that the DLP model predicted a MG for 0.5 < y <0.9.
Quillec et al (1982) performed a series of growths from the liquid phase
of InGaAsP inside the MG. For the same liquid composition and temperature,
they compared growth in nonepitaxial conditions (two phases were obtained)
to epitaxy of InP lattice matched alloy, where a single coherent phase was
obtained, as in many laboratories. This behaviour was interpreted in terms
of substrate induced strain stabilization. Near lattice matching, any deviation of the epilayer lattice parameter a_e from the substrate's a_s increases the solid energy by $\Delta G = \sigma [(a_s -a_e) / a_s]^2 V$ (V = volume of the
layer, σ is an elastic parameter). With this additional term, the stability
criterion is fulfilled (de Cremoux 1982). Stringfellow (1982b), following
Cahn (1961), noticed that the coherency strain energy makes coherent homogeneous alloys stable against small compositional fluctuations ; he found
that most of them would be stable at any temperature. He also pointed out
that taking into account the possibility of one or two dimensional composition non uniformity as developed by Khachaturyan (1969) does not modify
his conclusion. Actually, provided that the substrate has similar elastic
parameters as the epilayer, it can be easily verified that the above

expression is nothing but a particular case of equation 11 of Cahn (1961)
referring to any composition fluctuation in an alloy, in the isotropic approximation. In that approximation, the stability criterion is thus modified exactly like   the Cahn's coherent strain energy term,
so that the effective critical temperature drops also to $\sim 0°K$
(Stringfellow 1982b). The question remains however : why should that coherent metastable solid grow instead of the two solid phases minimizing
energy. The interfacial energy has to be taken into account : at the early
stage of growth, the substrate/layer interface costs less energy in the case
of coherent epitaxy, so that the energy is indeed minimized in that case if that
contribution is taken into account. A puzzling question was not raised by
de Cremoux (1982) and Stringfellow (1982b) : how come that the elastic
term almost exactly compensates the tendency to decomposition. According
to Bublik et al (1978), the interaction parameter involved in the regular
solution approximation may be written as $\Omega^S = \Omega^C + \Omega^E$, where Ω^C is the "chemical" part, directly determined from X-ray diffuse scattering experiments,
and Ω^E is the "elastic" part. If η is the relative difference between the
lattice parameters of the binaries involved in a ternary alloy, $\Omega^E = \sigma \eta^2$,
so that when $\Omega^C \ll \Omega^E$, Ω^S is equal to Cahn's additional elastic term,
in the isotropic approximation (equation 14 in Cahn (1961)). The DLP model
agrees with the above expression since it takes $\Omega^S \sim \eta^2$ (which is consistent with experiments). In the second derivative of the solid Gibbs free
energy, these terms cancel so that the solution behaves like an ideal one,
as noticed by de Cremoux (1982).

3. Microscopic effects

The coherency strain stabilization holds only for an homogeneous bulk
alloy, which remains metastable. On a lattice matched substrate, however,
growing a coherent epilayer costs less energy since it minimizes the interfacial contribution. During the growth process, spinodal decomposition
occurs as was first deduced from TEM observations and STEM measurements by
Henoc et al (1982), Glas et al (1982, 84) and Launois et al (1982). These
authors observed a long wavelength (100 to 300 nm) quasi-periodic contrast
modulation, propagating in the [100] and [010] directions, clearly related
to composition modulations (referred in the text as   CM)   measured by X-ray
microanalysis. They also observed a finer scale (5 to 15 nm) "speckle"
contrast, with not as well defined directions. Gowers (1983) attributed
this last observation to clustering without reporting the long wavelength
modulation. Mahajan et al (1983) interpreted the long wavelength structure
as due only to the strain fields arising at boundaries between spinodal
domains associated to the fine scale "speckle", rather than by C.M.
A close examination of the TEM long period contrast was performed by
Treacy et al (1984). Supporting the conclusions of Henoc et al (1982),
they interpret these fluctuations as a compositional modulation, which
follows the tie lines direction, with associated lattice parameter modulation of order 10^{-3}. This work emphasizes the importance of strain relaxation near the surface of the thinned foil which is the dominant source of
TEM contrast. Recently, Ueda et al (1984) published similar TEM observations obtained on InGaAsP and InGaP layers grown by LPE on (001) GaAs,
also within the MG.

4. New Developments

4.1. Stabilization by the substrate

$GaAs_ySb_{1-y}$ offers a clear case of instability, especially near y = 0.5.

This alloy should be lattice matched to InP, the stabilizing effect was proved by Quillec et al (1983a) who obtained $GaAs_{0.49}Sb_{0.51}$ on (001) InP, showing room temperature photoluminescence (see figure 1) comparable to

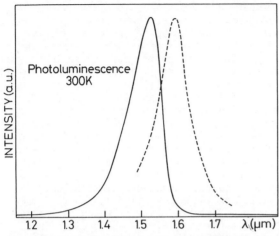

Fig. 1 - Room temperature photoluminescence obtained $GaAs_{0.49}Sb_{0.51}$ grown on (001) InP by LPE using substrate stabilization. Equivalent result were obtained by Stringfellow by MOCVD (dashed line).

that obtained by Cherng et al by MOCVD and reported by Stringfellow (1983b). A side effect, referred to as "latching" effect, previously reported in the GaInP system grown on GaAs by LPE (Stringfellow, 1972) is expected with other systems like GaInAs , when the growth temperature is not far from the critical one. This was verified (Quillec et al, 1983a, Joncour 1982) in that latter alloy.

4.2. Fundamental state of the stabilized alloy

In order to get a better understanding of the phenomenon, several annealing treatments (Launois et al, 1982) were tested on samples showing strong TEM modulation contrast. The conclusion was that the solid is quenched in a state resulting from interfacial spinodal decomposition. The important question is : what is the stable state of that alloy ? A series of new experiments proved that the stable (or metastable) state is the homogeneous one. Zinc is well known to enhance auto diffusion, on both sublattices in that alloy (Laidig, 1983 and Camras, 1983). After a 2 hours Zn diffusion at 550°C on a InGaAsP (λ = 1.3 µm) sample grown at 650°C, the contrast disappeared (figure 2) and compositional measurements show that similar epilayers were Zn doped ($P > 10^{18}$ cm^{-3}) by introducing the impurity into the bath ; they did not show C.M. These experiments support the concept of strain coherency stabilization of a single homogeneous phase. They confirm the previous assumption (Henoc et al, 1982) of a solid-liquid spinodal decomposition during the growth process which might be due to surface strain relaxation and enhanced atom mobility through the liquid phase. The composition modulations are quenched in the bulk solid due to low diffusion, but the system tends towards the homogeneous strainless alloy.

4.3. Practical consequences

Spinodal decomposition strengthen the matrix, it may thus have a positive effect, according to Mahajan (1983) : dislocations gliding and multiplying, as well as climbing, should be difficult in these epilayers. Whereas optical properties were not very affected by the C.M. since the

1000Å

As grown , at 650°C

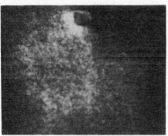

1000Å

with Zn annealing

Fig. 2 - STEM 220 dark field image obtained on a InGaAsP sample (λ=1.3μm) grown at 650°C by LPE : before and after 2 hours Zn diffusion at 550°C. The composition modulation disappears.

pseudo period is still small compared to the carrier diffusion length, low field electric transport proved very sensitive to the existence of the C.M. Electron Hall mobilities were measured on samples free of C.M. (grown at 740°C) in the whole compositional range, lattice matched to InP. Previously published data were obtained on samples grown at ∿ 650°C. The comparison shows clearly that in the substrate stabilized MG, the mobilities are much lower. In the extreme case (y ∿ 0.7, λ ∿ 1.3μm) the improvement is + 25% and + 140% at 300K and 77K respectively. The best result was μ_{77K} = 29400 cm^2/Vs for a λ = 1.35 μm InP lattice matched sample grown at 800°C. This was interpreted (Quillec et al 1983c, Benchimol et al 1983, Blood et al 1984) in terms of a new scattering process, previously included in the so-called "alloy scattering". It has also to be noted that high temperature LPE allows not only very pure layers (Benchimol et al 1983), but quite thick if desired (typically more than 10 μm from a 3g bath on 0.8 cm^2) thus not showing the thickness limitations previously reported (Takakei et al 1981) and associated to instability.

Summary

The existence of a large MG together with the substrate stabilization effect in the InGaAsP system has driven interest on the question of stability in the III-V alloys. It has been shown that most quaternary alloys show large MG, that "latching effect" is related to low stability and that strains play an essential role in epitaxy. The interface spinodal decomposition occurring in LPE was suppressed either by high temperature growth or by zinc diffusion, yielding improved electrical properties and better understanding of the phenomenon.

References

Belessa J, Gors C, Launois P., Quillec M and Launois H 1982 9th International
    Symposium on GaAs and related compounds, Alburquerque
Benchimol J L, Quillec M and Slempkes S 1983 J. of Crystal Growth 64 96
Blood P and Crassie A D C 1984 J. Appl. Phys. 56 1866
Bublik V T and Leikin V N 1978 Phys. Stat. Sol. 46 365
Cahn J W 1961 Acta Met. 9 795
Camras M D, Holonyak Jr N, Hess K, Ludowise M J, Dierze W T and Lewis C R
    1983 Appl.Phys. Lett. 42 185
de Cremoux B, Hirtz P and Ricciardi 1981 Inst. Phys. Conf. Ser. 56 115
de Cremoux B 1982 J. de Physique C5 43
Glas J F, Henoc P, Izrael A, Launois H, Quillec M and Treacy M M J 1982
    in "Proceedings of 10th Int. Congress on Electron Microscopy"

Glas F, Treacy M M J, Quillec M and Launois H 1982 J. Physique 43 C5-11
Glas F and Izrael A 1984 J. Micros. Spectrosc. Electron 9 237
Gowers J P 1983 Appl. Phys. A31 23
Gratton M F and Wooley J C 1973 J. of Electron Mat. 2 455
Henoc P, Izrael A, Quillec M and Launois H 1982 Appl. Phys. Letters 40 963
See for instance Hilliard J E 1970 "Spinodal decomposition" in Phase
    transformation, Aronson H I, ed. Am. Soc. for Metals, Metals Park, Ohio
Ilegems M and Panish M B 1974 J. Phys. Chem. Solids 35 409
Joncour M C 1982 Thesis, unpublished
Khachaturyan A G 1969 Phys. Stat. Sol. 35 119
Laidig W D, Lee J W, Chiang P K, Simpson L W and Bedair S M 1983 J. Appl.
    Phys. 54 6382
Launois H, Quillec M, Glas F and Treacy M M J 1982, 9th International
    Symposium on Gallium Arsenide and Related Compounds, Albuquerque
Mahajan S 1983 Inst. Phys. Conf. Ser. N° 67 259
Mikkelsen J C and Boyce J B 1982 Phys. Rev. Lett. 49 1412
Nahory R E, Pollack M.A., Beebe E.D. and De Winter J C 1978 J. Electrochem.
    Soc. 125 1053
Nakajima K, Osamura K, Yasuda K and Murakami Y 1977 J. of Crystal Growth
    41 87

Onabe K 1982a   Japan J. Appl. Phys. 21 797
Onabe K 1983 Japan J. Appl. Phys. 22 287
Onabe K 1982b Japan J. Appl. Phys. 21 L323
Quillec M, Daguet C, Benchimol J L and Launois H 1982 Appl. Phys. Lett. 40
    325
Quillec M, Launois H and Joncour M C 1983a J. Vac. Sci. Technol. B1 (2)238
Quillec M, Benchimol J L, Slempkes S and Launois H 1983 Appl. Phys. Lett.
    42 886
Springthorpe A J 1983 NATO InP Workshop, England
Stringfellow G B 1974 J. of Crystal Growth 27 21
Stringfellow G B 1982 J. of Crystal Growth 58 194
Stringfellow G B 1983a J. Appl. Phys. 54 404
Stringfellow G B 1972 J. Appl. Phys. 43 3455
Stringfellow G B 1983b J. of Crystal Growth 65 454
Stringfellow G B 1983c J.of Crystal Growth 62 225
Stringfellow G B 1982b J. Electron. Mat. 11 903
Stringfellow G B 1972 J. Appl. Phys. 43 3455
Takahei K and Nagai H 1981 Japan J. of Appl. Phys. 20 L313
Treacy M M J, Gibson J M and Howie A Submitted to Phil. Mag.
Ueda O, Isozumi S and Komiya S 1984 Japan J. of Appl. Phys. 23 L 241
Waho J, Ogawa S and Maruyama S 1977 Japan J. Appl. Phys. 16 1875

Inst. Phys. Conf. Ser. No. 74: Chapter 3
Paper presented at Int. Symp. GaAs and Related Compounds, Biarritz, 1984

175

Effects of MOCVD growth conditions on DC characteristics of GaAlAs–GaAs double heterojunction bipolar transistors

C. Dubon, R. Azoulay, M. Gauneau*, L. Dugrand, A. Sibille, J. Dangla, A.M. Duchenois and D. Ankri.

Centre National d'Etudes des Télécommunications - Laboratoire de Bagneux
196, Rue de Paris - 92220 BAGNEUX - FRANCE.

* LAB/ICM CNET-LANNION

Abstract: Double heterojunction bipolar transistors (DHBT's) grown by MOCVD have been fabricated and tested. An optimization of the heterointerface growth conditions has permitted to control the D.C. characteristics of DHBT's (β up to 5500) with a low recombination current and a reduced offset voltage ($\Delta Vce <$ 50 mV).

Reverse current gains of 15 have been measured on devices processed with a diffused technology.

1. INTRODUCTION

GaAs-GaAlAs heterojunction bipolar transistors (HBT's) have intensively been studied using liquid phase epitaxy (LPE). HBT's grown by molecular beam epitaxy (MBE) have recently shown impressive characteristics for integrated circuit applications (Ito et al 1984, Wang et al 1984). As compared to MBE, MOCVD can provide better interface quality structures (Drummond 1983) on large wafer size.

The advantages of double heterojunction bipolar transistors (DHBT's) have been pointed out by Kroemer (1982) : suppression of hole injection into the collector under saturated conditions and emitter collector interchangeability, which can simplify the design of ECL circuits. However, an accurate control of the two heterointerfaces quality is necessary in order to obtain an identical operation of the emitter base and base collector junctions and therefore to enhance the direct and reverse D.C. characteristics.

This paper presents a comprehensive study of the influence of the MOCVD growth parameters - grading in Al composition and in p type doping - on electrical characteristics of DHBT's.

In present MOCVD systems, the grading of the GaAs-GaAlAs heterojunction can be reduced to less than 20 Å and the layer thickness and doping level can be precisely controlled, hence this epitaxial technique appears suitable to build double heterojunction transistors.

2. MOCVD GROWTH CONDITIONS

The structures were grown in a classical vertical atmospheric pressure

MOCVD reactor. Diethylzinc (DEZn) was used for p type doping and H_2Se diluted to 50 ppm in H_2 for n type doping. The Al mole fraction was close to
0.4. The epitaxial structure has been designed to obtain high speed DHBT's
for ECL circuits (C. Dubon et al - 1983). The typical heterostructure is
shown in table 1.

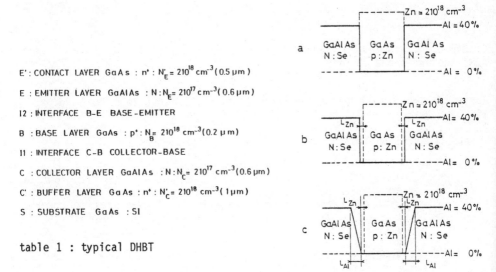

E' : CONTACT LAYER GaAs : n⁺ : $N_E' = 210^{18}$ cm⁻³ (0.5 μm)

E : EMITTER LAYER GaAlAs : $N:N_E = 210^{17}$ cm⁻³ (0.6 μm)

I2 : INTERFACE B-E BASE-EMITTER

B : BASE LAYER GaAs : p⁺ : $N_B = 210^{18}$ cm⁻³ (0.2 μm)

I1 : INTERFACE C-B COLLECTOR-BASE

C : COLLECTOR LAYER GaAlAs : $N:N_C = 210^{17}$ cm⁻³ (0.6 μm)

C' : BUFFER LAYER GaAs : n⁺ : $N_C' = 210^{18}$ cm⁻³ (1 μm)

S : SUBSTRATE GaAs : SI

table 1 : typical DHBT

fig. 1 : comparison of DHBT's interfaces.

We have compared the characteristics of DHBT's epilayer structures fabricated with three types of heterointerface , fig. 1.
Firstly, (case a), the heterointerfaces are abrupt in Al composition and in
p type doping.
Secondly (case b), the heterointerfaces are abrupt in Al composition but
the base doping is reduced and stopped at 300 Å from the heterointerfaces.
Finally, (case c), there is a gradient in Al composition and in p type
doping at 300 Å from the heterointerfaces.

3. INTERFACE CHARACTERIZATION

SIMS analysis was carried out to obtain the in-depth profile of the
different species. Dopant profiles were registered using the Cameca ims -
3f ion microanalyzer in the standard conditions to avoid signals arising
from the crater walls.
Selenium, Zinc and Oxygen were profiled by monitoring the $^{80}Se^-$,
$^{197}CsZn^+$ and $^{32}O_2^-$ secondary ions, respectively, under $^{133}Cs^+$ primary ion
bombardment.
To improve the absolute depth resolution as we were interested in the
true depth profile of the dopants near the heterointerface over 300 Å, some
samples were etched to remove about 6000 Å from the surface.
Some common failings of MOCVD reactors appeared on the first obtained
curves : the most impressive problem lies in the Al concentration spike
located at the E-B interface. It was due to an overpressure between the
reactor and the exhaust. But now, this problem is solved.
In case a), fig 2a, as the Zn diffuses into the N type GaAlAs layer, the

junction is no longer abrupt but presents a compensated zone which inter-
feres with the electron transit from the emitter to the base.

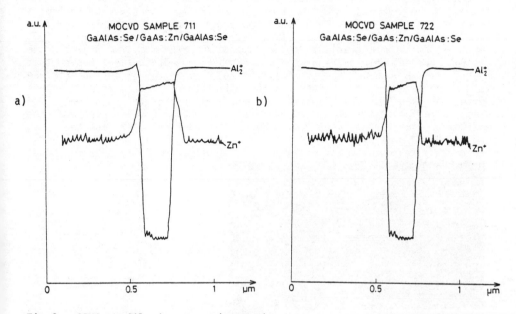

Fig 2 : SIMS profile in cases a) and b)

   The real Zn profile estimated from the SIMS curve has been included in
a simulation program (E. Caquot et al 1983). The band diagram is shown in
fig 3a, it proves that the compensated zone tends to widen the spike barrier impeding the electrons to cross it by tunnelling.
   In case b) and c) the Zn is contained inside the GaAs p$^+$ layer (fig. 2b)
and in case b) the E-B and C-B junctions are really abrupt heterojunctions.
The corresponding band diagram (fig 3b) presents a very sharp spike. The
diffusion coefficient of Zn inside GaAs at 750°C can be estimated from
these experiments at 2.10^{-15} cm^2/s.

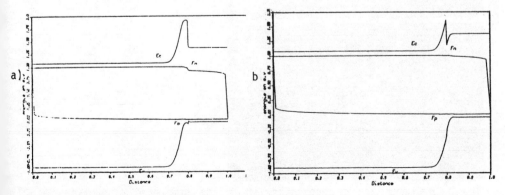

fig. 3 : Simulated band diagram in cases a) and b).

4. EXPERIMENTAL RESULTS

4.1 - DC characteristics
Common emitter characteristics of the three structures are shown in
fig 4. The maximum D.C. current gain for a structure with abrupt heterointerfaces (case a) is about 80, when the p type doping is stopped at 300 Å
from the interface (case b) the D.C. current gain is increased up to 400.
And when there is a gradient in Al composition and in p type doping (case
c), D.C. current gains as high as 5500 are obtained. These results show that
the current gain limiting factor in the case of an abrupt heterointerface
is not the base transfer factor but the injection efficiency Jn/Jp. Considering a p type doping of 2.10^{18} cm^{-3} and an injection efficiency of 1, an
electron diffusion length in the base of 7,6 μm can be calculated from the
maximum D.C. current gain of 5500. These values are the best reported up to
now for MOCVD growth.
The most impressive improvement due to this suspension of p type doping at

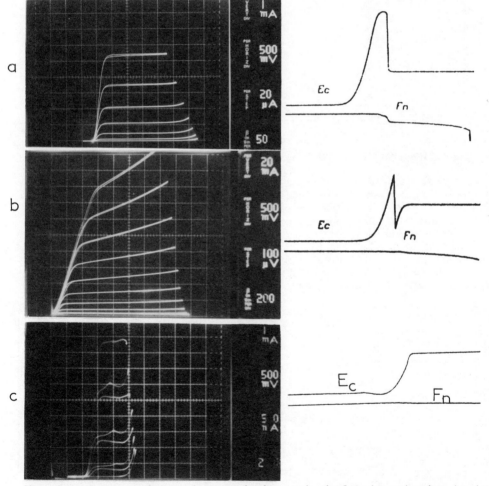

fig. 4 : Common emitter characteristics and simulated conduction band.

300 $\overset{\circ}{A}$ from the interface is the reduction of the recombination current.
In case a), when there is not any suspension of the doping, $\beta = 1$ for a
current density of 5.10^{-2} A/cm^2, but in case b) and c) it is decreased down
to $2 - 5$ 10^{-4} A/cm .

The ideality factor n_E obtained from I_C-V_{be} measurements is in good
agreement with the result of the equation [2] $n_E = 1 + (\varepsilon_E N_E / \varepsilon_B N_B)$

Another impressive improvement due to the p type doping suspension is the
reduction of the offset voltage ΔVce. As demonstrated by Dangla et al
(1984), the offset voltage equation can be reduced to [4].

[4] $\Delta Vce = V_T Ln (1/\alpha_I)^{n_2}$

when the reciprocity relation is achieved and when the EB and CB ideality
factors (n_1 and n_2) are equal, α_I being the reverse current gain.
In case a), these conditions are not realized and the offset voltage is
large (about 500mV). But in case b) and c), the good quality and the symmetry of the two heterointerfaces lead to identical values. In this case
the offset voltage is reduced to less than 50mV [fig 5].

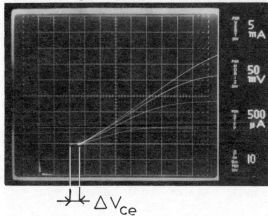

fig. 5 : The offset voltage is reduced to 40 mV

4.2 - CV and DLTS measurements
C-V measurements lead to the buit-in potential evaluation. ⌐c is higher in
case a) and b) than in case c), for the built in voltage is related to the
band structure by the equation [1] :

[1] $V_{bi} = E_{gb} + \Delta E_c + kT Ln \dfrac{N_E N_B}{N_{CE} N_{vB}}$

where ΔE_C is the conduction band discontinuity. As the graduality of
aluminium removes the spike barrier, the built in potential is lower in
case c). From the comparison between the obtained values in case a) or b)
and in case c), a value of 1V is deduced for ΔE_C .
C-V measurements have shown capacitance variations with frequency and temperature indicating the presence of deep levels in the layers. DLTS measurements [Lang 1974] were therefore performed on the structure presenting the

most prominent features (case b). A large concentration of DX centers [Lang 1979] was detected in the GaAlAs layers ; it was unambiguously correlated by admittance spectroscopy to the observed capacitance variations. The simila- rity of the two GaAlAs layers was assessed by the likeness of the EB and BC DLTS spectra.

5. CIRCUIT FABRICATION

These layers with good interface quality have been used for the fabrica- tion of ECL inverters. A scanning electron microscope photograph of the cir- cuit is shown in figure 6.

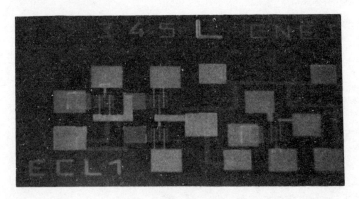

Fig. 6 : SEM of an inverter using DHBT's

It is a technological test circuit integrating 3 transistors and 3 resistors. A Zn diffusion is performed to contact the base ; as it enters into the GaAlAs N collector, it makes the E-B and B-C junction areas equal. The isola- tion is achieved by highly energetic   proton implantation. During this very step, the resistors are obtained by delimiting the active pads on the n^+ buffer layer. Layers treated with this process, because of the symmetry of the interfaces exhibited reverse and direct current gain in the same order of magnitude : 50 in direct mode and 15 in reverse mode.

6. CONCLUSION

We have carried out a study on the quality of heterointerfaces proving the influence of growth parameters on electrical characteristics. We have shown that an accurate control of the interface permits enhancing the D.C. current gain and to reduce the recombination current and the offset voltage.

References

Caquot E et al 1983 Proc. Nasecode III
Dangla J et al 1984 Proc. ESSDERC
Dubon C et al 1983 Proc. IEDM
Drummond T J et al 1983 Appl. Phys. Lett. 42 (7) 615
Ito H et al 1984 IEEE EDL-5(6) 214
Kroemer H 1982 Proc. IEEE 70 13
Lang D V J. Appl. Phys 45(7) 3023
Lang D V Phys. rev. B 19(2) 1015
Wang H et al 1984 Proc. ESSCIRC

Inst. Phys. Conf. Ser. No. 74: Chapter 3
Paper presented at Int. Symp. GaAs and Related Compounds, Biarritz, 1984

Low temperature growth of $Al_xGa_{1-x}As$ by MOCVD

T.F.Kuech, E.Veuhoff, D.J.Wolford, and J.A.Bradley

IBM Thomas J. Watson Research Center, P.O. Box 218, Yorktown Heights, NY 10598 USA

Abstract. We have demonstrated the growth of high quality MOCVD $Al_xGa_{1-x}As$ over the entire growth temperature range 600-800°C through suitable modifications to the growth system. These growths are characterized by both an increased photoluminescence intensity and a decreased background carrier concentration as the growth temperature is reduced. The maximum photoluminescence intensity was found at a growth temperature of ~650°C for $x \geq 0.2$ in $Al_xGa_{1-x}As$. The photoluminescence intensity also increased monotonically with AlAs mole fraction in the alloy. Controlled Si doping at moderate carrier concentrations ($\leq 10^{17}$ cm^{-3}) has been achieved at these temperatures.

Introduction

The application of metal-organic chemical vapor deposition (MOCVD) to the area of high-speed electronic devices has presented a new more stringent set of material requirements in order to achieve high performance devices. Two primary requirements are abrupt heterojunction interfaces and very high purity material. In order to minimize the interdiffusion at hetero-interfaces as well as dopant redistribution, lower growth temperatures are highly desirable, if not necessary. The growth of device structures at low temperatures is compatible with the growth of high purity GaAs since the electrical and optical properties of MOCVD GaAs generally improve with decreasing growth temperature (Dapkus P D, et al. 1981). Yet, while the growth of GaAs by MOCVD has been quite successful in producing very high purity materials, the growth of $Al_xGa_{1-x}As$ has presented particular problems (Stringfellow G B 1981). In both molecular beam epitaxy (MBE) and MOCVD, very high $Al_xGa_{1-x}As$ growth temperatures, relative to those found for the optimal growth of GaAs, are usually necessary in order to achieve low electrical compensation and good optical properties. The typical MOCVD growth temperatures employed for $Al_xGa_{1-x}As$ are in the range 750-800°C. At lower growth temperatures, the $Al_xGa_{1-x}As$ is usually found to be semi-insulating and possessing near-gap photoluminescence which is either weak or totally quenched (Hallais J, et al. 1981). The lower temperature limit for acceptable $Al_xGa_{1-x}As$ growth is, in part, determined by the background concentration of O_2 and H_2O in the growth ambient (Stringfellow G B 1981). Even small amounts of these oxidants in the growth system can severely degrade $Al_xGa_{1-x}As$ optical and electrical properties (Wallis R H 1980). By utilizing high growth temperatures, the oxygen incorporation into the $Al_xGa_{1-x}As$ is reduced. While high growth temperatures minimize the influence of oxygen in the $Al_xGa_{1-x}As$, the background level of impurities, most notably carbon, increases with temperature (Dapkus P D, et al. 1981 and Mohammed K, et al. 1983). In the absence of oxidants at the growing surface, it could be expected that the same improvements observed in the case of GaAs with decreasing growth temperature would be seen for $Al_xGa_{1-x}As$.

We have prepared $Al_xGa_{1-x}As$ in an optimized MOCVD system, whose features are directed towards the reduction of O_2 and H_2O at the surface of the growing layer. We have found that for the range of AlAs mole fractions, $0.16 < x < 0.5$, the material properties of $Al_xGa_{1-x}As$ improve with decreasing substrate temperature, T_s. This improvement is reflected in an increase in near-edge(2K) photoluminescence intensity, decreased impurity incorporation, and reduced electrical compensation with decreasing T_s. These trends are contrary to most previous reports in the literature and represent an advance in the application of MOCVD to the growth of $Al_xGa_{1-x}As$,

Growth and Characterization

The $Al_xGa_{1-x}As$ layers were grown through the co-pyrolysis of trimethyl gallium $(Ga(CH_3)_3)$, trimethyl aluminum $(Al(CH_3)_3)$, and arsine (AsH_3) in a hydrogen carrier gas. The range of growth temperatures investigated in this study was 600-800°C. The ratio of AsH_3 to metal species in the gas phase (V/III) was held constant at a value of 80. Silicon doping was achieved by the introduction of mixtures of H_2 with either silane (SiH_4) or disilane (Si_2H_6) to the growth ambient. The silane mixtures, as well as the AsH_3, were purified at the point of use by passage through a liquid-metal ternary melt prior to introduction into the growth ambient (Shealy J R, et al. 1982). The H_2 was purified by means of a Pd diffusion cell. The growth of $Al_xGa_{1-x}As$ was accomplished in a low pressure (76 torr) horizontal quartz reactor, possessing long entrance length before the susceptor. A quartz plate placed upstream of the susceptor insured a smooth gas-flow transition from the reactor walls to the leading edge of the susceptor. The linear flow velocity in the reactor was ~30 cm/sec. The <100> semi-insulating (Cr-doped) GaAs substrates were initially prepared by a standard procedure (Kuech T F, et al. 1984a). Growth of the $Al_xGa_{1-x}As$ layers was preceded by an *in-situ* thermal anneal of the substrate at the growth temperature in an AsH_3/H_2 ambient. A growth rate of 0.05 micron/min, corresponding to metal alkyl mole fraction of about $2x10^{-4}$, was used for all the layers. The layers in this study were all grown to a thickness of 3 microns. The electrical and optical properties of the layers were characterized by capacitance voltage profiling, Van der Pauw-Hall measurements, and low temperature photoluminescence. The alloy composition was determined by the photoluminescence peak positions (Dingle R A, et al.1977).

Photoluminescence (PL) was performed at 2K in a He immersion dewar. Samples were excited by the focussed 514.5nm output of a cw Ar^+ laser at power densities of ~$2.4x10^3$ W/cm^2 which leads to an average photoexcited carrier density of $\gtrsim 10^{17}$cm^{-3}. Emission was collected in a backscattering geometry and analyzed with a 0.75m double grating spectrometer, an S-20 photomultiplier, and photon counting. Particular care was given to obtaining reproducible and consistent spectra and luminescence intensities. Spectra presented below are therefore typical of at least four measurements from each sample for a given growth temperature or alloy composition. Relative integrated PL intensities derived from such studies, and plotted below, were reproducible to ±20% in all samples.

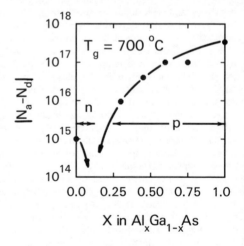

Figure 1. The influence of alloy composition on the background impurity concentration for a growth temperature of 700°C.

Experimental Results

The layers of $Al_xGa_{1-x}As$ grown in this study exhibited a smooth specular morphology under all growth conditions. Extensive electrical measurements were made on nominally undoped material, as well as Si-doped materials. Several systematic trends were seen in the electrical behavior of the undoped $Al_xGa_{1-x}As$, both as a function of alloy composition and T_s. Measurements were made over the entire range of alloy compositions, $0 \le x \le 1.0$. At AlAs mole fractions less than about x=0.20, there was an n-type to p-type transition as T_s was increased. As can be seen in Fig. 1, this transition occurred at $T_s \sim 700°C$ at a value of $x \simeq 0.15$. The hole concentration increases rapidly with AlAs mole fraction from $p = 8 \times 10^{15}$ cm^{-3} at x=0.3 to a value of 4×10^{17}cm^{-3} for pure AlAs. At AlAs mole fractions greater than about x=0.20, the layers were found to be always either p-type or non-conductive. The concentration of both donor and acceptor impurities increased strongly with T_s for all alloy compositions investigated. The hole concentration, particularly at higher AlAs mole fractions, decreased at $T_s > 750°C$ with a corresponding decrease in the hole mobility. Additional studies and results on the electrical behavior of undoped $Al_xGa_{1-x}As$ will be given elsewhere (Kuech T F, et al.).

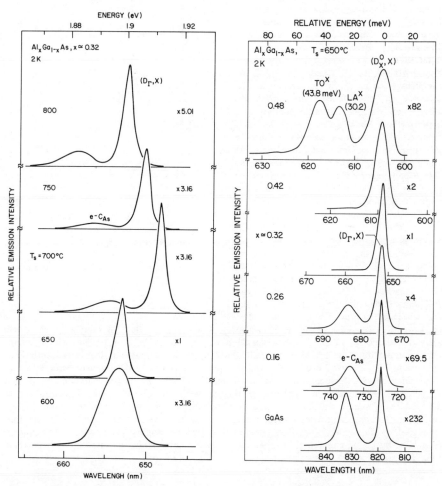

Figure 2. Photoluminescence spectra of $Al_xGa_{1-x}As$ with $x \simeq 0.32$ for material grown at differing substrate temperatures.

Figure 3. Photoluminescence spectra of $Al_xGa_{1-x}As$ of varying composition grown at 650°C.

Low temperature (2K) photoluminescence (PL) measurements were performed on all the undoped layers of $Al_xGa_{1-x}As$. Typical spectra at a value of $x \approx 0.32$ are given in Fig. 2 for the entire range of growth temperatures. The main features of these spectra are the sharp donor-bound-exciton structure seen at higher energies and the band-to-acceptor (carbon) recombination at lower energies. The shift in peak positions seen for different growth temperatures is due to minor differences in alloy composition. The relative intensities are noted at the right of the spectra. All growth temperatures result in intense near-edge emission, however, the most intense luminescence was obtained at $T_s \approx 650°C$. Broadening observed at $T_s = 600°C$ is not presently understood. Spectra taken from $Al_xGa_{1-x}As$ and GaAs samples grown at $650°C$ are shown in Fig. 3, again with their relative intensities indicated to the right. At all alloy compositions a sharp donor bound exciton peak is clearly resolved. The $x = 0.48$ sample exhibits a spectrum typical of indirect gap $Al_xGa_{1-x}As$ while the other spectra shown are in the direct bandgap region. The luminescence intensity monotonically increases with AlAs mole fraction (x).

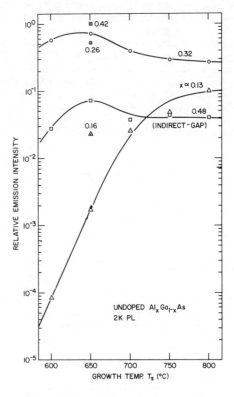

Figure 4. The integrated edge emission of $Al_xGa_{1-x}As$ over the entire range of alloy composition and growth temperature.

The dependence of the photoluminescence intensity on both alloy composition and growth temperature, T_s, is presented in Fig. 4. The relative integrated photoluminescence intensity of the near band edge emission is presented for both the direct and indirect bandgap materials. There are two main features of this figure: the increase in integrated intensity with increasing AlAs mole fraction for samples in the direct gap alloy region and the peaking of integrated intensity at $T_s \approx 650°C$ for samples with $x \gtrsim 0.20$. The alloy compositions which have $x \lesssim 0.20$ show a steep decrease in integrated intensity with decreasing T_s. At these lower AlAs mole fractions, the trends in the intensity with T_s are typical of previous MOCVD studies (Mohammed K, et al. 1983). The measurements made on Si doped $Al_xGa_{1-x}As$ indicated that controllable doping levels in the electron concentration in the 10^{16}-10^{17}cm^{-3} range could be achieved provided the Si concentration in the films was at least twice the background carrier concentration. While the carrier concentration in the films, as measured by capacitance profiling, indicated a uniform doping profile whose magnitude was controlled by the Si concentration, the Hall measurements yielded a much lower value (Kuech T F et al. 1984b).

Discussion

The degradation of the optical and electrical properties of $Al_xGa_{1-x}As$ due to the presence of O_2 and H_2O in the growth ambient has been observed by several authors (Mohammed K, et al. 1983, Wallis R H, et al. 1980, and Hersee S D, et al. 1981). The primary sources of O_2 and H_2O

are the gases used in the growth itself: H_2, AsH_3, and various dopant gases. Several researchers have reported on techniques for purifying these gases at the point of use. Layers grown from these purified gases have generally resulted in improved PL intensity (Shealy J R, et al. 1983). AsH_3 has been identified in several studies as the primary source of oxidants and the main determinant of the PL efficiency and properties in MOCVD $Al_xGa_{1-x}As$. Two approaches have emerged to remove oxidants from AsH_3. The first approach utilizes a desiccant such as molecular sieve or liquid ternary metal (Ga-In-Al) through which the AsH_3 is passed (Shealy J R, et al. 1983). The second approach getters the oxidants through a reaction involving the aluminum alkyl within the reactor vessel. By growing $Al_xGa_{1-x}As$ layers prior to the active region of the device, O_2 and H_2O are gettered for a period of time by the aluminum alkyls (e.g. $Al(CH_3)_3$) adsorbed or $Al_xGa_{1-x}As$ deposited on the reactor walls (Hersee S D, et al. 1981). Graphite baffles placed upstream of the hot susceptor have also been employed (Kisker D W,et.al. 1982). Pronounced gas phase transients have been observed in the case of the graphite baffles making this approach ill-suited to the growth of abrupt heterojunctions.

The growth system used in this study has employed several of these techniques in order to reduce the background levels of O_2 and H_2O. As mentioned above, the AsH_3 was dehydrated through the use of a ternary metal liquid bubbler (Kuech T F, et al. 1984b). It is known that the $Al(CH_3)_3$ used in the growth reacts readily with both O_2 and H_2O (Terao H, et al. 1984). The interaction of $Al(CH_3)_3$ and both O_2 and H_2O has been studied through the intentional introduction of these oxidants into the growth environment during the growth of $Al_xGa_{1-x}As$. The reaction of $Al(CH_3)_3$ and H_2O occurs rapidly in the gas phase producing $Al(OH)_3$ or Al_2O_3 which subsequently deposits onto the internal surfaces of the reactor. It has been proposed that O_2 and $Al(CH_3)_3$ react however to form a volatile specie, $(CH_3)_2AlOCH_3$ (Terao H, et al. 1984). This volatile specie is then transported to a hot surface where the compound is finally decomposed into Al_2O_3. We believe that the long entrance region (~30cm) of our system, through which the laminar flow profile is established, allows these gas phase reactions to occur with the subsequent deposition of the reaction products on the reactor walls prior to flowing over the growth surface. The efficiency of *in-situ* gettering of O_2 and H_2O should increase with the vapor phase concentration of $Al(CH_3)_3$ and the growth temperature. The increase in integrated PL intensity with increasing AlAs mole fraction seen in figures 3 and 4 may be attributed to an enhanced gettering of O_2 and H_2O due to the higher $Al(CH_3)_3$ concentrations in the growth ambient required when growing higher AlAs mole fraction $Al_xGa_{1-x}As$. The drop in PL intensity seen in the case of the $x\simeq0.13$ material as T_s is lowered is probably due to increased oxygen incorporation resulting from a low $Al(CH_3)_3$ concentration in the growth environment. Such oxygen incorporation is known to cause deep-state formation which competes with near-edge recombination, finally quenching it when oxygen contamination becomes severe (Stringfellow G B et al. 1979; Tsai M J et al. 1983,1984). In luminescence this principal competing process manifests itself in a 0.8eV band, which grows in as growth temperature or oxygen gettering is reduced (Tsai M J et al. 1983,1984). In deep level transient spectroscopy (DLTS) measurements oxygen injection during $Al_xGa_{1-x}As$ growth by MOCVD results in a new acceptor state located at E_c-0.41eV (Wallis R H et al. 1980) which most probably also competes with the near-edge PL. Both of these processes are most probably occuring in our material. We are currently involved in a more extensive study of the effect of oxygen on the growth and materials properties of $Al_xGa_{1-x}As$.

Since the gas phase containing the growth nutrients is further cleansed of oxidants *in-situ*, the growth of reduced-oxygen-content $Al_xGa_{1-x}As$ is then possible. The results presented here indicate trends in the material properties of the $Al_xGa_{1-x}As$ ($x\geq0.2$) which are, in many ways, quite similar to the MOCVD growth of GaAs. A primary effect of lower growth temperatures is the reduced incorporation of background impurities into the growing layer. In the particular case of carbon, its incorporation into GaAs decreases very rapidly as the growth temperature is reduced (Kuech T F, et al. 1984a). Carbon is also thought to be the dominant shallow acceptor in MOCVD $Al_xGa_{1-x}As$ (Stringfellow et al. 1980, Mircea-Roussel A, et al. 1982). The $Al_xGa_{1-x}As$ layers grown in this study exhibit trends with T_s similar to that of GaAs. Since the Al-C bond is considered much stronger than the Ga-C the carbon concentration is expected to increase with AlAs mole fraction in the film. At a given T_s the carbon content of the $Al_xGa_{1-x}As$

is always greater than that of GaAs grown under identical conditions as indicated by the high hole concentrations measured in our $Al_xGa_{1-x}As$. The increase in carbon concentration with x in $Al_xGa_{1-x}As$ is super-linear indicative a carbon incorporation reaction which is greater than first order in Al concentration. While at higher AlAs mole fractions the concentration of acceptors is greater than donors, the donor concentration also increases with increasing T_s. This is reflected in increased compensation at $T_s > 750°C$.

The growth of high quality $Al_xGa_{1-x}As$ at temperatures below 700°C will aid in the application of the MOCVD technique to high speed heterojunction devices. Devices relying on the formation of a two dimensional electron gas at a $Al_xGa_{1-x}As/GaAs$ interface require very high purity GaAs in order to achieve high electron mobilities and device performance. High purity GaAs is typically grown at temperatures of 650°C or less. The existing technology for $Al_xGa_{1-x}As$ required growth temperatures in excess of 750°C in order to obtain acceptable $Al_xGa_{1-x}As$ materials properties which is clearly incompatible with the growth requirements of GaAs. The results of this study indicate that with a growth system of appropriate design, the growth of both high purity GaAs and $Al_xGa_{1-x}As$ can be achieved at a single growth temperature.

Summary

We have demonstrated the growth of high purity $Al_xGa_{1-x}As$ over the growth range temperature of 600-800°C. The photoluminescence intensity for AlAs mole fractions greater than 0.2 increases monotonically with decreasing growth temperature exhibiting the most intense luminescence at $T_s \simeq 650°C$. The PL intensity also increased with AlAs mole fraction for alloy compositions in the direct bandgap region. These trends can be attributed to a reduced incorporation of oxygen in to the $Al_xGa_{1-x}As$ due to the gettering of O_2 and H_2O in the growth ambient by reaction with $Al(CH_3)_3$. These results indicate that both GaAs and $Al_xGa_{1-x}As$ can be optimally grown at the same temperature of ~650°C.

Acknowledgements This work was supported in part by ONR contract N00014-80-C-0376.

REFERENCES

Dapkus P D, Mansevit H M, Hess K L, Low T S, and Stillman G E 1981 J. Crystal Growth *55* 1.
Dingle R, Logan R A, and Arthur J R 1977 Inst. Phys. Conf. Ser. *A33* 210.
Hallais J, Andre J P, and Mircea-Roussel A 1981 J.Electron.Mat. *10* 665.
Hersee S D, di Forte-Poisson M A, Baldy M, and Duchemin J P 1981 J.Crystal Growth *55* 53.
Kisker D W, Miller J N, and Stringfellow G B 1982 Appl.Phys.Lett. *40* 614.
Kuech T F and Veuhoff E 1984a J.Crystal Growth (to be published) presented at the Second International Conference on Metal Organic Chemical Phase Epitaxy, April 10-12 Sheffield,England Post.31.
Kuech T F and Veuhoff E 1984b J.Crystal Growth (to be published) presented at the Second International Conference on Metal Organic Chemical Phase Epitaxy, April 10-12 Sheffield,England A1.4.
Mircea-Roussel A, Briere A, Hallais J, Vink A T, and Veenvliet H 1982 J.Appl.Phys. *53* 4351.
Mohammed K, Merz J P, and Kasemet D 1983 Materials Lett. *2* 1.
Shealy J R and Woodall J M 1982 Appl. Phys. Lett. *41,* 88.
Shealy J R, Claus C F, Wicks G W, and Eastman L F 1984 J.Crystal Growth (to be published) presented at the Second International Conference on Metal Organic Chemical Phase Epitaxy, April 10-12 Sheffield,England Post.2.
Stringfellow G B and Linnebach R 1980 J.Applied Physics *51* 2212.
Stringfellow G B 1981 J.Crystal Growth *55* 42.
Terao H and Sunakawa H April 10-12,1984; presented at the Second International Conference on Metal Organic Chemical Phase Epitaxy, Sheffield,England P2.13.
Tsai MJ, Tashima M M, and Moon R L 1983 unpublished.
Tsai MJ, Tashima M M, Twu B L, and Moon R L 1983 Inst. Phys. Conf. Ser. *65* 85.
Wallis R H, di Forte-Poisson M A, Bonnet M, Beuchet G, and Duchemin J P 1980 Proc. of the Int. Symp. on GaAs and Rel. Compounds (Vienna, Austria) 73.

Inst. Phys. Conf. Ser. No. 74: Chapter 3
Paper presented at Int. Symp. GaAs and Related Compounds, Biarritz, 1984

187

Ga$_{0.47}$In$_{0.53}$As–InP heterostructures grown by MOCVD

T.Fukui and H.Saito

Musashino Electrical Communication Laboratory,Nippon Telegraph & Telephone
Public Corporation, Musashino-shi, Tokyo 180 Japan

  Abstract.   High quality Ga$_{0.47}$In$_{0.53}$As and Ga$_{0.47}$In$_{0.53}$As-InP quantum
well heterostructures were grown by low-pressure metal organic chemical
vapor deposition(LP-MOCVD). Electron mobilities greater than 40,000
cm^2/V·s at 77K were obtained for Ga$_{0.47}$In$_{0.53}$As by precise gas flow
control using a leak-free gas system and introducing thin InP buffer
layers. A 0.998eV emission was obtained from an 8.8A thin quantum well.
Observed emission energy for well width,L$_z$<100A is slightly lower than
that calculated.

1. Introduction

Ga$_{0.47}$In$_{0.53}$As-InP heterostructures are attractive materials for optical
and microwave devices. Recently, the successful growth of GaInAs-InP
heterostructures have been reported using low-pressure metalorganic
chemical vapor deposition(LP-MOCVD)(Razeghi et al 1982,1983a). However,
the detailed growth conditions to obtain high purity GaInAs are not well
known. In the present paper, the effects of mole fraction ratios in the
vapor and introducing InP buffer layers on the electrical properties
are investigated for GaInAs grown by LP-MOCVD. Next, the photoluminescence spectra are measured for very thin GaInAs-InP quantum wells,
and the differences between measured and calculated emission energies
are discussed in connection with quantum well width.

2. Experimental Procedure

The epitaxial growth apparatus is constructed with a horizontal reactor
having a substrate loading chamber and a leak-free gas system, which is
helium leak-tested to a rate of less than 10^{-6}atom·cc/s. A leak-proof
gas manifold and reactor system are important to obtain reproducible
epitaxial wafers. A schematic diagram of the growth apparatus is shown
in Fig.1. The working pressure of 76 Torr is carefully controlled within
0.1 Torr during growth by adjusting the needle valve between the reactor
and the rotary pump. Triethylgallium(TEGa) and triethylindium(TEIn) are
used as the group III metal-alkyl sources, and are held in controlled
temperature baths at 40.0°C and 0°C(±0.01°C),respectively. Precisely
metered hydrogen gas is passed through each metal-alkyl bubbler. Arsine
(AsH$_3$) is used as the group V hydride source. The total hydrogen gas
flow rate is 3l/min, with no N$_2$ gas being used. The reactor and gas
system enables the rapid change (within 1 sec) of gas composition over
the substrate to obtain abrupt heterojunctions. Fe doped InP oriented
at (100)±0.1° is used as the substrate. The growth temperature is 550°C,

which is measured with an infrared radiation thermometer. Before the
growth of each GaInAs layer, a thin(1,000A) InP buffer layer is grown on
the InP substrate at 550°C. Typical growth conditions for $Ga_{0.47}In_{0.53}As$
were as follows: partial pressures for TEGa(P_{TEGa}),TEIn(P_{TEIn}) and AsH_3
(P_{AsH_3}) inside the reactor were 2.8×10^{-6}Torr,3.3×10^{-6}Torr and $3.0\times10^{-5} \sim$
4.9×10^{-4}Torr. At these conditions, the $[AsH_3]/([TEGa]+[TEIn])$ ratio was
varied from 5 to 80. For fixed conditions of P_{TEGa} and P_{TEIn},In compositions in solids were decreased about 2% as AsH_3 increased from 3×10^{-5}Torr
to 4.9×10^{-4}Torr. Therefore, partial pressure for TEGa had to be adjusted
to obtain lattice matching $Ga_{0.47}In_{0.53}As(\Delta a/a \leq 4\times10^{-4})$. Incorporation
ratios into solids for TEGa and TEIn were almost the same. The growth
rate for GaInAs was 2.2A/s, and was independent of [V]/[III] ratio.

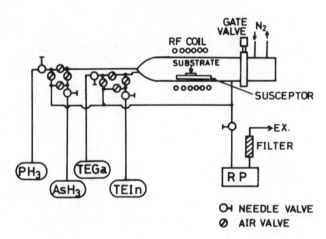

Fig.1   Schematic diagram of growth apparatus.

3. Results and Discussions

3.1 $Ga_{0.47}In_{0.53}As$ on InP

$Ga_{0.47}In_{0.53}As$ crystal properties were characterized by X-ray diffraction,
Hall measurements and photoluminescence. Double crystal X-ray rocking
curves were measured for $Ga_{0.47}In_{0.53}As$ using a $CuK\alpha_1$ line and InP(400)
reflection monoclometer. Figure 2 shows typical (400) reflection for the
epitaxial layer. Sharp X-ray diffraction was obtained. Full widths at
half the maximum(FWHM) were always in the region of $24''\sim60''$. Although
the value changed from run to run, it was independent of the [V]/[III]
ratio.   Surface morphology depends on [V]/[III] ratio. Mirror smooth
layers were obtained at mole fraction ratios,[V]/[III]$\geq$20, although at a
low mole fraction ratio, rough surfaces were often observed. This tendency
resembles previously reported InP MOCVD(Uwai et al 1984).

Figure 3 shows carrier concentrations and mobilities vs $[AsH_3]/([TEGa]+$
[TEIn]) for undoped $Ga_{0.47}In_{0.53}As$ grown on 1,000A InP buffer layers.
High electron mobilities greater than 40,000cm^2/V·s and a low carrier
concentration(n=4~5$\times10^{15}$cm^{-3}) at 77K were obtained when [V]/[III] ratios
were in the range of 20-40. Above and below this range, the background
increased as mobility decreased. All epitaxial layers were n-type, and

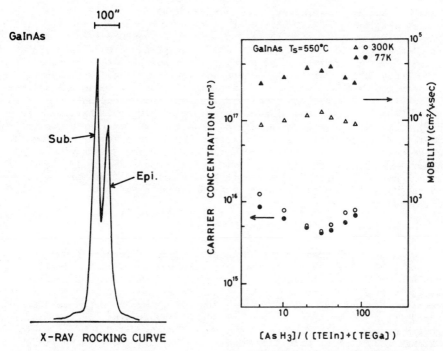

Fig.2   Typical X-ray rocking curve
for $Ga_{0.47}In_{0.53}As$.

Fig.3   Carrier concentrations and
mobilities vs $[AsH_3] / ([TEGa] + [TEIn])$
for undoped $Ga_{0.47}In_{0.53}As$.

p/n conversion was not observed when $[V]/[III]$ vapor ratios were in the
range of 5–80. GaInAs epitaxial layer grown directly on InP substrate
showed rather low electron mobility and poor surface morphology($n_{77K}=$
$4.7 \times 10^{15} cm^{-3}$ and $n_{77K}=13,000 cm^2/V \cdot s$). Figure 4 shows the temperature
dependence of carrier concentrations and electron mobilities for two types
of GaInAs i.e.,with and without the InP buffer layer. The electron
mobility for GaInAs grown directly on InP substrate was rather low and
showed almost the same temperature dependence of liquid phase or vapor

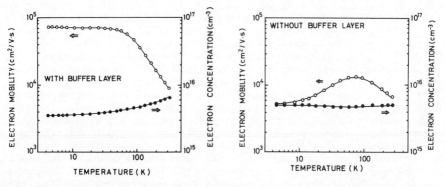

Fig.4   Temperature dependence of carrier concentrations and electron
mobilities for two type of GaInAs with and without InP buffer layers.

phase epitaxial grown layers. The maximum mobility was 13,000 cm^2/V·s
and was obtained at 77K. On the other hand, the electron mobility for
GaInAs with the InP buffer layer was more than 70,000cm^2/V·s at temperatures between 4.2K and 50K, and is almost constant in this temperature
region. This phenomenon suggests that 2-dimensional electron gas
accumulates at the GaInAs-InP heterojunction interface. The observations
of Shubnikov de Hass oscillations and magnetophonon resonances indicate
the coexistance of 2-dimensional and 3-dimensional electrons. Nicholas
et al(1982) also reported similar behavior for MOCVD GaInAs(1µm) with an
InP(1,500A) buffer layer. Therefore, a part of the carrier concentration
shown in Fig.3 is 2-dimensional electron gas which transferred from the
InP buffer layer. The highest mobility obtained for GaInAs with a InP
buffer layer is 12,100cm^2/V·s at 300K and 50,000cm^2/V·s at 77K(n_{77K}=
4.7x10^{15}cm^{-3}). These results show that the electrical properties of
undoped GaInAs are much improved introducing thin InP buffer layers between
GaInAs and InP substrate.

3.2 Ga$_{0.47}$In$_{0.53}$As-InP Quantum Well

Ga$_{0.47}$In$_{0.53}$As-InP quantum well heterostructures were also grown on
precise (100) oriented InP substrates. The quantum well layer thickness
was deduced from the steady-state growth rate, which was measured for
thick layers of InP and Ga$_{0.47}$In$_{0.53}$As. To minimize the thickness of the
interface transition layer, the gas flows of group III organometals were
stopped for a few seconds between InP and GaInAs growth. In that interval,
group V hydride gases in the reactor were exchanged(AsH$_3$/PH$_3$). Photoluminescence measurements were carried out at 77K, using a focused Kr ion
laser (6471A) which excites both the Ga$_{0.47}$In$_{0.53}$As and InP layers. The
photoluminescence was analyzed with a Jobin-Yvon H25 monochromator in
conjunction with a InSb detector.

For the thick Ga$_{0.47}$In$_{0.53}$As epitaxial layer(1.6µm), a narrow photoluminescence spectrum with a full width at half the maximum of 19meV was
obtained at 77K. The peak energy of 0.790eV is in good agreement with
the energy bandgap of Ga$_{0.47}$In$_{0.53}$As at 77K. The photoluminescence
spectra were measured for a Ga$_{0.47}$In$_{0.53}$As-InP quantum well having a well
thickness(L_z) of 8.8-250A. The luminescence intensity is at least
comparable to that of bulk Ga$_{0.47}$In$_{0.53}$As. Figure 5 shows the photoluminescence spectrum of the four-well structure and indicates the peak
assignments. The sample structure is illustrated in this figure. Figure
6 is a plot of luminescence peak energies as a function of the associated
well width,L_z. A 0.998eV emission was obtained from   an 8.8A thin
quantum well, which corresponds to only six monolayers. The photoluminescence line width at half the maximum(FWHM) was 55meV for L_z=8.8A.
This broad spectrum may be caused by thickness and/or composition
fluctuations in the quantum well. In this region, the difference of one
or two monolayers(1.5-2.9A) in the well width results in a large energy
difference in quantum size effect. FWHM decreased as L_z increased and is
28meV for L_z=52A. The theoretical curve calculated by Yamada et al(1984)
is also shown in Fig.6, using the envelope function approach introduced by
Bastard(1981). Emission energy vs well width agrees with the calculation
for $L_z \gtrsim$100A This result is in contrast with a previous report by Razeghi
et al(1983b) in which a large discrepancy was always observed between the
measured and calculated emission energy with Ga$_{0.47}$In$_{0.53}$As-InP quantum
wells(L_z=25-200A). For L_z<100A the observed emission energies are lower
than those calculated. It is probable that a number of factors contribute
to create the observed discrepancies between experiment and theory. The

layer thickness and composition of GaInAs are not well known because the GaInAs layer is formed within several seconds after gas flow switching. It is assumed that the steady-state growth rates and solid composition for bulk layers are valid for very thin layers. Layer thickness and interface abruptness are, however, fairly close in terms of the atomic order, because $(InAs)_1(GaAs)_1$ layered crystals have been successfully grown by a similar method recently(Fukui et al 1984). On the other hand, in the present calculation using the envelope function approximation it is assumed that the bulk band parameter is still valid for even only a few atomic layers. Therefore, the discrepancy discussed here may be caused partially by the epitaxial growth and partially by the calculation.

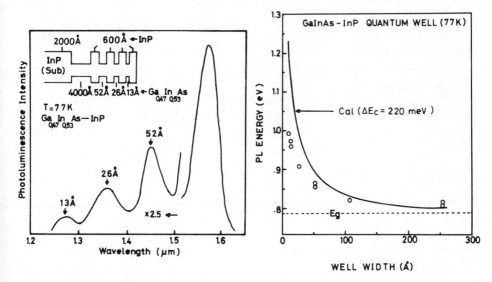

Fig.5 Photoluminescence spectrum of $Ga_{0.47}In_{0.53}As$-InP quantum wells measured at 77K. The inset schematically illustrates the sample structure.

Fig.6 Plot of measured photoluminescence peak energies as a function of associated well width, L_z. Theoretical curve was calculated using conduction band discontinuity, E_c=220meV.

4. Conclusion

High quality $Ga_{0.47}In_{0.53}As$ and $Ga_{0.47}In_{0.53}As$-InP quantum well heterostructures were grown by LP-MOCVD. A mobility of as high as 50,000cm^2/V·s was obtained at 77K for $Ga_{0.47}In_{0.53}As$ by means of a precise gas flow control using a leak-free gas system and introducing thin(1,000A) InP buffer layers. Furthermore, very sharp X-ray rocking curves and a narrow photoluminescence spectrum were observed. The photoluminescence spectrum measured for $Ga_{0.47}In_{0.53}As$-InP quantum well heterostructures showed a quantum size effect for well thickness of 8.8-250A. Emission energy vs well width agrees with the calculation for L_z>100A, and gradually decreases in comparison to the calculation for L_z<100A. The discrepancy may stem from both experiment and calculation.

Acknowledgements

The authors would like to thank Y.Kato and T.Kimura for their encouragement throughout the work. They would also like to thank K.Kumabe and N.Kobayashi for their many helpful discussions. Thanks are also due to A.Sugimura and S.Yamada for their calculations and K.Tsubaki for the Hall measurements.

References

Bastard G 1981 Phys.Rev.B24 5693
Fukui T and Saito H 1984 Jpn.J.Appl.Phys.23 L801
Nicholas R J, Brummell M A, Portal J C, Razeghi M and Poisson M A 1982 Solid state commun.43 825
Razeghi M, Poisson M A, Larivain J P, de Cremoux B, Duchemin J P and Voos M 1982 Electron.Lett.18 339
Razeghi M, Poisson M A, Larivain J P and Duchemin J P 1983a J.Electron. Matter.12 371
Razeghi M, Hirtz J P, Ziemelis U O, Delalande C, Etienne B and Voos M 1983b Appl.Phys.Lett.43 585
Uwai K, Susa N, Mikami O and Fukui T 1984 Jpn.J.Appl.Phys.23 L121
Yamada S, Taguchi A and Subimura A 1984 to be submitted.

Inst. Phys. Conf. Ser. No. 74: Chapter 3
Paper presented at Int. Symp. GaAs and Related Compounds, Biarritz, 1984

193

Low temperature epitaxial growth of device quality GaSb and GaAs by plasma-assisted epitaxy

T. Hariu, K. Matsushita, Y. Sato, Y. Sugiyama and Y. Shibata

Department of Electronic Engineering, Tohoku University, Sendai 980, Japan

Abstract. Undoped p-GaSb with carrier density of $4 \times 10^{16} cm^{-3}$ and mobility of $750 cm^2/Vs$, and undoped n-GaAs with carrier density of $3 \times 10^{17} cm^{-3}$ and mobility of $3500 cm^2/Vs$ were grown by plasma-assisted epitaxy (PAE) in hydrogen plasma at substrate temperatures around 400°C and 550°C, respectively. An optimum plasma power for the highest mobility and improved photoluminescence shifts to a higher value as the growth temperature is decreased.

1. Introduction

One of the important aspects in epitaxial growth of semiconductor crystals is a low temperature growth process for the fabrication of controlled device structures, as approached by relatively new technologies like MBE, IBE (Ion Beam Epitaxy), photoepitaxy etc. In IBE and photoepitaxy, external energy is added to the growing surface by increasing respectively the kinetic energy and internal energy of atoms, ions or molecules. In plasma-assisted epitaxy (PAE), the internal energy to activate chemical reaction and the momentum to enhance atomic migration on the growing surface of a crystal are simultaneously given to atoms and ions by using a discharging plasma. Both energies should be added to the surface, not to the bulk, in order to realize low temperature crystal growth. Furthermore, this PAE process has several other advantages such as (1) cleaning effect of the substrate and the growing surface (Hariu et al 1981), (2) high efficiency in impurity doping (Matsushita et al 1983), (3) high growth rate, if necessary and (4) good surface morphology across a large area.

In this paper, low temperature epitaxial growth of device quality GaAs and GaSb in hydrogen plasma is described, paying particular attention to the effect of discharging plasma on their crystallographic, electrical and optical properties.

2. Experimental

The experimental PAE apparatus and deposition procedure have already been described elsewhere (Hariu et al 1982). Electron density, electron temperature and plasma potential with respect to the earthed substrate are shown in Fig.1(a) as a function of applied RF power and their distributions across the reaction vessel are shown in Fig.1(b). It is seen here that the fractional content of ions is very small, and excited atoms or molecules are more important than ions in the present process, compared with IBE.

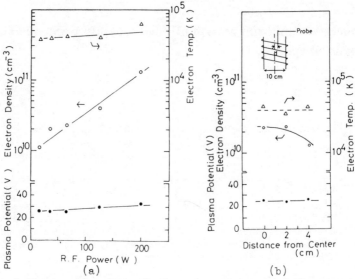

(a)   (b)

Fig. 1 Electron density, electron temperature and plasma potential
in the present PAE system (a) as a function of applied RF power
and (b) their distributions across the reaction chamber.

Electrical properties were measured by Van
der Pauw method on about 1μm thick GaSb
and GaAs layers, both grown on (100) semi-
insulating GaAs substrates. Photoluminescence (PL) spectra were measured at
4.2K by using a CW Ar-ion laser for the
excitation and a Si photodiode or a cooled
PbS for the detection of luminescence.

3. Results

Epitaxial layers of GaAs and GaSb were
grown on (100)GaAs substrates at 350°C
(Takenaka et al 1980, Sato et al 1984) in
the plasma of applied RF power 100W, while
only polycrystalline layers were grown
without plasma. However, it was found
that the epitaxial layers were deposited
only within a limited range of applied
plasma power and that polycrystalline
layers were produced at excess power
(Hariu et al 1981).

Electrical and optical measurements on PAE
layers also indicated that the applied
plasma power should be optimized to
improve these properties. Fig.2 shows the
variation of mobility in undoped GaSb
layers as a function of plasma power at
various substrate temperatures. An
optimum plasma power exists at each sub-

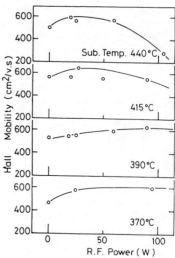

Fig. 2 Hall mobilities in
undoped p-type GaSb films
deposited on (100) semi-
insulating GaAs as a
function of plasma power
at various substrate
temperatures.

Table I Electrical properties of PAE GaAs and GaSb layers

	Discharging gas	Dopant	Conductivity	Carrier Density (cm^{-3})	Hall Mobility (cm^2/Vs)
GaAs	H_2	non	n	3×10^{17}	3500
	H_2	Te	n^+	2×10^{19}	900
	H_2	C	p^+	$10^{18} - 10^{20}$	110
GaSb	Ar	Te	n	$10^{18} - 10^{19}$	700
	H_2	non	p	6×10^{16}	750

strate temperature, and it shifts to a higher value as the substrate temperature is decreased.

Carrier densities and mobilities of some typical PAE GaAs and GaSb layers grown on (100) semi-insulating GaAs are summarized in Table I.

Fig.3 shows the mobilities of undoped GaSb films deposited by PAE in comparison with those by other methods like MBE,LPE, VPE and MOCVD as a function of substrate temperature. GaSb films by PAE have comparable mobilities in spite of lower substrate temperatures.

Photoluminescence spectra on PAE GaSb also indicated that the applied plasma power should be optimized to maximize the PL intensity due to bound exciton in relation to donor-acceptor or band-acceptor emission. Fig.4 shows a PL spectrum of a PAE GaSb on (100)GaAs and the variation of PL intensity at maximum peak near band edge as a function of plasma power. The maximum intensity is obtained for a layer deposited with plasma power of 20-30W at a substrate temperature of 440°C in the present deposition system. It is also shown that the PL intensity is always

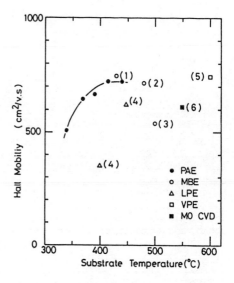

Fig. 3 Hall mobilities of un-doped p-type GaSb films deposited by PAE in comparison with those by other methods as a function of substrate temperature. (1)(Naganuma et al 1980), (2)(Gotoh et al 1981), (3)(Yano et al 1978), (4)(Miki et al 1974), (5)(Kakei et al 1970), (6)(Manasevit et al 1979).

higher for a layer grown on (100)GaSb substrate than (100)GaAs, probably due to the exact lattice matching in the former.

In applying PAE to the growth of some particular semiconductor crystal, proper discharging gas should be employed by considering its chemical reactivity and its physical effect like ion or atomic bombardment. Hydrogen plasma was found to be better than argon plasma in the present case in terms of higher carrier mobility, lower carrier density and higher

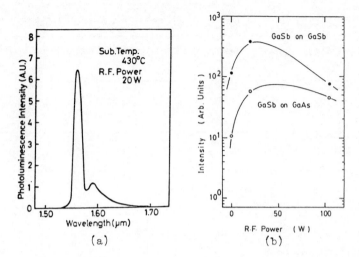

Fig. 4 (a) Photoluminescence spectrum of a GaSb film deposited
by PAE and (b) maximum luminescent intensity in GaSb layers
deposited on (100)GaSb and (100)GaAs as a function of plasma
power.

intensity of PL in the grown layers. Fig.5 compares the temperature
dependence of carrier densities and mobilities of undoped p-type GaSb
layers deposited in hydrogen and argon plasma, which have the similar
carrier mobilities at room temperature. This figure indicates less ionized
impurities in the former.

The PL intensity of layers grown in hydrogen plasma was much higher than
those in argon plasma (Matsushita et al 1984a).

Different crystal growth behaviours were observed in hydrogen and argon
plasmas. Fig.6 shows the variation of halfwidth of the (400) X-ray
diffraction lines of GaSb layers grown on (100)GaAs. The shift of lattice

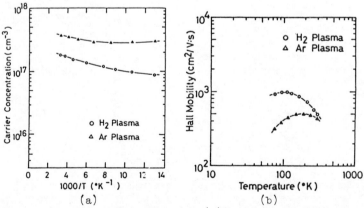

Fig. 5 Temperature dependence of (a) carrier density and
(b) Hall mobility in undoped p-type PAE GaSb layers deposited
in hydrogen and argon plasmas.

constant calculated from the shift of diffraction angles with respect to the (400) lines of GaAs substrates is also shown in the same figure. This shift indicates the elongation of lattice constant perpendicular to the substrate which is induced by the compressional stress along the substrate surface due to the larger lattice constant of GaSb than that of GaAs. The fact that the halfwidth for the layers grown in hydrogen plasma is larger than that in argon plasma does not necessarily mean inferior crystallographic properties of the former, as opposed to the superior electrical and optical properties described above. We should notice that the shift of lattice constant of GaSb layers grown in hydrogen plasma is larger than that in argon plasma. This is, we believe, caused by the different way to relax the lattice mismatch in the former, that is, by lattice distortion, while in the latter, higher densities of defects (like dislocation) are responsible for this relaxation.

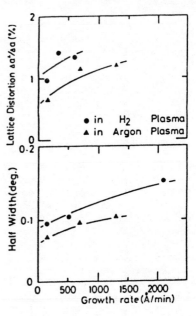

Fig.7 shows carrier density and Hall mobilities of undoped n-type PAE GaAs layers grown on semi-insulating (100) GaAs as a function of the supply ratio of As to Ga (not the flux ratio As_4/Ga as in MBE). An optimum supply ratio was around 2.5, much less than in MBE. Arsenic vapours produced by resistive heating are considered to be primarily composed of tetramic As_4 molecules and contain a dimer As_2 component, however, they are likely to be dissociated into

Fig. 6 The variation of half-width of (400) X-ray diffraction lines and the shift of lattice constant as a function of growth rate under constant plasma power 100W.

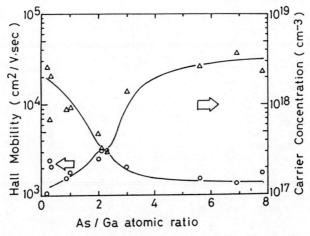

Fig. 7 Carrier density and Hall mobility of PAE GaAs layers as a function of the supply ratio of As to Ga.

lower molecular weight fragments such as As_3, As_2 or As by frequent collisions with a mean free path of the order of several millimeters in the present plasma. This dissociation will be favourable in improving crystal quality, as observed in MBE (Kuenzel and Ploog 1980).

Fig.8 shows the change of deep level PL spectra with the As source temperature. The 1.0eV band which is likely due to Ga-vacancy (Matsushita et al 1984a) and 0.67eV band which is likely due to As_{Ga}-related level or EL2 (Weber et al 1982) both decreases as the supply ratio of As decreases.

4. Concluding Remarks

PAE has been applied to grow epitaxial GaAs and GaSb at substrate temperatures as low as 350°C. Undoped p-GaSb layers with $p=6 \times 10^{16}$ cm^{-3} and $\mu_p=750 cm^2/Vs$ and undoped n-GaAs layers with $n=3 \times 10^{17} cm^{-3}$ and $\mu_n=3500 cm^2/Vs$ were grown at 410°C and 550°C, respectively, which are comparable to those obtained by other methods, in spite of a lower substrate temperature in PAE. As an example of possible application, the present PAE GaSb will be useful for the optoelectronic IC in 1.5µm wavelength region, monolithically combined with high-speed GaAs FET's. Heterojunction photodiodes, p-PAE GaSb on n-GaAs, which can reduce the dark current by several orders of magnitude compared with GaSb homojunctions, have been already demonstrated (Matsushita et al 1984b).

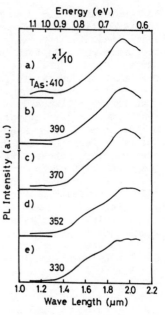

Fig. 8 Change of photo-luminescence spectra on PAE GaAs with As source temperature.

Acknowledgement

The authors would like to express their thanks to T. Sato, Y. Sasaki and K. Shida for their kind cooperation.

References

Gotoh H, Sasamoto K, Kuroda S et al 1981 Jpn. J. Appl. Phys.**20** L893
Hariu T, Takenaka, Shibuya S et al 1981 Thin Solid Films **80** 235
Hariu T, Matsushita K, Komatsu et al 1982 Inst. Phys. Conf. Ser.**65** 141
Kakei M, Shimokawa R and Arizumi T 1970 Jpn. J. Appl. Phys. **9** 1039
Kuenzel H and Ploog K 1980 Appl. Phys. Lett. **37** 416
Manasevit H M and Hess K L 1979 J. Electrochem. Soc. **126** 2031
Matsushita K, Sugiyama Y et al 1983 Jpn. J. Appl. Phys.**22** L602
Matsushita K, Sato T et al 1984a IEEE Trans. Electron Devices **ED-31** 1092
Matsushita K et al 1984b Ext. Abstract 16th Conf. Solid St. Dev. Mat.659
Miki H, Segawa K and Fujibayashi 1974 Jpn. J. Appl. Phys.**13** 20
Naganuma M and Miyazawa S 1980 Shinkuu **23** 326 (in Japanese)
Sato Y, Matsushita K, Hariu T et al 1984 Appl. Phys. Lett.**44** 592
Takenaka K, Hariu T et al 1980 Jpn. J. Appl. Phys. Suppl.**19-2** 183
Weber E R, Ennen H, Kaufman U et al 1982 Appl. Phys. Lett. **41** 6140
Yano M, Suzuki Y, Ishii T et al 1978 Jpn. J. Appl. Phys.**17** 2091

Inst. Phys. Conf. Ser. No. 74: Chapter 3
Paper presented at Int. Symp. GaAs and Related Compounds, Biarritz, 1984

199

Low temperature growth of high purity GaInAs for the preparation of GaInAs/InP multilayers using the hydride system

H Jürgensen, D Schmitz, M Heyen and P Balk

Institute of Semiconductor Electronics/SFB 202, Aachen Technical University, D-5100 Aachen, FRG

Abstract. The effects of arsenic and metal chloride vapor pressures on the growth of GaInAs have been investigated in order to optimize lattice matching to InP and the electrical properties of the material. These studies were carried out at temperatures necessary for the fabrication of high quality InP films (T_D = 900 K). In agreement with model calculations the Ga concentration in the film increases with the GaCl and AsH_3 pressures but decreases with the pressure of InCl. At optimum conditions lattice matching was reproducibly achieved within 5×10^{-4} lattice constants with background doping concentrations giving $n < 10^{14} cm^{-3}$. Donor concentrations between 10^{15} and $4 \times 10^{19} cm^{-3}$ were adjusted by injection of H_2S. The maximum room temperature mobility observed in our films was $12760 cm^2/Vs$.

1. Introduction

InP/GaInAs single and multiquantum well structures require layer sequences with extremely sharp interfaces. The successful preparation of such heterostructures demands their growth in one run at low deposition temperatures to suppress degradation of the substrate surface and interdiffusion between the layers.

In an earlier paper the feasibility of growing high quality InP at low temperatures in the hydride system was demonstrated (Jürgensen et al 1984a). Optimum deposition temperature was found to be 900 K. An important goal of the present investigation was to optimize the conditions for GaInAs deposition at the same temperature in order to be able to prepare complete heterostructures at one temperature.

In an earlier study of the epitaxial deposition of GaInAs Kordos et al (1982) explored the use of a system where the metal chlorides were formed by reaction of HCl with a single source containing a Ga/In alloy. It was shown that by this simple method reproducible results could be obtained over a number of runs. However, gradually a slight decrease of the Ga concentration in the source was obtained necessitating a small adjustment of the deposition temperature. Moreover, the preparation of multilayer structures would still require two sources: one for GaInAs, a second for InP. Therefore, in the present study two separate sources for Ga and In were used, as also done by other workers (Olsen et al 1983, Yamauchi et al 1982). It will be shown that exact lattice match and a wide range of impurity concentrations can be obtained by this method.

2. Experimental

The experiments were performed in a resistance heated quartz reactor 7 cm in diameter with two metal sources (figure 1). Four independently controlled temperature zones allowed adjustment of the required temperature profile with plateaus in the source and deposition regions, and a temperature peak in the mixing zone between source and substrate. In all experiments the source and deposition temperatures were kept at 970 K and 900 K respectively.

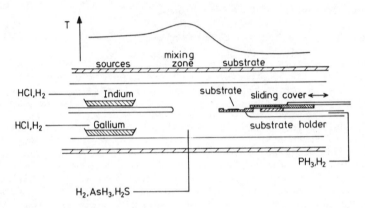

Figure 1. Schematic diagram of epitaxial reactor with substrate holder

The gas flows were adjusted using electronic flow controllers and pneumatic valves. The linear flow velocity in the deposition region was 20 cm/min, measured for a reactor temperature of 298 K. A total flow rate in the deposition zone of 800 ml/min and HCl + H_2 flow rates over the sources of 200 ml/min were utilized. All experiments were carried out with purified H_2, AsH_3 in H_2 and HCl (99.9999%). Substrates were Fe-doped semiinsulating InP wafers, cut 2° off the (100) plane. To avoid decomposition of the InP, a 10 Pa PH_3 pressure was maintained over the substrate during the heating and cooling cycles of the growth process (Jürgensen et al 1984a). To this end, the substrate holder was equipped with a sliding cover which could be flushed with this gas. Epitaxial growth was initiated by sliding the cover off the substrate area after turning off the PH_3 flow in the substrate holder. In contrast to a multiple chamber reactor, where the substrate has to be moved from one chamber to another (Olsen et al 1983, Beuchet et al 1982, DiGuiseppe et al 1983) our epitaxial system easily lends itself to automated growth of extremely thin films and multilayer structures under microprocessor control, which was used throughout the experiments.

3. Results and Discussions

At the same source and deposition temperatures (T_S=970 K; T_D=900 K) as used for the growth of InP the effects of the vapor pressures of arsenic and metal chlorides were studied. Results on the dependence of the composition of the ternary compound on the pressures of the metal chlorides are shown in figure 2. As expected, the Ga concentration in the film increases with the pressure of GaCl (figure 2a), but decreases with the InCl pressure (figure 2b). It should be noted that the conversion of HCl to metal chlorides at the sources is almost complete.

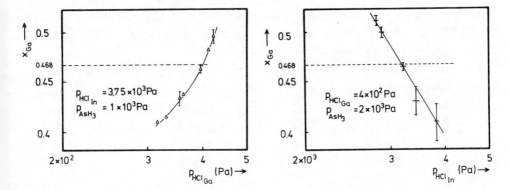

Figure 2. Dependence of Ga concentration (x_{Ga}) on the pressure of GaCl (a) and InCl (b).

Therefore, we assume that the pressures of InCl and GaCl in the reaction zone are equal to the corresponding HCl pressures. In order to obtain exact lattice matching the HCl flow rate over the In source must be larger by a factor of 6 to 10 than that over the Ga source. A 1% variation of the HCl_{Ga} flow rate causes a 0.6% change of the Ga concentration whereas the same variation of HCl_{In} flow gives rise to a 0.7 % change of the film composition. Since high quality electronic mass flow meters allow keeping the flow rates constant to better than 1%, reproducible fabrication of films with sufficient lattice match (within 1×10^{-3} lattice constants) can be achieved.

In figure 3 the effect of the AsH_3 pressure on the film composition is depicted. The observed decrease of the Ga concentration with increasing AsH_3 pressure can be explained from the difference in the heats of formation of GaAs and InAs. In this case a 1% variation of the flow rate causes only a 0.1 % change of the Ga concentration in the film.

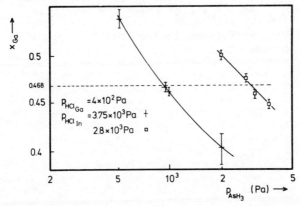

Figure 3. Effect of AsH_3 pressure on the composition of the ternary compound.

The GaInAs films grown in our system are of good lateral and vertical uniformity. As a measure for the depth uniformity a typical x-ray rocking curve of an epitaxial film exhibiting a FWHM of $\pm 1.5\times10^{-4}$ is given in figure 4. With $P_{HClGa}=4\times10^2$ Pa, $P_{HClIn}=3.4\times10^3$ Pa and $P_{AsH3}=2\times10^3$ Pa optimum lattice match ($\Delta a/a < 5\times10^{-4}$) and thus high mobilities (around 10000 cm^2/Vs) and low background doping concentrations ($n < 10^{14} cm^{-3}$) could be reproducibly achieved.

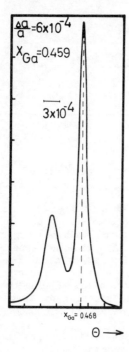

$\frac{\Delta a}{a}=6\times10^{-4}$

$X_{Ga}=0.459$

3×10^{-4}

$X_{Ga}=0.468$

$\Theta \longrightarrow$

Figure 4. Double crystal x-ray rocking curve of GaInAs film grown on InP.

Rocking curves and low temperature photoluminescence spectra obtained from points distributed over a substrate area of $1.5\times1.5cm^2$ indicate the excellent lateral uniformity of the film composition. For example, the variation of the lattice match over the sample was less than 1×10^4 lattice constants. However, due to a high gas phase supersaturation hillocks appeared on the grown surface. Additional experiments showed, that like in the case of InP, the surface morphology can be improved by additional injection of HCl in the mixing zone (Jürgensen et al 1984b).

At conditions close to those giving exact lattice matching the deposition rate of the ternary compound is almost linearly dependent on the pressures of both metal chlorides. An even weaker dependence was observed on AsH_3 pressure. Such a behaviour is expected from thermodynamic considerations.

Unintentionally doped samples were n-type with $n < 10^{14} cm^{-3}$. Higher electron concentrations were obtained by injection of H_2S into the deposition zone. In figure 5 the electron concentration has been plotted in dependence on the H_2S pressure. It can be seen that like in the case of the deposition of the binaries GaAs and InP a linear relationship between impurity and electron concentration pertains. This behavior indicates that dopant incorporation takes place at non-equilibrium conditions. It was proposed in the cases of GaAs and InP that the impurities incorporated in the surface of the layer are trapped by the fast growing film and incorporated in the bulk of the material (Korec et al 1984, Grundmann et al). It appears likely that a similar explanation holds in the case of the ternary. It can be seen from

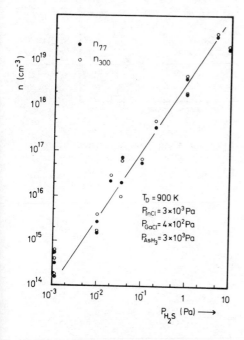

Figure 5. Dependence of electron concentration in GaInAs on H_2S pressure

the plot that donor concentrations from below 1×10^{15} to $4 \times 10^{19} cm^{-3}$ can be obtained by injection of H_2S without any indication of saturation.

The corresponding electron mobilities have been plotted in figure 6. As expected, the mobility decreases with increasing electron concentration similar to the cases of InP and GaAs. The solid lines have been drawn through the data points with the highest mobilities; these samples exhibit the optimum lattice match. Highest mobilities appear to be obtained at $n = 8 \times 10^{14} cm^{-3}$. This is possibly caused by carrier depletion effects at lower carrier concentrations in our relatively thin layers (d=1–3 µm). In the case of S-doped GaAs a drastic mobility drop occurs at $n = 4 \times 10^{18} cm^{-3}$, the concentration where the S uptake saturates. Such a saturation was not observed during growth of S-doped InP and GaInAs films; here only a weak decrease of mobility occurs at the highest concentrations. The dependence of the electron mobility on film composition measured on unintentionally and S doped films with low electron concentrations ($n < 10^{15} cm^{-3}$) is plotted in figure 7. These values are generally somewhat higher than those observed in an earlier study (Kordos et al 1982). The maximum mobility measured at RT was 12760 cm^2/Vs. The maximum appears to shift to smaller lattice constants for doped materials.

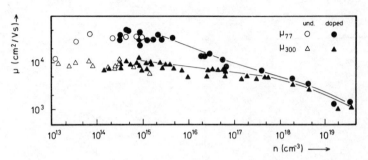

Figure 6. Dependence of electron mobility on electron concentration in S-doped and unintentionally doped films.

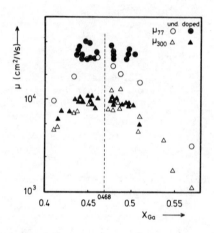

Figure 7. Dependence of electron concentration on GaInAs composition

References

Beuchet G, Bonnet M, Thebault P and Duchemin J P 1982 J. Crystal   Growth 57 379

DiGuiseppe M A, Temkin H, Peticolas L and Bonnmer W A 1983 Appl.   Phys. Lett. 43 906

Grundmann D, Jürgensen H, Heyen M and Balk P to be published

Jürgensen H, Korec J, Heyen M and Balk P 1984a J. Crystal Growth 66 73

Jürgensen H, Grundmann D, Heyen M and Balk P 1984b J. Crystal Growth accepted for publication

Kordos P, Schumbera P, Heyen M and Balk P 1982 Gallium Arsenide and Related Compounds 1981 (Inst. Phys. Conf. Ser. No 63) 131

Korec J, Grundmann D and Heyen M 1984 J. Electrochem. Soc. 131 1433

Olsen G H and Zamerowski T J 1983 RCA Review 44 270

Yamauchi Y, Susa N and Kanbe H 1982 J. Crystal Growth 56 402

Inst. Phys. Conf. Ser. No. 74: Chapter 3
Paper presented at Int. Symp. GaAs and Related Compounds, Biarritz, 1984

205

Growth of InP in the In$-$PCl$_3$$-H_2$$-PH_3$ system

I.H. Goodridge and N.B. Hasdell

Plessey Research (Caswell) Ltd., Allen Clark Research Centre, Caswell, Towcester, Northants NN12 8EQ, England

Abstract. Small additions of phosphine have been made to the In$-$PCl$_3$$-H_2$ system in an attempt to improve the deposition reaction for the production of InP for microwave applications. Our results indicate that the deposition reaction is strongly influenced by the presence of phosphine and that the formation of InP is kinetically controlled, the rate limiting step involving the catalytic decomposition of phosphine at the growing interface.

Optimisation of the phosphine concentration and other parameters has resulted in device quality structures being produced concomitant with a seven-fold conservation of reactants.

1. Introduction

In recent years, improvements in equipment and substrate quality have undoubtedly been instrumental in improving the yield of VPE grown InP. However, the yield and process economics of the In$-$PCl$_3$$-H_2$ system remain inferior to that of the comparable process for GaAs. This is somewhat surprising since the predicted (Shaw 1975) Group III element pick-up is thermodynamically more complete for InP than for GaAs. It would therefore seem that the problem lies in an inefficient deposition reaction.

2. Thermal Decomposition of Phosphine

In hydrogen, equation (1) could be written to describe the thermal decomposition of phosphine, the reaction thermodynamically being well to the right at $\sim$ 200°C. However, the reaction is a four body one, and in the gas phase, this is only likely at high pressure or high temperature where collision frequency or multi-collision frequency is high. Thus reaction (1) is kinetically (spatially) limited and the thermodynamic equilibrium is not established.

$$4PH_3 \rightleftarrows P_4 + 6H_2 \qquad\qquad (1)$$

The kinetic barriers can be overcome in a heterogeneous case where a solid or liquid surface is used to collect molecules or transition species. Obviously surfaces containing phosphorus would be best in this respect. In our case, such a surface could be InP itself $-$ the substrate, source, or even extraneous wall deposits. Although we do not present direct evidence of this behaviour, heterogeneous catalytic decomposition of PH$_3$ has been observed by other workers. Hinshelwood and Topley (1924), in an elegant study, demonstrated that powdered SiO_2 increased the reaction rate and more recently, Stringfellow (1984) showed that a better catalyst to be powdered GaP. By kind permission of G.B. Stringfellow, we reproduce his results in Fig.(1), to show that above ~ 600°C, in the presence of a suitable catalyst, phosphine is completely decomposed.

However these results would appear to be in direct conflict with those of Ban (1971) and Karlicek et al (1982). Ban (1971), by applying mass spectroscopy analysis techniques to a hydride system, established that at 850°C, ~ 10% of the phosphine remained undecomposed. Karlicek et al (1982), using laser spectroscopy analysis techniques, with a similar system, obtained a similar figure, but at the lower temperature of 700°C. We propose that the undecomposed phosphine seen by these workers, is a consequence of there being no suitable catalyst in their systems. We base this speculation on firstly, that it appears in both cases, that the phosphine was admitted downstream of

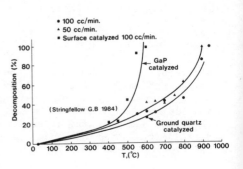

Fig.1.   Decomposition of PH_3 in H_2.

the source, and secondly, it would also appear, in the absence of substrates. If true, this means that Stringfellow's (1984) catalytic conditions for complete decomposition are not met, which could explain the presence of phosphine in their systems.

It is interesting to note from Ban's (1971) work, that he considers that reaction (2) better describes the decomposition of phosphine (where x is the mole fraction of PH_3 decomposing and α the degree of dissociation of P_4 molecules).

$$PH_3 \xrightarrow{\geqslant 200} (1-x) \ PH_3 + (1-\alpha) \ \frac{x}{4} \ P_4 + \alpha.\frac{x}{2} \ P_2 + \frac{3}{2} \ H_2 \qquad (2)$$

It is of further interest, that both Ban (1971) and Karlicek et al (1983) determined the presence of dimeric as well as tetrameric phosphorus as reaction products of the decomposition of PH_3. At a temperature of 850°C, Ban (1971) determined x and α to be 0.9 and 0.5 respectively.

3. Experimental

The reactor used in these experiments is schematically represented in Fig (2). It can be seen that selection of either hydrogen or nitrogen as the carrier gas can easily be made via a solenoid valve. Other solenoid valves and mass flow controllers route and control the flows as required. It can be seen that the reactor is fitted with two gaseous sources. One of these is a hydrogen sulphide in hydrogen mixture which Davies et al (1982) proved to be a useful dopant source for InP, and the other is the phosphine mixture whose properties and effect on such a system is the subject of this paper. Two phosphine mixtures were used in these experiments, firstly a 2-5% PH_3 in H_2 mixture and secondly, a 1% PH_3 in N_2 mixture. Both of these gas mixtures were obtained from BOC Special Gases. A more complete description of the reactor is given by Davies et al (1982).

In the experiments when hydrogen constituted the carrier gas, the PH_3/H_2 mixture was admitted, in separate experiments, to both source and deposition zones, whereas when nitrogen was used, the PH_3/N_2 mixture was added only to the substrate zone. In the latter case, growth was also attempted in the total absence of hydrogen and phosphine, and in another series, in the presence of hydrogen, when a hydrogen cylinder was fitted in lieu of the PH_3/N_2 mixture.

A 200g charge of indium at ~ 700°C constituted the indium source, and both iron and tin doped InP substrates were used. These were first vapour etched in-situ at ~ 750°C and then, by lowering the temperature to ~ 650°C, growth was initiated. Typical growth times were 60 mins.

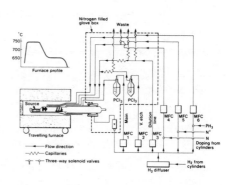

Fig.2.   Schematic of InP reactor.

Assessment of layers was made using a standard electrochemical C-V profiler (Polaron Equipment Ltd) and by standard Hall techniques. The latter were made at 298K and 77K, and using the calculations of Walukiewicz et al (1980), N_A and N_D were estimated.

4. RESULTS

The effect on growth rate of adding small amounts of PH_3/H_2 mixture to the In-PCl_3-H_2 system is shown in Fig.(3).
As can be seen, in all cases, growth rate increases as P_{PH_3} increases, and the extent of the increase differing over an order of magnitude dependent upon input point of the PH_3. Thus, at a PCl_3 mole fraction of 2.10^{-2} and a P_{PH_3} of 10^{-2} atm, a growth rate of ~ 140 μm/hr was obtained when the PH_3 was added downstream of the source, whereas this reduced to ~ 10 μm/hr when added upstream. It can also be seen, that lowering the PCl_3 mole fraction also reduced the growth rate. Thus, using a low PCl_3 molar fraction and a small PH_3 addition, growth rates of ~ 8 μm/hr were obtained. As such a low PCl_3 mole fraction would not normally be sufficient to produce InP, therefore these conditions offer the potential of a major conservation of both indium and

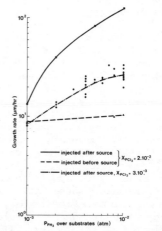

Fig.3.   Effect of PH_3 on growth rate.

phosphorus trichloride. However, in Fig(3), it can be seen that reproducibility of growth rate appears to worsen as P_{PH_3} increases. Possible reasons for this are discussed later.

In contrast, when growth was attempted in the $In-PCl_3-N_2$ system epitaxial growth could not be obtained until PH_3/N_2 was added downstream of the source. Surprisingly, when a hydrogen source was substituted for the PH_3/N_2 mixture, growth was only achieved with great difficulty even at a P_{H_2} of 2.10^{-1} and with a high degree of supersaturation!

In Fig(4), the relationship between phosphine partial pressure with N_D-N_A of

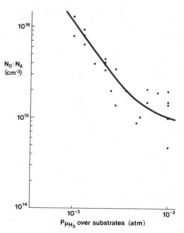

Fig.4. Effect of PH_3 on N_D-N_A.

layers grown in hydrogen are shown. It can be seen in Fig.(4), that increasing the P_{PH_3} results in a decrease in the residual[3] background doping level. Again there appears to be a large scatter in results and therefore the validity of the 'best fit' line could be questioned. However we believe the overall conclusion that N_D-N_A can be lowered by adding PH_3 is valid. By comparing Fig.(3) with Fig.(4), it can be shown that the results can be explained in terms of a growth rate dependence on N_D-N_A. Similar behaviour to this is often observed and there exist models, including those by Mullin (1977), Hurle (1976), and Giling and de Moor (1983), which claim to explain these observations. However analysis of these models is outside the scope of this paper.

From Hall data and the tables of Walukiewicz et al (1980), values of N_A and N_D have been estimated. We have found that,whereas N_A appears to be constant and unaffected by P_{PH_3} and growth rate changes, N_D is reduced by increasing the P_{PH_3}. This would therefore seem to explain the N_D-N_A vs P_{PH_3} relationship. Possible explanations for this behaviour could include a time dependence adsorption model, that is, the longer a growing surface is exposed, the greater the adsorption rate of impurities, or, a phosphorus vacancy model, which would claim that since donors are adsorbed at phosphorus vacancies, increasing the P_{PH_3} would decrease these and hence result in a lowering of N_D-N_A. Speculation of doping incorporation is however not the aim of this work.

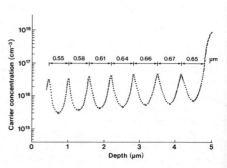

Fig.5. Periodic lo-hi-lo structure.

In an attempt to examine the origins of the non-reproducibility in growth rate (and hence growth rate related parameters), the test structure in Fig.(5) was grown. This is a C-V profile of a doped InP layer where the hydrogen sulphide was switched out for 18 secs every 120 secs. This resulted in formation of a periodic lo-hi-lo structure. It can be seen that

not only does the hi to lo peak distance change with time, but that the
carrier concentration of the lo regions also change with time. Thus there
is a reduction in growth rate with time. We believe that this effect is
related to extraneous wall deposits of InP between source and substrate.
Lowering the degree of supersaturation and vapour etching the reactor
between growth runs prevented such formation and resulted in a constant
growth rate and flat C-V profiles.

5. DISCUSSION

The results in both hydrogen and nitrogen ambients demonstrate the high
growth rate dependence upon phosphine concentration. The availability of
phosphine at the indium phosphide substrate would appear to be controlled by
the kinetics of the heterogeneous decomposition of phosphine. Clearly
kinetics will also play an important role at the growing interface itself.
In adding the PH_3 upstream of the indium source, we believe that the PH_3 is
probably completely dissociated on the surface of the source and hence has
little chance of affecting growth rate. Our results suggest that this may
be so. In contrast to this, when added downstream of the source, incomplete
decomposition is possible, which means that some PH_3 could be in the
deposition zone and available for reaction. Using the figures of Ban
(1971), if 10% of the PH_3 is undecomposed then this would represent, in the
case of small additions of PH_3, that a P_{PH_3} of $\sim 10^{-4}$ atm could be present
in the decomposition zone.

Thermodynamic analysis by Shaw (1975) and Jones (1982) of the equilibrium
partial pressures of phosphorus containing species in the $In-PCl_3-H_2$ system,
(i.e. without deliberately adding phosphine), indicates that P_{PH_3} is of the
order of 10^{-4} atm for an input PCl_3 molar fraction of $\sim 2.10^{-2}$.[3] Although
the PH_3 would appear to be by far the minor specie present, it is
interesting to note that, in our $In-PCl_3-H_2-PH_3$ system, our estimation of
P_{PH_3} is also of the order of 10^{-4} atm!

The importance of PH_3 in a conventional system was first suggested by
Fairhurst et al (1981). They claimed that it is the in situ formed PH_3
which provides phosphorus for the deposition reaction, and that the amount
of PH_3 available for that reaction depends upon the rate of competing
reactions. Although our experiments differ in that PH_3 has been
deliberately added, our overall conclusion would tend to support their
model.

Jürgensen et al (1984) have proposed that epitaxial growth is via a
multi-step adsorption process at phosphorus vacancies, and that the rate
determining step involves hydrogen as a reducing agent.

Our initial failure to form InP in the $In-PCl_3-N_2$ system does give initial
support to this theory. Although Giles et al (1983) confirmed the presence
of both InCl and P_4, we believed at that time, that our failure to form InP
was indeed related to the absence of hydrogen. However in this latest work,
this is refuted since addition of hydrogen, even at a partial pressure of $\sim$
2.10^{-1} atm, did not result in ready formation of InP. In contrast, when
small amounts of 1% PH_3/N_2 mixture were added, growth was readily achieved.
To us, this suggests that the role of PH_3 is more complex. We suggest that
PH_3 catalytically dissociates on the InP substrate and atomic hydrogen as
well as phosphorus are formed as transient species, and it is this which is
responsible for the enhanced growth rates. Although in the absence of
direct evidence, these ideas are somewhat speculative, our results suggest
that the formation of InP is by direct reaction of PH_3 with InCl at the

growing interface.

6. CONCLUSIONS

A new growth system has been assessed and found to offer advantages in conservation of reactants. A typical 200 g charge of indium should last for ~ 200 growth runs.

A second growth system, $In-PCl_3-N_2-PH_3$, has demonstrated that growth in a safer environment is possible.

The importance of phosphine as a reactive specie has been proposed.

ACKNOWLEDGEMENTS

The authors would like to thank P.L. Giles and P. Davies for their contribution to this paper, also D.Ashen and D. Anderson of RSRE-Malvern for useful discussions.

This work was carried out with the support of Procurement Executive, Ministry of Defence, sponsored by DCVD.

REFERENCES

Ban V.S, 1971, J. Electrochem. Soc. 118, 1473
Davies P, Hasdell N.B, Giles P.L, 1982 in: J. de Physique (France) Colloque C5, 43, 271.
Fairhurst K, Lee D, Robertson D.S, Parfitt H.T, Wilgoss W.H.E, 1981. J. Mat. Sci. 16, 1013.
Giles P.L, Davies P, Hasdell N.B, 1983, J. Cryst. Growth 61, 695.
Giling L.J, de Moor H.H.C, 1983 in: Proc. 4th European Conf. on CVD, Eindhoven, The Netherlands, 184.
Hinshelwood C.N, Topley B, 1924, J. Chem. Soc, 125, 393.
Hurle D.T.J, 1976 in: Proc. GaAs and Rel. Cmps, IOP London, 33a, 113.
Jones K, 1982 in: Proc. of SPIE Semiconductor Growth Tech. (Washington) 323, 81.
Jürgensen H, Korec, Heyen M, Balk P, 1984 J. Cryst. Growth 66, 73
Karlicek R.F, Donelly V.M, Johnson W.D, 1982 in: Proc. of SPIE Semiconductor Growth Tech. (Washington) 323, 62.
Mullin J.B, 1977 Cryst. Growth, 42, 77.
Shaw D.W, 1975, J. Phys. Chem. Solids, 36, 111.
Stringfellow G.B, 1984 in: Proc. of ICMOVPE-2 Sheffild, England, to be published.
Walukiewicz W, Lagowski T, Jastreblski L, Rava P, Lichtensteigen M, Gatos C.H, Gatos H.C, 1980, J. Appl. Phys. 51, 2659.

Inst. Phys. Conf. Ser. No. 74: Chapter 3
Paper presented at Int. Symp. GaAs and Related Compounds, Biarritz, 1984

211

Characterisation of high purity GaAs grown by hydride VPE

I A Dorrity, D Hewins and J D Medland

GEC Research Laboratories, Hirst Research Centre, Wembley, UK

Abstract. The influence of reactant mole fraction, growth temperature and source temperature on the purity of undoped GaAs grown by the hydride technique has been studied. Conditions have been established for the growth of high mobility layers (μ_{77} = 150–170,000 cm^{2}v^{-1}s^{-1}) with carrier concentrations 5–10x10^{13} cm^{-3}. Low temperature photoluminescence measurements were used to identify the residual acceptors, C, Zn and Ge. Comparison of these PL measurements with Hall measurements indicates that deep acceptors are dominant and shallow acceptor concentrations <10^{13} cm^{-3}.

Introduction

There is increasing interest in the growth of high purity VPE undoped GaAs as a buffer layer for ion implanted FETs and MMICs. Implants into buffer layers have produced FETs with higher gain and lower noise figures than equivalent structures fabricated on implanted substrates (Bartle et al 1984). Until recently the purest VPE material has been grown by the AsCl$_3$ technique since HCl purity problems apparently limited the performance of the related hydride method (Kennedy et al 1974). However, Abrokwah et al (1983) have now grown layers of comparable purity to the best AsCl$_3$ material.

In this study, we report the results of an investigation into the influence of reactant mole fraction, growth temperature T$_G$ and source temperature T$_s$ on the purity of undoped GaAs grown by the hydride technique. Optimisation of growth conditions has enabled growth of high purity layers with liquid nitrogen mobilities of 150–170,000 cm^{2}v^{-1}s^{-1} and carrier concentrations 5–10x10^{13} cm^{-3}. Incorporation of carbon, zinc and germanium acceptors, as determined by low temperature photoluminescence (PL) spectra, is also discussed.

Experimental

Growth was carried out in a Crystal Specialties 420 epitaxial reactor which has a vertical 85 mm OD quartz growth tube. Pd–diffused H$_2$ was used as a carrier gas. LEC undoped semi–insulating wafers cut 2° off (100) towards the (110) plane were used as substrates. These were solvent cleaned and etched in 3:1:1 H$_2$SO$_4$/H$_2$O$_2$/H$_2$O prior to loading the reactor and were in-situ etched with HCl immediately before growth.

Carrier concentrations and mobilities were measured by the van der Pauw method on 5 mm square samples with alloyed indium contacts. Compensation ratios and donor and acceptor concentrations were determined using the calculations of Walukiewicz et al (1982).

Photoluminescence measurements (4–20 K) were performed with the sample mounted in an Oxford Instruments cryostat. Luminescence was excited using the 5145 Å line of an argon ion laser, analysed using a SPEX 0.5 m monochromator and detected with an RCA 31035 A thermoelectrically cooled GaAs photocathode. Photon counting was used to produce spectra with a computer controlled 'SCADAS' instrument (Glen Creston).

Results

PL spectra

The PL spectra of the undoped layers were recorded in the excitonic region and the (lower energy) band-to-acceptor region. All layers with μ_{77} >100,000 $cm^2V^{-1}s^{-1}$ exhibit sharp well defined excitonic spectra and a typical spectrum is shown in Figure 1. Peaks due to free excitons (FE), donor bound excitons $(D^{\circ}X)$ and ionised donor bound excitons $(D^{+}X)$ are always observed. However, acceptor exciton $(A^{\circ}X)$ peaks are very weak and indeed are absent for many layers. Previous authors (Abrokwah et al 1983, Colter et al 1983) have related the low intensity or absence of $(A^{\circ}X)$ peaks to low compensation ratios in undoped layers. However, no such correlation was observed in this study. Moreover, as shown in Figure 1, $(A^{\circ}X)$ peaks were observed in the spectrum of layer with N_A/N_D of 0.1 but were absent for a sample with N_A/N_D of 0.5. $(D^{\circ}X)$ peaks were sharp with halfwidths in the range 0.15–0.25 meV confirming the excellent material quality but no well defined trend was apparent in the relationship between halfwidth and donor concentration.

PL spectra in the band-to-acceptor region showed broad bands due to donor-to-acceptor $(D^{\circ}-A^{\circ})$ and band-to-acceptor $(e-A^{\circ})$ transitions for various acceptors. Due to overlap of some peaks (i.e. $C(D^{\circ}-A^{\circ})$ and $Zn(e-A^{\circ})$) spectra were recorded over the range 4–20 K to ensure correct identification (Ashen et al 1975). Raising the temperature from 4 K increases the intensity of the $(e-A^{\circ})$ peaks and decreases the intensity of $(D^{\circ}-A^{\circ})$ peaks due to ionisation of shallow donors. At 16 K $(e-A^{\circ})$ transitions dominate the spectra and $(e-A^{\circ})/(D^{\circ}X)$ peak intensity ratios were used to compare acceptor concentrations.

Effect of GaCl mole fraction

The electrical properties of layers grown with different GaCl mole fractions are summarised in Figure 2. Impurity incorporation increases with decreasing mole fraction in accord with the earlier observations of Abrokwah et al (1983) and Kennedy et al (1974). At the highest mole fraction a layer with μ_{77} = 154,000 $cm^2V^{-1}s^{-1}$ and n = $6.2x10^{13}$ cm^{-3} and N_A/N_D = 0.25 was obtained. N_A increases and mobility decreases with decreasing mole fraction and a highly compensated layer with N_A/N_D of 0.9 and μ_{77} = 75,000 $cm^2V^{-1}s^{-1}$ resulted at the lowest mole fraction. The PL spectra showed the presence of dominant Zn and C acceptors and lower concentrations of Ge in all layers (Figure 3). However, in contrast to the N_A dependence, C acceptor concentrations decrease with decreasing mole fraction while the Zn and Ge concentrations remain constant.

Effect of T_G

The influence of T_G on impurity incorporation was studied over the range

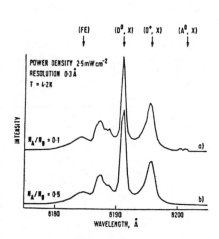

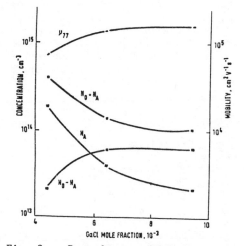

Fig 1: Excitonic region of PL spectrum for two undoped samples:
(a) $n=5.2 \times 10^{13}$ cm^{-3}, $\mu_{77}=170,000$ cm^2 V^{-1} s^{-1} $N_A/N_D=0.1$
(b) $n=4.3 \times 10^{13}$ cm^{-3}, $\mu_{77}=138,000$ cm^2 V^{-1} s^{-1} $N_A/N_D=0.5$

Fig 2: Dependence of N_D-N_A, μ_{77}, N_D+N_A and N_A on reactant mole fraction (Ga/As=0.6)

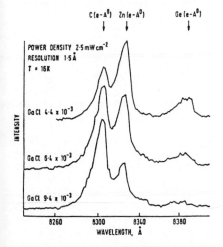

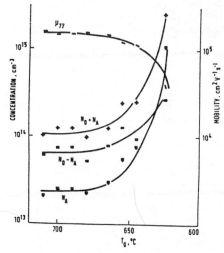

Fig 3: Band-to-acceptor region of PL spectrum for undoped GaAs layers grown with different GaCl mole fractions (Ts 850°C T$_G$ 710°C Ga/As=0.6)

Fig 4: Growth temperature (T$_G$) dependence of N_D+N_A, μ_{77}, N_A and N_D-N_A (Ts 850°C, GaCl mole fraction 9.4×10^{-3}, Ga/As=0.6)

710-635°C using high reactant mole fractions (Figure 4) N_D+N_A remains virtually constant 9.9-12.9x10^{13} cm^{-3} as T$_G$ decreased from 710-635°C with $\mu_{77} = 141-153,000$ cm^{2}V^{-1}s^{-1}. At T$_G$ K 665°C an increase in N_D+N_A is evident and a layer with $\mu_{77} = 40,000$ cm^{2}v^{-1}s^{-1} and $N_A/N_D = 0.8$ was obtained at 635°C. This compares with compensation ratios of 0.25-0.3 for layers grown at higher temperatures .

Only C and Zn acceptors were detected in the PL spectra of layers grown over the whole temperature range except at T_G 710°C when low concentrations of Ge were also detected. Between 710-665°C the concentration of Zn and C acceptors is almost independent of T_G with C/Zn ratio of 2-3. An increase in Zn and particularly C incorporation is apparent at T_G<665°C and C/Zn ratio of 10 obtained at 635°C. However, the magnitude of this increase in acceptor concentration is factor of 5 less than that indicated by increase in N_A.

Effect of Ts

High mobility layers (μ_{77} = 170-127,000 cm^{2}V^{-1}s^{-1}) were obtained throughout the range Ts 870-760°C. The highest mobility and least compensated layer N_A/N_D 0.1 was obtained at Ts 870°C. On decreasing Ts, N_A increases reaching a maximum (N_A/N_D = 0.5) at Ts ~800°C and then decreases as Ts is lowered further (Figure 5). This behaviour perhaps indicates that the Ga source getters acceptor impurities from H_2 or HCl. Impurity gettering of HCl by the In source has been proposed by Usui and Watanabe (1983) in the In/HCl/PH$_3$ system to account for increase in InP layer mobility with increasing source temperature.

However, this temperature dependence of acceptor incorporation is not reflected in the PL spectra. C and Zn are again dominant with lower levels of Ge acceptors present. Ge concentrations are independent of Ts and although a small increase in C acceptor concentrations is evident at Ts 800-760°C it is less than the increase observed in N_A. Furthermore, the highest Zn acceptor concentration was found for layer grown at T_S = 860°C whereas this layer had the lowest N_A value 0.6x10^{13} cm^{-3}.

Effect of secondary HCl

HCl is often injected directly into the growth zone in hydride growth of GaAs to lower the vapour phase supersaturation and hence suppress wall deposition and improve reproducibility of the growth rate (Mizutani and Watanabe 1982). However, all the layers described in the earlier sections were grown without HCl injection since as shown in Table 1 it was found to result in a marked increase in N_D+N_A (~2x10^{15} cm^{-3}), in agreement with observations of Kennedy et al (1984) and Abrokwah et al (1983). This behaviour is contrary to the model of Dilorenzo (1971) which predicts the suppression of impurity incorporation by HCl by stabilisation of the chlorides of elements such as Si, Zn and Ge. Furthermore, the PL spectra indicate a marked increase in Ge acceptor concentration in the presence of HCl whereas C+Zn concentrations are unaffected (Figure 6). This increase in Ge contamination may account for the increase in donor incorporation also observed with HCl injection since Ge donors are considerably in excess of Ge acceptors in VPE growth of GaAs (Low et al 1983, Skromme et al 1983).

(Ts=850, T_G=700°C, GaCl mole fraction 9.4x10^{-3}, Ga/As=0.6)

HCl mole fraction	μ_{77} cm^{2}V^{-1}s^{-1}	N_D-N_A (10^{13} cm^{-3})	N_D+N_A (10^{13} cm^{-3})
-	136,000	8.6	14.2
2x10^{-3}	54,000	110	184

Table 1: Electrical properties of undoped layers grown with and without HCl injection

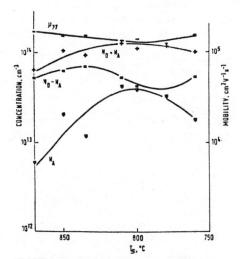

Fig 5: Source temperature (Ts) dependence of N_D+N_A, N_D-N_A μ_{77} and N_A (T_G 710°C, GaCl mole fraction 9.4x10^{-3}, Ga/As=0.6)

Fig 6: Band-to-acceptor region of PL spectrum for undoped GaAs grown with and without secondary HCl (Ts 850°C, T_G 700°C GaCl mole fraction 9.4x10^{-3}, Ga/As=0.6)

The origin of the Ge contamination is uncertain but it may be present in HCl and gettered by the Ga source under normal growth conditions (i.e. without secondary HCl). However, no variation in Ge acceptor concentration was apparent in layers grown at different source temperatures which would be expected to vary the gettering efficiency. Alternatively, Ge could be always present in the the vapour phase in the growth zone from HCl or AsH$_3$ and the kinetics of Ge incorporation enhanced in presence of HCl.

Discussion

Although growth conditions were observed to have a marked influence on N_A, this was not always apparent in the PL spectra. Inspection of Figure 7

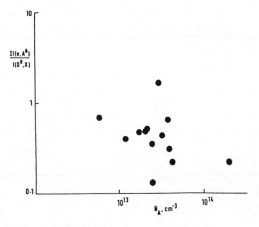

Fig 7: Total normalised (e-A°) peak intensities for Zn, C and Ge versus N_A for undoped GaAs

shows there is no consistent correlation between N_A and the total normalised (e-A°) peak intensities. These N_A values are probably overestimates for a number of reasons : 1). The mobility analysis assumes that deviations from the ideal theoretical mobility are solely due to additional ionised impurities whereas other mechanisms such as space charge and neutral impurity scattering could contribute : 2). The neglect of surface and interface depletion will lead to an underestimate of N_D-N_A since electrical thickness is less than metallurgical thickness : 3). The N_A value is an averaged value over the layer thickness and N_D+N_A may increase towards the substrate (Komiya and Wagner 1977) whereas PL measurements only give indication of acceptors in top 1 μm of the layer.

However, despite these inaccuracies the true N_A values are surely proportional to the derived N_A values for layers of similar doping and thickness and the use of corrected values is unlikely to improve the correlation with (e-A°) peak intensities. We believe that this lack of correlation may indicate that N_A is dominated by deep acceptors and that shallow acceptor concentrations $<10^{13}$ cm^{-3}. Thus variations in C, Zn and Ge acceptor concentrations apparent from PL spectra have no influence on N_A, typically $2-8 \times 10^{13}$ cm^{-3} for the majority of layers. Further evidence of the low levels of shallow acceptors is provided by the excitonic region of the PL spectrum (Figure 1) where very weak or non existent (A°X) peaks due to excitons bound to shallow acceptors are observed. DLTS and PL measurements of the deep acceptor region will be performed to examine this hypothesis and identify the deep acceptors.

Conclusions

High purity undoped GaAs (μ_{77} = 150-170,000 cm^{2}V^{-1}s^{-1}) has been grown under various growth conditions by hydride VPE. PL measurements have shown that C and Zn are the dominant shallow acceptors with lower concentrations of Ge. Injection of HCl into growth zone increases background doping and raises concentration of Ge acceptors. The weakness of (A°X) peaks in the PL spectra and the lack of correlation between N_A and (e-A°) peak intensities indicates that N_A may be dominated by deep acceptors with Zn, C and Ge concentrations $<10^{13}$ cm^{-3}.

References

Abrokwah J K, Peck T N, Walterson R A, Stillman G E, Low T S and Skromme B 1983 J. Electronic Mat. 12 681
Bartle D C, Tayrani R, Stewart C P and Grange J D 1984 GaAs and Related Compounds Inst. Phys. Conf. Ser. 74 521
Colter P C, Look D C and Reynolds D C 1983 Appl. Phys. Lett. 43 282
Dilorenzo J V and Moore G E 1971 J. Electrochem. Soc. 118 1823
Kamiya T and Wagner E 1977 J. Appl. Phys. 48 1928 1977
Kennedy J K. Potter W D and Davies D E 1974 J. Cryst. Growth 24/25 233
Low T S, Skromme B J and Stillman G E 1983 GaAs and related compounds 1982 (Inst. of Physics Conf. Ser. 65) 515
Mizutani T and Watanabe H 1982 J. Cryst. Growth 59 507
Skromme B J, Low T S, Roth T J, Stillman G E, Kennedy J K and Abrokwah J K 1983 J. Electron. Mater. 12 433
Usui A and Watanabe H 1983 J. Electronic Mat. 12 891
Walukiewicz W, Lagowski J and Gatos H C 1982 J. Appl. Phys. 53 769

Inst. Phys. Conf. Ser. No. 74: Chapter 3
Paper presented at Int. Symp. GaAs and Related Compounds, Biarritz, 1984

217

The influence of V:III flux ratio on unintentional impurity incorporation during molecular beam epitaxial growth

[a,c] S.C. Palmateer, [a] P.A. Maki, [b] W. Katz
[d] A.R. Calawa, [c] J.C.M. Hwang and [a] L.F. Eastman

Abstract. In this study, we show that surface stoichiometry is a dominant factor affecting unintentional impurity incorporation during molecular beam epitaxial growth. Epitaxial layers were grown under different V:III flux ratios and on various GaAs substrate types. Secondary
ion mass spectrometry measurements show that growth of GaAs at 600°C
with a 15:1 V:III flux (beam equivalent pressure - (P_{BE})) ratio results
in a surface accumulation of carbon, sulfur and manganese. This accumulation can be eliminated by reducing the V:III flux (P_{BE}) ratio to 10:1.
The influence of stoichiometry on GaAs/AlGaAs interface growth and the
growth of high purity GaAs has been investigated.

1. Introduction

A large body of experimental evidence has accumulated with regard to the
anomalous outdiffusion of impurities in compound semiconductors during high
temperature processing and epitaxial growth (Yu 1981, Klein et al 1980,
Huber et al 1982, Vasudev et al 1980, Tuck et al 1979). The accumulation
of these unintentional impurities at the surface of substrates can lead to
substrate epilayer interface buildup (Lum et al 1977, Palmateer 1982),
outdiffusion into epitaxial layers during growth (Anthony 1982, Huber et
al 1982) and surface accumulation (DeSimone 1980, DeSimone 1982) as observed for Cr in GaAs (Morkoc et al 1980) and AlGaAs, and Mn and Fe in GaAs
(Palmateer 1982, Covington et al 1980). This redistribution of unintentional substrate related impurities in epilayers during growth has been
shown to seriously influence device performance (Hyder et al 1982,
Palmateer et al 1982).

Crystal stoichiometry appears to be a dominant factor controlling the incorporation and redistribution of foreign impurities and intrinsic defects
in GaAs crystals. The semi-insulating properties of a GaAs ingot are determined by the melt stoichiometry during bulk liquid-encapsulated
Czochralski growth (Holmes et al 1982). The annealing of GaAs substrates
(Cr-doped and undoped) with a controlled surface stoichiometry produces a
surface accumulation of impurities, such as Mn (Palmateer et al 1983).
Control of surface stoichiometry during MBE growth has also been shown to
influence the incorporation of unintentional impurities at GaAs/AlGaAs
interfaces (Maki et al 1983).

An understanding of the diffusion mechanisms is needed to devise a way of
controlling atomic movement of impurities during growth. In this paper, we
show the influence of surface stoichiometry (by changing the V:III flux
ratio) on unintentional impurity redistribution in MBE layers. Results are
presented for epilayers grown on high pressure liquid encapsulated
Czochralski (HPLEC) and Cr-doped Bridgman substrates; as evaluated by Hall
measurements, photoluminescence, and Secondary Ion Mass Spectrometry
(SIMS). We also discuss the importance of growing MBE material at near
stoichiometric conditions.

2. Experimental

The epitaxial layers were grown in a Varian Gen II MBE system. The substrates used in this study were Cr-doped Bridgman GaAs and lightly Cr-doped HPLEC. The annealed and non-annealed substrates came from the same wafer.

Annealed substrates were prepared by heating at 750°C for 24 hours in palladium diffused H_2 in a slotted graphite block (Palmateer et al 1983). The surfaces were subsequently repolished using chlorox, removing approximately 1 - 3 mils of material.

Typical procedures for MBE growth were used as described elsewhere (Palmateer 1982). The holder was rotated during the growth of all layers (5 - 10 RPM). The doping and thickness uniformity has been measured using capacitance-voltage profiling techniques, and is ± 2% across a 2" wafer. GaAs epilayers were grown at 600°C with the V:III flux (P_{BE}) ratio specified throughout the text. Si was used as the n-type dopant. Elemental arsenic was used as the arsenic source. The undoped layers grown in this system are semi-insulating or low 10^{13} at/cm^3 p-type.

Secondary Ion Mass Spectrometry (SIMS) measurements were done on a Cameca IMS-3F system. Hall measurements were done by forming a clover leaf pattern and alloying indium or tin in H_2 at 450°C for 2 minutes. 4K photoluminescence studies were performed using the 6328Å line from an argon laser.

3. Results

3.1 The Effect of V:III Flux Ratio on Impurity Redistribution During MBE Growth

Undoped epilayers ~ 1.7 microns thick were grown under different V:III flux ratios, and on annealed and unannealed substrates. A 10:1 (P_{BE}) flux ratio corresponds to a Ga stable (3X1) RHEED pattern, while a 15:1 (P_{BE}) flux ratio corresponds to an As stable (4X2) pattern.

Figure 1 shows SIMS scans of undoped layers grown under a V:III flux (P_{BE}) ratio of 15:1. Scan 1(a) is of the epilayer grown on an unannealed substrate, whereas 1(b) is of an epilayer on an annealed substrate. There is a large surface accumulation of C,S and Mn at the surface of the epilayer grown on an unannealed substrate. This surface accumulation is eliminated when the epilayer is grown on an annealed substrate. Silicon is observed to accumulate both at the surface of epilayers grown on annealed and unannealed substrates. Other experiments indicate that this surface accumulation of Si is related to the surface electric field during epitaxial growth (Palmateer to be published).

Figure 2 shows SIMS scans of undoped epilayers grown under a (10:1) V:III flux (P_{BE}) ratio on unannealed and annealed substrates. In these scans, there is no large surface accumulation of C,S, or Mn. There is a Si accumulation as in the previous scans (Figure 1a, b).

3.2 The Effect of Arsenic Flux on the Growth of High Purity Gallium Arsenide.

The effect of arsenic flux and substrate type on the growth of high purity Si-doped epilayers has been investigated. Table I lists a set of layers which data is presented on.

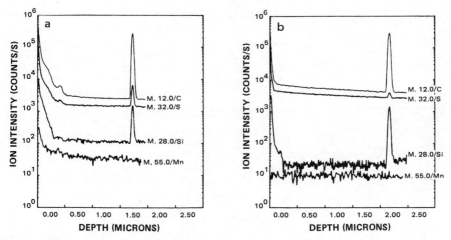

Fig. 1 SIMS scans of undoped MBE layers grown with a V:III (P_{BE}) flux ratio of 15:1; (a) epilayer grown on an unannealed substrate and (b) epilayer grown on an annealed substrate.

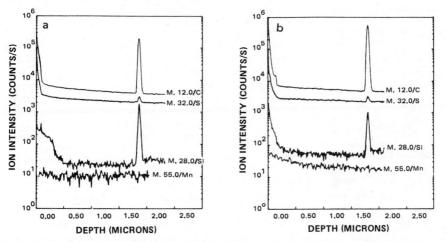

Fig. 2 SIMS scans of undoped MBE layers grown with a V:III (P_{BE}) flux ratio of 10:1; (a) epilayer grown on an unannealed substrate and (b) epilayer grown on an annealed substrate.

TABLE I - Si-doped GaAs Epilayers

Layer #	Thickness	300K (cm^2/V.sec)	Nd-Na (at/cm^3)	77K (cm^2/V.sec)	Nd-Na (at/cm^3)	k=Na/Nd
88HPLEC	19 μ m	high resistivity		high resistivity		
88CrB	19 μ m	7,639	1×10^{14}	131,185	1×10^{14}	.530
89HPLEC	10 μ m	9,800	1×10^{14}	147,000	1×10^{14}	.536
89CrB	19 μ m	7,834	1×10^{14}	146,000	1×10^{14}	.565

Layers 88 HPLEC/88 CrB and layers 89 HPLEC/89 CrB were each grown simultaneously. HPLEC is a lightly Cr-doped high pressure liquid encapsulated Czochralski grown substrate, while CrB is a Cr-doped Bridgman substrate.

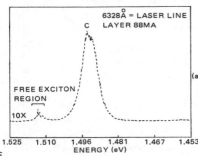

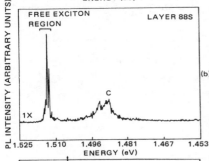

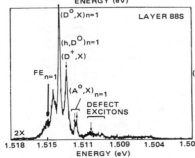

Fig. 3 4K photoluminescence spectra of Si-doped (N_d-N_a ~1×10^{14} at/cm³) GaAs layers (a) grown on a lightly Cr-doped HPLEC substrate, (b) grown on a Cr-doped Bridgman substrate, (c) grown on a Cr-doped Bridgman substrate (exciton region expanded).

Growth under a low As flux (V:III flux (P_{BE}) ratio of 10:1) results in increased incorporation of C as an acceptor (as seen in PL scan, Figure 3a). Layer 88 HPLEC was highly resistive, whereas 88 CrB grown simultaneously was n-type as determined by Hall measurements. The overall PL intensity for layer 88 HPLEC is reduced compared to 88 CrB. The ratio of the intensities of the C signal to exciton region is different in the two samples as seen in Figure 3(a) and 3(b). The insulating behavior of sample 88 HPLEC indicates that there may be a higher concentration of carbon in the LEC substrate which can outdiffuse and be incorporated as an acceptor during MBE growth. Figure 3(c) shows the exciton region of sample 88 CrB. This is a typical 4K PL scan of a high purity n-type (Si-doped) sample. The full width half maximum of the $(D^\circ,X)_{n=1}$ peak is 0.25 meV, indicative of high quality material.

Variations in mobility and carrier density were observed for epilayers grown on substrates which exhibited Mn and Cu in PL scans taken prior to growth. Epilayers grown on GaAs substrates with low levels of impurity contamination exhibit excellent uniformity in mobility and carrier concentration across a 2" wafer. Overall, in a high purity MBE system high mobility Si-doped layers can be grown reproducibly regardless of substrate type.

Figure 4 is a plot of mobility vs. carrier concentration for n-type GaAs. The solid line is a theoretical plot (Wolfe et al 1982) while the points represent experimental data for Si-doped layers grown on this Gen II MBE system. For Nd -Na ~ 1×10^{14} at/cm³ a 77K mobility of 147,000cm²/V.sec has been measured for Si-doped GaAs layers. This is the highest 77K bulk mobility measured for MBE grown material reported to date.

4. Discussion

Carbon, sulfur and manganese surface accumulate under increased V:III flux ratios in undoped MBE layers. The surface accumulation of certain impurity species used as intentional dopants during MBE growth have been reported previously (i.e., Si, Sn, Mn and Fe) (Palmateer to be published, DeSimone 1980, DeSimone 1982, Covington et al 1982). The mechanism or driving force for this surface accumulation is not known. Increased surface accumulation

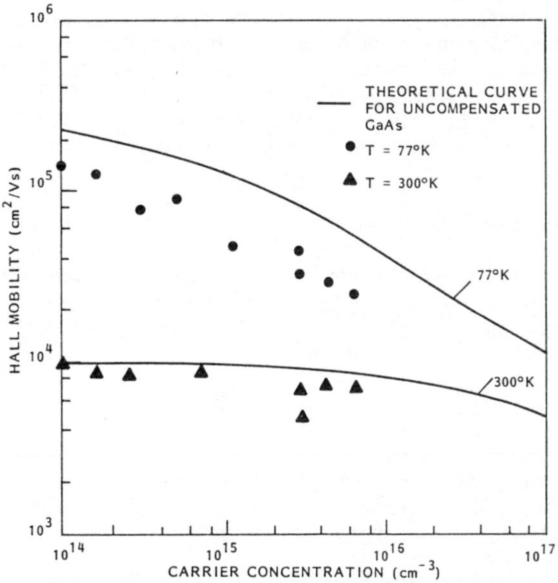

Fig.4 Plot of theoretical and experimental 300K and 77K mobility versus carrier concentration for n-type Si-doped GaAs.

observed with increased V:III flux ratio suggests that the diffusion is dependent on the Ga vacancy concentration during growth. To minimize the surface accumulation of unintentional impurities during growth, it is best to grow with a low V:III flux ratio. Another alternative to control surface accumulation is to preanneal substrates prior to growth.

The elimination of this surface accumulation is important for interface growth. We have shown in a previous paper that these impurities can degrade AlGaAs/GaAs interface quality (Maki et al 1983). The quality of the interface was found to be dependent upon the surface stoichiometry during interface growth. The present work indicates the importance of buffer layer growth at near stoichiometric conditions such that impurities available for incorporation are reduced.

The stoichiometry during the growth of low doped silicon ($\sim 1 \times 10^{14}$ at/cm^3) and undoped MBE GaAs layers strongly influences the resulting electrical properties of these layers. Substrate type and quality is also observed to affect the electrical properties of these layers. Poor substrate quality results in mobility and carrier density variations in low doped epilayers across a 2" wafer.

5. Conclusions

To grow the highest quality epitaxial layers, it is important to closely control stoichiometry during growth and preanneal the GaAs substrates. Previous work has shown that controlling impurity redistribution is necessary for the reproducible growth of certain epitaxial structures. We have demonstrated the importance of controlling surface stoichiometry during MBE growth. Impurity redistribution is influenced by the V:III flux ratio

possibly through changes in the vacancy concentration. This work indicates
that high quality buffer layers and active regions should be grown at near
stoichiometric conditions so as to minimize material degradation due to
impurity incorporation and redistribution. The growth of high mobility
modulation doped structures is essentially a problem of GaAs buffer layer
growth. We have reproducibly achieved extremely high mobilities (4K ~1.1
million cm^2/V.sec in the dark) by closely controlling growth stoichiometry
and substrate type.

6. Acknowledgements

The authors thank Dr. G. Wicks for photoluminescence measurements.

Parts of the work was supported by the Air Force Office of Scientific Research and the Office of Naval Research.

References

a Cornell University, Ithaca, New York
b General Electric, CRD, Schenectady, New York
c General Electric, Electronics Laboratory, Syracuse, New York
d Lincoln Laboratory, Massachusetts Institute of Technology, Lexington,
   Massachusetts
Anthony P J, 1982 Solid-State Electronics, 25 (10) 1003-1004.
Covington D W, Meeks E L and Hicklin W H, 1980
   Contract No. N00173-79-C-0033.
DeSimone D, 1980 Masters Thesis, Cornell University.
DeSimone D, 1982 Ph.D Thesis, Cornell University.
Holmes D E, Chen R T, Elliot K R, Kirkpatrtick C G and Yu P W. 1982 IEEE
   Transactions on Electron Devices, ED-29, (7), 1045.
Huber A M, Morillot G, Merenda P, Perrocheau J, Debrun J L, Valladon M and
   Koemnerer D, Institute of Physics, Conference Series No. 56, Chapter 8.
Huber A M, Morillot G, Bonnet M, Merenda P and Bessonneau G, 1982 Appl.
   Phys. Lett., 41, (7).
Hyder S B, 1982, J. Crystal Growth, 56, 369-375.
Klein P B, Nordquist P E R and Siebermann P G, 1980 J. Appl. Phys., 51 (9),
   4861.
Lum W Y, Weider H H, Koschel W H, Bishop S G and McCombe B D, 1977 Appl.
   Phys. Lett., 30, 1.
Maki P A, Palmateer S C, Wicks G W, Eastman L F and Calawa A R, 1983 J
   Electronic Mat., 12 (6), 1051.
Morkoc H, Hopkins C, Evans C A Jr., and Cho A Y, 1980 J. Appl. Phys., 51,
   (11), 5986.
Palmateer S C, 1982 Masters Thesis, Cornell University.
Palmateer S C, Maki P A, Hollis M, Eastman L F, Ward I and Hitzman C, 1982
   Proceedings 1982 Inter. Symp. on Gallium Arsenide and Related Compounds,
   pg. 149-156.
Palmateer S C, Schaff W J, Galuska A, Berry J and Eastman L F, 1983 Appl.
   Phys. Lett., 42 (2), 183.
Palmateer S C, to be published Ph. D Thesis, Cornell University.
Rumsby D, Ware R M, Smith B, Tyjberg M, Brozel M R and Foulkes E J, 1983
   GaAs IC Symp., Technical Digest p. 34.
Tuck B, Adegboyega G A, Jay P R and Cardwell M J, 1979 Institute of Physics
   Conference Series No. 45, Chapter 2.
Vasudev P K, Wilson R G and Evans C A Jr., 1980 Appl. Phys. Lett., 36 (10).
Wolfe C M and Stillman G E, 1982 Appl. Phys. Lett., 40, 6.
Yu P W, 1981 J. Appl. Phys., 52, (9), 5786.

Inst. Phys. Conf. Ser. No. 74: Chapter 3
Paper presented at Int. Symp. GaAs and Related Compounds, Biarritz, 1984

Study of the impurity redistribution in InP metalorganic chemical vapor deposited epilayers

A.M. HUBER, M. RAZEGHI, G. MORILLOT
THOMSON-CSF BP 10 - 91401 ORSAY - FRANCE -

Abstract. This paper reports new experimental results using secondary ion mass spectrometry analysis for quantitative determination of impurities in MOCVD InP layers. Mg, Mn, Fe, Cr and Si accumulation at the substrate-epilayer interface was observed. The origin of these elements was determined. The major source of impurities was the adsorption of atoms on the substrate surface during chemical etching prior to epitaxy.

1. Introduction.

In recent years InP has received much attention as a promising material for use in high-speed logic circuits and long wavelength (1,3 - 1,6 um) optical communication systems. Metalorganic chemical vapor deposition (MOCVD) is currently used for the preparation of this material. MOCVD has several potential advantages, namely, to grow layers uniform in thickness and doping level over a large area of substrate. In spite of great improvements in the purity of the organometallic sources, the material quality is not always reproducible.
Ogura et al (1984), by the deep level transient spectroscopy (DLTS) technique, have determined deep levels in undoped n-type InP layers grown by MOCVD on Fe doped semi-insulating InP substrates. They observed two kinds of electron trap, F_1 and F_2, with activation energies of 0,48 eV and 0,78 eV and they attributed these to a donor Fe complex center and to an Fe related defect respectively.
High resolution and low temperature photoluminescence, low temperature Hall effect measurements and DLTS are the currently used characterization techniques for MOCVD InP layers (Fukui et al 1980, Rao et al 1983, Razeghi et al 1983, Hsu et al 1983, Mircea et al 1984). Nevertheless, these are physical methods and the results are not always evident for the parasitic impurity determination necessary for material optimization.
For several years secondary ion mass spectrometry (SIMS) has been successfully applied in the electronics industry to the chemical measurement of impurity profiles in the III-V semiconductor materials.However, reports concerning the characterization of MOCVD InP layers have not yet been published. In this paper we present the results of SIMS studies of MOCVD InP epilayers which are : i) presence of impurities at epilayer-substrate interface ii) effect of some growth parameters on the impurity redistribution.

2. Experimental procedure.

2.1. SIMS measurement.

The analyses in our study are carried out by a modified CAMECA IMS 3F.

The replacement in the primary and secondary ion optics of the stainless steel apertures by tantalum, was one of the modifications. As a result of this the background of Fe and Cr diminished by a factor of 100 and an about 5×10^{12} at. cm^{-3} limit of detection was obtained in InP for these elements. This sensitivity is necessary for a better evaluation of InP because Fe and Cr are deep acceptor level elements in InP.

In our analysis of InP, the surface is scanned with a focused mass filtered oxygen ion beam (Ip ~ 1,5 uA at 10 keV). The scanned area is 250x250 um and the analysed region is 150 um in diameter. Offset target voltage -50V was applied to avoid possible interference with ionized hydrocarbons. Ion implanted samples were the standard for quantitative calibration of the instrument. The statistical results of various experiments show that the quantitative results of SIMS are given with an accuracy of $\pm 20\%$ above a concentration level of 1×10^{16} at. cm^{-3}. Below this level, results are less accurate, $\pm 50\%$ at 1×10^{14} at. cm^3. Talysurf measured depth precision is estimated at $\pm 10\%$.

The detection limit of the impurities which were measured are : Mg, Cr, Mn 5×10^{12} at. cm^{-3}, Fe 1×10^{13} at. cm^{-3}, Si 7×10^{13} at. cm^{-3}.

2.2. Growth condition of samples.

The epilayers used in our study were grown by the MOCVD low pressure method (Razeghi et al 1983). The growth apparatus and process have been reported in detail (Razeghi et al 1983). The chemical pretreatment of the substrates prior to epitaxial growth is as follows : dipping the wafer in H_2SO_4 for 3 minutes with ultrasonic agitation, rinsing in D.I. water and in hot methanol, dipping in 0,3% bromine methanol for 3 minutes and rinsing in hot methanol and finally dipping in H_2SO_4, rinsing in D.I. water and in hot methanol.

At a growth temperature of 550°C the surface of the layers is generally mirrorlike.

3. Experimental results.

3.1. Determination of impurities at the interface.

Each sample was analyzed in two different areas about 10 mm apart. Generally, analysis of two clean areas gives reproducible and representative results for the material. A complex profile situation is observed in MOCVD InP layers. Accumulation of impurities such as Mg, Fe, Cr, Mn and Si were detected by SIMS. A similar phenomenon was observed in MOCVD GaAs epitaxy (Huber et al 1982). In this case the major source of impurities was the adsorption of atoms during chemical etching prior to epitaxy (Huber et al 1984). The determination of the origin of this phenomenon was carried out in the same way as for GaAs.

The accumulation of impurities at the epilayer-substrate interface was believed to be due to one or more of three possible sources : i) outdiffusion from the substrate ii) incorporation from the ambient gas phase during preheating iii) absorption of impurities during chemical surface etching before epitaxy. It is possible that the phenomenon is partly related to the substrate (Kamada et al 1984). However a simple model of diffusion does not explain the observed high concentration of impurities.

The possibility of substrate contamination by exposure to air and heating in the MOCVD reactor was examined as follows : a 3 um thick layer of InP was grown and the wafer was removed from the reactor. After 30 minutes in the air the wafer was again placed in the reactor and a second epilayer

3,5 µm thick was grown. Figure 1 shows the " pile up " of impurities at the epilayer-substrate interface where peaks in the concentration of Mg, Cr, Mn, Fe, Si were analysed. At the epilayer-epilayer interface however, where the growth was stopped, the only concentration change was some local increase in Si level. These experiments indicate that exposure to air and heating under H_2/PH_3 pressure in the MOCVD reactor does not account for the accumulation of impurities.

The possibility of contamination of the substrate by the pre-growth chemical treatment was tested by growing a thick (10 µm)InP layer in the normal way. The pre-growth etch was then used to remove 2 µm of this first layer and a second epilayer was grown. The impurity profile of this sample is shown in figure 2. One can observe similar accumulations of Mg, Cr, Fe and Mn at the epi-substrate and epi-epi interfaces (Si was not analysed here). We conclude that the majority of impurities result from the chemical etching even though very pure reagents were used (e.g : " SUPRAPURE " MERCK H_2SO_4, Fe $(10^{-6}\%)$, Cr $(10^{-7}\%)$ and Mg $(10^{-6}\%)$conc).

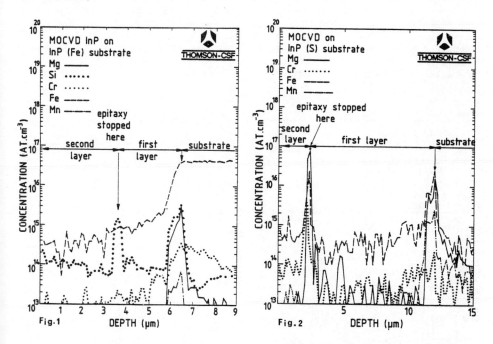

Fig 1. Depth profiles of Mg, Si, Cr, Fe and Mn in a MOCVD InP layer grown on InP (Fe) substrate. Epitaxy was stopped, the wafer was taken out of the reactor. A second layer was grown on the first.
Fig. 2. Depth profiles of Mg, Cr, Fe and Mn in a MOCVD InP layer grown on InP (S) substrate. The first layer was partly removed by controlled chemical etching. A second layer was grown on the chemically treated first epilayer.

3.2. The effect of some growth parameters on impurity redistribution.

3.2.1. In situ HCl etching prior to layer growth.
The impurities accumulated at the interface diffuse slightly during our " standard process" and layers are frequently very pure. Si and Fe about

1-2 $\times 10^{14}$at.cm^{-3}, Mg, Cr and Mn < 1×10^{13}at.cm^{-3}were analysed. However, impurity peaks at the interface in some cases ~ 10^{17}at.cm^{-3} could have some consequence on device quality (Huber et al 1984). Whiteley et al (1983) observe a fall in electron mobility in the vicinity of the interface GaInAs-InP grown by MOCVD. They propose using in situ HCl etching of InP prior to layer growth.Our experiment was carried out as follows : a 10/um thick layer of InP was grown and the wafer was removed from the reactor. A small part of the wafer was cleaved and kept for thickness measurement. The rest of the wafer was again placed in the reactor and in situ HCl etched during 1 minute at 550°C. (Volumetric flow of PH$_3$ 0,31/min, flow of HCl 40 cm^3/min in a flow of H$_2$ 21/min and N$_2$ 31/min).It was determined by comparing the thickness of etched and unetched layers, measured by SIMS, that a 1,7/um thick layer was removed.The peaks at the interface allowed accurate thickness determination of MOCVD layers by SIMS. The impurity profiles of this sample are presented in figure 3. The considerable peaks of Si, Fe, Mg, Mn and Cr at the epilayer-epilayer interface show elements incorporated from the HCl gas phase. In situ HCl etching without PH$_3$ flow created the same interface contamination but an enhanced impurity redistribution in the layer for Fe and Cr occurred as shown in figure 4. It is possible that in this case a surface with a large degree of non-stoichiometry could form and the layer could have a large density of phosphorus vacancies which accelerates impurity diffusion.

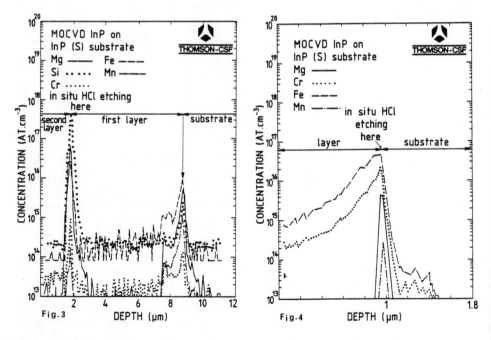

Fig.3. Depth profiles of Mg, Si, Cr, Fe and Mn in a MOCVD InP layer grown on InP (S) substrate. The first layer was partly removed by in situ HCl etching in PH$_3$/H$_2$ flow. A second layer was grown on the first.

Fig.4. Depth profiles of Mg, Cr, Fe and Mn in a MOCVD InP layer grown on InP (S) substrate. The substrate was in situ HCl etched without PH$_3$ in H$_2$.

3.2.2. Effect of doping elements of substrates.

InP layers on Fe doped semi-insulating, Sn and S (n $1x10^{18}$ at.cm^{-3})
doped substrates were deposited in the same run. Etch pit densities (EPD)
of substrates are $4x10^4$cm^{-2} $5x10^4$cm^{-2} and $9x10^3$cm^{-2} respectively. The
three (100) off 2° towards (110) oriented wafers were prepared
simultaneously side by side. Figure 5 shows Fe profiles. One can observe
that the Fe accumulation and the Fe concentration at the Fe doped substrate
epilayer interface and in the layer are about 10 times lower than in
the case of Sn or S doped substrates. Results are similar for Mg, Mn
and Si. The diverse levels of the same element in the layers grown in
the same run is most probably due to the substrate quality. The degree
of non-stoichiometry on the surface and in the substrates could influence
: i) the chemisorption ii) the crystal quality of layers, the formation
of vacancies and consequently the impurity redistribution. Layers grown
on Fe doped substrates often present lower impurity peaks at the interface
and lower level in layers. Nevertheless, we sometimes obtained similar
results on S and Sn doped substrates coming from different growths.

3.3.3. The effect of substrate orientation on the impurity redistribution.

 Growth of InP layers was carried out on Fe doped substrates oriented
(100) off 2°, 3° and 4° toward (110). Growth temperature here was 650°C
and the layers were doped by H$_2$S n 3.10^{17}at.cm^{-3}(300°K).The three crystals
were prepared simultaneously and layers were grown in the same run. Figure
6 shows the Fe depth profiles of layers. One can observe no significant
difference in Fe redistribution. Similar results were found concerning
Mg, Mn,Cr and Si.

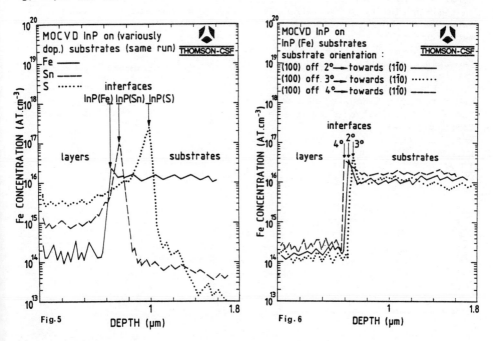

Fig. 5. Depth profiles of Fe on InP (Fe) InP (S) and InP (Sn) substrates.
Fig. 6. Depth profiles on InP (Fe) semi-insulating substrates oriented
(100) off 2°, 3° and 4° towards (110) .

4.Conclusion.

Using a very sensitive SIMS technique, the results presented here demonstrate the interface problem in the case of InP grown by MOCVD. It was shown that the adsorption of atoms during chemical etching before epitaxy represents a major source of impurities which accumulate at the epilayer-substrate interface. During in situ HCl etching the same phenomenon occurs. In some conditions, anomalous redistribution of these impurities in the InP layers was analysed. This is probably due to the non-stoichiometry of surfaces or substrates or both, rich in phosphorus vacancies, which enhance impurity diffusion. In the standard process, impurities accumulate at the interface, and diffuse very slightly into layers. In several cases, layers with Si, Fe $1-2 \times 10^{14}$at.cm^3 and Mg, Mn, Cr $< 1 \times 10^{13}$at.cm^{-3} were obtained. Further improvement in the material quality will require perfecting in situ HCl etching. The use of a buffer layer is not always a reliable solution.

5.Acknowledgements.

We would like to express our thanks to J.P. Hirtz and B. Lent for helpful discussions.

6.References

Fukui T, Horikoshi Y, 1980 Jpn. J. of Appl. Phys. 19 395.
Huber A M ,Morillot G,Hersee S.D.,Kazmierski K, Proc. of Third Conference on semi-insulating III-V materials. Oregon 1984 Edition Shiva 466.
Huber A M ,Morillot G,Bonnet M,Merenda P,Bessonneau G,1982 Appl. Phys. Lett. 41 638.
Hsu C C, Cohen R M, Stringfellow G B, 1983 J. Cryst. Growth 63 8
Kamada H, Shinoyama S, Katsui A, 1984 J. Appl. Phys. 55 2881
Mircea A, Azoulay R, Dugrand L, Mellet R, Rao E V K, Sacilotti M, 1984 J. of Electr. Mat. 13 603.
Ogura M,Mizuta M,Hase N,Kukimoto H,1984 Jpn. J. of Appl. Phys. 23 7.
Rao E V K ,Sibille A,Duhamel N, 1983 Physica (North-Holland Publishing Compagny) 116B 449.
Razeghi M,Duchemin J P, 1983 J. Cryst.Growth 64 76.
Razeghi M, Poisson M A , Larivan J P,Duchemin J P,1983,J.of Electr. Mat. 12 371.
Whiteley J S, Ghandi S K, 1983 Thin Solid Films 104 145.

Inst. Phys. Conf. Ser. No. 74: Chapter 4
Paper presented at Int. Symp. GaAs and Related Compounds, Biarritz, 1984

229

Experimental determination of velocity-field characteristic in N-GaAs and n GaInAs/InP using a new method based on the magnetoresistance effect

M.H. EVANNO, B. BOITTIAUX, A. SAFRIOUI and R. FAUQUEMBERGUE
Centre Hyperfrequences et Semiconducteurs, LA CNRS N° 287
UNIVERSITE DES SCIENCES ET TECHNIQUES DE LILLE
59655 VILLENEUVE D'ASCQ CEDEX – FRANCE

1. INTRODUCTION

The development of microwave devices at greaterand greater frequencies requires the utilisation of new semiconductor materials we need to characterize. In this way, the evolution of drift velocity versus electric field is
one of the most important criteria of their microwave performances, because
it determines the transit time of the carriers in the active layer and, indirectly, the quantity of transmissible informations. Since the discovery of
special velocity properties of Gallium Arsenide by J.B. Gunn (1963), III-V
direct gap semiconductor materials are expected to present greater carrier
velocity than silicon. Many attempts have been made for some years, especially by time of flight techniques which are based on the experimental determination of carrier transit time in a device (Ruch et al, 1967 – Houston
et al, 1977 – Evanno et al, 1982). However, these methods present some disadvantages such as special technology for devices and such as partial results for the technique ; particularly it is difficult to obtain the v(E)
characteristic at low fields.

Our purpose is therefore to present an experimental method using classical
F.E.T. planar structures and based on the classical magneto-resistance
effect, the validity of which is proved by Monte Carlo calculations:
relative variations of external current by application of a transverse
magnetic field is used to determine velocity field characteristic in
n GaAs and n GaInAs for various temperatures.

2. THE METHOD

In an infinite bias polarized semiconductor, a transverse magnetic field
induces the apparition of a Lorentz strength which bends the velocity vector v without changing its modulus, therefore the drift velocity is reduced
(fig 1).

We have investigated Monte Carlo calculations of this effect in GaAs and
the results show a reduction of drift velocity the greater as the magnetic
field increases for the same electric field and as the drift velocity increases for the same magnetic field (fig 2).

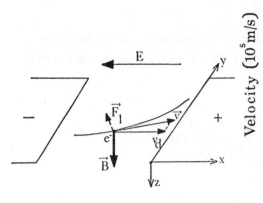

Fig. 1 : Classical magnetoresistance effect in the bulk

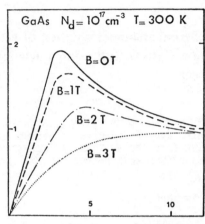

Electric field(kV/cm)

Fig. 2 : Effect of a transverse magnetic field on the v(E) characteristic

At low electric field, smaller than threshold field, that constitute the classical magnetoresistance effect and, as it is well known, the reduction of drift velocity v_d is given by

$$\frac{v_d(B)}{v_d(0)} = \frac{1}{1+\mu^2 B^2} \quad (1) \qquad \mu = \frac{v_d(0)}{E} \quad (2)$$

At higher field, Monte Carlo simulations allow us to generalize this formalism considering the secant mobility $\mu_s(E) = v_d(E)/E$ (3) and (1) becomes

$$\frac{v_d(B)}{v_d(0)} = \frac{1}{1+\mu_s^2 B^2} \quad (4)$$

Fig 2 shows that velocity field curves analytically obtained by (4) are in very good agreement with simulations for magnetic fields smaller than 3T.

In a finite bias polarized device, the free carrier concentration n, the electric field E and the carrier velocity v may depend on the position of the carriers in the active layer. Then, the value of external current is given by Ramo-Schockley law

$$I = \int_0^W \frac{e\,n(x)v(x)}{W}\,dx \quad (5)$$

where W is the length of the active layer. The application of a transverse magnetic field induces space variations of drift velocity in the active layer according to (4) and involves a reduction of the external current, the expression of which is given by combination of (4) and (5)

$$I(B) = \int_0^W \frac{e\,n(x)}{W}\,\frac{v_d(0)}{1+\mu_s^2 B^2}\,dx$$

Neglecting the diffusion current, the expression of relative variations of external current becomes, according to (6)

$$\frac{I(0)-I(B)}{I(B)} = <\mu_s^2>_x B^2 \quad (7)$$

$$< \mu_s^2 >_x = \frac{1}{w} \int \mu_s^2 \big(E(x)\big) dx \quad (8)$$

is the secant mobility mean square value in the active layer of the device.
Then, if we consider a development of $\mu_s^2(E)$ as a polynomial function of E

$$\mu^2(E) = \sum_{1}^{\infty} a_i E^i \quad (9)$$

the relative variations of current may be expressed as :

$$\frac{\Delta I}{I} = < \mu_s^2 >_x B^2 = \frac{B^2}{w} \int_o^w \sum_o^\infty a_i E^i(x)\, dx \quad (10)$$

$$\frac{\Delta I}{I} = \frac{B^2}{w} \left(a_o w + a_1 U + \int_o^w \sum_2^\infty a_i E^i dx \right) \quad (11)$$

Experimental results show that the variations of $< \mu_s^2 >_x$ versus applied voltage are piece-wise linear (see fig. 3). For each range of applied voltage labelled a, b, c, the terms of second and higher can be neglected in (11) so that the expression of relative order variations of external current may be approximated by:

$$\left(\frac{\Delta I}{I} \right)_{a,b,c} = \frac{B^2}{w} \left(a_{o_{a,b,c}} w + a_{1_{a,b,c}} U \right) \quad (12)$$

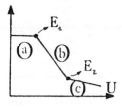 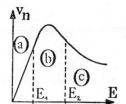

Fig 3 : Principle of the experimental method of $v_n(E)$ determination

The experimental determination of the constants a_o and a_1 for each zone of voltage variations allows us to obtain the secant mobility/electric field characteristic for electric fields corresponding to this voltage range and consequently the evolution of drift velocity

$$v_{a,b,c}(E) = \mu_{s_{a,b,c}}(E).E = \sqrt{a_{o_{a,b,c}} + a_{1_{a,b,c}} E}.E \quad (13)$$

3. THE SAMPLES AND THE EXPERIMENTAL SET-UP

Experimental determinations of v(E) have been performed by this method for n GaAs and n GaInAs/InP for temperatures ranging from 77 to 300 K.

For GaAs and GaInAs, the samples used in these measurements were both classical F.E.T. planar structures, the epitaxies of which were obtained by successive growth on a semi insulating substrate of a buffer and an active layer ; ohmic contacts to the substrate were made by electroplating and allowing Au.Ge eutectic (Fig 4).

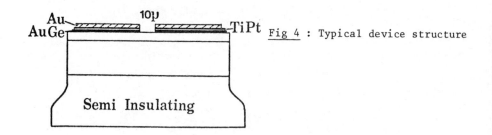

<u>Fig 4</u> : Typical device structure

For GaAs samples, the impurity concentration of active and buffer layer were respectively 10^{17} and 10^{14} cm^{-3} ; chemical etching of the active layer between the ohmic contacts up to the buffer layer allowed the determination of the $v_n(E)$ characteristic for the lowest impurity concentration.

The GaInAs/InP epitaxies so far   considered   are grown on an InP semi insulating substrate, have an InP buffer layer and a GaInAs active layer, the concentration of which is $2\ 10^{15}$ cm^{-3}.

All the devices were made by the technology team of the Centre Hyperfrequences et Semiconducteurs.

The experimental set up is really simple : it consists of an electromagnet in which the bias polarized device is mounted so that magnetic and electric field are orthogonal ; an amperemeter series connected in the polarization circuit of the device gives the current value.

4. <u>THE RESULTS</u>

Results are obtained for GaAs ($N_D = 10^{17}$ cm^{-3}) and GaInAs ($N_D = 2\ 10^{15}$cm^{-3}) at room temperature (300 K) and liquid nitrogen temperature (77 K).

All the velocity-field curves present the same shape as predicted by simulations obtained from band structure theory
 . a mobility zone at low field where $v = \mu E$
 . a maximum value of the velocity for the threshold field
 . a negative differential mobility zone where $dv/dE > 0$ up to the threshold field

They are presented on Fig 5 and 6 and the main parameters (low field mobility, maximum velocity value and threshold field) are recorded on Fig 7.

These results are compared with other experimental determinations (Ruch et al, 1967 - Gammel et al, 1981, Windhorn et al, 1982) and Monte Carlo calculations (Littlejohn et al, 1977 - Kaszynski, 1979) which are very close to them.

For GaInAs, they show one of the first complete experimental determination of $v_n(E)$ up to 10 kV/cm.

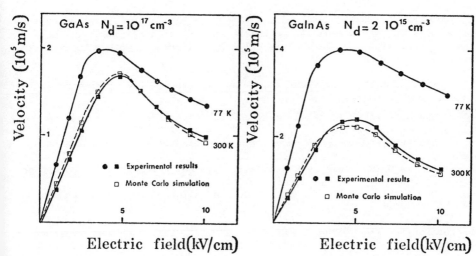

Fig 5 : Experimental $v_n(E)$ in GaAs

Fig 6 : Experimental $v_n(E)$ in GaInAs

	GaAs		GaInAs	
	77 K	300 K	77 K	300 K
μ (V/ms^2)	0.70	0.40	1.60	0.80
V_{max} (10^{5}m/s)	2.0	1.7	4.2	2.5
E_{th} (10^{5}V/m)	4.0	5.0	4.5	5.0

Fig 7 : Mean characteristic values of
         GaAs and GaInAs

5. CONCLUSION

The new method of velocity field characterization we have described has given results in GaAs and GaInAs which are in very good agreement to those previously obtained by Monte Carlo calculations and with other experimental results.

Moreover, as compared to other methods as for example time of flight techniques, some advantages can be pointed out :

. the working up is really easy and the experimental set-up needs only an electro-magnet
. the samples are photoconductors using classical FET's epitaxies and no peculiar structures are needed
. the magnetic field is the single parameter which is changing. Therefore the precision of measurement is better
. the method gives absolute results and allows the velocity determination at low fields

The interest of that method lies in the characterization of new semiconductor materials as III-V ternary and quaternary compounds the use of which depends the development of optical telecommunications and new structures as submicron devices which present theoretically predicted overshoot and ballistic properties (Ghis, 1983 - Fauquembergue, 1984).

REFERENCES

[1] J.B. Gunn, Solid State Comm. 1 p. 88-91 (1963)
[2] J.G. Ruch ; G.S. Kino, A.P.L. 10, 2, p 40-42 (1967)
[3] P.A. Houston ; A.G.R. Evans, S.S.E. 20 p 197.204 (1977)
[4] M.H. Evanno, J.L. Vaterkowski, Electronics Letters, 18, 10 p 417-418 (1982)
[5] A. Kaszynski, Thèse de Docteur Ingénieur, Lille, 1979
[6] M.A. Littlejohn, J.R. Hausen, T.H. Glisson, A.P.L., 30, 5, 1977
[7] J.C. Gammel Hideo Ohno, J.M. Ballantyne, IEEE QE 17, 2, 1981
[8] T.H. Windhorn, W. Cook, G.E Stillmann IEEE EDL.3, 1, 1982
[9] A. Ghis, E. Constant, B. Boittiaux, J. Appl. Phys., 54(1), p. 214, 1983
[10] R. Fauquembergue, M. Pernisek, A. Ghis, E. Constant, ESSDERC, Lille, 1984

Inst. Phys. Conf. Ser. No. 74: Chapter 4
Paper presented at Int. Symp. GaAs and Related Compounds, Biarritz, 1984

235

Characterisation of $Ga_{1-x}Al_xSb$ compound using high magnetic field and high pressure transport experiments

R.L. AULOMBARD*, L. KONCZEWICZ**[1], A. KADRI*, A. JOULLIE***, J.C. PORTAL****,

* Groupe d'Etude des Semiconducteurs, U.S.T.L., Montpellier, France.
** High Pressure Research Center, Warsaw, Poland.
*** E.M.M., U.S.T.L., Montpellier, France.
**** I.N.S.A., Département de Physique, Toulouse, France. S.N.C.I.,C.N.R.S., Grenoble, France.

Abstract. n-type $Ga_{1-x}Al_xSb$ alloys were prepared by a Bridgman method and by L.P.E. The Hall effect and resistivity as a function of magnetic field up to 30 T and hydrostatic pressure up to 20 kbar were measured in the temperature range 4.2 K - 300 K. The variation of the energetic distance between Γ and L minima was specified, and the values of the mobilities were determined.

1. Introduction

$Ga_{1-x}Al_xSb$ alloy is an attractive candidate for optoelectronic devices at long wavelengths in the spectral regions of 1.3 - 1.55 µm where the fused silica fibers exhibit lowest dispersion and loss. High performance avalanche photodiodes were prepared from this material (Law 1978-1981, Hildebrand 1981, Capasso 1981), as double heterostructure laser diodes with the GaAlSb/GaSb system (Dolginov 1981 - Tsang 1983). Recently Ohmori et al (1984) have demonstrated room temperature operation of $Ga_{0.83}Al_{0.17}Sb/GaSb$ multiquantum well lasers grown by MBE. Despite these encouraging published works, the $Ga_{1-x}Al_xSb$ alloy suffers from an insufficient knowledge of its band structure. A survey of the published literature indicates scattered results. One of the last published papers (Alibert, 1983) summarizes the situation : Γ and X band variations were quite well known, but no real direct determination of the L band variation was given in the alloy (the knowledge of the respective GaSb and AlSb is only known).

The precise knowledge of the band structure is necessary to evaluate its expected domain of application, so in order to define the transport characteristics of the central and satellite conduction band minima, we have prepared a series of Te doped $Ga_{1-x}Al_xSb$ crystals and studied the Hall effect and conductivity under high magnetic field (up to 30 T) and high hydrostatic pressure (up to 20 kbar) in the temperature range 4.2 K - 300 K.

2. Crystal growth techniques

$Ga_{1-x}Al_xSb$ n-type alloys were prepared by a Bridgman method and by liquid phase epitaxy. The Bridgman technique, described elsewhere (Aulombard, 1979)

(1) Present address : G.E.S., U.S.T.L., Montpellier, France.

furnished bulk polycrystalline ingots over the entire composition range. Homogeneous samples extracted from the first third of the ingot were selected for this study.

Single crystals of $Ga_{1-x}Al_xSb$ (x < 0.7) were grown by liquid phase epitaxy on undoped (111) B GaSb substrates. The n-type conductivity was obtained by tellurium doping by introducing Te doped GaSb crystals into the melt. The growth was performed using a conventional sliding graphite boat in a pure hydrogen atmosphere (Gautier, 1981). A graphite piston under a quartz rod maintained the flatness of the liquid solution of 3 mm height and prevented any Sb or Te loss during the heat treatment. The solution was saturated by a GaSb source (40 min) at 500°C (or 550°C); afterwards the source was withdrawn. After a period of another 40 min. the system was cooled (40°C/hour). This last operation, just before the cooling produced homogeneous liquid solutions. The substrate was introduced into the melt when the solution was 5°C super cooled. Large growth temperature intervals were used to obtain thick epilayers (50 μm).

The solid compositions were determined by quantitative electron microprobe analysis. The layer uniformity normal to the growth axis was controlled to be better than the accuracy of the microprobe analysis (± 1 atomic fraction). The effective segregation coefficient of tellurium was found to be dependent on the alloy composition and the growth temperature. As an example, Fig. 1 shows our measurements of the Te segregation coefficient for samples of GaSb and $Ga_{0.8}Al_{0.2}Sb$ grown at different temperatures.

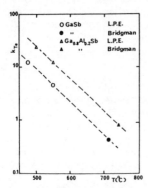

O GaSb	L.P.E.	
● "	Bridgman	
▲ $Ga_{0.8}Al_{0.2}Sb$	L.P.E.	
▲ "	Bridgman	

Fig. 1   Segregation coefficient of Te as a function of the growth temperature; k_{Te_S} was defined as $k_{Te} = X_{Te}^S/X_{Te}^L$ where X_{Te}^S and X_{Te}^L are the atomic fractions of Te in the solid and liquid phases.

3. Material characterization

We have made a variety of transport measurements on n-type $Ga_{1-x}Al_xSb$ bulk and epitaxial samples throughout the composition range to characterize the alloy system, and particularly the band parameters (crossover points of direct and indirect band - effective masses - mobilities).

The electrical characteristics of several studied samples are given in Table 1.

Sample	x	$R_H(cm^3/C)$		$\mu_H(cm^2/V.sec)$	
		T=300 K	T=77K	T=300 K	T=77 K
206 (B)	0.72	20.8	–	30	–
2 622 (B)	0.52	32.8	–	30	–
113 (B)	0.31	12.5	–	176	–
1 271 (E)	0.18	3.2	3.6	180	180
64 (B)	0.15	6.7	13.2	200	560
1 4 41 (E)	0.10	12.7	20.8	650	2350
1 412 (E)	0.03	8.5	8.3	1790	4800

Table 1. Hall coefficient R_H and Hall mobility $\mu_H(\mu_H=R_H/\rho)$ for several $Ga_{1-x}Al_xSb$ samples (B=Bridgman, E = L.P.E.).

Conventional bridge form samples with the sizes 5 x 0.1 x 0.5 mm³ have been used (for the epitaxial samples the non-insulating substrate has been eliminated). After chemical etching (2 % Br, 98 % methanol) ohmic contacts were soldered with pure tin or indium.

4. Experimental results

The Hall coefficient and the conductivity have been measured in the temperature range 4.2 K - 300 K using the standard D.C. method. For the investigation under pressure the samples were put inside a high pressure Be-Cu cell and helium gas was used as a pressure transmitting medium.

4.1 Low magnetic field results.

The variation of the apparent mobility $/\mu_H/$ for samples, having similar total carrier density ($n_T \approx 2.10^{18}$ cm^{-3}) is plotted versus Al content in Fig. 2.

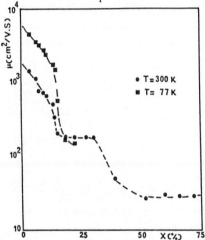

Though it is difficult to compare samples grown under different conditions, the results (Fig. 2) clearly show two plateau ranges in $\mu_H(x)$ dependence : the first one when Al content reaches 0.15 and the second one when it reaches 0.5. These results (μ_L #200 cm²/V.sec. and μ_X #30 cm²/V.sec. at T=300 K) may be supported by a band structure (Alibert, 1983), which gives two cross over points : from Γ to L at x=0.16 and from L to X at x=0.48.

Fig. 2. Apparent Hall mobility μ_H versus Al content for similar total carrier concentration samples.

The typical behaviours of the Hall effect under pressure up to 15 kbar for samples with x <0.2 are presented in Fig. 3. These results can be interpreted as a transfer of electrons from Γ to L minima. The pressure range P >7 kbar corresponds to the situation when all carriers are in the L minimum. The experimental pressure range was limited to 12 kbar because as we have noted in a previous paper (Konczewicz, 1982), for pressure higher than 15 kbar the deionization of Te donor impurity takes place. Moreover when the Al content reaches 0.3, these states are active at zero pressure.

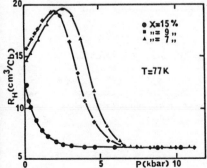

Fig. 3. Hall coefficient versus pressure for similar total carrier concentration samples.

4.2 High magnetic field experiments at low temperature.

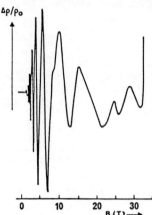

$\Delta\rho/\rho_0$

0   10   20   30
B (T) →

For highly doped samples with apparent high mobility (x < 0.15) the carriers of the Γ-direct band take an important part in the conduction and it makes possible to observe the quantized oscillations of magnetoresistance (Aulombard, 1978, 1980). As an example the second derivative longitudinal magnetoresistance up to 30 T is presented on Fig. 4. Two types of oscillation can be noticed : - first at low magnetic field the oscillations are due to Γ - band carriers - then at high field (B > 20 T) the appearance of new peaks in the signal can be attributed to the heavy L electrons.

Fig. 4   Typical longitudinal magnetoresistance oscillations for sample 1441 (T=4.2 K).

5. Analysis of the results

5.1 Contribution to the GaAlSb band structure

To analyze the experimental results of chapter 4.1, it has been assumed that the observed effects can be ascribed to the electrons in Γ and L bands only. The two band Hall coefficient is written : $R_H = (1/e).(r_\Gamma n_\Gamma b^2 + r_L n_L)/(n_\Gamma b + n_L)^2$ (1)

where $b = \mu_\Gamma/\mu_L$ is the mobility ratio, $n_\Gamma(n_L)$ is the carrier concentration and $r_\Gamma(r_L)$ is the Hall factor in Γ(L) bands respectively. Following Allgaier's analysis (1965) the mobility ratio b can be determined from the maximum of R_H (Fig. 3) through the formula : $(R_H/R_o)_{max} = (1+b)^2/4b$ (2)   where $R_o = r_L/n_T e$ and n_T is the total carrier concentration : $n_T = n_L + n_\Gamma = cte$. The formula (1) was solved using a three-band Kane's model for the Γ-conduction band and a parabolic shape for the L-band (the parameters used are listed in table II). The energetic separation $E_{\Gamma L}$ at the pressure corresponding to R_{Hmax} was determined and then assuming a linear variation of $E_{\Gamma L}$ with pressure ($dE_{\Gamma L}/dP = 9.3$ meV/kbar) (Aulombard, 1980) the zero pressure $E_{\Gamma L}$ was calculated. Taking the $E_\Gamma(X)$ variation (Alibert, 1983) and the determined values of $E_{\Gamma L}$, the following $E_L(X)$ variation law was found : $E_L = 0.86 + 1.45 x + 0.64 x (x-1)$ at T = 77 K.

The bowing parameter $c_L = 0.64$ determined in this way is the same as this one determined by Alibert (1983) for E_1 transition.

m^*_{dL}(GaSb)=0.57 m_o(Lee,1981)	m^*_{dL}(AlSb)=0.69 m_o(Pollack,1966)
E_p(GaSb)=22.4 eV (Lawaetz,1971)	E_p(AlSb) = 18.7 eV (Lawaetz,1971)

Table II : Parameters used in the calculation (linear variations have been taken for the alloy).

5.2 Carrier mobilities

From the knowledge of the apparent mobility ($\mu_H = R_H/\rho$), the respective n_Γ and n_L concentrations and the mobility ratio b at zero pressure, the mobility in each band can be deduced using the following formula (Allgaier, 1965) :

$$\mu_L = \mu_H\{(1-y)b+y\}/\{(1-y)b^2 + y\} \quad \text{where } y = n_L/(n_\Gamma + n_L)$$

The results we have obtained are presented in Fig. 5 where the respective μ_Γ and μ_L variations are reported in the Al content range up to 0.25 (at T=300 K and T = 77 K).

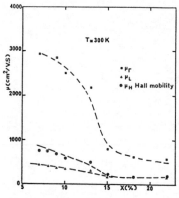
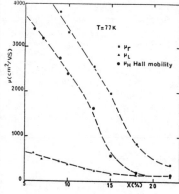

Fig. 5 μ_Γ and μ_L variations versus Al concentration.

5.3 Effective mass

From the analysis of the magnetoresistance oscillation amplitude (Aulombard, 1978) at different temperatures the conduction effective mass in Γ band at the Fermi level was determined. In order to determine the effective mass m^*_o at the bottom of the conduction band, the theoretical expression which holds for samples with the spin-orbit splitting Δ less than the energy gap (Aulombard, 1978) was used. The results we have obtained are plotted in Fig. 6. In contrast to the predictions of k-p theory (using a linear interpolation of the momentum energy E_p) the effective mass is found to be a linear function of alloy composition x (m^*_o=0.043 + 0.092 x).

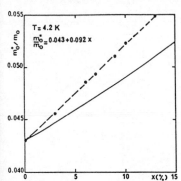

As suggested by Stradling (1979), the discrepancy is thought to originate from a decrease in E_p due to the alloy disorder.

Fig. 6   Variation of the electron effective mass in $Ga_{1-x}Al_xSb$. The full line was calculated using k-p theory. The dotted line is the linear variation of the mass with x.

6. Conclusions

6.1   The $E_L(x)$ variation law was determined, and the crossover between Γ and L-minima of conduction band was found to take place at Al composition x = 0.24 (T=77 K).

6.2   The conduction band minima mobilities were determined
- μ_Γ was found to be strongly composition dependent (from 3000 cm²/Vs at x=0.03 to 600 cm²/Vs at x=0.25 at T=300 K).
- in the same composition range μ_L was found to vary between $400 > \mu_L > 200$ cm²/Vs.

- μ_x was found to be of the order of 30 cm^2/Vs.

6.3 In contrast to the theoretical predictions the effective mass at the bottom of conduction band was found to be a linear function of alloy composition.

References

Alibert C., Joullié A.M. and Ance C., 1983, Phys. Rev. B <u>27</u>, 4946.

Allgaier R.S., 1965, J.A.P. <u>36</u>, 2429.

Aulombard R.L., Robert J.L., Raymond A. and Joullié A., 1978, Solid State Com. <u>26</u>, 697.

Aulombard R.L. and Joullié A., 1979 Mat. Res. Bull <u>14</u>, 349.

Aulombard R.L., Bousquet C., Robert J.L., Konczewicz L., Litwin E. and Porowski S. 1980, Int. Symp. on GaAs and related compounds (Vienna),pp. 651.

Capasso F., Hutchinson A.L., Foy P.W., Bethea C. and Bonner A., 1981 Appl. Physics Lett. <u>39</u>, 736.

Dolginov L.M., Brackin A.E., Druzhinina L.V., Elisew P.G., Milvidskii M.G., Skoipkin V.A. and Sverdlov B.N., 1981 IEEE, J. Quantum Electr. QE <u>17</u> 593.

Gautier P., Joullié A., Bougnot G. and Champness C.H., 1981, J. Crystal growth <u>51</u>, 336.

Hildebrand O., Kuebart W., Benz K.W. and Pilkuhn M.H., 1981, IEEE J. Quantum Electron. QE <u>17</u>, 284.

Konczewicz L., Litwin E., Porowski S., Iller A., Aulombard R.L., Robert J.L., Joullié A., 1982, Proc. of the 16th Int. Conf. on the Physics of S.C., Montpellier, pp. 92.

Law H.D., Tomasetta L.R., Nakano K. and Harris J.S., 1978, Appl. Phys. Letters <u>33</u>, 416.

Law H.D., Chin R., Nakano K. and Milano R., 1981 IEEE J. Quantum Electron. QE <u>17</u>, 275.

Lawaetz P., 1975, Phys. Rev. B <u>4</u>, 3460.

Lee H.J. and Wooley J.C., 1981, Can. J. Phys. <u>59</u>, 1844.

Ohmori Y., Tarucha S., Horikoshi Y. and Okamoto H., 1984, Japan J. Appl. Phys. <u>23</u>, L94.

Pollack F.H., Higginbotham C.W. and Cardona M., 1966, J. of the Physical Society of Japan, <u>21</u>, 20.

Stradling, 1979, J. of Magnetism and Magnetic Materials, <u>11</u>, 221.

Tsang W.T. and Olsson N.A., 1983, Appl. Phys. Lett. <u>43</u>, 8.

Inst. Phys. Conf. Ser. No. 74: Chapter 4
Paper presented at Int. Symp. GaAs and Related Compounds, Biarritz, 1984

Far-infrared spectroscopy of silicon donors in Al$_x$Ga$_{1-x}$As

T.N. Theis, T.F. Kuech, L.F. Palmateer and P.M. Mooney

I.B.M. Thomas J. Watson Research Center, P.O. Box 218, Yorktown Heights, NY 10598

Abstract. We report first measurements of the 1s-2p transition of shallow Si donors in Al$_x$Ga$_{1-x}$As. For AlAs mole fraction $x \gtrsim 0.25$, the transition is observed only after the material is briefly exposed to visible or near visible light. If we interpret this persistent photoeffect in terms of the D-X center model for deep donor states in Al$_x$Ga$_{1-x}$As, our results indicate that the photoionized D-X centers act as shallow donors.

Introduction

The electronic properties of n-type Al$_x$Ga$_{1-x}$As ($x \gtrsim 0.25$) are dominated by a deep donor level which, at low temperatures, gives rise to a large persistent photoconductivity (Nelson 1977). Large activation energies have been determined by Hall measurements. More importantly, the activation energy is just the difference between even larger thermal emission and capture energies observed in capacitance spectroscopy, and the photoionization energy is still larger (Lang et al. 1979, Künzel et al. 1983). Lang et al. (1979) have explained these features by postulating that the level is a donor-defect complex (D-X center) with a large lattice relaxation. The positively charged center is believed to give rise to a resonant state in the conduction band, which is associated with the X or L band minimum. The energy of this state is a sensitive function of the local lattice configuration. Local lattice deformation caused by thermal phonons can drive this state down toward the conduction band edge, leading to electron capture. Lattice relaxation then lowers the total energy, trapping the electron deep in the gap. Thermal emission is assumed to proceed by the inverse process, in which the local lattice deformation drives the trap state up toward the conduction band edge. If the electron is emitted, the lattice can go back to the resonant state configuration. The thermal energy required to initiate these charge state-controlled lattice relaxation processes represents a barrier for emission and capture.

The persistent photoconductivity occurs at low temperatures because, once the D-X center is ionized, the thermal barrier prevents electron recapture to the deep trap state. Here we show, for the first time, that the photoionized D-X center gives rise to shallow bound states associated with the Γ band. Far-infrared spectra, obtained before and after the sample is exposed to light, graphically demonstrate the transfer of electrons from deep to shallow levels. Electronic occupation of these states does not lead to large lattice relaxation and deep capture, as does occupation of the inferred (X or L band) resonant state.

Others have proposed that the persistent photoconductivity in n-Al$_x$Ga$_{1-x}$As is inherent in the conduction band structure of the alloy (Saxena 1982, Saxena and Sinha 1983, Chand et al. 1984). To our knowledge, this view cannot account for the fact that the thermal emission and capture energies are large compared to the thermal activation energy, but small compared to the photoionization energy. Thus, we shall refer to the deep donors in our material as D-X centers. However, our results place new constraints on any model of the deep donor level in n-Al$_x$Ga$_{1-x}$As.

Experimental Technique

The MOCVD growth of epitaxial $Al_xGa_{1-x}As$ layers with low acceptor concentrations is described elsewhere in this conference by Kuech et al. (1984). The electrical properties of the layers were characterized by capacitance voltage profiling and Van der Pauw-Hall measurements. The alloy composition was determined by electron micro-probe, and was also well correlated with low temperature (2K) photoluminescence peak positions. Si doped layers of varying AlAs mole fraction, x, were grown epitaxially on 100 GaAs substrates to a thickness of ~6 μm. Nominally undoped samples were also grown and systematically characterized to determine the net residual acceptor concentration, N_a, as a function of x. As x increased, N_a increased. The Si concentration, N_{Si}, was therefore adjusted for each value of x in order to obtain n-type material. Typically, $N_{Si}/N_a \gtrsim 4$.

The large lattice relaxation model assumes that Si is incorporated in $Al_xGa_{1-x}As$ either as part of a donor-defect (D-X) complex, or as an isolated shallow donor. Thus, $N_{Si}=N_{D-X}+N_d$, where N_{D-X} and N_d are concentrations of D-X centers and shallow Si donors, respectively. At room temperature both types of donors are readily ionized, so $N_{Si}-N_a$ could be determined from C-V measurements. Samples with $x \gtrsim 0.27$ showed pronounced carrier freeze-out to D-X centers upon cooling below 120 K. Indeed, the 6 μm thick epitaxial layers with $N_{Si}-N_a \approx 2-5 \times 10^{16}$ cm^{-3} became fully or nearly fully depleted, indicating $N_d \sim N_a$. A brief exposure to light ($\hbar\omega \gtrsim 0.8$ eV) persistently ionized the D-X centers (Lang et al. 1979), and made the material conductive. Below 60 K, electron recapture was negligible, and the density of ionized D-X centers could be extracted from C-V data. This was found to be nearly equal to the value of $N_{Si}-N_a$, consistent with $N_d \sim N_a$. Knowing N_a, we find typically, $N_{D-X}/N_d > 3$. To summarize these results, in our samples with $x \gtrsim 0.27$, $N_{D-X} > N_d \sim N_a$. Under these conditions, all or most of the shallow donor levels are compensated, and the Fermi level tends to be pinned by the D-X center.

Far-infrared (FIR) transmission spectra were obtained at 4.2 K. The substrates were wedged to reduce the effects of internal interference. Substrate absorption was accounted for by referencing the spectra to the transmission spectra of similarly wedged samples from which the $Al_xGa_{1-x}As$ layer had been removed. Three or four samples were usually stacked to increase the total absorption.

Experimental Results

We first discuss our FIR spectroscopic results on n-GaAs. Fig. 1 shows the transmission spectra obtained on layers with two different Si concentrations. Each spectrum features a prominent absorption line corresponding to the 1s-2p hydrogenic transition of the shallow Si donor. For the lower value of $N_{Si} \approx 6 \times 10^{15}$ cm^{-3} (total thickness = 32 μm) the well defined transmission minimum is at an energy of 4.5 meV, in good agreement with published values (Cooke et al. 1978). At $N_{Si} \approx 2.0 \times 10^{16}$ cm^{-3} (total thickness = 21 μm) the line is broadened and the minimum is shifted to slightly higher energy by Coulombic effects. Taking the differing total n-GaAs thicknesses of the samples into account, the broader line represents the absorption of about twice as many Si donors. Thus, the energy integrated absorption is correlated with the concentration of occupied donor levels.

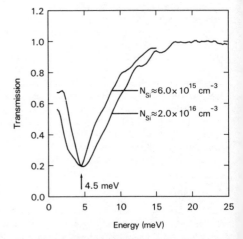

Fig. 1. Far-infrared transmission spectra of two n-GaAs samples ($N_a < 1 \times 10^{15}$ cm^{-3}) with differing Si dopant concentrations, showing the 1s-2p absorption line of the shallow donor.

The FIR transmission of $Al_xGa_{1-x}As$ ($x \gtrsim 0.27$) was quite different from GaAs. Fig. 2 contains several transmission spectra for a sample of n-$Al_{0.29}Ga_{0.71}As$. The first spectrum, showing little or no 1s-2p absorption, was obtained after the sample was cooled to 4.2 K in the dark. The lack of absorption is consistent with the C-V measurements, which indicate freeze-out to D-X centers and compensation of the remaining shallow donors. Before each of the remaining spectra were obtained, the sample was briefly exposed to light from a GaAs LED. An absorption line develops monotonically with the accumulated light exposure. The width of the line appears consistent with Coulombic broadening of the 1s-2p transition for $N_{Si}+N_a \simeq 4.2 \times 10^{16}$ cm^{-3}.

The energy of the transmission minimum for this and other samples is plotted versus x in Fig. 3. The smooth change in energy with composition shows that this is indeed the 1s-2p shallow donor transition. The dashed line gives the energy dependence calculated using the Γ band effective mass and dielectric constant dependence suggested by Casey and Panish (1978). For $x \lesssim 0.3$, the data are in fair agreement. For $x \gtrsim 0.3$, the energy rises rapidly, probably reflecting the influence of the X and L bands. However, we emphasize that even at x=0.34 the level is quite shallow and cannot explain the observed carrier freeze-out. Indeed, all the samples with $x \geq 0.27$ showed the freeze-out and the same general FIR transmission behavior upon exposure to light as that shown in Fig. 2.

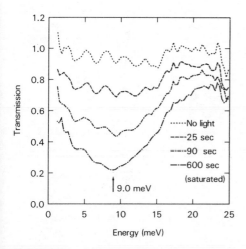

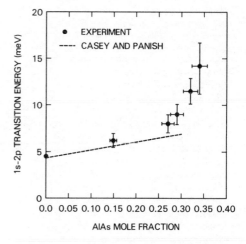

Fig. 2. Far infrared transmission spectra of an n-$Al_{0.29}Ga_{0.71}As$ sample ($N_{Si} \simeq 3.4 \times 10^{16}$ cm^{-3}, $N_a \simeq 8 \times 10^{15}$ cm^{-3}) after successively longer exposures to light from a GaAs LED. The LED current was 100 mA during each exposure. Each spectrum was obtained with the LED off. The 1s-2p absorption line develops as electrons are transferred from deep to shallow donor levels. The oscillatory structure evident in the spectra is caused by internal interference.

Fig. 3. The 1s-2p transition energy of shallow Si donors, determined from transmission minima of spectra like those of Figures 1 and 2, as a function of AlAs mole fraction, x. The horizontal error bars represent the relative accuracy of the electron microprobe measurement of x. The vertical error bars represent an estimate of the errors introduced by line broadening and internal interference effects in determining 1s-2p transition energies from the absorption lines.

Photoionization of a D-X center initially transfers an electron to the conduction band. Because the thermal barrier prevents recapture to the deep level, the electron can then be captured by one of the shallow donors, which would otherwise be compensated. This can cause an increase in the 1s-2p absorption upon exposure to light, but it cannot fully explain our results. After illumination, we find the energy integrated absorption to be well correlated with N_{Si}-N_a, and not correlated with N_d.

This is illustrated in Fig. 4 which compares the spectra of two samples of similar thickness, similar values of N_{Si}-N_a, and rather different values of N_d. One sample has a composition (x=0.15) below the range where a strong persistent photoeffect is observed, and there is little change in the 1s-2p absorption before (Fig. 4a) and after (Fig. 4b) exposure to light. The other sample has a composition (x=0.29) in this range and shows the dramatic increase in absorption discussed above. The former sample has a negligible concentration of D-X centers so that the concentration of uncompensated shallow donors which participate in the absorption is N_{Si}-$N_a \simeq 2.2 \times 10^{16}$ cm^{-3}. The latter sample has N_{Si}-$N_a \simeq 2.6 \times 10^{16}$ cm^{-3}, but with $N_{D-X} \simeq N_{Si}$-N_a, this leaves only $N_d \simeq N_a \simeq 8 \times 10^{15}$ cm^{-3} shallow donors to participate in the absorption after the D-X centers have been photoionized. Yet after illumination (Fig. 4b) the spectra are strikingly similar except for the expected shift in energy of the transmission minima. The somewhat broader absorption line of the $Al_{0.29}Ga_{0.71}As$ sample is consistent with the higher values of N_{Si} and N_a, but the energy integrated absorption can only be explained if essentially all the uncompensated Si atoms participate in the absorption.

In contrast to this lack of correlation with N_d, the absorption in samples of similar composition and thickness always showed strong correlation with N_{Si}-N_a. For instance, a sample with x=0.32, 24 μm total thickness, and N_{Si}-$N_a \simeq 1.1 \times 10^{16}$ cm^{-3} had a minimum transmission of 0.5 while a sample with x=0.34, identical thickness, and N_{Si}-$N_a \simeq 3.6 \times 10^{16}$ cm^{-3} had a minimum transmission of only 0.06 together with greater line broadening.

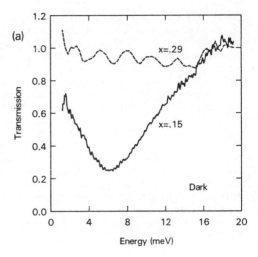

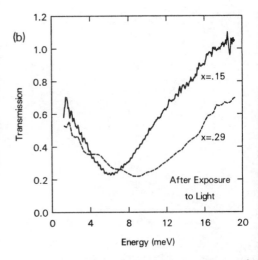

Fig. 4. Far-infrared transmission spectra of n-$Al_xGa_{1-x}As$ layers of similar total thickness (~24 μm) and similar N_{Si}-N_a for values of x below (x=0.15) and in (x=0.29) the composition range where strong persistent photoconductivity is observed. The sample with x=0.15 had $N_{Si} \simeq 2.4 \times 10^{16}$ cm^{-3} and $N_a \simeq 2 \times 10^{15}$ cm^{-3}. The sample with x=0.29 had $N_{Si} \simeq 3.4 \times 10^{16}$ cm^{-3} and $N_a \simeq 8 \times 10^{15}$ cm^{-3}. a) Spectra obtained before the samples were exposed to light. b) After exposure to light the two samples exhibit similar energy integrated absorptions, indicating participation by essentially all the uncompensated Si atoms in the shallow donor absorption.

Discussion

In order to reconcile these results with the large lattice relaxation model, we must assume that photoionization of D-X centers does not simply transfer electrons to otherwise compensated shallow donors. Rather, the D-X center itself can act as a shallow donor in one of its two metastable states. It is not surprising that a positively charged complex defect gives rise to a shallow Γ band bound state. The small effective mass ensures a small central cell perturbation, so the energy levels would be indistinguishable in our measurements from those of simple isolated Si donors. The electronic energy should also depend only weakly on the defect configuration, which may explain why this bound state does not lead to lattice deformation and deep capture of the electron. Deep capture must proceed only through the inferred X or L band resonant state.

Our results show that measurements of thermal activation energies of donors in n-$Al_xGa_{1-x}As$ must be interpreted with care. If samples are not exposed to light, then Hall measurements in the temperature range above ~120K will reflect the thermal activation of the deep state of the (numerically dominant) D-X center, and at lower temperatures, the activation of shallow donors if these are not compensated. Measurements conducted in the light, or at low temperatures after exposure to light, will reflect the thermal activation of the shallow donor state of the D-X center, as well as any uncompensated simple donors. For instance, donor binding energies in LPE material have been determined by thermally activated photoluminescence (i.e. under strong illumination) by Dingle et al. (1977). As expected, these results agree well with binding energies inferred from Fig. 3.

We emphasize that our results are not specific to MOCVD grown material. It appears that $N_{D-X} \gg N_d$ for typical MBE grown n-$Al_xGa_{1-x}As$ (Watanabe et al. 1984, Künzel et al. 1984,). N_d remained below 1×10^{16} cm^{-2} even at the highest doping concentrations. We suggest that it is also likely that $N_a > N_d$ in most MBE $Al_xGa_{1-x}As$ grown at relatively low temperatures for heterostructure fabrication. Concentrations of (possibly deep) acceptors ~5×10^{16} cm^{-3} have been inferred for material from various laboratories (Hwang et al. 1984, Hickmott and Solomon 1984, Batey et al. 1984) Under these conditions, the shallow donor levels are compensated, and the Fermi level is pinned by the D-X center, just as in much of the MOCVD material used in our study. In fact, the Fermi level must be pinned by a deep donor in order to explain the observed two-dimensional electron gas densities in modulation doped heterojunctions (Stern 1983, Vinter 1983, Hirakawa, et al. 1984).

Conclusion

We have presented the first measurements of the 1s-2p transition energy for shallow donors in $Al_xGa_{1-x}As$ over the range x=0.0 to x=0.34. A strong photoeffect observed for x$\gtrsim$0.27 graphically illustrates the transfer upon illumination of electrons from deep to shallow donor levels. The energy integrated absorption correlates well with the concentration of uncompensated Si atoms in the material, and not with the concentration of shallow donors deduced from temperature dependent C-V measurements. Interpreting these results within the large lattice relaxation model of Lang et al. (1979), we deduce that one of the metastable states of the D-X center acts as a shallow donor. Our results can also be interpreted in terms of a band structure model such as that of Saxena (1982), if it is assumed that essentially all of the electrons in deep levels are transferred to shallow levels upon exposing the material to light.

References

Batey J, Wright S L, DiMaria D J, and Theis T N (1984) to be published
Casey H C Jr., and Panish M B 1978 Heterostructure Lasers (New York:Academic Press) pp 188-94
Chand N, Henderson T, Klem J, Masselink W T, Fischer R, Chang Yia-Chung and Morkoç H 1984 submitted to Phys. Rev. B
Cooke R A, Hoult R A, Kirkman R F and Stradling R A 1978 J. Phys. D 11 945
Dingle R, Logan R A and Arthur J R Jr 1977 Inst. Phys. Conf. Ser. No. 33a 210
Hickmott T W, Solomon P M, Fischer R and Morkoç H 1984 J. Appl. Phys. to be published.
Hirakawa K, Sakaki H, and Yoshino J 1984 Appl. Phys. Lett. 45 253
Hwang J C M, Kastalsky A, Störmer H L, and Kerimidas V G 1984 Appl. Phys. Lett. 44 802
Kuech T F, Veuhoff E, and Wolford D J 1984 Inst.Phys.Conf.Ser. 74 181
Künzel H, Fischer A, Knecht J, and Ploog K 1983, Appl. Phys. A 32 69
Künzel H, Ploog K, Wünstel K, and Zhou B L 1984, J. Elec. Mat. 13 281
Lang D V and Logan R A and Jaros M 1979 Phys. Rev. B 19 1015
Nelson R J 1977 Appl. Phys. Lett. 31 351
Saxena A K 1982 Sol. State Elec. 25 127
Saxena A K and Sinha A K 1983 Indian J. Pure and Appl. Phys. 21 668
Stern, F 1983 Appl. Phys. Lett. 43 974
Vinter B 1983 Sol. State Comm. 48 151

Inst. Phys. Conf. Ser. No. 74: Chapter 4
Paper presented at Int. Symp. GaAs and Related Compounds, Biarritz, 1984

Measured ionization coefficients in $Ga_{1-x}Al_xAs$

J.P.R. David, J.S. Marsland, H.Y. Hall[*], G. Hill[*], N.J. Mason[*], M.A. Pate[*]
J.S. Roberts[*], P.N. Robson, J.E. Sitch[+], R.C. Woods.

Department of Electronic & Electrical Engineering, The University of
Sheffield, Mappin Street, Sheffield S1 3JD, UK.

[*]SERC Central Facility for III-V Semiconductors, University of Sheffield,
UK.

[+]Now at Bell Northern Research Ltd., Ottawa, Canada.

Abstract. The ionization coefficients α and β for electrons and holes
in $Ga_{1-x}Al_xAs$ have been measured for a range of Al fractions and doping
densities. Measurements show that both α and β decrease with increasing
Al composition. K ($=\alpha/\beta$) is >1 for GaAs and low values of x and becomes
<1 as x increases. The breakdown voltage of a $Ga_{1-x}Al_xAs$ diode normal-
ised to that of a similarly doped GaAs diode is a linear function of x.

1. Introduction

$Ga_{1-x}Al_xAs$ is increasingly being used in several novel device structures
such as HBTs , TEGFETs and APDs where the band-gap discontinuity and different material characteristics with respect to GaAs are exploited to give
improved performance over conventional device structures. In all these
applications knowledge of the breakdown characteristics of the material is
important, yet to the best of our knowledge there does not exist any information about the ionization coefficients of $Ga_{1-x}Al_xAs$ as a function of Al
fraction x.

We report here the ionization coefficients obtained by carrying out photomultiplication measurements on p-n heterojunction diodes. Our devices
cover a doping range of $3.5 \times 10^{15} cm^{-3}$ to $2 \times 10^{17} cm^{-3}$ and an Al fraction of
0 to 0.49. Throughout this variation we find that α and β, the electron
and hole ionization coefficients respectively, decrease with increasing x
for a given electric field. Less expectedly K ($=\alpha/\beta$) which is greater
than unity for GaAs becomes less than unity as the aluminium fraction is
increased.

2. Experimental Technique

The devices were fabricated from layers grown by LPE on $(100)n^+$ or p^+ GaAs
substrates. Table 1 describes the characteristics of some layers on which
measurements were carried out. All the layers were one-sided abrupt
uniformly doped p^-n^+ or n^-p^+ structures from which circular etched mesa
diodes were formed using standard lithographic techniques. Ge was used as
the p-type dopant and Sn for the n-type. Fig.1 is a schematic diagram of

the diodes and shows the contact metallization used. The top layer is
$\sim$2.0μm thick and the middle active layer $\sim$4.0μm; the mesa diameter is
$\sim$200μm.

Sample Number	Structure	Al fraction	$\lvert N_A-N_D\rvert$ (cm^{-3})	V_{BD} (volts)
1	p^+-n	0	3.5×10^{15}	132
2	n^+-p	0.13	8.6×10^{16}	18
3	n^+-p	0.18	5.5×10^{16}	24
4	n^+-p	0.25	7.0×10^{16}	24
5	p^+-n	0.26	3.5×10^{16}	39
6	n^+-p	0.26	7.0×10^{16}	24
7	n^+-p	0.45	2.5×10^{16}	55
8	n^+-p	0.47	2.0×10^{17}	16
9	n^+-p	0.49	6.0×10^{16}	32.5
10	n^+-p	0.76	2.0×10^{17}	19.5

Table 1. Details of devices used.

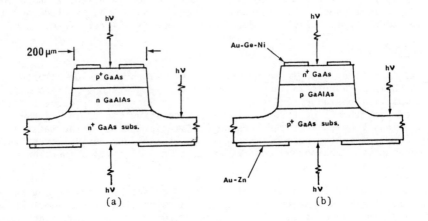

Fig.1   Schematic diagrams of device structures

A similar technique to that used for measuring photomultiplication
currents in InP diodes by Umebu et al (1980) was used. A chopped He-Ne
visible laser spot was focussed in turn on to each side of a reverse
biassed p-n diode thereby injecting electrons or holes and a lock-in
amplifier was used to detect and amplify the photo-current. An X-Y
recorder was finally used to record the variation of photocurrent with
reverse bias for both electron and hole injection.

Great care has to be exercised when obtaining the multiplication curves as
these determine the ionization rates, α and β. Only diodes with sharp
breakdown and low pre-breakdown dark currents were used for measurement.
A small laser spot (<5μm diameter) was focussed on top of the mesa diode
at several points to establish that the ionization rate is uniform and
that no local microplasmas exist. Ensuring pure injection of carriers
into the depletion region is essential for accurate determination of α
and β. In the case of the structure shown in Fig.1a pure electron
injection is readily achieved by focussing the spot onto the top of the
mesa. The minority carriers diffuse to the depletion region without
significant recombination. Hole injection can be achieved by either
focussing the spot on the floor of the mesa or by etching a hole in the
substrate to within a few microns of the junction and illuminating from
the substrate side. In most cases due to the difficulty of etching a hole
of carefully controlled depth in the substrate without a selective etch-
stop barrier, the former method was used. By increasing the distance of the
spot from the mesa-edge until two identically shaped multiplication curves
were obtained, the possibility of direct injection into the depletion
region was eliminated.

The complementary heterostructure shown in Fig.1b was also used. This
latter structure has the advantage that no barrier exists to the injection
of electrons from the GaAlAs region whereas this is not the case for the
structure of Fig.1a if the aluminium transition is abrupt. The effect of
the barrier could be observed through an initial steep increase in photocurrent with reverse bias voltage until the effect of the barrier was
overcome. At that stage the photocurrent temporarily saturated prior to
the onset of significant multiplication with further increase in voltage.
The saturated value of photocurrent was taken to correspond to unity
multiplication. Both types of structure gave essentially identical
results for α and β when the doping and aluminium fraction were comparable.

It is important to ensure that no mixed injection occurs in the high field
depletion region due to Franz-Keldysh electroabsorption of sub-band-gap
radiation. This is generated by the recombination of primary photoexcited electron-hole pairs in the heavily doped GaAs layers (Bulman et al
1982). The problem is greatly reduced in the present heterostructures due
to the large separation between the absorption edge of GaAlAs and that of
GaAs. The widening of the depletion region in the lower doped GaAlAs
layer with bias also distorts the multiplication by decreasing the length
photo-generated minority carriers have to diffuse before collection. This
was corrected for by using the method described by Woods et al (1973).
Finally a further factor which has to be allowed for is the dead-space
correction. To account for this the method described by Bulman et al
(1983) was used. The threshold ionization energies needed for this
correction were estimated for varying aluminium compositions as described
in section 4.

3. Experimental Results

The results for α and β versus reciprocal field are presented in Figs.2
and 3. In Fig.2 the data for x = 0 (sample #1) and x = 0.26 (sample #6)
are given together with the recent results of Bulman et al (1983) for GaAs,
whilst in Fig.3 the results for x = 0.18 (sample #3) and x = 0.48 (samples
#8 and 9) are shown. The aluminium fraction x was measured using an energy

dispersive X-ray spectrometer. The value of electric field is directly
dependent on the measured value of $|N_A-N_D|$ in the active layer. This was
obtained from C-V measurements but a 2% error in the measured radius of
the mesa leads to an approximate 8% error in $|N_A-N_D|$ and the corresponding
inferred field. This factor is responsible for some of the scatter
observed in the results in Figs.2 and 3 when they are taken from different
devices.

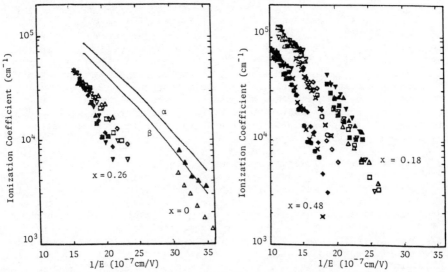

Fig.2 Experimental electron (closed symbols)
and hole (open symbols) ionization coefficients for GaAs and $Ga_{0.74}Al_{0.26}As$, and
recently reported ionization coefficients
(solid lines) for GaAs from Bulman et al(1983)

Fig.3 Experimental electron (closed symbols)
and hole (open symbols) ionization
coefficients for $Ga_{0.82}Al_{0.18}As$ and
$Ga_{0.52}Al_{0.48}As$.

Fig.4 shows the effect of aluminium fraction x on the breakdown voltage
(V_{BD}) normalised with respect to a GaAs abrupt one-sided junction of the
same doping density. These values for GaAs were taken from Sze (1981).
The appropriate value of doping density for each point in this figure is
given in Table 1.

4. Discussion

Fig.2 shows that our data agree with that of Bulman et al (1983), within
the bounds of experimental accuracy, over the field range considered. As
expected for the case x = 0.26, α and β are both lower than GaAs for a
given electric field but now β is seen to be slightly greater than α.
Fig.3 shows two sets of data, one for x = 0.18 and the other for x = 0.48.
Comparing Fig.3 with Fig.2 we see that the x = 0.18 data lies between that
of x = 0 and x = 0.26. Despite the scatter, each device with x = 0.18
shows an α slightly greater than β. For x = 0.48, α and β both continue
to decrease although now it is very obvious that β > α.

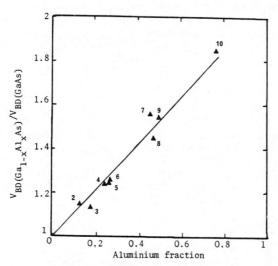

Fig.4 Breakdown voltage of GaAlAs normalised to GaAs
against aluminium fraction for various samples
(numbers refer to Table 1). Solid line is
$V_{BD(Ga_{1-x}Al_xAs)} = (1 + 1.1x)V_{BD(GaAs)}$

In order to interpret the multiplication results, estimates of the ionization thresholds were made following the method of Pearsall et al (1978).
The variation of the band minima (Saxena 1980) and effective masses (Hill
1979) with aluminium fraction are known, so that in the absence of detailed
band-structure calculations analytic estimates of the threshold energies
may be made using the parabolic band approximation. The threshold energies
calculated in this way are shown in Table 2.

x	Γ electron initiated (eV)	L electron initiated (eV)	Split-off hole initiated (eV)
0	1.61	1.68	1.63
0.1	1.75	1.75	1.78
0.2	1.90	1.84	1.94
0.3	2.07	1.93	2.12
0.4	2.25	2.04	2.31
0.5	2.44	2.16	2.51

Table 2. Ionization threshold energies versus x.

The increasing ionization threshold energies with x support the trends
shown in Figs.2 and 3. However, the differences between electron and hole
thresholds are insufficient and generally in the wrong sense to explain
the observed variation of K with x alone and the variation of mean
free path and phonon energy must be taken into account (Baraff 1962;
Ridley 1983). This is currently being investigated.

Fig.4 shows that the breakdown voltage of the heterojunction diodes
increases with increasing x as would be expected from Figs.2 and 3. The
normalised breakdown voltage increases linearly with x at least up to x =
0.76 and the simple empirical relationship

$$V_{BD(Ga_{1-x}Al_xAs)} = (1+1.1x)V_{BD(GaAs)}$$

appears to hold over the range of doping densities used.

5. Conclusions

The ionization coefficients in $Ga_{1-x}Al_xAs$ have been shown to decrease with
increasing Al fraction from photomultiplication measurements on heterojunction diodes. K which is >1 for GaAs, decreases with increasing x; at
x = 0.26, K <1 and it decreases even more for further increases in x. The
breakdown voltage for a $Ga_{1-x}Al_xAs$ junction is related to a similarly doped
GaAs junction by a simple empirical relationship.

6. Acknowledgements

This work was supported by the UK Science and Engineering Research Council.
The authors would like to thank Mr. M. Faulkner for his invaluable
assistance with the SEM analysis.

7. References

Baraff G A 1962 Phys. Rev. 128 2507
Bulman G E, Cook L W and Stillman G E 1982 Solid State Electron. 12 1189
Bulman G E, Robbins V M, Brennan K F, Hess K and Stillman G E 1983 IEEE
    Electron. Dev. Lett. EDL-4 181
Hill G 1979 Internal Report University of Sheffield (unpublished)
Pearsall T P, Capasso F, Nahory R E, Pollack M A and Chelikowsky J R 1978
    Solid State Electron. 21 297
Ridley B K 1983 J.Phys.C : Solid State Phys. 16 4733
Saxena A K 1980 J.Phys.C : Solid State Phys. 13 4323
Sze S M 1981 Physics of Semiconductor Devices (2nd Edition) (New York :
    Wiley) 101
Umebu I, Choudhury A N M M and Robson P N 1980 Appl. Phys. Lett. 36 302
Woods M H, Johnson W C and Lampert M A 1973 Solid State Electron. 16 381

Characterisation of thermal instability in GaAs–AlAs and GaAs–InAs superlattices with laser Raman spectroscopy

K. Kakimoto, H. Ohno*, R. Katsumi*, Y. Abe*, H. Hasegawa*,
and T. Katoda

Institute of Interdisciplinary Research, Faculty of Engineering,
The University of Tokyo

* Department of Electric Engineering, Faculty of Engineering,
Hokkaido University

Abstract. Thermal instability of GaAs–AlAs and GaAs–InAs superlattices was studied by Raman spectroscopy. Samples with various periods and grown at various temperatures were investigated to understand the stability of interface in superlattice and to obtain an activation energy of mixing by thermal annealing. Mixing occurred more easily in GaAs–InAs strained superlattice than GaAs–AlAs strain-free superlattice. Samples grown at a lower temperature are more stable against thermal annealing than those grown at a higher temperature.

1. Introduction

The superlattice (SL) is of great interest for opto-electronic and high speed devices. Phenomena occurring at an interface in a SL has not been made clear satisfactorily. To understand the properties at interface is a key to improve characteristics of devices. There are some works relating to characterization for the structure of interface in SL by transmission electron microscopy analysis (Petroff 1977, Petroff et al 1978, Okamoto et al 1983), Auger-electron-spectroscopy (Chang et al 1976), Rutherford backscattering (Chu et al 1984), X-ray diffraction study (Fleming et al 1980), and photoluminescence excitation spectroscopy (Weisbuch et al 1981).

In earlier work, the thickness of each layer in SL was measured by zone folding phonon on Raman spectroscopy from layered structure (Merz et al 1977, Sai-Halasz et al 1978, Colvard et al 1980, Sapriel et al 1983). In this paper, we report the thermal instability of SL by thermal annealing using Raman spectroscopy which is a non-destructive method for characterization. The disappearance of periodicity and formation of an alloy with a composition in midrange will be discussed in the following section.

2. Experimental procedure

The samples of SL such as GaAs–AlAs and GaAs–InAs used in this study were grown by MBE growth method. The orientation of the substrate is (100). Growth temperature of samples is in the range from 500 to 700°C for GaAs–AlAs and 500°C for GaAs–InAs. Thermal annealing of these SLs covered with SiO_2 or AlN cap whose thickness is about 1000 A was carried

out in the atmosphere of flowing N_2-H_2 mixed gas to prevent the
oxidization and/or evaporation of arsenic atoms at sample surface.

For Raman spectroscopy measurements, the sample was mounted in the quartz
box which was filled with N_2-H_2 mixed gas to prevent the oxidization of
the sample surface.   The typical excitation power was 100 mW with a
laser spot size of about 200 μm. The scattered light was analyzed by a
JRS 400T triple monochromater (JEOL) and detected by a cooled R943
photo-multiplier (Hamamatsu Photonics).

3. Results and discussions

3.1 Comparison of Raman spectra from annealed GaAs-AlAs and GaAs-InAs SLs

Zone folding optical and acoustic phonons are measured only from the
layered structure due to the destruction of a selection rule based on
the reduction of the symmetry in GaAs-$Ga_{0.7}Al_{0.3}As$ and GaAs-AlAs SLs
(Merz et al 1977). There is no report about Raman spectra from
strained-layered superlattices(SLS) such as GaAs-InAs though many
studies for GaAs-$Ga_{0.7}Al_{0.3}As$ and GaAs-AlAs have been already reported.
It is of   interest to study the thermal instability of SL which has
a shorter periodicity. However, only a few papers concerning
monolayered SL havebeen reported (Fleming et al 1980).

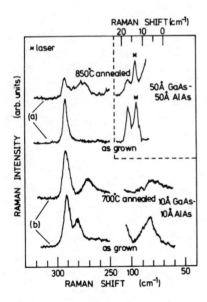
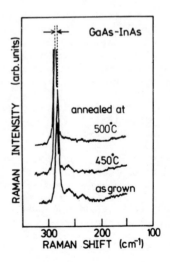

Fig. 1 Raman spectra from GaAs-
AlAs SLs
(a) 10 A GaAs-10 A AlAs
(b) 50 A GaAs-50 A AlAs

Fig. 2 Raman spectra from GaAs-
InAs SL for various
annealing temperatures

Figure 1 shows the Raman spectra from as-grown and thermally annealed GaAs-AlAs SLs whose periodicities are 20 or 100 A. Both optical and acoustic phonons were observed in the energy range from 5 to 105 cm^{-1} and from 250 to 320 cm^{-1} for various annealing temperatures, respectively. Peaks at about 290 and 282 cm^{-1} in the spectrum from 10 A GaAs-10 A AlAs SL annealed at 650°C for 2 hours correspond to the GaAs LO phonon and the zone folded phonon, respectively. The spectrum is almost identical to the spectrum obtained from as grown sample. In the case of 50 A GaAs-50 A AlAs SL annealed at 800°C for 2 hours, peaks at about 290 cm^{-1} is based on both GaAs LO and its zone folding phonons which can be also obtained from as grown sample. However, the zone folding optical and acoustic phonons disappeared after annealing at 700 and 850°C for the two samples respectively. The peak at about 272 cm^{-1} observed from both samples obtained from $Ga_{0.5}Al_{0.5}As$ random alloy at the same time. While the zone folding phonon is based on the periodicity of the sample disappeared by thermal annealing. In addition to this phenomenon, the zone folding acoustic phonon disappeared by thermal annealing. The results show that the random alloy semiconductor is produced from layered structure by thermal annealing.

Figure 2 shows the Raman spectra from GaAs-InAs SLS as grown and thermally annealed at 450 and 500°C in the spectral range from 150 to 310 cm^{-1} for various annealing temperatures. Peaks at 282 and 270 cm^{-1} obtained from as grown sample are based on GaAs LO and zone folding

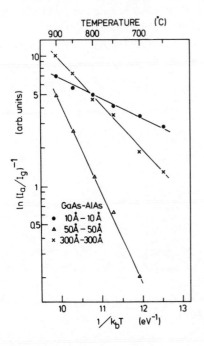

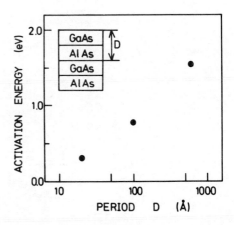

Fig. 3 Temperature dependence of variation for intensity ratio of $\ln(I_a/I_g)^{-1}$.

Fig. 4 Period dependence of activation energy for GaAs-AlAs SL.

phonons, respectively. Extra shift of GaAs mode from bulk GaAs which is
about 8 cm^{-1} was observed. The phenomenon is presumably due to stress
based on the difference of lattice constants. To characterize the
stress accumulated in the interface we adopt the model which is used in
the estimation of stress for SOS (Yamazaki et al 1984). In this model
the stress perpendicular to the interface is neglected. The estimated
stress which is tensile is between 2.7 and 6.4 x 10^{10} dyne/cm^2.

The zone folding optical phonon obtained from thermally annealed GaAs-
InAs SLS disappeared by thermal annealing at 500°C but we did not
observe any alloying effect. These results show that strained layered
superlattices such as GaAs-InAs SL are less stable than stress-free SL
such as GaAs-AlAs.

3.2 Exchange energy of atoms in alternative structure

There are a few works relating to the structure of interface in SL
theoretically and experimentally (Van Vechten 1980, Fleming 1980). We
have studied the activation energy of exchanging the GaAs to other
materials such as $Ga_{1-x}Al_xAs$ from relative intensity ratio of GaAs mode
from as grown sample against that mode from annealed sample.

Figure 3 shows the relation between annealing temperature and relative
intensity $\ln(I_a/I_g)^{-1}$. The intensity ratio $\ln(I_a/I_g)^{-1}$ is used to
estimate volume ratio between amorphous and crystalline in annealed Si
(Burke et al 1952). I_a and I_g are intensity GaAs LO modes from annealed
and as grown samples, respectively. If I_a and I_g correspond to the

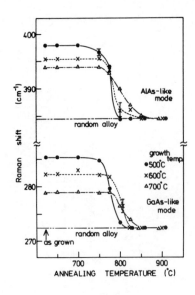

Fig. 5 Annealing temperature dependence of wavenumber
for zone folding optical phonons for GaAs-AlAs
grown at various temperatures.

volume fraction of crystallinity before and after annealing, the Avrami
expression (Burke et al 1952) for the volume fraction is given by

$$I_a = I_g \exp(-Bt), \tag{1}$$

where $B = \nu \exp(-E_a/k_b T)$. Annealing period was 1 hour. Samples used in
the experiment were x GaAs-y AlAs SLs where x=y=10, 50, or 300 A,
respectively. The slope of these lines shows the activation energy of
alloying from GaAs to other materials such as $Ga_{1-x}Al_xAs$ random alloy.
The relation between an activation energy and period of GaAs-AlAs SL
obtained from Fig. 3 is shown in Fig. 4. The result shows that the
sample having shorter periodicity has a lower activation energy. Van
Vechten (1980) and Fleming et al (1980) have reported the activation
energy of 0.25 and 0.201 eV, respectively. These values are almost
identical to that, 0.28 eV for 10 A GaAs- 10 A AlAs sample, obtained in
our experimental results. The reason that SL with small periodicity has
a lower activation energy is due to excess energy at the interface such
as an   entropy energy and/or strain energy based on stress.

3.3 Dependence of thermal instability in GaAs-AlAs SL on growth
     temperature

There is a critical temperature of alloying in a GaAs-AlAs SL as
reported by Petroff (1978). The result showed that SL grown at lower
substrate temperature is more stable than SL grown at higher substrate
temperature. We have studied dependence of thermal instability in GaAs-
AlAs SL by annealing on the growth of temperature.

Figure 5 shows the annealing temperature dependence of the wavenumber of
the zone folding optical phonon due to GaAs and AlAs like modes.
Samples used in the experiment are 10 A GaAs-10 A AlAs SL grown at
various temperatures.

The shifts in both modes are observed for as grown samples. The reason
for this phenomenon is considered that mixing occurred at the interface
in superlattices, that is, mixing causes the reduction of a period
forming $GaAs-Ga_{1-x}Al_xAs-AlAs$ layered structure. In other word,
variation in composition at the interface of a sample grown at a lower
substrate temperature is sharper than that of a sample grown at a higher
temperature. Therefore, mixing is apt to occur at interface grown at a
higher temperature because the alloyed layer between GaAs and AlAs
reduces the interface energy. While, the sample grown at a lower
temperature is apt to become $Ga_{0.5}Al_{0.5}As$ alloy by thermal annealing.
The result is obtained from the temperature dependence of a shift of
wavenumber based on zone folding optical phonons as shown in Fig. 5.

The critical temperature of mixing for 10 A GaAs-10 A AlAs SL by thermal
annealing was estimated in the range from 750 to 850°C from the change
of the wavenumber of zone folding optical phonon shown in Fig. 5. This
value is larger than the temperature (T=610°C) obtained by Petroff for
$(GaAs)_1-(AlAs)_1$.

4. Conclusions

Thermal instability in GaAs-AlAs and GaAs-InAs SL was studied by Raman
spectroscopy. Samples of shorter periodicity are less stable than those
with a longer periodicity for GaAs-AlAs. Strain free superlattices such

as GaAs–AlAs are more stable than GaAs–InAs SLS. Superlattices grown at a higher temperature are less stable than those prepared at a lower temperature. The critical temperature of alloying in 10 Å GaAs–10 Å AlAs superlattice is in the range from 750 to 850°C. This value is larger than that for $(GaAs)_1$–$(AlAs)_1$, that is, 610°C.

Acknowledgement

The authors thanks Drs S. Hiyamizu and S. Muto for supplying samples and fruitful discussions. Some of the samples used in this study were made by Fujitsu Limited as a part of the R&D Project of Basic Technology for Future Industries sponsored by Agency of Industrial Science and Technology, MITI.

References

Burke J E and Turnbull D, 1952 Prog. Metal Phys. **3**, 220.
Chang L L and Koma A, 1976 Appl. Phys. Lett. **29**, 138.
Chu W K, Ellison J A, Picraux S T, Biefeld R M, and Osbourn G C, 1984 Phys. Rev. Lett. **52**, 125.
Colvard C, Merlin R, Klein M V, and Gossard A C, 1980 Phys. Rev. Lett. **45** 298.
Fleming R M, McWhan D B, Gossard A C, Wiegmann W, and Logan R A, 1980 J. Appl. Phys. **51**, 357.
Merz J L, Barker A S Jr., and Gossard A C, 1977 Appl. Phys. Lett. **31**, 117.
Okamoto H, Seki M, and Horikoshi Y, 1983 Jap. J. Appl. Phys. **22**, L367.
Petroff P M, 1977 J. Vac. Sci. Technol. 14, 973.
Petroff P M, Gossard A C, Wiegmann W, and Savage A, 1978 J. Crystal growth **44**, 5.
Sai-Halasz G A, Pinczuk A, Yu P Y, and Esaki L, 1978 Surface Science **73**, 232.
Sapiel J, Michel J C, Toledano J C, Vacher R, Kervarec J, and Regreny A, 1983 Phys. Rev. **B28**, 2007.
Van Vechten J A, 1980 , Handbook on Semiconductors, edited by S. P. Keller North-Holland, Amsterdam.
Weisbuch C, Dingle R, Gossard A C, and Wiegmann W, 1981 Solid State Comm. **38**, 709.
Yamazaki K, Yamada M, Yamamoto K, and Abe K, 1984 Jap. J. Appl. Phys. **23**, 681.

Inst. Phys. Conf. Ser. No. 74: Chapter 4
Paper presented at Int. Symp. GaAs and Related Compounds, Biarritz, 1984

259

Characterization of sharp interfaces in III−V compound semiconductors

P. M. Petroff

AT&T Bell Laboratories, Murray Hill, NJ 07974 USA

Abstract. This paper reviews some of the methods available for the characterization of sharp interfaces in III-V compound semiconductor structures. Techniques aimed at the structural, chemical and electronic analysis of interfaces are presented along with examples of heterostructures and superlattice interfaces.

1. Introduction

Crystal growth techniques such as molecular beam epitaxy (MBE) and metalorganic chemical vapor deposition (MOCVD), are now producing epitaxial layers of III-V compound semiconductors with extremely sharp interfaces. These deposition techniques also allow to control the degree of sharpness of the interface from a structural and compositional point of view. Such control has permitted the fabrication of superlattices, epitaxial layers with small dimensions and made possible a range of structures with dimensions of the order of the electron wavelength (300−500Å). A large number of new devices using sharp interface structures have been proposed; e.g., the modulation doped field effect transistor,[1,2] the stair case avalanche photodetector,[3] heterojunction bipolar transistors,[4] etc.

The characterization of sharp interfaces has become essential for perfecting such devices since both the structure and composition of the interface are important to their characteristics. In this paper, we review some of the characterization techniques for this type of interface and illustrate their applications by using selected examples.

2. Structural Microcharacterization

The structural characterization establishes both the crystal structure of the epitaxial films at the interface and the crystal perfection of the interface, e.g., microdefects and steps. The main technique for this type of analysis is transmission electron microscopy (TEM). The preferred sample, electron beam configuration, is that shown in Figure 1, where the incident electron beam is parallel to the interface. Both diffraction information and real space imaging of the interface provide structural information.

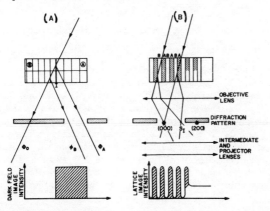

Fig. 1 Schematic of the sample−electron beam interactions for the analysis of interfaces by transmission electron microscopy.

(A) Single interface case, dark field imaging

(B) Superlattice case, lattice imaging

2.1 Diffraction Analysis

In modern electron microscopes, transmission electron diffraction (TED) of the epitaxial layers on both sides of the interface provides the crystallographic information from areas as small as 20Å in diameter. The lattice parameters for each epitaxial layer can be measured with a precision $\dfrac{\Delta a}{a} \cong 10^{-3}$. Tetragonal distortions in a plane perpendicular to the electron beam are also obtained from TED. The technique of convergent beam diffraction (CBD) which requires a convergent electron beam incident on a small area (20Å in diameter) establishes tetragonal distortions in a plane parallel to the interface. In addition, crystal space group symmetries are readily obtained from the CBD pattern.[5]

The precision in the lattice parameter and tetragonality measurements are strongly affected by the thin film nature of the sample. Indeed, the elastic relaxation at the film surfaces and the absence of a thick substrate in samples of the type shown in Figure 1 produce a different strain tensor than in the original structure. Thus, precision measurements of misfit strains using CBD and TED are rarely accurate.

Superlattice structures consisting of a periodic A-B-A-B epitaxial films are conveniently analyzed by TED.[6] The periodic crystal potential variation arising from the different epitaxial films behaves as a diffraction grating for the coherent parallel incident electron beam (Figure 1) and produces superlattice satellite spots on the TED pattern. The satellite spacing, Γ, is related to the superlattice periodicity, T, along the (hkl) growth direction by

$$T = d_{(hkl)}(m+n) \tag{1}$$

with

$$m+n = \frac{R_{(hkl)}}{\Gamma} \tag{2}$$

where $d_{(hkl)}$ is the interatomic distance along the (hkl) direction, m and n are the number of atomic layers of A and B respectively along <hkl> direction and $R_{(hkl)}$ is the distance between two adjacent [hkl] Bragg diffraction spots. For a superlattice period T smaller than the electron beam spatial coherency ($\approx$200–300Å), the TED analysis gives an accurate measurement of the periodicity ($\approx$5% error). It is also possible from quantitative measurements of the satellite intensities to establish the sharpness of the interface along the (hkl) direction (see Section 3). As an illustration of the diffraction method applications and limitations, we show the analysis of $(GaAs)_m–(AlAs)_n$ superlattices deposited by MBE on (100) misoriented GaAs surfaces. Here m and n are integer or fractional numbers $\geq$0.4 of monolayers.

A cross section through an idealized $(GaAs)_{0.50}–(AlAs)_{1.0}$ superlattice is shown in Figure 2A. The two periodicities in this superlattice are defined by $\overline{OA}$ and $\overline{OB}$. The computed TED pattern for an electron beam parallel to the $[0\overline{1}\overline{1}]$ direction is shown in Figure 2B. The misorientation angle α is measured from the TED pattern:

$$\alpha = (\overline{O'A'}, \overline{O'B'})$$

where the superlattice satellite spots are A', A'', A''' and the superlattice periodicity along the [100] direction is given by

$$(m+n)\, d_{(hkl)} = \frac{|\overline{O'B'}|}{|\overline{O'H}|} \tag{3}$$

In fact, such superlattices have been deposited by MBE on (100) GaAs vicinal surfaces, demonstrating the hyperfine control of the crystal growth that is achievable with this technique.[7] Figure 3 gives the TED pattern of two such superlattices deposited on GaAs (100) vicinal surfaces misoriented by 1° and 2°, respectively. The measured values of m+n in the superlattice are 2.33 and 2.06, and the expected value from the deposition rates and shutter sequence is 2.22. The measured angle of the deposited layers with the [100] axis are respectively 1° and 2°. These results indicate that the deposition sequence resulted in a superlattice similar to that shown in Figure 2A. We should note, however, that this interpretation is still not

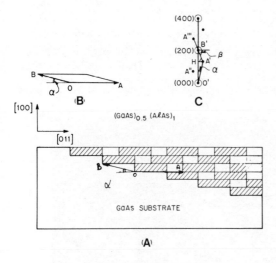

Fig. 2 (A) Cross section schematic of idealized $(GaAs)_{0.5}$-$(AlAs)_1$ super-lattice. The shaded and unshaded areas correspond to AlAs and GaAs, respectively.

(B) Primitive cell of this superlattice.

(C) Computed TED pattern for an electron-beam incident along the $[0\overline{1}\overline{1}]$ direction.

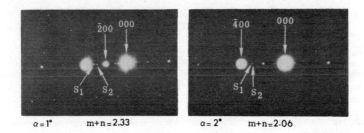

Fig. 3 TED pattern of a cross section through a $(GaAs)_{1.78}$-$(AlAs)_{0.44}$ superlattice deposited on (100) vicinal surfaces misoriented by 1° and 2°, respectively. The measured values of (m+n) are respectively 2.33 and 2.06.

unique. Indeed, a superlattice with composition $(GaAs)_m$-$(AlAs)_n$ with fluctuating m and n values from layer-to-layer would give a similar diffraction pattern providing their mean values are $\overline{m} = 1.78$ and $\overline{n} = 0.44$. To remove the nonuniqueness character of TED interpretation, the computed diffracted intensities of satellite spots using a model for the interface structure have to be compared and matched to the experimental ones. An alternative approach is the use of very high resolution TEM.

2.2 High Resolution Lattice Imaging

Lattice planes and atomic columns are directly imaged by this technique which requires very thin samples (50–500Å) and an electron microscope with a point-to-point resolution of $\cong 2\text{Å}$ or better. The sample geometry required for this type of interface analysis is that shown in Figure 1. Several diffracted beams are used for the image formation.

Under appropriate defocus and sample thickness conditions, the micrograph is a true representation of the atomic columns and interface positions in the sample. The main limitation of this method resides in the small volume of interface sampled. New, very high resolution instruments will eventually be able to distinguish the group III from group V elements in the images, thus opening up interesting new insights in the interface studies.

An example of lattice imaging is shown for the case of very small dimensions $(GaAs)_2-(AlAs)_2$ superlattices deposited by MBE on a (111) GaAs substrate. The high resolution micrograph in Figure 4 indicates that the superlattice is indeed ordered over a large volume (300–500Å in diameter). Atomic steps during growth produce antiphase boundaries as shown in Figure 4.

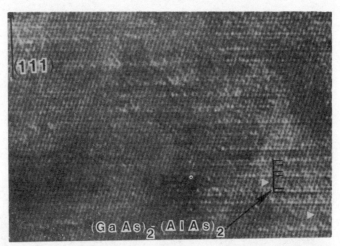

Fig. 4 High resolution TEM micrograph of a $(GaAs)_2-(AlAs)_2$ superlattice grown by MBE on a (111) GaAs substrate. The AlAs and GaAs bilayers appear as bright and dark atomic planes, respectively. Faults in the growth sequence are indicated by arrows. This figure should be viewed at grazing incidence along the direction of the white arrow.

2.3 Other Methods

A number of nondestructive methods are used in strain measurements at interfaces. Among these, Rutherford backscattering (RBS) and channeling methods[9–11] and X-ray methods based on double crystal diffraction[12] experiments have been applied to strained heterostructure interfaces and strained layer superlattices (SLS).

Strain at an heterostructure interface or in a SLS is detected by comparing its random RBS and channeling spectra with that of a strain-free epitaxial layer of the same crystallographic group. In the zinc blende structures, decreasing He^+ ion channeling yields are respectively observed for the <111>, <100> and <110> directions. A study of InAs-GaSb SLS[9] containing strained interfaces (lattice mismatch 0.7%) shows that the channeling yield along the <110> axis is always larger than along the <100> axis. The RBS analysis supports the existence of a strained interface in which the atom positions along the growth direction are aligned above each other across the interface. The misfit strain produces a uniform change in the SLS lattice parameter in a direction parallel to the interface and periodic variations of the lattice parameter perpendicular to the interfaces for each layer of the SLS.

With a more refined method using channeling angular scans,[11] it is possible to directly measure the interfacial strain with a high depth resolution. This method has been applied to measure the strain parallel and perpendicular to the interfaces of an InGaAs-GaAs SLS. The depth resolution allows to directly measure the strain for each layer of the SLS near the surface. This method is simple, nondestructive and applicable to heterostructures in which the signal of the top layer is resolved by ion beam technique.

More sensitive techniques based on double crystal X-ray measurements have been applied to superlattice structures. Superlattices with small interfacial strains, such as GaAs-GaAlAs, have been analyzed to measure the strain in a direction perpendicular to the crystal surface.[8] Recently, double crystal X-ray rocking curve methods[12] have been applied to measure the parallel and perpendicular strains in a GaAsP/GaP SLS. These measurements give an average value of the strain and show that most of the strain in this type of SLS is accommodated plastically by misfit dislocations at the buffer-substrate interface. This strain measuring technique is model dependent since it requires a matching of the experimental and computed rocking curve profiles for two crystallographic orientations of the sample with respect to the incident X-rays.

Both ion channeling and X-ray methods are macroscopic in nature and, thus, require uniform interfaces. In addition, the interpretation of the data is simple only for atomically sharp interfaces that are defect free.

3. Chemical Microcharacterization Methods

Compositional analysis at a sharp interface is difficult to achieve and only qualitative and semiquantitative information is usually attainable. The most direct microcharacterization techniques rely on the use of a scanning transmission electron microscope (STEM) with very small electron probes ($\cong 5\overset{\circ}{A}$) or TEM in the dark field imaging mode.

In the case of superlattices with small periods (20 to $200\overset{\circ}{A}$), an analysis of the superlattice satellite intensities and a comparison with computed intensities makes it possible to estimate the compositional roughness at an interface.[13]

3.1 Structure Factor Contrast and Compositional Analysis

The difference in atomic scattering factors between the group III and group V elements for certain Bragg reflections (e.g., (200), 420 ...) makes it possible to follow the chemical composition at an interface. This TEM method has been used to follow the chemical composition of sharp and diffused interfaces in GaAlAs-GaAs heterostructures produced by MBE.[6] An illustration of the method to evaluate compositionally graded GaAlAs-GaAs (MBE) interfaces is shown in Figure 5. The dark field transmitted intensity profile shows the sharp interface between the GaAs quantum wells and GaAlAs barrier layers and the expected parabolic variations for the linearly graded interfaces between the GaAlAs and GaAs. A deconvolution of this intensity profile taking into account the quadratic dependence of the (200) diffracted intensity as a function of Al content[6] indeed shows a linear compositional grading for the interface.[14] This technique is the only one presently available to evaluate compositionally graded and sharp interfaces with a high resolution ($\approx 5-8\overset{\circ}{A}$) and sensitivity (2%).

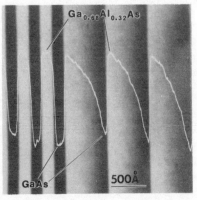

Fig. 5 Dark field TEM micrograph of a cross section through a structure containing GaAlAs (bright areas) barriers and GaAs triangular wells with one compositionally graded interface. The white trace is a densitometer scan of the (200) dark field diffracted intensity used for the image formation.

3.2 Very High Resolution Compositional Analysis

Earlier computations of very high resolution (TEM) images[15] indeed indicate that chemical analysis is achievable for GaAs and $Ga_{1-x}Al_xAs$ heterointerfaces by comparing computed and experimental images. Using dynamical image simulations, the choice of focus and sample thicknesses which make [110] axial lattice images of the interface most sensitive to variations in composition x have been investigated.[15] The atomic column separating the Ga and As <110> rows should be resolvable with the newest TEM instruments; however, the calculation was performed here for a microscope with point-to-point resolution larger than 2Å and thus only the GaAs and AlAs columns are resolved. There are optimal focusing conditions which make the contrast of the image a very sensitive function of x. Figure 6 shows the computed images of $Ga_{1-x}Al_xAs$—GaAs interfaces between x = o and x = 0.25 and x = 0.5 for two such defocus conditions. It is evident from this type of simulations that for each x value there is an optimal focusing for imaging the compositional variation across an interface. In the future this technique should take more importance because of its ultra-high spatial resolution and good sensitivity ($\frac{\Delta x}{x} \cong 0.1$).

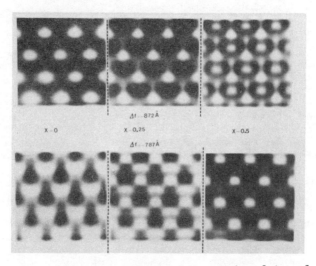

Fig. 6 Computed high resolution electron micrographs of interfaces between GaAs and $Ga_{0.75}Al_{0.25}As$ and $GaAs_{0.5}Al_{0.5}As$ interfaces for a sample thickness of 250Å and two values of defocusing. The electron beam energy is 200 KeV. The interfaces are indicated by the dotted vertical lines.

3.3 Other Methods

Numerous other methods for characterizing interfaces chemically are presently available. These methods all suffer from a lack of spatial resolution due to the probe beam spreading (STEM based techniques)[16] or ion knock on effects (SIMS, AES). The deconvolution procedure required to surmount these problems, in general limits these methods to a depth resolution of 30 to 50Å for compositional analysis of sharp interfaces.

Rutherford backscattering (RBS) is the only nondestructive technique which, when used in conjunction with ion channeling, will give a qualitative and quantitative chemical analysis of the interface[17] with a depth resolution of $\cong$30 to 50Å. For the amorphous-crystalline interface the spatial resolution is improved to a few Angstroms.[17]

4. Methods for the Electronic Microcharacterization of Interfaces

The electronic characterization of deep traps and defects at interfaces on a microscopic scale comes as a complement to the macroscopic methods such as capacitance-based measurements and deep level transient spectroscopy. The microscopic methods should provide, in addition to a spectroscopic information and depth resolution, a good spatial resolution. The methods are usually aimed at: (a) locating the depth of an interface and the doping profile as a function of distance from the interface or (b) characterizing deep levels or luminescence centers associated with impurity complexes at an interface. In fact, very few methods provide all these informations simultaneously. Both the electron beam induced current (EBIC) and scanning deep level transient spectroscopy (SDLTS),[16] when performed with the electron beam parallel to the interface, are limited in resolution by the minority carrier diffusion length. However, when the minority carrier diffusion path can be affected by an electric field or a potential barrier (such as in an heterostructure), the spatial resolution of the method can be greatly improved. The following example illustrates this point.

The new technique of field dependent cathodoluminescence[18] (FDCL) is an adaptation of the sensitive cathodoluminescence (CL) technique for the electronic characterization of interfaces. By moving a high electric field across a heterojunction or p-i or i-n homojunction interface, the diffusion path of the electron-beam induced minority carriers in each layer is affected. The high field region is produced at the edge of a depletion layer. By applying a bias voltage, V_R, the edge of the depletion layer is moved across the interfaces. The associated changes in the minority carrier diffusion and drift paths are then controlling the CL efficiency of each layer in the structure. A CL quenching is observed each time V_R reaches the punch-through voltage, V_{PT}, for an interface. By following the CL intensity of each layer as a function of V_R, the interfaces can be accurately located.[18] In addition, the high electric field will, in some cases, field ionize deep levels located inside the layer or at an interface. Introduction of these traps then controls (a) the minority carrier lifetimes and (b) the width of the depletion layer. The CL efficiency of each layer then reflects the presence of the ionized deep levels and the hysteresis in the CL versus bias voltage of each layer will permit their localization in the structure. For this method to be applicable, the doping and layer thicknesses must be chosen so that the low band-gap layers are undepleted at zero bias and the punch-through voltage is lower than the device break-down voltage.

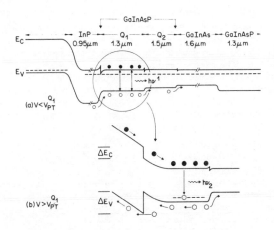

Fig. 7 Energy band diagram of an InP-InGaAsP-InGaAs APD and minority carriers behavior:

(a) $V_R = 0$

(b) $V_R > V_{PT}^{Q_1}$

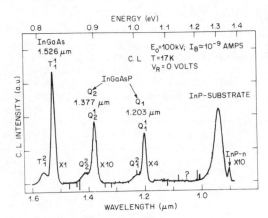

Fig. 8 Cathodoluminescence spectrum of the InP-InGaAsP-InGaAs APD shown in Figure 7 for V_R = 0 Volts.

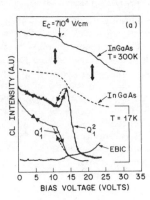

Fig. 9 Cathodoluminescence quenching versus reverse bias voltage for the InP-InGaAsP-InGaAs APD. The plotted CL emission lines are those shown in Figure 8. The punch-through voltage of the InP/Q_1 and Q_1/Q_2 interfaces indicated by arrows were determined from C-V measurements.

This technique has been applied to several heterostructures with abrupt or graded interfaces.[18] As an illustration of FDCL, we show the analysis of heterostructure interfaces in an avalanche photodiode (APD). The device structure consists of a large band-gap InP p$^+$-n junction supporting the avalanche adjacent to a narrow band-gap epilayers (InGaAsP and InGaAs n-type $1-4 \times 10^{16}$ cm^{-3}) absorbing the radiation. The energy band diagram and the CL spectrum for these devices are shown in Figures 7 and 8, respectively. The luminescence of the two quaternary layers, Q_1 and Q_2, consists of four lines, Q_1^1, Q_2^1, Q_1^2 and Q_2^2. The variations in the CL emission for the Q_1 and InGaAs (T_1^1 line) layers versus V_R at 17°K and 300°K are shown in Figure 9. The voltage dependent EBIC signal of this structure is also shown in Figure 9. A quenching of the CL intensity for the T_1^1 and Q_1^1 lines are detected for a value of V_R equal to $V_{PT}^{Q_1}$, the punch-through voltage for the InP-Q_1 interface. The arrows in Figure 9 indicate the interface position deduced from the capacitance-voltage technique. The increased Q_1^2 component has been related to field induced changes in the radiative path near the interface. A CL hysteresis (Figure 9) is observed after moving the depletion layer through and back across the interface. The size of the hysteresis loop has been observed to have a temperature dependence. This hysteresis effect has been ascribed to the presence of field ionized deep levels at the InP-Q_1 interface. The trap density estimated from the shift in V_{PT} is $\cong 10^{15}$ cm^{-3}. The trap distribution at an interface can then be mapped by monitoring the CL intensity of a given line at $V = V_{PT}$ and scanning the electron beam over the sample.

5. Conclusions

From this review which certainly does not cover all the methods available for the analysis of sharp interfaces, it should be evident that a combination of methods is needed to characterize their structure, chemistry and electronic properties. The most direct techniques for the structural analysis are based on transmission electron microscopy and transmission electron diffraction. On a macroscopic scale, ion channeling methods are now capable of measuring interfacial strain and composition with the advantage of depth resolution. The chemical composition of interfaces is best obtained nondestructively but with a limited depth resolution by ion channeling and Rutherford backscattering methods. TEM based techniques with the higher resolution still need further improvement to provide reliable quantitative chemical information. The electronic characterization of defect states at interfaces is still difficult to achieve with a high depth resolution and sensitivity. To this end, the recently introduced field dependent cathodoluminescence method comes as a useful complement to deep level transient spectroscopy and other capacitance based techniques.

REFERENCES

[1] T. Mimura, S. Hiyamizu, T. Fuji and K. Narbu, Jap. J. Appl. Phys. *19*, L225 (1980).

[2] D. Delagebeaudeuf, P. Delescluse, P. Etienne, M. Laviron, J. Chaplart and N. T. Linh, Electron. Lett. *16*, 667 (1980).

[3] F. Capasso, W. T. Tsang and G. F. Williams, IEEE Trans. Electron Dev. *ED-30*, 381 (1983).

[4] J. R. Hayes, F. Capasso, A. C. Gossard, R. J. Malik and W. Wiegmann, Electron. Lett. *13* 410 (1983).

[5] J. W Steeds, *Introduction to Analytical Electron Microscopy*, eds. J. Hren, J. Goldstein and D. Joy, Plenum Press, NY, Chapter 15 (1979).

[6] P. M. Petroff, J. Vac. Sci. Technol. *14*, 4, 973 (1977).

[7] P. M. Petroff, A. C. Gossard and W. Wiegmann, Appl. Phys. Lett. *45*, 620 (1984).

[8] R. M. Fleming, D. B. McWhan, A. C. Gossard, W. Wiegmann and R. A. Logan, J. Appl. Phys. *51*, 357 (1980).

[9] Chin-An Chang, J. Vac. Sci. Technol. *B1*, (2) 346 (1983).

[10] S. T. Picraux, L. R. Dawson, G. C. Osbourn, R. M. Biefeld and W. K. Chu, Appl. Phys. Lett. *43*, 11, 1020 (1983).

[11] S. T. Picraux, L. R. Dawson, G. C. Osbourn and W. K. Chu, Appl. Phys. Lett. *43*, 930 (1983).

[12] V. S. Speriosu, M. A. Nicolet, S. T. Picraux and R. M. Biefeld, Appl. Phys. Lett. *45*, 3, 223 (1984).

[13] T. S. Kuan, Mat. Res. Soc. Symp. Proceedings, eds. W. Krakow, D. A. Smith and L. W. Hobbs, North Holland, NY, Vol. 31, p. 143 (1984).

[14] D. Werder, P. M. Petroff, A. C. Gossard and W. Wiegmann (to be published).

[15] A. Olsen, J. C. Spence and P. M. Petroff, Proceedings of Electron Microscopy Society of American, ed. G. W. Bailey, Claitor Publishing Division, Baton Rouge, LA 70821, p. 318 (1980).

[16] P. M. Petroff, D. V. Lang, J. L. Strudel and R. A. Logan, *Scanning Electron microscopy*, ed. O. Johari, AMF O'Hare 60666, Vol. 1, p. 325 (1978).

[17] L. C. Feldman and J. M. Poate, Ann. Rev. Mat. Sci. *12*, 144 (1982).

[18] N. Magnea, P. M. Petroff, F. Capasso, R. A. Logan and P. W. Foy, *13th International Conference on Defects in Semiconductors*, ed. L. C. Kimerling (to be published) (1984).

[19] N. Magnea, P. M. Petroff, F. Capasso, R. A. Logan and P. W. Foy, Appl. Phys. Lett. (to be published).

Inst. Phys. Conf. Ser. No. 74: Chapter 4
Paper presented at Int. Symp. GaAs and Related Compounds, Biarritz, 1984

269

Contactless measurement of sheet carrier concentration and mobility of thin layers on semi-insulating GaAs

N. Braslau

IBM Watson Research Center P. O. Box 218, Yorktown Heights, New York 10598, USA

Abstract A method has been developed which measures the reflection coefficient of a thin conducting layer on a semi-insulating wafer and its change in a static magnetic field. From this the sheet resistance and magnetoconductive mobility are found. This method is contactless, non-destructive, absolute and fast. The sample is mounted across a waveguide so as to permit the wafer to physically exceed the guide dimensions. Layers have been measured at 300 and 85K with good agreement with Hall measurements. Heterojunction layers have also been measured.

1. Introduction

The electrical evaluation of epitaxial layers on semi-insulating III-V substrates is usually made with a Hall sample. This requires the sacrifice of a portion of the wafer, which must be fashioned into a Van der Pauw shape or a bar, with subsequent placement and alloying of ohmic contacts. It would be very convenient to have a simple, contactless method for determining sheet resistance and mobility, which could be applied to full wafers or portions of them prior to device processing. It is desirable to measure samples both at room and liquid nitrogen temperatures, so that heterojuntion layers can be evaluated.

For sheet resistance determination, commercial units are available which measure eddy current loading of a VHF oscillator tank (Miller et al 1978). This is convenient at room temperature but difficult to modify for use at 77K. By detecting the eddy current and Hall voltage in a very elegant and delicate manner, both sheet resistance and mobility can be determined (Horiguchi et al 1978).

There is a long history of the application of microwaves to the measurement of semiconductor materials. Most of these techniques are for thick bulk samples (Brand et al 1961, Bryant and Gunn 1965), or for epitaxial layers on heavily doped substrates (Bichara and Poitevin 1964). Techniques in which the open end of a waveguide are terminated with the semiconductor sample (Pozhela et al 1980) are not suitable for thin epitaxial layers because of the presence of shunt radiation conductance. For such layers, careful shaping of the sample for a close fit in the waveguide is necessary (Srivastava and Jain 1971, Dixon et al 1974).

The mobility μ of semiconducting layers can be determined from measurement of the magnetoconductance in an external field B, making use of the Drude expression

$$\frac{\sigma}{\sigma_0} = \frac{1}{1 + (\mu B)^2} \tag{1}$$

where the subscript o indicates the zero magnetic field value. Molnar and Kennedy (1978), using samples in the shape of a post inside a waveguide, determined the mobility of ion-implanted GaAs layers. Koch (1975), employed this method to determine carrier lifetime (proportional to mobility) of the two-dimensional electron gas in Si MOSFETs.

None of these techniques seem to be in very wide use, however. In this paper, we will describe how one can measure rapidly both sheet resistance and magnetoconductive mobility in the same apparatus without the need of calibrating samples, for wafers of arbitrary cross-section, so long as they fill the waveguide cross-section.

2. <u>Apparatus</u>

2.1 Principle of Measurement

We employ a conventional rectangular waveguide whose wave impedance (ohms) is given by

$$Z_o = \frac{377}{\sqrt{1 - (\frac{f_o}{f})^2}} \qquad (2)$$

where f is the operating frequency and f_o is the cutoff frequency for the TE_{01} mode (6.557 Ghz for WR-90 X band (1.01 x 2.29 cm.) waveguide). The voltage reflection coefficient for an impedance Z terminating the guide is given by

$$\Gamma = \frac{(Z - Z_o)}{(Z + Z_o)} \qquad (3)$$

If a uniform resistive film of sheet resistance R (ohm square^{-1}) is backed by an adjustable short circuit, the admittance Z^{-1} is the sum of the film conductance R^{-1} and the susceptance of the shorted line. If the length of the shorted line is adjusted to give zero susceptance, corresponding to $(2n+1)$ quarter guide wavelengths from the short, the line is then terminated by the impedance $Z = R$ (Huxley 1947, pg. 143).

The epitaxial sample we wish to measure is this resistive film on a somewhat thicker dielectric substrate. It can be shown theoretically (Ragan 1948, pg. 218) and experimentally that the substrate behaves as a series reactance in the guide. Thus the short position can be adjusted to compensate for this additional reactance and one finds that the reflection coefficient as a function of short position shows a distinct minimum when the terminating impedance is purely resistive. This requires that the skin depth in the resistive film be large compared to its thickness, which is easily satisfied for layers of practical interest at X band.

2.2 Sample Mount

If one cuts the broad wall of the waveguide to insert a wafer in front of the short, a significant discontinuity is introduced which has a severe effect on the reflection coefficient. The so-called 'choke flange' (Ragan 1948, pg. 193) was introduced to make mechanical coupling of waveguide components with minimal discontinuity. This flange introduces a half-wavelength series branching line which presents zero series impedance to the main guide, and is widely used. If the axis of the waveguide is preserved, we have seen that it is possible to separate this choke flange from its mating flat flange by almost 2 mm. before significant effects on wave transmission and reflection are seen. This is more than enough for the insertion of semiconductor wafers.

When the wafer, larger than the guide cross-section, is held against the flat flange with a gap of this magnitude, the propagating mode is preserved across the section, the portion of the wafer within the guide is sampled, and that part within the radial series branching line acts to slightly modify the guide wavelength in that section. Near midband of the dominant mode propagation, this is almost undetectable.

Fig. 1 shows a drawing of our wafer mount. The wafer is held by a vacuum chuck against a brass plate with a section of guide milled in it, and it is, in turn, inserted into place and held with a second vacuum chuck. This assures preservation of alignment. The flanges are extended to provide mechanical alignment away from the 'choke' area and to provide cooling for the sample as described below. For samples too small for the vacuum chuck, two spring clips are provided, located so as to remain near the edges of the narrow section of guide where the tangential electric field is zero. An optical fiber penetrates the center of the broad guide wall at one bend and is positioned to transmit illumination from a tungsten lamp to the conducting channel of the sample when required. The mount is placed in the magnet gap so that B is normal to the wafer surface.

2.3 Waveguide Arrangement

The apparatus, with the exception of the sample mount itself, is assembled from standard microwave components. The magnitude of the reflection coefficient is measured with a conventional reflectometer, consisting of 20 db directional couplers sensing the reflected and forward waves, with their ratio displayed on a Hewlett-Packard Model 416B Ratiometer. The phase is sensed with the slotted line. A water-cooled oil bath provides a low drift environment for the

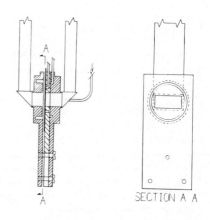

Fig. 1. Waveguide sample mount showing position of choke flange with respect to the sample.

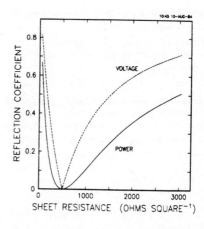

Fig. 2. Reflection coefficient vs sheet resistance calculated from eq. 3. The power reflection coefficient is the square of the voltage reflection coefficient.

Varian X-13 klystron, which is square-wave amplitude modulated at 1 Khz, as required for the Ratiometer. The frequency is set conveniently at 10.1 Ghz, where Z_0 is 507 ohms.

3. Measurement Procedure

The magnitude of the reflection coefficient (eq. 3) as a function of the sheet resistance of the sample is shown in Fig. 2. The response is double valued. The initialization of the measurement is made by replacing the sample with a metal plate and the reflection coefficient set to 1. The slotted line is moved to a standing wave minimum, which defines the phase of the short . Points on the branch of Fig. 2 below Z_0 will have this phase and the slotted line signal will be a local minimum. Points on the higher branch will be 90° out of phase and the slotted line position will be a local maximum. This removes the ambiguity.

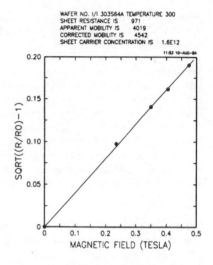

Fig. 3. Typical measurement showing least squares fit to the Drude expression (the 'apparent mobility'). R0 is the sheet resistance at zero field. The corrected mobility incorporates the calculation of Rees (1970) and Boardman (1971) to obtain Hall mobility.

The sample is put into place and the short circuit is adjusted to give a minimum in the reflection coefficient. Its value at this point gives the sheet resistance. The reflection coefficient as a function of static magnetic field is measured and the sheet resistance as a function of magnetic field is fit to the Drude expression, eq. 1, by the method of least squares.

A typical measurement result is shown in Fig. 3. The slope of this line gives the magnetoconductive mobility. In order to compare these measurements with the mobility determined from Hall measurements, it is necessary to use the results of a magnetotransport calculation by Rees (1970) and Boardman (1971). For room temperature, the Hall mobility is approximately 13% higher, and at 77K about 6% higher. This correction is included in the results shown in Fig. 3. The sheet carrier concentration is then

$$n_s = \frac{1}{R\mu q} \tag{4}$$

where q is the electronic charge. The measurement and computer data reduction takes only a few minutes; there is no mechanical stress on the sample.

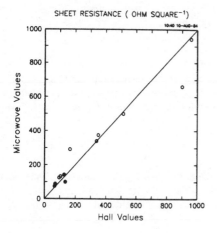

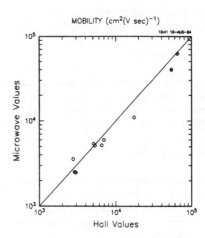

Fig. 4. Comparison of microwave measurement of sample sheet resistance to values obtained by Hall measurements. Dark dots are heterojunction samples at liquid nitrogen temperature.

Fig. 5. Comparison of microwave measurements of mobility to values obtained by Hall measurements. Dark dots are heterojunction samples at low temperature.

The waveguide sample holder fills most of the 7 cm magnet gap. In order to cool the sample, a simple stainless steel cup of square cross-section, covered by foam insulation, is filled with liquid nitrogen and raised so the level is slightly below the waveguide cross-section. The sample is cooled by conduction and convection. Thin walled stainless waveguide is used above the sample holder to minimize conduction losses. There is a slight temperature gradient (4K) across a mounted sample, and its mean temperature under equilibrium conditions is about 85K. For GaAlAs-GaAs heterojunctions, the temperature dependence of electron mobility is less than 1% deg^{-1} (Mendez 1984), so this measurement underestimates the two-dimensional electron gas (2DEG) mobility by less than 10%. Cooling of a mounted sample takes about 4 minutes.

4. Results

In Figs. 4 and 5, we compare microwave and Hall measurements on several wafers made by a variety of growth methods. Sheet resistance in the range 50-3000 ohm square^{-1} can be determined. As we are limited to magnetic fields of .5 T, the smallest mobility that can be measured is about 3000 cm^2(V sec)$^{-1}$. The value obtained is, of course, an average over the guide cross-section.

The dark dots in figs. 5 and 6 represent measurements on some heterojunction layers at liquid nitrogen temperature. The complication here is that shunt conductance in the doped AlGaAs is also sensed by the microwave signal. At low temperatures, the zero magnetic field sheet resistance of the 2DEG is low enough that the parallel combination is close to this lower value. With applied field, the resistance of the high mobility 2DEG increases rapidly so that the sample's sheet resistance tends to saturate at the value of the shunt layer, and the experimental data does not fit the Drude expression. In these cases, restricting the fit to very low field values is sufficient if the 2DEG mobility is high.

5. Conclusions

A simple microwave system has been described which measures sheet resistance and mobility of epitaxial layers on semi-insulating substrates. It is non-destructive, contactless, absolute, and fast. It should be useful for electrical evaluation of wafers as a replacement for conventional Hall measurements.

Acknowledgments

The author appreciates the free access to their collection of waveguide components provided by N. S. Shiren and T. G. Kazyaka. W. A. Kahn furnished imaginative machining. Encouragement, together with samples, were provided by S. L. Wright and T. F. Kuech.

References

Bichara M R E and Poitevin S P R 1964 IEEE Trans. IM-13 323
Boardman A D Fawcett W and Ruch J G 1971 Phys. Stat. Sol. (A) 4 133
Bryant C A and Gunn J B 1965 Rev. Sci. Inst. 36 1614
Dixon Jr S Giordano P F and Jacobs H 1974 J. Appl. Phys. 45 2570
Horiguchi F Matsumura H Furukawa S and Ishiwara H 1979 Jap. J. Appl. Phys. 18 Suppl 18-1 165
Huxley L G H 1947 A Survey of the Principles and Practice of Waveguides (Cambridge)
Jacobs H Brand F A Mendl J D Benanti M and Benjamin R 1961 Proc. I R E 49 928
Koch J F 1979 Festkorperprobleme XV ed H J Queisser (Pergamon/Vieweg) pp 79-112
Mendez E E Price P J and Heiblum M 1984 App. Phys. Lett. 45 294
Miller G L Robinson D A H and Ferris S D 1978 Semiconductor Characterization Techniques ed P A Barnes and G A Razgonyi The Electrochemical Society Proceedings 78-3 1
Molnar B and Kennedy T A 1978 J. Electrochem. Soc. 125 1318
Pozhela Y K Tolutis R B and Laurinavichyus A K 1980 Defektoskopiya (USSR) 9 35
Ragan G L ed 1948 Microwave Transmission Circuits Radiation Laboratory Series 9 (McGraw Hill,New York)
Rees H D 1970 J. phys. C 3 965
Srivastava G P and Jain A K 1971 Rev. Sci. Inst. 41 1793

Inst. Phys. Conf. Ser. No. 74: Chapter 4
Paper presented at Int. Symp. GaAs and Related Compounds, Biarritz, 1984

Pressure dependence of near-edge (5 K) recombination in high-purity III−V materials

D.J. Wolford, H. Mariette+, and J.A. Bradley
IBM Thomas J. Watson Research Center, Yorktown Heights, NY 10598 USA

Abstract. PL, PLES, and time-resolved PL have been studied in high-purity n-type GaAs and InP at 5 K and hydrostatic pressures of up to 80 kbar. Fast, sharp direct-edge PL arises from free and bound excitons and BA transitions. Above 41.5 kbar GaAs band structure becomes indirect; slower X_1-gap transitions resemble n-type GaP and reveal multiple donors. Excitons remain shallow and follow the gaps, thus giving precise Γ_1 and X_1 dependences on pressure (10.74 and -1.34 meV/kbar) and accurate band and level crossings. New indirect band gaps are established for GaAs.

1. Introduction

In semiconducting materials energy gaps and electronic states associated with them depend on external perturbations. Foremost among these perturbations is pressure which, if hydrostatic, preserves crystal symmetries while altering crystal properties through changes in lattice parameter. In this work we report new luminescence spectroscopy studies (Wolford et al. 1984a, 1984b), at *both* liquid He temperatures and large hydrostatic pressures, of undoped high-purity n-type GaAs and InP.

To accomplish this, we obtain pressures which are hydrostatic at 5 K to better than 1 part in 400 at up to 80 kbar in diamond anvil cells. By extensive photoluminescence (PL), PL excitation spectroscopy (PLES), and fast time-resolved PL, we identify and trace versus pressure shallow intrinsic and impurity bound states surrounding *both* the direct and indirect band gaps. Because of high spectral resolution and hydrostatic conditions, we obtain precise Γ_1- and X_1-gap variations with pressure, and accurate band and level crossings which conflict with the most recent pressure studies (Welber et al. 1975, Yu et al. 1978, Olego et al. 1980). In addition, we establish new indirect band gaps (Aspnes 1976) for GaAs at atmospheric pressure and prove that, in contrast to recent claims (Kobayashi et al. 1983), direct-gap donor excitons do not become anomalously deep near Γ_1-X_1 crossover. Results presented here represent the most comprehensive and spectroscopically detailed account of the influence of pressure on shallow electronic states yet reported for direct-gap III-V materials.

2. Methods

The GaAs was prepared by VPE on (100) substrates. Layers were undoped high-purity n-type ($\sim 2\times10^{14}$ cm^{-3}) and 15 μm thick. The InP was undoped n-type ($\lesssim 10^{15}$ cm^{-3}) prepared by VPE on (100) substrates. Samples were mechanically lapped and etched to thicknesses of $\sim$ 40 μm and cleaved into 100-μm-square die. Single die were loaded into diamond anvil pressure cells (Jayaraman 1983), together with a 10-μm ruby chip and a pressure transmitting fluid. Measurements were made at 5 K in flowing He. Pressures were deduced from calibrated ruby (R1) emission (Jayaraman 1983).

Photoluminescence (PL) was excited by the focussed 488.0 nm line of an Ar^+ laser at power densities of $1\text{-}6\times10^3$ W/cm^2. PL excitation spectra (PLES) were obtained with a tunable cw dye laser (R110). Time-resolved PL was obtained with a mode-locked, cavity-dumped, dye laser (R6G) of average power 2 mW and pulse width $\lesssim 20$ psec. PL was collected in backscattering and analyzed with a 0.75 m double-grating spectrometer, an RCA 31034 or a Varian 154A photomultiplier, and photon counting. Time decays were measured by time-correlated single-photon-counting with a resolution of 0.6 nsec.

3. Results

Low temperature PL spectra of the high-purity GaAs studied under pressure are shown in Fig. 1 at atmospheric pressure. To illustrate sample purity and help identify observed transitions, data for our usual 5 K operation and 1.8 K are compared, and presented together with 1.8 K PL from a more compensated high-purity MOCVD sample. At 1.8 K the VPE GaAs shows the dominant transitions of excitons bound to neutral (D_X^0) and ionized (D_Γ^+) direct donors, together with the intrinsic free exciton-polariton, and 2-electron Auger recombination at donors (Heim et al. 1974). In the comparison MOCVD sample, because of greater residual acceptor (C) doping, bound exciton recombination at the neutral acceptors (A°) is also detected; we present this to later illustrate the basis for possible influence of acceptors in PLES absorption data. The spectrum at 5 K may be identified predominantly as exciton recombination at the D_Γ^+ donors (indistinguishable from $h-D_X^0$), with weak free

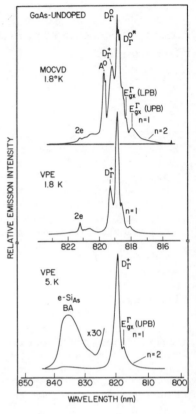

Fig. 1 PL of high-purity GaAs at atmospheric pressure.

exciton emission, and still weaker band-to-acceptor (BA) emission involving residual Si acceptors (Ashen et al. 1975).

Results from the VPE GaAs in a diamond cell are shown in Fig. 2 versus applied pressures from 4.07 to 60.3 kbar. For convenience, the spectra are aligned vertically according to the principal exciton transition. As in Fig. 1, pressures corresponding to direct gaps yield spectra composed of a strong D_Γ^+ exciton line whose intensity is nearly independent of pressure (note scales), and weaker free excitons (n = 1 and 2) and BA transitions. Above ~ 40 kbar a new set of weak, sharp lines appear which resemble donor PL in n-type GaP (Dean 1967) and we attribute to excitons bound to neutral indirect donors (D_X^0) and their weak momentum-conserving phonon replicas. These represent the first clear spectra in GaAs showing optical transitions involving impurity states attached to indirect minima (Wolford et al. 1984a, 1984b). As the scale factors indicate, this new emission rapidly weakens as pressure is increased and band structure becomes more indirect.

We note that because of the large binding energy of X_1 donors (Dean 1967), at least two residual donor species (unresolved as direct donors) appear as D_X^0 lines in these indirect spectra above ~40 kbar. Although chemical identity of these donors is unknown, amphoteric Si is likely, as are the common residual group-VI donors (e.g., S, Se, or Te) (Wolfe et al. 1977, Cook et al. 1978). Note also in Fig. 2 an ~ 0.6 meV splitting of these D_X^0 lines. This may arise from a non-hydrostatic pressure component which lifts the degeneracy of the valence band edge. The D_X^0 excitons themselves thus provide a sensitive calibrated measure of sample stress. From detailed stress results for D_X^0 states in GaP (Mathieu et al. 1979), we deduce a non-hydrostatic component of at most 1 part in 400, for our purposes, a virtually negligible stress.

Results of similar PL studies at more than 60 pressures between 0 and 80 kbar are plotted in Fig. 3. These indicate shallow "effective-mass" states cling to their respective band edges, thus giving the most precise and comprehensive measure of the GaAs band structure yet reported (Wolford et al. 1984a, 1984b). The D_Γ states located 6-8 meV below Γ_1 (Nam et al. 1976) shift at an apparent linear rate of <u>10.74 ± 0.05</u> meV/kbar, as do the BA transitions lying ~ 35 meV below the gap, and the free excitons located 4-6 meV (Nam et al. 1976) below the gap. These detailed results for direct states in GaAs (Fig. 3) agree remarkably well (in fact, within error) with our recent report of 10.7 meV/kbar for D_Γ PL in GaAs$_{1-x}$P$_x$:N (x = 0.03, 0.115) (Wolford et al. 1983, 1984c, 1984d) and with recent theoretical results of 10.5 meV/kbar (Lee et al. 1984).

At ~ 40 kbar Fig. 3 shows the D_X^0 levels abruptly appear in the gap and coexist (e.g., Fig. 2) with the D_Γ emission until ~ 43 kbar. This is the first such observation under pressure of simultaneous recombination of both direct and indirect impurity states near band crossing. The D_X^0 levels also shift linearly, again reflecting the pressure-shift of the corresponding X$_1$-edge, and yielding a value of

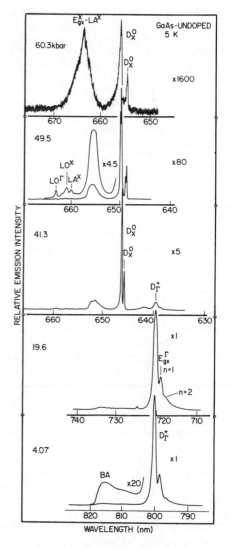

Fig. 2 PL of VPE GaAs (5 K) in a diamond cell.

<u>-1.34 ± 0.05</u> meV/kbar which agrees very well with our recent value for D_X^0 of -1.4 meV/kbar in GaAs$_{1-x}$P$_x$:N (Wolford 1983, 1984c) and with -1.49 meV/kbar measured for D_X^0 in GaP (Gil et al. 1983). Assuming these levels are bound by ~ 40 meV in GaAs as they are in GaP (Dean 1967, Sturge et al. 1978), a result confirmed in recent PLES studies (Wolford et al 1984a, 1984b, 1984d), we may also plot the X$_1$ energies in GaAs versus pressure shown in Fig. 3. From these results the precise Γ_1-X$_1$ band crossing of <u>41.5 kbar</u> at 5 K may be obtained which places the middle spectrum of Fig. 2 at this valley degeneracy. The remaining L$_1$ dependence shown in Fig. 3 is that reported by Aspnes (1976).

The dramatic difference in k-composition of the D_Γ and D$_X$ wavefunctions is illustrated in Fig. 4. Here are shown time decays of the D_Γ exciton at 39.2 kbar, and the D_X^0

exciton at 43.1 and 61.3 kbar. D_Γ emission is rapid, thus reflecting a zone-center "allowed" transition (and possible surface recombination), and giving a lifetime of $\lesssim 0.6$ nsec, our instrumental response time. In contrast, the D_X^0 line at 61.3 kbar shows an ~ 10 nsec buildup and a long 94 nsec exponential decay. This transient is consistent with a relatively forbidden indirect-gap transition. D_X^0 decay at 43.1 kbar is non-exponential and intermediate in time between the others, thus showing possible mixing of Γ_1 and X_1 states and band structure enhancement of D_X^0 recombination near crossover.

Figure 5 presents PL and PLES (upper) data at 33.7 and 40.3 kbar which, taken together, reveal the dominant absorption processes occurring near the GaAs band edges under pressure. In the top (PLES) spectrum BA intensity is monitored while the dye laser energy is varied. Here absorption peaks at the free exciton, the A° exciton resonance which is unresolved in the above-gap-excited PL (Fig. 1), and the D_Γ exciton resonance. At 40.3 kbar D_X^0 intensity has been recorded and the same direct exciton resonances dominate. In additional PLES we have also traced both the Γ_1 and X_1 edges above 40 kbar; results are reported elsewhere (Wolford et al. 1984a, 1984b, 1984c).

We have examined near-edge PL under pressure from other direct-gap high-purity III-V's. Figure 6 shows results from undoped n-type ($\sim 10^{15}$ cm^{-3}) InP. Identical direct-gap transitions occur as in GaAs (Fig. 2), except in this material D_Γ^0 -donor excitons dominate at 5 K (Dean et al. 1983). Again, all edge emission shifts smoothly with pressure, both that of the excitons and BA transitions, thus again marking the Γ_1 gap. Here the derived gap-shift amounts to ~ 7.9 meV/kbar, a value in good agreement with recent 300 K measurements (Muller et al. 1980). Detailed results will be reported elsewhere (Mariette et al. 1984).

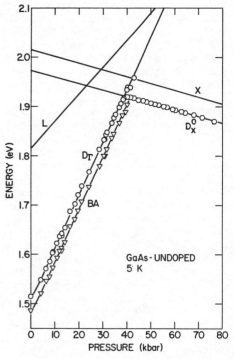

Fig. 3 PL energies of VPE GaAs (5 K) under pressure.

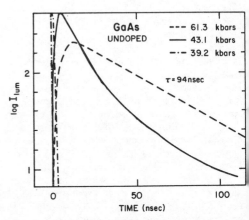

Fig. 4 PL time-decay of VPE GaAs (5 K) under pressure.

4. Discussion and Conclusions

Recent studies of GaAs under pressure have found decidedly different variations of the Γ_1 gap than reported here. Examples are a *nonlinear* dependence at 300 K such as $E_g(P)$

$\sim E_0(300\ K) + 12.6\times10^{-3}P - 3.77\times10^{-5}P^2$ eV (Welber et al. 1975, Yu et al. 1978, Olego et al. 1980) and, in contrast, a *linear* dependence of 8.5 meV/kbar at 120°K (Olego et al. 1980). Neither of these agree with the 10.74 meV/kbar shift we find for 0–40 kbar in GaAs or $GaAs_{1-x}P_x:N$ (x = 0.03, 0.115) (Wolford 1984a, 1984b, 1984d). Instead, our data agree more closely with certain earlier stress and pressure measurement (77–300 K) giving results of 10.5–11.3 meV/kbar (Welber et al. 1975, references therein). In determining gap variations from the shallow exciton and impurity states, we have neglected possible pressure dependences of their binding energies. Although such effects should, in principle, be considered, they will not contribute sufficiently in the 0–40 kbar range to account for the present discrepancy. Instead, we suggest it may result from difficulties in (1) applying gap fits at high pressures to data below 40 kbar (2) identifying transitions in broad optical absorption or PL of heavily doped materials and (3) obtaining sufficiently hydrostatic pressure media at low temperatures or high pressures (op. cit. 1975, 1978, 1980). Similar considerations may apply to the discrepancy between our derivative of -1.34 meV/kbar for the X_1 edge and literature estimates ranging from -1 to -3 meV/kbar (op. cit. 1978, 1980).

Based on the data in Fig. 3, we conclude the accepted indirect band gaps of GaAs at atmospheric pressure should be revised (Wolford 1984a, 1984b, 1984d). Aspnes (1976) quotes these as 1.981 eV for X_1 and 1.815 eV for L_1 at 0 K. However, extrapolation of the D_X^0 levels in Fig. 3 to P = 0 leads to the unacceptable result of degeneracy of these bound excitons with this X_1 energy. Instead, our data provide a new low temperature X_1 gap of $\underline{2.015 \pm 0.010}$ eV. Because Aspnes accurately deduces from core-level spectroscopy the X_1-L_1 separation as 170 ± 30 meV, not the respective band energies with respect to the valence band edge, we must also therefore reinterpret the L_1-gap as $\underline{1.845 \pm 0.040}$ eV. These new P = 0 band

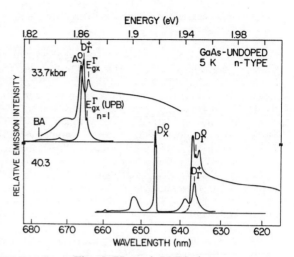

Fig. 5 PL and PLES (upper spectra) of VPE GaAs (5 K) under pressure.

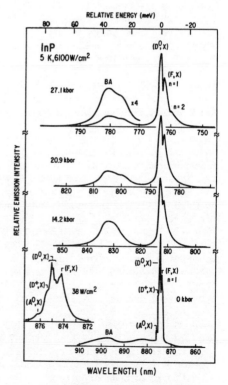

Fig. 6 PL of LPE InP (5 K) in a diamond cell.

gaps will necessarily influence existing composition dependences of band structure in GaAs-based alloys.

Our data also places in question recent report of pressure-induced deepening of donors in GaAs. Kobayashi et al. (1983) assert that in PL the D_Γ exciton abruptly deepens with respect to Γ_1 above 30 kbar and that such optical data supports the conclusion from Hall measurement in $Al_xGa_{1-x}As$ that direct donor activation energies necessarily increase near the Γ_1-X_1 crossover. Figure 3 demonstrate the D_Γ excitons are not detectibly perturbed above 30 kbar but instead smoothly follow the intrinsic states to and beyond crossover. Further, Fig. 3 proves that even the symmetry-forbidden level crossing (Altarelli et al. 1971) with the D_X^0 exciton (A_1) near 40 kbar does not induce detectible deepening of the D_Γ exciton. We conclude the dominant direct donor-exciton in our GaAs remains shallow as expected by effective mass theory. D_XA -pair PL near crossover may, in their more impure doped material, account for the anomalous results of Kobayashi et al. (1983).

Acknowledgements

We thank T. Kuech for supplying the MOCVD GaAs and V. Thierry-Mieg for supplying the InP. This work was supported in part by ONR under contract N00014-80-C-0376. +Permanent address: Laboratoire de Physiques des Solides CNRS, Meudon-Bellevue, France.

References

Altarelli M and Iadonisi G 1971 Nuovo Cim. 5B 36
Ashen D J, Dean P J, Hurle D T J, Mullin J B, White A M and Greene P D 1975 J. Phys. Chem. Solids 36 1041
Aspnes D E 1976 Phys. Rev. B14 5331
Cooke R A, Hoult R A, Kirkman R F and Stradling R A 1978 J. Phys. Appl. Phys. D11 945
Dean P J 1967 Phys. Rev. 157 655
Dean P J and Skolnick s 1983 J. Appl. Phys. 54 346
Gil B, Baj J, Camassel J, Mathieu H, Benoit a lu Guillaume C, Mestres N and Pascual J 1983 Phys. Rev. 28
Heim U and Heisinger P 1974 Phys. Stat. Sol. 66 461
Jayaraman A 1983 Rev. Mod. Phys. 55 65
Kobayashi M, Yokoyama T and Narita S 1983 Jpn. J. Appl. Phys. 22 L612
Lee S, Sanchez-Dehesa J and Dow J D 1984 to be published
Mariette H, Wolford D J and Bradley J A 1984 to be published
Mathieu H, Archilla B, Merle P and Camassel J 1979 Phys. Rev. 20 4268
Muller H, Trommer R, Cardona M and Vogl P 1980 Phys. Rev. B21 4879
Nam S B, Reynolds D C and Litton C 1976 J. Lumin. 12/13 277
Olego D, Cardona M and Muller H 1980 Phys. Rev. 22 894
Sturge M D, Vink A T and Kuijpers F P J 1978 Appl. Phys. Lett. 32 49
Welber B, Cardona M, Kim C K and Rodriguez S 1975 Phys. Rev. B1 5729
Wolfe C M, Stillman G E and Korn D M 1977 Inst. Phys. Conf. Ser. 33b 120
Wolford D J, Bradley J A, Fry K, Thompson J and King H E 1983 Inst. Phys. Conf. Ser. 65 477
Wolford D J and Bradley J A 1984a Bull. Am. Phys. Soc. 29 291
Wolford D J and Bradley J A 1984b Proc. 1st Inter. Conf. on Spectroscopy of Shallow Centers in Semiconductors, Solid State Commun.
Wolford D J, Bradley J A, Fry K and Thompson J 1984c Proc. 17th Inter. Conf. on the Physics of Semiconductors
Wolford D J and Bradley J A 1984d, to be published
Yu P Y and Welber B 1978 Solid State Commun. 25 209

Inst. Phys. Conf. Ser. No. 74: Chapter 4
Paper presented at Int. Symp. GaAs and Related Compounds, Biarritz, 1984

281

Defect clustering phenomena and related hopping type conduction in GaAs and near its surface

P. Langlade, S. Makram-Ebeid

Laboratoires d'Electronique et de Physique Appliquée
3, avenue Descartes - 94450 Limeil Brévannes (France)

Abstract - Electronic defects introduced by boron ion implantation have
been investigated. Differential Deep Level Transient Spectroscopic (DDLTS)
measurements have been performed to study the thermal electron emission
properties of these defects under different annealing and electric field
conditions. The leakage current induced by boron ion implantation is observed to anneal at the same temperature as the DLTS peaks. We interpret our
leakage current and DLTS observations by invoking a mechanism where a key
role is played by electronic hopping processes between mid-gap and shallower levels within defects clusters.

1. Introduction

Ion implantation induced defects play an important role in GaAs discrete
and I.C. planar technologies. These defects are used for the electron
carrier removal needed to delineate the active device layers and to isolate different devices sharing a common semi-insulating substrate. The
most frequently implanted species are oxygen and boron ions as well as
protons. We concentrate in this paper on boron ions.

Apart from carrier removal, ion implantation also gives rise to unwanted
leakage conductions. A recent study of Deng et al (1984) has demonstrated
that this current is due to a hopping mechanism between localized states
in the band-gap. Implanting GaAs with boron or heavier ions is known to
induce defects which are distributed in space and tend to be concentrated
in a few hundred angströms size clusters (Lang 1977). The defect density
within these clusters is so high that electron hopping between the different defects is probable. The small electron effective mass enhancesproba-
bility for multiphonon assisted tunnelling type of hopping between defects.

In this paper, we show that such hopping processes occurring within defect
clusters may help us interpreting the DLTS spectra observed in n-GaAs boron
implanted samples. The same type of process may also explain the annealing
behaviour of the ion implantation induced leakage current.

2. Experimental details
2.1. - Sample preparation

We use, as starting material, Bridgman n-GaAs with n = 6 to 10×10^{16} cm^{-3}.
These electron densities are of the same order of magnitude as those used
for the active GaAs MESFET layers (n = 1 to 2×10^{17} cm^{-3}). The native
electron trap concentration is made to decrease by capping the polished
surface with a 1,000 Å LPCVD Si_3N_4 layer and then annealing at 870 °C
for 15 mn. This treatment (Makram-Ebeid et al 1982) greatly reduces the
native electron trap concentration in a 1 µm thick layer near the surface.
After removing the Si_3N_4 capping layer, the n-GaAs surface is implanted
with different doses of boron ions ($\emptyset = 10^{10}$ to 10^{14} cm^{-3}) having 100 keV

kinetic energy. This is followed by a 15 mn anneal at temperatures ranging from 300 to 800 °C. Annealing is achieved in an open furnace by the close-contact technique under ion H_2 flux. This avoids surface As losses. Schottky diodes are then fabricated by evaporating gold electrodes (area = 6×10^{-2} cm^{-2}) on the surface.

2.2. Experimental techniques

We have made capacitive Differential Deep Level Transient Spectroscopy (DDLTS) measurements on the samples for which the average defect concentration was small compared to the initial free electron concentration. This condition was always met with boron ion doses smaller than 10^{11}cm^{-2}. For sample annealed above 500 °C, the boron ion doses could be taken higher. DDLTS results gave us information on the defect profiles and their sensitivity to the electric field.

3. Experimental results and interpretation

Fig. 1 shows typical DLTS spectra for boron implanted n-GaAs. The level EL2 was present in the sample prior to implantation. The peaks induced by ion implantation are EL5 and EL6 and a broad peak appearing below room temperature and labelled U-peak in conformity with the nomenclature of G.M. Martin et al (1982). Direct comparison with the literature is difficult because of the peak sensitivity to the electric field.

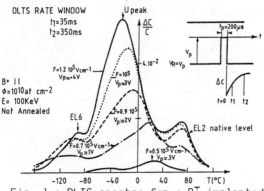

Fig. 1 : DLTS spectra for a B$^-$ implanted GaAs as a function of the electric field

3.1. Annealing behaviour

Fig. 2 shows the effect of a 15 mn anneal on the U-peak position. This position shifts to higher temperature for increasing annealing temperature (T_A). For T_A near or above 600 °C, the U-peak coïncides with EL2, the main native electron trap in bulk and VPE GaAs. Evidence of the creation of EL2 by such ion implantation and anneal has been first reported by Taniguchi and Ikoma (1982).

We are thus led to believe that U-peak is related to the mid-gap level EL2. We suggest that the U-peak can be attributed to electron emission from the mid gap level EL2 assisted by a prior hopping step to the shallower level EL6. Such a multiphonon assisted tunnelling hopping process is probable when the defect separations are of the order of (or smaller than) 100 A as in the case within defect

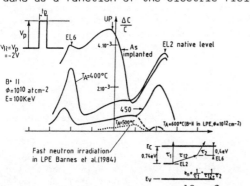

Fig. 2- DLTS spectra for the 10^{10}cm^{-2} implanted sample as a function of the annealing temperature

clusters. The electron emission rate e_n can be put in the form :

$$e_n = \frac{1}{\tau_1} + \frac{1}{\tau_2 + \tau_{12}} \qquad (1)$$

where $1/\tau_1$ and $1/\tau_2$ are the emission rates from isolated EL2 and EL6 defects respectively. The hopping transition between the two types of defects per unit time is $1/\tau_{12}$. The quantity τ_{12} is exponentially related to the defect separation. The broad DLTS and DDLTS peaks are in fact due to the wide statistical distribution of the τ_{12} time constants which give rise to very non-exponential electron emission transients. For high annealing temperature, the average defect separation increases and so do the τ_{12} values. This results in isolated defect behaviour.

Fig. 3a-3b show the annealing behaviour of the concentration profiles of the EL6 and U-peaks. Although some changes in the profiles are observable for annealing temperatures as low as 300 °C, the main annealing step, where the defect concentration sharply decreases, occurs above 400 °C. Different annealing behaviour for different distances from the surface can also be observed. Complex interaction between defect and defect migration may be responsible for this behaviour. A more detailed study will be published elsewhere (P. Langlade, S. Makram-Ebeid).

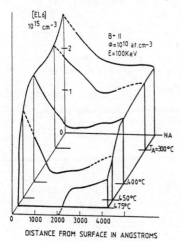 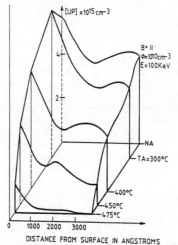

Fig.3 : Annealing behaviour of the concentration profiles
of a) EL6 ; b) U-peak

Fig. 4 shows the annealing behaviour of the low field Schottky diodes leakage current induced by the ion implantations. The main annealing step occurs above 400 °C and coincides with that of the DLTS peaks.

3.2. Electric field sensitivity of the U-peak

Fig. 5 shows the DDLTS peak position versus the corresponding electric field. The slope of the curve is seen to decrease when the annealing temperature T_A increases. Referring back to equation (1), the time constant τ_{12} was attributed to tunnelling between defects and expected to play an important role when the defect separation is small. Since τ_{12} is expected to vary with the applied external field, it is not surprising to observe a large electric field sensitivity before annealing (high defect densities in cluster) and a very small electric field induced change for high annealing temperatures (small defect densities).

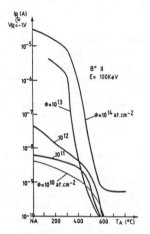

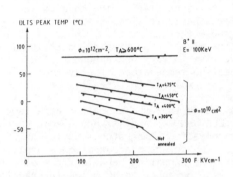

Fig. 4 - Annealing behaviour of the low field Schottky diode leakage current as a function of the implantation dose.

Fig. 5 - Electric field sensitivity of the U-peak with the annealing temperature.

4. Theoretical model

We have built a computer model to account for the electron emission proper-ties of the defect clusters. In this model, an electron trapped in the fundamental EL2 state (labelled 1 in Fig. 2) can be either directly emit-ted to the conduction band at a rate $1/\tau_1$ or via a shallower level 2 (EL5 or EL6) at a rate $1/(\tau_2 + \tau_{12})$. The inter-defect distance r_{12} is assumed to be small enough to permit multiphonon tunnelling transitions. In calculating the rate of transition $1/\tau_{12}$ between the two defects label-led 1 and 2, we had to include a large back transition probability from level 2 to level 1. We have dealt with the kinetic rate equation in a manner inspired from that of Rhees et al. (1980) who made a thorough investigation of the similar situation of a defect having a fundamental and an excited state. The rate $1/\tau_{12}$ is in fact significantly larger than $1/\tau_1$ only if the electron, after hopping to defect 2, is driven away from defect 1 at a sufficiently high rate $1/\tau_2$. One mechanism which can achieve this is again multiphonon electron tunnelling transition between neigh-bouring type 2 defects (with shallow energy levels).

The presence of an electric field can greatly enhance the rate $1/\tau_2$ and thus also the measured emission rates in accordance to observations. All the basic electron transition probability rates are evaluated by means of a generalization of the multiphonon tunnelling ionization quantum model of Makram-Ebeid and Lannoo (1982).

In addition to the preceding remarks, our model takes into account statist-cal aspects related to the random distribution of inter defect distances r_{12} . We have assumed a three dimensional Poisson distribution for inter-defect separation within the clusters and a reduced uniform concentration in regions connecting the clusters together. Among other things, the model accounts for the very non-exponential electron emission transients we observe (wide and distorted DLTS peaks). The change in DLTS peak shape and position with electric field and with the reduction of defect

densities after annealing is also correctly accounted for. The details of
the model will be published elsewhere. Example of the type of agreement
obtained with experiments can be seen in Fig. 6a-b.

5. Similarity with near surface defects

Fig. 7a-7b show DLTS spectra for unimplanted n-GaAs which has been annealed at 870 °C for 15 mn under Si_3N_4 cap. Although the native EL2 sharply
decreases near the surface, another broad peak appears. This peak is similar to the U-peak reported above and corresponds to a large defect density
($>10^{16}$ cm^{-3}) in a 600 Å thick layer near the GaAs surface prior to etchings
(curves a and b of Fig. 7).

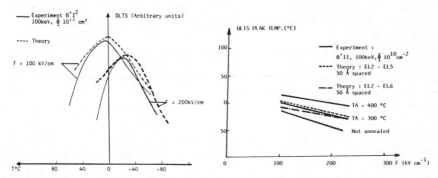

Fig. 6 a-b - Comparison between experimental and theoretical :
a) - DLTS spectra for the U-peak - b) - DLTS U peak temperature
sensitivity as a function of the electric field.

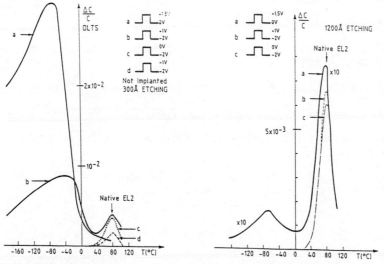

Fig. 7 - DLTS spectra for a n-GaAs sample annealed at 870 °C for
15 mn under an Si_3N_4 cap and then etched for a) 300 Å, b) 1200Å

6. Conclusion

We have studied in detail the deep level defects introduced by boron ion implantation in n-GaAs. The DDLTS peak position of the main defect is observed to depend on the post implantation   annealing temperature. The electric field dependence of the peak position also differs for different annealing temperatures. Both these observations suggest an electron hopping process between a mid gap electron trap and shallower levels. Detailed modelling of such a process and comparison with experimental results will be published elsewhere (S. Makram-Ebeid, P. Langlade).

The annealing behaviour of the DDLTS profile suggests complex defect recombination and long range migrations. The boron ion induced leakage current conduction is observed to have the same main annealing stage as the DLTS peaks, thus suggesting a correlation between defect densities and leakage currents.

We have also reported a process induced defects, which occur in great densities, near the GaAs surface after an 870 °C anneal under an Si_3N_4 cap. The similarity of the DLTS spectra with the U-peak suggests a common nature of the defects involved.

References

Deng X.C., Liu X.H., Boehringer K., Kalbitzer S., Appl. Phys. A 33 (1984), p. 29

Lang D.V., Kimerling L.C., Lattice defects in Semiconductors (1977), p. 581

Langlade P., Makram-Ebeid S., to be published

Makram-Ebeid S., Langlade P., to be published

Makram-Ebeid S., Gautard D., Devillard P., Martin G.M., Appl. Phys. Lett. 40, 2 (1982), p. 161

Makram-Ebeid S., Lannoo M., Phys. Rev. B., 25 n° 10 (1982) p. 6406

Martin G.M., Secordel P., Venger C., J. Appl. Phys. 53 (1981), p. 8706

Rees G.J., Grimmein H.G., Jenzen E., Skarstam B., J. Phys. C., 13 (1980), p. 6157

Taniguchi M., Ikoma T., Semi-insulating III-V materials, Evian (1982), p. 283. Edited by S. Makram-Ebeid and B. Tuck (Shiva).

Inst. Phys. Conf. Ser. No. 74: Chapter 4
Paper presented at Int. Symp. GaAs and Related Compounds, Biarritz, 1984

287

Surface and bulk traps in GaAs MESFETs

R.H. Wallis[*], A. Faucher, D. Pons[**] and P.R. Jay[***]
Thomson-CSF, Laboratoire Central de Recherches, B.P. 10, 91401 Orsay,- France
*Present address : GEC plc, Hirst Research Centre, East Lane, Wembley, England
**Present address : Enertec Schlumberger, 1 chemin de la Roseraie, 67380 Lingolsheim, France.
***Present address: Thomson-CSF, DHM, BP10, 91401 Orsay, France.

Abstract. The existence of hole traps in GaAs microwave FET devices, proposed to explain effects seen in conductance DLTS on short gatelength devices, is interpreted in terms of a mechanism related rather to surface depletion than bulk traps. Experiments on different gatelength FETs confirm the model, which also explains other effects such as "lagging" looping and frequency dispersion of transconductance.

1. Introduction

Gallium arsenide MESFETs frequently exhibit time dependent effects such as drain current drift and looping of the characteristics. One origin of such behaviour is deep traps located within the active layer of the device. Such traps may in principle be detected by DLTS ; although the capacitance of a short-gate FET is too small for observing capacitive transients, the transient response of the drain current may be used instead to monitor changes in trap occupation, as first suggested by Adlerstein (1976). A common feature of such conductance-mode DLTS experiments on short-gate devices has been the observation of peaks in the DLTS spectra whose sign corresponds to hole traps, frequently in surprisingly high concentrations (Adlerstein (1976), Zylbersztejn et al (1979), Itoh and Yanai (1980), Mitonneau (1981), Meignant et al (1979)). This is a surprising result, since hole injection would not normally be expected in a majority carrier device, although a tentative explanation was proposed by Zylbersztejn et al (1979).

A second origin of time dependent behaviour is changes in the population of states on the free surface of the device in the access regions between the gate metallisation and the source and drain ohmic contacts. These changes will induce changes in the thickness of the surface depletion region and so modify the total parasitic resistance R_p which is in series with the gated channel. This mechanism has been suggested by Itoh et al (1981) to explain drain current drift. Of particular relevance is the work of Ozeki et al (1982), who have observed dispersion of the transconductance at low frequencies and shown that the magnitude of this effect is related to gate leakage currents and can be modified by surface treatment. However, dispersion of the transconductance can also be caused by deep levels under the gate (Wada et al(1975)).

We report here a series of measurements designed to distinguish between these two mechanisms for time dependent behaviour. We show that in general the dominant mechanism is trapping by states on the free surface, and that this mechanism leads to a natural explanation of the "hole traps" previously observed in conductance-mode DLTS experiments. We further show that the effect of the surface states can be interpreted very simply by supposing that the gate behaves as if it were electrically larger than its metallurgical length, and we demonstrate that this electrical length can be easily obtained from d-c measurements.

2. Theory

Before presenting our results it is convenient for later discussion to re-state elementary expressions describing the operation of a MESFET in the linear regime (i.e. with the drain-source voltage V_{ds} sufficiently small that the depletion depth is constant along the channel length, so that the channel may be treated as a resistance). We assume an active layer of thickness a and uniform donor density N_D and mobility μ , ignore band-bending at the substrate, and use the depletion approximation : these gross over-simplifications are adequate for present purposes. The device will be pinched off when the applied negative gate bias V_{gs} reaches the pinch-off voltage V_p, where :

$$V_b - V_p = a^2 q N_D / 2\epsilon\epsilon_o \qquad (1)$$

where V_b is the built-in voltage of the Schottky barrier. Writing η for $[(V_b - V_{gs})/(V_b - V_p)]$, then the width of the conducting channel at a gate bias of V_{gs} is $a(1 - \eta^{1/2})$, and so the resistance of the channel $R_{ch}(V_{gs})$ is :

$$R_{ch}(V_{gs}) = R_o/(1 - \eta^{1/2}) \qquad (2)$$

where
$$R_o = L/ZqN_D a\mu \qquad (3)$$

(L and Z being the gate width and length respectively), and the intrinsic transconductance g_{mo} is :

$$g_{mo} = (\delta i_D/\delta V_{gs})_{V_{DS}} = V_{DS}/2R_o(V_b - V_p)\eta^{1/2} \qquad (4)$$

In a real device the resistance of the source and drain ohmic contacts and the ungated access regions gives a parasitic resistance R_p in series with $R_{ch}(V_{gs})$ so that the resistance measured between source and drain R_{SD} is :

$$R_{SD} = R_p + R_{ch}(V_{gs}) = R_p + R_o/(1 - \eta^{1/2}) \qquad (5)$$

while the measured transconductance $g_m^* = g_{mo}(1 + R_p/R_{ch})^{-2} \qquad (6)$

3. Transconductance Dispersion Results

Measurements of g_m^* were carried out over the range 5-100 kHz, using a lock-in technique and keeping both V_{DS} and the superimposed gate modulation signal V_{gs} small ($\sim$ 50 mV). Results for the variation of g_m^* with V_{gs} at three different frequencies measured on a 1 μm gate-length power FET fabricated on a VPE active layer are shown in Fig 1. The overall shape of these curves is as expected from Eqs (4) and (6), since although g_{mo} decreases monotonically as V_{gs} is increased towards pinchoff, the term in Eq (6) correcting for the presence of R_p increases more rapidly until V_{gs} approaches V_p, leading to the characteristic peaked curve. What is significant in Fig.1, however, is that g_m is frequency-dependent, most noticeably close to zero gate bias. As stated earlier, this could be due either to deep traps under the gate (Wada et al 1975), effectively making a frequency-dependent contribution to N_D and hence to g_{mo}, or to surface state trapping (Ozeki et al 1982) which would influence g_m^* through the correction term in Eq (6) for the parasitic resistance. However, the former explanation would require that the density of deep traps be greatest just beneath the gate, and much less deep in the channel, since the frequency dispersion is greatest near $V_{gs} = 0$. This is illustrated more clearly in Fig. 2 where data from a second device, this time a small-signal FET on an MOCVD active layer, are presented as curves of g_m^* against frequency for different gate biases (all curves being normalised to 1 at the lowest measurement frequency). Further, the magnitude of the observed dispersion, about 20% at $V_{gs} = 0$ in both Figs 1 and 2, requires the trap density to be about this fraction of the donor density, while the fact that g_m^* <u>decreases</u> with increasing f implies that the traps are hole traps. None of these requirements appears very probable. Moreover this same behaviour has been observed in a wide variety of devices.

If, however, surface states are invoked, then the results of Figs 1 and 2 can be easily explained. At low frequencies occupation of these states will follow changes in the gate potential and so the gate will control the channel width over a length L_e greater than its metallurgical length L_m; at higher frequencies the surface states will not be able to follow the gate potential and so the electrical length L_e will fall to be equal to L_m. At first sight this would seem to imply that the transconductance should rise rather than fall at high frequency, since $g_{mo} \propto 1/L$ according to Eqs (3) and (6). However L also enters the parasitic correction term in Eq (6), both because $R_{ch} \propto L$ and because any increase in the effective gate length must imply a corresponding decrease in the access regions and so in R_p. It is easy to see that inasmuch as away from pinch-off this correction term makes g^* increase rather than decrease, with increasingly negative V_{gs}, so it will also invert the dependence of g_m^* on L. This is exactly what is observed in Fig 2 ; the decrease in g_m^* with increasing frequency is greatest for $V_{gs} = 0$, when the correction term in Eq (6) is greatest, and it becomes smaller as V_{gs} is increased until at $V_{gs} = -2.5V$, g_m actually rises slightly at higher frequency rather than falling. These conclusions are only valid when the device is operated in the linear part of its characteristics : in saturation Eqs (2) to (6) do not apply.

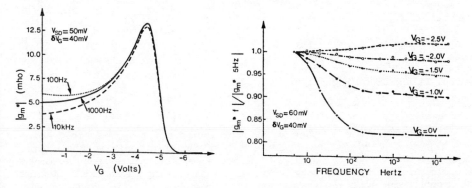

Fig. 1 Measured g_m vs gate voltage   Fig. 2 Frequency dispersion of g_m as
        as a function of frequency.   a function of gate voltage.

Experimentally the dispersion in g_m^* is much smaller when measured under saturation conditions as expected. The effect is nevertheless still important since both the noise and small signal gain of FETs are particularly sensitive to the values of parasitic resistance in series with the channel.

A consequence of the above model is that the transconductance dispersion should be much smaller on a long-gate device, since the relative difference between L_e and L_m will be much smaller. To test this prediction we have compared two test devices with gate lengths of 2 and 43 μm fabricated side-by-side on the same wafer. The results are shown in Fig 3, and clearly confirm that surface states rather than deep levels (for which no dependence on gate length would be expected) are responsible. Finally, in agreement with Ozeki et al (1982) we also have observed that the magnitude of the transconductance dispersion can be modified by different surface treatments during processing.

4. Measurement of Effective Gate Length

If our model of the surface states acting as an extension to the gate is correct, then it should be possible to deduce this effective electrical length L_e by quantitative analysis of the electrical characteristics. For power devices Ozeki et al (1982) have shown that at low frequencies the gate capacitance is anomalously high, but for small-signal devices stray capacitances preclude the use of this technique. Instead we have obtained L_e from the I_D-V_{gs} characteristics in the linear regime ($V_{DS} \approx$ 50 mV), using Eqs (3) and (5). As first suggested by Grebene (1967), plotting R_{SD} against $(1-n)^{-\frac{1}{2}}$ should according to Eq (5), give a straight line of slope R_0 and intercept R_p. This is done by digitising the I_D-V_{gs} curve and then optimising the choice of V_p in fitting to Eq (5), as proposed by Fukui (1979). The values of V_p and R_0 so obtained are then inserted into Eqs (1) and (3), together with V_b, which is known from Schottky barrier measurements, the mobility, which can be determined by magneto-transconductance measurements (Jay and Wallis 1981) and the gate width Z and length L. If L_m is taken for the value of L, then the only remaining unknown quantities are a and N_D, so these may be obtained by solving Eqs (1) and (3) simultaneously. However, this method gives values not consistent with the known parameters of the epilayer, exaggerating a and understimating N_D. Conversely, the use of this method to deduce the value of μ from R_0 by introducing known values of N_D, a, Z and L_g frequently gives rise to abnormally low values of mobility as a consequence of the apparent value of L_g (Fukui 1979). If instead we use the value of N_D obtained from C-V measurements on the epilayer, then Eq (1) gives a value for a which is entirely reasonable. These values of N_D and a can now be inserted into Eq (3) to obtain a value for the effective gate length L_e. The values we obtain are consistently greater than the metallurgical length, typically by 0.6 -0.8μm for devices with 0.7μm gates. It should be noted that this discrepancy is much greater than that expected for simple fringing at the depletion region at the ends of the gate.

To confirm this point further we have performed a fit as described above for a series of six adjacent test devices differing only in their gate lengths, which were 2, 5, 8, 12, 20 and 43μm. The values of R_0 obtained are plotted against gate length in Fig. 4. According to Eq (3) R_0 should be proportional to L and so the points in Fig 4 should lie on a straight line passing through the origin. The points in Fig 4 do indeed lie on a straight line, but one which passes approximately 1μm to the left of the origin, implying that electrically the gates are this amount longer than their metallurgical lengths.

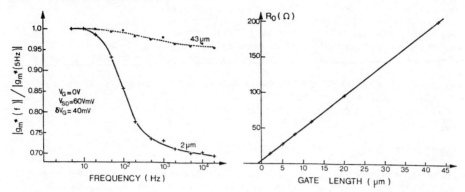

Fig.3 Frequency dispersion of g_m as a function of gate length.

Fig.4 R_o as a function of metallurgical gate length.

5. DLTS Results

The response of the free surface to changes in gate potential provides a natural explanation for the previously puzzling appearance of "hole traps" in conductance-mode DLTS experiments on short-gate devices. These DLTS signals originate not from transient changes in the depletion depth, as conventionally interpreted, but in the channel length. When, following a positive "filling pulse", the gate is returned to its quiescent value of negative bias, the surface states will respond with a characteristic time τ, contributing a transient increase to R_{SD}. Since τ may be expected to be thermally activated (especially if the surface state population is changed by a surface leakage current, as found by Ozeki et al (1982)) then a peak in the DLTS spectrum resembling hole emission results. A spectrum containing more than one peak would require only that the surface response cannot be represented by a single activation energy.

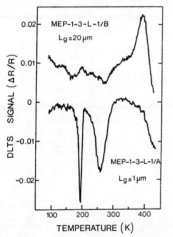

Fig.5 Conductance DLTS spectra for 2 different gate length devices on the same wafer.

As in the case of the transconductance dispersion, the relative importance of these surface-related effects should be smaller in a long-gate device and so there should be a difference in the DLTS spectra of devices from the same wafer. In the long gate device (Fig 5a) the spectrum contains only peaks corresponding to electron traps. We interpret these peaks as arising from true deep levels beneath the gate. The dominant peak at 390 K corresponds to the ubiquitous EL2 trap. The same spectrum, albeit with reduced signal-to-noise, was obtained by capacitance-mode DLTS on this long-gate device. In contrast the spectrum for the short-gate device (Fig 5b) shows only "hole traps", which we ascribe to the response of the surface as discussed above.

One noteworthy aspect of the surface states is that their filling and emptying should proceed at the same rate, in contrast to bulk deep levels for which capture is a rapid process and re-emission thermally activated.

A consequence is that their population will not be appreciably changed by narrow pulses applied to the gate and so "hole traps" will only be observed if long pulses are used. This is exactly the behaviour observed by Zylbersztejn et al (1979).

Conclusions

By analysing the frequency dependence of transconductance measurements on microwave FETs, and on transistors of varying gate length, we have shown that effects previously attributed to hole traps in DLTS conductance spectroscopy can be better explained in terms of a depletion region due to surface states. The contribution of this region to the effective gate length of the device leads to the notion of an electrical gate length being greater than the metallurgical gate length at lower frequencies, and thereby explains the erroneous deduction of mobility from static measurements in the linear region. The behaviour of these surface states also helps to explain the sensitivity of microwave FETs to surface preparation and passivation techniques, and transient effects such as "looping", "lagging", etc.

Acknowledgments

We wish to thank those colleagues that provided devices for these studies, D. Kaplan for numerous discussions, and the Direction de Recherches, Etudes et Techniques for financial support under contracts n° 83.34.050.00.470.75.01.

References

Adlerstein M G 1976 Electronics Letters 12 297
Fukui H 1979 Bell Syst. Tech. Journal 58 771
Grebene A B 1967 Proc. IEEE 2031
Itoh H, Ohata K and Hasegawa F 1981 IEEE Trans. Electron Devices ED-27 10.
Jay P R and Wallis R H 1981 IEEE Electron Device Letters EDL-2 265
Meignant D, Boccon-Gibod D and Bourgeois J M 1979, Electron Letters 15 779
Mitonneau A, Meignant D, Baudet P, Berth M, Hollan L 1981 "GaAs and Related
    Compounds 1980" pp 445
Ozeki M, Kodama K and Shibatomi A 1982 "GaAs and Related Compounds 1981"
    (Inst. Phys. Conf. Ser. 63) pp 323
Wada O, Yanagisawa S and Takanashi H 1975 Japan J. Appl. Phys. 14 157
Zylbersztejn A, Bert G and Nuzillat G 1979 "GaAs and Related Compounds
    1978" (Inst. Phys. Conf. Ser. 45) 315

Inst. Phys. Conf. Ser. No. 74: Chapter 4
Paper presented at Int. Symp. GaAs and Related Compounds, Biarritz, 1984

Electron velocity vs electric field in n-type $Ga_{0.47}In_{0.53}As$ short samples-evidence for ballistic transport

T. Ohashi, M.I. Nathan[*], S.D. Mukherjee, G.W. Wicks,
G. Rubino[**] and L.F. Eastman
School of Electrical Engineering, Phillips Hall
Cornell University, Ithaca, NY 19453 USA

Introduction

Current developments in modern semiconductor technology have led to the realization of high speed devices in submicron dimensions. These dimensions are becoming comparable to the mean free path for electron scattering, particularly for III-V compounds. At these dimensions electrons move through the device making no or at most a few collisions. Transport in this regime is known as ballistic or near ballistic transport. It is expected that the velocity will be higher in the ballistic regime than in the collision dominated case which will lead to higher device speeds. Measurements of current-voltage characteristics should give some information on this effect. Such measurements have been made by Hollis on GaAs. In this paper we report similar measurments on $Ga_{0.47}In_{0.53}As$. This material has a higher low field mobility than GaAs. Evidence for ballistic transport is found.

Experimental Results

Lattice matched multilayer $Ga_{0.47}In_{0.53}As$ $n^+n^-n^+$ structures were grown on (100) oriented n^+ InP substrates by MBE. The n^+ layers were Si doped with thickness of 0.25μ, and the n^- layers unintentionally doped with $1.0 \pm 0.2 \times 10^{16}$ cm^{-3} with thickness of 0.25μ, 0.5μ and 1.0μ. In addition a structure with no n^- layer was used to be able to account for contact resistance.

Since the impedance of $n^+n^-n^+$ diode samples with n^- layer thickness of less than 1μ would be toosmall to measure accurately, the $n^+n^-n^+$ structure was repetitively grown on top of each other in order to increase sample impedance to an accurately measurable amount. Five stacks of $n^+n^-n^+$ structure were grown for 0.25μ samples, three stacks for 0.5μ and two stacks for 1μ samples. n^+ layers were grown thick enough to prevent hot electrons from being injected from one n^- region to next n^- region, in this case 0.25μ. After the layers were mesa-etched, with 100μ diameters down to the n^+ substrates, AuGeNi/Ag/Au ohmic contacts were evaporated on top of the mesas, the areas surrounding the mesas and the backside of substrates and then alloyed. 25μ-diameter Au wires were bonded on the contacts after cleaved devices were mounted on TO-5 headers.

[*]IBM Research Center, Yorktown Heights, NY 10598, USA.

[**]IBM Corporation, Endicott, NY USA

Three terminal pulsed I-V measurements were made at 300K and 77K up to current densities of 1.2×10^5 A/cm^2. The direction of electron transport was perpendicular to the layers, or <100> direction. Pulse width and duty cycle were varied to check that there was no Joule heating effect. Measurements were carried out using pulse width of 50 ~ 100 nsec and repetition rate of 50 ~ 500 sec^{-1}.

The highest current density was limited by the driving capability of the pulse power supplies, not by burning-out of the samples. For lower voltage measurements (Figs. 1-3), HP 214B pulse generator was used mainly because this pulser gives easy and reproducible control of pulse width, repetition rate and pulse height. A mercury reed pulser was used for the higher voltage measurements (Figs. 4 and 5) which required higher driving capability than the former pulser could provide.

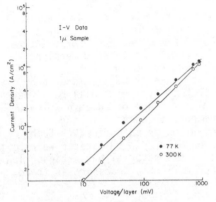

Fig. 1. Log-log plot of current density vs voltage drop across a 1μm n⁻ layer. Voltage drop across the contacts has been subtracted out. The straight lines are drawn to guide the eye.

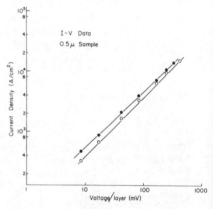

Fig. 2. Log-log plot of current density vs voltage dropped across a 0.5μm n⁻ layer. Voltage drop across the contacts has been subtracted out. The straight lines are drawn to guide the eye.

Results and Discussions

Results of the I-V measurements are shown in Figs. 1 - 5. Since potential difference is more essential than electric field when calculating the energy of ballistic electrons, the plots are shown with respect to voltage rather than electric field. Figs. 1-3 show I-V data at lower voltage at 300K and 77K, and Figs. 4 and 5 at higher voltage at only 77K since slight Joule heating effect was observed at higher current level at 300K. Data shown here for lower and higher voltage were intentionally taken from different samples in order to demonstrate the extent of reproducibility from one device to another. Variation observed among the many diodes tested for this experiment was within ± 20% which could be explained by fluctuation of background doping from growth to growth and non uniformity of contact resistance across wafers. The voltage drop across the contact resistance (0.2 Ω per diode) was extracted from the I-V data of samples with no n⁻ layer. The contact voltage has been subtracted from the raw data to give the

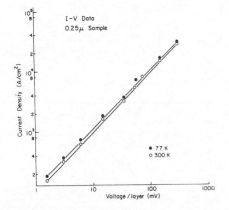

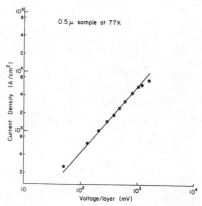

Fig. 3. Log-log plot of current density vs voltage dropped across a 0.25μm n⁻ layer. Voltage drop across the contacts has been sub-tracted out. The straight lines are drawn to guide the eye.

Fig. 4. Log-log plot of current density vs voltage dropped across a 0.5μm n⁻ layer. Voltage drop across the contacts has been sub-tracted out. The straight line is drawn to guide the eye.

data shown in Figs. 1-5.

The I-V curves for the 0.25 μ sample are almost temperature independent. Collision dominated transport should have a temperature dependence because of intrinsic temperature dependence of the various collision mechanisms. Ballistic transport, on the other hand, should be temperature independent. The I-V curves become progressively more temperature dependent as the length is increased. This implies that electrons have more collisions in a longer sample, which is expected from ballistic theory.

Next, the slopes of the logI-logV curves are different depending on the length of the devices. The 77K curves for 1 μm sample in Fig. 1 shows an upward bowing and has a slope of 1 below 45 mV and 0.8 above 45 mV. The 300K curve also has a slight upward bowing, but the slope over the entire voltage range is about unity. The 300K curve for the 0.5 μ sample has the same characteristics, slight upward bowing and unity slope, same as the one for the 1 μ sample, whereas the 77K curves show a slope of 0.9 below 150 mV and 1 above 150 mV in Fig. 2, and Fig. 4. Above 1V in Fig. 1, the slope becomes smaller than 1, which is due to upper valley transition of electrons accelerated by electric field. The 0.25 μ sample has similar characteristics as the 0.5 μ sample. Both 77K and 300K curves have about same slope of 0.95 (Fig. 3) below 300 mV. Above that voltage, the 77K slope increases to 1.1, then shows a tendency toward leveling off at the highest voltage applied (Fig. 5). This tendency is also because of upper valley transition of electrons.

The velocity-field curve cannot be deduced from the data directly be-cause the electron concentration is not uniform in the n-region, since the Debye length, L_d is an appreciable fraction of the sample length even at 77K. For example, Debye length of this material at 77K is: L_d = 220 Å. Furthermore, the electron spillover will increase with field. In other words, space charge injection is important. Thefore we can only discuss the data semi-quantitatively and make inferences about

Fig. 5. Log-log plot of current density vs voltage drop across a 0.25 μm n⁻ layer. Voltage drop across the contacts has been subtracted out. The straight lines are drawn to guide the eye.

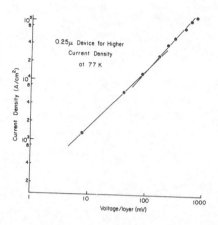

ballistic effects.

The simple ballistic model considered by Shur and Eastman gives I α $V^{1/2}$ relation at low voltage and I α $V^{3/2}$ dependence at high voltage. The transition point between the two regions shifts to higher voltage with increasing doping level in n⁻ region. This explains why a I α $V^{3/2}$ characteristics is not observed in 1 μ samples. For doping level of 10^{16} cm⁻³, the transition voltage from I α $V^{1/2}$ dependence region to I α $V^{3/2}$ region is higher than the point where intervalley saturation occurs. The 0.25 μ sample has a slope of 1.1 for voltages above 100 mV. This could be indicative of a I α $V^{3/2}$ dependence. The 0.5 μ sample also shows increasing slope with increasing voltage. But the slope doesn't become larger than 1. This could be explained by a higher chance of scattering compared to 0.25 μ sample simply because the n⁻ region is longer.

Shur and Eastman proposed a near-ballistic model taking account of momentum and energy relaxations due to weak scattering. In this model, electrons gain their kinetic energy through electric field more slowly than in the case of simple ballistic model because of energy loss due to weak scattering. This leads to a I α V^{α} dependence with α smaller than 3/2, hence in better agreement with the measured data.

Finally, when current densities are compared at the same voltage, or at the same electric field, shorter samples have higher current densities. This could be explained by two effects: space charge injection or higher electron velocity. It is likely that both mechanisms contribute at the same time. So, in order to estimate electron velocity, the amount of space charge, $\Delta n = M_{sc}$, should be determined. As in a depletion region we assume that space charge is proportional to applied voltage and inversely proportional to length squared, or:

$$\Delta n = N_{sc} = K \frac{V(\text{volt})}{\ell^2 (\mu^2)}; \quad (K = \text{const.})$$

K can be obtained from the depletion width vs. voltage relationship; K ≈ 1.25×10^{15} (μ²/volt.cm³). For example, when data for 0.25 μ and 1 μ samples at the same electric field are compared (say, 100 mV for 0.25 μ diode and 400 mV for 1 μ diode), the 0.25 μ sample has twice as large a current density as the 1 μ sample even after compensating for space change injection using the assumptions made above. This also suggests that shorter devices suffer fewer collisions, approaching to ballistic transport. Furthermore, with the assumptions made above for space charge, approximate electron velocity could be estimated. The 0.25 μ sample had the highest current density of 1.2×10^5 A/cm^2 at 0.77 volts. The density of space charge for this current density is:

$$\Delta n = N_{sc} \tilde{=} 1.5 \times 10^{16}/cm^3,$$

which gives $v_{elec} = 3 \times 10^7$ cm/s. This value is larger than the peak electron velocity observed in bulk InGaAs, which indicates that velocity overshoot occurs in a short dimension device, giving shorter transit time of electrons than can be expected by conventional collision dominated transport.

Conclusion

I-V measurements on short $n^+n^-n^+$ Ga$_{0.47}$In$_{0.53}$As samples reveal that shorter samples are less temperature sensitive, in agreement with ballistic transport theory.

The slopes of the I-V curves for 0.5 μ and 0.25 μ long samples showed a vestige of simple ballistic theory or near-ballistic theory.

Average electron velocity was estimated, with compensation for the effect of space charge injection. Shorter samples had higher electron velocities for identical voltage. Also, the highest electron velocity was found to be ~ 3×10^7 cm/sec.

Since electron mobility in Ga$_{0.47}$In$_{0.53}$As does not decrease as much as that of GaAs when temperature is increased from 77°K to 300°K, this material could be a good candidate for fabricating ballistic devices operating at room temperature.

Acknowledgements
We would like to thank J. Berry for his technical support. This work was supported by the Office of Naval Research under contract no. N00014-75-C-0739 and the Air Force Office of Scientific Research under contract no. F49620-81-C-0082.

References
Hollis M A, M.S. Thesis, May 1981, Cornell.
Shur M S and Eastman L F, IEEE Trans. Electron Dev. ED-**26** 1677 (1979).
Shur M S and Eastman L F, Solid State Elect. **24** 11-18 (1981).
Marsh J H, Houston P A and Robson P N, GaAs and Related Compounds **58** 621 (1980).

Inst. Phys. Conf. Ser. No. 74: Chapter 4
Paper presented at Int. Symp. GaAs and Related Compounds, Biarritz, 1984

299

Anomalous behaviour of Mn-doped LPE $Ga_xIn_{1-x}As_yP_{1-y}$ revealed by transport and high pressure measurements

L.G. Shantharama and A.R. Adams
Department of Physics, University of Surrey
Guildford, Surrey, GU2 5XH, U.K.

E.M. Allen and P.D. Greene
Standard Telecommunication Laboratories Ltd
London Road, Harlow, Essex CM17 9NA, U.K.

ABSTRACT

The Hall constant and resistivity of Mn doped $Ga_xIn_{1-x}As_yP_{1-y}$ has been measured as a function of temperature and pressure. Analysis using the quantum dielectric theory indicates that the Mn acceptor level is not pinned to the vacuum level as has been suggested by Ledebo and Ridley for other deep levels. Under certain conditions holes are not scattered by ionized impurities in heavily compensated Mn and Ge doped material. This is believed to be due to the creation of small dipoles formed by the diffusion of neutral Mn.

1. INTRODUCTION

Because of its relatively low vapour pressure, low affinity for oxygen and low toxicity, Mn is a convenient p-type dopant for InP and the commercially important alloy $Ga_xIn_{1-x}As_yP_{1-y}$ grown lattice matched to InP ($y = 2.1x$). It as been shown (Smith et al. 1983) that, E_A, the activation energy from the top of the valence band to the Mn acceptor level, varies linearly across the alloy range. It is expressed in meV by $E_A = 230-185y$. Measurements have now been made of the Hall constant and resistivity as a function of temperature, and pressure for different alloy compositions. Two interesting aspects of the work are considered here. Firstly, measurements of the Hall constant have been used to determine E_A as a function of pressure in order to investigate the theory of Ledebo and Ridley (1982), which suggests that deep levels are fixed in energy with respect to the vacuum level. Secondly, the hole and electron mobilities and their temperature dependence have been used to study the carrier scattering mechanisms which are occurring. These yield the surprising result that, under certain circumstances, material heavily doped with Mn and Ge can have high hole mobilities limited only by phonon scattering as in high purity material.

2. EXPERIMENTAL PROCEDURE

The material used was grown by liquid phase epitaxy on semi-insulating (100) doped InP substrates. Van der Pauw clover-leaf shaped samples were used and p-type material contacted by evaporating and alloying Au-Zn electrodes. These were found to show good ohmic characteristics down to

140K on material with resistivities up to 10^4 ohm-cm. In order to extend the measurements of the Hall constant and resistivity previously reported (Smith et al. 1983) to lower temperatures and lower y values, the experimental arrangement of Hemenger (1973) was adopted. The voltage contacts were interfaced with high input impedence, unity gain amplifiers and, to minimise the effects of cable capacitance and leakage currents, the inner shield of the cables was connected to the amplifier outputs. With this system accurate measurements could be made on samples with resistances up to $10^{10}\Omega$. The pressure dependence measurements were made in a piston-and-cylinder apparatus with castor oil as the pressure transmitting medium. These measurements were made at room temperature in the pressure range 1 bar to 8 kbar.

The carrier concentration in the p-type samples varied from 6×10^{15} to 4×10^{16} cm^{-3} at 300K and in general N_A, N_D and E_A were obtained from a computer fit to the standard expression

$$\frac{p(p+N_D)}{N_A-N_D-p} = \frac{N_v}{g} e^{-E_A/kT} \tag{1}$$

where the symbols have their usual meaning. In the composition range $0 \leqslant y \leqslant 0.4$ a plot of $\ell n \; pT^{-3/2}$ vs T^{-1} yielded a straight line over the temperature range available and only E_A could be determined.

RESULTS AND DISCUSSION

a) Carrier Concentration

Nearly all measurements of the band structure of semiconductors give only values for the relative energy separations of, for example, the band extrema and rarely give any absolute energy values. This causes considerable difficulty when trying to determine important parameters such as the barrier heights of heterojunctions. Also it makes it difficult to test theories which claim to give absolute energy values or their temperature and pressure dependencies. Therefore the observation by Ledebo and Ridley that deep levels such as those associated with the transition metals appear to be fixed relative to the vacuum level, is very exciting since such a dopant would produce a marker against which different semiconductor band structures could be measured. Since Ledebo and Ridley (1982) used the band structure model of Phillips and Van Vechten (Phillips 1968, Van Vechten 1969 a, b) to show that the impurity level is constant, this model should also predict the composition and pressure dependences of the activation energy. Figure 1 shows a comparison between theory and experiment for the present results. In 1(a) the theoretical variation of the top of the valence band, I, as a function of composition is shown by the dashed line. The experimental points show the position of the Mn level, E_{Mn}, obtained by determining E_A from the temperature dependence of p fitted using equation 1. The Mn level appears to move up somewhat towards the vacuum level as the top of the valence band also moves up, however the change is probably well within the error in the theoretical determination of the variation of I with composition. Figure 2 shows the pressure dependence of the hole density at different alloy compositions. As can be seen, the hole density decreases more swiftly with pressure as the composition parameter y decreases. Fitting these curves using equation 1 gives $dE_A/d\pi$, the pressure coefficient of the activation energy, which is plotted as a function of y in Figure 1b. According to the Ledebo

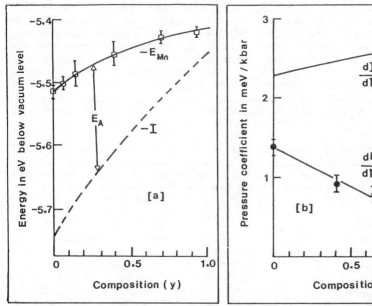

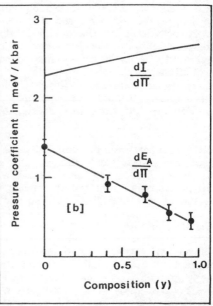

Figure 1. The composition dependence of (a) the ionization potential I and the energy of the Mn acceptor level $E_{Mn} = I - E_A$; (b) the pressure coefficients $dI/d\pi$ and $dE_A/d\pi$.

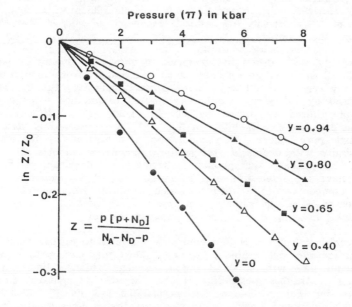

Figure 2. The pressure dependence of the hole density parameter Z for different alloy compositions.

and Ridley model, $dE_A/d\pi$ should be equal to the absolute pressure dependence of the top of the valence band $dI/d\pi$. This is shown by the dashed line in Figure 1b calculated on the model of Phillips and Van Vechten assuming that the compressibility of the alloy can be obtained by linear interpolation of the compressibility of its binary compound constituents. Not only is the predicted value of $dI/d\pi$ much larger than $dE_A/d\pi$, it increases with increasing y instead of decreasing as measured. Results, to be discussed elsewhere, indicate that this discrepancy cannot be explained in terms of a pressure dependent Franck-Condon shift since this is expected to be too small (Barnes and Samara 1983).

It can therefore be concluded that, within the terms of the model proposed by Ledebo and Ridley, the absolute energy of the Mn level in $Ga_xIn_{1-x}As_yP_{1-y}$ does not remain fixed with respect to the vacuum level as the pressure is changed. Instead it appears that the closer the Mn level is to the top of the valence band the more strongly it is influenced by it.

b)   Carrier Mobility

Since Mn forms a relatively deep acceptor in InP and alloys of low y composition, it is a suitable compensating dopant for the production of high resistivity material. We have therefore studied compensation in samples of the alloy $Ga_{0.03}In_{0.97}As_{0.07}P_{0.93}$ concurrently doped with both Mn and the shallow donor Ge. This alloy was preferred to InP because it was easier to grow with good morphology and freedom from surface terracing. The results for four samples are listed in Table 1. Consider first sample A1 which was grown from a melt containing 0.00344mg of Mn and 3.792mg of Ge per gram of In melt. Since Ge has a much lower segregation coefficient than Mn this produced a sample in which the donor concentration N_D slightly exceeded the acceptor concentration N_A, leading to a free electron concentration of $1x10^{17}cm^{-3}$. However, the measured mobility of $500cm^2V^{-1}s^{-1}$ would, in InP, correspond to a density of ionized impurities of almost $4x10^{18}cm^{-3}$ indicating that the sample is heavily compensated and the mobility is dominated by ionized impurity scattering. This conclusion is confirmed by the decreasing mobility with decreasing temperature shown in Figure 3 (sample A1). Samples B1 and C1, grown from melts containing a lower concentration of Ge showed p-type conductivity. Considering sample C1, the Ge content has been reduced by 50% which would be expected to produce a reduction in the density of ionized impurities by at the most, 25%. However, the measured value for the hole mobility of $136cm^2V^{-1}s^{-1}$ observed is very close to that of high purity InP. This conclusion is confirmed by the strongly increasing mobility with decreasing temperature shown for samples C1 and D in Figure 3. This indicates that phonon scattering is the dominant mechanism. We therefore have the interesting and potentially important phenomenon that Mn and Ge compensate in a manner which produces little or no ionized impurity scattering when the sample is p-type. It is interesting to note that below 200K $\mu_p > \mu_n$ in samples with similar impurity content.

One possible explanation is that neutral Mn is able to diffuse much more freely than ionized Mn and is able to approach close to donors before forming a donor acceptor pair with a small dipole moment which is ineffective in scattering holes. Another possibility is that Mn-Ge precipitates form in p-type material but not in n-type. However, the low density of holes in sample D, which contains no Ge, indicates that this sample is also heavily compensated. Nevertheless the high mobility and its temperature dependence indicates that Mn is equally effective in neutralising the scattering effects of the donors in this sample which are normally present in nominally undoped

TABLE 1   Material: $Ga_{0.03}In_{0.97}As_{0.07}P_{0.93}$

Sample	Dopant in mg/g In melt		Majority carrier	Carrier concentration	Mobility
	Mn	Ge		cm^{-3}	$cm^2V^{-1}s^{-1}$
A1	0.00344	3.792	n	1.0×10^{17}	500
B1	0.00344	2.668	p	6.0×10^{15}	121
C1	0.00344	1.91	p	2.5×10^{15}	136
D	0.00344	0	p	6.3×10^{15}	139

TABLE 2   Material: $Ga_{0.47}In_{0.53}As$

A2	0.0196	15	n	1.0×10^{18}	82
B2	0.0196	13	p	6.0×10^{16}	53
C2	0.0196	11	p	6.0×10^{16}	73
E	0	11	n	2.0×10^{17}	1366

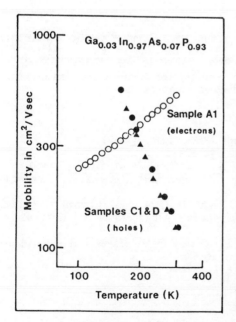
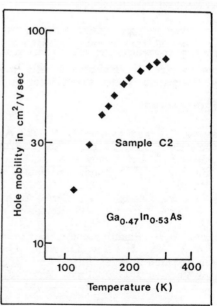

Figure 3.   The temperature dependence of the mobility of electrons (A1) and holes in Mn doped material showing the dependence of the hole mobility on composition.

LPE material. SIMS measurements by Silberg et al. (1982) have already shown that the diffusion of Mn is inhibited by the presence of Ge and we will show elsewhere that sulphur also inhibits Mn diffusion.

In order to test the hypothesis that the diffusion of neutral Mn plays a central rôle, a similar range of $Ga_{0.47}In_{0.53}As$ was grown. The results are given in Table 2. As can be seen, in this material both the hole and the electron mobilities are considerably reduced below that of high purity material. Curve C2 in Figure 3 confirms that the mobility in the p-type material is now dominated by ionized impurity scattering. This is to be expected since, in $Ga_{0.47}In_{0.53}As$, the activation energy E_A is only 45meV and, at the growth temperature, there will be relatively few neutral Mn atoms free to diffuse.

CONCLUSIONS

Although the composition dependence of the activation energy of Mn in the alloy $Ga_xIn_{1-x}As_yP_{1-y}$ does not disagree with the suggestion by Ledebo and Ridley that 'deep' levels are fixed with respect to the vacuum level, its variation with pressure is quite contrary to that predicted by this model. This may be due to the proximity of the level to the valence band which appears to have some influence on its pressure behaviour.

Mn compensates Ge in such a manner that, when the material is p-type, the ionized impurities have little effect on the scattering of holes. This may be due to the diffusion of neutral Mn which allows the formation of small dipoles or multipoles. The work is being extended to a study of Mn with other donor species and to a search for Mn-Ge precipitates using transmission electron microscopy.

ACKNOWLEDGEMENTS

One of us (LGS) wishes to thank the Commonwealth Scholarship Commission for financial assistance. Thanks are due to Dr. J. Marsh for growing the (GaIn)As samples at the SERC central growth facility in Sheffield and to Mrs. V. Hinton and Mr. B.J. Gunney for technical assistance. Mr. D. Lancefield is also thanked for his help in computation.

REFERENCES

Barnes C E and Samara G A 1983 Appl. Phys. Lett. **43** 677.
Hemenger P M 1973 Rev. Sci. Instrum. **44** 698.
Ledebo L-Å and Ridley B K 1982 J. Phys. C: Solid State Phys. **15** L961.
Phillips J C 1973 Bonds and Bands in Semiconductors (New York: Academic Press).
Silberg E, Chang T Y, Caridi E A, Evans Jr C A and Hitzman C J 1982 Gallium Arsenide and Related Compounds (Albuquerque) in Inst. Phys. Conf. Ser. No. **65** 187.
Smith A W, Shantharama L G Eaves L, Greene P D, Hayes J R and Adams A.R. 1983 J. Phys. D: Appl. Phys. **16** 679.
Van Vechten J 1969a Phys. Rev. **182** 891.
Van Vechten J 1969b Phys. Rev. **187** 1007.

Inst. Phys. Conf. Ser. No. 74: Chapter 4
Paper presented at Int. Symp. GaAs and Related Compounds, Biarritz, 1984

Investigation of deep levels at interfaces by means of FET structures and optical excitation

F.J.Tegude[*], K.Heime
Universität Duisburg, D 4100 Duisburg, FRG

Abstract: Deep level distributions in actual FETs are studied using optical rather than thermal excitation in the FET itself. It is demonstrated that by this method the sensitivity to traps near the layerSI substrate interface is increased, the lateral resolution is improved and the concentration even of very deep levels near the interface is determined correctly, which is very difficult with thermally activated measurements. A simple yet correct analytical model for the evaluation of near interface traps and experiments performed on GaAs LPE, MBE and diffused layers on different substrates (Cr-doped and undoped SI, p^+ and n^+) are presented.

1. Introduction

The influence of semi-insulating (SI) GaAs substrates on FET performance has been investigated by many authors /e.g.1/ and they established this negative influence of not properly prepared substrates. Considerable effort was necessary during the last years for GaAs IC-technology to appear. Especially the advent of undoped SI substrates was an important milestone on this way. Nevertheless, many unsolved problems remain, most of which are related to deep levels introduced during the FET fabrication process. Therefore, methods for the characterisation of these deep levels are requested, which are particularly adopted to FETs. Consequently, we used the FET device itself for deep level measurements. But furthermore it is very important to choose the adequate excitation of the deep levels. As will be demonstrated optical is superior to thermal excitation, convenient for the very popular DLTS technique, when FET structures are to be studied. In our work we investigated LPE, MBE and diffused layers on Cr-doped and undoped SI, n^+ and p^+ GaAs substrates.

2. Advantages of optical compared to thermal excitation

Because GaAs technology nearly exclusively uses MESFETs, the Schottky contact space-charge region (SCR) is appropriate for the detection of deep level transitions, when in depth profiling is demanded. With thermal excitation in this case only majority carrier traps, i.e. deep donors in the n-FET channels, are detectable. Optical excitation induces

transitions between deep levels and both band edges, so deep
donors and acceptors can be measured (fig.1). A second point
is particularly important for the characterisation of the
thin FET layers on SI substrates: thermal excitation occurs
only at a distance $X > w(V)$ from the SCR edge. Especially
for very deep levels this causes an error in determination
of activation energy, E_T, due to high electric fields, and
another in determination of concentration N_T, when the SCR
edge reaches the SI substrate. In the latter case levels
within a distance w (fig.1) from the interface are not detected. When for this reason the reverse voltage is further
increased, the SCR region is pushed into the SI substrate,
so for example in CV measurements the deep level concentration is determined relative to the N_D^+ in the SI substrate
and not relative to the N_D^+ layer concentration. Both errors
are eliminated using optical excitation, because transitions
in the whole SCR are induced. Furthermore, optical excitation yields E_T correctly as Gibbs free energy G, whereas
evaluation of an Arrhenius plot (DLTS) yields E_T as enthalpy
H thus causing considerable error by the entropy term $Tx\Delta S$,
especially for very deep levels in wide bandgap materials
/2/. Another temperature dependence arises with the thermally activated capture crossection $\sigma(T)$. The above mentioned
arguments demonstrate clear advantages of optical compared
to thermal excitation from theoretical as well as device
related point of view.

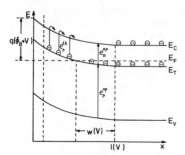
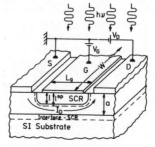

Fig.1. Deep level transitions
inside a Schottky contact
space-charge region

Fig.2. Crossection of a
MESFET under optical
excitation

3. Photo-FET technique

With this technique /3/ optically induced drain current variations ΔI_D are measured. The FET is illuminated with mono-
chromatic light, the quantum energy of which is increased
thus giving the ΔI_D (hv) spectrum. Theoretical arguments and
comparison of photo-FET and photocapacitance spectra indicate that SCR-width variations rather than photoconductivity
effects are responsible for the photo-FET signal /4/. E_T is
achieved from thresholds in the spectrum, the sign of ΔI_D
indicating the deep level type, donor or acceptor. N_T is
evaluated from the amount ΔI_D in the way to be described
now. The condition for high in depth resolution is

the SCR-edge to be parallel to the surface (fig.2), that is

(1) $V_{DS} \ll V_{sat}$, $V_{DS} \ll V_G$; V_{DS} drain source voltage, V_{sat} saturation value of V_{DS}, $V_G = \Phi_B$ + gate voltage

For the normalized signal we get

(2) $\dfrac{\Delta I_D}{I_D} = \dfrac{l - l^{op}}{a - l}$ l, l^{op} SCR width with and without ionized deep levels (fig.2)

Assuming $N_T \ll n$, as convenient, and using Shockleys formula for the I-V characteristics of MESFETs /5/ one finds /4/

(3) $\dfrac{\Delta I_D}{I_D} = \dfrac{\pm \dfrac{N_T^+}{2n}}{\sqrt{(V_P/V_G)} - 1}$ V_P pinch-off voltage

Solving for N_T/n yields

(4) $\dfrac{N_T^+}{n} = 2 \dfrac{\pm \Delta I_D}{I_D} (\sqrt{\dfrac{V_P}{V_G}} - 1)$

Analogous formula to (3) and (4) containing V_{DS} instead of I_D exist for constant current instead of constant voltage measurement, but with a change in sign. Eq.(3) demonstrates the distinct increase of sensitivity for $V_G \longrightarrow V_P$, i.e. for interface near levels, in contrast to capacitance measurements. $N_T/n < 10^{-5}$ is estimated. Further advantages are (a) very high lateral resolution of the order of 10 μm^2, when short gate FETs are used, (b) the guarantee of correlation with FET performance. It should be noted, however, that the interface-SCR may complicate the evaluation of photo-FET measurements.

4. Simple model for near interface traps

The model we propose and which is verified by our experimental results is
shown in fig. 3. The concentrations of
shallow and deep levels are assumed to
be abrupt and deep levels exist only
for $X > l_0$. For this configuration
Poissons equation reads

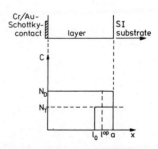

Fig.3. Model for near interface traps

(5) $\varepsilon \dfrac{\partial^2 \psi}{\partial x^2} = -\rho = -qn$, $0 < x < l_0$

$\qquad\qquad = -q(n + N_T^+)$, $l_0 < x < l$

Integration and solving for l^{op} yields

(6) $l^{op} = \left[qN_T^+ l_0 + \sqrt{(qN_T^+ l_0)^2 + 2q(n + N_T^+)\varepsilon V} \right] \left[q(n + N_T^+) \right]^{-1}$

V = reverse voltage including Φ_B

Eq.(6) can be used in eq.(2) for photo-FET measurements.
Because of the above mentioned complication due to the
interface-SCR this model has been adopted to photocapacitance measurements. The result is /4/:

$$V_T = V(l=l_0)$$

$$(7) \quad \frac{N_T^+}{n} = \left[1 \pm \sqrt{1 - 4\frac{V_T}{V}(\frac{C}{C_o})^2\left[(\frac{C}{C_o})^2 - 1\right]}\right] \left[2\frac{V_T}{V}(\frac{C}{C_o})^2\right]^{-1}$$

$$C_o = \sqrt{\frac{qn\epsilon}{2V}}$$

V is the parameter for the detection depth, C, C_o and V_T
(threshold voltage) are measured values. For very exact measurements without an error introduced by the interface-SCR,
eq.(7) can be used. Photo-FET may be calibrated by photocapacitance and then used to give the advantages of high sensitivity near the interface to the SI substrate and very
high lateral resolution.

5. Experimental results

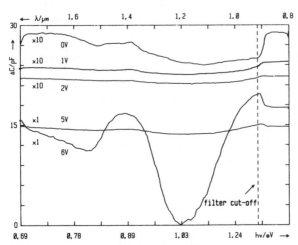

Fig.4. Photocapacitance spectra of
a LPE layer grown on SI GaAs:Cr

Fig.4-6 show some
selected experi-
mental results. In
fig.4 photocapaci-
tance spectra of a
LPE layer grown on
a SI GaAs:Cr sub-
strate are presented
for different bias
voltages. Near con-
tact (V=0V) and near
interface (V=6V)
traps are detected,
but no deep levels in
the layer bulk. Re-
garding the different
scales for the 0V - 2V
and the 5V - 6V spec-
tra, a high deep level
concentration near the
interface and a low
near the Schottky con-
tact is read. In most samples solely the interface levels exist
In fig.5 photo-FET and photocapacitance spectra of a highly Sn-
doped layer diffused into SI GaAs:Cr is shown, biased to detect
near interface traps. Nearly the same spectra are obtained for
both photo-FET and photocapacitance (traces 1 and 2). Comparison
of traces 2 and 3 shows the high lateral resolution, for these
spectra are taken from two different short gate FETs (Lg=
1μm, W= 300μm), situated side by side on the same sample. In
fig. 6 the two FETs are compared with respect to the dc-
characteristics: it is demonstrated that the FET with high
deep level content (trace 2, fig.5) shows loops in its dc-
characteristic, the one with low deep level concentration
does not. In table 1 our investigations performed on LPE,
MBE and diffused layers on Cr-doped and undoped SI GaAs, and

on p^+ and n^+ GaAs substrates, are summarized.

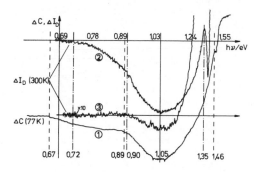

Fig.5. Photo-FET and photocapacitance spectra of a Sn-doped layer diffused into SI GaAs:Cr

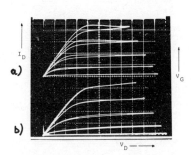

Fig.6. FET dc-characteristics:a) high, b) low deep level concentration in the channel

level type	E_T/eV	N_T/cm^{-3}							
		Sn-Diffusion in SI-GaAs:Cr	LPE on SI-GaAs:Cr		MBE on SI-GaAs:Cr	LPE on un-dop.SI-GaAs	LPE on n$^+$GaAs	LPE on p$^+$GaAs	
		interface	contact	interface	interface	interface	contact		
deep acc.	0,62(Fe)	—	—	~3×10^{15}	—	1÷10×10^{15}	—		measurements performed with pn-junctions instead of Schottky-contacts, NO DEEP LEVELS DETECTED, $N_T \lesssim 5\times10^{11}$cm^{-3}
	0,7(HL9Cr)	~2×10^{16}	0÷10^{14}	0,5÷20×10^{16}	—	—	~2×10^{14}		
	0,76	—	0÷10^{14}	—	—	—	—		
	0,79	—	—	—	—	~10^{17}	—		
	0,83	—	—	—	~10^{15}	~3×10^{15}	—		
	0,9(HL1Cr)	~5×10^{16}	0÷2×10^{14}	1÷20×10^{16}	—	—	5×10^{14}		
	0,93(HL1?)	—	—	—	~2×10^{14}	~5×10^{14}	—		
deep don.	0,7	—	—	~10^{16}	—	~5×10^{16}	—		
	0,76	—	—	—	—	~10^{15}	—		
	0,82(EL2)	} no EL2!	0÷10^{14}	0,8÷30×10^{15}	—	—	—		
	0,88(EL2)		—	—	~5×10^{14}	0,3÷10×10^{16}	~10^{14}		
	1,05	~5×10^{16}	—	3÷10×10^{16}	~2×10^{14}	~5×10^{14}	—		
	1,13	—	0÷10^{13}	—	—	—	~5×10^{14}		
	1,26	—	—	~5×10^{16}	—	—	—		
Resonance	1,35	X	—	—	—	—	—		

Tab.1. Summary of experimental results

6. Discussion

The partially very strong accumulation of Cr and Cr-related acceptor levels, 0.9eV (HL1) and 0.7eV (HL9), has been discussed by the authors previously /6,7/. Concerning the near contact Cr-related levels our investigations clearly show that the Cr/Au Schottky contact is responsible, for we never

found these levels in samples grown on p^+ substrates, where the pn junction SCR was used. In this paper we like to concentrate on EL2, because of its extreme importance in undoped SI GaAs substrates. We identify the two donor levels at 0.82eV and 0.88eV as EL2, because in literature the activation energy of EL2 ranges from 0.75eV to 0.88eV. Furthermore, Zou /8/ has proposed three different EL2 levels, so the existence of more than one EL2 level is most likely. Table 1 shows that for different layer-substrate combinations we find different EL2 levels: 0.82eV in LPE samples on SI GaAs:Cr, and 0.88eV in MBE on SI GaAs:Cr, LPE on undoped SI GaAs and LPE on n^+ GaAs samples. This fact combined with the following findings confirm the assignment of EL2 to intrinsic defects: no EL2 in the bulk of the Ga-rich grown LPE layers; no EL2 in the highly Sn-doped layers diffused into SI GaAs:Cr; no EL2 near the interface to n^+ and p^+ substrates. These results indicate a correlation to Ga vacancies, antisite defects As_{Ga} and interstitial As as has been discussed by other authors /8,9/. Identifying EL2 with As_{Ga} its absence in the highly Sn-doped diffused layers and in layers on n^+ and especially on Zn-doped p^+ substrates may be explained by a mechanism, where the As is "kicked out" of the Ga-site by Sn and Zn atoms, respectively /4/.
To summarize, we found that all deep levels detected were introduced by the SI GaAs substrates, Cr-doped and undoped, to partially rather high concentrations, or, to a minor extent, by the formation of the Cr/Au Schottky contact.

Acknowledgements

This work was supported by the Deutsche Forschungsgemeinschaft.

References

/1/ T. Itoh, H. Yanai (1980), Inst. Phys. Conf. Ser. 56
    London, 537
/2/ J.A. van Vechten, C.D. Thurmond (1976)
    Phys. Rev. B 14(8), 3539
/3/ F.J. Tegude, K. Heime (1980),
    Electronics Lett. 16, 22
/4/ F.J. Tegude (1983)
    Dissertation, Universität Duisburg, FRG
/5/ S.M. Sze (1969),
    Physics of semiconductor devices, Wiley, New York
/6/ F.J. Tegude, K. Heime (1980),
    Semi-insulating III-V materials, Nottingham, Shiva Publ.
    Ltd.,Orpington, 321
/7/ F.J. Tegude, J. Baston, K. Heime (1982)
    Semi-insulating III-V materials, Evian, Shiva Publ. Ltd.
    Nantwich, 291
/8/ Zou Yuanxi (1982),
    Inst. Phys. Conf. Ser. 63, London, 185
/9/ J. Lagowski et al (1982)
    Appl. Phys. Lett. 40, 342

Inst. Phys. Conf. Ser. No. 74: Chapter 4
Paper presented at Int. Symp. GaAs and Related Compounds, Biarritz, 1984   311

Carrier dynamics in GaInAs

M E Prise, M R Taghizadeh, S D Smith and B S Wherrett

Heriot-Watt University, Riccarton, Edinburgh EH14 4AS

1. Introduction

Carrier recombination processes in GaInAsP have been suggested as one of
the possible mechanisms causing the large temperature dependence of
threshold current seen in lasers made from these materials (Horikoshi,
1982). Here we describe a direct optical measurement of photo-excited
carrier relaxation in GaInAs. Using an optical parametric amplifier as a
source of picosecond pulses tunable near the band-edge of GaInAs, we
carried out excite-probe measurements of absorption bleaching and recovery.
Because the initial excited carrier density is uniform through the part
of the sample the probe pulse passes through, we could then use a simple
model to obtain the carrier recombination rate, and in particular the
Auger rate.

2. Experimental Set-Up

Fig. (1) illustrates the experimental
configuration used. The picosecond
tunable radiation is provided by an
angle tuned Optical Parametric
Amplifier consisting of a Lithium
Niobate crystal pumped by a 35 pS
pulse from a flashlamp pumped passiv-
ely modelocked Nd:YAG laser and operating at 1.06μm. A grating feedback
arrangement was used to obtain
tunable radiation from 1.45 to
2.1μm. For an input pulse energy of
2mJ between 5 and 50μJ is obtained
depending on the operating wave-

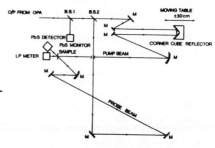

Fig (1) Experimental Layout

length (Prise 1983). The output pulse from the optical parametric
amplifier is split to provide one intense excite pulse and second weak
(< 1%) probe pulse. A relative delay of up to 4ns between the excite
pulse and probe pulse is introduced by using a stepper motor driven
translator and a corner cube reflector. The excite beam is focussed to
have a $(1/e^2)$ diameter of 700μm on the sample. In order to reduce
complication due to the Gaussian spatial profile of the beam, the probe
beam was focussed to half its diameter. The probe and excite beams
were at an angle of ∿5° to enable detection.

The input energy, the excite pulse transmission and the probe pulse
transmission were measured using PbS photodetectors and neutral density
filters. Calibration was performed using a Laser Precision Energy

ratiometer. All the detectors and the stepper motor driving the delay
line were interfaced to a microcomputer for ease of data collection.

The sample used in these experiments was a 4.4μm thick epitaxially grown
layer of InGaAs (n-type 10^{16} cm^{-3}) grown by vapour phase epitaxy on InP.
The composition of this sample was determined using double crystal x-ray
diffraction (Moss & Ritchie 1983). Two wavelengths are used in these
experiments, 775meV and 806meV, which are respectively 35 and 66meV above
the initial room temperature gap.

3. Experimental Results

The transmission of the excite pulse as a function of the pulse energy
was measured, and was found to increase to a maximum of 28%. This high
energy limit (with peak energy   densities of 1mJ/cm^2) corresponds to the
reflectivities of this sample and the absorption of the InP substrate.
In this limit the absorption of the InGaAs epilayer is completely
bleached, and a uniform carrier density is induced through the sample.
We then went on to examine the time dependence of this absorption bleaching using our excite-probe configuration. The results are shown in
Figure (2).

These measurements were carried out
in the high energy limit. This high
excite energy limit provides an ideal
regime for the measurements since the
probe beam then traverses a sample
region in which the photoexcited
carrier density generated by the
excite pulse is completely uniform.
The initial carrier density can be
obtained, since the quasi-fermi
levels in the conduction and valence
bands are equal under completely
bleached conditions.

4. Interpretation

Transmission recovery after the
initial bleaching is very rapid,
taking several hundred picoseconds.
This recovery is due to photoexcited carrier recombination.
In order to obtain the recombination rates, a number of assumptions have to be made. The dependence of sample transmission on
carrier density at a given wavelength must be known.

Fig.(2) Probe transmission as a
function of time-delay and wavelength. The solid lines are the
best theoretical fit including
Auger recombination. The dashed
line, for comparison is a fit
accounting for only extrinsic and
radiative recombination.

We have assumed that the photoexcited carriers thermalize very rapidly
within the conduction and valence bands, so their distribution can be
characterized by a constant representative temperature. The carriers
will scatter into a thermal distribution in less than 1ps which is much
less than the pulse length of 35ps. Since the carriers are excited near
the band edge with little excess kinetic energy we have taken this
constant representative temperature to be the lattice temperature T=300K.
This will be discussed later. We have also assumed only valence and

conduction band states with the same wavevectors are optically coupled, since our sample has a low impurity concentration. We can therefore write the absorption as a function of frequency and carrier density as

$$\alpha(\hbar\omega,N) = \alpha_o \; (\hbar\omega) \; (f_v(E_v,N,T) - f_c(E_c,N,T))$$

where $\alpha_o(\hbar\omega)$ is the linear absorption in the absence of excitation, E_c and E_v are the energies of the optically coupled states, T is the temperature and f_c and f_v are the electron fermi occupancies of the optically coupled states.

We can now curve fit by estimating the induced carrier density at zero time delay, N_0, and then by modelling the time-dependence of the carrier density N. Since we are working under initial bleaching conditions N_0 is determined as the density for which $f_c = f_v$. We do not need a value for $\alpha_o(\hbar\omega)$. The fermi energies as a function of carrier densities are evaluated using band edge effective masses $m_c = 0.041m_e$, $m_h = -0.50m_e$ (Pearsall, 1982) and Kane's four band model setting the spin-orbit splitting as zero to estimate the conduction band non-parabolicity (Kane,1957). The Joyce-Dixon approximation for fermi energy as a function of carrier density, which is valid under these experimental conditions, was used to reduce the amount of computation required (Joyce & Dixon,1977). The carrier-density dependence of the band-gap due to exchange and correlation effects was assumed to be the same as that of GaAs (Dutta,1981).

We represent the carrier recombination rate by

$$\frac{dN}{dt} = - AN^3 - BN^2 - CN$$

where A, B and C are the Auger, radiative and extrinsic rate coefficients respectively. Using the momentum matrix element $(2m_e \, P^2/\hbar^2 = 23.2eV)$ (Sugimura,1982) one can calculate $B = 2.3\times10^{-10} \mathrm{cm}^3\mathrm{s}^{-1}$ (Yariv,1975).

The solid lines in figure (2) indicate the results of fitting to the experimental data at the two frequencies, with the same temperature and Auger rate. Best fits of both sets of data are obtained using a temperature T=300K and $A = 2.5 \pm 0.5 \times 10^{-28}\mathrm{cm}^6\mathrm{s}^{-1}$. An upper limit on the extrinsic recombination rate of $2 \times 10^8\mathrm{s}^{-1}$ is obtained. It has been suggested that the radiative recombination rate calculated by our simple model may be too high (Su et al 1982). In order to evaluate the importance of this effect we fitted our results assuming no radiative recombination. In this case a best fit was obtained with $A = 3.2 \times 10^{-28}\mathrm{cm}^6\mathrm{s}^{-1}$, indicating our results are not very critical on the value we chose for the radiative recombination rate.

Another effect we must consider is carrier temperature. Some carrier heating may occur by the excitation of electrons high into the conduction band by the Auger process. This results in the electron distribution being described by a higher temperature than that of the lattice. Experimental measurements of carrier heating in 1.3µm InGaAsP indicate an increase in carrier temperature of much less than 100K (Etienne et al 1982). Fitting our data to a carrier temperature of 400K yields an Auger coefficient of $1.8\pm0.3 \times 10^{-28}\mathrm{cm}^6\mathrm{s}^{-1}$.

CONCLUSIONS

We have determined the Auger recombination rate in $Ga_{0.47}In_{0.53}P$ to be $2.5(\pm 0.5) \times 10^{-28} cm^6 s^{-1}$ at carrier densities in the range 1.0 to 2.5 x $10^{18} cm^{-3}$, similar to those at threshold in semiconductor lasers. This result is slightly higher than that obtained from a similar experiment using only 1.06μm radiation and a more complicated model taking into account spatial carrier distributions   (Wintner and Ippen, 1984). It is substantially higher than measurements of the Auger coefficient in 1.3μm InGaAsP (Sermage et al 1983, Mozer et al 1982, Henry et al 1983). This is in agreement with scaling arguments put forward by Sugimura, 1982 and Nelson and Dutta 1983. The magnitude of the result is greater than that predicted by these results, which is in agreement with the suggestion of Burt and Smith, 1984 that the overlap integral used by previous authors in Auger calculations may be too high.

ACKNOWLEDGEMENTS

We would like to thank British Telecom Research Laboratory for the provision of samples, and the U.S.A.F. Rome Air Development Centre at Griffiss Air Force Base for partial financial support under Grant No. AFOSR-83-151.

REFERENCES

Burt M G and Smith C 1984 J.Phys.C.Solid State Phys., 17,247
Dutta N K 1983 Appl.Phys.Lett.42,259
Etienne B, Shah J, Leheny R F and Nahory R E 1982 Appl.Phys.Lett.41,1019
Henry C H, Levine B F, Logan R A and Bethea C G 1983 IEEE J.Quantum
   Electron.QU19,905
Horikoshi Y 1982 GaInAsP Alloy Semiconductors ed T.P.Pearsall
   (New York: Wiley)pp379
Joyce W B and Dixon R W 1977 Appl.Phys.Lett.31,354
Kane E O 1957 J.Phys.Chem.Solids,1,249
Moss R H and Ritchie S 1983 British Telecom Tech.J.1,7
Mozer A, Romanek K M, Schmid W, Piluhn M H and Schlosser E 1982
   Appl.Phys.Lett.41,964
Nelson R J and Dutta N K 1983 J.Appl.Phys.54,2923
Prise M E 1983 PhD Thesis, Heriot-Watt Univ.Edinburgh
Sermage B, Eichler H J, Heritage J P, Nelson R J and Dutta N K 198=
   J.Appl.Phys;52,55(1981)
Su C B,Olshansky R,Manning J and Polwazinik 1984 Appl.Phys.Lett.44,732
Sugimura A 1982 IEEE J.Quantum Electron QE 18,352
Wintner E and Ippen E P 1984 Appl.Phys.Lett.44,999
Yariv A 1975 Quantum Electronics (Wiley: New York)pp160

Inst. Phys. Conf. Ser. No. 74: Chapter 5
Paper presented at Int. Symp. GaAs and Related Compounds, Biarritz, 1984

Magneto-optical study of hydrogenic donors confined in GaAs/Al$_{0.25}$Ga$_{0.75}$As quantum wells

R. J. Wagner, B. V. Shanabrook, J. E. Furneaux and J. Comas
Naval Research Laboratory, Washington, D. C. 20375

N. C. Jarosik and B. D. McCombe
SUNY at Buffalo, Amherst, N.Y. 14260

Abstract. Far-infrared magneto-absorption experiments have been performed on donors confined in GaAs/Al$_{0.25}$Ga$_{0.75}$As multi-quantum wells. The structures had GaAs layer thicknesses of 80-450 Å and were selectively doped either at the edge or center of the quantum wells. The experiments show that the 1s - 2p (m = +1) absorption lines attributed to donors at the center of the well shift to higher energies as quantum well widths are reduced. However, the experimentally determined transition energies are somewhat smaller than those predicted theoretically.

1. Introduction

Recent innovations in epitaxial growth techniques have resulted in detailed physical studies of semiconductor heterostructures and superlattices. Most of these studies have focused on the effect of reduced dimensionality on the conduction and valence band continua. In contrast, this paper is concerned with the effect of reduced dimensionality on the shallow donor in GaAs. A donor in bulk GaAs is well described by the effective mass approximation with a binding energy of 5.8 meV and a Bohr radius of about 100 Å. Theoretical calculations have indicated that this weakly bound impurity should be relatively strongly affected by confinement between Al$_x$Ga$_{1-x}$As layers. Specifically, it has been suggested (Bastard 1982, Greene and Bajaj 1984) that the binding energy of a shallow donor is sensitive to its location in the quantum well. Furthermore, the binding energy of a donor at the center of the well is expected to be significantly larger than that observed in bulk material. It will be shown that while some of the characteristics of the experimentally-observed donor-related features are compatible with these theoretical predictions others remain unexplained.

2. Experimental Considerations

The samples used in this study were GaAs/Al$_x$Ga$_{1-x}$As multiple quantum well structures prepared in a commercial MBE system. The width of each of the Al$_{0.25}$Ga$_{0.75}$As layers was approximately 150 Å for all samples examined. The width of the GaAs layers, L$_z$, was varied between 80 Å and 450 Å. The overall superlattice thickness was about 1 µm for each sample. The Si donors were 'spike doped' at either the center or the edge of the GaAs wells. The samples had a designed Si donor density from 5 x 10^{15} cm^{-3} to 1 x 10^{17} cm^{-3}. In most samples, the width of the doping profile, as

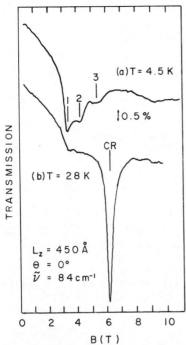

Fig. 1 —.The magneto-transmission
of 84 cm^{-1} radiation versus
magnetic field for a quantum well
of width 450 Å that has been
selectively doped with donors at
the center of the well.

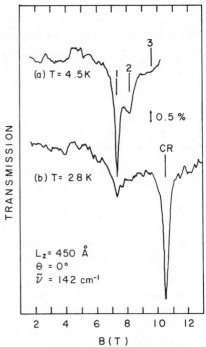

Fig. 2 —.The magneto-transmission
of 142 cm^{-1} radiation versus
magnetic field for a quantum well
of width 450 Å that has been
selectively doped with donors at
the center of the well.

determined by the growth rate, was 1/3 L_z. After these samples were
characterized via photoluminescence (Shanabrook and Comas 1984) and Raman
scattering (Shanabrook et al 1984) two types of far infrared
magneto-absorption experiments were performed. In one case, a far infrared
interferometer and superconducting solenoid was employed. Alternatively, a
far infrared laser was used with a Bitter solenoid magnet which provided
magnetic fields to 14 T. These approaches are complementary with each
having particular strengths and weaknesses. Since some of the
interferometer data has already been reported by Jarosik et al (1984) ,this
report will focus on the laser-Bitter magnet data. All of the data
presented below was taken by measuring the change of the transmission of
far infrared laser light by the sample as a function of magnetic field. A
detailed discussion of the experimental apparatus can be found elsewhere
(Wagner and Prinz 1970, Wagner et al 1973).

Under some experimental conditions (i. e. if the samples were cooled to
4.2K in the dark) and for some samples, the absorption lines were either
very weak or even non-existent. Furthermore, differences in line intensity
were apparent when interferometer and laser data were compared. Guided by
experience of workers who have studied persistent photoconductivity in
selectively doped heterostructure transistors (Stormer et al 1981), we
applied visible radiation to some samples. This procedure had a profound

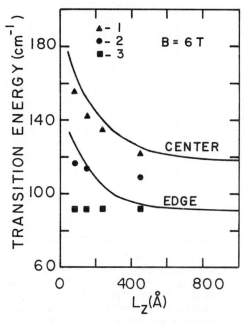

Fig. 3 — The transition energies for the peaks labeled 1, 2 and 3 at 6 T versus quantum well width. The theoretically determined 1s − 2p (m = +1) transitions for donors at the center and edge of the quantum well are indicated with solid lines (Greene and Bajaj, 1984).

effect on the absorption line intensity, sometimes increasing it by a factor of 10. Since the cause of this phenomenon has not been explored, this report will focus on the binding energies of the shallow donors as determined by the position of the absorption line.

3. Data Presentation and Discussion

Shown in figures 1 and 2 are transmission data as a function of magnetic field obtained from a sample with a quantum well width of 450 Å with far infrared laser energies of 84 cm^{-1} and 142 cm^{-1} respectively. This sample was selectively doped with Si donors in the central 150 Å of the quantum well. In the 4.5K data shown in figure 1(a) and 2(a), three distinct absorption lines are observed and labeled 1, 2 and 3. If the sample were bulk GaAs containing Si donors, only one feature due to the 1s–2p (m=+1) transition would be observed in this field range. These peaks due to intracenter transitions in the bulk material would have occurred at 3.7 T and 7.7 T for infrared laser energies of 84 cm^{-1} and 142 cm^{-1} respectively. The peaks observed in figures 1 and 2 do not occur at these magnetic fields and therefore are not related to bulk donors in GaAs. As is shown in figures 1(b) and 2(b), we observe cyclotron resonance due to free electrons when these samples are heated to 25K. In addition, we observed a decrease in the intensity of the lines labeled 1, 2 and 3 as the temperature of the sample was increased from 4.5K to 28K. These observations suggest that the peaks labeled 1, 2 and 3 are probably related to defects, presumably the intentionally incorporated Si donors, that can weakly bind an electron. Similar experiments were performed on samples with four different well widths, L_z. These samples had a variety of donor densities and in each case three absorption lines of varying width and intensity were observed. The spectral features were sufficiently well resolved to select a line position in each case. The positions of each of the absorption peaks as a function of quantum well width for a field

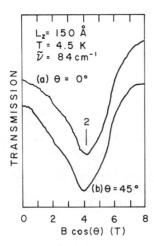

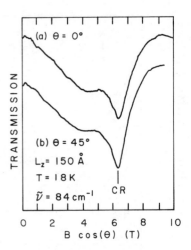

Fig. 4 — The 4.5 K magneto-transmission of 84 cm^{-1} radiation versus magnetic field of a quantum well of width 150 Å that has been tilted by (a) 0° and (b) 45° with respect to the field normal.

Fig. 5 — The 18 K magneto-transmission of 84 cm^{-1} radiation versus magnetic field of a quantum well of width 150 Å that has been tilted by (a) 0° and (b) 45° with respect to the field normal.

of 6 T is shown in figure 3. Because we measure field dependent absorption peaks, it was necessary to transform our field-dependent peaks into frequency-dependent positions by assuming a linear dependence of frequency and field. (This assumption can only be unambiguously tested with the interferometric spectrometer.) Figure 3 also includes the calculated transition energy at 6 T for donors at the center and edge of a GaAs well of width L_z (Greene and Bajaj 1984). The excellent agreement between the position of peak 1 and the theoretical calculation strongly suggests that line 1 is associated with donors at the center of the well. The interferometer data taken on the same samples (Jarosik et. al. 1984) also confirms this conclusion. On the other hand, the transitions responsible for lines 2 and 3 are not as yet clear. One possible explanation of the origin of line 2 in the samples with quantum well widths of 80 Å and 150 Å is that the GaAs buffer or cap layers are contaminated with donors. In order to test this hypothesis, the magneto-absorption of the sample with L_z = 150 Å was studied with the sample tilted at 45°. One would expect that the wave function of the donor would be mainly two-dimensional for L_z values comparable to the its Bohr diameter. Under such circumstances, we expect that the transition energies measured in magneto-absorption measurements would only be sensitive to the magnitude of the magnetic field that is normal to the layers. This indeed happened for a sample with L_z = 150 Å (Figs. 4, 5). Here the data for the (b) trace in each figure was scaled by cos(Θ). Within the limits imposed by system noise, traces (a) and (b) are identical. Figure 5 also confirms that the free electron gas has 2-D character. Similar behaviour was observed for peaks labeled 1, 2 and 3 at laser energies of 142 cm^{-1}. However, it was not possible to

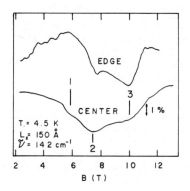

Fig. 6 — The comparison between the magneto-transmission observed for samples exhibiting quantum well widths of 150 Å and have either been doped at the center or edge of the well.

observe CR in this case because it was above the total field limit of the magnet. The donor related features of the samples with quantum well widths of 240 Å and 450 Å behaved in a somewhat similar, although more complicated, manner when tilted with respect to the magnetic field direction. Because the magnetic fields at which the peaks labeled 1, 2 and 3 are sensitive to the direction of the magnetic field relative to the growth axis, we conclude that these peaks arise from transitions involving defects in the multi-quantum wells. Therefore, they are not related to donors in the buffer or cap layers of GaAs.

On a sample with L_z = 150 Å and a well-center doping density of 1×10^{16} cm^{-3}, the temperature dependence of lines 1, 2, 3 and CR has been studied. This experiment showed that the line intensity of 3 went down with respect to 1 and 2. Over the same temperature range the intensity of CR increased. This observation shows that line 3 represents a distinct, very shallow, donor-like center and does not arise from a transition involving the same ground state that gives rise to either line 1 or 2. It is also interesting to note, that the magnitude of the transition energy for line 3 is not sensitive to the width of the quantum well.

Additional experimental information regarding the origin of the transitions that give rise to peaks 2 and 3 is displayed in Fig. 6. Here a comparison of absorption spectra for two samples is made. In the CENTER case, the sample was Si-doped at the center of the well. In the EDGE case, the sample was doped for 25 Å in the GaAs before starting Al$_{0.25}$Ga$_{0.75}$As growth. While the issue of line intensity for lines 2 and 3 is somewhat puzzling, it appears that line 1 at 6 T is almost totally absent from the edge doped sample. While this observation is consistent with the association of line 1 with the donors at the center of the quantum well, it also suggests that either line 2 or line 3 arises from donors at the edge of the quantum well. As indicated in figure 3, theoretical calculations would suggest that line 2 for this sample is to be associated with the 1s – 2p (m=+1) for donors at the edge of the quantum well. If this assignment is correct, then the data shown in figure 6 for the sample that was doped at the center of the well exhibit a significant number of donors at the edge of the well. This implies that either donor diffusion or segregation

during growth over a distance scale of 50 Å is significant for
'spike-doped' superlattices grown at $650^{\circ}C$. This point of view may also
explain why the transition energy of line 2 (figure 3) for L_z = 450 Å is
larger than expected from theory. If in this wide well, donors did not
reach the well edge, absorption structure might occur at an intermediate
energy, based on a density of states argument (Bastard 1981).
Additional measurements are necessary before the nature of the transitions
that give rise to lines 2 and 3 can be firmly established.

4. Conclusions

These experiments indicate that donors located at the center of a quantum
well exhibit significantly larger binding energies than that characteristic
of shallow donors in bulk GaAs. In addition, the magnitude of the increase
in binding energy versus quantum well width is in substantial agreement
with theoretical predictions. These measurements also reveal the existence
of two interesting additional absorption features. While the feature
labeled line 2 and observed in the narrower quantum wells (i. e. < 150 Å)
is possibly related to donors at the edge of the quantum well, the origin
of line 3 remains unexplained. However, we have discussed that the state
responsible for line 3 exhibits a significantly smaller binding energy than
that of the states responsible for lines 1 and 2. Furthermore, the
position of this line does not appear to be sensitive to the width of the
quantum well. In conclusion, we have also presented evidence suggesting
that either donor diffusion or segregation over a distance scale of 50 A
has taken place during the growth process. If this suggestion is true,
then donor motion may have important consequences for materials growth
and/or device design.

Acknowledgements

The authors wish to thank R. L. Greene and K. K. Bajaj for allowing us to
present the results of their calculation prior to publication and the staff
of the NRL High Magnetic Field Facility for their technical support.

References

Bastard G 1981 Phys. Rev. B24 4714
Greene R L and Bajaj K K 1984 Phys. Rev. B (to be published)
Jarosik N C, McCombe B D, Shanabrook B V, Wagner R J, Comas J and Wicks G
    1984 Proc. 17th Intl. Conf. on the Phys. of Semiconductors (to be
    published)
Shanabrook B V and Comas J 1984 Surface Science 142 504
Shanabrook B V, Comas J, Perry T A and Merlin R 1984 Phys.Rev.B29 7096
Stormer H L, Gossard A C, Wiegmann W and Baldwin K 1981 Appl.Phys.Lett.39
    912
Wagner R J and Prinz G A 1971 Appl. Opt. 10 2060
Wagner R J, Zelano A J, and Ngai L H 1973 Opt. Commun. 8 46

Optimized GaAs/(Al,Ga)As modulation doped heterostructures

H. Lee, W.J. Schaff, G.W. Wicks, L.F. Eastman and A R Calawa*
School of Electrical Engineering, Phillips Hall
Cornell University, Ithaca, NY 14853
* Cornell lecturer - permanent address:
MIT Lincoln Laboratories, Lexington, MA 02173, USA

Abstract

Modulation doped GaAs/(Al,Ga)As heterostructures using an AlAs spacer
layer and a Si atomic plane doping technique have been grown by MBE.
Reduced electron mobility degradation due to high temperature
annealing and high mobility at relatively large electron sheet
concentration have been achieved when AlAs is substituted for
$Al_{0.3}Ga_{0.7}As$ as a spacer layer material. For the atomic plane
doping, a thin (~ 0.1% monolayer) Si atomic plane was used to dope
the heterostructures. Electron mobilities of 7160 cm^2/V-s at 300K
and 122,500 cm^2/V-s at 77K have been obtained.

1.   Modulation Doped Heterostructures with AlAs Spacer

A GaAs/AlAs modulation doped heterostructure aimed at making use of the
large bandgap and refractory nature of AlAs was studied by Wang (1982).
However, the device performance was not satisfactory, presumably due to
difficulty in making good ohmic contact to the two dimensional electron
gas (2DEG) through thick AlAs.

In this paper, GaAs/(Al,Ga)As modulation doped heterostructures using
thin, undoped AlAs only as spacer layer at the heterojunction to
separate undoped GaAs and doped $Al_xGa_{1-x}As$ (x typically between 0.25 to
0.3) are discussed, taking full advantage of the large bandgap and
refractory nature offered by AlAs, while avoiding the ohmic contact
problem through thick AlAs.

1.1 Introduction

The bandgap of AlAs is 0.73 eV larger than that of GaAs, and is 0.37 eV
larger than that of $Al_{0.3}Ga_{0.7}As$ (Casey and Panish, 1969) which has been
typically used in GaAs/(Al,Ga)As modulation doped heterostructures. A
larger conduction band discontinuity by a factor of two is obtained by
replacing $Al_{0.3}Ga_{0.7}As$ with AlAs at the heterojunction interfacing with
GaAs.
A typical GaAs/(Al,Ga)As heterostructure using AlAs as a spacer layer is
shown in Fig. 1.   Both GaAs with a lattice constant of 5.6533 Å and
AlAs with a lattice constant of 5.6605 Å at 300K have zincblende crystal
structure. Despite the 1.7% lattice mismatch, the structure in Fig. 1

200 Å GaAs Undoped	

200 Å Graded Al$_x$Ga$_{1-x}$As
100 Å Al$_{0.3}$Ga$_{0.7}$As Undoped
200 Å Al$_{0.3}$Ga$_{0.7}$As:Si N$_d$ = 2×10^{18} cm^{-3}
100 Å AlAs Undoped
1.0 µm GaAs Undoped

SI GaAs

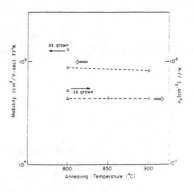

Fig.1   MD heterostructure   Fig.2   Annealing characteristics
        with AlAs spacer   of the structure shown in
                                          Fig.1

has shown 2DEG Hall mobilities of 7,100 cm^2/V-s at 300K 135,000 cm^2/V-s at 77K and 380,000 cm^2/V-s at 4K indicating good quality of the heterojunction interface comparable to that of GaAs/Al$_{0.3}$Ga$_{0.7}$As. The corresponding 2DEG sheet density at these temperatures (from 300K to 4K) are 4.17 x 10^{11} cm^{-2}, 4.1 x 10^{11} cm^{-2} and 4.0 x 10^{11} cm^{-2} respectively.

The substrate temperature during growth was 640°C. The Ga flux was set to grow GaAs with growth rate of 1.0 µm/hr, and Al flux was adjusted to obtain Al$_{0.3}$Ga$_{0.7}$As under the same Ga flux. The growth rate of AlAs was determined as follows:

G.R. of AlAs = G.R. of Al$_{0.3}$Ga$_{0.7}$As - G.R. of GaAs

$$= \frac{1.0 \ \mu m/hr}{0.7} - 1.0 \ \mu m/hr$$

$$= 0.4285 \ \mu m/hr$$

1.2. High Temperature Annealing Characteristics

Structures with AlAs spacer layers were annealed at 800°C, 850°C and 900°C respectively for 15 minutes in a capless anneal system using InAs as a source of As overpressure (Woodall et al 1981).

The 2DEG mobilities and sheet electron densities at 77K characterized by Van der Pauw measurements before and after annealing are plotted in Fig. 2. The 2DEG mobility suffered a 27% decrease as a result of the 800°C anneal. The mobility remained relatively constant at this value after annealing at temperatures up to 900°C indicating there was no further degradation of the heterostructure. The 2DEG sheet density showed the same trend. Sheet density decreased by 17% with annealing at 800°C and did not degrade further at annealing temperatures up to 900°C.

These results are an improvement over earlier work which used a conventional AlGaAs spacer layer (Lee et al 1984). In that work, the magnitude of mobility degradation resulting from high temperature annealing was strongly dependent on the annealing temperature. The fact that the high temperature annealing characteristics of the AlAs spacer MD heterostructures are superior to those of the $Al_{0.3}Ga_{0.7}As$ spacer MD heterostructures confirms the refractory nature of AlAs. This well-behaved annealing characteristic of the AlAs spacer MD heterostructure is important for its potential application in the self-aligned gate ion implanted MODFET technology.

1.3. Low Field 2DEG Mobility Characteristics

A typical plot of 2DEG mobility versus electron sheet concentration at low temperatures is shown in Fig. 3 for $Al_{0.3}Ga_{0.7}As$ spacer MD heterostructures (Hiyamizu et al. 1981). The dotted line represents the theoretical study done by Mori and Ando(1980), which indicates that the electron mobility starts to drop as the sheet electron concentration goes beyond a threshold of about 7.5×10^{11} cm^{-2}. The reason is that at this value the ground subband is filled and electrons start to fill the first excited subband, hence intersubband electron scattering drives the mobility down.

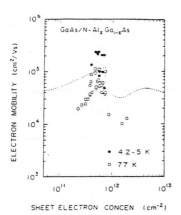

Fig.3

2 DEG mobility in a GaAs/$Al_xGa_{1-x}As$ heterostructure with $Al_xGa_{1-x}As$(x=0.3) spacer at low temperatures versus sheet electron concentration

A schematic diagram showing subband energy levels can be seen in Fig. 4. The obvious advantage by replacing $Al_{0.3}Ga_{0.7}As$ spacer with AlAs is the higher energy level of the first excited subband, hence more electrons can be accomodated before the onset of intersubband scattering. The result is that high electron mobility can be achieved even at high electron sheet concentration as shown by Fig. 5, which demonstrated that for the same 20 Å spacer layer thickness, the electron mobilities of AlAs spacer MD heterostructures are higher by a factor of two than its $Al_{0.3}Ga_{0.7}As$ spacer counterpart for sheet electron concentration over 1×10^{12} cm^{-2}. The growth and material parameters were kept the same for the two heterostructures compared here except for different spacer layer material.

This result should be of great merit to the cryogenic operation of MODFET to reduce the source series resistance, since low field mobility still plays an important role between source and gate. Thus higher extrinsic transconductance can be expected.

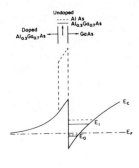

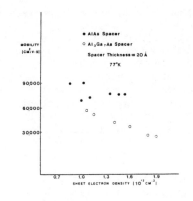

Fig.4   Schematic conduction
        band diagram showing
        subband levels

Fig.5   2 DEG mobility at 77K
        with different spacers
        of 20Å

1.4   Summary

By the use of AlAs only as a spacer layer to the GaAs/(Al,Ga)As MD
heterostructures, the more refractory nature and higher conduction band
discontinuity at the heterojunction interface provided by AlAs compared
to $Al_xGa_{1-x}As$ (x ~ 0.3) normally used offers (a) better annealing
characteristics, i.e., maintaining high 2DEG mobility and stable
electron sheet concentration after high temperature annealing up to
900°C, which is crucial for the fabrication of self-aligned gate ion-
implanted MODFETs and (b) raised first excited subband level, hence high
electron mobility at low temperatures can be achieved even at high
electron sheet concentration, which is helpful to reduce the source
resistance of cryogenic MODFETs and higher extrinsic transconductance is
expected.

2.   Modulation Doped Heterostructures Using Atomic Plane Doping

2.1   Introduction

Atomic plane doping, which was first introduced to synthesize complex
free-carrier profiles in MBE grown GaAs using Ge as a dopant by Wood et
al. (1980) was done by interrupting GaAs growth by stopping the Ga flux,
while Ge deposition onto the surface was used to form an atomic plane of
dopants. Recently, this technique using Si as a dopant has been studied
by Ohno et al. (1984) to synthesize free-carrier profiles in
metalorganic chemical vapor deposition (MOCVD). C-V free carrier
profiling showed that the sheet carrier concentration of a dopant plane
was proportional to the amount of Si introduced.

In an attempt to minimize the thickness of the doped $Al_xGa_{1-x}As$ layer,
which is crucial to optimize the transconductance achievable for FET
applications, modulation doped heterostructures using Si atomic doping
technique in $Al_{0.3}Ga_{0.7}As$ have been successfully grown by molecular beam
epitaxy (MBE).

2.2 Structure and Growth Parameters

A typical modulation doped heterostructure used to study the effect of

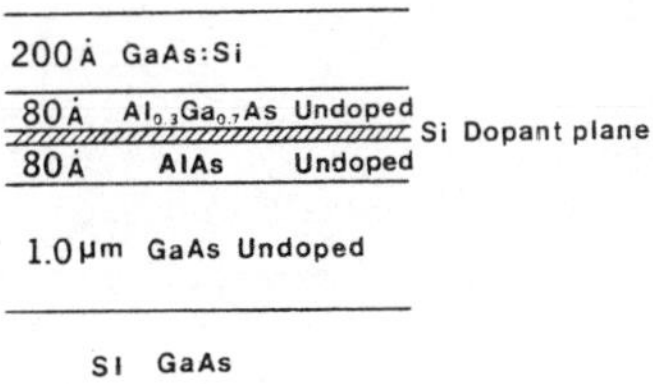

Fig. 6. A typical modulation doped heterostructure used to study the effect of atomic plane doping

atomic plane doping is shown in Fig. 6. Undoped GaAs with thickness of 1.0 µm was first grown on (100) undoped LEC semi-insulating GaAs substrate with growth rate typically at 1.0 µm/hr. After a spacer layer of 80 Å $Al_{0.3}Ga_{0.7}As$ was grown on top of the undoped GaAs, the sequence of the atomic plane doping followed: (A) shuttering Ga and Al flux to suspend growth, (B) introducing Si flux to form an atomic plane of dopant on the surface, (C) restart the $Al_{0.3}Ga_{0.7}As$ growth to incorporate the dopant layer. The Si flux introduced would result in 5 x 10^{17} cm^{-3} doping level were the layer grown at 1.0 µm/hr and doped uniformly.

2.3 Results and Discussion

Van der Pauw measurements made on the sample in Fig. 6 yielded electron mobility of 7820 cm^2/V-s at 300K and 70,325 cm^2/V-s at 77K. The sheet electron concentration was 8.7 x 10^{11} cm^{-2} at 300K and 7.4 x 10^{11} cm^{-2} at 77K. Electron concentration profiles of the atomic doping plane at 300K and 77K in GaAs made by C-V measurements are shown in Fig. 7 and Fig.8. An estimate of the width of the real doping profile can be determined as follows: assuming the width is the sum of two factors, real width A and Debye length B $\sqrt{T}$ which is proportional to the square root of the absolute temperature T.

$$FWHM(T) = A + B\sqrt{T}$$

with FWHM(300) = 306 Å and FWHM(77) = 216 Å two equations can be solved and A = 126 Å is obtained, probably due to diffusion of Si in GaAs. The spread of the doping profile instead of maintaining an atomically abrupt dopant plane results in an effectively thinner spacer layer thickness which also explains the lower electron mobility at 77K than what would be expected in a uniformly doped modulation doped heterostructure with comparable spacer thickness. The same structure in Fig. 6 with AlAs as a spacer was also grown. Electron mobilities of 7160 cm^2/V-s at 300K and 122,500 cm^2/V-s at 77K with sheet electron concentrations of 4.4 x 10^{11} cm^{-2} at 300K and 4.2 x 10^{11} cm^{-2} at 77K were obtained.

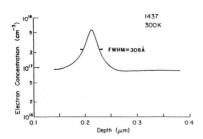

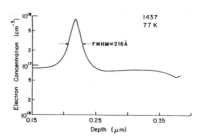

Fig.7 C-V profile of Si atomic doping plane in GaAs at 300K with back-ground doping of 1×10^{17}cm^{-3}

Fig.8 C-V profile of Si atomic doping plane in GaAs at 77K with back-ground doping of 1×10^{17}cm^{-3}

2.4 Summary

Thin modulation doped heterostructures using Si atomic plane doping has been successfully grown by MBE. Electron mobilities of 7160 cm^2/V-s at 300K and 122,500 cm^2/V-s at 77K has been achieved. This structure has potential use for MODFETs with reduced thickness between gate and 2DEG channel, thereby enhancing the transconductance which is important for microwave and high speed logic applications.

Acknowledgement

The authors would like to thank J. Berry for his excellent technical assistance. Special thanks go to Mrs. E. Wever for preparation of the manuscript. H. Lee was supported by a Honeywell Fellowship. The work was supported by the U.S. Air Force Office of Scientific Research under contract no. F49620-81-C-0082.

References

Casey H C and Panish, M 1969 J. Appl. Phys. **40** 4910
Hiamizu S, Minmura T, Fujii T, Nanbu K and Hasimoto H, 1981
    Jpn. J. Appl. Phys. **20** L245
Lee H, Wicks G and Eastman L, 1984, Proc. of IEEE/1983 Cornell
    Conference on High Speed Semiconductor Devices and Circuits 204
Mori S and Ando J, 1980 J. Phys. Soc. Jpn. **48** 865
Ohno H, Ikeda E and Hasegawa, H., 1984 Jpn. J. Appl. Phys. **23** L369
Wang W, 1982 Appl. Phys. Lett., **41** (6), 540
Wood C, Metze G, Berry J and Eastman L, 1980 J. Appl. Phys. **51**
    383
Woodall J, Rupprecht H, Chicokta R and Wicks G, 1981 Appl. Phys.
    Lett., **38** 639

Inst. Phys. Conf. Ser. No. 74: Chapter 5
Paper presented at Int. Symp. GaAs and Related Compounds, Biarritz, 1984

327

Electronic properties of MOVPE grown GaAs/GaAlAs heterostructures and quantum wells using electroreflectance and spectroscopic ellipsometry

M. Erman, P. Frijlink, J.B. Theeten
Laboratoires d'Electronique et de Physique Appliquée,
3, avenue Descartes - 94450 Limeil Brévannes (France)
and
C. Alibert, S. Gaillard - Equipe de Micro-opto-électronique de Montpellier
Unité   associée au C.N.R.S. n° 392 - USTL - F34 060 Montpellier (France)

Abstract. MOVPE grown GaAs-Ga$_{1-x}$Al$_x$As heterostructures and quantum wells have been analyzed using electroreflectance (E.R.) and spectroscopic ellipsometry (S.E.). For electroreflectance, E.R. gives the electronic levels for GaAs and Ga$_{1-x}$Al$_x$As layers. S.E. gives an analysis of the interface region in terms of roughness and chemical composition. For quasi two dimensional electron gas E.R. gives all allowed transitions around the Γ and L points together with a quantitative analysis of Stark shifts of the fundamental levels. S.E. gives both the thickness and the dielectric function of the GaAs well. The two techniques have been used to analyse the E_1 and $E_1 + \Delta_1$ transitions of quantum wells.

1. Introduction

Quantum well (QW) structures are of great technological interest especially for low threshold quantum well lasers. The understanding of their physical properties, as well as a complete characterization of the structures (thicknesses, interface quality, 2 dim electronic levels) is therefore needed. Usually, electronic states in the QW are measured with photoluminescence [1] or luminescence under excitation conditions [2] where transitions other than the fundamental one can be observed. We show that similar information can be obtained using spectroscopic ellipsometry (SE) and electroreflectance (ER). Furthermore, SE gives an accurate - at the monolayer scale - determination of the QW thickness, while ER is capable of checking the electro-optic behaviour of the QW (Stark effect). We have studied single GaAs QW embedded in a GaAlAs medium. The structures have been characterized around the fundamental gap (Eo transition), and around the L point (E1 transition corresponding to L4,5 v → L6c), a region which is not usually explored.

2. Experimental

Three samples, with QW thicknesses of respectively 4,5 and 7 nm, were grown using an OM-CVD reactor (growth temperature 650°C, growth rate 5 A/sec) with optimized geometry in order to achieve sharp transitions. The gas transport in the VPE reactor was arranged in such a way that gas composition over the wafer could be changed within 0.1 s. The QWs have been grown near the surface (10 nm, in order to allow the analysis of the E1 optical transition), on top of a 1 μm Al$_{54}$Ga$_{46}$As buffer layer. All epitaxial layers are not intentionally doped with the residual doping in the low 10^{15}cm^{-3} region.

The spectroscopic ellipsometer operates in ambient air. It is of the rotating polarizer type and covers the 1.2 to 5.4 eV range [3]. SE measures the ratio ρ between the reflection coefficients rp and rs for a light polarized parallel (p) and perpendicular (s) to the plane of incidence. The ratio is expressed as follows : $\rho = r_p/r_s = \tan \psi \exp (i\Delta)$. On bulk, homogeneous materials, one can directly measure the dielectric function of the material [4]. This has been done for GaAs, and thick GaAlAs epitaxial layers. Such reference dielectric functions are used for the multilayer analysis of the QWs.

For ER, parts of the wafer have been covered with 80 nm of Cu_2S evaporated film as a top electrode, the back side of the n+ substrate being contacted using silver paste. The experimental set up has been described in [5]. In the ER experiment, a modulated voltage is applied to a Schottky barrier. The induced electric field changes the dielectric function of the material and therefore the reflection coefficient.The AC component of the reflected light is measured which gives the ratio $\Delta R/Ro$. In the low field approximation, $\Delta R/Ro$ is proportional to the third derivative $(\frac{\partial^3 \epsilon}{\partial E^3})$ of the dielectric function vs energy.

3. Thickness determination

Thicknesses of all layers (except the buffer GaAlAs layer) have been determined using ES data taken in the 2.6 to 3.6 eV region. In this region one can observe the E_1, $E_1 + \Delta_1$ optical transitions of both GaAs (QW) and GaAlAs. Because of the thickness of the buffer layer (1 μm) and the penetration depth of the light in this region (<.1μm) no contribution to the experimental spectra from the substrate can be seen. The position of the E_1,$E_1 + \Delta_1$ peaks of GaAlAs is a direct measure of the Al composition. Furthermore, due to the strong variation of the optical absorption near E_1 for both GaAs and GaAlAs, the overall lineshape of tan (ψ) and cos (Δ) curves contains also information upon the thicknesses of the layers. Fig. 1 shows a typical experimental cos Δ curve, as well as the first derivative, which enhance the E_1,$E_1 +\Delta_1$ structures. Using reference dielectric functions, such experimental curves can be fitted with a theoretical multilayer model. The parameters of the model (thicknesses) are determined using linear regression. The dielectric function of the QW is to a first approximation derived from the bulk GaAs one with an overall energy shift because of the quantum localization energy. A detailed description of this fitting procedure can be found in [6]. The results are given in Table I. (see page 3).

The interface between GaAs and GaAlAs have been assumed to be sharp in our model since a previous SE analysis of heterojunctions grown in the same MO-VPE reactor had shown that the GaAs-GaAlAs transitions are within one monolayer [7]. Thickness of the buffer layer can be obtained performing measurements in the near IR region as discussed in the next section.

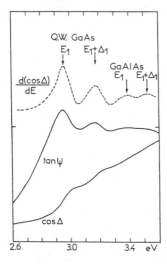

Fig.1 - SE spectra in the 2.6-3.6 eV range (tan ψ, cos Δ vs E curves) for a 7 nm thick QW. Dotted line is the first derivative of the cos Δ.

		SHF 382	SHF 383	SHF 384
Top oxide layer	density	$120 \pm 13\%$	$179 \pm 18\%$	$216 \pm 24\%$
	d_{ox}	2.57 ± 0.07 nm	2.69 ± 0.07 nm	2.18 ± 0.06 nm
AlGaAs overlayer	d_{ov}	7.08 ± 0.22 nm	6.43 ± 0.16 nm	4.46 ± 0.15 nm
GaAs QW	d_{QW}	3.73 ± 0.13 nm	5.00 ± 0.12 nm	7.00 ± 0.16 nm
Mean square deviation model and data		2.9×10^{-3}	3.7×10^{-3}	4.6×10^{-3}

Table I

4. The fundamental gap (Eo region)

The results obtained around Eo by ER are reported on Fig. 2. For comparison the data measured on bulk GaAs are also shown. In this case the Eo transition ($\Gamma 6c \rightarrow \Gamma 8$ v) and Eo+Δo ($\Gamma 6c \rightarrow \Gamma 7$ v) are clearly seen. On the QW's data, the contribution of the substrate can not be observed because the modulated electric field does not penetrate into the GaAs n+ substrate. Therefore all the structures below 2.1 eV (which is the Eo transition of GaAlAs) are originating from the QW. The dependence of the energy position on the QW thickness is clearly seen. Comparing the peak position with a theoretical modeling of the QW [6] one can identify the structure labeled 1 as being the transition from the first level of heavy holes to the first level of electrons, and 1' as the transition from first level of light holes to first level of electrons. Other transitions (2,2'...) correspond to higher levels of electrons and holes. It is clear that in the tentative

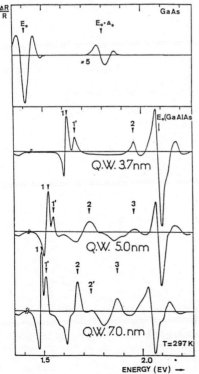

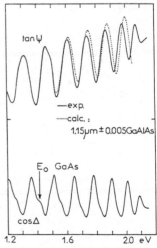

Fig. 3 - SE spectra in the near IR region for a 3.7nm thick QW. Dotted line is calculated tan ψ curve for a 1.5μm thick GaAlAs layer on top of GaAs substrate, as determined by linear regression.

Fig. 2(left) - ER spectra obtained from the QWs as well as reference GaAs sample in the Eo energy region.

labeling of the optical transitions, the QW thickness determination is of prime importance.ES and ER appear therefore as complementary optical methods

In the same energy region, ES data exhibit strong interferences due to the GaAlAs buffer layer. This can be used to calculate accurately the thickness of this layer as seen in Fig. 3. In contrast with ER, there is a strong contribution of the substrate, and Eo transition of the substrate can be seen on the tan (ψ) and cos (Δ) curves. The transitions due to QW are not easily seen and a mathematical subtraction of the interference regime would be compulsory in order to observe them with the additional complication of possible thickness and composition inhomogeneities since the light penetrates through the all wafer.

5. The E₁ (L point) region

The discontinuities between GaAs and GaAlAs valence and conductance bands around the L point give rise to a QW similar to the one around the Γ point. Thus, localization energies for $E_1, E_1 + \Delta_1$ are also associated with transitions while different from those observed at the direct gap Γ. Both ER and SE have been used to measure the E_1, $E_1 + \Delta_1$ positions in the QW. The position can be determined by SE using derivatives of the experimental curves (Fig. 1), which is quite similar to the ER, but the results are obtained without applying an electrical field. Another way is to subtract mathematically from the SE data the contribution of the surface overlayers and buffer layer, then calculating the dielectric function of the QW itself. Ref.[6] explains the mathematical procedure. We emphasise that such a "deconvolution" is only possible when the thickness of the QW is known. Fig. 4 gives the dielectric functions for GaAs, 3.7 and 7 nm thick QW. The shift towards higher energies due to the localization of electrons and holes is observed.

Using the effective masses around the L point, in which the contribution

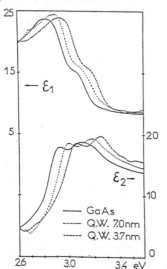

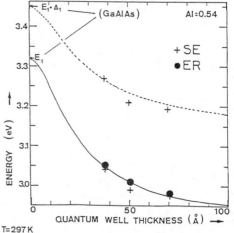

Fig. 5 - Modelling of the optical transitions around E_1 and $E_1+\Delta_1$. Only the first level transition is represented (n=1). The experimental values from ER and SE are given for comparison.

Fig. 4 - Dielectric functions of the GaAs QW layers obtained from the SE data using the thicknesses deduced from the multilayer model fitting. The dielectric function of bulk GaAs is given for comparison.

from the Λ lines has been incorporated, we have computed the optical transition energies corresponding to E_1 and $E_1 + \Delta_1$, as a function of QW thickness. The results are shown in Fig. 5 and are compared with experimental data. The agreement is reasonably good.

6. Stark effect in QW

Fig. 6 gives the ER spectra for two samples near the excitonic transitions $HH_1 \rightarrow E_1$ and $LH_1 \rightarrow E_1$ for various electric fields. The peak positions clearly shift towards lower energies as the electric field is increased. An approximate value of the electric field in the AlGaAs layers can be obtained [8] from the Franz-Keldysh oscillations above the AlGaAs E_0 excitonic transition [9]. We have interpreted those shifts in terms of Stark effect, using a model based on a variational method [10] and neglecting excitonic effects as well as tunneling through finite height barriers. Fig. 7 gives the comparison between the experimental shifts and model. A satisfactory agreement is obtained when applying a scaling factor of 0.67 to the estimated electric field.It is to be noted in particular that the prediction for a quadratic shift as a function of the electric field is confirmed.

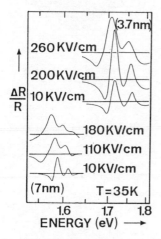

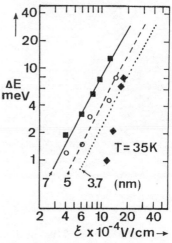

Fig. 6 - 35 K E.R spectra from 3.7 and 7 nm thick single GaAs quantum well near the $HH_1 \rightarrow E_1$ transition for various longitudinal electric fields.

Fig. 7 - Experimental and theoretical shifts of the $HH_1 \rightarrow E_1$ transition vs the longitudinal electric field. Full, dashed and dotted lines are theoretical values, ■, O and ◆ experimental values for 7nm, 5nm and 3.7nm thick single quantum wells. Experimental data are shifted with a C-scaled electric field ($\mathcal{E} = 0.67 \ \mathcal{E}_{exp}$).

7. Conclusions

In summary, ER and SE have proven to be complementary for the analysis of electronic states in AlGaAs / GaAs quantum wells. In fact, the methodology developed here can be extended to other heterostructures such as InGaAs / InP quantum wells. In this case, the sensitivity of SE is even improved as illuminated in Fig. 8 which gives the computed difference $\Delta\rho$ between SE measurements on a stack with and without the quantum well layer.

$\Delta\rho$ is computed as a function of the photon energy and overlayer thickness d_{ov} both for AlGaAs / GaAs and InP / InGaAs systems. In both cases, most of the information is seen to originate from the 3 eV region because of the difference in E_1 peak positions for the QW layer and overlayer material. As shown in Fig. 8, a InGaAs QW is more visible since InP is less absorbing in the InGaAs E_1 transition region as compared with the corresponding situation for AlGaAs/GaAs system.

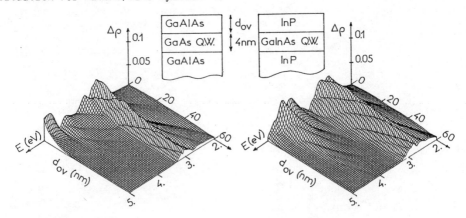

Fig. 8 - Sensitivity of ES to 4 nm QW of GaAs embedded in GaAlAs (left) and 4 nm QW of GaInAs embedded in InP (right). See text.

References

1 - C. Weisbuch, R.C. Miller, R. Dingle, A.C. Gossard and W. Wiegmann, Solid State Commun, 37, 219 (1981)

2 - Z.Y. Xu, V.G. Kreismanis and C.L. Tang, Appl. Phys. Lett. 43, 415 (1983)

3 - Our system is comparable to the one described in : D.E. Aspnes and A.A. Studna, Rev. Sci. Inst. 49, 291 (1972)

4 - D.E. Aspnes, A.A. Studna, Phys. Rev. B. 27, 985 (1983)

5 - See for instance, C. Alibert, Fan Jia Hua, M. Erman, P. Frijling, P. Jarry and J.B. Theeten, Rev. Phys. Appl., 18, 709 (1983)

6 - M. Erman, J.B. Theeten, P. Frijlink, S. Gaillard, Fan Jia Hia, C. Alibert, to be published in J.A.P.

7 - M. Erman, J.B. Theeten, N. Vodjdani and Y. Demay, J. Vac. Sci. Technol., B1 (2), 328 (1983)

8 - D.E. Aspnes, Phys. Rev. 12, 2297 (1975)

9 - C. Alibert, S. Gaillard, J.A. Brum, G. Bastard, P. Frijlink and M. Erman, to be published.

10 - G. Bastard, E.E. Mendez, L.L. Chang and L. Esaki, Phys. Rev. B28, 3241 (1983)

Inst. Phys. Conf. Ser. No. 74: Chapter 5
Paper presented at Int. Symp. GaAs and Related Compounds, Biarritz, 1984

333

Sub-micrometer quantum well HEMT with $Al_{0.3}Ga_{0.7}As$ buffer layer

L.H. Camnitz, P.A. Maki, P.J. Tasker, and L.F. Eastman
School of Electrical Engineering, Phillips Hall
Cornell University, Ithaca, NY 14853   USA

Abstract
The first modulation doped AlGaAs/GaAs heterostructure field effect
transistors have been fabricated utilizing an undoped $Al_{0.3}Ga_{0.7}As$
buffer layer. Transconductances (g_m) of 125 mS/mm and 210 mS/mm have
been measured at 300K and 77K, respectively for 0.35 μm length
enhancement mode devices. The devices exhibit this high g_m at low
drain current with excellent pinchoff characteristics, attributed to
a high quality GaAs quantum well channel and AlGaAs buffer layer. A
state of the art 8.3 GHz noise temperature of 10.5K was measured at a
physical temperature of 12.5K, demonstrating great potential of these
devices for low noise amplification.

1)   Introduction
The use of an AlGaAs buffer layer to reduce the parasitic buffer current
or substrate conduction in GaAs MESFET's has been proposed by Eastman et
al (1979) and investigated by many laboratories (Hallias et al 1979).
Improvements in output resistance have been noted (Ghosh et al 1984,
Wang et al 1981) and some improvement in microwave power efficiency has
been found (Schaff et al 1983). However, very little direct evidence of
significantly improved high frequency behavior has yet been observed.
An important reason for this lack of evidence may be that most of the
devices investigated either had gate lengths longer than one micrometer
or were not tested at microwave frequencies. One would expect the
parasitic substrate conduction to become dominant for very short gate
lengths, for it scales inversely as the gate length squared (Eastman et
al 1979).

A problem which exists in MESFET's with AlGaAs buffer layers is the
difficulty of growing high quality GaAs on AlGaAs. Such devices may
show degradation of g_m at low I_d or poor pinchoff characteristics,
indicating a poor heterojunction interface or buffer layer. The use of
smoothing layers (Kopp et al 1982, Arnold et al 1984) or multi-layer
buffers (Schaff et al 1983) results in improved pinchoff
characteristics.

Recent modeling work (Widiger et al 1984) has shown that in High
Electron Mobility Transistor (HEMT) structures the excited electron
energy states of the two dimensional electron gas (2DEG) are not
confined by the potential well, rather they are bulk-like states similar
to those in the MESFET. Since hot electrons are prevalent in the
channel of an operating short channel HEMT, one would expect a large

substrate conduction component. Indeed, state of the art 0.35 micrometer HEMT's exhibit values of output resistance as low as 43 Ω-mm (Berenz et al 1984). An inverted HEMT or double heterostructure HEMT which incorporates a thick AlGaAs buffer layer might then outperform a short channel conventional HEMT.

In this paper, we report on the successful fabrication of a submicron low noise quantum well HEMT. A conventional HEMT structure with the addition of an undoped $Al_{0.3}Ga_{0.7}As$ buffer layer was chosen over the inverted HEMT or double heterostructure HEMT in order to avoid the problems of Si-doping in the AlGaAs below the GaAs channel (Drummond et al 1982). This work reflects the application of previously reported MBE growth techniques for undoped single quantum wells (Maki et al 1983a) and modulation doped single quantum wells (Maki et al 1983b) on thick (1 μm) AlGaAs buffers. The important aspects are high temperature growth (680°C) at conventional growth rates (1-1.4 μm/hr) with careful control of surface stoichiometry through the V:III flux ratio. These conditions result in a quantum well structure exhibiting narrow photoluminescence linewidth and high electron mobility, indicating smooth well interfaces and low defect and impurity concentration. In addition, a high quality AlGaAs buffer is obtained which is important to the device results presented here.

Epilayer Growth

The structure was grown by MBE in a Varian Gen II machine on an undoped SI LEC substrate. The AlGaAs layers and the GaAs quantum well were grown at a substrate temperature of 680°C at a GaAs growth rate of 1 μm/hr under metal stable growth conditions. The device structure grown is shown in Fig. 1. First, a 2100 Å GaAs buffer layer was grown, followed by a graded composition 1000 Å thick AlGaAs layer to preclude the formation of a second 2DEG. The 1 μm thickness of the unintentionally doped AlGaAs buffer was chosen as this is a depth over which the fringing fields are high due to the 0.35 μm gate geometry (Eastman, 1979). The undoped 130 Å thick GaAs quantum well channel is followed by an 85 Å undoped AlGaAs spacer layer and 350 Å Si-doped AlGaAs at a Si atom concentration of 1.5 $\times 10^{18}$ cm^{-3}. Finally, a 350 Å n$^+$ contact layer was grown. Van der Pauw Hall measurements yielded a mobility

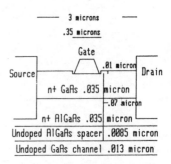

Fig. 1. Device structure.

of 6,400 cm^2/V-sec and 70,000 cm^2/V-sec with sheet carrier densities of 7.6 x 10^{11} cm^{-2} and 6.1 x 10^{11} cm^{-2} at 300K and 77K, respectively. The 77K sheet carrier density was ~ 4 x 10^{11} cm^{-2} when measured in the dark, indicating a significant persistent photoconductivity (PPC) effect in the Si-doped AlGaAs.

Device Fabrication

Low noise sub-micrometer gate length HEMT's were fabricated using mesa isolation, Ni/AuGe/Ag/Au alloyed ohmic contacts, and recessed gates formed by direct write EBL. The ohmic contact is a 100 Å Ni, 900 Å AuGe, 1000 Å Ag, and 1000 Å Au metallization deeply alloyed at a peak temperature of 570°C in a 3 minute transient cycle (Zwicknagl et al 1984, Jones). The ohmic contact resistivity was measured to be 0.5 Ω-mm at 77K and at 300 K. Wet chemical etching was used to etch off ~ 100 Å of the n^+ GaAs layer before the Ti/Pt/Au gate metal deposition. The edges of the recessed gate trench coincided with the footprint of the 0.35 μm long gate. The 300 μm width gate was centered in the 3 μm channel formed by the source and drain ohmic contacts, and fed by 2 gate pads in the standard π configuration.

DC Characteristics

The drain characteristics at 300K (Figure 2) showed an enhancement-mode like behavior, with a threshold of voltage of -0.2 V and a maximum transconductance of 38 mS, or 125 mS/mm. The recessed Schottky gate metal to GaAs quantum well spacing, a, was 700 Å, as confirmed by the gate capacitance. This explains the relatively low transconductance compared to HEMT's with spacings of 350-400 Å. Considering that the channel aspect ratio, L/a, is only about 5:1, the drain output resistance value of 500 Ω or 150 Ω-mm is rather high.

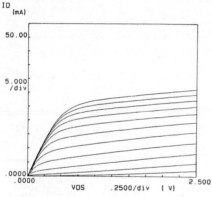

Fig. 2: 300K enhancement mode drain characteristic (Vds 0.1V/step)

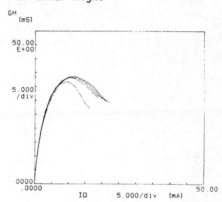

Fig. 3: 300K g_m vs I_d

A very important characteristic for low noise FET's is to have high transconductance at low current in order to reduce the noise generated by velocity saturated electrons (Pucell et al, 1975). A plot of g_m vs I_d (Fig. 3) reveals that the AlGaAs buffered HEMT achieves its maximum transconductance at only 11 mA. In addition, it does not exhibit g_m degradation with increasing drain bias. This behavior is much different than that seen for an equal gate length HEMT with a GaAs buffer layer and a channel aspect ratio, L/a, of 10:1 (Fig. 4) (Camnitz et al 1984). With the GaAs buffer, increasing the drain voltage degrades the g_m at low current, indicating an increase in substrate current which is not well controlled by the gate voltage.

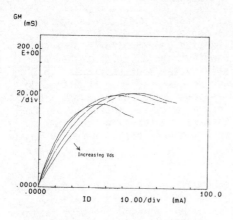

Fig. 4: 300K g_m vs. I_d for 0.33
μm GaAs buffered HEMT

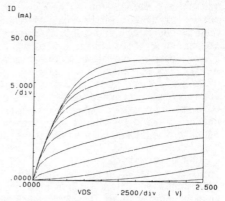

Fig. 5: 77K enhancement mode drain
characteristic (Vds 0.1V/step)

At 77K in light, the threshold voltage and transconductance increased to
0V and 210 mS/mm, respectively (Fig. 5). In the dark, the drain curves
collapsed when V_{ds} exceeded 1V, due to the deep traps in the doped
AlGaAs (Rochette, 1982). The plot of g_m vs I_d at 77K measured in light
(Fig. 6) shows that the maximum transconductance is reached at only 8
mA, as compared to 30-50 mA for conventional MESFET's. The variation in
g_m at higher current is caused by the onset of drain current collapse.

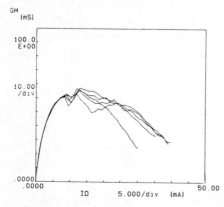

Fig. 6: 77K g_m vs I_d

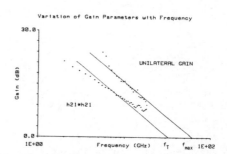

Fig. 7: Gain vs freq. at I_d=15mA

Microwave Characterization

S-parameters were measured at 300K from 2-18 GHz at varying bias
conditions. The maximum microwave gain was observed at 15 mA drain
current and 2V drain bias. The maximum stable gain was 15.4 dB at 8
GHz. Fig. 7 shows a plot of current gain and unilateral gain vs
frequency. Extrapolation at 6 dB/octave yields 32 GHz for f_T and a very
good 60 GHz f_{max} for this operating current.

Noise Measurements

Some chips were packaged and evaluated for noise temperature in a slug tuned 8 GHz test set by M. Pospieszalski of the National Radio Astronomy Observatory. The minimum 300K noise temperature was 104K (1.33 dB) with an associated gain of 8.4 dB at a drain bias of 5 mA and 2V (Fig. 8). The noise figure is fairly low between 2 mA and 10 mA while the gain degrades only below 5 mA. This optimum current is significantly lower than in GaAs MESFET's, whose optimum current is usually 8 to 15 mA at a bias of 3-3.5V. This noise figure is limited by the parasitic gate and source resistances. Using the Fukui equation and the measured values $R_g = 7\ \Omega$, $R_s = 9\ \Omega$, $gm_0 = 47$ mS and $C_{gs} = .2$ pF one obtains a fitting factor $K_f = 1.82$. This value compares closely to that obtained for conventional HEMT structures by Laviron et al (1984) and is lower than normally seen for GaAs FET's.

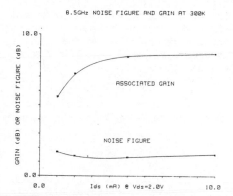

Fig. 8: 300K noise parameters vs I_d

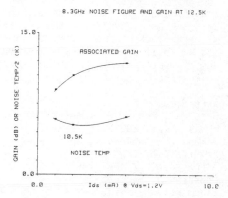

Fig. 9: 12.5K noise parameters vs I_d

The test set and device were cooled to 12.5K physical temperature and the minimum noise temperature of 10.5 K with an associated gain of 10.5 dB was measured for the amplifier at a drain bias of 1.5 mA and 1.2 V. This is a state of the art value of noise temperature while the optimum bias current is far lower than normally measured on similarly sized GaAs MESFET's (~ 8 mA). See Fig. 9.

Discussion

The ability for this HEMT to keep high gain at very low current is aided by the high electron mobility and confinement of the electrons to the quantum well channel. This helps to lower the noise figure and also lower the optimum bias current for minimum noise figure.

The advantage of an AlGaAs buffer layer becomes more apparent when GaAs buffer layer conduction becomes significant at very short gate lengths. A major improvement is seen in the sharper pinchoff characteristic, and the increased output resistance. A high quality AlGaAs buffer layer and quantum well channel are essential to high performance.

Conclusion

We have fabricated the first submicrometer gate HEMT with an undoped $Al_{0.3}Ga_{0.7}As$ buffer layer. We have demonstrated its superior pinchoff characteristics and high gain at low current. Its noise performance is comparable to conventional HEMT's at 300K and state of the art at 12.5K physical temperature. The HEMT's fabricated in this work were limited by high gate resistance, low sheet carrier density, and lower than desired doping in the n^+ AlGaAs. High source resistance and lower than optimum transconductance resulted. By increasing the doping density and adjusting layer thicknesses, the transconductance and parasitic resistances can be made equivalent to GaAs buffered HEMT structures. Without these limitations, the AlGaAs buffered HEMT may surpass the conventional HEMT in noise performance.

Acknowledgements

The authors gratefully acknowledge helpful discussions with H. Lee, W. Schaff, and D.W. Van der Merwe, and technical assistance from J. Berry. We thank Dr. E. Wolf for the use of the NRRFSS electron beam microfabricator, Prof. J.P. Krusius for assistance in DC measurements, and M. Pospieszalski of NRAO for noise figure measurements. This work was supported by NRAO, IBM Corporation, and the Air Force Office of Scientific Research under contract no. F49620-81-C-0082.

References

Arnold D, Kopp W, Fischer R, Henderson T, and Morkoc H 1984 IEEE Elect. Dev. Lett. **EDL-5** pp 82-84.
Berenz J J, Nakano K, and Weller K P 1984 IEEE MTT-S Digest pp 98-101.
Camnitz L H, Lee H, Tasker P J, Van der Merwe D, and Eastman L F to be presented at 1984 Int. Elect. Dev. Mtg.
Chevrier J, and Linh N T 1984 Dev. Res. Conf., Santa Barbara, California
Drummond T J, Fisher R, Miller P, Morkoc,H and Cho A Y 1982 J. Vac. Sci. Tech. **21** pp 684-688.
Eastman L F and Shur M S 1979 IEEE Trans. Electron Dev. **ED-26** 1359-1361.
Ghosh C L and Layman R L 1984 IEEE Elec. Dev. Lett. **EDL-5** pp 3-5.
Hallais J, Andre J P, Baudet P, and Boccon-Gibod D 1979 Inst. Phys. Conf. Ser. **45** p. 361.
Jones W L private communication.
Kopp W, Su S L, Fischer R, Lyons W G, Thorne R E, Drummond T J, Morkoc H, and Cho A Y 1982 Appl. Phys. Lett. **41** pp 563-565.
Laviron M, Delagebeaudeuf, D, Rochette J F, Tung P N, Delescluese P, Chevrier J and Linh N T 1984 Dev. Res Conf. Santa Barbara, CA USA
Maki P A, Palmateer S C, Wicks G W, Eastman L F and Calawa A R 1983a Jour. Electron. Mtrls. **12** pp 1051-1063.
Maki P A, Wicks G W, and Eastman L F 1983b Proc. of IEEE/Cornell Conf. on High Speed Semi. Dev. and Circuits, 209-217.
Pucel R A, Haus H A, and Statz H 1975 Adv. in Electr. and Elect. Phys. **38** pp 195-265.
Rochette J F, Delescluse P, Laviron M, Delagebeaudeuf D, Cheurier J, and Linh N T 1982 Inst Phys Conf Ser **65** 385-392.
Schaff W J and Eastman L F 1983 Proc of IEEE/Cornell Conf on High Speed Semi. Dev. and Circuits pp 226-233.
Wang, W I, Judaprawira S,Wood C E C, and Eastman L F 1981 Appl. Phys. Lett. **38** pp 708-710.
Widiger D, Hess K, and Coleman J J 1984 IEEE Elect Dev Lett **5** 266-269.
Zwicknagl P, Mukherjee S D, Jones W L, Lee H, Capani P M, Griem, H T, Berry J D, Rathbun L, and Eastman L F 1984 Inst.Phys.Conf.Ser. <u>74</u> 575

Inst. Phys. Conf. Ser. No. 74: Chapter 5
Paper presented at Int. Symp. GaAs and Related Compounds, Biarritz, 1984

339

Resonant tunneling in MBE-grown pnp-GaAs quantum well structures

U. Prechtel, Ch. Zeller, and G. Abstreiter

Physik-Department, Technische Universität München, D-8046 Garching, FRG

and

K. Ploog

Max-Planck-Institut für Festkörperforschung, D-7000 Stuttgart, FRG

Abstract. The resonant tunneling of electrons between valence band
states and discrete subbands in the conduction band is studied in pnp-
GaAs quantum well structures. The I(V) characteristics for direct tunneling processes are analyzed theoretically. Experimental results are
obtained with pnp-triode structures with specially designed selective
contacts. The $k_{||}$-dependence of the tunneling probability is important
for doping concentrations of the order of 2×10^{18} cm^{-3}.

1. Introduction

In this contribution we present a
study of tunneling of electrons between
valence band states and discrete subbands in the conduction band in pnp-
GaAs quantum well structures. The samples consist of one, very thin n-doped
layer of GaAs which is imbedded between two, rather thick p-doped layers.
Charge transfer between donor states
in the n-type layer and acceptor states
in the p-type material causes a space
charge potential, similar as discussed for doping superlattices (see
for example, Ploog and Döhler (1983)).
The variation of the band edges in
such a pnp-GaAs structure is shown in
Fig. 1 for zero applied bias voltage.
The total well depth V_o depends on the
doping concentration N_D and N_A and on
the thickness d_n. It can be obtained
easily by integrating Poisson's equation. For thin enough n-type layers
($d_n \lesssim 1000$ Å) the lower half of the
potential well is nearly parabolic.
It leads to quantization of the motion
of carriers normal to the layers with
a nearly equidistant energy separation
of the electric subbands. Rather high

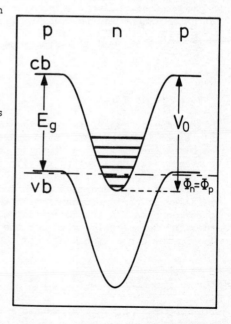

Fig. 1: Variation of band edges
in a pnp-triode structure

doping concentrations are required in order to achieve occupation of sub-bands in the n-type layers at zero bias ($N_D \simeq N_A \gtrsim 10^{18}$ cm^{-3}). The pnp-triode structures are very similar to a bipolar transistor. Due to the high doping concentrations, however, tunneling of carriers can occur through the forbidden energy gap from conduction to valence band states and vice versa. It therefore also resembles very closely the Esaki tunnel diode (Esaki 1958). In our ultrathin structures, however, the tunneling involves subband states in the n-type layer which have two-dimensional character. The density of states is constant for each subband. The total density of states in the conduction band is like a staircase. The effect of this density of states function on the p-n tunnel current is discussed in the following.

2. Theoretical considerations

A voltage applied in backward direction tends to deplete the n-type layer as the well depth is increasing (Zeller et al. 1984). The quasi Fermi level in the conduction band Φ_n is moved below Φ_p, the Fermi energy in the p-type layers. Tunneling of occupied valence band states into empty subband states becomes possible. The tunneling probability depends on the barrier (overlap of wavefunctions) and on the density of filled and empty states. It therefore reflects the staircase-like density of states of the two-dimensional system in the conduction band. Forward bias increases the free carrier concentration, Φ_n is above Φ_p. Tunneling of subband electrons into hole states of the p-type layers may occur. If the bottom of the lowest subband lies above the top of the valence band no direct tunneling is possible anymore. However, tunneling via deep levels can contribute to the np-current for voltages $eU_{np} > \Phi_n - E(O) + \Phi_p$, where $E(O)$ is the bottom energy of the lowest subband above the conduction band edge. Only direct tunneling, where the wave vector $k_{\parallel}$ is conserved in the process is con-sidered in our theoretical considerations. The transmission coefficient for direct tunneling is given by (Kane 1961)

$$T = \frac{\pi^2}{g} \; \exp \; (-E_g/2\bar{E}_{\parallel}) \cdot \exp \; (-2E_{\parallel}/\bar{E}_{\parallel}) \tag{1}$$

with

$$\bar{E}_{\parallel} = \sqrt{2} \; \hbar \; F/(\pi \sqrt{m^* E_g}) \tag{2}$$

A large energy gap E_g is reducing the tunneling probability; a large elec-tric field F on the other hand increases the transmission coefficient. The second exponential in eq. 1 shows that T is also reduced with increasing wavevector $k_{\parallel}$, parallel to the layers. The reason is the larger energy gap for carriers with finite parallel energy $E_{\parallel}$. $\bar{E}_{\parallel}$ is the average value of parallel electron energies which contribute to the tunneling process. The decay of the wavefunction in the energy gap is determined by the effective mass. In the following we neglect the contribution of the heavy holes and for simplicity we set the light hole mass equal to the electron mass in the conduction band. For the tunnel current one obtains (Kane 1961)

$$j = \frac{em^*}{18\hbar^3} \cdot \exp \; (-E_g/2\bar{E}_{\parallel}) \int\int \exp \; (-2E_{\parallel}/\bar{E}_{\parallel}) \cdot (f_c(E) - f_v(E)) \; dE_{\parallel} dE \tag{3}$$

where $E = E_{\perp} + E_{\parallel}$, $f_c(E)$ is the Fermi distribution function in the con-duction band of the n-type layer and $f_v(E)$ in the valence band of the p-type layers, respectively. Tunneling is only possible from occupied to

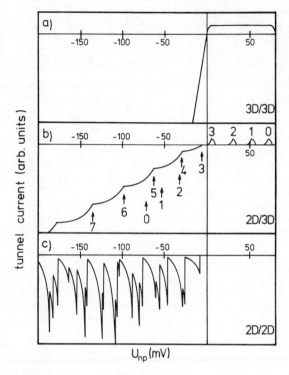

Fig. 2:
Calculated I(V) characteristics for three different
tunnel diodes
a) three-dimensional density
   of states on both sides
   (Esaki-diode)
b) three-dimensional/two-
   dimensional (pnp-structure)
c) two-dimensional/two-dimen-
   sional (doping super-
   lattice)

unoccupied states. If both sides are three-dimensional, the integration
within the allowed energy region leads to the usual Esaki tunneling characteristics (Fig. 2a). If one side is two-dimensional, one integration in
eq. 3 is reduced to a summation over all subbands. For each $k_\parallel$, tunneling
from only one fixed valence band state is possible. In the case of two-
dimensional systems on both sides tunneling between electron and hole subbands is only possible for one discrete $k_\parallel$. Eq. 3 reduces to two summa-
tions. In Fig. 2 we show the calculated tunnel characteristics considering
only direct tunneling processes for the three situations. The parameters
used are Φ_n = 80 meV, Φ_p = 5 meV, and $\bar{E}$ = 10 meV, d_n = 700 Å, $N_D = N_A$ =
2 x 10^{18} cm^{-3} for the 3d-2d case. In the following we concentrate only on
Fig. 2b which is representative for our pnp-triode structures. For the
given parameters, 4 subbands (0 to 3) are occupied at zero bias. With reverse bias one clearly recognizes the onset of subbands 4, 5, 6, 7, ...
The partial saturation of the tunnel current observed for each subband is
caused by the decreasing tunneling with increasing $E_\parallel$. For subband 3 the
saturation value is smaller because of its partial occupation. The onset
of subbands 0, 1, 2 is also marked by arrows. Their contribution to the
tunnel current is negligibly small due to the large $E_\parallel$. Fig. 2b shows
clearly that the two-dimensional density of states is reflected in the tunnel current. The onset of subbands should be observable as peaks already
in the conductance curve. This is a direct consequence of the dependence
of the tunneling probability on $E_\parallel$. With increasing electric field F this

dependence gets weaker and weaker. The tunnel current with forward bias voltage is also shown in Fig. 2. Tunneling into valence band continuum states is assumed. Acceptor impurities and deeper levels are neglected. The small width of Φ_p and the $k_{\parallel}$ conservation lead to peaks in the tunnel current for each subband already in the direct I(V) characteristics. This is different if $k_{\parallel}$ non-conservation is included. However, even then one expects some structure whenever the bottom of the subbands moves through the energy of the hole states of the p-type layers.

3. Experimental aspects

The samples used in our experiments have been grown by molecular beam epitaxy. Si and Be were used as n- and p-type dopants, respectively. The thickness of the n-type layers was varied between 400 Å and 1000 Å. The doping concentrations are $N_D = N_A \gtrsim 1 \times 10^{18}$. Two types of contact technologies have been applied. In one set of samples the layers were grown either on p+- or semiinsulating (100) oriented GaAs substrates. Selective ohmic contacts were formed to the different layers by alloying small Sn and Sn/Zn balls as n^+ and p^+ electrodes, respectively. The arrangement is shown schematically in Fig. 3a. To avoid the problems related to the diffusion of n^+-contacts through the p-layers and opposite, we used also different types of samples, as shown in Fig. 3b. A special masking technique was incorporated into the MBE system in order to facilitate the fabrication of selective contacts to the individual layers. Difficulties arising from this masking technique are the uncertainty in the change of growth conditions and mainly shadowing problems close to the corner of the mask. To reduce the latter effect a second arsenic source was incorporated in the system. Contacts to these samples have been fabricated with In or also with Sn and Sn/Zn balls. The samples were then mounted on a special sample holder and immersed in liquid He. Both direct current-voltage (I(V)) curves and the first and second derivatives (dI/dV, d^2I/dV^2) of the I(V) curves were measured using a modulation technique.

4. Results and discussion

The measured I(V) characteristics are found to be very sensitively dependent on the doping concentration in the layers. The current is strongly increasing with N_D and N_A. For lower carrier concentrations ($N_D = N_A \lesssim 2 \times 10^{18} cm^{-3}$) no negative resistance region is observed in the forward bias region. A careful examination of I(V) curves, however, shows that there exist some small kinks both for positive and negative applied voltages. These structures become very clear in the first and second derivative curves. In Fig. 4a we show the conductance curve of a pnp-sample with $d_n = 700$ Å

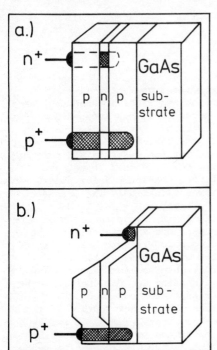

Fig. 3:   Sample arrangements

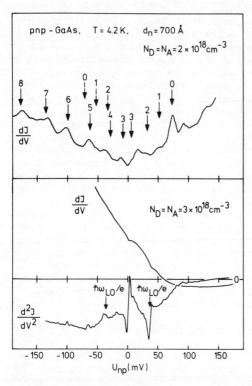

Fig. 4:
First and second derivatives
of the I(V) characteristics of
pnp-triode structures. The
arrows mark the positions of
the onset of tunneling currents
via new subbands as expected
from the theory.

and $N_D = N_A = 2 \times 10^{18} cm^{-3}$. This curve shows a series of nearly periodic
peaks with a separation of approximately 30 mV. The peak positions are
compared to the calculated structures as discussed above. The arrows from
Fig. 2b coincide very well with the structures observed in the (dI/dV)
versus voltage characteristics. Around zero bias, an additional dip
appears which causes some distortion in this region. There are only very
small additional structures around $\pm$ 37 mV which corresponds to the energy
of the LO-phonon in GaAs. With increasing carrier concentration the total
tunnel current is growing drastically. The relative size of the subband
related structures, however, is decreasing. The dip at V = O V as well as
LO-phonon related peaks are more pronounced. In Fig. 4b we show the first
and second derivative of the I(V) curve obtained with a sample with
d_n = 700 Å and $N_D = N_A = 3 \times 10^{18}$ cm^{-3}. This sample has been grown with
the special masking technique where the sample arrangement as shown in
Fig. 3b is achieved. The conductance is negative for a certain forward
bias region. Besides   the zero bias and the LO-phonon related structures
there exist only very small additional peaks which are more pronounced in
the second derivative curves. In the reverse bias region they are probably
related to subbands in the conduction band. In forward bias also additional peaks are observed which are caused by deep levels both in the n- and
p-type layers and in the interface region. Because of these complications
no direct relation to subbands in the conduction band is possible.

The observed effects of the two-dimensional subbands on the I(V) characteristics are much less pronounced than expected from the calculations.
Several reasons may account for this reduction. The theory considers only

direct tunneling with perfect $k_\parallel$ conservation, it does not include the heavy mass valence band, and it assumes that subband energies in the conduction band are sharp. Taking these effects into account, one finds a strong washout of the calculated structures. In addition there exist other channels for the current due to experimental reasons. The diffused-in contacts cause leakage current directly between the individual layers and the highly doped regions. In highly doped samples this leakage current usually dominates the I(V) characteristics. Therefore we included a masking technique into the MBE system. This, however, causes other problems which are related to the properties of the edge where the p-layer stops.One also has to consider the large sample size (several mm^2) used in the present experiments, which might cause spatial potential fluctuations due to inhomogeneities in thickness and doping concentration. With all these uncertainties we believe that it is not surprising that subband related structures are only observed in the derivative curves and not in the direct I(V) characteristics. The fact that direct tunneling into electric subbands is measured more easily in the lower doped samples ($N_D = N_A = 2 \times 10^{18}$ cm^{-3}) is probably related to the $k_\parallel$ dependence of the transmission coefficient. However, this dependence is decreasing with the strength of the electric field. It should be noted that in tunneling experiments through rectangular potential barriers, e.g. thin oxides, such a $k_\parallel$ dependence has not been observed (see for example Kunze and Lautz, 1982). In our case one can simulate a rectangular potential barrier by increasing the electric field. Then one loses the $k_\parallel$-dependence because $\bar{E}_\parallel$ goes to infinite.

In summary, we have studied the tunneling characteristics of thin pnp-triode structures both theoretically and experimentally. Special features due to the two-dimensional nature of the conduction band states in the n-type layers are observed in the derivative curves. The lineshape for lower doped samples indicates some evidence for the $k_\parallel$-dependence of the tunneling transmission coefficient. For future studies as well as for possible applications a more sophisticated sample preparation technology is needed in order to avoid leakage currents which partially mask the desired effects of two-dimensional systems.

We wish to thank A. Fischer for the expert help in sample preparation.

References:

Esaki L 1958, Phys. Rev. 109, 603-604
Kane E O 1961, J. Appl. Phys. 32, 83-91
Kunze U and Lautz G 1982, Surface Sci. 113, 55-68
Ploog K and Döhler G H 1983, Adv. in Physics 32, 285-359
Zeller Ch, Abstreiter G, and Ploog K 1984, Surface Sci. 142, 456-459

Inst. Phys. Conf. Ser. No. 74: Chapter 5
Paper presented at Int. Symp. GaAs and Related Compounds, Biarritz, 1984

345

Interface recombination in GaAs–Ga$_{1-x}$Al$_x$As quantum wells

B. SERMAGE, F. ALEXANDRE, J.L. LIEVIN, R. AZOULAY, M. EL KAIM,
H. LE PERSON and J.Y. MARZIN
Centre National d'Etudes des Télécommunications
Laboratoire de Bagneux
196 rue de Paris – 92220 BAGNEUX – FRANCE

Abstract. GaAs–Ga$_{1-x}$Al$_x$As interface recombination velocity has been investigated at room temperature by measuring the GaAs band edge decay time after short laser pulse excitation at 0.53μm. The GaAs layers were confined by two Ga$_{1-x}$Al$_x$As layers and their thickness was varied between 20Å and 5μm. The structures were grown by molecular beam epitaxy onto GaAs substrates. The recombination velocity at the interface Ga$_{1-x}$Al$_x$As layer to GaAs layer was found to be large, but is drastically improved by the insertion of a GaAs quantum well in the Ga$_{1-x}$Al$_x$As layer. For thicknesses smaller than 1000Å, the effective interface recombination was found to decrease. This improvement is attributed to quantum effects.

1. Introduction

Interface recombination is an important parameter for the performance of GaAs–Ga$_{1-x}$Al$_x$As heterostructure devices such as double heterostructure lasers (DH) and double heterojunction bipolar transistors (DHBT). Indeed, the very high threshold current of the first GaAs–Ga$_{1-x}$Al$_x$As DH lasers prepared by molecular beam epitaxy (MBE) was attributed to a large interface recombination velocity (Tsang 1978). Since that time, substantial progress has been made in reducing the threshold current by increasing the MBE growth temperature up to 700°C. Presently, there is a widespread interest in quantum well lasers because of the possibility of extremely low threshold current operation.

However, very few interface recombination velocity measurements are available on MBE GaAs–Ga$_{1-x}$Al$_x$As heterostructures (Duggan et al. 1980) and on GaAs quantum wells. We report here such results obtained on a series of different GaAs–Ga$_{1-x}$Al$_x$As MBE structures. This study points out that : (i) the interface recombination velocity can be determined from the investigation of the carrier lifetime in the GaAs layer as a function of its thickness, (ii) the interface recombination velocity is large for the MBE heterointerface Ga$_{1-x}$Al$_x$As to GaAs layer but can be improved by the insertion of a GaAs quantum well at this interface, (iii) for very thin GaAs layers, the carrier lifetime saturates and the apparent interface recombination decreases.

2. Experimental procedure

The carrier lifetime was determined at room temperature by recording the decay of the luminescence of the sample after excitation by an ultra short laser pulse as shown on fig. 1. The laser pulses were provided by a 1.06μm modelocked YAG laser beam doubled in a KTP crystal. In order to avoid the heating of the sample, one pulse out of sixteen was selected with an extinction ratio better than 20, by focusing the beam on an acoustooptic cell driven at 400MHz in synchronism with the modelocked laser. The pulses were 70ps long and repeated every 160ns. The pulse energy was varied between 10^{-9} and 10^{11} joules by using neutral filters. By means of a microscope objective, the beam was focused on the sample to a spot 20μm in diameter.

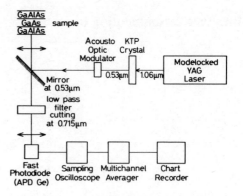

The luminescence of the sample is focused through the same objecti-ve on a fast Ge avalanche photo-diode. The electrical signal is processed in a sampling oscillos-cope followed by a multichannel

Fig. 1. Diagram of the luminescence measurements set up.

analyzer and an X-Y recorder. The overall time resolution of the system is about 200ps.

The investigated structures are a GaAs active layer with a thickness d, confined by a 0.1μm thick $Ga_{0.7}Al_{0.3}As$ buffer layer and cap layer. These structures were grown ont (100) GaAs substrates, in a second generation MBE system (Riber 2300). The growth conditions used (high substrate temperature $T_S=680°C$ and low flux ratio $R=J_{As_4}/J_{Ga}\leq5$) are known to improve the optical properties of $GaAs/Ga_{1-x}Al_xAs$ double heterostructures. The value of the investigated parameter d is determined with a precision of 2% from the GaAs growth rate. This precision is obtained by deducing the GaAs growth rate from the periodicity of a $GaAs/Ga_{0.7}Al_{0.3}As$ superlattice characterized by X-ray diffraction.

In order to select only the luminescence of the GaAs layer, the luminescence due to $Ga_{1-x}Al_xAs$ is filtered out. The decay time is determined from the slope at the origin of the luminescence decay curve. The purpose of the $Ga_{1-x}Al_xAs$ cap layer was to prevent recombination at the free surface of the GaAs layer which can be as high as $3\times10^5cm\ s^{-1}$. It has been checked that for the lowest exciting pulse energy useable, the lifetime is nearly independent of excitation. The carrier concentration in the GaAs layer is estimated to be smaller than $5\times10^{17}cm^{-3}$.

3. Results and discussion

3.1 Thick GaAs active layers (d>1000Å)

The carrier lifetime τ in a layer of thickness d is expressed in a simple manner as the sum of bulk and interface contributions :

$$\frac{1}{\tau} = \frac{1}{\tau_b} + \frac{(S_1+S_2)}{d} \quad (1)$$

where τ_b is the bulk lifetime in the active layer for the same conditions of excitation, S_1 and S_2 are the recombination velocities at both interfaces (Ettenberg et al. 1976). Eq. 1 is valid for d<<L and S_1, S_2<<L/τ, L being the diffusion length of the carriers in the GaAs layer. As we shall see, these conditions are fulfilled in our experiment, since L and L/τ are respectively larger than 1μm and 10^5cm.s^{-1}.

From eq. 1, (S_1+S_2) is determined by investigating the variation of the carrier lifetime τ for samples with different GaAs thicknesses d. Two series of samples have been analyzed ; they correspond respectively to an undoped GaAs layer(background doping, $2\times10^{14}\text{cm}^{-3}$) and a p type doped one (Be, $p=8\times10^{17}\text{cm}^{-3}$). The fig. 2 shows that the experimental points follow quite well eq. 1 within the dispersion limits due to the precision of the lifetime measurements ($\sim 10\%$) and to the non perfect reproducibility of the growth conditions. It has been previously reported that the first interface (GaAs grown on $Ga_{1-x}Al_xAs$) is of lower quality than the second one ($Ga_{1-x}Al_xAs$ grown on GaAs) (Morkoç 1982). So the recombination velocity deduced from fig. 2 is mainly due to the first interface. The values of (S_1+S_2) obtained (see table 1) are relatively high (6×10^3 and $1.1\times10^{14}\text{cm.s}^{-1}$) as compared to that measured by the same technique for samples grown by metal organic chemical vapor deposition (MOCVD) in our laboratory ($7\times10^2\text{cm.s}^{-1}$).

However, we have found that the insertion in the structures grown by MBE of a thin GaAs well or a few periods of GaAs-$Ga_{1-x}Al_xAs$ superlattice instead of the first low quality interface of the GaAs active layer could improve drastically the carrier lifetime. This is shown in fig. 2 for the case of an undoped and a Be doped 1000Å thick GaAs active layer, with and without an added single 20Å thick GaAs quantum well below the active layer.

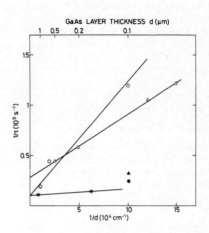

Fig. 2. Variation of the inverse carrier lifetime ($1/\tau$) as a function of the inverse GaAs layer thickness ($1/d$) for MBE structures. The experimental points follow nearly $1/\tau = 1/\tau_b + (S_1+S_2)/d$. The open circles and triangles correspond respectively to an undoped and a Be doped GaAs layer. The full circles and triangles refer to the same structure with an added 20Å thick GaAs well below the active layer. The stars correspond to structures grown by MOCVD, with an undoped GaAs layer.

Epitaxy technique	GaAs doping	τ_b (ns)	$(S_1+S_2)(\text{cms}^{-1})$
MBE	$p=8\times10^{17}\text{cm}^{-3}$ (Be doped)	3.6	6 000
MBE	undoped	10	11 000
MBE(with an added GaAs well)	undoped		$\leq$ 1 500
MOCVD	undoped	10	700

Table 1. Bulk lifetime and interface recombination velocity calculated from eq.1 and the data of Fig. 2.

The carrier lifetime is increased by a factor of three in the second case. It is difficult to separate the respective influence of this added quantum well on τ_b and S_1. However since τ_b could not be larger than the radiative lifetime (6ns) in p type GaAs with a concentration of $8 \times 10^{17} cm^{-3}$ (Nelson 1978), the recombination velocity $(S_1 + S_2)$ for the improved structure must be lower than $1.5 \ 10^3 cm.s^{-1}$. This value corresponds to good interfaces and is compatible with reported low threshold lasers. This result also shows that the first interface $Ga_{1-x}Al_xAs$ to GaAs layer has a much larger recombination velocity than the second one.

3.2 Thin GaAs active layers (d<1000Å)

In order to investigate the interface recombination in quantum wells, carrier lifetime measurements have been performed on structures with an undoped GaAs active layer with a thickness d lower than 1000Å. Results reported in fig. 3 show that the experimental points do not follow eq.1 for d<1000Å and that $1/\tau$ saturates at a value of about $2.5 \times 10^9 s^{-1}$. If we assume that eq.1 is still valid in that case and since any variation of $1/\tau_b$ would not be sufficient to explain the results, a decrease of the apparent interface recombination velocity as shown in fig. 4 must be supposed. The interface recombination velocity is calculated assuming a constant value of 10ns for τ_b in undoped GaAs (see table 1). Values as low as $300 cm.s^{-1}$ are obtained for structures with quantum wells.

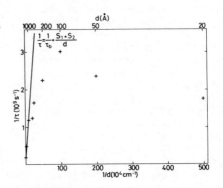

Fig. 3. Variation of the inverse carrier lifetime $(1/\tau)$ as a function of the inverse thickness $(1/d)$ for MBE structures with an undoped GaAs layer.

This phenomenon can be related to the wave function of the electrons and the holes in the fundamental level of a quantum well. The wavefunctions have their maximum at the center of the well and decrease notably at the interfaces. This leads to a low probability of presence for electrons and holes at the interfaces and thus to a small interface recombination velocity. In the case of thick GaAs layers, quantum well level description is no more valid since the free path of electrons and holes is smaller than the thickness of the well. In this case, we consider that the density of carriers is constant over all the GaAs layer

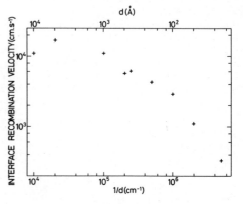

Fig. 4. Interface recombination velocity (calculated from data of fig. 3 using eq. 1 with τ_b=10ns) as a function of the inverse thickness $(1/d)$ for MBE structures with an undoped GaAs layer.

thickness so that the interface recombination will be larger than in a quantum well. Finally the carrier lifetime has been measured in a multi-quantum well (MQW) structure consisting of 10 GaAs wells 100Å thick separated by the 1000Å thick $Ga_{0.5}Al_{0.5}As$ barriers. The measured carrier lifetime (3.8 ns) is larger than that obtained for a 100Å thick GaAs single quantum well (0.33 ns). This gives an apparent interface recombination velocity smaller than $260 cm \cdot s^{-1}$. This relatively low value is due to the fact that the first well of the MQW structure improves the quality of the subsequent interfaces as it has been observed in fig. 2.

4. Conclusion

The carrier lifetime in GaAs layer have been measured at room temperature from the GaAs luminescence decay time after short pulse excitation in $GaAs-Ga_{1-x}Al_xAs$ double heterostructures. The $Ga_{1-x}Al_xAs/GaAs$ interface recombination velocity has been deduced from the carrier lifetime measurements as a function of the thickness of the GaAs active layer. The MBE hetero-interface $Ga_{1-x}Al_xAs$ to GaAs layer has a large interface recombination velocity. However, the insertion of a thin GaAs well at this interface improves drastically the effective carrier lifetime in the GaAs layer and the interface recombination velocity. Moreover, in the case of GaAs quantum wells the apparent interface recombination decreases. This improvement is attributed to the shape of the wave function of electrons and holes in the n=1 quantum level which decreases the importance of interface recombination. This is a favorable feature for single or multiquantum well lasers.

Acknowledgments

The authors would like to thank particularly G. Bastard and J.P. Noblanc for useful discussions and support.

References

Duggan G, Scott G B , Foxon CT and Harris JJ, 1980 Appl. Phys. Lett. 38, 246
Ettenberg M and Kressel H, 1976 J. Appl. Phys. 47, 1538
Morkoç H, Drummond TJ and Fischer R, 1982 J. Appl. Phys. 53, 1030
Nelson RJ and Sohers RG, 1978 J. Appl. Phys. 49, 6103
Tsang WT 1978 Appl. Phys. Lett. 33, 246

Inst. Phys. Conf. Ser. No. 74: Chapter 5
Paper presented at Int. Symp. GaAs and Related Compounds, Biarritz, 1984

351

Luminescence properties of GaAs single quantum wells confined by short-period all-binary GaAs/AlAs superlattices

Kenzo Fujiwara[*], Helmut Oppolzer[**] and Klaus Ploog

Max-Planck-Institut für Festkörperforschung, Heisenbergstrasse 1, D-7000 Stuttgart 80, Federal Republic of Germany

Abstract. New GaAs single quantum well heterostructures (SQWH) composed of all-binary AlAs/GaAs heterostructures by using AlAs/GaAs short-period superlattices (SPS) instead of the ternary $Al_xGa_{1-x}As$ alloy as cladding layers were grown by molecular beam epitaxy. Detailed photoluminescence and excitation spectroscopy measurements at 2K revealed that these SPS confined SQWH exhibit superior luminescence properties due to a specific vertical transport of photoexcited carriers. The structures are there- fore very attractive for application in new optoelectronic devices.

1. Introduction

Recent investigations on $Al_xGa_{1-x}As$/GaAs single quantum well hetero- structures (SQWH) grown by molecular beam epitaxy (MBE) revealed a strong influence of growth conditions and of the specific heterostructure configuration used on the QW luminescence efficiency (Gossard et al 1982 and Petroff et al 1984). This effect is attributed to the inferior crystal quality of the ternary $Al_xGa_{1-x}As$ alloy, and the resulting poor interface properties. In this paper, we have studied two series of MBE grown GaAs SQWH with well widths of L_z = 9.2 and 6.4 nm which are entirely composed of all-binary GaAs/AlAs heterostructures by using GaAs/AlAs short-period superlattices (SPS) with well widths l_z and barrier layer thicknesses l_B, instead of the ternary alloy as cladding layers. The new SQWH are complete- ly free from any alloy problems at the peculiar $Al_xGa_{1-x}As$/GaAs hetero- interface. For the electronic properties of this attractive heterostructure system only two-dimensional quantum states (levels) are involved which replace the three-dimensional energy bands of $Al_xGa_{1-x}As$ and of monolayer- superlattices (Gossard et al 1976). We show that the use of GaAs/AlAs SPS cladding layers significantly improves the photoluminescence properties of GaAs SQWH. In addition, we provide first experimental evidence for an efficient trapping of photoexcited carriers into the SQWH by tunneling through the SPS.

2. Experimental

The GaAs SQWH samples with GaAs/AlAs-SPS of systematically varied barrier and well widths and with the ternary $Al_xGa_{1-x}As$ alloy of similar Al content were grown at $570 \leqslant T \leqslant 600°C$ from As_4 molecular beams on semi-insulating (100) GaAs substrates. The MBE apparatus used has been described in details previously by Jung et al (1984). The application of a continuously azimuth- ally rotating substrate holder kept the thickness and composition varia-

---

[*] on leave of absence from Central Res. Lab., Mitsubishi Electric Corp., Amagasaki, Hyogo 661, Japan

[**]Siemens AG, Otto-Hahn-Ring 6, 8000 München 83, Germany F. R.

tions of the epilayers to less than 1% over 2 cm lateral dimension. The rotation speed of 6rpm provided a minimum shutter control time of 10 s, which limits our minimum layer thickness to 1.3 nm in this study. Before growing a set of heterostructures, thick GaAs and AlAs epilayers were grown to determine the growth rates precisely by measuring the layer thicknesses with a scanning electron microscope. The lower growth rates of 0.44 μm/h for GaAs and 0.46 μm/h for AlAs is favorable to minimize effects of flux fluctuations which may cause layer thickness variations.

The heterostructure configurations of the SQWH were investigated by transmission electron microscopy (TEM) on (110) cleaved and thinned specimen. In Fig. 1 we show a TEM picture obtained from the SQWH sample with L_z = 9.2 nm and (l_z, l_B) = (2.4,2.6) nm. The directly measured L_z is 10.4 ± 0.5 nm and agrees within the experimental accuracy of ± 9 % with the nominal value. This result provides an independent support for the high degree of control for growth of our heterostructure samples. A summary of the design parameters for two series of studied samples is given in Table 1. In these samples the average Al composition of the confinement layers was kept constant at x = 0.5 (= $l_B/[l_z + l_B]$) and only the periodicity l_z and l_B of the SPS were varied. The photoluminescence (PL) measurements were performed at 2 K and 300 K. The luminescence light excited by a cw Kr$^+$ laser of power density 10^{-2}-10^2 W/cm^2 was analyzed with a 1 m monochromator (SPEX 1704)

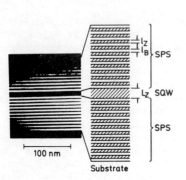

Fig.1 TEM picture of GaAs SQW confined by GaAs/AlAs SPS. Design parameters are L_z = 9.2 nm, l_z = 2.4 nm and l_B = 2.6 nm.

Table 1.   Layer thicknesses, QW PL peak energies and calculated eigenvalues of GaAs SQWH samples. Symmetrical cladding layers are used for both sides of the QW.

Samples	SQW	Cladding layers GaAs/AlAs SPS	Al$_{0.5}$Ga$_{0.5}$As	E_{1hh}^{SQW} observed		E_{1hh}^{SQW} calculated	
	L_z	(l_z, l_B)	L_B	300K	2K	300K	2K
	(nm)	(nm)	(nm)	(eV)		(eV)	
#4419	9.2	(2.4,2.6) 10periods		1.472	1.553	---	---
#4418	9.2	(1.2,1.3) 10periods		1.477	1.549	---	---
#4446	9.2		107	1.472	1.551	1.470	1.565
#4451	6.4	(2.5,2.6) 20periods		1.525	1.603	---	---
#4448	6.4	(1.3,1.3) 40periods		1.516	1.593	---	---
#4447	6.4		107	1.510	1.587	1.504	1.599

and detected by a cooled GaAs photomultiplier (RCA C31034 A) attached to a photon counting system. Excitation spectra at 2 K were obtained with a tunable cw dye laser using DCM dye.

3. Results and Discussion

In Fig. 2 we show PL spectra obtained at 300 K from two SQWH with L_z = 6.4 (a) and 9.2(b) nm cladded by SPS with l_z = 2.4-2.5 nm and l_B = 2.6 nm. The main peaks from the QW observed at 813.0 nm (1.525 eV) in (a) and 842.3 nm (1.472 eV) in (b) are attributed to the emission based on confined par-ticle transitions, the n = 1 electron to heavy-hole transitions, E_{1hh}^{SQW}, of the GaAs single quantum well. In Table 1 these energies are compared with the energy eigenvalues calculated for the single quantum wells. A detailed discussion for the energy variations will be given later. The shoulder at 798 nm in the spectrum (a) is associated with the light-hole transitions, E_{1lh}^{SQW}. The double peaks observed at 704 nm (1.76 eV) and 675 nm (1.84 eV) both in (a) and (b), however, are related to the SPS cladding layers. The calculated energy eigenvalue of E_{1hh} for the SPS well with l_z = 2.4 nm is 1.79 eV, assuming the 85/15 rule (Dingle 1975). The observed red shift of 30 meV for the E_{1hh} energy is due to the coupling between adjacent wells within the SPS. These room-temperature PL spectra, which exhibit a narrow line width of 42 meV for the SQWH with L_z = 6.4 nm, exemplify the excellent quality of the samples and confirm the designed configurations of the hetero-structures.

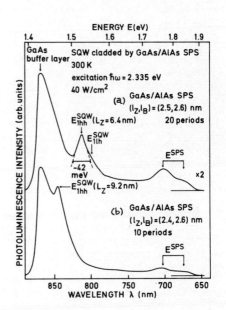

Fig. 2 Photoluminescence spectra at 300 K for the SQWH samples with L_z = 6.4 (a) and 9.2 (b) nm confined by GaAs/AlAs SPS with l_z = 2.4-2.5 nm and l_B = 2.6 nm.

In Fig. 3 we show the PL spectra at 2 K of the SQWH with L_z = 9.2 nm cladd-ed by the SPS (a,c) and the ternary $Al_{0.5}Ga_{0.5}As$ (b,d), taken at two dif-ferent excitation energies of 2.604 eV and 1.916 eV. The sharp peak observed at 789.3 nm (1.553 eV) in (a) and (c) and 799.9 nm (1.550 eV) in (b) and (d) is attributed to the E_{1hh}^{SQW} transition. Peaks and shoulders at 1.491 eV and 1.51 eV are assigned to GaAs related emission from the 200 nm undoped GaAs buffer layers. The expected energy position of bulk GaAs band to band transitions (B.B.) is indicated in the figure. Inspection of the PL spectra shown in Fig. 3 clearly reveals that the SQWH confined by the SPS with l_z = 2.4 nm and l_B = 2.6 nm exhibits superior luminescence prop-erties as compared to the SQWH using the ternary $Al_{0.5}Ga_{0.5}As$ alloy as cladding layers. If we take into account the low growth temperature and the large lattice mismatch of $\Delta a/a$ = 1.6 x 10^{-3} for the GaAs/AlAs system (Casey and Panish 1978), the linewidth (FWHM) of 10.7 meV for the dominating luminescence line at 1.553 eV is reasonable for a SQWH. It should be noted that as yet no special care was taken to optimize the growth conditions,especially with respect to background impurities during growth.

Fig. 3 Low temperature photoluminescence spectra for two types
of SQWH samples with GaAs/AlAs SPS
(a)and (c) and $Al_{0.5}Ga_{0.5}As$ (b)
and (d) cladding layers excited by
the Kr^+ blue line at 2.604 eV and
the red line at 1.916 eV.

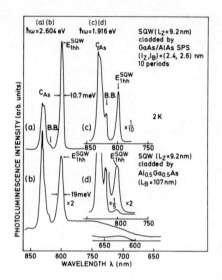

The QW luminescence line obtained from the SQWH confined by $Al_{0.5}Ga_{0.5}As$ layers, on the other hand, shows a broader FWHM of 19 meV, which indicates inferior interface properties of the ternary $Al_xGa_{1-x}As$ cladding layers, as reported by Masselink et al (1984). In addition, we observe a broad emission band from 1.5 eV to 2.0 eV originating from the $Al_{0.5}Ga_{0.5}As$ (indirect gap E_g = 2 eV). A similar improvement of the PL spectra was observed for the SQWH with L_z = 6.4 nm confined by the SPS. When we use the SPS cladding layers with shorter l_z and l_B (= 1.3 nm), however, the improvement of the PL spectra as compared to the ternary alloy cladding layer is less pronounced.

The peak intensity of the QW luminescence from the two types of SQWH samples shows a different dependence on the photoexcitation energy as shown in Fig. 3. Using the blue line of the Kr^+ laser at $\hbar\omega$ = 2.604 eV for excitation, we observe an intensity ratio of 2.6. When the red line of the Kr^+ laser at $\hbar\omega$ = 1.916 eV is used, however, this ratio increases to 21. This enhanced intensity ratio is caused by the difference of the photo-excitation process in the cladding layers using the Kr^+ red line because $E_{1hh}^{SPS} < \hbar\omega < E_g^{Al_{0.5}Ga_{0.5}As}$. The major contribution to the enhanced luminescence intensity of the SQW arises from photoexcited-carrier diffusion in the SPS and trapping within the wider well at $\hbar\omega > E_{1hh}^{SPS}$. The carriers generated in the SPS cladding layers may also recombine in the narrow quantum wells of the SPS as indicated in Fig. 2. This emission can also contribute to the excitation of the SQWH. However, photon recycling effects seem to be unimportant in this case because the PL intensity from the SQWH is much more intense than that from the SPS. Therefore, the transport of these carriers to the SQWH by tunneling through the thin AlAs barriers dominates the radiative recombination process, and this leads to the observed enhanced PL intensity ratio for the SQWH samples cladded by SPS.

In Fig. 4 we provide further convincing evidence for the efficient trapping of the photoexcited carriers by the SPS cladded SQWH. The inset of the figure shows the excitation spectrum of the SPS with l_z = 2.5 nm and l_B = 2.6 nm which reveals a peak around 655 nm (1.89 eV) due to the E_{1hh}^{SPS} transition. When the excitation energy of the tunable dye laser passes the

Fig. 4 Low temperature photoluminescence spectra for the SPS
confined SQWH sample at the excitation energy below and above the
threshold to excite the SPS cladding
layers. The inset is the excitation
spectrum of the SPS at the monochromator wavelength of 772.5 nm which
corresponds to the E_{1hh}^{SQW} transition
energy.

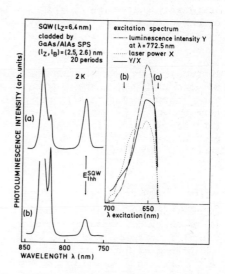

threshold value to excite the carriers
in the SPS cladding layers, a significantly enhanced luminescence of the
SQWH is observed, as compared with the
PL spectra (a) and (b) in Fig. 4. For
clarity, two tuning energies at 1.814
eV (683.5 nm) and 1.941 eV (638.7 nm)
as indicated in the figure were
selected to give the same power of 45
mW, and only the photoexcitation
energy was varied. When the excitation
energy of 1.814 eV is used, only the
photoexcited carriers generated within
the single quantum well contribute to the PL intensity of the E_{1hh}^{SQW} transition. With increasing the photoexcition energy to 1.941 eV above the
threshold value E_{1hh}^{SPS} which is observed in the excitation spectrum, however,
more carriers are created both in the SQWH and the SPS cladding layers.
The observed enhancement of the E_{1hh}^{SQW} transition are therefore attributed
to the excess carriers generated in the SPS cladding layers. These experimental results demonstrate for the first time that vertical transport is
indeed possible through thin AlAs barrier layers. This efficient trapping
of the photoexcited carriers and the resulting superior luminescence properties make the SPS confined SQWH very attractive for application in newly
designed heterostructure lasers with separated superlattice waveguide and
superlattice barriers.

The final important result of our investigation is that the PL peak energy
of the L_z = 6.4 nm SQWH shifts markedly when the heterostructure configuration of the cladding layers is varied, as shown in Table 1. The
$Al_{0.5}Ga_{0.5}As$ cladding layers correspond to the limiting case of SPS
configurations with $l_z = l_B = 0$, i.e. a random mixture of the Al and Ga
sublattice in the direction of growth. The other extreme case with $l_B = \infty$
represents a single uncoupled quantum well separated by thick AlAs barrier
layers. The SQWH with wider L_z of 9.2 nm exhibits only a minor shift of the
PL peak energy with heterostructure configuration both at 2 K and 300 K.
For the narrower well width of L_z = 6.4 nm, however, the PL peak energy of
the single quantum well is shifting by~15 meV at 2 K and 300 K. In this
case the wave functions associated with the confined particle states spread
considerably into the cladding layers as L_z decreases. We therefore assume
that the observed shift of the SQWH emission is due to enhanced coupling
between the SQWH and the SPS when advancing from $Al_{0.5}Ga_{0.5}As$ cladding
layers to the SPS with $l_z = l_B$ = 2.5 nm. Comparison of the calculated SQWH
transition energy with the PL results indicates that the agreement is
satisfactory at room temperature in the limit of $l_z = l_B = 0$. This agree-

ment is attributed to thermalization effects of photoexcited carriers which cause the carrier population more closely to approach the actual two-dimensional density of states. A blue-shift of the QW emission energy with increasing the SPS periods is due to modified physical constants such as effective masses of the SPS barrier layers, because the new SQWH confined by SPS studied here represent a fully interacting quantum well system.

In summary, we have studied MBE grown GaAs single quantum well hetero-structures (SQWH) which are composed of all-binary GaAs/AlAs heterostruc-tures by using GaAs/AlAs short-period superlattices (SPS) instead of the ternary $Al_xGa_{1-x}As$ alloy as cladding layer. These SPS confined SQWH exhibit superior trapping of photoexcited carriers leading to improved luminescence properties. The structures are thus highly attractive for application in novel optoelectronic devices.

Acknowledgements

The authors would like to thank Y. Horikoshi for stimulating discussions, E. O. Göbel for critical reading of the manuscript, A. Fischer and J. Knecht for expert help with the sample preparation, W. Heinz for technical assistance with the luminescence measurements , A. Jonitz for setting up the dye laser and S. Schild of Siemens AG for assistance with the TEM measure-ments. Part of this work was sponsored by the Bundesministerium für Forschung und Technologie of the Federal Republic of Germany.

References

Casey Jr. H C and Panish M B 1978 Heterostructure Lasers B (New York: Academic) pp8
Dingle R 1975 Festkörperprobleme XV (Advances in Solid State Physics) ed H.J Queisser (Braunschweig: Pergamon-Vieweg) pp21
Gossard A C, Petroff P M, Wiegmann W, Dingle R and Savage A 1976 Appl. Phys. Lett. 29 323
Gossard A C, Wiegmann W, Miller R C, Petroff P M and Tsang W T 1982 Proc. 2nd Int. Conf. on MBE (Tokyo) pp39
Jung H, Fischer A and Ploog K 1984 Appl. Phys. A33 97
Masselink W T, Klein M V, Sun Y L, Chang Y C, Fischer R, Drummond T J and Morkoc H 1984 Appl. Phys. Lett 44 435
Petroff P M, Miller R C, Gossard A C and Wiegmann W 1984 Appl. Phys. Lett. 44 217

Vertical transport in GaAs–GaAlAs: photoluminescence and decay time measurements

B. Lambert[*], R. Romestain[**], D. Miller[**], A. Chomette[*], A. Regreny[*], and B. Deveaud[*]

[*] CNET/LAB/ICM, BP 40, 22301 Lannion, France
[**] Laboratoire de spectroscopie physique, Université de Grenoble, BP 68, 38402 St Martin d'Heres Cedex, France

Abstract. Photoluminescence (PL) and excitation (PLE) experiments support the evidence of vertical transport in superlattices (SL) : photo excited carriers created by the laser move in the conduction and valence bands of the SL and are trapped in purposely introduced enlarged wells. We show that photo excited carriers are able to diffuse over more than 500 nm in a SL with $L_Z = L_B = 3$ nm.

Although L. Esaki and R. Tsu (1970) did predict remarkable properties for the transport along the growth axis (direction Z) in superlattices (SL), very few experimental results have been published in this field. Furthermore, two mechanisms have been invoked to explain transport results along Z : band conduction (Esaki and Chang - 1974) or hopping between localized states (Palmier et al. - 1984). As far as optical experiments are concerned, most results were obtained on isolated quantum wells (LL. Chang - 1983, B. Lambert and B. Deveaud - 1984). Despite this lack of experimental results about SLs, transport parallel to the growth direction Z (vertical transport) is thought (L. Esaki - 1984) to be one of the new research areas in the next years.

We report the first observations by luminescence of vertical transport in GaAs-GaAlAs SLs. Wells larger by some monolayers have been purposely introduced as probes inside regular SLs. Electrons and holes are transferred from the SL to the larger wells and the transfer efficiency increases with decreasing SL period. As a consequence the intensity ratio between luminescence peaks due to the enlarged well (EW) and luminescence peaks due to the SL is greater than predicted from the mere proportion of EW to regular wells.

Our samples were grown in a home modified MBE 500 Riber system. The growth conditions have been adjusted to obtain for multi quantum wells, luminescence spectra corresponding to the best results. The growth process and the photoluminescence experimental conditions have been described by B. Deveaud et al. (1984).

For the present study we have grown three series of samples. For all of them the Al concentration (x) in the barrier is about 30 %. Firstly we have grown regular SLs ; the barrier (L_B) and well (L_Z) widths are equal, ranging between 7 and 3 nm (the samples are labelled 3/3, 4/4, 5/5 and 7/7).

Secondly we have grown SLs with the same parameters as above in which a few wells were enlarged by three monolayers (3a/2 = 0,85 nm), the mean distance between two enlarged wells (EW) being kept constant and equal to 140 nm (symmetrical structure).

Finally we have grown SLs with one enlarged well (L_Z + 5 a/2) : starting from the surface we have a 900 nm thick SL followed with the EW then by 100 nm of SL (asymmetrical structure).

Typical luminescence results are displayed on figure 1. The spectrum 1.a is the luminescence of a 4/4 regular SL. It shows a split peak that corresponds to heavy hole exciton recombination (at 1.631 and 1.635 eV). The observed splitting is induced by one monolayer steps at the interfaces giving rise to different possible sizes in each well (B. Deveaud et al.). The exciton linewidth is 3 meV both in luminescence and in PLE, the peak positions and the splitting both correspond to the predicted values.

Results obtained on a symmetrical structure consisting of 4,8 nm EWs (L_Z + 3 monolayers) embedded in a 4/4 SL are shown in figure 1.c. In that case, the luminescence mainly comes from the EWs (1,612 eV peak, 6 to 1 ratio) although carriers are mostly created in the SL due to the volume ratio of standard wells to EWs : Z motion of the photo excited carriers is evident.

From the steady-state rate equations governing the dynamics of carrier recombination in the SL and the EWs, we derive relation (1). In the case of symmetrical structures the laser generation rate can be assumed to be uniform in the sample so that :

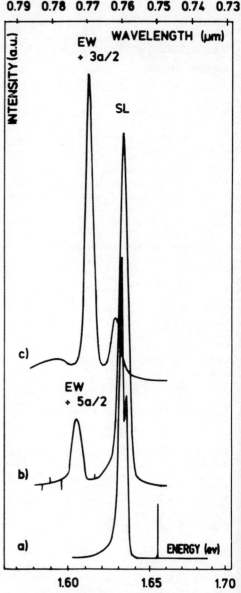

Fig. 1 Photoluminescence spectra at 4K :

a) Regular 4/4 SL,

b) Non symmetrical structure consisting of 900 nm of 4/4 SL followed by one 5,4 nm EW

c) Symmetrical structure : 4/4 SL with 1/18 4,8 nm EW

$$\frac{\tau_{lW}}{\tau_{tr}} = \frac{N}{N+1} R - \frac{1}{N+1} \frac{\tau_{lW}}{\tau_{lSL}} \quad (1)$$

where τ_{tr}, τ_{lW}, τ_{lSL} are the mean trapping time into the EWs, and the carrier lifetimes in the EW and the SL. R is the intensity ratio between excitonic intensities in the EW and in the SL and 1/N is the proportion of EWs to nominal wells. The second term on the right hand side of (1) is smaller than the first   because of the $\frac{1}{N+1}$ factor. The $\frac{\tau_{tr}}{\tau_{IW}}$ ratio can easily be obtained from luminescence experiments.

Decay time measurements of the luminescence signals have been performed under picosecond excitation for 7/7 samples. The decay time of the regular SL exciton is found to be about 350 ps.   This value is in good agreement with the results obtained by R. Höger et al (1984) on MQW. As expected, the decay time is slightly shorter in the enlarged well and it is measured to be 270 ps. Decay times have not yet been measured on the other structures. However it seems reasonable to assume a decay time in the enlarged wells between 200 ps and 300 ps.

If the carriers can be described by Bloch type wave functions it is possible to calculate the trapping times in our structures. The most efficient mechanism would be ionized impurity scattering, the only adjustable parameter is then the impurity concentration. Taking 3.10^{15} cm^{-3} in GaAs wells and 5.10^{16} cm^{-3} in GaAlAs barriers (ternary alloys are known to trap more impurities), we obtain very good agreement between the calculated and experimental values (see A. Chomette et al.)

Luminescence of a non symmetrical structure, consisting of a 900 nm thick 4/4 SL followed by an enlarged well (EW) equal to 4 nm + 5 monolayers (5,4 nm) then by 100 nm of 4/4 SL, is shown on fig. 1.b. In that case, due to slightly lower growth quality, the SL exciton peak linewidth is larger (6 meV) than the monolayer splitting (4 meV) and thus only one exciton peak is observed. A second peak appears at 1.605 eV and corresponds to recombination in the EW. The distance between the two peaks (27 meV) corresponds quite well to the calculated value (29 meV). Due to the asymmetry of the structure, exciting photons are typically absorbed within the first 500 nm so that almost no exciting light reaches the EW. Furthermore, the EW only represents less than a hundredth of the volume of the structure. The ratio between the two peaks is five to one and can only be explained by diffusion of the photoexcited carriers throughout the structure, from near the surface   where they are created up to the EW. The same type of diffusion was observed by Göbel et al. (1983) in a GaAlAs/GaAs well structure and is indicative of easy carriers motion in the growth direction, through the GaAlAs barriers of the SL. As a matter of fact, our experimental structure closely corresponds to one of those studied by Göbel et al (1983) except   that in our case the surface layer is a SL instead of a GaAlAs layer. A first rough analysis of our results indicates that the average drift velocity in our SL is at least 10^5 cm/s. Such a high velocity is in favor of band conduction for the carriers.

PLE spectra of symmetrical structures were presented by Chomette et al. In such a case, as the mean trapping distance is small, excitons are observed to move through the GaAlAs barriers : exciton absorption in the SL gives rise to efficient luminescence in the EW. This is more and more true as the SL period decreases. On the contrary, PLE studies of asymmetrical structures (see figure 2) reveal that exciton motion is not possible over 900 nm before recombination. This is   shown   by the striking differences between the two spectra of figure 2 obtained on an   asymmetrical structure consisting of 900 nm of a 3/3 SL followed by one 4,4 nm EW.

i) Fig. 2.b : PLE when detection is set on the SL exciton : this is the usual spectrum of a SL showing heavy hole and light hole exciton resonances. Contrary to what is observed in MQW structures of equivalent quality, the PLE signal does not cancel between the two excitons as a result of the reduction of exciton binding energy in a SL.

ii) Fig. 2.a : PLE when set on the EW exciton peak. Heavy hole and light hole transitions for the EW are observed : exciting light can reach the EW when below the SL gap energy. A broad absorption band is then observed above 1,640 eV, corresponding to electron-hole pair formation in the SL. No exciton resonances but rather small anti resonances appear as a result of larger absorption coefficient and thus smaller penetration depth of the exciting light.

As a summary, we have experimental arguments for band transport of photo excited carriers along the growth direction of superlattices. Luminescence studies allow us to observe a ratio of excitonic transition in the superlattices and in purposely introduced enlarged wells that does not correspond to the volume ratio. The explanation lies in the ability for the carriers to move in the growth direction. This motion is observed in two types of structure :

i) symmetrical structures where the mean distance is of the order of 100 nm and where the experimental times compare well with the calculated times.

ii) Asymmetrical structures where it is shown that photo excited carriers are able to diffuse over more than 500 nm .

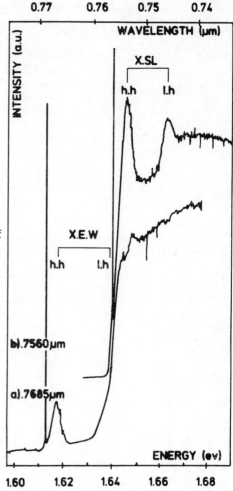

Fig. 2 PLE spectra of a non-symmetrical structure.a) Detection on the EW exciton peak (0.7685 μm). b) Detection on the SL exciton peak (0.7560 μm)

Chang L.L. 1983 J. Vac. Sci. Technol. B1 120

Chomette A., Deveaud B., Emery J.Y., Lambert B., Regreny A., To be published

Deveaud B., Emery J.Y., Chomette A., Lambert B. and Baudet M. 1984 Appl. Phys. Lett. To be published

Dingle R. 1975 Advances in Solid State Physics 15 21

Esaki L. and Tsu R. 1970 I.B.M. J. Res. Dev. 14 61

Esaki L. and Chang L.L. 1974 Phys. Rev. Letters 33 495

Esaki L. 1984 Proceeding of the 17th International Conference on the Physics of Semiconductors San Francisco (U.S.A.)

Göbel E.O., Jung H., Kuhl J., Ploog K. 1983 Phys. Rev. Lett. 51 1588

Höger R., Göbel E.O., Kuhl J. Ploog K. and Weimann G. 1984 Proceeding of the 17th International Conference on the Physics of Semiconductors San-Francisco (U.S.A.)

Lambert B. and Deveaud B. 1984 A travers la Physique (Paris - les Editions de Physique) pp 219-224

Palmier J.F., Le Person M., Minot C., Chomette A., Regreny A. and Calecki D 1984 First International Conference on Superlattices, Microstructure and Microdevices Champaign - Urbana (U.S.A.)

Inst. Phys. Conf. Ser. No. 74: Chapter 5
Paper presented at Int. Symp. GaAs and Related Compounds, Biarritz, 1984

Lateral electron transport measurements using very low resistance ohmic contacts to two dimensional electron gas at $Al_{0.48}In_{0.52}As/Ga_{0.47}In_{0.53}As$ heterointerfaces

P.M. Capani,[*] S.D. Mukherjee, H.T. Griem, G.W. Wicks and L.F. Eastman

School of Electrical Engineering and National Research and Resource
Facility for Submicron Structures, Phillips Hall, Cornell University,
Ithaca, New York 14853, USA

Abstract. Very low resistance alloyed ohmic contacts with transfer resistances ρ_t = 0.19±0.05 Ωmm have been made to the two
dimensional electron gas (2DEG) in lattice matched MBE grown
$Al_{0.48}In_{0.52}As/Ga_{0.47}In_{0.53}As$ modulation doped heterostructures.
Measured sheet carrier concentrations and Hall mobilities were
n_s = 1.64 x 10^{12}/cm^2, μ_H = 10,500 cm^2/V.s for 300K and n_s =
1.56 x 10^{12}/cm^2, μ_H = 54,100 cm^2/V.s for 77K. Preliminary experiments indicate electron velocities in the 2DEG exceed 2.5 x 10^7 cm/
sec at 300K. Velocity saturation, however, was not observed due to
observed charge injection currents at high fields.

1. Introduction

The $Al_{0.48}In_{0.52}As/Ga_{0.47}In_{0.53}As$ modulation doped structure is
extremely attractive for MESFET applications since the sheet electrons
are capable of experiencing higher mobilities and therefore higher
velocities due to the significant reduction in ionized impurity
scattering as compared with bulk GaAs. Sheet carrier density is four times
that of AlGaAs/GaAs. In short channels, electron saturation velocity
determines performance, whereas electron mobility governs the performance in long channel devices. In this paper, we report on the
fabrication of low resistance alloyed ohmic contacts to the Fermi gas
at the heterointerface and present data obtained from preliminary
velocity-field measurements.

2. Experimental

Heteroepitaxial layers were grown by MBE on the ⟨100⟩ surface of insulating InP in the following sequence: Insulating $Al_{0.48}In_{0.52}As$ (3800 Å)/
unintentionally doped n-$Ga_{0.47}In_{0.53}As$ (800 Å)/ insulating $Al_{0.48}In_{0.52}As$ (130 Å)/ n-$Al_{0.48}In_{0.52}As$: Si(65 Å)/ insulating $Al_{0.48}In_{0.52}As$ (320 Å)/
unintentionally doped n-$Ga_{0.47}In_{0.53}As$ (160 Å). The insulating 3800 Å
$Al_{0.48}In_{0.52}As$ layer serves as a buffer between the insulating substrate

and the active layer to prevent parasitic current flow between the epitaxy-substrate interface. The thin 160 Å $Ga_{0.47}In_{0.53}As$ top layer acts as passivation for the epi layer and prevents oxide formation on the $Al_{0.48}In_{0.52}As$ layer beneath. Such oxides are very difficult to remove and would prevent the liquified ohmic contact metallization from reaching the semiconductor during alloying, thus resulting in poor ohmic contacts. The epitaxial layer was characterized by Van der Pauw-Hall measurements at room and liquid nitrogen temperatures with the following Hall mobilities and sheet carrier concentrations: $\mu_{300} = 10,500$ cm^2/V.s, $n_{s300} = 1.64 \times 10^{12}/cm^2$ and $\mu_{77} = 54,100$ cm^2/V.s, $n_{s77} = 1.56 \times 10^{12}/cm^2$. The corresponding sheet resistivities of the two dimensional electron gas (2DEG) were 362 $\Omega/\square$ at 300K and 74 $\Omega/\square$ at 77K. Carrier freeze-out is less than 5 percent indicating a high quality modulation doped layer. Following layer growth and characterization, mesa-etched electrical isolation and metal evaporation lift-off masks were defined by 10x optical projection lithography on AZ-1350J positive photoresist. The photomasks were designed to include transmission line measurement (TLM) patterns adjacent to the velocity-field measurement patterns, thus allowing evaluation of the ohmic contacts to the 2DEG.

The formation of a high quality low resistance ohmic contact to the electron gas below a thick $Al_{0.48}In_{0.52}As$ layer posed a number of difficulties. These included the need to maintain the n-$Ga_{0.47}In_{0.53}As$ stoichiometry, limiting the amount of Ga removed from the semiconductor by the interaction with excess Au metallization, reducing In outdiffusion caused by the Ag-In interaction, and preserving the metal/semiconductor morphology. In earlier experiments, we succeeded in forming extremely low resistance ohmic contacts to 1000 Å n+-GaInAs:Si ($N_D = 2 \times 10^{18}/cm^3$) through 600 Å of insulating AlInAs with the resulting transfer resistance of 0.06 ± 0.02 Ω.mm (Capani (1984)). The metallization system, as evaporated, was Ni/Ge/Au/Ag/Au. Alloying was carried out in a transient alloy furnace where the sample was heated to a desired maximum temperature T_{max}, then allowed to cool. Care was exercised to adjust the amount of Ge to ~24 atomic percent of the total Au used. Ag used was reduced to ~15 atomic percent to avoid gross Ag-In interaction which, as mentioned earlier, causes extreme interdiffusions. Some Ag was retained as a solute for Au-Ag alloy formation which is known to reduce Au-Ge affinity. This ohmic metallization was used to obtain ohmic contacts to the 2DEG system. The metallization consisted of Ni (100 Å)/ Ge (500 Å)/ Au (1000 Å)/ Ag (220 Å)/ Au (1000 Å) and was thermally evaporated onto the sample as shown in the inset in Fig. 1. Metal lift-off was carried out in acetone. Samples containing TLM patterns were cleaved from the processed layer before alloying and were used to characterize ohmic contact electrical performance as a function of the maximum sample alloy temperature.

After the ohmic contacts had been evaluated, samples containing velocity-field measurement patterns were cleaved from the processed layer, alloyed and then epoxied to multiple pin headers where electrical connections were made through Au wires attached by thermal compression bonding. Velocity-field patterns consisted of a 40 μm wide active region with ohmic contacts at either end, thus defining a channel of predetermined length. Current-voltage characteristics were verified with a curve tracer. The velocity-

field measurements were made at 300K and 77K by applying a rectangular
current pulse train having a pulse width of 0.10 μsec and pulse repetition rate of 5 msec to give a 0.002 percent duty cycle. This
extremely short duty cycle was chosen to minimize sample heating which
would alter current-voltage characteristics of the device. Voltage
across the ohmic contacts was measured by a high impedance voltage
probe. Both the input current and the output voltage waveforms were
displayed on an oscilloscope. All velocity-field measurements were
made in complete darkness to avoid photoconductivity effects.

3. Results

Specific transfer resistance ρ_t in Ω.mm and sheet resistivity of the semiconductor in $\Omega/\square$ were determined by the transmission line method after
Reeves (1982). Data were recorded at both 300K and 77K. The behavior of
transfer resistance as a function of maximum sample alloy temperature T_{max}
is shown in Fig. 1 for a constant furnace temperature T_f = 600°C. A
single sample progressively raised
to higher peak temperatures is shown
by open circles (300K) and open
triangles (77K). Samples heated to
particular maximum temperatures in
single alloy cycles are indicated
by filled circles (300K) and filled
triangles (77K). Excellent ohmic
contacts with transfer resistances
$\rho_t \lesssim 0.20\pm0.05$ Ω.mm are realized
over the broad temperature range
of $400 < T_{max} < 490$°C, demonstrating
that good ohmic contact forma-
tion (e.g., ρ_t = 0.19±0.05 Ω.mm
for T_{max} = 410°C) does occur at
low alloy temperatures. As T_{max}
is increased above 440°C, data point
deviation increases. SEM measurements
show changes in the interelectrode
spacings probably due to liquifi-
cation of the metal along the
semiconductor surface. For

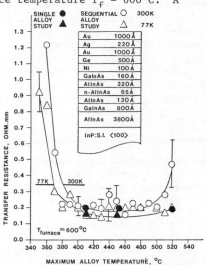

TRANSFER RESISTANCE VS MAXIMUM ALLOY TEMPERATURE
Fig. 1

temperatures exceeding 520°C, severe interelectrode separation changes
are observed. This is caused by surface diffusion or dendrite formation
due to large surface spreading of the liquid alloy. Above 500°C, the
metal surface morphology degenerates rapidly. Contact to the 2DEG is verified since the sheet resistivity of the interelectrode semiconductor
layer as obtained from TLM measurements is to within 5 percent of the
resistivity calculated from room temperature Hall data and is within 1
percent of the 77K Hall data resistivity for alloys heated to T_{max} = 410°C.

Velocity-field characteristics have been obtained from pulsed current-
voltage measurements. The electric field applied to the 2DEG is calculated after first subtracting the voltage drop across the ohmic contacts.
Contact resistance, as measured on a neighboring TLM pattern, was 12 Ω.
Our best data are shown in Fig. 2 for 300K and 77K with the sample in

complete darkness. The active device region was 40 μm wide x 4.5 μm long. Electron velocities were calculated from the relationship

$$\upsilon = I(n_s eW)^{-1}$$ (1)

where I = applied current, n_s = sheet carrier concentration, e = 1.6 x 10^{-19}C and W = device width.

4. Discussion

The electron velocities depicted in Fig. 2 fail to saturate at high applied electric fields. With low fields, band deformation is negligible and all of the conduction electrons are confined to the 2DEG. As the electric field is increased, band bending becomes more severe and causes electrons to spill out of the potential well and flow into the AlInAs buffer layer. This charge injection current increases with increasing field strength, preventing saturation of the device. Furthermore, the injection current may cause the measured velocities to be 10 - 15 percent larger than they actually are under high fields. Hence, electron velocities in the 2DEG are measured to be 2.86 x 10^7 cm/sec at 300K (Field ∿4.75kV/cm) and 3.01 x 10^7 cm/sec at 77K (Field ∿

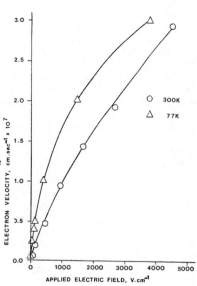

VELOCITY—FIELD MEASUREMENT DATA, GaInAs
MODULATION DOPED STRUCTURE
Fig. 2

3.75kV/cm). If the actual velocity at 300K is reduced to 2.46 x 10^7 cm/sec due to contributions from charge injection currents, this velocity exceeds the saturation velocity of modulation doped (Al, Ga)As/ GaAs reported by Su (1982) to be 1.7 x 10^7 cm/sec.

5. Conclusion

We conclude that we have fabricated high quality low resistance ohmic contacts with ρ_t≈0.19 Ω.mm at low alloy temperatures T_{max} = 410°C to the AlInAs/GaInAs 2DEG heterosystem. These contacts allow measurement of lateral electron velocities of 2.86 x 10^7 cm/sec for applied electric field strength of 4.75 kV/cm (300K) and 3.01 x 10^7 cm/sec at 3.75 kV/cm (77K). However, due to the lack of saturation within the device, charge injection current is thought to be present at high electric fields, causing the measured velocity to exceed the actual velocity within the 2DEG by as much as 10 - 15 percent. Low field velocity-field data are in excellent agreement with Hall data.

Further work is planned to investigate lateral electron transport and high field charge injection current in the AlInAs/GaInAs modulation doped structure. Although ohmic contact performance is good, the contact metallurgy could be improved upon with continued effort.

This work was supported by the Army Research Office through contract DAAG29-82-K-011 monitored by Horst Wittman. We acknowledge J.D. Berry, E. Van Gieson, M.I. Nathan, T. Ohashi, S. Seidman, W. Schaff and P. Zwicknagl for their assistance. One of the authors (P.M. Capani) acknowledges the IBM Corporation for Fellowship support.

References

Capani PM, Mukherjee SD, Zwicknagl P, Berry JD, Griem HT, Rathbun L and
    Eastman LF 1984 Electron Lett vol 20 no 11 p446 - 449
Reeves GK and Harrison HB 1982 IEEE Elect Dev Lett EDL-3 p111 - 113
Su SL, Fischer R, Drummond TJ, Lyons WG, Thorne RE, Kopp W and
    Morkoç H 1982 Electron Lett vol 18 no 18 p794 - 796

*IBM Corporation, 1701 North Street, Endicott, New York 13760, USA

Inst. Phys. Conf. Ser. No. 74: Chapter 5
Paper presented at Int. Symp. GaAs and Related Compounds, Biarritz, 1984

High mobility modulation doped Ga$_{.47}$In$_{.53}$As/Al$_{.48}$In$_{.52}$As structures

T. Griem, M. Nathan*, G.W. Wicks and L.F. Eastman
School of Electrical Engineering, Phillips Hall,
Cornell University, Ithaca, NY 14853

Abstract
Two dimensional electron gas (2DEG) mobilities of 12,000 cm^2/V-s at
room temperature and 63,000 cm^2/V-sec at 77K have been obtained for an
AlInAs/GaInAs modulation doped structure with a room temperature sheet
concentration of 1.4 x 10^{12} cm^{-2}. This is the highest room temperature mobility reported to date for a 2DEG. The combined high
mobility and high sheet concentration results in a room temperature
conductivity 3 times higher than that for similar structures in the
AlGaAs/GaAs material system. In comparison to the AlGaAs/GaAs system
the electron mobilities show a similar but less pronounced dependence
on spacer thickness and sheet concentration, the difference possibly
being due to alloy scattering and additional impurity scattering
resulting from the higher background impurity concentration in the
GaInAs. For a low sheet concentration of 4.1 x 10^{11} cm^{-2} a persistent
photo-conductivity (PPC) of 300% was observed. At high carrier
concentrations, however the PPC becomes negligible ($<$1%) (Griem et
al).

Introduction
A modulation doped GaInAs/AlInAs structure has been previously reported
with a sheet concentration of 2.1 x 10^{12} cm^{-2} and a 300K mobility of
10,000 cm^2/V-sec, giving a conductance 3.5 times higher than the best
GaAs/AlGaAs results (Griem et al). Because of the higher mobility of
electrons in GaInAs and of the larger conduction band discontinuity
between AlInAs and GaInAs which allows higher 2DEG sheet concentrations,
the low field room-temperature conductance is inherently higher for this
new material system than for GaAs/AlGaAs. At high electric fields where
FET's operate, electron saturation velocities in excess of 3.0 x 10^7
cm/sec have been obtained (Capani et al) compared to 2.0 x 10^7 cm/sec for
the GaAs/AlGaAs structure (Banerjee et al). The GaInAs/AlInAs modulation
doped FET's should therefore operate at higher microwave frequencies. In
this paper we report a room temperature mobility of 12,000 cm^2/V-sec
obtained in a GaInAs/AlInAs modulation doped structure, which is the
highest reported 300K mobility obtained for any 2DEG. The 77K mobility
is 63,000 cm^2/V-sec. These mobilities were attained after light
exposure.

Experimental Method
The samples were grown by molecular beam epitaxy (MBE) using a Varian 360
machine. Horizontal Bridgman-grown <100> semi-insulating InP was used
for the substrate material. The wafers were polished on one side using a
*IBM Corporation, Yorktown Heights, NY 10598

simple polishing wheel apparatus and a 0.5% bromine methanol solution.
Following a solvent degrease the substrates were rinsed in deionized
water. Immediately before loading into the growth chamber the substrates
were etched in a room temperature 6:1:1 solution of $H_2SO_4:H_2O:H_2O_2$ for 25
seconds which removed about 1 μm of material. The fluxes of the group
III elements were set on the MBE machine to give the correct lattice
matched compositions of $Ga_{0.47}In_{0.53}As$ and $Al_{0.48}In_{0.52}As$ on InP. This
condition has been predetermined with photoluminescence (PL) spectra from
thick epitaxial GaInAs and AlInAs. The lattice matched composition was
assumed to be that which gave the minimum 4K PL linewidth. The growth
rate used was 1.15 μm/hr. The GaInAs was grown at 505°C, the AlInAs
buffer was grown at 515°C and the AlInAs above the 2DEG was grown at
535°C. The low temperature used to grow the AlInAs buffer is in accord
with recent work by Welch which shows that improved structural quality is
attained for undoped AlInAs grown at lower temperatures (Welch, 1985).
The structures were grown under an As_4 beam equivalent pressure of ~ 5 x
10^{-6} Torr.

Photoluminescence linewidths were also used as an indication of the
quality of the grown layers. An unintentionally doped GaInAs epilayer
grown prior to growing the modulation doped structures had a 4K
photoluminescence spectral linewidth with a full width at half maximum
(FWHM) value of 6 meV and a net electron concentration of ~ 8 x 10^{15} cm-
3, with a 300K mobility of ~ 8000 cm²/V-sec. The AlInAs has a 4K PL
linewidth of ~ 19 meV attained from a PL scan of a modulation doped
sample. A comparison with PL linewidths of < 1 meV (Ashen et al) for
GaAs and ~ 3 meV (Ballingall et al) for AlGaAs show that neither GaInAs
nor AlInAs can yet be grown of comparable quality.

All structures in this work have AlInAs buffers (except one with a
superlattice buffer) which effectively bypasses the problems associated
with the GaInAs/InP interface. Earlier modulation doped structures grown
in this material system which were fabricated into FET's could not be
pinched off because their 2DEG was not isolated from this trouble prone
interface with the substrate (Hsieh). AlInAs buffers were found to be
equivalent to superlattice (GaInAs/AlInAs) buffers. Mobilities improved
less than 5% with the superlattice buffer which consisted of 24 periods
of 30 Å GaInAs and 130 Å AlInAs.

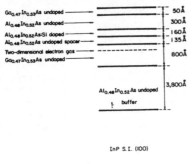 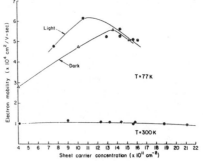

Fig.1. Schematic of 2 DEG structure
giving a room temperature mobility
   of 12,000 cm²/V-sec.

Fig. 2.   Electron mobility vs.
sheet concentration for Hall
measurements done in the dark and
light.

Fig. 1 is a schematic of the structure with the highest room temperature mobility, referred to earlier. The structure is analogous to that often grown for GaAs/AlGaAs modulation doped structures, with two differences: the buffer layer and the relatively thin GaInAs layer which contains the 2DEG. The GaInAs layer is thin to minimize parallel conduction (Hsieh). The 3,800 Å undoped AlInAs buffer is followed by 800 Å of undoped GaInAs which contains the 2DEG. The 135 Å undoped AlInAs spacer is followed by 160 Å Si doped ~ 2.6 x 10^{18} cm^{-3} AlInAs. The spacer layer was varied to study the effect of changing the separation between the Si ions and the 2DEG. In another experiment the width of the doped AlInAs layer was varied to change the sheet concentration of the 2DEG.

Results

The dependence of carrier mobility on sheet electron concentration is shown in Fig. 2. Sheet electron concentrations are varied from 4 x 10^{11} cm^{-2} to 2.1 x 10^{12} cm^{-2} by varying the thickness of the AlInAs:Si from 30 to 190 Å. Hall measurements were made under two conditions, in the light and in the dark. The 300K electron mobility is independent of the sheet concentration and whether or not the sample had been exposed to light, while at 77K a distinct maxima in the 2DEG mobility is observed. After exposure to light an electron mobility maximum of > 60,000 cm^2/V-sec is observed at a sheet concentration of 1.2 x 10^{12} cm^{-2}. In the dark, the peak mobility drops to 56,000 cm^2/V-sec but the corresponding sheet concentration increases to 1.4 x 10^{12} cm^{-2}. At sheet concentrations above 1.4 x 10^{12} cm^{-2} the mobilities are independent of the incident radiation. The change in sheet carrier concentration observed at 77K when a sample is exposed to light is shown in Fig. 3. The dashed line represents no change and the solid line follows the observed concentration changes. The data indicates that the light generates as much as 4.5 x 10^{11} cm^{-2} sheet electrons at a dark sheet concentration of 3 x 10^{11} cm^{-2}. This matches with the persistent photoconductivity (PPC) data taken which showed PPC increasing with decreasing sheet concentration. A PPC effect of 300% (increase from equilibrium dark value conductivity) was seen for the lowest sheet concentration sample grown (4.1 x 10^{11} cm^{-2}). The PPC decreased to an insignificant value (< 1%) for the sample with a sheet concentration of 2.1 x 10^{12} cm^{-2} (Griem et al).

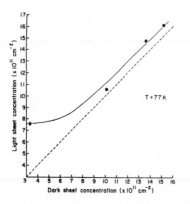

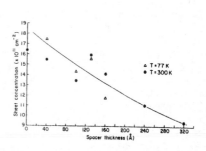

Fig 3. Light sheet concentration vs. dark sheet concentration at 77K.

Fig. 4. Sheet concentration vs. spacer thickness.

In Fig. 4 the sheet concentration is observed to decrease monotonically with increasing spacer thickness both at 300K and at 77K. Fig. 5 shows that the 300K mobility is essentially independent of spacer layer thickness. At 77K however a mobility maximum (after light exposure)of 63,000 cm^2/V-sec with a spacer layer thickness of about 160 Å is observed. The mobility decreases sharply at smaller spacer thicknesses and moderately at larger spacer thicknesses.

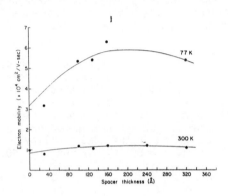

Fig. 5. Electron mobility versus spacer thickness.

Discussion

Higher 300K sheet conductance can be obtained in GaInAs/AlInAs modulation doped structures than have been achieved for the GaAs/AlGaAs system. This results from the higher intrinsic electron mobility in GaInAs and a higher sheet electron concentration for the 2DEG. The higher mobility results from the smaller electron effective mass in GaInAs. The higher attainable sheet electron concentration is believed due to a larger conduction band discontinuity (> 0.5 eV compared to < 0.3 eV) in the GaInAs/AlInAs system (Welch, 1984).

The dependence of 77K 2DEG electron mobility on sheet concentration shows a peak mobility at a concentration of about 1.2 x 10^{12} cm^{-2} in the light and 1.4 x 10^{12} cm^{-2} in the dark. At the lower sheet concentrations the increasing mobility with sheet concentration is likely due to the increased screening effect of the added carriers which results in a decrease in ionized impurity scattering. The drop in mobility for high sheet concentrations is a trend also seen in the GaAs/AlGaAs system (Hiyamizu et al). There are two common explanations given: parallel conduction in the heavily doped large bandgap region or intersubband scattering. Neither of these explanations are very satisfactory as the first explanation is a very small effect at 77K and the second explanation's effect should occur at a lower sheet concentration (~ 4.5 x 10^{11} cm^{-2}) for the GaInAs/AlInAs system (Kastalsky et al). It is not understood why there is an offset of the 77K peaks between the dark and light measurement.

A peak in 77K 2DEG electron mobility is also observed with a change in the undoped AlInAs spacer layer thickness. The mobility maximum occurs for a spacer layer thickness of about 160 Å. The increase in mobility with increasing spacers up to 160 Å is most likely due to the reduced coulombic interaction between the 2DEG and the donors in the AlInAs. As the spacer layer thickness is increased beyond 160 Å electron transfer into the 2DEG is decreased reducing its sheet concentration and mobility. The effect of increased parallel conduction in a low mobility region or intersubband scattering may again prevail. As explained earlier these are weak explanations for the 77K mobility decrease.

The results reported here are not believed to be optimized. By analogy with the GaAs/AlGaAs modulation doped structure where the highest mobilities are obtained with the highest purity MBE grown GaAs, higher 2DEG mobilities for the GaInAs/AlInAs system should be possible by increasing the GaInAs purity. Our nominally undoped MBE grown GaInAs has an electron concentration of about 8×10^{15} cm^{-3} with a 77K mobility of about 20,000 cm^2/V-sec. The highest reported mobility for undoped MBE grown GaInAs is 11,000 cm^2/V-sec and 43,000 cm^2/V-sec at 300K and 77K respectively with an electron concentration of 2×10^{15} cm^{-3} (Bonnevie et al). The highest 77K mobility ever reported for GaInAs (grown by LPE) is 70,000 cm^2/V-sec with an electron concentration of 3.5×10^{14} cm^{-3} (Oliver). The highest 300K mobility also attained by Oliver using LPE was 13,800 cm^2/V-sec. GaInAs 2DEG mobilities should approach these values once the purity of the GaInAs used for the 2DEG structures is improved. The high background doping is probably due to either the In source or outdiffusion from the substrates. Steps are being taken to reduce this problem (Brown et al).

Acknowledgements
My gratitude goes to A.R. Calawa who helped in the editing, to J. Berry for his knowledgeable technical assistance, and to E. Weaver for typing the manuscript. This work was supported by the Army Research Office under contract no. DAAG29-82-K-0011.

References
Ando T, 1982 J Phys Soc Jap **51** 3900
Ashen D J, Dean P J, Hurle D T J, Mullin J B, and White A M, 1975 J Phys
    Chem Solids **36** 1041
Ballingall J M and Collins D M, 1983 J Appl Phys **54**
Banerjee P, Bhattacharya P K, Ludowise M J, Dietz W T, 1983 IEEE Electron
    Dev Lett **EDL-4**
Bonnevie D and Huet D 1982 J. De Physique **5**
Brown A S, Palmateer S C, Wicks G W, Eastman L F, Calawa A R and Hitzman
    C, 3rd Conf. of Semi-insulating
Capani P 1984 Symp on GaAs and Related Compounds Inst.Phys.Conf.Ser. <u>**74**</u> 361
Griem, T. 1984 Int. MBE Conf to be published
Hiyamizu S, Saito J, Nanbu K, and Ishikawa T, 1983 Jap Jour Appl Phys 22
    pp L609-L611
Hsieh K, Ph.D. Thesis, Cornell University 1984
Oliver J D Jr. Ph.D. Thesis, Cornell University 1980
Welch D F, Wicks G W and Eastman L F 1984 J Appl Phys **55**
Welch D F 1985 to be published

Optical properties of strained $In_xGa_{1-x}As$–GaAs superlattices

J.Y. MARZIN, M.N. CHARASSE, G. LE ROUX, L. GOLDSTEIN and M. QUILLEC
Laboratoire de Bagneux* C.N.E.T., 196 rue de Paris, 92220 Bagneux, FRANCE

* Laboratoire associé au C.N.R.S. (LA 250)

Abstract. We present optical data obtained on $In_{0.15}Ga_{0.85}As$-GaAs strained superlattices grown by Molecular Beam Epitaxy (M.B.E.). Clear excitonic transitions are observed in absorption spectra. An original method has been used to display the optical selection rules. An envelope function model, including the stress effects, yields a good agreement with the experimental results. It suggests a new configuration for the valence bands in the superlattice.

1. Introduction

Since the first studies on $GaAs$-$GaAs_xP_{1-x}$ by Matthews (76), strained layer superlattices have regained interest in the last few years. Several systems have been examined and, among them, $In_xGa_{1-x}As$-GaAs has been thoroughly studied. After it was successfully grown (Goldstein 82, Fritz 82), structural (Picraux 83), transport (Fritz 83), crystallographic (Quillec 84) and optical (Marzin 83, Camras 83, Nakayoma 84 and Laidig 84a) properties have been investigated. This system is well suited to optical applications and several groups have used it to realize devices such as lasers (Ludowise 83, Laidig 84b), photodetectors (Myers 84), and Field Effect Transistors (Zipperian 83).

2. Growth procedure

The samples, grown by M.B.E., consist in ten periods of $In_{0.15}Ga_{0.85}As$ layers which thickness has been varied from 50 to 300 Å alternating with 200 Å thick GaAs layers. This superlattice is grown on 5000 Å GaAs buffer layer deposited on a (100) GaAs substrate. We want to point out that we do not use any graded buffer layer to match the superlattice equilibrium parameter.

The standard substrate preparation has been described elsewhere (Quillec 84). The growth rate was about 1 μm/hr and the growth temperature 550 to 560°C. Indium desorption was found to be negligible from the recording of the RHEED oscillations (Charasse 84). Moreover, this techni-

que allows a precise determination of the GaAs growth rate. This measurement was made during the buffer layer growth to prevent damage of the
superlattice since it needs to interrupt the growth and the substrate
rotation. The thickness of the GaAs sublayers can thus be directly obtained. Besides, double X ray diffraction profiles (Quillec 84) give an
accurate determination of sublayer thicknesses as well as of the
$In_xGa_{1-x}As$ composition.

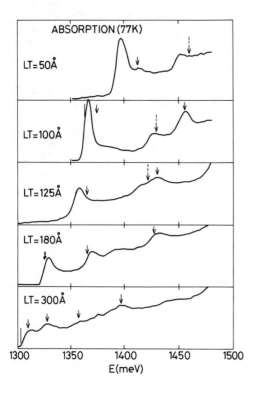

Fig.1:77K absorption spectra
for five samples with different
ternary layer thicknesses
L_T.($\rightarrow$) and ($--\rightarrow$) indicate the
calculated heavy and light hole
related transition energies,
respectively.

3. Absorption experiments and discussion

Fig.1 shows the 77K absorption spectra obtained on 5 samples. Their
excellent optical quality is indicated by the observation of clear excitonic transitions (even at room temperature for the samples with a ternary layer thickness smaller than 200 Å). These exciton peaks are appearing
on a steplike background which is characteristic of these two dimensional
systems. The sharpest peaks are obtained for L_T=100 Å, for L_T=50 Å, small
thickness fluctuations may broaden the peaks, whereas, for L_T 100 Å,
broadening can be due to the decreasing of the exciton binding energies,
as well as to stress relaxation.

To assign the different transitions, we have to take into account the stress effects. As shown by X-ray diffraction and T.E.M. analysis, for $L_T < 180$ Å, the 1 % mismatch between $In_{0.15}Ga_{0.85}As$ and GaAs is totally accommodated by elastic strain. In this range of thicknesses, $In_{0.15}Ga_{0.85}As$ only is under biaxial compression in the plane of the layers, this stress being independent of L_T, while the GaAs barriers remain unstrained. For $L_T=180$ Å, we observe a degradation of the double crystal X-ray diffraction pattern, and T.E.M. study reveals the presence of dislocations which tend to be confined in the first layer near the superlattice-buffer layer interface. For $L_T=300$ Å, crosshatch can be observed on the surface of the sample under optical microscope examination (x500).

The valence band of the strained $In_{0.15}Ga_{0.85}As$ is split , with the heavy hole band top lying 80meV above the light hole one (for the "well" strained samples with $L_T<180$ Å, and using GaAs deformation potentials given by Chandrasekhar, 77). The heavy hole states remain uncoupled to the light particle ones. If this valence band splitting is small compared to the spin orbit splitting, we can neglect the stress induced coupling between light and spin orbit split off bands : we can then use the same type of calculations as in unstrained superlattices except that, due to this stress induced valence band splitting, the wells are different for the heavy and light holes.

Therefore, the lowest energy transition is arising from the first heavy hole (HH_1) and electron (E_1) levels. The existence of an associated exciton implies that the electron and heavy holes are both confined in the ternary layers. As the band gap difference is equal to 162meV (Leu, 79), using interpolated effective masses, we can deduce that there can be only one electronic level for $L_T=50$ Å. The second transition observed for this sample and the 1.43eV transition in the spectrum of the sample with $L_T=100$ Å are then thought to be related to the first light hole level (LH_1). The 1.46eV transition in the latter spectrum is assigned to (HH_2, E_2) exciton.

To obtain the band configuration in the superlattice from the position of these transitions, we have to be sure of these tentative assignments, so we tried to verify them in the experiments we describe below.

4. Complementary experiment : photoluminescence excitation

As already pointed out by Dingle (75), light and heavy hole associated transitions in such superlattices have different selection rules : for a linear light polarization along the growth axis z, the heavy hole ones are forbidden, while the light hole ones are allowed. In our structures, the substrate is transparent in the energy region of the superlattice transitions. This fact can be used to excite its photoluminescence with a dye laser focused on the cleaved edge of the sample, as shown in the inset of Fig.2. The light polarization can then be set nearly along or perpendicular to z, inside the sample. This is not possible in standard surface excitation, because of the high refractive index of these materials. This new technique should be usefully employed in structures with "transparent" substrates.

Fig.2 shows such excitation of the photoluminescence performed at 77 K on the sample with $L_T=100$ Å. The luminescence at the energy of the first excitonic transition α is plotted versus the dye (oxazine 750) laser energy. Fig.2b spectrum results from an on-edge perpendicular to z light polarization experiment and is comparable to the c spectrum, corresponding to "surface" excitation, as expected. On the contrary, the a spectrum, obtained in the on-edge, parallel to z polarization configuration is different : heavy hole related transitions are forbidden in this polarization, so the HH_1-E_1 continuum contribution is much weaker, β line is still observed while γ one is not. These results support our previous interpretation : β transition is likely to be related to light holes.

5. Calculation results and discussion

Using an envelope function model (Bastard 81), we have calculated the transition energies in our superlattices, with the conduction band discontinuity as an adjustable parameter.

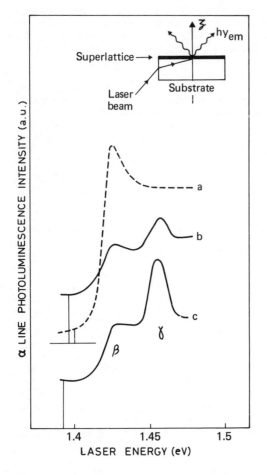

Fig.2:77K excitation spectra of the sample with L_T=100 Å
a) on-edge light polarized linearly // z.
b) on-edge light polarized linearly ⊥ z.
c) surface unpolarized.

For the samples with $L_T < 180$ Å, we assumed that $In_{0.15}Ga_{0.85}As$ layers only are strained. The best fit, which results are indicated by arrows on Fig.1 is obtained for a 55 meV deep heavy hole well. For this value, the electron well is 107 meV deep and the $In_{0.15}Ga_{0.85}As$ light hole valence band top is at 25 meV under the GaAs valence band top. For light holes, GaAs layers then are the wells and $In_{0.15}Ga_{0.85}As$ layers the barriers. Because of their small masses, coupling between successive GaAs wells through the 25 meV high barriers is very efficient when L_T is small enough. The corresponding wave functions are strongly leaking in $In_{0.15}Ga_{0.85}As$ layers, which explains that we can however observe the light hole-electron transitions β. As our calculation does not include the exciton binding energies, the calculated transitions lie at higher energies than the experimental ones.

For thicknesses $L_T \geqslant 180$ Å, we have assumed that the superlattice has taken its equilibrium parameter, which seems to be more in agreement with the experimental data, as shown in Fig.1.

6. Conclusion

As a summary, we have grown strained $In_{0.15}Ga_{0.85}As$_GaAs superlattices of excellent optical quality. Our results suggest a new valence band configuration in this system, due to strain induced splittings, i.e. light and heavy hole levels are confined in GaAs and $In_{0.15}Ga_{0.85}As$ layers, respectively.

Acknowledgments

The authors wish to thank P. HENOC for T.E.M. observations and A.M. JEAN-LOUIS for some P.L. measurements.

REFERENCES

Bastard G., 1981, Phys. Rev. B24, 5693
Camras M.D., Brown J.M., Holonyak N., Nixon M.A., Kaliski R.W., Ludowise M.J., Dietze W.T., and Lewis C.R., 1983, J. Appl. Phys.54, 6183
Chandrasekhar M. and Pollack F.H., 1977, Phys. Rev. B15, 2127
Charasse M.N. and Golstein L., to be published
Dingle R., 1975, Festkorper probleme, H.J. Queisser, ed., Pergamon-Viewig, XV, 21
Fritz I.J., Dawson L.R. and Zipperian T.E., 1983, Appl. Phys. Lett.43, 846
Goldstein L., Quillec M., Rao E.V.K., Henoc P., Masson J.M. and Marzin J.Y., 1982, Journal de Physique 12, C5, 201
Laidig W.D., Caldwell P.J., Lin Y.F. and Peng C.K., 1984a, Appl. Phys. Lett.44, 653
Laidig W.D., Blanks D.K. and Scherzina J.F., 1984b, J. Appl. Phys.56, 1791
Leu Y.T., Thiel F.A., Scheiber H., Robin J.J., Miller B.I. and Bachmann J., 1979, J. Electron. Mater., 8, 663
Ludowise M.J., Dietze W.T., Lewis C.R., Camras M.D., Holonyak N., Fuller B.K. and Nixon M.A., 1983, Appl. Phys. Lett.42, 487

Marzin J.Y. and Rao E.V.K., 1983, Appl. Phys. Lett.43, 560
Matthews J.W. and Blakeslee A.E., 1976, J. Cryst. Growth 32, 216
Myers D.R., Zipperian T.E., Biefeld R.M. and Wiczer J.J., 1983,
    International Electronic Device Meeting 83, 700
Nakayama M., Kubota K., Kato H. and Sano N., 1984, Solid State Comm.51,
    343
Picraux S.T., Dawson L.R., Osbourn G.C. and Chu W.K., 1983, Appl. Phys.
    Lett.43, 930
Quillec M., Goldstein L., Le Roux G., Burgeat J. and Primot J., 1984, J.
    Appl. Phys. 55, 2904
Zipperian T.E., Dawson L.R., Osbourn G.C. and Fritz I.J., 1983,
    International Electronic. Device Meeting 83, 696.

Inst. Phys. Conf. Ser. No. 74: Chapter 5
Paper presented at Int. Symp. GaAs and Related Compounds, Biarritz, 1984

379

Optical studies of GaInAs/InP quantum wells

M. Razeghi, J. Nagle and C. Weisbuch
Thomson-CSF, Laboratoire Central de Recherches, B.P. N°10, 91401 Orsay, France

Abstract. The highly-efficient photoluminescence of GaInAs/InP quantum wells grown by low-pressure MOCVD has been investigated. Narrow lines are observed down to nominal thicknesses of $\approx$ 8 Å. Various excited states of quantum wells have been measured by photoluminescence excitation spectroscopy, allowing a preliminary determination of the band-gap offsets at the GaInAs-InP interface. Carrier capture is shown to be extremely efficient even for the narrowest wells studied.

1. Introduction

Quantum-well (QW) luminescence has been by now extensively studied in the GaAs-GaAlAs system and has provided significant information about the structural properties of heterostructure devices and about the 2-D nature of the elementary excitations. QW's made of ternary or quaternary materials such as GaInAs or GaInAsP imbedded in InP (Razeghi et al 1983) or InAlAs(Welch et al 1983) barrier material have been up to now much less studied. We present here some recent results concerning the photoluminescent properties of GaInAs/InP QW's grown by low-pressure MOCVD. Standard conditions were used, leading to a growth rate of $\approx$ 2.5 Å/s and well-defined interfaces by rapid switching of gas composition. We study multiple single-well samples, i.e. samples with a stack of quantum wells with different thicknesses grown under the same conditions. Such structures were introduced for GaAs/GaAlAs by Frijlink and Maluenda (1982). We address three main subjects in this paper : (i) the origin and properties of luminescence lines ; (ii) the observation of the excited states of the quantum wells by photoluminescence excitation spectroscopy (PLE) and the determination of QW parameters from these data ; (iii) the carrier-capture properties of thin quantum-wells.

2. Luminescence results

Photoluminescence spectra were obtained at various temperatures, using a Helium gas-flow variable-temperature cryostat. Luminescence was excited using a He-Ne laser (excitation of the surface InP barrier material), a YAG laser (direct excitation of the GaInAs or GaInAsP QW) or a spectrometer filtered lamp (used as tunable source for PLE spectra). Luminescence was analyzed in a 60 cm grating-spectrometer and detected with a high-sensitivity N -cooled Ge-photodiode. The high quantum efficiency of the material and the careful optimization of the experiment for maximum sensitivity allow the detection of luminescence signals with excitation powers well below the microwatt range as in PLE conditions.

Typical spectrum for a five-well structure is shown in figure 1. The well-thicknesses indicated ("nominal") are those deduced from the growth-rate as extrapolated for thin layers from the measured growth rate

of thick layers. The remarkable features are the sharpness of lines, even
for QW with an 8 Å nominal thickness ($\Delta E \simeq$ 29 meV) at 5K. The progress
in sample quality can be traced by comparing these spectra with those
obtained a little more than one year ago (Razeghi et al 1983). Extremely
large shifts of luminescence due to the quantum confinement effect of
QW's are observed, up to 0.320 eV for the nominal 8 Å QW.

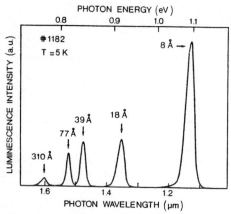

Figure 1 : Photoluminescence
spectrum of a multi-well
sample at 5K excited by an
He-Ne laser in the InP
overlayer.

The growth of layers with different thicknesses has allowed us to assess
the origin of the variation in energy while scanning across the sample
(Fig.2). The recombination energy in an alloy QW can be approximated by :

$$E = h\nu = E_o(x,y) + E^e_{conf} + E^h_{conf} \qquad (1)$$

Figure 2 : Luminescence of a
multi-well sample for different
spots, 5 mm away on the surface.
He-Ne laser excitation.

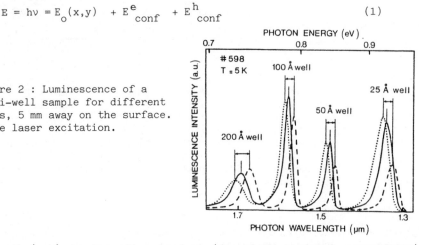

where $E_o(x,y)$ is the alloy bandgap (dependent on local composition),
E^e_{conf} and E^h_{conf} are the confining energies for the electron and hole
ground states respectively. We have neglected here the small shift between the quantized electron-hole levels and luminescence line position
due to the relaxed nature of the luminescent state and stemming from
exciton binding, carrier trapping, energy relaxation of the excitations
in the disordered alloy band or in the quantum-well disordered energy
band (Weisbuch et al 1981b). Variations of h according to eq. (1) can
have two main origins : (i) a variation in $E_o(x,y)$ due to alloy macroscopic variations of composition across the sample ; (ii) a change of the
confining energies in a quantum well due to a macroscopic variation of QW

thickness (originating in a non-uniform growth rate across the wafer). In case (i), one expects a similar shifting of the QW energies across the wafer, independent of the thickness of the quantum well under observation. In case (ii), if v_1 and v_2 are the two growth rates at two points 1 and 2, and remembering that $E \sim L^{-2}$ in the infinitely deep well limit, we find :

$$\Delta E \approx L^{-3} \Delta L = L^{-2}(v_2 - v_1)/v_2$$

In this case the energy shift between two points on the wafer depends on the well thickness. The observation of a constant shift for all the QW in those samples which display energy shifts across the wafer allows us to conclude that this is due to the macroscopic variations of the alloy composition across the sample.

We are not able with our present level of understanding to decide the nature of the luminescent level at low temperatures. Comparing with GaAs-GaAlAs quantum wells where the dominance of free-exciton-mediated recombination has been established through spin-orientation and lineshape analysis (Weisbuch et al 1981a), we would tentatively ascribe the same origin to the luminescence line in the GaInAs/InP system : the 2D exciton binding energy should be $\gtrsim$ 4 times larger than in 3D (as was recently well-established in the GaAs/GaAlAs system by Maan et al 1984), i.e. should be 10 meV.

The observed linewidth of the luminescence lines (and excitation spectra) is surprisingly small. A model of PLE spectra in GaAs/GaAlAs has described this linewidth as due to the spatially-varying well-thickness due to the microscopic island-like topology of the interface due to a layer-growth mode (Weisbuch et al 1981b) ; (Singh et al 1983) : assuming a half-monolayer height for the islands and a correlation length $\geq$ the exciton Bohr radius a_B, one expects variations of the exciton absorption or recombination energy given by $\Delta E \approx L^{-3} a/2$. The usually-observed continuum of energy fluctuations is due to the distributions of the lateral sizes of the islands, which lead to a distribution of E's (Bastard et al 1984). If we try to apply this island-disorder model in the present case, it yields numbers for E far astray from the experimental results. An a/2 fluctuation in interface position yields $\Delta E \simeq 42$ meV for a 20 Å thick sample, much above the observed linewidth. We then are led to consider that the previous model, although very successful in the case of GaAs/GaAlAs, does not apply on our GaInAs/InP case. We therefore tend to describe the interface as rather fuzzy, both in the growth direction where we believe from our growth condition that the interface occurs by compositional grading over several atomic layers and along the interface plane where the compositional correlation length should be quite small, well below a , i.e. at most tens of Angstroms. However the observation of a narrow, well-defined, intense luminescence peak, even for thicknesses down to 8 Å (nominal), leads us to emphasize the very good uniformity of this fuzzy well, i.e. very small deviations from the average thickness are allowed over macroscopic (here larger than a_B) distances.

2. Excitation spectra

Excitation spectra have been recorded for a number of GaInAs/InP quantum wells. These spectra are quite comparable to those reported at this conference in the GaInAs/InAlAs system or the strained GaInAs/GaAs system grown by MBE. A remarkable feature as compared to the GaAs/GaAlAs case is that we do not observe pairs of absorption peaks due to each confined electron level associated with either the heavy or light-hole level. We rather observe two series of peaks, one at lower energies and rather sharp, the other quite broader and more rightly described as a bump (figure 3). We ascribe tentatively these two series as due to heavy-hole to electron and light-hole to electron transitions. The different shapes of these transitions might be due to the different non-parabolicities of the two hole bands due to the strong valence-band mixing recently evidenced in the GaAs/GaAlAs case (Sooryakumar et al 1984).

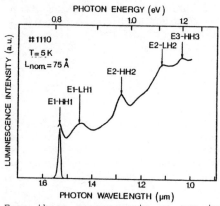

Figure 3 : Excitation spectrum for a 75 Å quantum-well.

From these measurements we can try to evaluate the various parameters entering equation (1). It should be remarked that the situation is much more complicated that in the GaAs/GaAlAs case, as E_o is not so precisely known (we are not fully informed of the well chemical composition) and the various band parameters (mainly the hole masses) are also not too well determined. We also do not have at present a reliable measurement of the well thicknesses. We therefore only produce here preliminary fit of our experimental results taking all the unknown as fitting parameter (figure 4). We use a standard quantum-well calculation of energy levels,

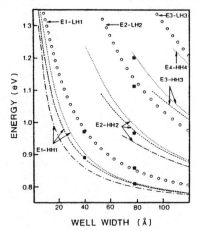

Figure 4 : Fit of the transitions observed in excitation spectra for two wells with 75 and 35 Å nominal thickness. Fitting parameters :
$E_o = 0.742$ eV
$m_e = 0.041\ m_o$, $m_{hh} = 0.5\ m_o$,
$m_{lh} = 0.042\ m_o$.
$L = 39$ Å ; $L = 78$ Å
The various E-HH transition curves correspond to $Q = \Delta E_c / \Delta E_g = 0.3(- -)$
0.5 (——) and 0.7 (.....). Only one curve is traced for E-LH transitions as this parameter has negligible influence on transition energies.

introducing in a simple iteration form the non-parabolicity of the energy bands. No provision is taken for the mixing of the valence bands which are considered fully uncoupled. We neglect exciton effects small in our scale of confining energies. As dramatically pointed out by Miller et al

(1984) in the GaAs/GaAlAs case, we also observe that interband transitions do not lead to an accurate determination of Q : the change in the confining energies induced by a change in Q can easily be corrected by a change in other fitting parameters, here mainly E_o and L. A more detailed knowledge of sample structure and band parameters is therefore needed for a better precision in Q.

4. Capture process

The multiple single-well configuration offers an interesting means to study the perpendicular carrier diffusion and capture as the various wells act as probes of the local carrier concentration away from the surface when the photoluminescence is excited by light absorption near the sample surface. A similar probing of carrier transport in multilayer material has been used by Lambert et al. (1984) in an attempt to evidence superlattice perpendicular transport. The luminescence of a given quantum well can be approximately expressed as the product of three factors :

$$I_{lum} = N(z) \times R \times \eta$$

where $N(z)$ is the excitation density at the well, R is the capture rate and η the luminescence quantum efficiency of that given well.

In order to evaluate η, excitation by light <u>below</u> the barrier material bandgap allows a direct, uniform excitation of the various quantum wells as each well absorbs only a small fraction of the incident light. The various wells do not however absorb exactly the same amount of incident light, as the absorption coefficient at a given energy increases with the energy difference to the quantum-well ground state (roughly proportionally to the number of quantum states excited by light, neglecting excitonic effects and non-parabolicities). We thus expect the absorption probability to be increasing with increasing quantum-well thickness. If η is constant with well thickness, one therefore expects the <u>YAG-excited</u> luminescence to be larger for larger wells which is quite often observed but not always. This shows the uncontrolled variation of η from sample to sample for wells with similar widths. Excitation with the <u>He-Ne laser</u> however systematically shows a strongest emission feature by the narrowest well, situated near the surface in all our samples, which evidences the large fraction of carriers captured by the first well, in detrimental action to the luminescence of the other wells. This efficient carrier capture was already noted by several experiments dealing with time-resolved carrier dynamics (Göbel et al 1983) or from the ratio of QW to barrier luminescence intensities (Christen et al 1984). We extend it here to the competition between wells and to the GaInAs/InP system. Even the very narrow well of 8 Å (nominal) has a very efficient capture evidencing that the eventual inefficient capture by very narrow wells is not a strong issue for design rules of quantum-well lasers (Tang et al 1982).

Finally the temperature dependence of quantum well luminescence efficiency and capture rate has been studied (fig. 5). The most remarkable feature is the rather constant I_{lum} up to 40 K, above which I_{lum} varies by about two orders of magnitude, <u>independently</u> of quantum well thickness and excitation wavelength. This shows that the capture rate is rather constant and that the main varying parameter is η. We believe that the change occuring at $T \simeq 40K$ is due to the switching of the main recombination process from excitonic to electron-hole pair recombination. The latter process leads to a variation of radiative lifetime as T^{-2}, insufficient to explain the observed $T^{-5/2}$ or $T^{-7/2}$ variation. There might be some contribution of Auger recombination and further transfer in the barrier material, although this hypothesis remains to be evaluated more quantitatively.

Figure 5 : Integrated luminescence intensity as a function of temperature for He-Ne and YAG laser excitation.

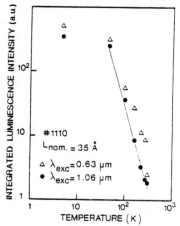

5. Conclusion

We have evidenced the very good properties of narrow (down to 8 Å) GaInAs/InP QW's grown by MOCVD : high quantum efficiency, narrow lineshapes. A preliminary evaluation of the data yields $Q=0.5\pm0.1$ for the repartition of the bandgap discontinuity between conduction and valence bands. Capture is shown to be very efficient even in the narrowest wells, independently of temperature.

6. Acknowledgments

We thank R. Cordeau for expert technical assistance. Useful discussions with T. Weil are gratefully acknowledged.

7. References
Bastard G, Delalande C, Meynadier MH, Frijlink PM and Voos M 1984, Phys. Rev. B29
Christen J, Bimberg D, Steckenborn A and Weimann G 1984, Appl. Phys. Lett. 44 84
Frijlink PM and Maluenda J 1982, Jap. J. Appl. Phys. 21 L574
Göbel EO, Jung H, Kuhl J and Ploog K 1983, Phys. Rev. Lett. 51 1588
Lambert B, Romestain R, Miller D, Chomette A, Regreny A and Deveaud B 1984, These Proceeding p. 357.
Maan JC, Belle G, Fasolino A, Altarelli M and Ploog K 1984, Phys. Rev. B30 2253
Miller RC, Kleinman DA and Gossard AC 1984, Phys. Rev. B29 7085
Razeghi M, Hirtz JP, Ziemelis VO, Delalande C, Etienne B and Voos M 1983, Appl. Phys. Lett. 43 585
Singh J and Madhukar A 1983, J. Vac. Sci. Technol. B1 305
Sooryakumar R, Chemla DS, Pinczuk A, Gossard A and Wiegmann W 1984, J. Vac. Sci. Technol. B2 349.
Tang JY, Hess K, Holonyak Jr N, Coleman JJ and Dapkus PD 1982, J. Appl. Phys. 53 6044.
Weisbuch C, Miller RC, Dingle R, Gossard AC and Wiegmann W 1981a, Solid State Comm. 37 219
Weisbuch C, Dingle R, Gossard AC and Wiegmann W 1981b, Solid State Comm. 38 709
Welch DF, Wicks GW and Eastmann LF 1983, Appl. Phys. Lett. 43 762.

Inst. Phys. Conf. Ser. No. 74: Chapter 5
Paper presented at Int. Symp. GaAs and Related Compounds, Biarritz, 1984

Alloy scattering in InP−Ga$_{1-x}$In$_x$As$_y$P$_{1-y}$ modulation-doped quantum wells

J.A. Brum and G. Bastard

Groupe de Physique des Solides de l'Ecole Normale Supérieure,
24 rue Lhomond, F-75005 Paris, France

Abstract. We present the results of self consistent calculations of charge transfer and alloy scattering-limited mobility in modulation-doped InP-Ga$_{1-x}$Ir$_x$As$_y$P$_{1-y}$ quantum wells at T = 0.

1. Introduction

The quaternary quantum wells (Q.W.) of Ga$_{1-x}$In$_x$As$_y$P$_{1-y}$ clad between InP barriers and lattice-matched to InP have potential applications in opto-electronics since their band gap can be adjusted to fit the 1.5 μm transparent window of optical fibers. The present communication explores the influence of an unavoilable scattering mechanism on the electronic mobility in these systems : the alloy scattering. Calculations of alloy scattering in InP-Ga$_{0.47}$In$_{0.53}$As single heterojunctions have already been reported by Bastard (1983) and Basu and Nag (1983). Here we will show that modulation-doped Q.W.'s display a lower electronic mobility than single heterojunctions, the two-dimensional (2D) charge density n$_e$ being the same.

2. Self-consistent calculations of charge transfer

To be specific we consider a single modulation-doped Q.W. (fig.1). The barrier is assumed to be uniformly doped with donors (volume concentration N$_d$) and the doping is the same in left-hand side and right-hand side barriers. The barrier height is V$_b$. As a result of electron transfer each of the barriers is depleted over a length L$_d$. The electrical neutrality implies

$$2 N_d L_d = n_e \tag{1}$$

The spatial separation between the electrons and their parent donors gives rise to a band bending - e$\phi(z)$ where $\phi(z)$ is the electrostatic potential. We have to solve simultaneously Poisson and Schrödinger equations. If we restrict our considerations to the electric quantum limit (E.Q.L.) we obtain :

$$\phi''(z) = \frac{4\pi e}{\kappa} [- N_d^+(z) + n_e \chi_{E_1}^2(z)] \tag{2}$$

$$[- \frac{\hbar^2}{2m^*} \frac{d^2}{dz^2} + V_b Y(z^2 - \frac{L^2}{4}) - e\phi(z)] \chi_{E_1}(z) = E_1 \chi_{E_1}(z) \tag{3}$$

In addition to eqs (1-3) one should ensure that the chemical potential μ is the same in the well :

$$\mu = E_1 + \frac{\pi\hbar^2 n_e}{m^*}$$

(4)

and in the barrier :

$$\mu = V_b - \varepsilon_b - \frac{2\pi e^2}{\kappa} N_d L_d^2$$

(5)

In eqs (2-5) κ is the relative dielectric permittivity of the structure; E_1 the confinement energy of the first electron subband, $\chi_{E_1}(z)$ the asso-ciated wave-function; m^* the carrier effective mass (assumed to be position-independent); ε_b the donor binding energy in the barrier and $Y(x)$ the step function. Finally the assumption of a zero temperature is implicit in eqs (2,4,5).

The energy levels of modulation-doped GaAs-Ga$_{1-x}$Al$_x$As multiple Q.W.'s have been variationally calculated by Ando and Mori (1979). More recently Stern and Schulmann (1984) have proposed numerical solutions which include many-body corrections to the Hartree contribution $-e\phi(z)$. Fishman (1983) has shown that the band bending can be accurately treated in perturbation for Q.W. structures. His calculated subband energies are in good agreement with optical experiments (Pinczuk and Worlock 1982). The accuracy of a perturbative treatment of the effects of $\phi(z)$ is inherent to the Q.W. structure : even if $\phi(z) = 0$ E_1 is bound. This is in marked contrast with single heterojunctions where the carrier confinement is entirely due to a non-vanishing $\phi(z)$. We adopt here a reduced version of Fishman's approach. Namely we treat $\phi(z)$ by first order perturbation theory on each Q.W. level. Then the eqs (2,3) decouple since the wavefunction appearing in eq(2) is the unperturbed (by $-e\phi(z)$) solution of the square well problem.

The calculated thickness (L) dependence of the transferred charge is pre-sented in fig.(2) for modulation-doped InP-Ga$_{1-x}$In$_x$As Q.W.'s. The para-meters used in the calculations are N_d = 3×10^{16} cm^{-3} (residual doping in InP); x = 0.53; m^* = 0.041 m_0; κ = 13.2. Three different conduction band offsets have been considered : $V_b - \varepsilon_b$ = 0.2 eV; 0.35 eV; 0.5 eV. As seen in fig.(2) n_e vanishes at vanishing L. This has to be expected since in this limit E_1 approaches the top of the Q.W. Beyond $\sim$ 100 Å the transfer saturates as the confinement energy E_1, which is the only quantity depending on L in eqs (4,5), becomes a small fraction of $V_b - \varepsilon_b$. The curves shown in fig.(2) have been interrupted near the thicknesses corresponding to the population of the second electronic subband E_2. For wider wells our calcu-lations are no longer self-consistent.

3. Alloy scattering-limited mobility

At T = 0 there exists a linear relationship between the velocity relaxation time $\tau_i(E_F)$ of the ith subband at the Fermi energy E_F and $E_F - E_i$ (see e.g. Ando, Fowler, Stern 1983) :

$$E_F - E_i = \sum_j K_{ij} \tau_j(E_F)$$

(6)

We evaluate the electronic mobility in the E.Q.L. accounting for the carrier scattering by the short-range fluctuating alloy potential. This scattering mechanism is likely to be dominant at low temperature when the 2D electron gas is mostly confined within a ternary or a quaternary alloy material. On the other hand, the scattering by ionized impurities is strongly reduced over the bulk value by the spatial separation between the donors and the electrons (see e.g. Stormer 1984). We consider that the Q.W. energy levels (section 2) have been calculated in the virtual crystal approximation (V.C.A.) for the alloy part of the structure. The difference between the actual and the V.C.A. potential is then treated in the Born approximation and $\tau_1(E_F)$ is evaluated. For InP-Ga$_{1-x}$In$_x$As we obtain for the electron mobility the expression :

$$\mu_{alloy}(E_F) = \frac{e\hbar^3}{(m^*)^2 x(1-x)\,\Omega_0\,[\delta V]^2} \times \frac{1}{\int_{-\frac{L}{2}}^{+\frac{L}{2}} |\chi_{E_1}(z)|^4 \, dz} \tag{7}$$

where Ω_0 is the volume of the V.C.A. unit cell; $[\delta V]$ the strength of the fluctuating alloy potential averaged over Ω_0 and $x = 0.53$.

In marked contrast with single heterojunctions, μ_{alloy} is independent of n_e in Q.W. structures. To a large extent the shape of $\chi_{E_1}(z)$ is fixed by the Q.W. potential and not by the band bending $-e\phi(z)$. In single heterojunction $-e\phi(z)$ determines $\chi_{E_1}(z)$ whose shape is therefore strongly n_e-dependent. This, in turn, implies the decrease of μ_{alloy} with n_e in single hetero-junctions.

When a second subband just becomes populated the mobility of the first subband drops due to intersubband scattering (see e.g. Ando, Fowler, Stern 1983). If the alloy scattering is the dominant scattering mechanism the relative mobility drop at the onset of population of the second subband is given by

$$\frac{\Delta\mu_{alloy}}{\mu_{alloy}(E_F=E_2)} = \frac{-\int_{-L/2}^{L/2} \chi^2_{E_1}(z)\,\chi^2_{E_2}(z)\,dz}{\int_{-L/2}^{+L/2} \chi^2_{E_1}(z)\,[\,\chi^2_{E_1}(z) + \chi^2_{E_2}(z)\,]\,dz} \tag{8}$$

where $\chi_{E_2}(z)$ is the wavefunction of the E_2 subband. We present in fig.(3) the thickness dependence of the alloy scattering-limited mobility of InP-Ga$_{0.47}$In$_{0.53}$As modulation-doped Q.W.'s. The parameter $[\delta V]$ has been taken equal to its bulk value : 0.6 eV (Hayes et al 1982); Ω_0 is equal to 1/4 (5.87 Å)^3 and for each L the transferred charge has been evaluated as in fig.(2). The mobility diverges near L = 0. This is due to the increasing leakage of the χ_{E_1} wavefunction in the InP barrier where the alloy scattering does not take place. (Note that the mobility increase would be severely limited in Al$_{0.47}$In$_{0.53}$As-Ga$_{0.47}$In$_{0.53}$As Q.W.'s by the alloy scattering occurring in the Al$_{0.47}$In$_{0.53}$As barriers). At large L and in the E.Q.L. the mobility also increases. This reflects the increasing delocalization of the χ_{E_1} wavefunction over the well whereas the alloy fluctuations occur on a fixed scale (the V.C.A. unit cell). Finally we remark that the mobility figures are smaller in Q.W. structures than in single heterojunctions n_e and V_b being kept the same. This again reflects the stronger spatial localization of $\chi_{E_1}(z)$ wavefunction in Q.W.'s for actual n_e.

It is very difficult to calculate reliable mobility figures in InP-Ga$_{1-x}$In$_x$As$_y$P$_{1-y}$ Q.W.'s. The calculations can be done along the same line as followed in eqs (1-8). However the lack of knowledge of V_b or $[\delta V]$ hampers numerical estimates. The mobility should be smaller than in InP-Ga$_{0.47}$In$_{0.53}$As Q.W.'s, all parameters being kept constant. Taking L = 150 Å, y = 0.5, x = 0.77, V_b = 0.5 eV we have obtained $\mu = 1.4 \times 10^4$ cm^2/V/s if $[\delta V]$

is taken from the energy gap difference or $\mu = 3.8 \times 10^4$ cm^2/V/s if [δV] is taken from the electron affinity difference (Littlejohn et al 1978).

4. Conclusion

We have performed selfconsistent calculations of charge transfer in InP-Ga$_{0.47}$In$_{0.53}$As modulation-doped Q.W.'s treating the band bending effects in perturbation. The calculated results are rather insensitive to the assumed conduction band discontinuity between InP and Ga$_{0.47}$In$_{0.53}$As. We have shown that the alloy scattering is an efficient scattering mechanism in InP-Ga$_{0.47}$In$_{0.53}$As Q.W.'s at low temperature and found that it affects more adversely Q.W.'s than single heterojunctions. The lack of knowledge of several parameters has precluded the obtaining of precise mobility figures in InP-Ga$_{1-x}$In$_x$As$_y$P$_{1-y}$ al hough they appear to be quite low.

Acknowledgements

The "Groupe de Physi.ue des Solides de l'Ecole Normale Supérieure" is "Laboratoire associé au CNRS". This work has been partly supported by the "Centre d'Etude des Télécommunications". One of us (J.A.B.) expresses his gratitude to the "Conselho Nacional de Desenvolvimento Cientifico e Tecnologico" (CNPq, Brazil) for financial support.

References

Ando T. and Mori S., 1979, J. Phys. Soc. Japan 47, 1518.
Ando T., Fowler A. and Stern F., 1982, Rev. Mod. Phys. 54, 437.
Bastard G., 1983, Appl. Phys. Lett. 43, 591.
Basu P.K. and Nag B.R., 1983, Appl. Phys. Lett. 43, 689.
Fishman G., 1983, Phys. Rev. B 27, 7611.
Hayes J.R., Adams A.R. and Greene P.D., 1982, GaInAsP alloy Semiconductors
    ed. T.P. Pearsall (New York, Wiley) p.189.
Littlejohn M.A., Hauser J.R., Glisson T.H., Ferry D.K. and Harrison J.W.,
    1978, Solid State Electron. 21, 107.
Pinczuk A. and Worlock J.M., 1982, Surf. Sci. 113, 69.
Stern F. and Schulman J.N., 1984, Proc. 1st Int. Conf. on Superlattices
    and Microstructures, to be published.
Störmer H.L., 1984, Surf. Sci. 142, 130.

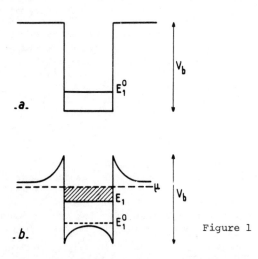

Figure 1

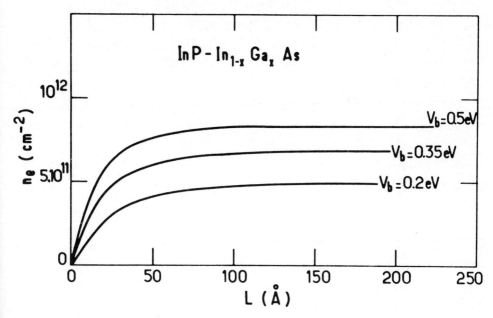

Figure 2.

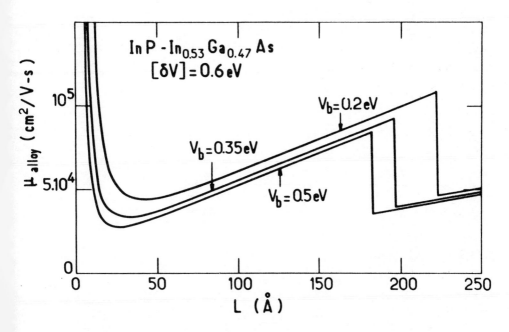

Figure 3.

Inst. Phys. Conf. Ser. No. 74: Chapter 5
Paper presented at Int. Symp. GaAs and Related Compounds, Biarritz, 1984

Energies of sub-band minima in GaAs–(AlGa)As quantum well heterostructures

P.Dawson, G.Duggan, H.I.Ralph and K.Woodbridge
        Philips Research Laboratories
    Redhill, Surrey, RH1 5HA, England

We have shown in a previous publication that a simple implementation of Kane's model for the band edges in a III-V semiconductor is able to give a better account of the energies of optical transitions than other calculations normally used to date. In this paper we show that the primary reason for the improved agreement is the inclusion of the non-parabolicity in the conduction band.

INTRODUCTION

The first attempts to calculate the sub-band energies of a quantum well assumed that the carriers behaved like fixed mass quasi-particles and moved in a square well potential. Using this calculation, Dingle et al. (1974) were able to describe the optical absorption of layers of GaAs between layers of (AlGa)As sufficiently well that it was quite clear that the layers formed quantum wells. Using the same calculation, Frijlink and Maluenda (1982) and Kawai et al. (1984) have attempted to show that the boundaries between the barrier and well regions of quantum well samples prepared by organo-metallic chemical vapour deposition did not exhibit a grading in composition by comparing calculated and measured transition energies. This and other conclusions based on the precise calculation of the optical transition energies rely on our knowledge of the errors incurred in such a calculation. We have, therefore, addressed the problem of how well various methods of calculation can be made to fit experimental data.

Among the assumptions whose effects should be investigated are that the light hole and the electron bands are non-parabolic, that the close-by electron bands at the X and L points of the Brillouin zone can be neglected and the shapes of the band minima are different in the (AlGa)As from those in the GaAs. Miller et al. (1984) have performed calculations which take account of the different band edge effective masses in the well and barrier regions. We have made a calculation which takes the non-parabolicity as well as the different effective masses into account and find that the experimental transition energies can be fitted very closely for a number of different samples.(Dawson et al. 1984a) Only a less good fit can be obtained by using either of the other approaches. In this paper we investigate the significant differences between the different calculations.

EXPERIMENTAL

The sample used for comparison in this paper is the single quantum well sample #261 reported by Dawson et al. (1984a). The GaAs well

thickness is known to be 26±1 monolayers (=73.5±3Å) from RHEED
oscillations observed during the growth. The 5K photoluminescence
excitation spectra (PLE) were recorded by the method described by Dawson
et al. (1984b). The Al concentration of the barrier region (0.26) was
derived from the position of the edge in the PLE spectrum corresponding to
absorption in the barriers.

THEORY

     The three calculations of confined particle states which we have
compared are all envelope function approximations. In all cases we
calculated the energies of the confined states in the independent particle
approximation and assumed that the exciton binding energy was given
correctly by the calculation of Greene et al. (1984). The one particle
Hamiltonian separates into motion parallel and perpendicular to the plane
of the layers. If the momentum in the plane is zero, the light and heavy
holes do not interact and can be treated separately. We assumed that the
minimum energy gaps occur for zero momentum in the plane.
     The first method, which is essentially that used by Dingle (1975),
assumes that the electrons and holes are quasi-particles of fixed mass
which move in a finite square well potential. The masses of the particles
were assumed to be those appropriate to the zone centre band edges in
GaAs. The wave functions were obtained from the Schroedinger equation
inside and outside the well and were matched by making the wave function
and its derivative continuous at the boundary.
     The second method was that used by Miller et al. (1984) and is the
same as the first except that the masses are taken to be different inside
and outside the well. To match the wave functions at the edge of the well,
the function and the ratio of the derivative of the function and the mass
have to be made continuous.
     The third method is Bastard's (1982) implementation of the Kane
model. The band energies are calculated by the k.p method. This is a
simple method of taking some account of the non-parabolicity of the
electron and light hole bands. Only the atomic s and p states near the
band gap are included and the strengths of the k.p matrix elements are
derived from the electron and light hole band edge masses in the bulk
material. The heavy hole band becomes decoupled from all the others in
this approximation and is treated separately by the second method. The
spin-orbit split off band can be included in this approach but we found
that its inclusion in this case made very little difference and it has
not, therefore, been included in deriving any of the results presented
here. The main difficulty with the simple Kane model is that only one
matrix element is retained which has to take different values for the
electron and light hole bands.
     The electron mass in the GaAs has been taken as $0.0665m_0$ throughout.
This value has been obtained by Chamberlain et al. (1972) from cyclotron
resonance and by Stillman et al. (1971) from the infra-red spectra of
shallow donors in GaAs. There is a certain amount of discussion in the
literature about which value to use for the hole masses in GaAs,
especially the heavy hole.(Dingle 1975, Kleinman 1983, Miller et al.
1984) The GaAs hole masses have been measured rather thoroughly using
cyclotron resonance by Mears and Stradling (1970) and by Skolnick et al.
(1976) and the most recent of these quotes A=6.98, B=4.5 and C=6.1 in the
expression

$$E = -Ak^2 \pm \sqrt{[B^2k^4+C^2(k_x^2k_y^2+k_y^2k_z^2+k_z^2k_x^2)]}$$

where the values of A, B and C are given in units of $h^2/2m_o$. Following
Dingle et al. (1974), we have assumed that the mass in the direction
perpendicular to the plane is the correct one to use. In our samples this
is the <100> direction where the heavy hole mass is $m_o/(A-B)$ and the light
hole mass is $m_o/(A+B)$ giving $0.40m_o$ and $0.087m_o$ respectively. These
values, which we prefer, are a little different from those recently proposed
by Kleinman (1983) and by Miller et al. (1984) but we have tested the effect
of using different values for the hole masses. The effective masses in
(AlGa)As were obtained by linearly interpolating between the masses in the
two binary limits. The effective masses in AlAs are not very well known.
We have normally used $0.15m_o$ for the electron mass and $0.75m_o$ and $0.16m_o$ for
the heavy and light hole masses respectively but the effect of changing
these values is considered below.

    We have tried to display the results as clearly as possible by
plotting contours such as those in Fig.1. We have calculated the allowed
electron-heavy hole and electron-light hole transitions as a function of
the well width, L_z, and the fraction of the band gap difference between
the well and barrier regions that is assigned to the conduction band, Q_e.
The pairs of values of L_z and Q_e which give rise to a particular
transition energy define a contour in the L_z-Q_e plane. From the
excitation spectrum, we have determined, for each transition, a range
within which the transition energy must lie. The pairs of contours
plotted in the figure correspond to the lower and upper limits of these
ranges.

RESULTS AND DISCUSSION

    The transition energies determined for sample #261 are reproduced in
table I. Fig 1. shows the contours obtained from the simple Kane model
calculation using the GaAs hole masses in the <100> direction derived from
the cyclotron resonance experiments (above). There is a small area which
lies within all the bounds implying that the experimental transitions can
be fitted within the assumed experimental error using these masses and
appropriate values of well width, L_z, and band offset parameter, Q_e. The
values derived are $71.5\pm.5$Å for L_z and $0.75\pm.03$ for Q_e. The range of L_z
lies within the range obtained from the RHEED oscillation measurements.
The range obtained for Q_e is about the same for all the samples
investigated.(Dawson et al. 1984a)

    Miller et al. (1984) have suggested that values of $0.34m_o$ and $0.094m_o$
for the heavy and light holes respectively should be used. Putting these
values into our calculation also enables the experimental energies to be
fitted, this time giving $L_z = 71\pm1$Å and $Q_e = 0.66\pm.04$.

    In Figs.2 and 3 we have plotted the contours obtained from
calculating the transition energies using the first and second methods
respectively. The <100> hole masses were used for GaAs. Both the methods
can be seen to give a less good fit to the experiments whatever the values
chosen for L_z and Q_e.

    The transition energies are likely to be more sensitive to the
electron mass than to the hole mass in the barrier region because the
electron energy gives the largest contribution to the total confinement
energy. There are no measurements of the electron mass in AlAs but
calculations of Hess et al. (1973) give $0.15m_o$ and those of Lawaetz (1971)
give $0.22m_o$. The AlAs mass used in fig. 1 is $0.15m_o$. The result of the
same calculation using $0.22m_o$ for the electron mass is shown in Fig.4.

The fit is even better than in Fig.1 but a slightly wider well ($73\pm1\overset{\circ}{A}$) has to be assumed. Other than this improved agreement, however, we have no reason to prefer the higher electron mass. Other workers, who have derived values of Q_e different from ours, have normally used the spectra obtained from quantum wells wider than those used by us. We have found that for wider wells the transition energies depend only weakly on the value of Q_e. We found that only if narrow wells are used ($<100\overset{\circ}{A}$) is it possible to obtain enough dependence of the transition energies on Q_e to fix its value with any precision.

CONCLUSIONS

We have already shown that the energies of the transitions in GaAs-(AlGa)As quantum wells can be predicted using Bastard's implementation of the Kane model and the well widths obtained from the fit always lie inside the range obtained from RHEED oscillation measurements.(Dawson et al. 1984a) Within the Kane model, the correct heavy hole effective mass to use for GaAs is $0.40m_o$ but changing this effective mass a little does not alter the ability of the calculation to fit the results. The more simple calculations using parabolic bands can only be made to give less good fits. The primary difference between the two types of calculation is the inclusion of the effect of non-parabolicity in the Kane model and we conclude that the non-parabolicity of the conduction band is the most important omission from the simpler calculations.

REFERENCES

Bastard G 1982 Phys. Rev. B 12, 7584
Chamberlain J M, Simmonds P E, Stradling R A and Bradley G G 1972 Proc. 11th Int. Conf. on Physics of Semiconductors, Warsaw, ed M.Miasek, PWN-Polish Scientific, Warsaw p.1016
Dawson P, Duggan G, Ralph H I, Woodbridge K and 't Hooft G W 1984a International Conference on Superlattices, Microstructures and Microdevices, Champaign, Illinois
Dawson P, Duggan G, Ralph H I and Woodbridge K 1984b Phys. Rev. B 28, 7381
Dingle R, 1975 Festkorperprobleme 15, 21
Dingle R, Wiegmann W and Henry C H 1974 Phys. Rev. Letts. 33, 827
Frijlink P M and Maluenda J 1982 J. de Physique C5, 185
Greene R L, Bajaj K K and Phelps D E, 1984 Phys. Rev. B 29, 1807
Hess E, Neumann H and Topol I 1973 Phys. Status Solidi (b) 55, 187
Kawai H, Kaneko K and Watanabe N 1984 J. Appl. Phys. 56, 463
Kleinman D A 1983 Phys. Rev. B 28, 871
Lawaetz P 1971 Phys. Rev. B 4, 3460
Mears A L and Stradling R A 1970 J. Phys. C, 3, L21
Miller R C, Kleinman D A and Gossard A C 1984 Phys. Rev. B 29, 7085
Skolnick M S, Jain A K, Stradling R A, Leotin J, Ousset J C and Askenasy S 1976 J. Phys. C, 9, 2809
Stillman G E, Larsen D M, Wolfe C M and Brant R C 1971 Solid State Commun, 9, 2245

Table I

transition assignment	transition energy (eV)
n=1 electron-heavy hole	1.5807 ± 0.001
n=1 electron-light hole	1.5972 ± 0.001
n=2 electron-heavy hole	1.7329 ± 0.001

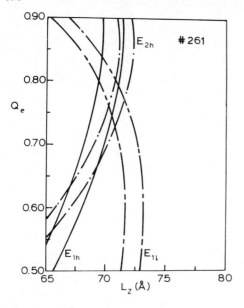

Fig.1 Transition energy contours
for the simple Kane model.

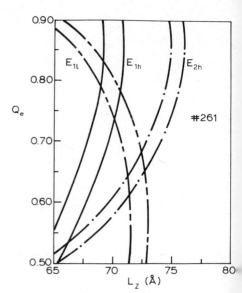

Fig.3 Contours for the square
well model accounting for the
discontinuity in mass at the
interfaces.

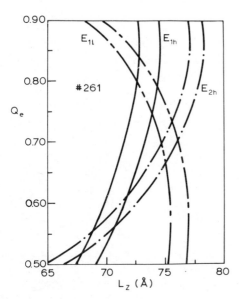

Fig.2 Contours for the finite
square well model.

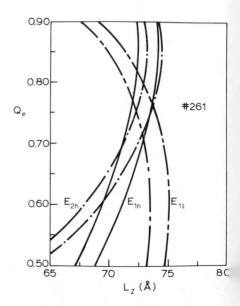

Fig.4 Contours for the simple
Kane model but with 0.22m for
the electron mass in the barrier

Inst. Phys. Conf. Ser. No. 74: Chapter 6
Paper presented at Int. Symp. GaAs and Related Compounds, Biarritz, 1984

397

III−V compound semiconductor optical detectors

G.E. Stillman, N. Tabatabaie, V.M. Robbins, and J.A. Aguilar
Electrical Engineering Research Laboratory,
Coordinated Science Laboratory, and Materials Research Laboratory
University of Illinois at Urbana-Champaign, Urbana, Illinois 61801 USA

1. Introduction

Detectors for the 1.3 μm–1.6 μm spectral range are very important for application in wide-bandwidth, long-distance fiber optical communications systems, and III-V compound semiconductor alloys are of particular interest because of the ability to adjust the bandgap to obtain high quantum efficiency in this range. Quaternary III-V compounds permit independent variation of the lattice constant and the bandgap over a significant range so that heterostructures are possible with these lattice matched alloy systems. The most extensively studied quaternary alloy system is InGaAsP/InP which permits the 0.91-1.72 μm wavelength range to be covered while the lattice constant is maintained at that of InP. The growth and properties of this alloy system have recently been reviewed by Stillman et al. (1983). High quality InGaAs/InP structures can now be prepared routinely by several different growth techniques, but it is clear that this important material system has several fundamental limitations which influence the ultimate performance of the devices fabricated from these semiconductor alloys. In this paper we review limitations imposed by the properties of the material and by the structures that are utilized to minimize the degrading effects of these properties. Within the limitations imposed by the material properties, we calculate the required gain and allowable dark current necessary to obtain the best performance possible with InGaAsP/InP avalanche photodiodes, and we estimate the improvement in performance that can be obtained by using avalanche photodiodes rather than PIN diodes without internal gain in fiber optical transmission systems.

2. Material Limitations

It is well known that it is important to minimize the leakage current in avalanche photodiodes because of both the saturation of the available gain and the excess noise of the multiplication process. In the narrow bandgap lattice-matched alloy systems the band-to-band or defect-assisted tunneling current becomes important for electric fields greater than about 2×10^5 V/cm, much too low to obtain significant avalanche gain at room temperature. A structure first proposed by Susa et al. (1980) that is widely used to minimize the tunneling dark current is shown in Fig. 1. In this structure, the optical absorption occurs in InGaAs where the electric field is kept below the value for tunneling to occur ($\sim 2 \times 10^5$ V/cm) while the multiplication occurs in the high field region in the InP. In this separate absorption and multiplication (SAM) avalanche photodiode structure the quantum efficiency is determined by the properties of the InGaAs and by the interface and valence band discontinuity between the

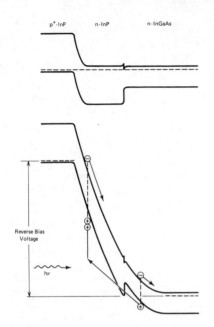

Fig. 1 SAM avalanche photodiode structure.

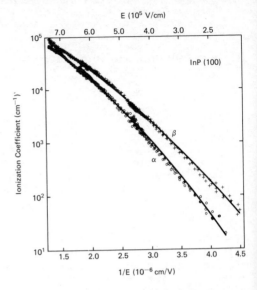

Fig. 2 Impact ionization coefficients for (100) InP.

InGaAs and InP, while the excess noise properties are determined by the avalanche multiplication process in the InP. The excess noise of the multiplication process is governed by the ratio of the electron and hole impact ionizaton coefficients, α and β respectively, and recently, reliable values for these parameters in (100) InP have been reported (Stillman et al, 1982). The variation of α and β with electric field is shown in Fig. 2. These data were obtained from photocurrent multiplication measurements on devices over a doping range that permitted reliable measurements to be made over the electric field range from less than 2.5×10^5 V/cm to greater than 7×10^5 V/cm. The solid lines in the figure are least square fits to the experimental data given by

$$\alpha = 3.454 \times 10^5 \exp\left\{-(1.041 \times 10^6/E)^{1.54}\right\} \tag{1}$$

and

$$\beta = 3.795 \times 10^5 \exp\left\{-(1.007 \times 10^6/E)^{1.46}\right\}. \tag{2}$$

Since the excess noise due to the multiplication process is smallest when the values of α and β are significantly different from each other it is rather disappointing that the ratio of α and β is close to unity over the entire electric field range where reasonable values of avalanche gain can be obtained.

To evaluate the performance potential of InGaAsP/InP avalanche photodiodes, we have calculated the minimum incident optical power (per unit bandwidth) necessary to produce a given signal to noise power ratio for several different InGaAsP/InP SAM avalanche photodiode configurations.

The minimum average 100% modulated optical power per unit bandwidth is given by

$$\frac{P_o}{B} = \frac{2h\nu F_p}{\eta} \left(\frac{S}{N}\right) \left\{ 1 + \left[1 + \frac{I_{eq}}{qBF_p^2(S/N)} \right]^{1/2} \right\},$$ (3)

where F_p is the excess noise factor for the photogenerated current,

$$I_{eq} = I_B R_B + I_{DB} F_D + \frac{I_{DS}}{M^2} + \frac{2kT_{eff}}{qR_{eq}M^2},$$ (4)

and the other terms are defined by Stillman and Wolfe (1977). The excess noise factor used for these calculations is given by McIntyre et al. (1972) for the general case in which $k = \alpha(x)/\beta(x)$ is a function of electric field or position in the depletion region. For pure hole injection, the avalanche gain is

$$M = \frac{\exp[-\int_o^w (\alpha-\beta)dx]}{1 - \int_o^w \alpha \, \exp[-\int_o^x (\alpha-\beta)dx']dx} .$$ (5)

The excess noise factor is given by

$$F_p(M) = k'_{eff} M + (2 - \frac{1}{M})(1 - k'_{eff})$$ (6)

where

$$k'_{eff} = \frac{k_{eff}}{k_1^2}, \quad k_{eff} = \frac{(k_2-k_1^2)}{1-k_2},$$ (7)

$$k_1 = \int_o^w \beta(x)M(x)dx / \int_o^w \alpha(x)M(x)dx, \text{ and}$$ (8)

$$k_2 = \int_o^w \beta(x)M^2(x)dx / \int_o^w \alpha(x)M^2(x)dx.$$

To calculate the minimum value of P_o/B for a given S/N ratio, four specific SAM device configurations were considered. Since the avalanche gain and noise properties of the SAM avalanche photodiode structure are determined primarily by the electric field profile in the InP depletion region, the four configurations differed in the width and doping concentration of the n-type InP layer shown in Fig. 1. For each configuration the doping concentration was adjusted for the selected width so that avalanche breakdown occurs when the electric field at the InP/InGaAs interface is about 1×10^5 V/cm, well below the electric field where the tunneling current in the small bandgap InGaAs becomes significant. The numerical integrations were performed for the electric field profiles determined by these assumed structures. For the calculations presented here a value of $h\nu/\eta = 2.8$ was assumed which corresponds to a quantum efficiency of 0.3 at a wavelength of 1.55 μm. The scale for P_o/B can be adjusted as needed for other values of $h\nu$ and η. For each value of the S/N ratio there is a value of multiplication or gain, M_{opt}, which minimizes the value of P_o/B, and the variation of M_{opt} with S/N for the four different configurations and three different capacitance values is shown in Fig. 3. For the device configurations given, the different values of capacitance translate into different device active areas. The resistance R_{eq} in Eq. (4) is directly related to the capacitance and the bandwidth, and here we have simply assumed that $R_{eq} = 1/(2\pi CB)$. The minimum detectable power can probably be slightly reduced from that calculated here by using an integrating preamplifier with a larger value of R_{eq} and with equalization. The variation of the excess noise factor calculated for these configurations is given in Fig. 4 and shows that the configuration with the widest active region in the InP and the lowest doping level has the smallest excess noise factor, but there is little difference between the two configurations with the widest n-type InP layers. The values of M_{opt} for a given value of S/N as shown in Fig. 3, are larger for the higher capacitance values since the thermal noise of R_{eq} for a given bandwidth is larger when the capacitance is higher.

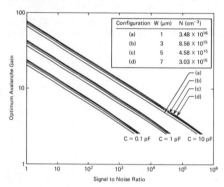

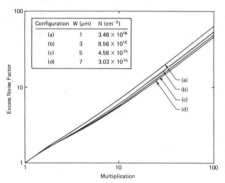

Fig. 3 Optimum avalanche gain vs. signal–noise ratio.

Fig. 4 Excess noise factor vs. multiplication for SAM APDs.

Using M=1 and M=M_{opt}, the variation of P_o/B with S/N for a PIN detector or for a SAM avalanche photodiode, respectively, can be calculated and the results of such a calculation for configuration (c) are shown in Fig. 5. At low values of S/N, over a 17 dB improvement in P_o/B is obtained by using an avalanche photodiode, while at high S/N ratios there is little difference between the PIN and APD detector performance. For Gaussian noise, an S/N ratio of about 144 is required to achieve a bit error rate of less than 10^{-9} (Stillman, 1980). At this signal to noise ratio there is still more than a 10 dB improvement in P_o/B when an APD is utilized.

The calculations shown in Fig. 5 assumed that I_{eq} in Eq. (4) was due entirely to the Johnson noise current on R_{eq} at T_{eff} = 300 K. Assuming that the excess noise factor for multiplied dark current I_{DB} is the same as that for the photocurrent it is easy to calculate the value of I_{DB} and I_{DS} per unit bandwidth that will cause a 3 dB increase in P_o/B, and the results of this calculation are shown in Fig. 6. Thus, for a 0.1pF capacitance PIN detector with a 1 GHz bandwidth and a S/N of 144, the dark current can be as high as 0.1 mA and only cause a 3 dB increase in P_o, while for the avalanche photodiode, a bulk dark current of 1 μA will cause a similar degradation of P_o. With the SAM structure it is possible to obtain dark currents significantly lower than these values so that leakage currents do not limit the performance of avalanche photodiodes for the 1.3–1.5 μm wavelength range.

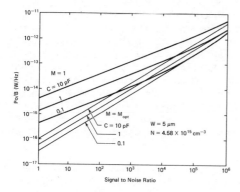

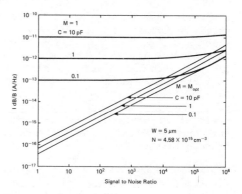

Fig. 5 Minimum average optical power per unit bandwidth for given S/N ratios.

Fig. 6. Multiplied dark current which causes a 3 dB increase in P_0 at a given bandwidth.

3. SAM Structure Limitations

The SAM structure effectively eliminates the tunneling dark current in the small bandgap alloys, and even though the ratio α/β is not much different from unity in InP, these devices still permit a significant improvement in performance over PIN diodes with no internal gain. However, this struc-ture introduces another problem which can strongly influence the high fre-quency performance of these devices—they have a slow transient response to long-wavelength optical pulses (Kim et al., 1981). This slow response is due to the trapping of holes at the valence band discontinuity evident at the InGaAs/InP interface in Fig. 1 (Forrest et al., 1982) and it was suggested that this slow response could be eliminated by intentional com-positional grading at the interface or by growing additional fixed composition intermediate bandgap layers between the InP multiplication regions and the narrow-bandgap absorption region. By applying these tech-niques, Campbell et al. (1983) have demonstrated minimum average powers for 10^{-9} bit error rates at 1.55 μm of −41.5 dBm at 420 M bit/s and −37.5 dBm at 1G bit/s and at 1.3 μm, minimum average powers of −38 dBm at 420 M bit/s and −38 dBm at 1 G bit/s.

The nearly equal values of electron and hole ionization coefficients in InP and most other compound semiconductor materials suitable for long wavelength detectors have led to attempts to devise device structures that will somehow artificially enhance the ionization coefficient ratio. One such structure designed to enhance the ratio of α to β was proposed by Chin et al. (1980). The avalanche region consists of a superlattice of alternating layers of wide and narrow bandgap materials. In addition, the majority of the bandgap discontinuity occurs in the conduction band. In this structure an electron is accelerated by the electric field through the wide bandgap materal until it reaches the small bandgap material, where it effectively gains on additional energy of ΔE_c, the band edge discontinuity in the conduction band. The threshold for impact ionization of electrons in the small bandgap material is therefore effectively lower-ed by ΔE_c. Holes experience the same effect, but because the impact ionization rate depends exponentially on the impact ionization threshold, the electron impact ionization rate is enhanced much more than that of holes and therefore α/β increases.

A superlattice diode similar to the one proposed by Chin et al. was fabricated by Capasso et al. (1982). Measurements of the ionization coefficients using both hole and electron injection into the high$_s$ field superlattice region indicated that for an electric field of 2.5×10^5 V/cm, $\alpha/\beta = 8$ with an electron initiated multiplication of 10. This does indicate an enhancement of the α/β ratio from the value of 2.2 in bulk GaAs at the same electric field. Unfortunately, there have been no published results of noise measurements on these devices so it is not clear if in fact a low excess noise factor is obtainable in this sort of structure. Clearly, further work is needed in investigating the behavior of superlattice devices.

There have been other device structures proposed to enhance the α/β ratio (Capasso, 1983). Also, the graded-gap APD (Capasso, 1982b) and the channeling photodiode (Capasso, 1982c and 1982d) are structures designed to enhance the α/β ratio by tailoring the bandgap. These various devices have been recently reviewed by Capasso (1984).

In conclusion, avalanche photodiodes fabricated in the InP/InGaAsP material system have demonstrated improved performance over PIN diodes for wide-bandwidth, long-wavelength fiber optical applications. Unfortunately, the useable gain of the APDs is limited by the small ratio of the ionization coefficients. This fundamental material limitation may possibly be overcome by innovative device structures.

Acknowledgments.   This work was supported by the National Science Foundation under Contracts NSF ECS 92-09090, and NSF DMR 83-16981 and by the Joint Services Electronics Program, U.S. Air Force, U.S. Army, and U.S. Navy under Contract N00014-76-C-0806. The authors thank G. Morris and B. Payne for their assistance in the preparation of this manuscript.

References

Campbell J C, Dentai A G, Holder W.S, Kasper B L 1983 Electronics Letters Vol. 19, pp. 818-820
Capasso F, Tsang W T, Hutchinson A L, and Williams G F 1982a Appl. Phys. Lett. 40, 38-40
Capasso F, Tsang W T, Hutchinson A L, and Foy P W 1982b Proc. 1981 Symp. on GaAs and Related Compounds Inst. Phys. Conf. Ser 63 pp 473-478
Capasso F 1982c Electron Lett. 18 12-13
Capasso F 1982d IEEE Trans. Electron Devices ED-29 1388-1395
Capasso F, Tsang W T, and Williams G F 1983 IEEE Trans. on Electron Devices ED-30 381-390
Chin R, Holonyak N, Stillman G E, Tang Y Y, and Hess K 1980 Electron Lett. 17, 467-469
Forrest S R, Kim O K, and Smith R C 1982 Appl. Phys. Lett. 41, 95-98
Kim O K, Forrest S R, Bonner W A, and Smith R G 1981 Appl. Phys. Lett. 39, 492
McIntyre R J 1972 IEEE Trans. Electron Devices ED-19, 703
Stillman G E and Wolfe C M 1977 Semiconductors and Semimetals, vol. 12 (Academic Press, New York) pp. 291-393
Stillman G E 1980 Proc. SPIE, vol. 239 pp. 52-52
Stillman G E, Cook L W, Tabatabaie N, Bulman G E, and Robbins V M 1983 IEEE Trans. on Electron Devices ED-30 364-381
Susa N, Nakagome H, Mikami O, Ando H, and Kanbe H 1980 IEEE J. Quantum Electron. QE-16 864-870

Inst. Phys. Conf. Ser. No. 74: Chapter 6
Paper presented at Int. Symp. GaAs and Related Compounds, Biarritz, 1984

403

Ga₁₋ₓAlₓAs–GaAs heterojunction bipolar phototransistor grown by MBE for monolithic photoreceivers

J.L. Liévin, H. Wang, C. Dubon, F. Alexandre, B. Sermage and D. Ankri

Centre National d'Etudes des Télécommunications

Laboratoire de Bagneux

196 Rue de Paris - 92220 BAGNEUX - FRANCE

Abstract : We have optimized the MBE growth conditions of (Ga,Al)Al/GaAs heterojunction bipolar phototransistors (HPT's). Devices with current-gains higher than 3000 and low input capacitances have been obtained using a fabrication process suitable for IC's. A photoreceiver has been designed with predicted sensitivity of -39.5 dBm at 140 Mbits/s.

I - INTRODUCTION

The increasing demand for low-noise photodetector amplifiers, due to the advent of optical fiber telecommunications, has induced a large amount of work in the field of the new III-V photodetectors. The non avalanching nature of the heterojunction bipolar phototransistor (HPT) as well as its simple structure for integration are quite attractive for local networks in optical transmission links. On the other hand, the comparison between APD's PIN-FET's and HPT's (Tabatabaie- Alavi 1981, Brain 1983) has demonstrated that HPT's can have high sensitivity at high bit-rates (>1 GBits/s). The use of diffused epitaxial structures allows the reduction of capacitances and low-noise characteristics, so that impressive performances have already been demonstrated on LPE grown HPT's (Scavennec 1983) even at moderate bit-rate (~140 Mbits/s).

The purpose of this work is to optimize the MBE growth conditions for the (Ga,Al)As-GaAs system in order to obtain high current gain low input capacitance HPT's. Molecular Beam Epitaxy (MBE) is a suitable technique for IC's, since it permits a good uniformity on large area wafers. These performances can be used to build a monolithic photoreceiver for $0,85\mu m$ multimode links operating at 140 Mbits/s with a sensitivity close to -40dBm.

II - DESIGN OF THE EPITAXIAL STRUCTURE

The basic advantage of HPT's resides in the possibility to obtain high current-gains h_{FE} independently of the emitter and base doping levels because of the wide-gap emitter. Following the analysis of Brain and Smith (1983), the figure of merit to be improved for moderate bit-rate transmissions is : $h_{FE}^{1/2}$ / C_E (C_E = emitter capacitance).

Therefore, a typical HPT epitaxial structure has to meet the following requirements :

a) low n-type doped emitter in order to reduce the input capacitance of the photodetector.

b) a high current-gain must be obtained at the same time and requires an injection efficiency and a base transfer factor close to unity. Thus, we must grow a graded emitter-base heterointerface to avoid the "spike" limitations on the injected electron current. Recombination currents at low levels must be reduced at the heterointerface and in the E-B space charge region. High base transfer factor necessitates the reduction of the recombination current in the bulk (GaAs : p-type). Finally, a compromise between base doping level and base thickness must be found in order to ensure a minority carrier diffusion length L_{eB} much greater than the base width W_B.
c) the need for high-quantum efficiency requires a large base-collector depletion region, where the electron-hole pairs created by the incident light are collected. Therefore, a thick lightly doped collector is grown ($W_C \simeq 2$ μm).

Previous MBE results did not compare favorably with those obtained on LPE grown structures which demonstrate current-gains up to 20 000. Therefore, electron life-time measurements in p-type GaAs have been carried out to characterize the quality of the HPT's base layers.

III - LUMINESCENCE TIME-DECAY MEASUREMENTS

III.1 - Experimental procedure

The measurements were performed on double heterostructures (DHS's) which consisted of the following layers : an undoped 5000Å-thick GaAs buffer layer, a Be-doped GaAs "active layer" (p=8.10^{17} at. cm^{-3}) of variable thickness d, sandwiched between two intrinsic 1000 Å-thick $Ga_{0.6}Al_{0.4}As$ layers. The active layer thickness d ranged from 1μm to 600 Å. Similar growth conditions in the RIBER 2300 growth chamber were used for HPT's and DHS's.

The electron life-time in the p-type GaAs was measured at room temperature at low excitation conditions. The $Ga_{0.6}Al_{0.4}As$ cap layer was excited by a 0.53μm radiation of a 1,06μm modelocked YAG laser beam doubled in a KTP crystal. Pulse width at half height was 70ps and pulse selection by an acousto-optic crystal was employed. The luminescence of the sample is detected on a fast Ge-APD. The electrical signal is sent to a sampling oscilloscope, a multichannel analyser, and is finally plotted on a X-Y recorder. The luminescence of the (Ga,Al)As is eliminated by a low pass filter. A presentation of the experiment is given by Sermage (1984).

III.2 - Results

The results are summarized in fig. 1. Assuming the relevance of the relation : $1/\tau = 1/\tau_b + (S_1+S_2)/d$, where τ_b is the bulk lifetime of the active layer S_1 and S_2 the recombination velocities at the heterointerfaces (Ettenberg 1976), we can deduce τ_b and $(S_1 + S_2)$ with varying d values.

MBE is known to produce (Ga,Al)As on top of GaAs heterointerface (S_2) of better quality than the GaAs on top of (Ga,Al)As heterointerface (S_1^2) (Morkoç 1982). As a consequence, we can expect $S_1 > S_2$. From fig. 2, we obtain $(S_1 + S_2) \simeq 6.10^3 \text{cm.s}^{-1}$. One 1000Å-thick sample has been grown with a thin GaAs layer below the active layer in order to improve the quality of the "bad" interface. It results in a drastic reduction of $(S_1 + S_2)$ which confirms that the recombination was mainly due to the GaAs on top of (Ga,Al)As heterointerface ($S_1 \gg S_2$). From these measurements, we obtain $\tau_b \simeq 3.5\text{ns}$ in the p-type GaAs : $p = 8.10^{17} \text{ at.cm}^{-3}$. By means of an electron mobility estimation at this doping level (Walukiewicz 1979) we find $L_{eB} \approx 4.5\mu\text{m}$, which is slightly lower than the reported LPE diffusion lengths. This value will be compared to the HPT's DC characteristics in the next part.

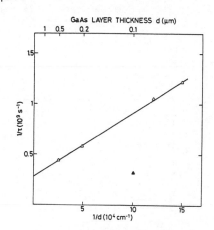

Fig. 1 : $1/\tau$ versus $1/d$ plot for p-type GaAs : $p = 8.10^{17} \text{cm}^{-3}$

IV - HPT's GROWTH

MBE is reputed to easily grow abrupt heterostructures as well as abrupt doping profiles. On the contrary, the growth of gradual heterointerfaces necessitates computer facilities to monitor the cell temperatures. This is achieved by coupling the RIBER 2300 growth chamber with a HP 9816 computer which controls the cell temperatures, the cell shutters operations and can be extended to arsenic overpressure monitoring or pyrometric substrate temperature corrections.

Different HPT's structures were grown and tested. The base doping level was $p = 8.10^{17} \text{ at.cm}^{-3}$. The substrate temperature was kept at 600°C during the growth, with a flux ratio : $J_{As} / J_{Ga} = 6$. These conditions correspond to a good morphology of the Si-doped $Ga_{1-x}Al_xAs$ emitter (Alexandre 1984).

The Al-gradient included at the E-B heterointerfaces starts from 7 % up to 26 %. The transition shape (deduced from the cell temperature measurements during the growth) can be approximated to a parabolic transition for simulation. A comparison between this graded heterointerface and an abrupt transition is given in fig. 2 for two bias voltages. It demonstrates that the spike has been considerably reduced and smoothed, leading to a higher injection efficiency.

Using these growth conditions, two high current-gain HPT's were obtained : $h_{FE} = 1000$ ($W_B = 1500$ Å) and $h_{FE} = 3000$ ($W_B = 500$ Å), which is the highest reported value to date for MBE grown structures (Fig. 3). The emitter doping level was $N_E = 4.10^{17} \text{ at.cm}^{-3}$ for both structures. These values are close to the LPE results, but the MBE grown HPT's are working at higher collector currents (Fig. 4). This suggests that the recombination currents have still to be reduced to make full use of the MBE advantages.

Using the relation : $h_{FE\ max} = 2\ L_{eB}^2\ /\ W_B^2$ which assumes an injection efficiency equal to one, we can compare the "apparent" diffusion lengths of the tested HPT's to the value obtained by the luminescence measurements of III-2. Since the latter is higher, we shall conclude that the injection efficiency is less than unity.

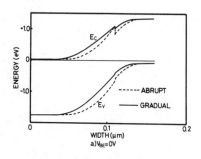 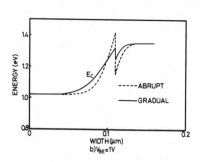

Fig. 2 : Energy-band diagram of the Emitter-Base heterostructure for abrupt and gradual cases : a) at V_{BE} = 0V., b) at V_{BE} = 1V

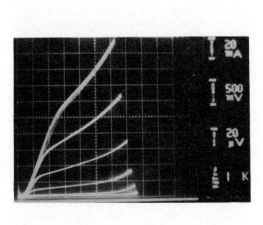 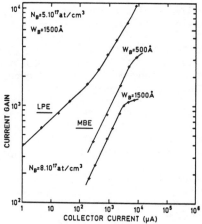

Fig. 3 : Common emitter characteristics of high-gain MBE
grown HPT.

Fig. 4 : Current-gain versus
collector current for LPE
and MBE grown HPT's.

V - FABRICATION TECHNOLOGY OF THE PHOTOTRANSISTOR AND PREAMPLIFIER

The device processing has been reported elsewhere for phototransistors and high-speed transistors (Ankri 1982). Localized Zn diffusion process to convert the sensitive area of the phototransistor into p-type is used. The emitter base capacitance is then reduced and the critical emitter layer etching step for the two preamplifier HBT's is avoided. The emitter-base area of the transistors is reduced to 5.6 10^{-7} cm^2 and with a four micron design rule the input capacitances are in the range of 0.15 pF at 0V. The process includes the fabrication of resistors for the preamplifier using ion implantation and metallic deposition.

The electrical circuit and the layout of the photoreceiver are shown in Fig. 5. Taking into account the previous device parameters, the simulation results on the monolithic transimpedance photoreceiver have demonstrated that a bandwidth of 300 MHz is compatible with a sensitivity of - 39.5 dBm at 140 Mbits/s for 10^{-9} BER. The optimized value of the feedback resistor R_F can be set from 16 KΩ to 30 KΩ (see Fig. 6 (a) and (b)).

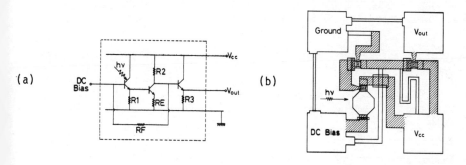

Fig. 5 : Electrical circuit (a) and layout (b) of the photoreceiver using a transimpedance common collector configuration with an HPT in front end.

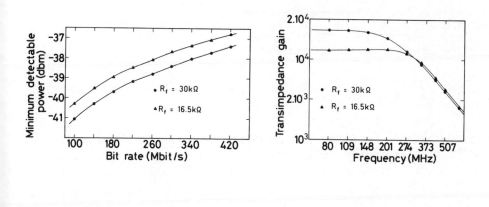

(a)   (b)

Fig. 6 : a) Minimum detectable power of the integrated photoreceiver
         b) Magnitude of the transimpedance gain versus frequency.

In conclusion, we have presented HPT's growth condition improvements using MBE technique. High-current gain and low input capacitance HPT's were obtained, which are suitable for IC's photoreceivers. These preliminary results indicate that heterojunction bipolar technology is promising for monolithic photoreceivers operating at 140 Mbits/s with a sensitivity of - 39.5 Mbits/s.

VI - <u>AKNOWLEDGEMENTS</u>

The authors gratefully aknowledge Dr. A. Mircea for encouragements and discussions during the course of this work, and Dr. Dangla for heterostructure simulations.

F. ALEXANDRE, L. GOLDSTEIN, G. LEROUX, M.C. JONCOUR, H. THIBIERGE and EVK RAO, Proc. third Int. Conf. on MBE (San Francisco) 1984.

D. ANKRI, A. SCAVENNEC, C. BESOMBES, C. COURBET, F. HELIOT, J. RIOU, Appl. Phys. Lett., <u>40</u>, 816, (1982).

M.C. BRAIN and D.R. SMITH, IEEE Trans. Electron Devices, 30 (4), 390, April 1983.

M. ETTENBERG and H. KRESSEL, J. Appl. Phys. <u>47</u>, 1538, 1976.

H. MOKOC, T.J. DRUMMOND and R. FISHER, J. Appl. Phys. 53(2), Feb. 1982, 1030.

A. SCAVENNEC, D. ANKRI, C. BESOMBES, C. COURBET, J. RIOU, F. HELIOT, Electron. Lett., 19(10) 394, May 1983.

B. SERMAGE, F. ALEXANDRE, J.L. LIEVIN, R. AZOULAY, M. EL KAIM, H. LE PERSON and J.Y. MARZIN. Proc. 11th. Int. Symp. on GaAs and Related Compounds (Biarritz) 1984. Inst.Phys.Conf.Ser. <u>74</u> 345.

K. TABATABAIE - ALAVI and C.G. FONSTAD, JR., IEEE J. Quantum Electron., vol. QE-17, n° 12, 2259-2261, December 1981.

W. WALUKIEWICZ, J. LAGOWSKI, L. JASTRZEBSKI and H.C. GATOS, J. Appl. Phys. 50(7), July 1979, 5040.

Inst. Phys. Conf. Ser. No. 74: Chapter 6
Paper presented at Int. Symp. GaAs and Related Compounds, Biarritz, 1984

409

Time-response of InGaAs/InP avalanche photodiodes

Reiner Trommer and Ludwig Hoffmann

Siemens Research Laboratories, Otto-Hahn-Ring 6,
D-8000 München 83, Fed. Rep. of Germany

Abstract. The time-response of InGaAs/InP avalanche photodiodes with separated regions of absorption (InGaAs) and multiplication (InP) and a thin intermediate InGaAsP layer was investigated. At low reverse bias voltages the bandwidth is limited by accumulation of holes at the heterointerfaces, close to breakdown a gain·bandwidth product is observed. A bandwidth up to 3 GHz and a gain·bandwidth product of 30 GHz were achieved. These properties make the devices well suited for use in high speed optical receivers.

1. Introduction

InGaAs/InP SAM avalanche photodiodes with separated regions of absorption and multiplication have recently proven to be most promising detectors in modern optical communication systems which operate at light wavelengths between 1 and 1.6 μm. Their principal layer sequence consists of an InP p^+n junction followed by a n-InGaAs absorption layer with a cutoff wavelength of 1.65 μm (Kanbe et al 1980). Photoresponse sets in at a reverse bias voltage at which the space charge layer starts to extend into the InGaAs. A high electric field in the n-InP leads to avalanche multiplication whereas a moderate electric field in InGaAs yields a low dark current and serves only to separate the photon-generated electron-hole pairs. In addition only holes, which possess higher ionization probability in InP than electrons, are injected into the multiplication region and the excess noise associated with the avalanche gain process becomes low. One problem of such semiconductor layer structures is the discontinuity in the valence bands present at the heterojunction. A potential well is formed there in which holes may be accumulated which then leads to a slow photodiode response (Forrest et al 1983). However, the heterointerface can be graded or the valence band discontinuity may stepwise be reduced by insertion of a thin InGaAsP layer.

In this paper we report experimental studies of the time-response of such optimized SAM-type InGaAs/InP avalanche photodiodes. Devices with different layer thicknesses and doping profiles have been investigated. Both time- and frequency-domain measurements have been performed as a function of reverse bias voltage, incident light power or temperature. The mechanisms which lead to bandwidth limitations are discussed in connection with the application of the diodes in optical receivers.

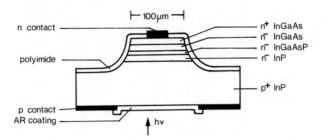

Fig. 1   Schematical cross section of the avalanche photodiodes

2. Avalanche Photodiode Structure

A schematical cross section of our devices is shown in Fig. 1. On the basis of a p^+-InP substrate a layer sequence of n^--InP, n^--InGaAsP, n^--InGaAs and n^+-InGaAs was deposited by liquid phase epitaxy. The active area was defined by mesa etching and the diodes are illuminated through the substrate. Details about the fabrication process and the extremely good performance concerning dark current (primary dark currents as low as 0.1 nA, total dark current < 10 nA at a gain of 10) and multiplication (up to 10 000) have been reported before (Trommer 1983, 1984). Various doping levels of the n^--InP multiplication layer have been realized. The corresponding layer thickness was adjusted, such that the electric field at the heterointerface was $(1-2) \cdot 10^5$ V/cm for voltages close to breakdown. In addition "low-high-low" structures were prepared, which possess a n^-n sequence in InP with a n^- layer thickness of about 1 µm and $n \approx 2 \cdot 10^{16}$ cm^{-3}. In all cases the background doping level in the n^--InGaAs was below $4 \cdot 10^{15}$ cm^{-3}.

3. Time-Response Measurements and Discussion

The time-response of the avalanche photodiodes was studied by illuminating the central part of the active area either with short light pulses ($\approx$ 100 ps halfwidth) or with sinusoidally modulated light from a semiconductor laser emitting at a wavelength of 1.3 µm. Light power levels between 30 nW and 1 mW were used. The principal behaviour of the photocurrent as observed in the pulse measurements can be seen in Fig. 2. At voltages slightly above the onset of photoresponse, that is when the space charge region just extends into the InGaAs absorption layer (by about 1 µm) and the electric field at the heterointerfaces is low ($< 5 \cdot 10^4$ V/cm), a slow risetime and exponential decay of the photocurrent are measured (upper curve). By increasing the reverse voltage the photoresponse gets fast (central curve) and remains constant up close to breakdown. From the shortest measured pulse-halfwidths of 110 ps a bandwidth of at least 3 GHz is estimated. This is close to the value expected for a response limited by the transit time of holes through the space charge region. Finally, at high multiplication factors close to breakdown, a slow exponential decay is observed again (lower curve). The same principal behaviour can be seen in the frequency domain measurements. In Fig. 3 the ac-photocurrent amplitude is shown as a function of modulation frequency of the light. As a parameter the reverse bias voltage and the multiplication factor are used. Increasing the gain

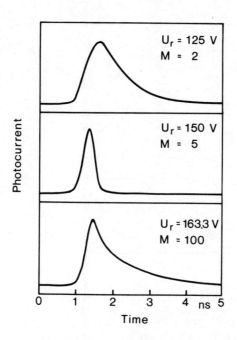

Fig. 2 Pulse-response of a photodiode; measured at three reverse bias voltages: 125 V (slightly above the onset of photoresponse), 150 V, and 163.3 V (close to breakdown)

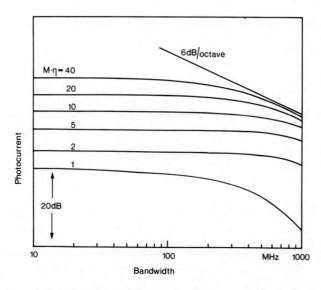

Fig. 3 ac-photocurrent versus modulation frequency of the
light at various reverse bias voltages and multiplication
factors M (the quantum efficiency of the device is $\eta = 0.6$)

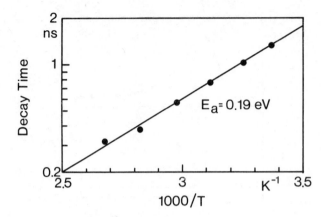

Fig. 4 Arrhenius plot of the decay time constant, measured at low light power and low voltage

first gives an increasing bandwidth. At high multiplication factors and frequencies all curves tend towards one limiting line, which gives a gain bandwidth $G \cdot B$ type of behaviour. In the following we will now discuss in detail the bandwidth limiting mechanisms at low voltages and the dependence of $G \cdot B$ on the diode doping profile.

Two processes may contribute to the slow response observed at lower voltages: accumulation and slow emission of holes at the heterojunctions and/or diffusion of photon-generated minority carriers from the neutral region. We therefore investigated the temperature dependence of the decay time constant at low illumination intensities. As can be seen in Fig. 4 the process is thermally activated, similar as observed by Forrest et al (1983) for diodes without an intermediate InGaAsP layer. We thus conclude that for the low light power levels detected in optical receivers the low voltage bandwidth limitation is due to slow thermal emission of holes accumulated in the potential wells at the heterointerfaces. The activation energy of 0.19 eV is in close agreement with the values for the valence band discontinuities which distribute into about equal amounts over the InGaAs/InGaAsP and the InGaAsP/InP heterojunctions (Nakao et al 1984). Contrary to Forrest et al (1983), however, we observe that with increasing light intensity the decay gets slower and non exponential. This result can only be explained by a shortening of the space charge layer width whereby more light is absorbed in the neutral region and the diffusion current becomes dominant. By increasing the voltage the low power pulse shape can be retained. There are two contributions to the additional positive charge in the space charge region w which accounts for this shortening. First there are the trapped holes at the interfaces and second the photocurrent itself gives an unbalanced charge of holes drifting through the multiplication layer. Correspondingly the deviation of w from its value without illumination w_0 can be taken from the following equation:

$$w = \left(w_0^2 - \frac{d^2}{qNv} j - \frac{2d}{N} p \right)^{1/2} \tag{1}$$

Here j is the photocurrent density, p the sheet carrier concentration of the accumulated holes, N the doping of the InGaAs layer, q the elementary

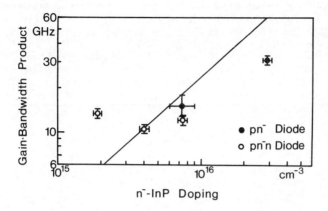

Fig. 5 Gain · bandwidth product versus the n⁻ doping level of a p⁺n⁻ or a p⁺n⁻n InP layer sequence. The solid line is according to the theory of Emmons (1967) (see text)

charge, v the hole drift velocity and d the thickness of the n⁻-InP layer. We can only roughly estimate that in our experiments both terms may contribute to the observed increasing pulse tail. The fact that, unlike Forrest et al (1983) no barrier lowering with increasing light power is measured can be attributed to the abrupt interfaces in our devices. Transmission electron microscope investigations on similar layer sequences grown by our liquid phase epitaxy show an interface grading ≤ 1 nm (Oppolzer 1984). It has to be pointed out that the disappearance of the slow decay at higher voltages cannot only be explained by barrier lowering, which occurs only at a graded heterojunction. Instead, tunneling may become the dominant fast emission process of trapped holes at higher electric fields. From the theory given by Moll (1964) we calculate that at 10^5 V/cm the tunneling current flowing from the InGaAs to the InGaAsP light hole valence band through an energy barrier of 0.19 eV is as high as 1000 A/cm^2 per Volt voltage drop across the tunneling layer.

As described above the bandwidth of our diodes is at high voltages close to breakdown limited by a gain · bandwidth product G · B. This is caused by the avalanche buildup time. From frequency domain measurements G · B could directly be evaluated as seen in Fig. 3. From pulse measurements the integral pulse area and the exponential decay were evaluated for determining multiplication and bandwidth. The resulting value for G · B is in good agreement with that from the modulation experiments. At higher light intensities the maximum obtainable gain was found to be limited by a series resistance. At a fixed applied bias voltage the multiplication decreased with increasing light power, but simultaneously the bandwidth increased so that G · B remained unaltered. The series resistance is attributed to a reduction of the electric field across the avalanche region, as described by Shockley (1961). In Fig. 5 G · B is plotted as a function of doping level in the n⁻-InP layer of our devices. The "low-high-low" structures all give about the same value for G · B, which means it is limited by the constant thickness of the n⁻-InP layer in these structures (≈ 1 μm). The devices with uniform doping profile in the multiplication layer show an increase of G · B with doping. Values up to 30 GHz have been measured. Both diode structures give the same $G \cdot B$ at a doping of around $7 \cdot 10^{15}$ cm^{-3}, which means the effective avalanche width is about 1/4 of the total space charge width w_b of a p^+n^--InP junction at breakdown, as given by Sze (1981). The solid line is calculated from Emmons (1967) theory for a pin junction, where the carrier drift velocity was taken to be 10^7 cm/s and the avalanche width was identified with w_b. For a shorter avalanche width e.g. 1/4 w_b the curve would even be shifted up. So obviously Emmons theory cannot be applied to an avalanche photodiode with a nonconstant electric field profile.

4. Conclusion

Our experiments show that in SAM-type InGaAs/InP avalanche photodiodes with an intermediate InGaAsP layer the bandwidth can at low voltages still be limited by accumulation of holes at the heterointerfaces. At higher voltages the response gets fast and bandwidths of 3 GHz are possible. Close to breakdown limitation due to a gain · bandwidth product is observed, which depends on the doping profile of the InP multiplication layer. Values up to 30 GHz have been achieved. These results show that such photodiodes have high potential for application in high speed optical receivers up to about 3 GHz.

Acknowledgements

The authors wish to thank H. Huber, W. Kunkel and G. Meier for epitaxial growth and preparation of the photodiodes, to R. Freytag for performing the modulation measurements, and to M. Plihal for a critical reading of the manuscript. This work has been supported by the Ministry of Research and Technology of the Fed. Rep. of Germany. The authors alone are responsible for the contents.

References

Emmons RB 1967 J.Appl.Phys. 38 3705
Forrest SR, Kim OK, Smith RG 1983 Sol.State Electr. 26 951
Kanbe H, Susa N, Nakagome H, Ando H 1980 Electronics Lett. 16 163
Moll JL 1964 Physics of Semiconductors (New York: McGraw-Hill) p 253
Nakao M, Yoshida S, Gonda S 1984 Sol.State Comm. 49 663
Oppolzer H 1984 private communication
Shockley W 1961 Solid-State Electr. 2 35
Sze SM 1981 Physics of Semiconductor Devices (New York: Wiley) pp 77 and 104
Trommer R 1983 Proc. 9th Europ. Conf. on Opt. Comm. ed. H Melchior and A Sollberger (North-Holland: Elsevier) pp 159-162
Trommer R 1984 Frequenz 38 212

Inst. Phys. Conf. Ser. No. 74: Chapter 6
Paper presented at Int. Symp. GaAs and Related Compounds, Biarritz, 1984

415

Strained-layer superlattice emitter and detector structures using modulation-doped active regions

L. R. Dawson, T. E. Zipperian, C. E. Barnes, J. J. Wiczer, and G. C. Osbourn

Sandia National Laboratories, Albuquerque, NM 87185

Abstract. Double-heterostructure p^+n diodes have been fabricated from strained-layer superlattice (SLS) material in which the emitting and absorbing region consisted of alternate 120 Å thick layers of GaAs and $In_{.2}Ga_{.8}As$. This region was modulation doped by a Si impurity spike in only the central 30 Å of the GaAs layers. When operated as detectors (0 volts bias), these structures exhibited internal quantum efficiencies of 75%. As emitters they produced output at 1.02 μm greater than or comparable to both uniformly-doped SLS structures and commercially available 1.06μm devices. The achievement of such high performance in modulation-doped structures identical to those required for optimizing parallel transport properties (horizontal electron mobilities) implies that the InGaAs/GaAs SLS system is an attractive candidate for integrated optoelectronic applications.

1. Introduction

Strained-layer superlattice (SLS) structures allow the use of mismatched semiconductor alloys as layer materials without the accumulation of strain energy sufficient to cause the generation of misfit dislocations. This freedom from the need for precise lattice matching of the superlattice components greatly broadens the choice of compatible superlattice alloys, and increases the ability to tailor electronic and optical properties in such structures. For optoelectronic applications, two attractive features of the $In_xGa_{1-x}As$ SLS system are the ability to tune the energy gap to suitable values (Osbourn 1983) and the ability to enhance carrier transport properties by suitable doping profiles (Fritz, et al. 1983). Recent trends in device applications favor monolithic integration over the interconnection of discrete elements. This fact suggests that if the full potential of SLS materials is to be realized, the same structures will be required to perform emitter, detector, and gain (carrier transport) functions. Since carrier transport parallel to the SLS interfaces is greatly enhanced by modulation doping (Fritz, et al. 1983) (similar to that used in selectively doped single-well AlGaAs/GaAs MODFETs), it is important to demonstrate good emitter and detector performance in similar (modulation doped) SLS material. Modulation doping of the n-type material in the SLS structures reported here was not specifically intended as a means of improving the emitter or detector characteristics, but as a

---

*This work was performed at Sandia National Laboratories and supported by the U. S. Department of Energy under contract number DE-AC04-76DP00789.

demonstration that SLS materials can simultaneously be optimized for both optical (energy gap) and transport (horizontal electron mobility) properties.

2.   Device Structure

For this study, double heterostructure p^+n diodes were fabricated using modulation-doped (MD), n-type superlattices as the emitting or absorbing medium for LED or detector operation, respectively. As shown in Figure 1, the ~1.5 μm thick SLS region was composed of equally thick GaAs and In$_{.2}$Ga$_{.8}$As layers (repeat distance = 240 Å), confined below (buffer) and above (window) by 1 μm of the quaternary alloy In$_{.1}$Al$_{.3}$Ga$_{.6}$As. The p^+n junctions were imbedded within the SLS material approximately 0.1 μm below the window layer. The p-type SLS region was uniformly doped with Be at 2.5×10^{18} cm^{-3}, and the n-type SLS region was modulation-doped by a Si impurity spike varied between 3×10^{17} and 2×10^{18} cm^{-3} in the central 30 Å of the 120 Å GaAs barrier layers (leading to average carrier populations in the narrow-bandgap wells of 8×10^{16} to 5×10^{17} cm^{-3}). The energy gap of the In$_{.1}$Al$_{.3}$Ga$_{.6}$As buffer and window material has been determined by photoluminescence measurements to be 1.65 eV (Jones and Dawson), substantially higher than the 1.22 eV energy gap of the SLS material (Dawson, et al. 1984). Since the In content of the buffer/window material, x_{InAs} = 0.1, is equal to that of the average In content in the SLS region, no net strain (lattice mismatch) exists in the entire struc-

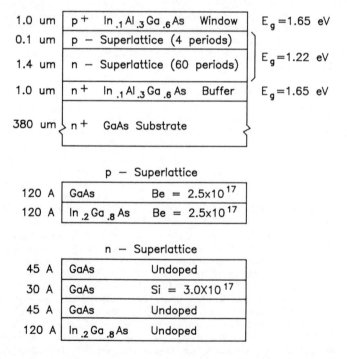

Fig. 1   Double heterostructure p–n diodes using modulation-doped strained-layer superlattice material.

ture, except at the substrate-buffer interface, where the lattice mismatch is ~0.7%. Misfit dislocations generated due to accumulated strain energy within the thick buffer layer are removed by the alternate tensile and compressive strains at the first few SLS interfaces (Matthews and Blakeslee, 1974, Matthews and Blakeslee, 1976).

3. MBE Growth

These structures were grown in a Varian 360 MBE system modified for computerized shutter control. The entire structure was grown at a substrate temperature of 500°C at a growth rate of 0.7 μm/hr and a V/III ratio of 5-7. Be and Si were used as p and n type dopants, respectively. These are the same growth conditions used for the growth of enhanced mobility structures (Fritz, et al. 1983). While this growth temperature is considerably below the optimum for GaAs growth, it is near the optimum for $In_{.2}Ga_{.8}As$ growth, and represents an acceptable compromise (Wood, et al. 1984). RHEED patterns remain distinct and well streaked during the growth of the SLS structure. No change in surface smoothness is apparent during the nucleation of each layer in spite of the fact that the layer materials differ in bulk lattice constant by 1.4%.

4. Emitter Results

For evaluation of LED performance, diodes made from these structures were compared with other SLS structures identical in every way except that the entire SLS region was uniformly doped (UD). Comparison was also made with commercially available RCA InGaAs devices (C30116) grown by vapor phase epitaxy, using a graded buffer layer to accommodate the lattice mismatch. These commerical devices were supplied on headers with a well in the center designed to reflect some of the side emission out of the top of the package. The SLS diodes were mounted in similar packages. A gel-like substance was removed from the RCA devices for purpose of comparison.

Electroluminescence (EL) intensity and spectra were taken at room temperature and 85K using a Spex 1-meter monochromator with corrected S-1 photomultiplier detection. The LEDs were pulsed at 4.0 KHz with a 100 μs pulse width (40% duty cycle). The configuration used strongly favors the sampling of front surface emission, and is likely to be indicative of power available for optical fiber transmission.

Typical room temperature EL spectra for the three types of devices are shown in Figure 2. The SLS emission peaks at 1.02 μm, very near the wavelength corresponding to the energy gap calculated for such structures (~ 1.22eV)(Dawson, et al. 1984). Both SLS structures produce room temperature spectra with shoulders on the short wavelength side of the band, while for 85K emission the shoulders disappear. This behavior is not understood at present. The peak intensity for both SLS devices is somewhat greater than that of the RCA device, with the MD device substantially better than the UD device. Room temperature peak EL intensities as a function of current density are shown in Figure 3. Both SLS types have intensities greater than or equal to the RCA LEDs over the entire current density range, with the MD device again substantially better than the UD device. The better relative behavior of the SLS devices at lower current density may indicate the increased effect of non-radiative recombination due to dislocations in the RCA non-SLS material.

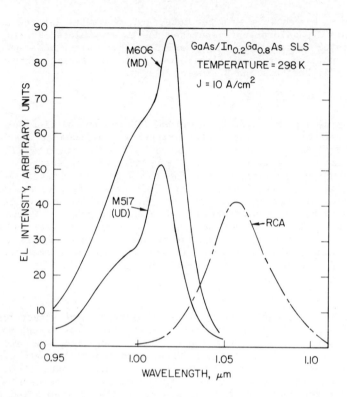

Fig. 2   Electroluminescence spectra at 298K for modulation doped (MD) and uniformly-doped (UD) SLS LEDs, and graded-buffer RCA InGaAs LEDs.

Total power output was measured at room temperature with a Photodyne Model 88XL radiometer/photometer system. This integrating sphere technique should not severely discriminate against the source of photons, although the degree to which edge-emitted photons are reflected into the sphere by the sides of the package is difficult to predict. Power output was comparable for all device types for current ranging from 2 to 50 mA. Specific values were 100 μW at 30 mA, 200 μW at 50 mA. It is clear from the above results not only that SLS emitters can perform well compared to commercial devices operating at similar wavelength, but also that modulation doping of the SLS active region does not degrade performance.

5.   Detector Results

The same MD devices were also operated as unbiased photodetectors. The external quantum efficiency of these uncoated devices was measured as a function of wavelength and appears in Figure 4. The cutoff in spectral response at 740 nm and 1030 nm corresponds closely to the energy gaps of the SLS and window regions, 1.22 and 1.65 eV, respectively. The peak response of 54% at 760 nm (uncorrected for surface reflection of ~ 30%) suggests a peak internal quantum efficiency in excess of 75%. The wavelength dependence of the response is probably due to the increased collection efficiency for minority carriers created close to the p-n junction by virtue of the higher absorption coefficient of the higher energy pho-

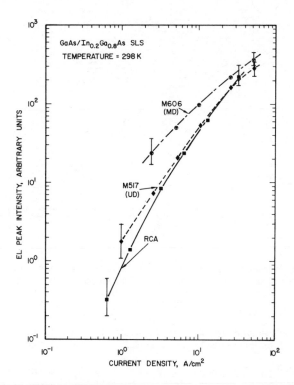

Fig. 3   Peak EL intensity versus current density at 298K for modulation-doped (MD) and uniformly-doped (UD) SLS LEDs, and RCA InGaAs LEDs.

tons. The absorption depth is especially important since SLS diffusion lengths perpendicular to the interfaces can be considerably less than the in-plane values. By modeling the photoresponse of similar (UD) SLS detectors (Dawson, et al. 1984), a perpendicular diffusion length of 0.1 μm for holes in an n type SLS has been deduced. The UD devices were also measured for comparison. Their performance was similar to that of the MD devices, as shown in Figure 4.

6. Discussion

In addition to the results described above for MD SLS structures with relatively light Si doping (3 x 10^{17} cm^{-3} in the doping spike, 8 x 10^{16} cm^{-3} in the wells), diodes with heavier doping (2 x 10^{18} cm^{-3} in the spike, 2 x 10^{17} cm^{-3} in the wells) were also fabricated and tested. The emitter characteristics were virtually identical to those shown in Figures 2 and 3. Detector characteristics were somewhat inferior, with peak external quantum efficiency of 38% (~54% internal quantum efficiency) compared to 54% for the lightly doped devices. This difference is probably due to the wider depletion region in the lightly doped material, which would enhance collection of minority carriers over a larger volume adjacent to the junction. This suggests that even lower doping might further enhance the quantum efficiency of such SLS structures.   The

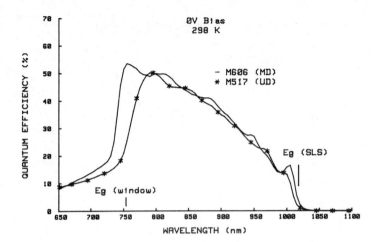

Fig. 4   External quantum efficiency as a function of wavelength for un-coated, modulation-doped (MD) and uniformly-doped (UD) unbiased photo-detectors.

doping levels represented by the devices of Figures 2 and 3 are quite compatible with enhanced-mobility structures (Fritz, et al. 1983).

In conclusion, the results shown above clearly indicate that good emitter and detector performance can be achieved in material essentially idential to that required for enhanced mobility structures. Taken with previous transport (Fritz, et al. 1983) and transistor studies (Zipperian, et al. 1983), these very encouraging LED and photodetector results suggest that SLS materials can be simultaneously optimized for a variety of optoelec-tronic applications. The InGaAs SLS materials system is thus an attractive candidate for integrated optoelectronic structures.

7.   Acknowledgments

It is a pleasure to acknowledge the expert technical assistance of R. E. Hibray, T. A. Plut, R. G. Keefe, and L. V. Hansen.

8.   References

Dawson L R, Osbourn G C, Zipperian T E, Wiczer J J, Barnes C E, Fritz I J, and Biefeld R M 1984 J. Vac. Sci. Technol. B 2 179
Fritz I J, Dawson L R and Zipperian T E 1983 Appl. Phys. Lett. 43 846
Jones E C and Dawson L R unpublished
Matthews J W and Blakeslee A E 1974 J. Cryst. Growth 27 118
Matthews J W and Blakeslee A E 1976 J. Cryst. Growth 32 265
Osbourn G C 1983 Phys Rev B 27 5126
Wood CEC, Singer K, Ohashi T, Dawson L R and Noreika A J 1983 J. Appl. Phys. 54 2732
Zipperian T E, Dawson L R, Osbourn G C and Fritz I J Proc. 1983 Internat. Electron Dev. Mtg. 696

Inst. Phys. Conf. Ser. No. 74: Chapter 6
Paper presented at Int. Symp. GaAs and Related Compounds, Biarritz, 1984

421

In$_x$Ga$_{1-x}$As photodiodes for the 1.0 to 2.4 micron spectral region prepared by low-pressure MOCVD

P. Poulain, M.A. Di Forte-Poisson, K. Kazmierski and B. de Cremoux
Thomson-C.S.F., Laboratoire Central de Recherches,
Domaine de Corbeville, Boîte Postale n° 10, 91401 ORSAY (France)
and
J.P. Moy and P. Radisson
Thomson-C.S.F., Division Tubes Electroniques
38, rue Vauthier, Boîte Postale n° 305, 92192 BOULOGNE BILLANCOURT
Cedex (France)

ABSTRACT :

Photodiodes with a cut-off wavelength extended in the infra-red region up to 2.4 µm have been made in a mismatched In$_{.8}$Ga$_{.2}$As/InP double heterostructure prepared by low-pressure MOCVD on InP substrates.

Initial devices have exhibited good performance although grown without InAsP graded layers to accomodate the mismatch between the substrate and the active region.

1. INTRODUCTION

The development of high-purity silica fibers with low-loss and low-dispersion in the 1.0 - 1.6 µm wavelength region has widely promoted the use of the quaternary alloy InGaAsP lattice matched to InP for the fabrication of light sources and detectors in this wavelength region. The maximum wavelength accomodated in this system is 1.65 µm corresponding to the bandgap of the ternary solution In$_{.53}$Ga$_{.47}$As.

As the intrinsic Rayleigh scattering loss varies as the inverse fourth power of the wavelength, fluoride and chalcogenide optical fibers suitable for use at longer wavelength are considered promising ultra-low loss transmission media. Recently, a fluoride glass fiber has been reported by Mitachi et al. (1984) exhibiting the lowest loss 8.5 dB/km at 2.12 µm and a useful range up to 2.6 µm. Thus, optoelectronic devices operating in this wavelength region are of interest for future use.

In-rich InGaAs lattice mismatched to InP has already been used by Gregory et al. (1980) to produce by VPE photocathodes reaching the 2.1 µm region, and by Morrison et al. (1983) to grow by LPE photodiodes reaching 2.0 µm.

More recently, Lee et al. (1984) has reported InAsP homojunction
detectors by MBE with a cut-off also at 2.0 μm.

In this paper, we present $In_{.8} Ga_{.2} As$ photodiodes prepared by low
pressure MOCVD and showing reasonable responsivity up to 2.4 μm.

2. DEVICE FABRICATION

The InGaAs/InP photodiode structures were grown in a horizontal-
geometry low pressure MOCVD reactor from standard grade sources of
triethylgallium (TEG), triethylindium (TEI), pure arsine and pure
phosphine, at a temperature of 550° C. Three nominally undoped layers
were deposited on an N type InP substrate. The growth started with an
InP buffer layer, followed by the mismatched InGaAs absorbing layer
and another InP "window" layer. The thicknesses were respectively 1 μm,
2.6 μm and 1 μm. Capacitance – voltage measurements indicated that the
background carrier concentration was in the 10^{15} cm^{-3} range (N type),
and light transmission experiments indicated a bandgap, for the ternary
layer, at a wavelength close to 2.3 μm.

As InAsP can be grown lattice matched to $In_{.8} Ga_{.2} As$ and has in this
case a higher bandgap, it would be possible to improve the device
by using an InAsP buffer layer of graded composition to accomodate the
mismatch between the substrate and the active region, as done by
Morrison et al. (1983) by LPE. Also by growing a matched InAsP window
layer, only the short wavelength part of the photoresponse would be
sacrificed. Nevertheless, in our devices which did not contain InAsP
the wafers exhibited a smooth mirror-like surface after growth.

The p-n junction was then located in the active layer at a depth of
1.1 μm from the surface by Zn diffusion, using the semi-closed box
technique reported by KAZMIERSKI et al. (1984). The ohmic contacts
were made by sputtering Au-Zn and Au on p and n sides respectively.
Finally mesas were chemically etched with areas ranging from 1 to
4.10^{-4} cm^2.

It should be noted that no cross-hatched pattern was revealed outside
the mesas at this etching step. Figure 1 is a schematic representation
of the photodiode.

3. EXPERIMENTAL RESULTS

Packaged devices were characterized by measuring the I-V characteristics,
spectral response, capacitance-voltage curve and response time.

Figure 2 is a typical room temperature I-V curve measured on a broad
area device. The characteristic is soft with a forward conduction of
1 mA at 0.5 V and a reverse conduction (also 1 mA) at 2.5 V. The
reverse current of these devices was also measured as a function of the
temperature and at .5 V bias was found to decrease from 40 μA at room
temperature to 2 nA at 170 K. Room temperature leakage currents were
also measured before any packaging operation and values as low as 300 nA
were found on small mesas at .1 V.

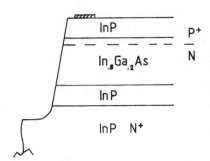

Fig. 1   Schematic cross-section of
         the photodiode.

Fig. 2   Room temperature

         Forward and reverse I-V
         characteristic

         Horiz. .5V/div –
         Vert 1 mA/div

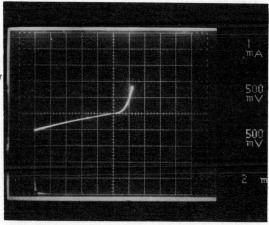

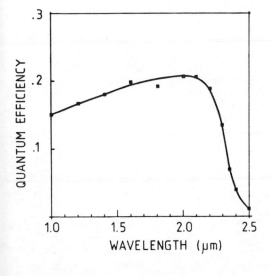

Fig. 3   Quantum efficiency of
         the long wavelength
         photodiode under front
         illumination.

Figure 3 represents the quantum efficiency measured for front illumination and at .5 V bias. It gradually increases from 15% at 1.0 μm to 20% at 2.2 μm and then exhibits a sharp decrease between 2.3 and 2.5 μm. The quantum efficiency of these devices is much lower than the 70% value measured on double heterostructure photodiodes also grown by MOCVD but with a lattice matched InGaAs active region (1.65 μm cut off wavelength) and reported by Poulain et al (1984). We believe that this effect mainly arises from a loss of photocarriers owing to a probable increase of the recombination velocity at the interfaces between the InP layers and the mismatched active region.

The speed of response was measured at a reverse bias of 0.5 V using 100 ps pulses from an InGaAsP laser diode which emitted at 1.3 μm. The fall time was about 1 nS and mainly limited by the combination of the 8 pF of the packaged device with the 50 Ω impedance of the sampling scope.

4. CONCLUSION

Using low-pressure MOCVD, we have made long - wavelength photodiodes in a mismatched InGaAs/InP double heterostructure. We believe that the 2.4 μm cut-off is the longest one reached to date in this alloy family.

Although good performance was obtained on these initial devices, we believe that further improvements could be obtained by using the InAsP ternary alloy instead of InP during the growth, in order to accomodate the lattice mismatch between the InP substrate and the InGaAs active layer.

Therefore, the ternary system InGaAs already appears as a suitable material for detectors at wavelengths higher than 1.7 μm. It offers some advantages when compared to InAsP recently proposed by Lee et al. (1984) owing to its smaller bandgap at a given lattice constant. It has also the advantage over the quaternary system InGaAsb or II-VI compounds, that InP and related compounds have now reached a high degree of technology and are readily available with device quality.

5. ACKNOWLEDGEMENTS

The authors wish to acknowledge R. Blondeau, G. Vilain, P. Hirtz and L. Noel for technical assistance during this work, and also P. Philippe for the low temperature measurements on these devices.

6. REFERENCES

Gregory P.E., Escher J.S., Saxena R.R., and Hyder S.B. 1980
    Appl. Phys. Lett., 36, 639.

Kazmierski K., Huber A.M., Morillot G., and de Cremoux B. 1984
    Jap. J. Appl. Phys., 23, 628.

Lee T.P., Burrus C.A., Sessa W.B., and Tsang W.T. 1984
    Electron. Lett., 20, 363.

Mitachi S., Ohishi Y., and Takahashi S. 1984
    Rev. Elec. Comm. Lab., 32, 461.

Morrison C.B., Bedair S.M., and Bachman K.J. 1983 presented at
    "Spring Meeting of the Electrochemical Society", paper n° 333.

Poulain P., Razeghi M., Hirtz P., Kazmierski K., and de Cremoux B. 1984
    presented at "Ninth IEEE international semiconductor laser conference"
    Rio de Janeiro.

Inst. Phys. Conf. Ser. No. 74: Chapter 6
Paper presented at Int. Symp. GaAs and Related Compounds, Biarritz, 1984

427

Short wavelength quantum well lasers grown by molecular beam epitaxy

P Blood, E D Fletcher, K Woodbridge, P Dawson and P J Hulyer

Philips Research Laboratories, Redhill, Surrey, RH1 5HA, England.

Abstract. We have fabricated AlGaAs multiple quantum well lasers from
a variety of structures grown using molecular beam epitaxy. Devices
with GaAs wells from 13Å to 55Å wide operate at wavelengths of 707nm
to 837nm respectively, though these are longer than calculated for the
n=1(e-hh) transition using the measured well width and barrier
composition. Electroluminescence spectra as a function of drive
current suggest this is not due to reduction in sub band separation at
high injection. Devices with 25Å wide wells in a 4000Å wide optical
waveguide show a systematic increase in threshold current with
increasing number of wells from 2 to 40, and this cannot be explained
by the calculated variation in the confinement factor. These devices
operate at about 760nm with a broad area threshold current density of
about 1.7kA cm^{-2} for the 2 and 4 well structures.

1. Introduction

There is widespread interest in the use of the quantum size effect to
modify the operating wavelength of semiconductor lasers, particularly in
the AlGaAs system. High quality current injection lasers have been
demonstrated by molecular beam epitaxy (MBE) (Tsang 1981a, 1981b) and
metalorganic chemical vapour deposition (MOCVD) (Hersee et al 1982,
Kasemset et al 1982) using GaAs well widths down to about 60Å producing
emission at wavelengths as short as 820nm. Hersee et al (1984) have
reported injection structures with wells nominally as thin as 20Å but
emission wavelengths apparently suggest the wells are much thicker than
the nominal values. Further reductions in wavelength have been achieved
by adding Al in relatively wide wells (Burnham et al 1983, Tsang and
Ditzenberger 1981) and using external cavities (Epler et al 1984), and in
these structures the wavelength shortening due to the quantum size effect
is small. We have investigated the growth by MBE and the device
characteristics of current injection multiple quantum well lasers with
narrow (<55Å) GaAs wells in which we have achieved laser emission at
wavelengths as short as 707nm using quantum size effects alone. In this
paper we describe the characterisation of the MQW material with
particular reference to the barrier height and well width, and the
emission wavelengths and threshold currents of devices with various well
widths and devices containing different numbers of wells.

2. Layer Growth and Characterisation

The structures were grown in a laboratory built MBE system equipped with
facilities for monitoring RHEED intensity oscillations. The substrate
temperature was 700°C for all of the structure except the buffer layer,

which was grown at 620°C, and the growth rate for GaAs was about 1μm hr⁻¹. The substrate was rotated at 120 rpm giving at least one rotation in the time to deposit one monolayer and thereby ensuring good lateral uniformity for very thin layers.

A variety of device structures, all with GaAs wells, have been grown using Be and Si as the p and n dopants respectively. In all cases the multiple quantum well active regions were not intentionally doped, which for the particular growth system used implied a residual doping of about $10^{16}cm^{-3}$ p type. Some devices had graded refractive index regions on each side of the active layer. These regions were about 1000Å wide with an Al profile determined by the thermal response of the cell. The growth rate for each structure was determined from measurement of the Ga flux by an ion gauge rotated into the substrate position immediately before growth, and these flux measurements were calibrated in terms of a GaAs growth rate using the RHEED oscillation method (Neave and Joyce 1983).

Successful fabrication of working lasers requires material of high optical efficiency, and in the case of quantum well lasers employing a large number of thin layers this implies that the interface recombination rate must be as low as possible. The decay of photoluminescence following pulse excitation of a 60 period 55Å GaAs quantum well structure with 30% Al in the barriers and grown in the system used for the lasers is shown in Figure 1. This exponential decay corresponds to a lifetime of 14.2ns. If we assume that the recombination process in such thin layers is dominated by the interfaces, then for purposes of comparison we can express the lifetime τ wholly in terms of a recombination velocity S, although it is probable that this concept is not strictly valid in these samples (Dawson et al 1984). Using the relation

$$\tau^{-1} = \frac{2S}{d}$$

where d is the well width, we find the data in Figure 1 corresponds to

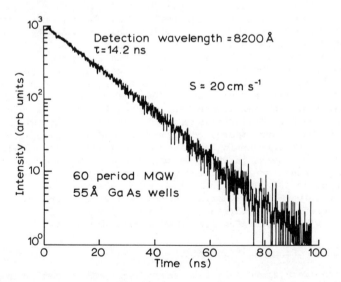

Fig.1   Decay of the room temperature photoluminescence from a 60 period MQW with GaAs wells 55 Å wide.

$S \sim 20\mathrm{cm}\ \mathrm{sec}^{-1}$. The best reported values for conventional samples in the AlGaAs/GaAs system are around $500\mathrm{cm}\ \mathrm{sec}^{-1}$. Therefore, insofar as the comparison is valid, the decay measurements indicate that the quantum well material is of high optical quality.

Further characterisation of the laser material is required for a meaningful analysis of the device performance. In particular we need accurate values of the height and width of the wells. We determined the thickness of the upper GaAs contact layer of the device structure using an electrochemical C-V profiler with the depth scale calibrated for each sample by means of a measurement of the depth of the etched crater. From the timing of the computer controlled cell shutter sequence for the growth we then calculated the well width. The results agree within the uncertainty of $\sim 10\%$ with the determination based on RHEED oscillations referred to above. Electrochemical profiling was also used to record surface photo-voltage spectra to determine the band edges of the cladding layers, and in some instances of the barrier layers also (Stagg et al 1982), from which we could infer the composition. Figure 2 shows spectra recorded in the cladding region (A) and in the waveguide region (B) of a separate confinement quantum well device. When the etching reaches the vicinity of the quantum well structure sometimes it is possible to observe a spectrum similar to that shown in Figure 2C which is characteristic of the MQW structure. Because the etching is inherently non-uniform over a depth range which is large compared with the quantum well period the etch penetrates the wells and the voltage is probably set up by photoexcited carriers being extracted along the wells. The spectrum shown in Figure 2C shows well resolved n=1(e-hh) and (e-lh) exciton absorption peaks similar to conventional absorption spectra (Miller et al 1982).

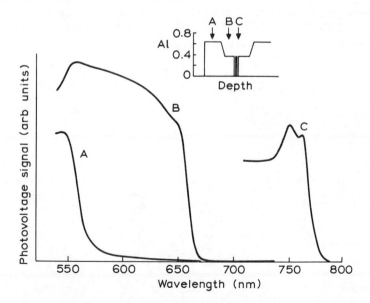

Fig. 2   Surface photovoltage spectra recorded at the three depths indicated, while electrochemically profiling a separate confinement MQW laser structure.

3. Laser Results

Figure 3 shows the observed emission wavelengths of a series of 50μm oxide stripe MQW lasers with GaAs well widths from 55Å to 13Å; full details of the structures and growth conditions have been reported previously (Woodbridge et al 1984). We have achieved laser emission in the visible part of the spectrum at 707nm with GaAs wells only 13Å wide. As a result of the material characterisation described above we have been able to calculate the wavelength of the n=1(e-hh) and (e-lh) transitions for the well width and barrier composition of each structure (assuming the conduction band discontinuity is 85% of the band gap difference). These values are in good agreement with the wavelengths of the observed absorption peaks (e.g. Figure 2) when correction has been made for the exciton binding energy (Dawson et al 1983, Green and Bajaj 1983). The results in the figure show that all the devices operate at longer wavelengths than that of the n=1(e-hh) transition. This effect cannot be due to error in assignment of the well width, graded barriers, or fluctuations in composition as suggested by Hersee et al (1984), because we observe differences between laser emission wavelengths and absorption data measured on the same material. One possibility is that the sub-band separation shrinks as the injection level is increased, but spectra of the electroluminescence from the 10 well device with 55Å wide wells show emission at the laser wavelength down to currents as low as 7% threshold. Using spot excitation on the front surface we have observed a wavelength difference of 18nm between the edge and front surface photoluminescence of a similar MQW structure which we interpret as being due to absorption of light emitted in the plane of the wells. Reabsorption effects may have an influence on the laser wavelength. Similar phenomena have been reported by Tarucha et al (1983).

We have also investigated the influence of the number of wells on the threshold current using a separate confinement structure to provide an

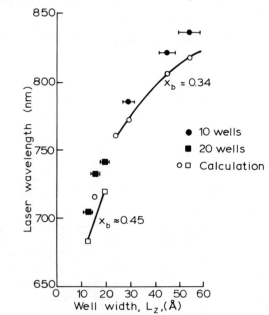

Fig. 3 Emission wavelengths as a function of well width for a series of MQW lasers with GaAs wells, and 80 Å wide barriers. The wavelengths of the n=1 (e-hh) transition calculated for each device is shown by open symbols.

• 10 wells
■ 20 wells
○□ Calculation

$x_b \approx 0.34$

$x_b \approx 0.45$

optical waveguide of fixed width. The outer cladding layers contained 65% Al and were graded down to a 4000Å wide waveguide containing ~ 35% Al. Samples were grown with different numbers of GaAs wells in the waveguide, these being symmetrically disposed about its centre and having a width of 25Å, separated by 75Å barriers. We calculate that the confinement factor of the total waveguide changes by only 5% from having no wells to being filled with 40 wells. The samples were grown in random order and the threshold currents increased systematically from 2 wells to 40 wells. The devices operated at wavelengths around 760 nm and the minimum threshold current measured on 2 well 50 μm oxide stripe devices was equivalent to a broad area current density of about 1.6 kA cm^{-2}.

The threshold current density J may be written in the following form (Casey and Panish 1978)

$$\frac{J}{d} = \frac{J_o}{\eta_i} + \frac{1}{\Gamma} \left\{ \frac{1}{\eta_i \beta} \left[\alpha_i + \frac{1}{L} \ln R^{-1} \right] \right\}$$

where d is the total thickness of GaAs (number of wells multiplied by well width), η_i and α_i are the internal efficiency and optical losses respectively, L is the device length, R the facet power reflectivity, J_o and β define the intercept and slope of the gain plotted as a function of nominal current density and Γ is the average confinement factor for the wells. Since the wells are not coupled we assume J_o and β are independent of the number of wells so that if η_i and α_i are also constant then for devices of the same length a plot of J/d versus Γ^{-1} should be a straight line with positive intercept of J_o/η_i on the vertical axis. Since the wells have only a small effect on the electric field distribution in the waveguide, Γ can be calculated by integrating the square of the field across the wells for each structure. Due to the decrease in field away from the centre of the waveguide the increase in Γ is sublinear with respect to the number of wells.

Figure 4 shows values of J/d for 250μm long, 50μm wide oxide stripe devices with no correction for current spreading plotted as a function of

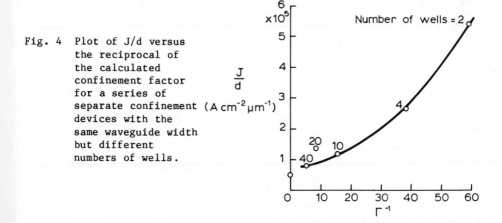

Fig. 4   Plot of J/d versus the reciprocal of the calculated confinement factor for a series of separate confinement devices with the same waveguide width but different numbers of wells.

Γ^{-1} obtained by calculation. This plot is not linear and does not have
a positive threshold current intercept except for structures with a large
number of wells. We infer from this data that Γ alone cannot account for
the variation of J with number of wells and that other parameters may
also vary in these structures.

Acknowledgements

We thank G. Duggan for the theoretical calculation of the transition
wavelengths, and P. Kershaw and J. Bellchambers for help with
measurements. We thank G. Duggan and C.T. Foxon for helpful discussions
on this work.

References

Burnham R D, Lindstöm C, Paoli T L, Scrifes D R, Streifer W and
   Holonyak N Jr 1983 Appl. Phys. Lett. 42 937
Casey H C Jr, Panish M B 1978 "Heterostructure Lasers" (Academic Press,
   New York) Part A p183
Dawson P, Duggan G, Ralph H I and Woodbridge K 1983 Phys. Rev. B 28 7381
Dawson P, Duggan G, Ralph H I and Woodbridge K 1984 Intl. Conf. on
   Physics of Semiconductors, San Francisco
Epler J E, Holonyak N Jr, Brown J M, Burnham R D, Streifer W and
   Paoli T L 1984 Journ. Appl. Phys. 56 670
Greene R L and Bajaj K K 1983 Sol. State Comm. 45 831
Hersee S D, Baldy M, Assenat P, de Cremoux B and Duchemin J P 1982 Elec.
   Lett. 18 870
Hersee S D, de Cremoux B and Duchemin J P 1984 Appl. Phys. Lett. 44 476
Kasemset D, Hong C S, Patel N B and Dapkus P D 1982 Appl. Phys. Lett. 41
   912
Miller D A B, Chelma D S, Eilenberger D J, Smith P W, Gossard A C and
   Tsang W T 1982 Appl. Phys. Lett. 41 679
Neave J H and Joyce B A 1983 Appl. Phys. A.31 1
Stagg J P, Hulyer P J, Foxon C T and Ashenford D 1982 Journ. de Phys. 43
   suppl. 12, C5-377
Tarucha S, Horikoshi Y and Okamoto H 1983 Jap. J. Appl. Phys 22 L482
Tsang W T 1981a Appl. Phys. Lett. 38 204
Tsang W T 1981b Appl. Phys. Lett. 39 786
Tsang W T and Ditzenberger J A 1981 Appl. Phys. Lett. 39 193
Woodbridge K, Blood P, Fletcher E D and Hulyer P J 1984 Appl. Phys.
   Letts. 45 16

Inst. Phys. Conf. Ser. No. 74: Chapter 6
Paper presented at Int. Symp. GaAs and Related Compounds, Biarritz, 1984

433

Ga$_{1-x}$Al$_x$As power lasers made by MO-VPE

F.P.J. Kuijpers*, G.A. Acket and H.G. Kock

Philips Research Laboratories
P.O.Box 80.000, 5600 JA Eindhoven, The Netherlands

Abstract

Threshold current (ith), output power (P) and far field distributions were investigated on MO-VPE grown gain guided lasers, having a planar proton bombarded structure, in their respective relation to the thickness of the active layer (d) in the wavelength regime 810-870 nm. For d $\approx$ 1000 Å: i$^{th}_{CW-30°C}$ $\approx$ 55 mA and P$_{CW-30°C}$ $\approx$ 65 mW kink free. The lateral mode behaviour is non-Gaussian and consistent with a cosh^{-2} gain profile even for thin active layers. Multi-mode fibre coupled devices with output powers as high as 40 mW from the fibre (CW 30°C) were obtained.

1. Introduction

Ga$_{1-x}$Al$_x$As high power laser structures have been grown by LPE and MO-VPE. In both growth techniques the quintessence is the thickness control of a thin active layer. For LPE this has been achieved by growing on profiled substrates so that thin active layers arise (Botez et al.1982, Wada et al. 1983). However, good thickness control can easily be obtained by using MO-VPE.

Semiconductor lasers having output powers of about 40 mW are of great interest especially for digital optical recording. So far this has only been achieved on profiled substrates with LPE as well as with MO-VPE (Scifres et al. 1981, Ackley 1983, Ackley and Hom 1983) or by a two step zinc diffusion process in a planar n-type structure grown by MO-VPE (Kumabe et al. 1984). In this respect also the more complex structures consisting of quantum well-lasers and phase locked arrays (e.g. Burnham et al. 1982, Scifres et al. 1983) should be mentioned as possible future candidates.

In this paper we report on simple and successful manufacturing of kink-free 65 mW CW power lasers from planar proton bombarded structures grown by MO-VPE (see fig. 1). The wavelength regime is 810-870 nm.

As the (thin) active layer thickness d is the key parameter for high power lasers we studied the dependence of ith, P and lateral mode behaviour on d for 550 Å $\leq$ d $\leq$ 1350 Å. Referring to the latter relation we like to extend the work of J. Biesterbos et al. (1983) on the lateral mode behaviour for gain guided Ga$_{1-x}$Al$_x$As lasers with 1200 Å $\leq$ d $\leq$ 2600 Å, which were grown by LPE. We will also compare our results with some recently published data by Mamine (1984). All lasers discussed have a stripe width of about 5 μm, defined by a shallow proton bombardment into the p-GaAlAs cladding layer.

Additionally we will present the longitudinal mode spectra of the lasers which were coupled to multimode optical fibres with a core diameter of only 35 μm yielding a fibre CW output power of 40 mW. Also some CW lifetest data at 60°C of these lasers will be given.

---

* Now with semiconductor laser development of Philips Main Industry Group Electronic Components and Materials, Eindhoven, The Netherlands.

2. Experimental

2.1 MO-VPE

All lasers were grown according to the same recipe which was optimized beforehand with respect to i^{th} and laser degradation. The only parameters varied were:
a. 550 Å $\lesssim$ d $\lesssim$ 1350 Å
b. the aluminium content in the active layer (x) was $0 \leq x \leq 0.10$, the difference in Al content between the active and cladding layers being 0.30.

The growth temperature was 775°C, the V/III ratio in the vapour phase was 9 and the molar fraction of TMG was 10^{-4}. Hydrogen selenide and diethylzinc as dopants were used for n-and p-doping respectively.

The wafers were grown on (100) n-GaAs substrates, doped with Si up to 2.10^{18} cm^{-3}. Buffer layers consisting of n-Ga$_{0.97}$Al$_{0.03}$As and n-Ga$_{1-y}$Al$_y$As, where y ramps gradually from 0.03 towards 0.3 + x, were always incorporated (Hersee et al. 1981). The laser structure is given in fig. 1. The wafers were grown in an MR-100 apparatus from Cambridge Instruments Ltd.

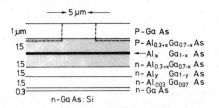

Figure 1.
Structure of proton bombarded MO-VPE laser; the n-type layers are doped with Se, whereas the p-type layers are doped with Zn.

2.2 Laser manufacturing, fibre coupling and measurements

The wafers grown were subjected to a shallow proton bombardment penetrating into the p-GaAlAs cladding layer at a distance $\gtrsim$ 1 µm away from the active layer. The stripe width was 5 µm. Processing involved standard AuGeNi and CrPtAu metallization.

For mirror coating Al$_2$O$_3$ was used, the front mirrors were provided with a 1/4λ coating while the rear mirrors were 1/2λ coated. Uncoated lasers were studied as well. The lasers were mounted into TO-5 or SOT-148 headers.

The ends of the multimode fibres used in coupling were first thinned down to a diameter of 70 µm by etching and then provided with a half-spherical lens by flaming (Khoe et al. 1984).

Layer thickness measurements were performed on a Philips PSEM 505 and are accurate to 50 Å for the active layer. Far field measurements at 5 mW and 60 mW were performed at room temperature using 100 ns pulses in a 10% duty cycle. Output powers are quoted as measured from the front.

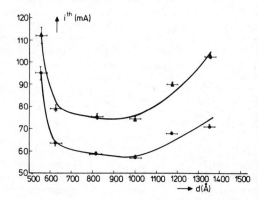

3. Laser performance

Figure 2.
Dependence of threshold current of 1/4λ - 1/2λ coated lasers on layer thickness: ● pulse measurements at 30°C, ▲ CW measurements at 60°C. Cavity length = 250 µm.

The dependence of ith on d is given in fig. 2 for pulse measurements at 30°C and CW measurements at 60°C. The lowest value of ith is obtained in the regime 800-1000 Å. Coated lasers with 550 Å $\lesssim$ d $\lesssim$ 1000 Å show kinkfree output powers at 30°C of 65 mW CW (see fig. 3), independent on their wavelength in the regime investigated. The kinkfree output power in a 100 ns pulse at 30°C is 160 mW.

To our best knowledge these power values are the highest reported so far for a non-quantum well planar proton bombarded laser structure with 5 µm stripe width. The driving currents at these CW output powers being ~ 230 mA are low.

At output powers of 60 mW CW the lasers did not show any degradation during 120 hrs CW operation at room temperature. Lifetest results at 60°C for CW operation during 1000 hrs at 15 mW show that the averaged degradation at 15 mW is less than 10%/1000 hrs, indicating a rather low susceptibility to mirror erosion, which is probably caused by the thin active layer. Indeed, no mirror erosion at all was found after a 2000 hrs lifetest at 60°C, 15 mW CW. Noteworthy is the fact that the spread in degradation rate over the various lasers batches from 10 different growth runs is less than 7%, indicating the high reproducibility of the MO-VPE process.

In fig. 4 typical longitudinal mode spectra are shown at CW output powers of 2 and 15 mW. The half width at 50 mW CW still corresponds to 4 modes. The large number of modes even at high output power is caused by the antireflex coated front mirror of the laser. This was concluded from a comparison with lasers with both facets 1/2λ coated.

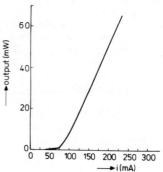

Figure 3.
CW output power at 30°C as a function of driving current for 1/4λ - 1/2λ coated lasers with d = 1000Å.

Lasers from different batches were coupled to multimode optical fibres with a core diameter of only 35 μm. CW output powers from the fibre of 40 mW were very reproducibly obtained at 30°C. This means a coupling efficiency of about 65%, whereas for thicker active layers this efficiency is about 50% (see also chapter 4).

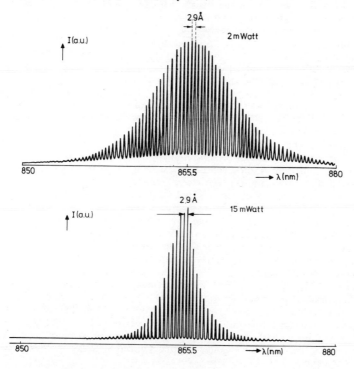

Figure 4. Longitudinal mode spectra of 1/4λ - 1/2λ coated lasers at 2 and 15 mW CW output power.

4. Lateral mode behaviour at various output powers

Thin active layers are essential for high-power semiconductor lasers. The fact that the optical flux of the mode spreads out into the cladding reduces the power density, furthermore this is accompanied in narrow gain guided diodes by a decrease of both far-field width perpendicular to the active layer and parallel to it (Biesterbos et al. 1983, Mamine 1984); both far-field widths depending on the vertical confinement factor. This decrease of the far-field width in both directions considerably enhances the coupling efficiency into many optical systems. However, quantitatively and qualitatively there is a discrepancy between the results reported. Mamine (1984) claims that for active layers thinner than 1500 Å the far-field parallel to the active layer becomes single-lobed and can be described in terms of a parabolic gain profile (Gaussian far-field) whereas Biesterbos et al. described their data over the whole range between 0.12 μm and 0.26 μm with the $\cosh^{-2}$ gain profile leading to the well-known non-Gaussian far-field distributions (Asbeck et al. 1979). The lasers investigated by Biesterbos et al. (1983) were LPE-grown devices, the results by Mamine were obtained on MO-VPE-grown samples. In either case the lasers were fabricated by using a shallow proton-bombardment sufficiently far away from the active layer ($\geqslant$ 1 μm). Therefore it is especially interesting to investigate the lateral mode behaviour in our present MO-VPE grown shallow proton-isolated lasers as a function of active layer thickness, both for low (5 mW) and high (60 mW) pulsed power outputs.

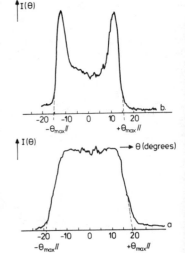

Results of a typical lateral far-field distribution both at 5 mW and 60 mW for a MO-VPE-grown stripe laser with an active layer thickness of 0.10 μm are shown in fig. 5. The shape of the far-field at 5 mW is clearly non-Gaussian and limiting angles $\pm 2\theta_{max\!/\!/}$ can be defined by extra-polating the slopes. The non-Gaussian character of the mode is most strikingly revealed at high output power where the side maxima dominate and hence a $\cosh^{-2}$ gain profile is much more appropriate than a parabolic one. In fact these far-fields are not appreciably different from the LPE ones except from the decreasing width due to the reduced antiguiding for thinner active layers.

We have also investigated the maximum far-field angles $2\theta_{max\!/\!/}$ as a function of the vertical confinement factor $\Gamma_\perp$ for both low and high output power. The results are plotted in fig. 6. Apart from some spread which was also present in the data of Biesterbos et al., the data show the continuation of the decrease of far-field width observed earlier down into the region of $\Gamma_\perp$-values of 0.1 − 0.2. The only difference obtained is that now the slopes of the $2\theta_{max\!/\!/}$ as a function of $\Gamma_\perp$ are nearly equal, the difference in $2\theta_{max\!/\!/}$ being about 5° − 6° whereas at higher $\Gamma_\perp$-values the high power data show a significantly lower slope in agreement with Biesterbos et al. Qualitatively, this may indicate that at very low $\Gamma_\perp$-values the variation in antiguiding becomes less significant.

Figure 5.

Typical FF$/\!/$ distributions for uncoated and 1/4λ - 1/2λ coated lasers with d = 1000 Å emitting 5 mW (a) and 60 mW (b) from the front facet. The pulse width used is 100 ns in a 10% duty cycle.

Van Eck (1983) reported far-field wavefront measurements on some of the LPE grown diodes reported earlier. The wavefront indicated the presence of a non-Gaussian mode, the near field wavefront farther away from the axis tending towards a straight line corresponds to the limiting angle in the far-field. Recently van Eck (1984) performed some identical measurements on MO-VPE-grown diodes with active layers around 0.1 μm. Also these measurements show similar wavefronts as for the LPE lasers albeit with a reduced limiting angle due to the reduction of active layer thickness.

5. Conclusions

Results have been presented confirming the usefulness of thin active layers for obtaining high output power laser diodes.

Proton bombarded, gain guided lasers with an active layer thickness of about 600-1000 Å, manufactured out of simple, planar structures grown by MO-VPE, yield low threshold currents, kink-free output powers up to 65 mW CW in the wavelength regime 810-870 nm. For the active layer regime of 550-1350 Å investigated the far-field distributions parallel to the active layer were of high symmetry. The penetration of optical flux into the cladding layers as a result of thin active layers reduces facet degradation. The reduced beam divergence both parallel and perpendicular to the active layer yields good fibre coupling; even for fibres with a core diameter reduced to 35 μm a kinkfree output power of 40 mW from the fibre was reproducibly obtained.

It has been found that even for very thin active layers down to 500 Å, in contrast to a recent report (Mamine 1984), the lateral mode is essentially non-Gaussian and more consistent with a cosh^{-2} gain profile than with a parabolic one. The results can be regarded as an "extrapolation" of the relation $2\theta_{max\parallel}$ versus $\Gamma_\perp$ for earlier LPE-grown proton bombarded, gain guided diodes (Biesterbos et al. 1983) towards lower $\Gamma_\perp$.

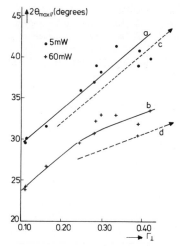

Figure 6.
Dependence of FF$_\parallel$ width $2\theta_{max\parallel}$ on the confinement factor $\Gamma_\perp$ for uncoated lasers emitting 5 mW (curve a) and 60 mW (curve b) per facet. $\Gamma_\perp$ was calculated according to H. Kressel and J.K. Butler in "Semiconductor lasers and heterojunction LED's", p.215, 216. The broken lines are taken from fig. 2 of Biesterbos et al. (1983), being the results for LPE grown proton bombarded lasers with a stripe width of 5 μm emitting 5 mW (curve c) and 60 mW (curve d) per facet.

Acknowledgments

The authors want to express their thanks to H.J.M. van der Laak and H.J.A. Roberts for providing the far-field data, to J.A. de Poorter for the accurate layer thickness measurements and to D.C. van Eck for performing wave front measurements on some of the MO-VPE grown samples. The discussions with H. Koelmans, B.H. Verbeek and P.J. de Waard were very much appreciated.

REFERENCES

D.E. Ackley 1983, Appl. Phys. Lett. **42**, 152.
D.E. Ackley and G. Hom 1983, Appl. Phys. Lett. **42**, 653.
P.M. Asbeck, D.A. Cammack, J.J. Daniele and V. Klebanoff 1979, IEEE J. Quantum Electron. **QE-15**, 727.
D. Botez, D.J. Channin, M. Ettenberg 1982, Optic. Engin. **21**, 1066.
J.W.M. Biesterbos, R.P. Brouwer, A. Valster, J.A. de Poorter and G.A. Acket 1983, IEEE J. Quantum Electron. **QE-19**, 961.
R.D. Burnham, W. Streifer, D.R. Scifres, C. Lindström, T.L. Paoli 1982, Electron. Lett. **18**, 1095.
D.C. van Eck 1983, IEEE J. Quantum Electron. **QE-19**, 986.
D.C. van Eck 1984, Private communication.
S.D. Hersee, M.A. Di Forte-Poisson, M. Baldy and J.P. Duchemin 1981, J. Cryst. Growth **55**, 53.
G.D. Khoe, H.G. Kock, D. Küppers, J.H.F.M. Poulissen and H.M. de Vrieze 1984, J. Lightwave Techn. **LT-2**, 217.
H. Kumabe, J. Ohsawa, K. Isshiki, N. Kaneno, H. Namizaki, K. Ikeda, S. Takamiya and W. Susaki 1984, paper presented at the 9th IEEE semiconductor laser conference, Rio de Janeiro, August 7-10.

T. Mamine 1984, Appl. Phys. Lett. **44,** 1109.
D.R. Scifres, R.D. Burnham and W. Streifer 1981, Appl. Phys. Lett. **38,** 915.
D.R. Scifres, R.D. Burnham, C. Lindström, W. Streifer and T.L. Paoli 1983, Appl. Phys. Lett. **42,** 645.
M. Wada, K. Hamada, H. Shimizu, T. Sugino, F. Tajiri, K. Itoh, G. Kano and I. Teramoto 1983, Appl. Phys. Lett. **42,** 853.

Inst. Phys. Conf. Ser. No. 74: Chapter 6
Paper presented at Int. Symp. GaAs and Related Compounds, Biarritz, 1984

A new device for high brightness AlGaAs red LED

Ming-Jong Tsai and Chen H. Wu

Optoelectronics Division, Hewlett Packard, Palo Alto, California 94304

Abstract. A new device is described in this paper to completely
eliminate contact shadowing due to current crowding in AlGaAs
grown-in p-n junction devices. A comparison between this new
device and the conventional device indicates that the light
output of the new device is about a factor of two higher than
that of the conventional device on the same wafer. Low current
red LED's made from this new AlGaAs device structure yield the
highest light output (2.6 lm/A at 2 mA without encapsulation)
reported to date. Furthermore, this new device design can also
be used in other LED material systems.

1. Introduction

AlGaAs has recently been demonstrated to be a very promising material
for red LED (650-665 nm) applications (Nishizawa, et al, 1983 and 1977;
Ishiguro, et al 1983; and Varon, et al, 1981). Brightness twice that of
existing red GaAsP:N LED has been reported. Because of this higher
brightness than other visible LED's, this AlGaAs red LED has generated
much interest. New applications such as automobile tail lights and cross-
walk signals are conceivable.

Most AlGaAs red emitting devices (both single and double heterojunctions)
reported so far use an n-side up structure and have a p-type active layer
which has been reported to yield much higher efficiency than an n-type
active layer (Varon, et al, 1981 and Alferov, et al, 1973). It is known
that a double heterojunction LED usually has higher efficiency than a
single heterojunction LED. Efficiency of 8% for AlGaAs double heterojunction red LED on transparent AlGaAs substrate has been reported (Ishiguro
et al, 1983).

In these AlGaAs grown-in p-n junction devices, however, current crowding
is often a problem and thus contact shadowing can severely reduce the external light output efficiency. In this paper, a novel device structure
is described, which completely eliminates contact shadowing while still
maintaining the internal quantum efficiency of a grown-in p-n junction.

2. Device Structure

The new device described in this paper uses a grown-in p-side up double
heterojunction structure with a Zn-diffused region directly underneath

the metal contact, as shown in Figure 1. The double heterojunction structure was grown by liquid phase epitaxy on (100)- oriented n–GaAs substrate by ramp cooling from 830 °C to 580 °C at a rate of ~1.2 °C/ minute. Zn diffusion was carried out typically at 655 °C in a sealed ampoule with Zn and As sources. As can be seen in Figure 1, the diff- usion penetrates through the p–active layer and extends into the bottom n-type confining layer. Because of the lower turn-on voltage of the grown-in p–35%/n–70% AlGaAs junction, as compared to the diffused p–70%/ n–70% junction, the current flow and thus the light emitting area is out- side of the diffusion area and away from the metal contact. Light output is therefore enhanced by eliminating contact shadowing. Light output of these new diffused devices is compared with that of the conventional non- diffused devices made from the same wafer and the results are presented in the next section.

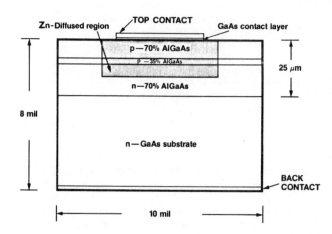

Fig. 1. New device structure: a p-side up double heterojunction structure with a Zn- diffused region dir- ectly underneath the metal contact.

In addition, due to the high surface doping concentration in the contact region obtained from Zn diffusion, aluminum top contact can be used in this new device for excellent bondability. On the other hand, the commer- cially available n-side up device uses gold based metal as top contact and its bondability is inferior to that of aluminum. This advantage of using aluminum top contact coupled with the higher external light output makes the new device very attractive and practical for mass production.

3. Device Results

3.1 Light Emitting Pattern

Metal contact shadowing reduces external light output efficiency of all conventional LED devices. Depending on the severity of the current crowd- ing, the contact shadowing can be a very serious problem that greatly re- duces the external light output of a LED. In the new device design, how- ever, the problem of contact shadowing is eliminated as expected. Figure 2 shows the photograph of a typical light emitting pattern from one of the new devices. The round diffusion area is dark and the light emitting region is indeed outside of the diffusion area. The top contact metal near the center of the diffusion area can also be seen from this photograph. Elimination of top contact shadowing is therefore evident.

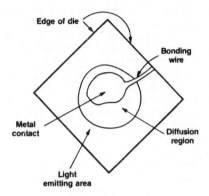

Fig. 2. Photograph of light emitting pattern of the new device.

3.2 Light Output Efficiency

Figure 3 shows the light output comparison (tested at 20 mA on bare chips) between the new diffused devices and the conventional non-diffused devices made from 22 wafers. Each data point represents an average light output measured from 40 chips (20 from the new device and 20 from the non-diffused device) processed on the same wafer. As expected, the new devices with Zn-diffusion directly underneath the metal contact yield much higher light output than that of non-diffused devices. Light output improvement factor between 1.3 and 4 has been obtained for this new diffused device. A typical improvement factor of ~2 is obtained for the new device as indicated in Figure 3.

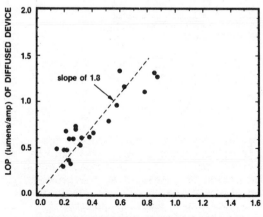

Fig. 3. Light output (lm/A) comparison between the new diffused devices and conventional non-diffused devices made from the same wafer (tested at 20 mA on bare chip).

3.3 Electroluminescence Spectra

The electroluminescence (EL) spectra of both diffused and non-diffused
devices are shown in Figure 4 at two driving currents. For a conventional
non-diffused device, the EL has a major peak corresponding to the bandgap
of the active layer, and a second peak about 14-18 nm longer in wavelength.
At the higher driving current, the second peak tends to be enhanced. This
second peak is determined to be due to the light emitting from the four
sides of the chip. Because majority of the light generated underneath the
contact needs to travel through the absorbing active layer before reaching the sides, photons with energies near the bandgap are strongly absorbed
while photons at lower energies are only weakly absorbed. The net effect
is the shift of peak wavelength toward the long side. On the other hand,
the diffused device usually shows only one EL peak even at high driving
current. This is because that the main peak in the diffused device is
much enhanced due to the elimination of contact shadowing and the light
emitting from four sides of the chip is only a small portion of the total
light output.

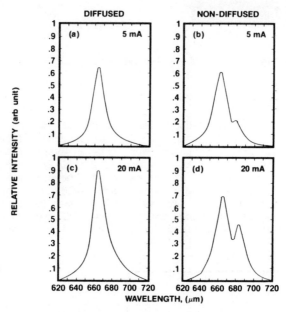

Fig. 4. Typical EL spectra of AlGaAs red LEDs for a new diffused
device: (a) and (c) and a non-diffused device: (b) and (d) at 5 mA
and 20 mA, respectively.

3.4 Minority Carrier Lifetime

EL decay times of these two types of devices were also measured. The
experimental data show that both the diffused and non-diffused devices
have the same average minority carrier lifetime. This indicates that
light emitted from the new diffused device is generated at the grown-
in p-35%/n-70% junction rather than at the diffused p-70%/n-70% junction. This also indicates that the grown-in p-n junction quality of
the diffused device is not affected by the diffusion process used to
make this new device.

3.5 Low Current Performance

A remarkable property of the new diffused device is its low current performance. The new device usually has a maximum external light output
efficiency at low driving current (1-4 mA) while the conventional non-
diffused device has a maximum efficiency at a higher driving current (7
-20 mA). Figure 5 shows the typical light output performance of both
devices as a function of driving current. The drop-off of the light
output efficiency at high currents for both devices is probably due to
thermal heating caused by the high current density. As mentioned early
in this section, current crowding underneath the contact is observed in a
conventional non-diffused device, while in the new device current is
crowding around the outer periphery of the diffusion region. This makes
the current density higher in the light generating region for the new
device compared with the conventional device under the same driving
current. This current density effect is responsible for the excellent
performance at low driving current level. As shown in Figure 5, the light
output efficiency at 2 mA for the new device is higher than that at 20 mA.
Light output as high as 2.6 lm/A without encapsulation has been obtained
for the new device at 2 mA which corresponds to about 6-7% external quantum
efficiency at 655 nm when encapsulated. This is believed to be the highest
light output efficiency reported to date for the AlGaAs red LED with GaAs
absorbing substrate.

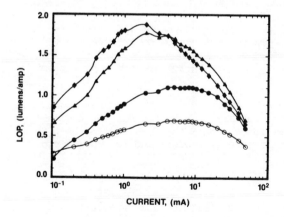

Fig. 5. Light output
(lm/A) as a function
of driving current
for two new diffused
devices (♦ and ▲) and
two non-diffused de-
vices (● and o).

4. Discussion

In conventional non-diffused AlGaAs device and most of other LED's, there
are two major problems that limit their light output performance. The
first is that their maximum light output efficiency is not obtained until
high driving currents (usually > 10 mA) are reached. The second is that
metal contact shadowing severely limits the external light output efficiency of these devices. Both of these problems are significantly reduced
by using the new device design, as described in sections 3.1 and 3.5.
The reduction of these two problems makes the new device extremely
attractive for low current applications.

Although the new device design is illustrated here using an AlGaAs red LED, the idea can also apply to other heterojunction material systems. The general idea is to confine the "active" region where the photons are generated, into a region where light extraction from the device can be most efficient. Zn diffusion used in the new device is made to guide the current flow path in such a way that the active p-n junction is confined to a region where contact shadowing is eliminated. In addition, because the diffusion region is not active, the diffusion process is not as critical as in the other cases where the diffused junction is the light generating region.

Furthermore, when different diffusion patterns are used, various devices can also be made to fit specific applications. For a device which requires a small light emitting area in a specific region, e.g. light source coupling to fiber, this new device idea can easily be applied to obtain such a device by using a proper diffusion pattern.

5. Conclusion

A new device is described to completely eliminate contact shadowing due to current crowding in AlGaAs red LED devices. The light output of this new device is generally a factor of two higher than that of the conventional device on the same wafer. This AlGaAs new device performs particularly well at low current and yields about 6-7% external quantum efficiency (at 655 nm) at 2 mA, which is believed to be the highest efficiency reported to date for the AlGaAs red LED with GaAs absorbing substrate.

This new device design can also be applied to other LED material systems. Besides eliminating the contact shadowing problem due to current crowding, this device design idea can also be used to obtain devices with various light emitting pattern for various applications.

Acknowledgements

The authors would like to thank Pat Tang, Charlotte Garvey, and Ann Davis for their technical assistance, and Betty Harrison for help in preparing the manuscript, and Dr. George Craford and Dr. Wayne Snyder for their continuing support.

References

Alferov Z I, Amosov V I, Garbuzov D Z, Zhilyaev Y V, Konnikov S G, Kopev P S, and Trofim V G, 1973 Soviet Phys.-Semicond 6 1620

Ishiguro H, Sawa K, Nagao S, Yamanaks H, and Koike S, 1983 Appl. Phys. Lett. 43 1034

Nishizawa J, Koike M and Jin C C 1983 J. Appl. Phys. 54 2807

Nishizawa J, Suto K, and Teshima T 1977 J. Appl. Phys. 48 3484

Varon J, Mahieu M, Vandenberg P, Boissy M-C and Lebailly J 1981 IEEE ED-28 416

Inst. Phys. Conf. Ser. No. 74: Chapter 6
Paper presented at Int. Symp. GaAs and Related Compounds, Biarritz, 1984

445

Leakage currents in 1.3 μm buried heterojunction laser diodes

P.I. Kuindersma, W. Dijksterhuis, A.H.E. Willekes and H.P.M. Ambrosius

Philips Research Laboratories, 5600 JA, Eindhoven, The Netherlands

Abstract

An experimental investigation into the leakage current behaviour of 1.3 μm DCPBH laser diodes, both under low and high electrical stress, is reported. Leakage currents are identified by electro- and photo-luminescence and we demonstrate the novel phenomenon of leakage enhancement due to transistor action. Thyristor breakdown is shown to be absent in DCPBH- diodes, primarily due to lateral conduction in the blocking layers. Stress test aging of diodes, which typically strongly levels off in time, is shown not to be caused by an increase in leakage current via the blocking layers.

1. Introduction

Stability in the long term as well as operation at high temperature are major issues in the field of GaInAsP laser diodes. Recently a method has been reported in which 1.3 μm laser diodes to be used in field systems are selected on the base of electrical stress test (Nakano et al 1984, Higuchi et al 1983). In these tests the diodes are subjected to high current at elevated temperature. We studied the behaviour of 1.3 μm laser diodes under such high electrical stress (currents up to 500 mA). In this paper we describe the observed leakage current behaviour of DCPBH-diodes at (low and) high electrical stress and its consequences for device performance. It is generally believed that an appreciable amount if not all of the degradation of GaInAsP buried heterostructure type lasers is related to degradation of the leakage paths, i.e. yielding an enhanced leakage after stress testing. In this paper we specifically try to find experimental proof for the deterioration of the current blocking layers in DCPBH diodes during stress testing.

2. Experimental

Investigations were carried out on DCPBH diodes made at our laboratories. Diodes of about 250 μm length have room temperature threshold currents in the range of 12 - 20 mA, usual T_0 values around 70° C (but depending on doping levels and layer thicknesses) and differential efficiencies of

the order of 0.3 to 0.4 mW/mA/facet at room temperature. The
cross section of the active layer in the mesa is about
2×0.12 (μm)^2 whereas the channel widths are about 7 μm.
Doping levels for both p and n InP layer ranged from
5×10^{17}/cm^3 up to 3×10^{18}/cm^3. The diodes were provided
with an oxide stripe, with widths from ten to fifty micron.

3. Low Currents

Leakage phenomena in DCPBH diodes are substantially different
for high and low currents. At low currents the blocking
layers as such work perfectly and leakage is found to be
dominated by a residual leakage path (LCL). The amount of LCL
follows from measuring the intensities of the quaternary
spontaneous emissions from both the mesa and the outside
channel regions via a window in the backside metallization.
An example is given in fig. 2 (left), whereas fig. 2 (right)
gives a calculated example of leakage and mesa currents,
using a simple transmission line model of distributed InP and
quaternary diodes with series resistances. In- and outside
channel sheet resistivities of the p-InP blocking layer are
input parameters together with the ratio of InP and
quaternary diodes saturation current densities and the mesa
and channel width. As shown in fig. 2, at a few mA the major
portion of the current is leakage to the outside channel
regions, but this leakage levels off at high current.

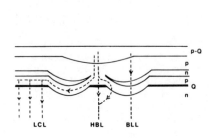

Fig. 1. Leakage paths in DCPBH
diodes.

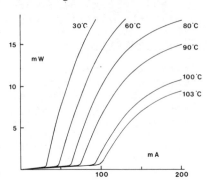

Fig. 3. Power per facet vs.
current.

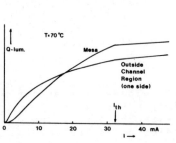

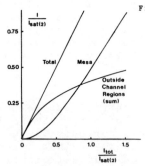

Fig. 2. Measured LCL
(luminescence and
calculated LCL
(currents);
$I_{sat}(2) \equiv 2nkT/(eR_{ch})$
where R_{ch} is the
lateral channel
resistance.

4. High Currents

High temperature operation (see e.g. fig. 3) of 1.3 μm laser diodes is limited by saturation of lasing power at high currents. This power saturation extends beyond the well-known threshold T_0-problem. The current I_{max} for which the power output is at its maximum, is usually 150 to 250 mA. Power saturation at these relatively low currents I_{max} is very unlikely to be due to thermal runaway of threshold only, despite the relatively low T_0 values of GaInAsP. Thermal runaway of threshold would give an I_{max} given by

$$I_{max} = I_t(cw) + (T_0/(VR_{th}))\ln\left[T_0/(I_t(cw)VR_{th})\right]$$

Substituting $T_0 = 60^{\circ}C$, $V = 2$ Volt, $R_{th} = 100^{\circ}C/W$ and the cw threshold current $I_t(cw) = 50$ mA, yields an I_{max} of about 500 mA. Below we will show that saturation of lasing power occurs primarily due to leakage, both across the buried mesa active layer (HBL) and parallel via the blocking layers (BLL). These high current leakage paths are localized by measuring topography and strength of their corresponding InP electroluminescences; both from the front mirror (fig. 4) and from the window in the backside metallization. Selective photoluminescence of n-InP (both the floating layer and the bufferlayer) is of help in locating the electroluminescences.

The leakage current HBL is found to consist of two parts: homobarrier leakage as well as heterobarrier leakage were observed (as indicated in fig. 1). The (field assisted) heterobarrier leakage is found to be either hole or electron leakage, or both, depending on doping levels. Blocking layer leakage (BLL) leads to electroluminescence of the top p-InP layer. Figure 5 gives an example of the increase of HBL and BLL at the expense of the quaternary spontaneous emission (QL) at increasing current levels.

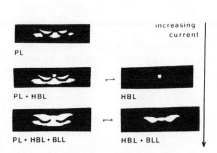

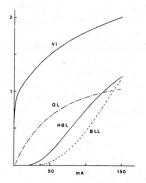

Fig. 4. Front mirror InP photo- and electroluminescence diode with electron heterobarrier leakage and blocking layer leakage.

Fig. 5. Quaternary (QL) and InP (HBL and BLL) spontaneous emission vs. current; voltage (V) vs. current.

5. Transistor Action

BLL is strongly enhanced by transistor action of the lower npn
transistor of the blocking layers. Experimentally this is
demonstrated on lasers with an additional (collector) contact
to the n-InP floating layer, while using the top contact as a
(rather remote) base contact (fig. 6). Figure 7 gives, in a
common emitter configuration, the collector current I_C as a
function of the collector-emitter voltage V_{CE} for increasing
values of the top contact laser diode current I_L; the
component I_B of I_L is acting as base current. At low V_{CE}
the collector current is dominated by the forward current I_F
in the upper pn, whereas at higher V_{CE} the collector current
equals the blocking np leakage current I_R. Moreover fig. 7
clearly demonstrates the (increasing) transistor action: a
strong enhancement of the blocking np leakage current with
increasing top laser diode current. The forward current I_F
leads to (BLL-) luminescence of the top p-InP.

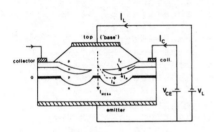

Fig. 6. Laser with additional
collector contact to
n-InP floating layer in
a common emitter
configuration.

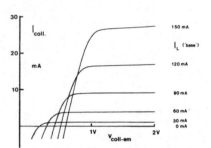

Fig. 7. Collector current I_C
vs. collector emitter
voltage V_{CE} at
increasing top contact
laser currents I_L.

Fig. 8 gives an example of a measured differential external gain,
(dI_C/dI_E) at V_{CE} constant, versus the emitter current I_E,
showing the increasing effect of blocking layer leakage with
increasing current. A gain dI_C/dI_E of 0.5 implies a reduction
of the differential quantum efficiency with an equal amount
(corresponding to 0.25 mW/mA/facet) due to BLL only. Based on
electroluminescence we estimate the HBL contribution to lowering
the laser differential efficiency to be about the same as the BLL
contribution. Furthermore there will of course be a (smaller)
thermal contribution.

Fig. 8. Measured external gain
(dI_C/dI_E) at
constant V_{CE} of 2
volts vs. the emitter
current I_E. (This
gain cannot exceed
unity.)

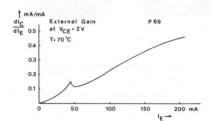

Having demonstrated the annoying phenomenon of transistor action in the lower npn, an important question remains whether or not the blocking layers pnpn does show breakover from the off- to the on-state. For diodes without the additional collector contacts we used the potential contrast in SEM to follow the voltage drops across the various junctions as a function of current. We found that up to the highest currents (500 mA) the central np in the blocking layers remained reversely biased (off-state).

6. Equivalent Circuit

Observed leakage phenomena can be described using the thyristor-ladder equivalent circuit of fig. 9. Two features are important:
a) the first npn transistor representing the channel region has a higher gain than the subsequent npn (representing the outside channel region with heterobarrier emitters) and b) lateral conduction is present in a thyristor-ladder, in contrast to the saturation one would have in a single thyristor. Due to lateral conduction only part of the collector current of the first npn transistor is available as base current to the first pnp transistor. Hence the gain of the latter is kept low and the regenerative process leading to thyristor breakover is strongly suppressed.

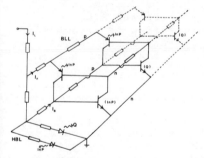

Fig. 9. Thyristor ladder equivalent circuit.

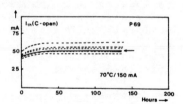

Fig. 10. Threshold vs. current during stress testing at 70°C/150 mA of devices with the additional collector contacts.

Fig. 11.

Common emitter and common base transistor characteristics before and after one week stress testing at 70°C/ 150 mA. The increase in open collector threshold is also shown.

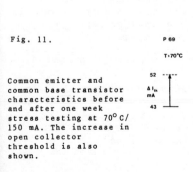

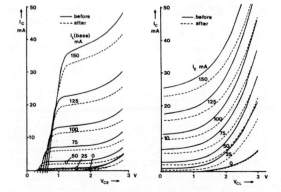

7. Leakage and Aging

It is generally believed that the degradation of InGaAsP buried heterostructure type laser diodes is related to the degradation of the leakage paths, in the sense that the junctions around the active region are less effective in confining the current to the active region after high temperature high current stress tests. We checked this point experimentally and came to opposite conclusions: laser diode devices with the additional collector contacts were stress tested at 70° C/150 mA (for one week) with open collector and we observed the well known aging which strongly levels off in time. Examples are given in fig. 10. During the stress test we measured periodically the transistor characteristics and found that leakage currents through the blocking layers did not increase. Actually we consistently found small decreases in leakage current. All diodes showed very similar behaviour. A typical example is given in fig. 11, which shows the common emitter and common base collector currents vs. voltage for various top contact currents c.q. emitter currents; together with the open collector threshold current, before and after stress testing at 70° C/150 mA. The results clearly demonstrate that neither the initial aging during the stress test nor the subsequent aging is due to increased leakage via the blocking layers. The results imply that aging is located in or around the mesa, either by increased non-radiative recombination and/or by increased HBL. Obviously such degradation in the mesa region is accompanied by an increased blocking np leakage current, when measured at constant lasing power, simply due to the transistor action described above.

8. Conclusions

Leakage in DCPBH laser diodes at low currents was found to be dominated by a residual leakage path via lateral conduction in the p-InP in the channels. At high currents/temperature DCPBH diodes suffer from leakage via the blocking layers and across the mesa, leading to power output saturation. We demonstrated that blocking layer leakage is dominated by transistor action in the lower InP npn. Initial and final aging during stress tests, of 1 week 70° C and 150 mA, were shown not to be caused by an increased leakage via the blocking layers. Thyristor breakover is shown to be absent in DCPBH diodes up to very high currents, primarily due to lateral conduction in the n-InP blocking layer.

References

Nakano Y, Iwane G and Ikegami T 1984 Electron.Lett. 20, pp 397-398

Higuchi H, Oomura E, Hirano R, Sakakibara Y, Namizaki H, Susaki W and Fujikawa K 1983 Techn.Digest of 4th Int.Conf. on IOOC, Tokyo, pp 204-205

Inst. Phys. Conf. Ser. No. 74: Chapter 6
Paper presented at Int. Symp. GaAs and Related Compounds, Biarritz, 1984

451

LP-MOCVD growth and cw operation of high quality SLM and DFB semiconductor $Ga_xIn_{1-x}As_yP_{1-y}$–InP lasers

M. RAZEGHI, R. BLONDEAU, J.C. BOULAY*, B. de CREMOUX, J.P. DUCHEMIN
THOMSON-CSF, Domaine de Corbeville, BP n°10, 91401 Orsay Cedex (FRANCE).
*CNET, 196 Rue de Paris, 92220 Bagneux, France.

SUMMARY. Room temperature CW operation of single longitudinal mode (SLM) and distributed feed back (DFB) semiconductor lasers have been achieved in $Ga_xIn_{1-x}As_yP_{1-y}$-InP (1.2-1.6 µm region) DH lasers fabricated on material grown by two-step low pressure metalorganic chemical vapor deposition (LP-MOCVD), over large area of InP substrate (10 cm^2).

Room temperature CW threshold currents as low as 25 mA have been measured for stripe buried (SB) devices with cavity length of 300 µm and stripe width of 2 µm.

1.57 µm low threshold current (60 mA CW) stable SLM (DFB) operation at fixed mode was obtained under the CW condition, in the temperature range from 9°C to 90°C, with the wavelength shift of 0.9 A/°C.

1. Introduction

The choice of a radiation source suitable for optical cable telecommunication systems is connected with two typical parameters of the transmission medium : attenuation as a function of wavelength and material dispersion. Optical glass fibres have their lowest loss at a wavelength around 1.5 µm and the miniumm dispersion around 1.3 µm.

The $Ga_xIn_{1-x}As_yP_{1-y}$ alloys lattice matched to InP with resulting band gap of 0.75 to 1.35 eV cover the region of low loss and minimum dispersion of silica optical fibres. This makes the material a suitable active layer in double heterostructure lasers which may be used as light sources in fibre communication systems.

Room temperature pulse operation and continuous wave (CW) operation in the 1.2-1.6 µm region have been achieved in GaInAsP-InP DH lasers fabricated on material grown by LP-MOCVD (Razeghi et al., 1984). Threshold current densities as low as 430 A/cm^2 have been obtained for devices emitting at 1.3 µm (Razeghi et al., 1984), and 960 A/cm^2 for devices emitting at 1.55 µm (Razeghi et al., 1984).

In this paper we report very low threshold current Buried Ridge Structure (BRS) GaInAsP-InP lasers emitting at 1.3 µm and 1.55 µm and also the growth of Distributed Feedback (DFB) lasers emitting at 1.57 µm fabricated on material grown completely by LP-MOCVD.

2. BRS lasers emitting at 1.3 µm

Fig.1 shows the schematic diagram of the cross-section of a BRS laser grown by two-step low pressure metalorganic chemical vapor deposition (LP-MOCVD). This laser was fabricated as follows.

First the following layers were grown successively on InP(S) substrate with orientation (100) exact or (100) 2° off.

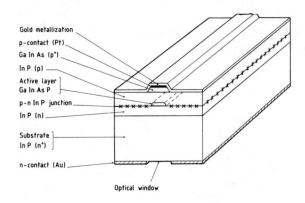

Gold metallization
p-contact (Pt)
Ga In As (p⁺)
In P (p)
Active layer
Ga In As P
p-n In P junction
In P (n)
Substrate
In P (n⁺)
n-contact (Au)
Optical window

Fig.1 : Schematic dia-
gram of the Buried
Ridge Structure of
1.3 μm laser.

- 1 μm InP sulphur doped for confinement layer,
- 0.2 μm GaInAsP (λ = 1.3 μm) undoped active layer,
- 0.2 μm InP, Zn doped confinement layer.
Next, a ridge of about 2 μm width and 0.4 μm depth was etched in the InP
and GaInAsP layers through a photolithographic resist mask, using a selective etchant composed of $SO_4H_2-H_2O_2-H_2O$ (1:8:40).
After removing the resist mask, the ridge was then covered by a 1 μm thick
Zn doped InP confinement layer and 0.5 μm Zn doped $Ga_{0.47}In_{0.53}As$ contact
layer with N_A-N_D = 1×10^{19} cm^{-3} (using again the LP-MOCVD growth technique).
In order to localize the injection only in the buried ridge active region,
a deep proton implantation was realized through a 5 μm wide photoresist
mask after the metallisation of the contacts. Further localisation of the
current in the buried ridge is achieved by the built-in potential difference between the P-N InP homojunction on each side of the active region and
the N-P InP-GaInAsP heterojunction of the active region.
After cleaving and scribing into chips 300 μm and 350 μm wide, the lasers
were mounted epi-side down, onto nickel-plated copper heatsinks using indium.
Fig.2 shows the light-current (P-I) and voltage-current characteristics
under CW operation at 20°C. A threshold current as low as 15 mA with a
linear output power of more than 10 mW have been measured. External quantum efficiencies of 60 % have been achieved. The standard deviation of
threshold current is less than 5 % over 8 cm^2.
Single transverse mode operation up to 15 mW has been obtained with multi
longitudinal mode.
Fig.3 presents the lasing characteristics of this BRS laser under CW operation at 20°C.
A few diodes randomly selected were life-tested.
Fig.4 shows that these lasers operate at 70°C with output power of 5 mW
per facet for more than 1000 hours without significant degradation.

3. BRS lasers emitting at 1.5 μm

Fig.5 presents the scanning electron microscope cross-section of BRS lasers
emitting at 1.5 μm, fabricated on material grown by two-step LP-MOCVD. The
fabrication process is similar to the 1.3 μm BRS laser mentioned before. A
threshold of 25 mA with a linear output of 10 mW at 20°C has been measured.
Fig.6 shows the variation of threshold current for different temperatures.
The value of T_0 is 65° for this device.

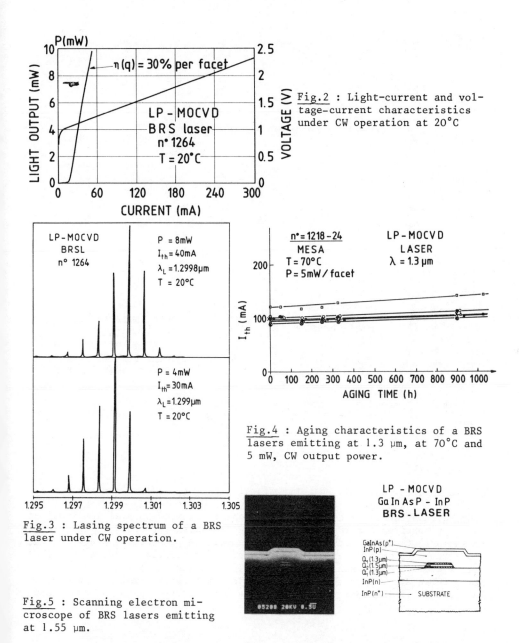

Fig.2 : Light-current and voltage-current characteristics under CW operation at 20°C

Fig.4 : Aging characteristics of a BRS lasers emitting at 1.3 μm, at 70°C and 5 mW, CW output power.

Fig.3 : Lasing spectrum of a BRS laser under CW operation.

Fig.5 : Scanning electron microscope of BRS lasers emitting at 1.55 μm.

The far field patterns parallel to the junction plane of these lasers are shown in fig.7. These patterns were measured under CW operation for different optical output powers. A stable operation in the fundamental transverse mode without significant distortion was observed at least up to 10 mW. Far field full width at half power, both parallel and perpendicular to the junction plane, were about 25 and 38 degrees, respectively.

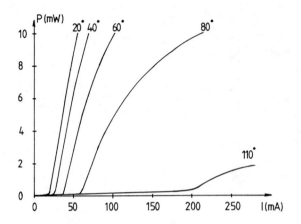

Fig.6 : Light-current
characteristics for dif-
ferent temperature
heatsink, for a laser
emitting at 1.5 μm.

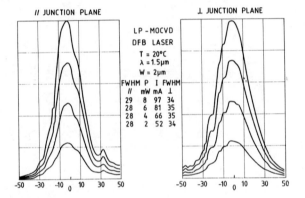

Fig.7 : Far-field pat-
terns parallel and per-
pendicular to the
junction of a DFB laser
for different output
powers under CW opera-
tion.

4. DFB lasers

Single longitudinal mode selection is difficult to obtain in a semiconductor laser with a Fabry-Perot optical cavity. Owing to the broad gain spectrum, even at moderate current drive along threshold, several longitudinal
modes are excited ; a typical lasing spectrum with reference to this case
is shown in fig.3. Spectral selectivity could be provided by Bragg diffraction from a periodic spatial variation of the refractive index within the
gain medium of the laser, as suggested by Kogelink and Shank (1972), and
may provide a feedback mechanism to replace the Fabry-Perot cavity mirrors.
This occurs if the variation period Λ is fixed according to the Bragg condition for a back (180°) reflection as

$$\Lambda = \frac{m\lambda_L}{2n_{ep}} \qquad m = 1,2, \ \dots$$

where m is the grating order, λ_L, lasing wavelength and n_{ep} the equivalent
refractive index of the medium.
In this case, possible resonance frequencies have approximately the same
spacing in wavelength as the Fabry-Perot cavity, but the threshold gain
increases sharply with deviations from the wavelength selected from the
Bragg condition. A single longitudinal mode operation can be obtained by
means of this feedback mechanism with a spectral bandwidth $\Delta\lambda \leq 1$ Å.

For a semiconductor laser, a periodic variation of the refractive index may be caused by a corrugation grating on a dielectric waveguide near the active layer of the device.

Such DFB lasers with second order corrugation grating have been fabricated on material (GaInAsP-InP) grown for the first time by the LP-MOCVD growth technique (Razeghi et al., 1984).

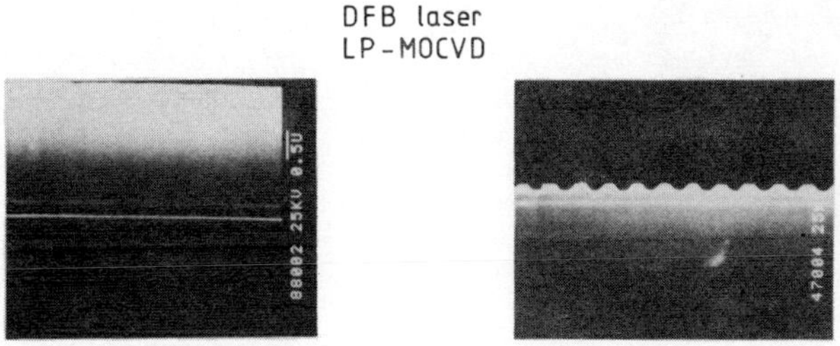

DFB laser
LP-MOCVD

Fig.8 shows scanning electron microscope cross-sections of the DFB laser structure before and after LP-MOCVD regrowth. No significant surface deformation has occurred and the final depth of the corrugation is similar to the pregrowth height (see M. Razeghi et al., 1984).

Fig.8 : SEM cross-section of the corrugated structure before and after LP-MOCVD overgrowth.

Typical light-current characteristics and lasing spectrum of the DFB laser at 20°C are shown in Fig.9.

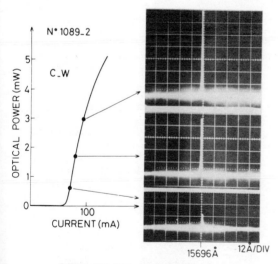

N° 1089-2

C_W

OPTICAL POWER (mW)

CURRENT (mA)
100

15696Å   12Å/DIV

RIDGE MO-CVD DFB LASER

Fig.9 : Light current and lasing spectrum characteristics of DFB laser at 20°C under CW operation.

This DFB laser operated in single longitudinal mode in the temperature range between 10 and 90°C, with a temperature coefficient of the lasing wavelength $d\lambda_L/dT$ of 0.9 Å/C.

The CW threshold current density of 60 mA at 20°C has been obtained with a linear output power of 6 mA. An external quantum efficiency of 15 % per facet has been measured.

Stable SLM operation under high speed direct modulation up to 500 P sec and under sinusoïdal modulation at 1 GHz have been obtained (fig.10).

In conclusion, we have shown that the LP-MOCVD growth technique is promising for large scale production of GaInAsP-InP DH lasers emitting at 1.3 μm and 1.55 μm. We have also demonstrated that high quality DFB lasers may be fabricated on material grown completely by LP-MOCVD.

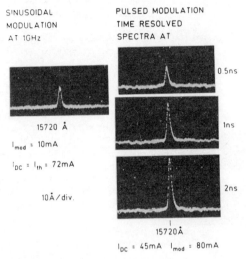

SPECTRA OF MODULATED LIGHT

SINUSOIDAL
MODULATION
AT 1GHz

PULSED MODULATION
TIME RESOLVED
SPECTRA AT

0.5ns

15720 Å

1ns

I_{mod} = 10mA

I_{DC} = I_{th} = 72mA

2ns

10Å/div.

15720Å

I_{DC} = 45mA I_{mod} = 80mA

Fig.10 : Lasing spectrum under rapid direct modulation at 1 GHz.

1.57 μm RIDGE OM-CVD DFB LASER

Acknowledgments

We want to thank D. Leguen, L. Noel, G. Vilain and J. Ricciardi for technical assistance, K. Kazmierski for preparation of corrugation grating, M. Krakwiski, P. Hirtz and B. Lent for useful discussion and characterisation.

References

1. M. Razeghi, B. de Cremoux, J.P. Duchemin. J. Crystal Growth (1984) (to be published).
2. M. Razeghi. "Light wave technology for communication". Edited by W.T. Tsang and A.C. Beer (1984). (Academic Press).
3. M. Razeghi and J.P. Duchemin. J. Crystal Growth (1984), to be published.
4. H. Kogelink and C.V. Shank. J. Appl. Phys. 43 (1972) 2327.
5. M. Razeghi, R. Blondeau, M. Krakowski, J.P. Boulay, B. de Cremoux and J.P. Duchemin. Appl. Phys. Lett. (1984), to be published.

Inst. Phys. Conf. Ser. No. 74: Chapter 6
Paper presented at Int. Symp. GaAs and Related Compounds, Biarritz, 1984

457

Properties of semi-insulating $In_{0.53}Ga_{0.47}As$ epitaxial layers and photoconductive detectors

M. V. Rao and P. K. Bhattacharya

Solid State Electronics Laboratory, Department of Electrical
Engineering and Computer Science, University of Michigan, Ann Arbor,
Michigan 48109, USA

Abstract. Growth of semi-insulating $In_{0.53}Ga_{0.47}As$/InP by liquid phase epitaxy has been investigated. Fe and Cr were used as dopants and added to the growth melt. Fe-doped layers with $\rho{\sim}500\Omega$-cm, $n{\sim}2{\times}10^{12}$ cm^{-3}, and high carrier mobilities could be grown consistently. Cr doping under identical growth conditions failed to produce high resistivity layers. Photoconductive detectors fabricated on the Fe-doped layers exhibit dark currents in the μA range, optical gains ~10 and photo-response to pulsed excitation with FWHM~400ps. These figures can be further improved with more optimal design.

I. Introduction

$In_{0.53}Ga_{0.47}$As lattice-matched to InP is important for microwave and opto-electronic device applications. In particular, its room-temperature bandgap of 0.74eV makes it useful as a photodetector material in fiber optical communication systems operating at wavelengths between 1.1 and 1.7 μm.

Liquid phase epitaxial (LPE) $In_{0.53}Ga_{0.47}As$ with free-electron concentration $n{\sim}10^{14}cm^{-3}$ has been grown by adopting extended melt-baking schemes (Oliver and Eastman 1980, Cook et al 1982, Amano et al 1981, Kuphal and Pöcker 1982, Bhattacharya et al 1983) and by compensating the residual shallow donors with controlled addition of Zn shallow acceptors (Pearsall et al 1981). In order to reduce n further to the near-intrinsic limit, an alternate method is to introduce impurities which produce compensating deep levels in the energy bandgap. Suitable dopant species are Fe and Cr. This technique has been used to grow SI GaAs by LPE (Mattes et al 1975, Woodward et al 1981, Kojima and Hasegawa 1980, P. A. Houston 1980).

Photoconductive detectors have advantages over photodiodes because of simplicity of fabrication, uniformity, and their potential for achieving photoconductive gains greater than unity. Carrier recombination and transit time in these devices are independent, so that speed, gain and device dimensions can be traded off against each other depending on the desired application. The performance of photoconductive detectors made on n- and p-type $In_{0.53}Ga_{0.47}As$ and modulation doped $In_{0.53}Ga_{0.47}As$/$In_{0.52}Al_{0.48}As$ heterostructures has been reported (Degani et al 1981, Gammel et al 1981, Klein et al 1981, Chen et al 1983, 1984a, 1984b, 1984c). However, the performance of devices made with the high-resistivity ternary compound has not been demonstrated. A useful application is the integration of such detectors with FETs made with the same high-resistivity material (Weider et al 1983).   It is therefore essential that the carrier mobilities in   the high-resistivity material be as high as possible.

We report here our experiments on Fe and Cr doping of LPE $In_{0.53}Ga_{0.47}As$ and the performance of photoconductive detectors made with the Fe-doped material. Comparisons have been made with identical devices made on Zn-doped p-type $In_{0.53}Ga_{0.47}As$.   It is observed that the internal optical gains in the devices are amongst the highest measured and the optical pulse response characteristics are favorable for high-speed applications.

II.   Simplified Theory of Photoconductive Detectors

The gain of a photoconductor (defined as the number of carriers collected in the external circuit per incident photon) is given by:

$$\Gamma = \frac{h\nu I_{ph}}{q \, n_i P_L} \tag{1}$$

where $h\nu$, n_i, P_L, and I_{ph} are photon energy, internal quantum efficiency, incident optical power on the active area, and photocurrent, respectively. The internal optical gain of the device is given by:

$$G_{opt} = \frac{h\nu I_{ph}}{q \, P_L} = n_i \Gamma \tag{2}$$

In terms of material and device parameters the internal gain can be expressed as (Beneking 1982):

$$G_{opt} = n_i \tau_{eff}(\mu_n + \mu_p) \frac{V}{L^2} \tag{3}$$

where τ_{eff}, L and V are effective carrier lifetime, length of the active area and applied bias, respectively. Since $\mu_p \ll \mu_n$,

$$G_{opt} \cong n_i \, \tau_{eff}\mu_n \frac{V}{L^2} \tag{4}$$

$$= n_i \frac{\tau_{eff}}{t_r} \tag{5}$$

where t_r is the carrier transit time. The gain-bandwidth product of a photoconductive detector is given by (Beneking 1982):

$$\Gamma B = \frac{1}{2\pi}(\mu_n + \mu_p) \frac{V}{L^2} \tag{6}$$

From Eqns. (4) and (6) it is evident that both the internal gain and the (ΓB) product can be enhanced by reducing L. Taking account of surface reflection by the reflection coefficient R and incomplete absorption in the active region, the external photoconductive gain is given by:

$$\Gamma_{ext} = (1-R)(1-e^{-W/a})\Gamma \tag{7}$$

where W is the thickness of the active layer and a is the absorption length of the incident light.

III.   Material Growth and Properties

Epitaxial layers, 5-10 μm thick, were grown typically at 650°C using the standard step-cooling technique. Extended melt-baking was utilized (Bhattacharya et al 1983) to lower the residual donor and acceptor concentrations in the epitaxial layers to $\sim (1-5) \times 10^{15} cm^{-3}$. In of 6N purity was baked at 730°C for 10 hr. followed by the addition of polycrystalline GaAs and InAs and a further bake at 680°C for 10 hr.

Controlled amounts of elemental Fe or Cr were added to the melt at the time of loading the substrate. The upper limit on $X^{\chi}_{Fe/Cr}$ was dictated by melt carry-over problems and/or degradation in surface morphology. Figure 1 depicts the measured variation of net donor density with X^{χ}_{Fe} at room temperature in representative epitaxial layers, as obtained from Hall measurements. For $X^{\chi}_{Fe} \simeq 3 \times 10^{-4}$ we have consistently achieved a low free-carrier concentration, contrary to the observation of Clawson et al (1983). The 300K Hall mobility varied from 10,200 $cm^2/V \cdot s$ in undoped layers to $\sim 6000\ cm^2/V \cdot s$ in layers doped heavily with Fe. From the analysis of temperature-dependent Hall data on conducting Fe-doped samples the Fe acceptor ionization energy is found to be 0.46 eV. On the contrary, the free-electron concentration in Cr-doped layers was found to increase with increased addition of Cr. This could be due to residual impurities like S in Cr metal which act as donors or a donor-like complex involving Cr and a native-defect being formed in the solid phase.

Photoluminescence spectra of undoped and Fe-doped epitaxial layers at 4.2K are shown in Fig. 2. It is clear that no new transition due to Fe is present in the edge luminescence. The sharp peak centered at 1531 nm has been studied by us earlier (Bhattacharya et al 1983) and has been attributed to donor-bound exciton (D-X) transitions.

IV. Device Fabrication
Photoconductive detectors were fabricated with Fe-doped $In_{0.53}Ga_{0.47}As$ having $(N_D-N_A) \sim 2 \times 10^{12} cm^{-3}$. The measured Hall mobility at 300K in this material is $\sim 6000\ cm^2/V \cdot s$. For comparison, devices were also fabricated with Zn-doped

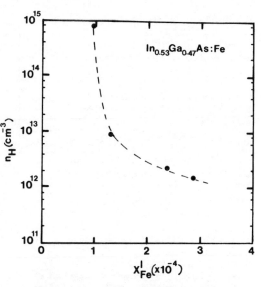

Figure 1.   Variation of net donor density at room temperature with X^{χ}_{Fe} in Fe-doped LPE $In_{0.53}Ga_{0.47}As$. The growth temperature is 650°C.

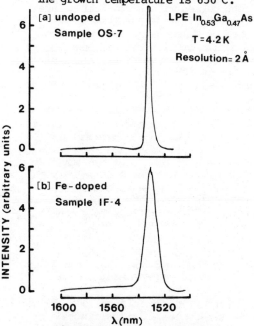

Figure 2. Photoluminescence spectra recorded at 4.2K with undoped and Fe-doped LPE $In_{0.53}Ga_{0.47}As$. The peak centered at 1531 nm is attributed to donor-bound exciton transitions.

p-type $In_{0.53}Ga_{0.47}As$ having
$p \sim (2-5) \times 10^{15}cm^{-3}$. The
devices are of source–drain
and interdigitated configurations with Ni/Au–Ge electrodes having 7 μm spacings.
Typical dark current–voltage
characteristics of non-
interdigitated devices on
$In_{0.53}Ga_{0.47}As$:Fe are shown in
Fig. 3. The measured currents
are about three times larger
for interdigitated devices. The
peak response of the devices is
at 1.68 μm at room temperature.

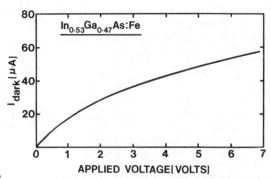

Figure 3. Dark current–voltage
characteristics for a planar
source–drain photoconductive
detector.

V. Device Results and Discussion

The response speed for a
non-interdigitated device mounted on
a coaxial stripline was measured using
a (GaIn)(AsP) laser emitting at 1.29 μm.
The laser was driven by a comb generator
resulting in 300 ps or 11 ps (FWHM)
light pulses with a repetition rate
of 100 MHz. Fig. 4 (a) shows the
detector response for 300 ps pulses
at a bias of 6V. The response is
characterized by a rise time t_r of
200 ps (which includes light pulse
and oscilloscope rise times), FWHM of
400 ps, and a fall time t_f of 1.4 ns.
The large t_f is a result of a 3 ns
decay in the turn-off characteristics
and limits the 3dB bandwidth to $\sim$ 114
MHz. However, fall times almost twice
as large have been reported (Chen et
al 1984a) in devices made on n-type
$In_{0.53}Ga_{0.47}As$ even with shorter
channel lengths. The results
indicate that faster devices can be
fabricated with more optimal design.
Figure 4(b) shows that for 11 ps
excitation pulses at 6V bias t_r, t_f
and the FWHM are 6ns, 10ns, and 11 ns,
respectively, in the same device.

The variations of internal optical gain
with bias voltage in devices made on
Fe- and Zn-doped alloys are shown in
Fig. 5. Excitation was provided with
a 0.63 μm He–Ne laser. The Zn-doped
device exhibits higher gain due to
higher minority carrier (electron)
mobility and possibly higher carrier
life time. The measured gains of $\sim$ 8

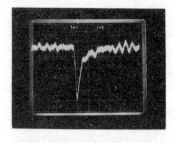

[a]

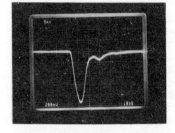

[b]

Figure 4. Photoresponse of
$In_{0.53}Ga_{0.47}As$ photoconductor
at 6V bias to (a) 300 ps and
(b) 11 ps pulses provided by
1.29 μm InGaAsP laser.

and 11 in these devices at
6V are amongst the highest
values reported. Still
higher values can be expected
with smaller channel lengths.
At very low bias values the
injected electron-hole pair
lifetime is determined mainly
by bulk and surface
recombination so that
photoexcited pairs recombine
before reaching the contact.
With increasing bias the
minority-carrier transit time
decreases and the gain increases.
The initial linear increase of
gain with bias is in accordance
with Eqn. 3. Values of $\tau_{n,p}$
= 350 ps and ($\mu_n + \mu_p$) = 6400
and 8200 $cm^2 V \cdot s$ for Fe and Zn-
doped material, respectively,
give reasonable fits to the
initial linear regions of the
data. The subsequent small
increase of G_{opt} with bias
is probably caused by a
combination of carrier velocity
saturation and amplification

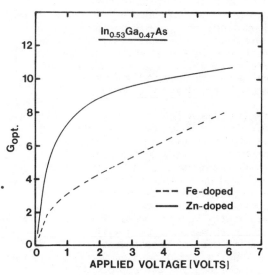

Figure 5. Variation of
internal optical gain with
applied bias.

induced by avalanching. The latter may be facilitated by the relatively
small bandgap (Klein, et al 1981). The sensitivity of Fe and Zn-doped
devices are 2.2 and 3.2 A/W, respectively, at a bias of 2V. The
variation of G_{opt} with temperature at different incident power levels
is depicted in Fig. 6. The
measurements were performed
with a 0.63 μm He-Ne laser.
It is apparent that the
variation of carrier mobility
and lifetime with temperature
is responsible for the observed
trend in the gain. The solid
lines in Fig. 7 indicate
theoretical values of G_{opt}
calculated from estimated
values of carrier mobility
and lifetime. The material
parameters used in the analysis
are identical to those reported
earlier by Bhattacharya et al
(1983). The decrease in gain
with increasing light power is
due to the decrease in the value
of τ_{eff} with increasing light
intensity (Rose 1956). It has
been assumed that τ_{eff} varies
in the range 300-400 ps
with increasing light intensity
under conditions in which the

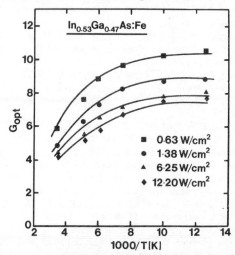

Figure 6. Variation of
internal optical gain with
temperature at various levels
of excitation intensity.

data of Fig. 6 were obtained. However, the temperature dependence of τ_{eff} has not been considered. Similar trends are reflected in Fig. 7 which depicts the variation of G_{opt} with bias voltage in a Fe-doped device under varying excitation levels.

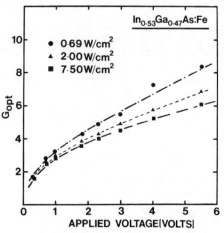

Figure 7. Variation of internal optical gain with applied bias at various levels of excitation intensity.

Acknowledgements
The authors thank Dr. C. Y. Chen of AT&T Bell Laboratories, Murray Hill, for performing the response speed measurements. The help provided by Professor E. Leith and other members of the Electro-optics Laboratory and L. Shive is gratefully acknowledged. Part of this work was performed while the authors were at Oregon State University, Department of Electrical Engineering, and was supported by the National Science Foundation.

References
Amano T, Takahei K and Nagai H 1981 Japan. J. Appl. Phys. 20 2105
Beneking H 1982 IEEE Trans. Electron Dev. ED-29 1420
Bhattacharya P K, Rao M V and Tsai M-J 1983 J. Appl. Phys. 54 5096
Chen C Y, Pang Y M, Garbinski P A, Cho A Y and Alavi K 1983 Appl Phys. Lett. 43 308
Chen C Y, Pang Y M, Alavi K, Cho A Y and Garbinski P A 1984a Appl. Phys. Lett. 44 99
Chen C Y, Kasper B L and Cox H M 1984b Appl. Phys. Lett. 44. 1142
Chen C Y, Pang Y M, Cho A Y, Alavi K and Garbinski P A 1984c J. Vac. Sci. Technol. B 2 262
Clawson A R, Mullin D P and Elder D J 1983 J. Cryst. Growth 64 90
Cook L W, Tashima M M, Tabatabaie N, Low T A, and Stillman G E 1982 J. Cryst. Growth 56 475
Degani J, Leheny R F, Nahory R E, Pollack M A, Heritage J P and DeWinter J C 1981 Appl. Phys. Lett. 38 27
Gammel J C, Ohno H and Ballantyne J M 1981 IEEE J. Quantum Electron. QE-17 269
Houston P A 1980 J. Electron. Mater. 9 79
Klein H J, Kaumanns R and Beneking H 1981 Electron. Lett. 17 421
Kojima K and Hasegawa H 1980 Phys. Stat. Solidi(a) 62 673
Kuphal E and Pocker A 1982 J. Cryst. Growth 58 133
Mattes B L, Houng Y and Pearson G L 1975 J Vac. Sci. Technol. 12 869
Oliver J D and Eastman L F 1980 J Electron. Mater. 9 693
Pearsall T P, Beuchet G, Hirtz J P, Visentin N and Bonnet M 1981 Inst. Phys. Conf. Ser. 56 639
Rose A 1956 Proc. Photoconductivity Conf. eds R G Breckenridge, B R Russell and E E Hahn (New York:Wiley) pp 3–4
Weider H H, Veteran J L, Clawson A R and Mullin D P 1983 Appl. Phys. Lett. 43 287
Woodward D W, Kirchner P D, Schaff W J, Tiwari S, Stall R and Eastman L F 1981 Inst. Phys. Conf. Ser. 56 83

Inst. Phys. Conf. Ser. No. 74: Chapter 6
Paper presented at Int. Symp. GaAs and Related Compounds, Biarritz, 1984

463

2 DEG III-V photoconductive detectors: an attractive device for photoreceiver applications

D. Decoster, J.P. Vilcot, A. Boyodi, J. Ramdani and M. Constant*

Centre Hyperfréquences et Semiconducteurs - L.A. C.N.R.S. n° 287 - Bât. P4
Université des Sciences et Techniques de Lille I - 59655 VILLENEUVE D'ASCQ
CEDEX - FRANCE.

* Laboratoire de Spectrochimie Infrarouge et Raman. Université des Sciences
  et Techniques de Lille I - 59655 VILLENEUVE D'ASCQ CEDEX - FRANCE.

Abstract : Gain and noise properties of GaAs, GaInAs, GaInAs-InP,
GaAlAs-GaAs planar photoconductive detectors are studied. The influence
of the 2 DEG on the internal gain is considered. The noise due to
the illumination of these devices is found to be lower than the noise
of avalanche photodiodes.

1. Introduction

Several experimental and theoretical investigations have reported the
high performance of planar photoresistors in III-V materials (Gammel et al
1981, Klein et al 1982, Pang et al 1983). More precisely, high steady-
state gains (up to 10^9) and large gain-bandwidth products (up to 30 GHz)
can be obtained using N-type GaAs planar photoconductive detectors with
short electrode spacing (Vaterkowski et al 1982). Moreover it has been
shown that, for the same value of the gain, the noise due to the illumination of such devices is lower than that of an avalanche photodiode (Vilcot
et al 1984 b). The high-gain performances have been associated with trapping effects at the surface (Vilcot et al 1984 a). As a consequence,
an improved device can be obtained if : (i) the trapping effects are
well controlled, and (ii) the transit time is reduced. Thus, heterojunction
structures would appear to be very promising. In particular, a control
of trapping effects at the interface could be expected. Furthermore,
the reduction of Coulombian scattering in the modulation-doped 2 DEG
gives rise to extremely high mobility (Linh 1983). For these reasons,
GaAlAs-GaAs and GaInAs-InP photoresistors have been studied, and comparisons with GaAs and GaInAs planar photoresistors have been made.

2. GaAs planar photoresistors

N-type GaAs planar photoconductive detectors have been fabricated in
our laboratory. The basic device is constituted of two ohmic contacts
on layers suitable for F.E.T. (see insert fig. 1). This structure allows
monolithic integration. Buffer-type devices have also been fabricated
by etching the GaAs layer.

Examples of the high performance obtained with GaAs planar photoresistors
are given in figures 1, 2 and 3. Figure 1 presents typical steady-state
gains obtained for N-type GaAs planar photoresistors. These experiments

shows that the sensitivity can reach 10^7 A/W for a 15 µm electrode spacing. The high gain values have been interpreted using the relation

$$G = \tau_v/\tau_t \qquad (1)$$

where τ_v is the electron-hole pair lifetime connected with trapping effects at the surface and τ_t is the electron transit time (Vilcot et al 1984 a).

In figure 2 is plotted the dynamic gain of a notched N-type GaAs planar photoresistor. It decreases as the frequency increases with a gain-bandwidth product higher than 10 GHz. These experimental results can be described by the expression (Vilcot et al 1984 c).

$$G(\omega) = \frac{\tau_v}{\tau_t} \frac{1}{\sqrt{1+\tau_v^2\omega^2}} \qquad (2)$$

corresponding to the assumption of a generation-recombination process. In figure 3 is reported a typical fast response obtained with a buffer-type GaAs photoresistor. In this case the steady-state gain is, obviously, lower than for the N-type device.

All of these results show that the performance of planar photoresistors depends strongly upon the electron-hole pair lifetime and the electron transit time. In the next part of this paper, we present the influence of the 2 DEG on the performance of the devices.

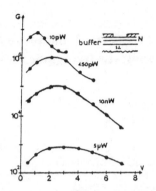

Fig. 1 : Steady state gain of a N-type GaAs planar photoresistor versus the bias voltage V for various light power. Electrode spacing 15 µm. Measurements performed using an HeNe laser.

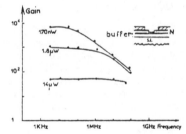

Fig. 2 : Dynamic gain of a notched N-type GaAs planar photoresistor versus the frequency for various light powers. Measurements performed using a sinusoïdaly modulated laser diode (0.8 µm wavelength). Electrode spacing 10 µm. Bias voltage 5V.

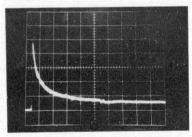

Fig. 3 : Picosecond response of a buffer type GaAs photoresistor. Measurements performed using a synchronously pumped mode-locked cavity-dumped dye laser. Time base 2ns/div.

3. Heterojunction structure planar photoresistors

The heterojunction structure planar photoresistors are presented in figure 4 (see insert). The GaAlAs-GaAs heterostructure was made by M.B.E. It is constituted of a 1000 Å thick and highly doped (10^{18} at/cm³) GaAlAs layer matched to a 1 μm thick and undoped GaAs layer with a 20 Å thick undoped GaAlAs spacer. The GaInAs-InP heterostructure is constituted of a 400 Å thick undoped ($< 10^{15}$ at/cm³) InP layer matched to a 1.25 μm thick undoped

GaInAs layer which was grown by MOCVD on a InP substrate. Under such conditions GaAlAs and InP thin layers are windows for the 0.8 μm and 1.3 μm (or 1.5 μm) wavelengths, respectively. We have also made GaInAs planar photoresistors by etching the InP layer of the GaInAs-InP heterojunction structure.

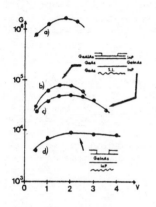

Fig. 4 : Steady-state gain versus the bias voltage of N-type GaAs (a), GaAlAs-GaAs (b), GaInAs-InP (c) and GaInAs (d) planar photoresistors. Electrode spacing 5 μm. For the GaAs and GaAlAs-GaAs photoresistors, the measurements have been performed using a 0.8 μm wavelength CW laser diode. For the GaInAs and GaInAs-InP photoresistors, a 1.3 μm wavelength CW laser diode has been used. P_L = 3 nW.

We present (figure 4) a comparison between the steady-state gain of N-type GaAs and GaAlAs-GaAs photoresistors. These experimental results clearly show that the heterojunction reduces the gain. The same order of magnitude has been found for the gain of the GaInAs-InP heterojunction structure. The steady-state gain of GaInAs planar photoresistors is even lower. As we can deduce from equation 1, the heterostructure decreases the lifetime τ_V for GaAs and increases τ_V for GaInAs. These results can be explained by the high value of the surface potential of GaAs (Vilcot et al 1984 a), by the low value of the surface potential of GaInAs, and by interface effects in the heterojunction structures. The influence of the interface on the electron-hole pair lifetime can also be observed on the dynamic gains of N-type GaAs and GaAlAs-GaAs photoresistors (figure 5). The dynamic

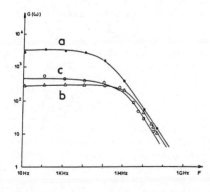

Fig. 5 : Dynamic gain of N-type (a), GaAlAs-GaAs (b), GaInAs-InP (c) planar photoresistors. Electrode spacing 20 μm. The measurements have been performed using a 0.8 μm wavelength sinusoïdaly modulated laser diode. Light power 4.5 μW. Bias voltage 2.5 V.

gain of GaAlAs-GaAs photoresistors can be described by the relation (2). But in the low frequency range, it is lower than the GaAs photoresistor gain. In this case, the 2 DEG increases the high-frequency cut-off. It can be noted that the GaInAs-InP photoresistor gain presents the same behaviour as the GaAlAs-GaAs photoresistor.

The noise figures of these components have been measured over the 10 MHz-

1.5 GHz frequency range using a HP 8970 A noise figure meter. The noise measurements have been performed in darkness and under an HeNe illumination. Figure 6 shows typical results. For all of the devices, the noise

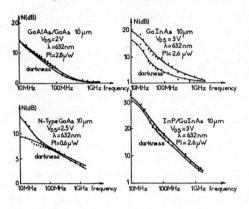

Fig. 6 : Noise power against frequency of N-type GaAs, GaAlAs-GaAs, GaInAs-InP planar photoresistors measured in darkness and under illumination. The 0 dB level corresponds to the noise power of a 50 Ω resistor at 293 K.

values decrease as the frequency increases. The noise due to illumination can be measured by subtracting the noise in darkness from the noise under illumination. Under an assumption of a generation-recombination process, this noise detected in a bandwidth B centered at ω can be expressed as a function of the dynamical gain (Vilcot et al 1984 b).

$$\overline{i^2} = 4q\,I_{ph}\,\frac{G^2(\omega)}{G}\,B \qquad (3)$$

where G is the steady-state gain. This expression can be experimentally verified by comparing (figure 7) the gain values determined from noise measurements to those obtained using sinusoïdaly modulated laser diode (0.8 μm wavelength). This comparison is presented in figure 7 for the 2 DEG photoresistors. The good agreement, within experimental errors,

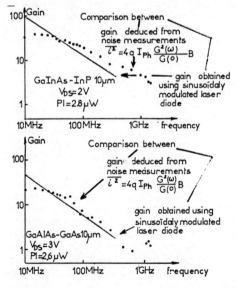

Fig. 7 : Comparison between gain values deduced from noise measurements to those obtained using a sinusoïdaly modulated laser diode (0.8 μm wavelength).

between the data obtained by the two methods shows that equation (3) represents, to a good approximation, the noise due to illumination for the 2 DEG photoresistors as well as for the N-type GaAs photoresistors (Vilcot et al 1984 b). In the case of an avalanche photodiode (APD) the noise due to illumination is given by

$$\overline{i^2} = 2q\,I_{ph}\,\frac{G^2(\omega)}{G}\,FB \qquad (4)$$

In this relation, F is the excess noise ratio due to the hole ionisation mechanism whose minimum value is 2 when the hole ionisation ratio β is equal to zero. By comparing equations (3) and (4) it is apparent that, for the same value of the internal gain, the noise due to the illumination of the 2 DEG planar photoresistors is always lower than that of an APD. This noise is equal to that of an ideal APD for which β is reduced to zero.

4. Conclusion

It has been experimentally demonstrated that III-V planar photoconductive detectors are characterized by an internal gain which can be controlled by the 2 DEG. Moreover, the noise due to the illumination of these devices is lower than the noise of APD. As a consequence, 2 DEG planar photoresistors could be useful for short wavelength gigabit photoreceiver applications, in particular in the 1.3 μm (or 1.5 μm) wavelength range.

Acknowledgements

The authors would like to thank M. RAZEGHI and N.T. LINH for supplying them the wafers of GaAlAs-GaAs and GaInAs-InP materials, as well as S. MARICOT for her technical assistance and R. FAUQUEMBERGUE for many stimulating discussions.

References

Gammel J.C., Metze G.M. and Ballantyne J.M., 1981, IEEE trans. ED-28, 841.

Klein H.J., Beneking H. and Bimberg D., 1982, Thin. Sol. Films, 90, 371.

Linh N.T., 1983, Proc. ESSDERC Canterbury (Inst. of Phys.)

Pang Y.M., Chen C.Y. and Garbinski P.A., 1983, Elec. Lett. 19, 717.

Vaterkowski J.L., Constant M., Berdai M. and Hombert B., 1982, ESSDERC Munich.

Vilcot J.P., Vaterkowski J.L., Decoster D. and Constant M., 1984 a, Elect. Lett. 20, 86.

Vilcot J.P., Decoster D., Raczy L. and Constant M., 1984 b, Elect. Lett., 20, 274.

Vilcot J.P., Constant M., Decoster D. and Fauquembergue R., 1984 c, Proc. ESSDERC Lille (North Holland).

Inst. Phys. Conf. Ser. No. 74: Chapter 6
Paper presented at Int. Symp. GaAs and Related Compounds, Biarritz, 1984

469

Graded band gap heterostructures for high efficiency GaAs solar cells made by a modified isothermal liquid phase epitaxy

M. GAVAND, L. MAYET, A. LAUGIER

Laboratoire de Physique de la Matière (LA CNRS n° 358)
Institut National des Sciences Appliquées de Lyon
20, Avenue Albert Einstein 69621 VILLEURBANNE Cédex (France)

Abstract. A modified liquid phase epitaxial growth technique is used to make high efficiency p-GaAlAs:Be/p-GaAs:Be/n-GaAs:Si heterostructure solar cells. A thin graded band gap window is grown by isothermally soaking a n-GaAs substrate in a saturated Be doped GaAlAs melt. The Al composition increases toward the surface and Be diffuses in the n-GaAs substrate, forming a GaAs p/n junction. This isothermal process allows to use the same melt for several structures.

1. Introduction

Liquid phase epitaxy (LPE) has been used successfully to grow a wide variety of opto-electronic devices such as double heterostructure lasers, light emitting diodes and solar cells. At present time the highest efficiency reported for a GaAlAs-GaAs heterojunction solar cell has been achieved by LPE process (Sahai et al. 1979). This technique remains the easiest and the cheapest method for growing good quality GaAlAs layers. Nevertheless it does not allow the production of large surfaces as it is possible when using new thin layers growth techniques, MBE or MOCVD. That is an important drawback in the case of large size devices as solar cells even for high concentration use.

In this paper we present a simple and fast LPE growth process easy to automatize. In the case of $Ga_{1-x}Al_xAs$-GaAs solar cells many authors have treated theoretically the influence of a built-in electric field in the window layer on the improvement of conversion efficiency (Hutchby 1975, Konagai and Takahashi 1976, Therez et al. 1979). The major effect is to reduce surface and bulk recombination losses. Konagai et al. (1975), Sahai et al. (1977), Woodall and Hovel (1977), Imenkov et al. (1978) have fabricated graded heterostructures producing a built-in electric field in the surface layer. This field (E=1/q dEg/dx, Eg : band-gap) can be increased only by reducing the window thickness because the band-gap variation is limited to 0.7 eV (AlAs : 2.26 eV ; GaAs : 1.4 eV). So, we use the isothermal procedure reported by Kordos et al. (1979, 1980) allowing the growth of thin graded band-gap $Ga_{1-x}Al_xAs$ layers (w < 100 nm) from a saturated melt. Non equilibrium conditions between binary solid GaAs and ternary GaAlAs liquid lead in a first time to the dissolution of GaAs, and then a graded GaAlAs layer appears. The Al concentration in the window increases from zero at the interface to the value in the solid in equilibrium with the melt at the surface.

A n-GaAs substrate is soaked in a saturated GaAlAs melt. In our case, this melt is Be doped and during the soaking period Be diffuses in the n-GaAs substrate, forming a GaAs p-n junction. The p-side of the junction presents a graded carrier concentration. A schematic energy band diagram of the cell structure is shown in Fig.1. The built-in electric field produced in the window is estimated to more than

5.10^4 V/cm and allows improvement of the spectral response in the short wavelength region.

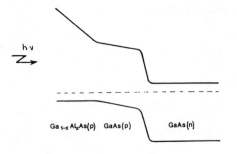

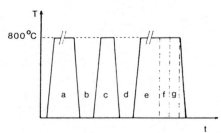

Fig. 1 Schematic energy band diagram of the p-n device

Fig. 2 Temperature vs time cycle. a : Ga baking ; b : Be loading ; c : Be incorporation ; d : Al and GaAs loading ; e : saturation; f : first substrate-melt contact ; g : second substrate-melt contact

2. Experimental process

The $Ga_{1-x}Al_xAs$-GaAs heterostructure solar cells are fabricated in a classical horizontal liquid phase epitaxial growth apparatus. The substrate are < 100 >-oriented Si-doped GaAs wafers with n = $1.2\,10^{18}cm^{-3}$ and $2.5\,10^5\,cm^{-2}$ EPD. The GaAlAs melt is saturated with As from undoped polycrystalline GaAs wafer floating on the top. In a first step, the Ga charge is baked in Pd-purified hydrogen, then an accurately weighted amount of Be is loaded and a new baking takes place. This procedure is necessary because Be has a very strong oxygen affinity. Secondly, GaAs substrate and GaAs source are added with Al and the temperature is raised to 800°C for 2 hours (saturation time). The GaAs substrate is located in a recess of the graphite boat to reduce As losses. Then, the substrate is brought into contact with the saturated melt during a 5 to 30 minute period at constant temperature. This allows the growth of the graded p-GaAlAs layer and the formation of a p-n junction in the GaAs substrate by Be diffusion. A typical time-temperature cycle of the growth procedure is shown in Fig.2.

For some experiments a second n-GaAs substrate is added and pushed under the melt in the same conditions as the first one. Indeed, the isothermal process, the small consumption of Be and Al, the weak vapor pressure of Be (10^{-6}Torr for Be, 200 Torr for Zn at 800°C) lead to consider that the GaAlAs:Be melt is not affected by one growth and can be reused. After a rough estimate the process of many tens of 2 cm^2 substrates for a 5 g Ga melt seems possible.

Generally, the Al concentration in the melt is fixed for equilibrium with $Ga_{0.15}Al_{0.85}As$ solid, but some experiments are carried out with melts in equilibrium with $Ga_{0.4}Al_{0.6}As$ and $Ga_{0.7}Al_{0.3}As$ solids. The Be concentration in the melt, x_{Be} is in the range 4-8.10^{-4}.

The back contact of solar cells is made by a classical Au-Ge-Ni evaporation and the front contact grid is obtained by evaporating on the GaAlAs a 200 nm Au-Be (99-1)

layer. The ohmic contacts are then annealed at 450°C for 3 mn in dry H_2. The p-contact measured resistivity is 5.10^{-4} ohm.cm^2.

3. Growth results

3.1. GaAlAs window

The Fig.3 shows a typical surface of a sample which was in contact with a melt in equilibrium with $Ga_{0.15}Al_{0.85}As$ solid. On the left the surface has been exposed to HCl which is a selective etch for $Ga_{1-x}Al_xAs$ when x >0.4, 0.5. The right side shows the eliminated layer. We observe the same morphology as Small et al. (1979a) which is a characteristic of substrate dissolution. The etched surface is rougher in the Small case, probably resulting from a larger amount of dissolution induced by undersaturated solutions. On the other hand, Kordos et al. (1979 a) have not observed that type of morphology and the layers prepared by these authors from isothermal contact of a GaAs substrate with a GaAlAs saturated solution were smooth, without surface undulation. The major difference in the experimental process is the absence in our case of a GaAs layer grown on the substrate to eliminate possible effects of surface decomposition during the saturation period of the melt.

 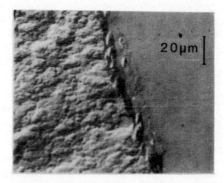

Fig. 3 Typical surface of a wafer after contact with the melt in equilibrium with $Ga_{0.15}Al_{0.85}As$ solid (see text)

Fig. 4 Same conditions as Fig. 3. But after contact a small cooling ramp is generated (see text)

Small et al. (1979 b, 1980 a) have largely treated theoretically the problem of non-equilibrium between GaAs solid and GaAlAs melt, and report many experiments (1980 b, 1981). They show that the graded GaAlAs layer is formed by Al solid state diffusion through the moving interface, which implies a large diffusion coefficient of Al in GaAs (5.10^{-12} cm^2/s) due to defect generation in the dissolving substrate. They get a good agreement between calculations and experimental results. There is perhaps only one inconsistency with this model : the smoother morphology of the upper GaAlAs layer shown in Fig.3 is characteristic of a growth mechanism. Moreover, if the growth goes on after isothermal contact by generating a small ramp cooling (0.5°C/mn during 5 mn), the surface of the layer becomes flat, without undulation as shown in Fig.4. The layer thickness is about 700 nm.

3.2. Be diffusion

The isothermal contact duration does not modify largely the Al profile of the graded window thickness (Kordos et al. 1979 a). So, it is possible to vary it for a good control of Be diffusion in the substrate. The junction depth is measured with an optical microscope on a cleaved cross section stained with $HF:HNO_3$ mixture (Fig.5). SEM beam-induced current technique allows to check the location of the electrical junction at the chemically revealed line.

Many authors have used Be as dopant in GaAs solar cells (Ewan et al. 1978, Flores et al. 1980, Fanetti et al. 1981). An extensive study of Be diffusion during LPE growth has been made by Masu et al. (1980, 1983). They report a Be diffusion coefficient of about $2.5 \cdot 10^{-12}$ cm^2/s at 800° C. For $x_{Be} \simeq 4 \cdot 10^{-4}$, we find $6 \cdot 10^{-12}$ cm^2/s. This value is larger, this may be due to different growth procedure. Masu uses a thick GaAlAs window layer grown by equilibrium cooling or step-cooling. For $x_{Be} = 8 \cdot 10^{-4}$ we find $3 \cdot 10^{-11}$ cm^2/s. This large value is in agreement with the anomalous behaviour observed by Masu (1983) for $x_{Be} > 6 \cdot 10^{-4}$. In this case, Be diffusion cannot be represented by a constant diffusion coefficient and involves a more complex process.

Fig. 5 Photomicrograph of a stained cleaved face showing the junction line

4. Photovoltaic results

Normalized quantum efficiency curves vs wavelength are shown in Fig.6. We can observe the improvement of the spectral response due to the thin graded band-gap GaAlAs window by comparison with a thick constant band-gap $Ga_{0.1}Al_{0.9}As$ window (Yoshida 1980). Normalized quantum efficiency curves of the $Ga_{1-x}Al_xAs$-GaAs heterostructure solar cells are shown for several x values (x = 0, 0.3, 0.6, 0.85). All the growth conditions (contact duration, temperature, x_{Be}) are the same. Only the Al composition is varied to obtain melts in equilibrium with the different solids. The influence of the Al composition at the surface of the graded GaAlAs window is obvious.

The performances of our devices measured using a solar simulator (Xenon light, AM1, output intensity : 107 mW.cm^{-2}) are the following : open-circuit voltage of 970 mV, short circuit current : 22 mA cm^{-2} and a fill factor of 0.73. The resulting power efficiency is 14.5 % without antireflection coating.

The results reported above concern the case where only one wafer is processed during the growth run. Now, we have to consider the case of two wafers. A schematic drawing of the graphite boat is shown in Fig.7. The wafers are called Ia in the first case and IIa and IIb in the second case. Wafer IIa is the first to be brought into contact with the melt and remains at growth temperature during the process of wafer IIb. The performances of solar cells prepared from wafers IIb and Ia with the same conditions are very similar and reproducible. However, the IIa electrical properties were very different from IIb ones. Solar cells IIa present catastrophic performances and sometimes have no rectifying characteristics. We attribute this result to the alteration of the Be diffusion profile in GaAs by subsequent diffusion during the process of the wafer. Several things corroborate this interpretation. First the p-GaAs layer resistivity of IIa is higher than the one of IIb and Ia (from $3 \cdot 10^{-2}$ ohm.cm for IIb and Ia with junction depth of 3 microns, to $6-9 \cdot 10^{-2}$ ohm.cm for IIa). Secondly, we

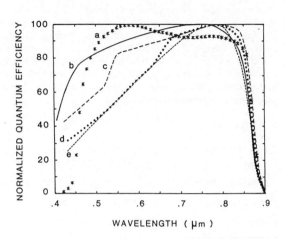

Fig. 6 Normalized quantum efficiency vs wavelength. Curve a : structure with a thick constant band-gap window (x_{AlAs} = 0.9) (Yoshida et al. 1980). Curves b, c, d, e : this work. The Al concentrations of the solids in equilibrium with the different melts are respectively : 0.85, 0.60, 0.30, 0.00. Other conditions remain the same

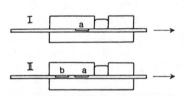

Fig. 7 Schematic view of the graphite boat. I, one substrate ; II, two substrates

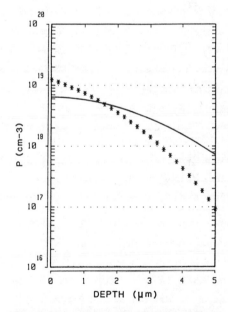

have estimated the evolution of the Be profile after contact with the melt at growth temperature. According to the data of Masu et al. (1983), we suppose an initial Be profile given by $p(z) = p_0 \, \mathrm{erfc}(z/2(Dt)1/2)$ where p_0 is the surface free-carrier concentration, z the distance from the surface, t the diffusion time (contact duration) and D the diffusion coefficient. Fig. 8 shows the evolution of this profile after a period t corresponding to the growth of IIb. We can observe the decrease of Na-Nd in the first microns of the substrate. This could suppress the junction.

Fig. 8 Example of Be diffusion profile evolution. ∗ : Initial profile ;——: Profile at time t at growth temperature outside the melt.

The alteration of the Be diffusion profile is the major limitation with a view to increase the processed surface and to automatize the LPE system. This limitation can be avoided. It is possible and very useful to put the wafer IIa (and eventually

others) at low temperature immediately after the growth period. This requires so
modifications of the graphite boat. Mukai et al. (1980) have reported a process to
shorten the substrate heating time during saturation period : the SSI (substrate -
slide-in) method. The substrate is pushed into the high temperature zone of the
furnace just before the growth. A similar process can be used for the present method
to lead the wafers in a cold zone outside the high temperature zone of the furnace
just after the growth. Then it becomes entirely compatible with continuous
automatized production run.

5. References

Ewan J, Knechtli R C, Loo R and Kamath G S 1978 Proc. 13th IEEE Photovoltaic
    Specialists Conf. (New York : IEEE) pp 941-945

Fanetti E, Flores C, Guarini G, Paletta F and Passoni D 1981 Solar Cells 3 187

Flores C and Passoni D 1980 Solid-State Electron. 23 911

Hutchby J A 1975 Appl.Phys.Lett. 26 457

Imenkov A N, Stamkulov A A, Taurbaev T I, Tsarenkov B V, Shorin V F and
    Yakovlev Y P 1978 Sov.Phys.Semicond. 12 559

Konagai M and Takahashi K 1975 J.Appl.Phys. 46 3542

Konagai M and Takahashi K 1976 Solid-State Electron. 19 259

Kordos P, Pearson G L and Panish M B 1979 J.Appl.Phys. 50 6902

Kordos P and Pearson G L 1980 Solid-State Electron. 23 399

Masu K, Konagai M and Takahashi 1980 Appl.Phys.Lett. 37 182

Masu K, Konagai M and Takahashi K 1983 J.Appl.Phys. 54 1574

Mukai S, Gonda S, Shindo M, Yamashita T and Fujimoto A 1980 Jpn.J.Appl.Phys. 19
    1571

Sahai R, Harris J S, Edwall D D and Eisen F H 1977 J.Electronic Materials 6 645

Sahai R, Edwall D D and Harris J S 1979 Appl.Phys.Lett. 34 147

Small M B, Ghez R, Potemski R M and Woodall J M 1979a Appl.Phys.Lett. 35 209

Small M B and Ghez R 1979b J.Appl.Phys. 50 5322

Small M B and Ghez R 1980a J.Appl.Phys. 51 1589

Small M B, Ghez R, Potemski R M and Reuter W 1980b J.Electrochem.Soc. 127 1177

Small M B, Ghez R, Reuter W and Potemski R M 1981 J.Appl.Phys. 52 814

Therez F, Martinot H and Esteve D 1979 Proc. 2nd E.C. Photovoltaic Solar Energy
    Conf. (Dordrecht : Reidel) pp 926-937

Woodall J M and Hovel H J 1977 Appl.Phys.Lett. 30 492

Yoshida S, Mitsui K, Oda T, Sogo T, Yukimoto Y and Shiramata K 1980 14th IEEE
    Photovoltaic Specialists Conf. (New York : IEEE) pp 1508

Analysis of electrical performance of GaInAs PIN photodiodes for telecommunication applications

J.N. PATILLON, S. MAKRAM-EBEID, J.P. CHANE, B.G. MARTIN, G.M. MARTIN

Laboratoires d'Electronique et de Physique Appliquée
3, avenue Descartes, 94450 LIMEIL-BREVANNES, France

Abstract : The ultimate sensitivity of $Ga_{0.47}In_{0.53}As$ PIN diodes for telecommunication applications is directly related to the minimum achievable value of dark reverse current. The analysis of parameters which affect the dark leakage current is of key importance. This study has been realised on GaInAs homojunction grown on InP vapor phase epitaxy. The growth technique makes use of an $AsCl_3 + H_2$ gaseous mixture flowing over GaAs and InAs solid sources. The InP substrate is introduced in the hot zone inside a quartz box hermetically closed with a sliding cover to prevent decomposition before growth of the GaInAs active layer. The size of the wafer is 2.2 x 3.2 cm^2. This technique is very close to that already proposed by Yanochi and Susa (1981). PIN diodes are then prepared by Zn diffusion, from a spray deposited ZnO layer covered by a SiO_2 film and mesa etching. An AuBe front layer on the p$^+$ diffused top region is used as ohmic contact.

1. Current analysis

In the PIN diodes made on $Ga_{0.47}In_{0.53}As$ lattice matched to InP, most of the authors in the literature (Trommer 1983, Takanashi 1981, Forrest 1980) have noticed, as we have (see fig. 1) an exponential dependence of the dark

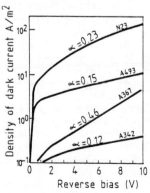

Figure 1

current I_R on the reverse bias voltage in the form $I_R = I_0 exp(+ \alpha V_R)$. The origin of such a variation has never been firmly established and the aim of this paper is to shed some light on the physical mechanisms which seem to be responsible for that effect. This may help us finding remedies to these parasitic effects.

Figure 1 displays types of different dark current behaviours. There appears no relevant correlation between variations of α and of I_0 pre-exponential factor. This exponential behaviour cannot be explained by the well-known thermal generation-recombination mechanism. Furthermore, a pure band to band electron tunnelling would have the following bias dependence (Sze 1969) :

$$I_R \propto V_R \ exp(- A/Em) \backsim V_R \ exp(- B/V_k{}^t) \qquad (1)$$

where A and B are constants, Em is the maximum electrical field, and k a constant equal to 0.5 or 2/3 if the doping level in the depleted region is constant or linearly varying. Both the overall variation of the I-V law and the value of B cannot fit at all the experimental data. It must also be noted that equation (1) predicts temperature independent currents which are not observed in practice.

The observed dependence can be actually taken into account by a more complex current generation mechanism for which the electron emission rate e_n in a mid-gap level is drastically increased under electrical field and to a larger extent than hole emission rate e_p. The expression of the generation current is given by :

$$I_G = N_T e_n e_p/(e_n + e_p) \qquad (2)$$

where e_n and e_p are usually different by orders of magnitude. In the case of a trap for which $e_n \ll e_p$, $I_G = N_T e_n$. Electron emission from that level may be strongly enhanced by a phonon assisted electrical tunnelling effect, as it was observed for many deep levels in III-V's and in GaAs in particular (Makram-Ebeid 1980 a, 1980 b). Hole emissions are less sensitive to the electrical field because of the effective mass as compared to that of electrons. Then I_G increases proportionally to e_n as a function of the electrical field, i.e. the applied voltage before saturating when $e_n \simeq e_p$.

The emission rate $e_n(x)$ has been shown to be expressed by $\beta \exp(\gamma E(x))$ and depends on the value of the electrical field $E(x)$, at the spatial position x inside the depleted layer W (Makram-Ebeid 1982). Then equation (2) can be re-written :

$$I_G = \beta \int_0^\omega \exp(\gamma E(x)) dx \qquad (3)$$

The integrand varies very quickly with x so that I_G can be seen to be roughly proportional to $N_T \exp(\quad E_m)$. The doping level in the side of the PIN diode is usually not constant but increases away from the p$^+$ transition. The maximum value E_m of E is proportional to V^k with k : 2/3 in case of a linear increase of $n(x)$ for instance, and k = 1 for a true PIN structure. Thus a simplified expression for I_G would be :

$$I_G \propto \beta N_T \exp(\quad (V + \phi_B)^t) \qquad (4)$$

with k close to unity and ϕ_B the built-in junction potential; this variation fits the observed curves reported in fig. 1.

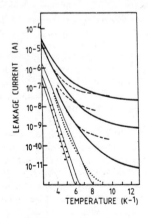

LEAKAGE CURRENT (A)

10^{-4}, 10^{-5}, 10^{-6}, 10^{-7}, 10^{-8}, 10^{-9}, 10^{-10}, 10^{-11}

4   6   8   10   12
TEMPERATURE (K^{-1})

Figure 2 : Measured (dotted lines) and theorical (full lines) current voltage curves

A computer calculation based on the non simplified equation (3) has led to the theoretical I-V curves plotted in figure 2 as a function of temperature. These calculations correspond to a deep level, close to mid-gap with a thermal ionisation energy E_A = 0.35 eV, present in a concentration of 10^{15} cm^{-3}. Variations of $n(x)$ have been supposed to be linear, varying between 2 10^{15} cm^{-3} and 8 10^{15} cm^{-3}. A rather good fit of experimental data is obtained. As expected, there is a dependence with temperature, because the electron tunnelling emission from the level can be enhanced by thermal phonon absorption which provide energy and then give a lower apparent tunnelling barrier height. This effect is, of course, particularly sensitive for low values of E, and V, while for large values of V, saturation is obtained in any case whatever the temperature. The value of α , obtained by simple fit of experimental curves with equation (4), is shown in figure 3 for various layers. In figure 3, it has been plotted as a function of the capacitance for the same value of the dark current density. Some parts are distinctly apart from the overall linear tendency. They have been obtained

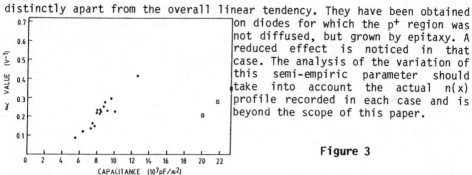

on diodes for which the p+ region was not diffused, but grown by epitaxy. A reduced effect is noticed in that case. The analysis of the variation of this semi-empiric parameter should take into account the actual n(x) profile recorded in each case and is beyond the scope of this paper.

Figure 3

2. DLTS results

DLTS (Deep Level Transient Spectroscopy) measurements have been performed in order to characterize the deep levels on the studied diodes. DLTS spectra are shown on figure 4 in the case of N23 material. Two electron traps TE1 and TE2 and one hole trap TH1 are detected : then e_n versus $1000/T$ characteristics (signature) are displayed in figure 5 and their corresponding apparent activation energy E_A and apparent capture cross section given in table I. The values reported for TH1 may not be very accurate, because of TE2 in the spectrum prevents a precise analysis of emission rates.

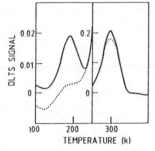
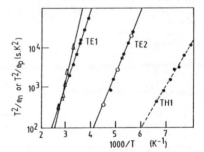

Figure 4 : Typical DLTS Spectra
$e_n = 6.4$ s^{-1}, reverse bias = 0.1 V
(dashed curve) and 4 V (full curve)

Figure 5

Recently reported DLTS data (Trommer 1983, Bhattachaya 1979) obtained on LPE materials are also given. It is clear that the same electron traps are detected on both VPE (this study) and LPE, which would suggest that they are related to impurities rather than to stoichiometry related defects. Their concentrations may be very different in each case, as seen on table I, the lowest values being measured in LPE materials. All the data of table I comes from materials which have been Zn doped for the p+ region. It may be relevant that none of these deep levels has been detected in a recent study of Ga$_{0.47}$In$_{0.53}$As grown by LPE where ternary was located far from the Zn p+ region (Forrest 1982). Can some of these 3 levels be related to a Zn complex ?

Our observations suggest that isolated mid-gap traps may not be directly responsible for the reverse leakage current. The most probable candidate could be mid-gap trap having a hole trap behaviour low field ($e_n \ll e_p$) and becoming an electron trap at higher field ($e_n \gg e_p$). This has in fact been seen for Cr level in GaAs (Makram-Ebeid 1980 a).

	TE1	TE2	TH1	Density of dark current at -2 V (A m²)	ND (cm⁻³)	α (v⁻¹)
VPE N23	ET = 0.42 eV σ_∞ = 6 10^{-18}cm² NT = 8 10^{14}cm⁻³	ET = 0.35 eV σ_∞ = 9 10^{-6}cm² NT = 8 10^{13}cm⁻³	ET = 0.23 eV σ_∞ = 3 10^{16}cm² NT = 8 10^{12}cm⁻³	15	1.5 10^{15}	0.23
VPE A493	ET = 0.42 eV σ_∞ = 6 10^{-18}cm² NT = 6 10^{14} cm⁻³	NT < 10^{12} cm⁻³	NT < 10^{12}cm⁻³	3.5	2 10^{15}	0.15
VPE A342	NT = 5 10^{14} cm⁻³	NT < 10^{12} cm⁻³		0.13	4 10^{15}	0.12
LPE (Trommer et al 1979)	ET = 0.57 eV 10^{13} < NT < 10^{14}cm⁻³	NT < 10^{11}cm⁻³	NT < 10^{11}cm⁻³	0.1	3-4 10^{15}	0.1
LPE (Bhattachaya et al 1983)	unobs.	ET = 0.35 eV σ_∞ = 9 10^{-16}cm⁻² NT = 7 10^{13}cm⁻³	unobs.			

Table I : Parameter for trap levels identified in VPE and LPE n type
$In_{0.53}Ga_{0.47}As$

Our DLTS results do not directly support such a hypothesis. Instead we observe the simultaneous occurrence of hole and electron traps. We suggest an other native model where a coupling between electron and hole traps may be involved as sketched in figure 6. Equation 2 should then be replaced by

$$I_G \propto N_T \; 1/(1/e_n + 1/e_p + 1/e_{hop}) \qquad (5)$$

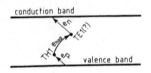

Figure 6

This model cannot be completely quantitative because we lack confirmation about the statistic of electron trap/hole trap distance. It is not surprising to find that good agreement with theory in fig. 2 is observed for low fields and high temperatures. Both equations (2) and (5) predict a current proportional to e_n. More detailed modelling would require statistical information in deep level spatial distribution. One possible assumption is the formation of a donor-acceptor pair during the Zn diffusion process.

References

Bhattacharaya P K, Ku J W, Owen S, Chao S and Yeat R, 1979, Elect. Lett. 15 753

Forrest S R, Lehery M F, Nahory N F and Pollak M A, 1980 Appl. Phys. Lett. 37 322

Forrest S R and Kim O K, 1982 J. Appl. Phys. 53 5738

Makram-Ebeid S, Martin G M and Woodward P N, 1980 S. Phys. Soc. Jpn 49 Suppl A, 287

Makram-Ebeid S, 1980 Defect on Semiconductors, MRS Meeting Proc. Vol. 2, Edit. by Narayan and T.Y. Tan (North Holland, New York) 495

Makram-Ebeid S and Lanno M, 1982, Phys. Rev. B 25 6406.

Sze S M, 1969, Physics of Semiconductors Devices (New York, Wiley) 100

Takanashi Y and Horikoshi Y, 1981 Jap. J. Appl. Phys. 20 1907

Trommer P and Albrecht H, 1983 Jap. J. Appl. Phys. 22 L 634

Weber E R, 1984 Concluding remarks of 13th Conf. on Defects in Semiconductors Inst. Phys. Conf. Ser. to be published

Yanachi Y, Susa N, 1981 J. Electrochem. Soc. 128 210

Inst. Phys. Conf. Ser. No. 74: Chapter 6
Paper presented at Int. Symp. GaAs and Related Compounds, Biarritz, 1984

Computer aided design and realization of VSIS AlGaAs laser diodes

F. LOZES-DUPUY, G. VASSILIEFF, A. BENSOUSSAN, H. MARTINOT, B. SAINT-CRICQ

Laboratoire d'Automatique et d'Analyse des Systèmes du C.N.R.S.
7, avenue du Colonel Roche - 31077 TOULOUSE CEDEX (France)

Abstract. A theoretical and experimental investigation is presented on the visible AlGaAs laser diodes. Theoretical calculations on the gain coefficient in the active layer and the electrical behaviour of the double heterojunction show that the threshold current density is very sensitive to dopings and aluminium content in the confining layers. Computer aided design of V groove channeled substrate inner stripe yields a set of technological parameters for devices emitting at wavelength below 700 nm.

1. Introduction

Visible laser diodes are of increasing interest for various applications such as video and audio disc playback systems, laser printers, plastic optical fiber communications. AlGaAs double heterojunction lasers have been the most widely studied using channeled substrate planar laser (T. Kajimura et al, 1979), terraced substrate laser (M. Wada et al, 1981), ridged substrate internally diffused stripe laser (T.G.J. Van Oirschot et al, 1983). The shortest CW wavelength has been reported by Yamamoto et al (1982-1983) and Hayakawa et al (1982) where internal current confinement and built in optical waveguide was achieved with a V-channeled substrate inner stripe laser (VSIS laser).

This paper reports on computer aided design and fabrication of visible AlGaAs V-channeled substrate inner stripe laser. The threshold current densities of AlGaAs DH lasers are calculated as a function of the aluminium content in the active layer, using gain spectra model in AlGaAs alloy system and numerical simulation of the electrical behaviour of the double heterojunction. The technological parameters of visible VSIS lasers are deduced from these results and from the analysis of the optical properties of the inner stripe geometry. Then the performance limits of optimized structures are presented. Finally, the fabrication procedure, consisting of a two-step LPE process is described, and some inherent difficulties are set up.

2. Threshold current densities of visible broad-area AlGaAs DH lasers

The achievement of AlGaAs laser diodes with emission below 770 nm involves more difficulties than for devices emitting in the infrared wavelength range :
 a) the aluminium composition of the active layer is near the cross-over point between the direct and indirect conduction band minima
 b) the doping levels in the confining layers and the heterobarrier height between the active and the confining layers may not be large enough to prevent the leakage currents due to carriers not confined in the active layer.

Therefore, we have set up a global model of visible DH AlGaAs laser in order
to determine the influence of the aluminium composition of the active layer
on the threshold current density. First, we introduce a gain spectra model
for the AlGaAs alloy system in order to calculate the relation between the
gain coefficient and the injection level in the active layer. Second, a
numerical solution of the charge transport equations in the double heterojunction gives the relation between the gain coefficient and the total current density through the device. Then the threshold current density is
determined by the threshold gain of the optical cavity.

The dependence of the gain coefficient on excitation has been calculated by
Stern for GaAs (Stern, 1976 ; Casey and Stern, 1976), but accurate calculations for the whole direct AlGaAs alloy system would be a lengthy work .
Gain spectra can be usefully obtained from a simplified model. If one considers the electron-hole pair generation coefficient defined by :

$$\alpha'(E) = \frac{\hbar \pi q^2}{\varepsilon_0 \hbar c E m^2} \int_{-\infty}^{+\infty} |M(E,E')|^2 \, \rho_c(E') \, \rho_v(E') \, f_v(E') \, (1 - f_c(E')) \, dE$$

which is related to the gain coefficient as follows :

$$g(E) = \alpha'(E) \left[\exp \frac{(E_{FN} - E_{FP}) - E}{kT} - 1 \right]$$

The electron-hole pair generation coefficient has been measured by Henry et
al (1981) in the case of GaAs. In the case of the alloy system, these results can be extrapolated, as it is shown in Figure 1, by translation along
the energy axis with an amplitude equal to the difference between the direct
energy gaps of GaAs and AlGaAs. An example of calculated gain spectra for
$Al_{0.35}Ga_{0.65}As$ is shown in Figure 2.

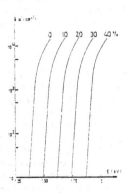
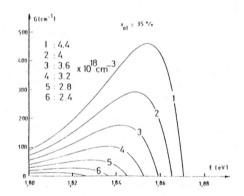

Figure 1. Electron-hole pair generation coefficient versus energy
for GaAs (experimental data given
by Henry et al 1981) and for
$Al_xGa_{1-x}As$ (extrapolated values).

Figure 2. Calculated gain spectra
for $Al_{0.35}Ga_{0.65}As$.

The more complete description of the electrical behaviour of the double
heterostructure laser is given by the numerical solution of the one-dimensional system of the charge transport equations. The HETLASF software has
been developed to numerically solve the general equations using finite
difference discretization and NEWTON's method. FERMI statistics and distri-

bution of electrons between direct and indirect conduction band minima are taken into account and the form of the spontaneous recombination term is following the gain spectra model.

Typical solutions to the transport equations are presented for a DH AlGaAs laser emitting at $\lambda = 670$ nm, with aluminium concentration equal to 0.35 in the active region. The influence of the aluminium concentration and the doping levels in the confining layers on the total current density through the device are illustrated in Figures 3 and 4. These figures highlight the amount of leakage current due to the weak heterobarrier height and the low efficiencies of the heterojunctions. They show that the aluminium composition and the doping level of the confining layers must be chosen greater than 0.75 and 10^{18} cm^{-3} respectively.

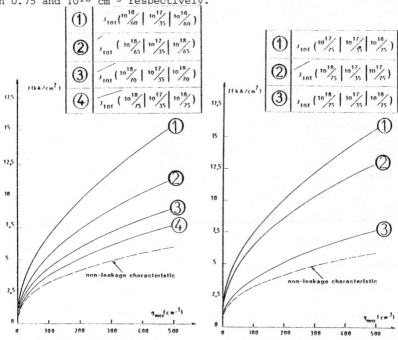

Figure 3. Influence of the aluminium concentration in the confining layers on the total current density. The most important device parameters are Al$_x$Ga$_{1-x}$As active layer : x=35%; N_A=10^{17}cm^{-3}/Al$_y$Ga$_{1-y}$As confining layers : N_A,N_D=10^{18} cm^{-3} ; 60% < y < 75% .

Figure 4. Influence of the doping levels in the confining layers on the total current density. The most important device parameters are : Al$_x$Ga$_{1-x}$As active layer : x=35% ; N_A=10^{17} cm^{-3}, Al$_y$Ga$_{1-y}$As confining layers : y=75% ; 10^{17}cm^{-3} < N_A, N_D < 10^{18}cm^{-3} .

These results are finally combined with the classical calculation of the optical confinement factor in a symmetric three-layer slab waveguide in order to determine the threshold current density of visible double hetero-junction lasers. Figures 5 and 6 illustrate the influence of the aluminium composition of the active layer on the threshold current density when several values of cavity losses are considered. Figure 5 shows that physical limitation of the wavelength range in AlGaAs DH laser is about 650 nm for devices with aluminium composition and doping level in the confining layers equal to 0.75 and 10^{18} cm^{-3} respectively. This wavelength range is reduced to 680 nm, as shown in Figure 6 if the aluminium composition of the confining layers is only equal to 0.60.

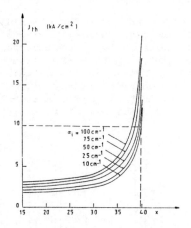

Figure 5. Threshold current density of DH $Al_{0.75}Ga_{0.25}As/Al_xGa_{1-x}As/Al_{0.75}Ga_{0.25}As$ laser, versus the Al composition in the active layer. Dopings in the confining layers are $10^{18}cm^{-3}$. The cavity length is 500 μm, and several cavity losses values are considered.

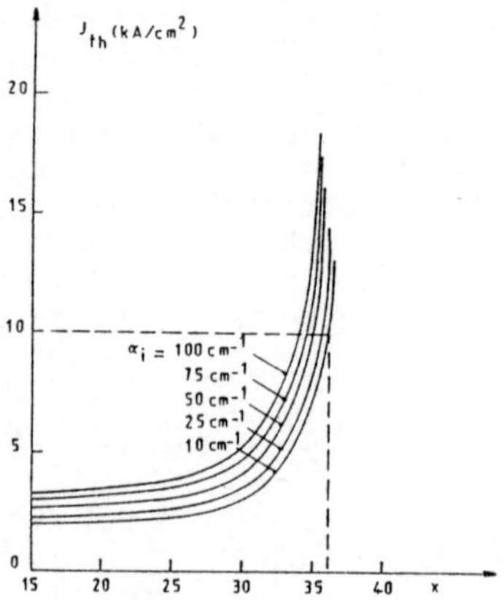

Figure 6. Threshold current density of DH $Al_{0.6}Ga_{0.4}As/Al_xGa_{1-x}As/Al_{0.6}Ga_{0.4}As$ laser, versus the Al composition in the active layer. Dopings in the confining layers are $10^{18}cm^{-3}$. The cavity length is 500 μm, and several cavity losses are considered.

3. Computer aided design of visible V channeled substrate inner stripe

The VSIS laser is schematicaly depicted in the Figure 7. In this structure,away from the channel, the n GaAs current-blocking layer has to be heavily doped with a thickness greater than the hole-diffusion length. This will force the electron-hole pairs to recombine in this layer where light absorption takes place. On the other hand, lateral optical confinement is obtained by a built-in waveguide self-aligned with the current channel. The optical loss into the n GaAs layer provides effective complex refractive index in the outside regions which is different from that in the channeled region.

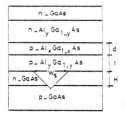

Figure 7. Schematic diagram of AlGaAs VSIS laser.

The lateral mode behaviour has been calculated using a general method of lateral mode analysis with boundary matching technique reported by CHINN et al (1982). The pump gain of each lateral mode is plotted in Figure 9 as a function of the stripe width W_s, for a typical structure emitting at $\lambda \sim$ 695 nm, with threshold gain cavity equal to 50 cm^{-1}. The parameters used are given in Figure 8.

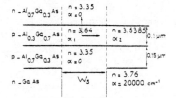

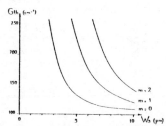

Figure 8. Effective refractive indexes and thicknesses of the slab-waveguide used in the analysis of a typical VSIS structure emitting at $\lambda \sim 695$ nm

Figure 9. Calculated pump gains of each lateral mode as a function of the stripe width.

Figure 9 shows that a stripe width in the range of 4-8 µm is suitable for selecting fundamental mode oscillation without reaching strong pump gains. The thickness of the p-confining layer is found to be thinner than 0.2 µm in order to ensure stable mode confinement.

Therefore, threshold current intensities can be determined for optimized VSIS structures, following the methodology given above for the DH broad-area laser. Starting from the threshold gain of the cavity, the pump gain of the fundamental mode is deduced from the optical model. Injecting and leakage currents are then obtained from the gain spectra model and the electrical behaviour of the heterostructure. A lateral distribution of diodes and resistances along the p-p junction is used to evaluate the spreading current in the p confining layer. Resulting threshold currents are reported in Figure 10 for three optimized structures. These theoretical predictions seem to indicate that CW operation is feasible below 680 nm.

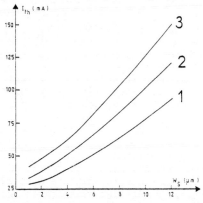

Figure 10. Calculated threshold currents in 3 VSIS structures as a function of the strip width. The parameters of the $Ga_{1-y}Al_yAs/Ga_{1-x}Al_xAs$ heterostructure are :
curve 1. x=0.30 ; y=0.70
curve 2. x=0.35 ; y=0.75
curve 3. x=0.40 ; y=0.80
Dopings in confining layers are 10^{18} cm^{-3} ; d=0.1µm ; t=0.15µm ; L(cavity)=250µm ; g_{th}(cavity)=50cm^{-1} .

4. Fabrication process and experimental results

The fabrication of conventional broad-area AlGaAs visible lasers has formed the earlier part of the work. Pulsed lasers were elaborated in order to verify the theoretical results concerning the influence of the aluminium content of the confining layers on the threshold current density.The Al concentration in the active layer was varying from x=0.15 to x=0.35 while the Al concentration in the confining layers were y=0.6. p-type confining layer were Zn doped (10^{18} cm^{-3}) while the n-type were Te doped (10^{18}cm^{-3}). Curves 1,2,3 in Figure 11 shows the P/J characteristics of 3 devices grown in a sequence of runs in which only the aluminium composition of the active layer is varied. Threshold current density increases with the x Al concen-

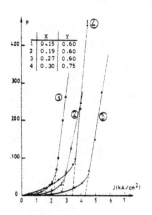

tration and the lowest wavelength we obtained at room temperature is λ =705 nm. As shown by theoretical predictions, shorter wavelength with lower threshold current was achieved by   increasing the aluminium composition of the confining layer from y=0.6 to y=0.75 (curve n°4 in Figure 11).

Figure 11. P/J characteristics of broad-area DH lasers

Curve 1. x=0.15 and y=0.60
Curve 2. x=0.19   y=0.60
Curve 3. x=0.27   y=0.60
Curve 4. x=0.30   y=0.75

In addition to the work on broad-area lasers, the effort is now applied to the growth of VSIS CW devices. The fabrication process is given by a two-step LPE process. First the n-GaAs blocking layer is grown at the cooling rate of 0.2°C/mn from 800°C, then V-shaped channel is grooved by chemical etching. Further, preliminary results have been obtained concerning the second step of the epitaxy. We have determined that step cooling of 3°C is adequate to fill the groove and obtain a flat surface on the confining layer. Moreover, experimental results and theoretical calculations using SPELLAS software (Vassilief et al 1984) have shown that it is necessary to include a precursor substrate in the process in order to obtain sufficiently low growth rates for achiving the p-layers defined by the theoretical design.

5. Summary

A global methodology for computer aided design of visible AlGaAs lasers has been applied to broad-area DH lasers and V-channeled substrate inner stripe lasers. It is based on a gain spectra model for the active region,on a complete description of the electrical behaviour of the heterostructure and an accurate model of the device. The critical properties of the theoretical analysis have been experimentally verified on fabrication of broad-area laser The main features of the VSIS layer fabrication process have been defined and are in investigation.

References

Casey H C and Stern F 1976, J. Appl. Phys. 47, 631-642
Chinn S R and Spiers R J 1982, IEEE J. on Qu. Electron. QE-18,n°6,984-991
Hayakawa T, Miyauchi N, Yamamoto S, Hayashi H, Yano S and Hijikata T, 1982, J. Appl. Phys. 53, 7224-7234
Henry CH, Logan RA and Bertness KA 1981, J. Appl. Phys. 52, 4457-4461
Kajimura T, Kuroda T, Yamashita S, Nakamura M and Umeda J, 1979, Appl. Optics 18, 1812-1815
Stern F, 1976, J. Appl. Phys. 47, 5382-5386
Van Oirschot TGJ, Valster A and de Poorter JA, 1983, Appl. Phys. Lett. 43, 809-811
Vassilieff G and Saint-Cricq B, 1984, J. Appl. Phys. 55, 743-751
Wada M, Itoh K, Shimizu H, Sugino T and Teramoto I, 1981, IEEE J. of Quantum Electron. QE-17, n°5, 776-780
Yamamoto S, Hayashi H, Yano S, Sakurai T and Hijikata T, 1982, Appl. Phys. Lett. 40, 373-375
Yamamoto S, Hayashi H, Hayakawa T, Miyauchi N, Yano S and Hijikata T, 1983, IEEE J. of Quantum Electron. QE-19, n°6, 1009-1015.

Inst. Phys. Conf. Ser. No. 74: Chapter 6
Paper presented at Int. Symp. GaAs and Related Compounds, Biarritz, 1984

485

Integrated buried crescent GaAs–(AlGa)As lasers operating in single-longitudinal-mode

Pan Huizhen, Yang Yi, Xiao Zongyao, Wang Dening

Shanghai Institute of Metallurgy, Academia Sinica

Abstract. An integrated buried crescent GaAs–(AlGa)As laser, which has two active cavities coupled with a convex passive waveguide, has been fabricated by liquid phase epitaxial growth on a channeled-substrate. The gradually changing thickness of the active and waveguide layer provides an index guiding for maintaining a fundamental-transverse-mode. Controllable single-longitudinal-mode operation due to optical injection has been achieved. The lasing threshold is about 150~250mA. The device lased in a single-longitudinal-mode at the drive current just above the threshold to about 1.3 times the threshold.

1. Introduction

For high-capacity and long-distance multi-mode and single-mode optical fiber communication systems, it is very desirable to have a semiconductor laser that emits in one single frequency even under high bit-rate direct modulation, i.e., the so-called dynamic-single-mode laser. However, it has been shown that conventional transverse-mode stabilized Fabry-Perot laser diodes tend to exhibit multi-longitudinal-mode oscillation when directly modulated (Kishino et al., 1982). In order to stabilize the axial-mode, a dynamic-single-mode laser must consist of tightly mode-controlled resonator which helps in maintaining fixed axial as well as transverse-mode under certain operation condition.

Several device structures have been demonstrated to date that have potential for satisfying the above criteria. Many of these incorporate critically dimensioned very fine gratings, over which good crystal growth must be done, such as the distributed feedback and distributed Bragg reflector-buried-heterostructure lasers (Abe et al., 1982; Utaka et al., 1981). Others lose the advantages of a monolithic structure by requiring external cavities (Preston et al., 1981) or injection locking sources (Smith and Malyon, 1982). A more attractive candidate is the coupled-cavity structure. This structure has the advantage of simplicity, compatibility with many state-of-the-art index-guided laser structures, and electronic controllability via the three terminal electrical connections. Very clean cw single-mode operation under 1 Gb/s modulation has been demonstrated by Tsang et al. (1983) with their newly developed cleaved-coupled-cavity lasers. Continuous operation of dynamic-single-mode two-section coupled-cavity laser has also been obtained by Coldren et al. (1984). Nevertheless, these devices are not capable of integration with other optical devices.

Within the past few years we have reported a number of experiments with coupled-cavity lasers that is integrated by means of passive waveguide (Xiao Zongyao et al., 1982; Pan Huizhen et al., 1983; Xiao Zongyao et al., 1984). Since there is a waveguide coupled to the active region, this device has the advantage of easy integration with other optical devices, and if two dimensional waveguides with curved portions and single-mode characteristics are made, a degree of design flexibility for developing sophisticated optical circuits with a high packing density. Moreover, the fabrication processes are simple, only a single step LPE growth and selective chemical etching will be involved.

In this paper, we report significant advances achieved over the early broad-area twin-mesa type integrated lasers. By adopting index-guided buried-crescent structure, the threshold currents have been reduced to about 200mA, and single-longitudinal-mode operation has been extended to about 1.3 times the threshold current.

2. Device Structure

Evanescent field coupling was adopted for the waveguide structure of the integrated laser under investigation.

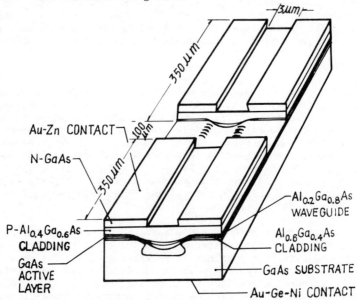

Fig. 1: Schematic view of the integrated device.

Fig.1 shows the schematic view of the device. The active cavities consist of a four layer composite waveguide structure. Between the GaAs active layer and the $Ga_{0.4}Al_{0.6}As$ lower confinement layer, there is a $Ga_{0.8}Al_{0.2}As$ waveguide layer. The layer parameters are chosen so that an appreciable fraction of the optical energy is in the passive waveguide layer by means of the evanescent field. The Al content for the waveguide layer was chosen to be 0.2 since the optical confinement factor will decrease more rapidly, causing a much higher threshold, if the Al content is lower, as shown in Fig.2. Taking account of the easy fabrication of the ohmic contact, the Al content of the cladding layer was chosen to be 0.35~0.4. The output passive waveguide that interconnects the two active cavities is simply an

asymmetric three layer waveguide with air as the cladding. The Al content
of the lower confining layer was chosen to be about 0.6 so that the propagating optical energy is fairly restricted within the waveguide layer.

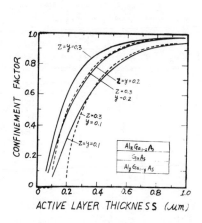

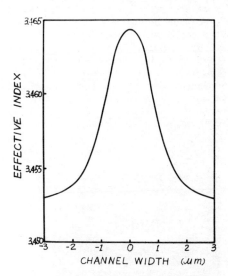

Fig. 2: Optical confinement factor   Fig. 3: Effective guide index
versus active layer thickness for   profile along the lateral direction.
different Al content profile.

The channeled-substrate crescent configuration was adopted to obtain lateral mode confinement. The gradually changing thickness of the active and
waveguide layer gives a corresponding effective guide index profile along
the lateral direction of the p-n junction, as shown in Fig.3. It guides
the light to the center of the structure, and can maintain a stable fundamental-transverse-mode if the channel width is narrow enough. An additional n-GaAs layer on the top serves as the current-confining layer, injection current is confined within the crescent region by the reversed p-n
junction.

3. Device Fabrication

Standard photolithographic techniques were used to etch 6μm wide channels
along [1$\bar{1}$0] direction in {100}-oriented, n-GaAs substrate. $H_2SO_4:H_2O_2:H_2O$
in the ratio 1:8:2 was used as the etchant. The channel depth is approximately 2.5μm. Liquid phase epitaxy was then carried out over the channeled-substrate. Five layers as shown in Fig.1 were grown successively in
a single step. Partial melt-back sometimes occurred at the shoulder of
the channels, although the growth temperature was lowered to 760°C and a
supersaturation of 6°C was used. Fig.4 shows the scanning electron micrograph of the epitaxial wafer cross-section.

The smoothing effect of LPE automatically results in thick growth inside
the channel and thin growth elsewhere, and the as-grown wafer has a flat
surface that facilitates the subsequent processing of the 3μm stripe contact. The n-GaAs current-confining layer was first etched with non-

selective $H_2SO_4:H_2O_2:H_2O$ solution and then with the selective superoxol etchant to thoroughly expose the $p-Al_{0.4}Ga_{0.6}As$ within the contact stripe and the interconnecting-waveguide region. After Zn diffusion to provide a high doping level at the surface, gold was evaporated over the entire surface of the wafer. A second photolithographic step was used to define the contact pads on top of the active cavities, and the exposed gold between resist pads was removed with the iodine etch. The $Al_{0.4}Ga_{0.6}As$ layer exposed by this process was then removed with hot hydrofluoric acid. Finally, the etched laser mirrors were formed by removing the exposed GaAs active layer with superoxol. The wafer was lapped and Au, Ge and Ni was evaporated over the substrate. Alloying of the p and n-contact was carried out simultaneously.

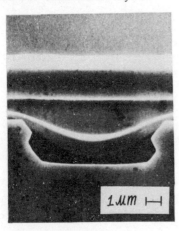

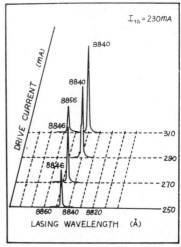

Fig. 4: SEM photo of the epitaxial wafer cross-section.

Fig. 5: Lasing spectra of the integrated device at different drive currents.

4. Mode Characteristics

The experimental setup for the measurement of mode behavior consists of an infrared TV microscope and a grating monochromator system to simultaneously display the near-field patterns and lasing spectra of the integrated lasers. The device was operated at room temperature under pulsed condition with the width of 1μsec and the repetition rate of 1 KHz. The injection currents into the two active cavities can be controlled separately. For convenience, we applied a synchronous pulse current of the same value on both cavities. Threshold currents measured were generally 150 to 250mA.

Mode characteristics were measured from one of the cleaved-facet of the device. It has been shown that the lasing spectra measured from the other cleaved-facet are just the same. This phenomenon demonstrates the wavelength selection mechanism resulting from the mutual optical injection locking. Fig.5 gives the lasing spectra of a typical device at different drive currents. The threshold current measured was 230mA. Single-longitudinal-mode operation was observed up to an injection current 1.3 times the threshold. As the discrete longitudinal modes of a cavity can be shifted by changing the refractive index of the active region through variation of the two injection currents, the coincidence-resonant-modes of the coupled cavity are shifted, causing a mode hopping from 8846Å to 8840Å as shown in Fig.5.

5. Evaluation of Threshold Current

For further improvement of the device performance, we have evaluated the effect of active layer central thickness d_o and channel width w on threshold current density according to the model that described by Ooumra (1981) and Wang (1984). The influence of lateral current spreading in layers other than the active layer on the threshold current for stripe-geometry lasers was also included in the evaluation (Tsang, 1978). The results are plotted in Fig.6 and Fig.7. It indicates that the optimal d_o is about 0.15 to 0.2μm, and the threshold current will reduce to below 100mA if w is less than 4μm. It can also be seen from Fig.7 that the estimated threshold current for the actual device shown above is in good agreement with the experimental result.

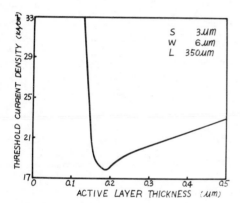
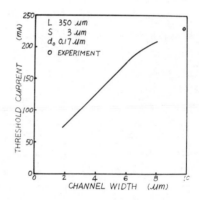

Fig. 6: The effect of active layer thickness on threshold current density.

Fig. 7: The effect of channel width on threshold current.

6. Conclusions

An integrated waveguide-coupled-cavity laser with buried crescent structure has been devised. Single-longitudinal-mode operation due to wavelength selection has been demonstrated. Evaluation of the threshold current indicates that further improvement can be achieved through optimizing the active layer thickness and the channel width. The monolithic structure should be more compatible with integration.

References

Abe Y, Kishino K, Tanbun-ek T, Arai S, Koyama F and Suematsu Y 1982 Electron. Lett. 18 410-411

Coldren LA, Ebeling K J, Rentschler J A and Burrus C A 1984 Appl. Phys. Lett. 44 368–370

Kishino K, Aoki S and Suematsu Y 1982 IEEE J. Quantum Electron. QE–18 343–351

Oomura E, Murotani T, Higuchi H, Namizaki H and Susaki W 1981 IEEE J. Quantum Electron. QE–17 646–649

Preston K R, Woolard K C and Cameron K H 1981 Electron. Lett. 17 931–932

Pan Huizhen, Xu Guohua, Xiao Zongyao and Wang Dening 1983 Tech. Digest pp.180–181, Fourth International Conference on Integrated Optics and Optical Fiber Communication, Tokyo, Japan; 1983 Acta Optica Sinica 3, 786–791

Smith D W and Malyon D J 1982 Electron. Lett. 18 43–45

Tsang W T, Olsson N A and Logan R A 1983 Appl. Phys. Lett. 42 650–652

Tsang W T 1981 J. Appl. Phys. 49 1031–1044

Utaka K, Akiba S, Sakai K and Matsushima Y 1981 Electron. Lett. 17 961–963

Wang Dening and Pan Huizhen 1984 Acta Optica Sinica 4 126–134

Xiao Zongyao, Shen Pengnian and Pan Huizhen 1982 Tech. Digest, paper THB6, Sixth Topical Meeting on Integrated and Guided-wave Optics, Pacific Grove, USA; 1983 Chinese Physics 3 716–720

Xiao Zongyao, Pan Huizhen and Chen Lianyong 1984 Acta Optica Sinica 4 499–506

Inst. Phys. Conf. Ser. No. 74: Chapter 6
Paper presented at Int. Symp. GaAs and Related Compounds, Biarritz, 1984

Industrial development and reliability evaluation of 1.3 μm buried heterostructure in GaAsP/InP lasers

E. VETU - C. RUINEAU - C. CARRIERE - C. CANNEAU

CIT ALCATEL - Division "CSO" Route de Nozay - 91620 LA VILLE DU BOIS

Abstract. Using studies made at the "CGE research laboratory",
CIT-ALCATEL is developing a technological process aiming at the industrial production of 1.3μm In Ga As P/InP buried heterostructure lasers.
A good control of the LPE growth technique and characterization tests
at given steps of the fabrication lead to the obtaining of reproducible
structure with a 1 500 Å-thick, 2.8 ± 0.5μm wide active stripe.
We present the statistical results obtained on the first 1500 lasers :
mean threshold current = 28mA, mean optical output power at 200mA drive
current = 30mW - and the main optical characteristics.
We also expose preliminary results concerning the ruggedness of lasers
after application of various stress steps. The good thermal characteristic of the lasers and the intrinsic quality of their mirror facets lead
to a much better reliability than for previous laser structures.

1. Introduction

In Ga As P/InP double heterostructure laser diodes are mainly developed as
light sources for long-haul optical fiber telecommunications.
At the Laboratoires de Marcoussis (CGE Company), a process has been developed for the obtaining of 1.3μm. In Ga As P/InP buried heterostructure (BH)
lasers (Landreaux and al 1982).
This index guided structure exhibits better performances than gain-guided
lasers : the lower threshold current allows operation at higher temperatures;
the absence of astigmatism and the stronger tendency to operate stably in
the fundamental transverse mode provide a better coupling to monomode optical fibers.
From the studies done at the Laboratoires de Marcoussis, we at CIT-ALCATEL
are developing a reproducible technology which leads, as a first step, to
the production of 300 BH laser diodes per month.

2. Laser fabrication

Fig.1 shows the buried stripe structure to be realized : the refractive
index difference between InP and the In Ga As P active layer provides optical confinement, and current confinement is due to the reverse biased
In/P, P-N junction.

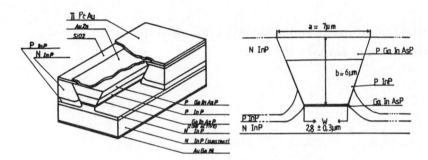

FIGURE 1 : BURIED HETEROSTRUCTURE LASER

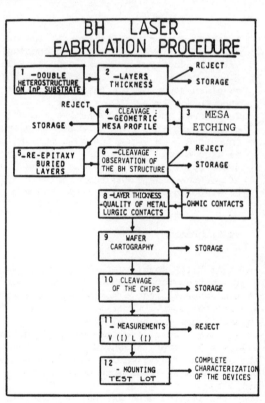

Fig.2 presents the fabrication procedure, with the required control tests ; the possibility of storing the wafers after certain steps is also indicated. This procedure has been designed for an industrial production of the devices.

The techniques involved are : liquid phase epitaxy on a monocrystalline substrate using the two-phase technique ; chemical etching of the reverse mesa ; electrical contacting and insulating.

- FIGURE 2 -

3. Results

In this section, we present results obtained on a series of 1500 BH lasers, mounted P side up with In solder.
A large area (∅ = 12mm) Ge photodetector is used for recording the L (I) curves.
Fig.3 shows the statistical distribution of the C.W. threshold current.
More than 60% of the lasers have a threshold current of 28 ± 9mA, lasing is still obtained at 60°C   Typical L (I) curves at different temperatures are presented in Fig.4 .

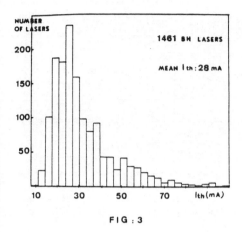

FIG : 3

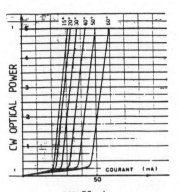

FIGURE 4

TYPICAL L(I) CHARACTERISTICS AT DIFFERENT TEMPERATURES

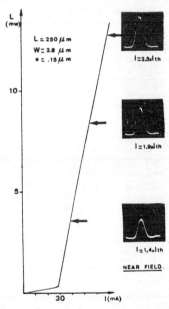

The near-field width $\Delta NF_{1/e}$ of our lasers is 2 - 4µm.
Fig.5 presents a near-field pattern at different light output powers.
No change in the near-field pattern is observed as the output power increases, which is a proof of the good stability of the fundamental transverse mode.

4. Preliminary step-stress and reliability tests

Step-stress tests have been performed to evaluate the ruggedness of our lasers.

4.1 Automatic power output control (APC) mode aging

Such tests are currently performed on our standard BH laser production at levels of 3,5 and 10mW. 3 types of behaviour are observed on unscreened lasers (Fig.6).
- a quick variation exceeding 50% change of the driving current within less than 2000h (type 1).
- a saturable   variation. A saturation is appearing after 200 to 1000 hours (type 1).
- a slower   variation without saturation (type 3).
These three types of behaviours were observed by Mizuishi et al (1981 and

1983). The type of variation and the slope are characteristic from a wafer. As the production control improves we observe a slower variation in type 2 and 3.

4.2 Electroluminescent mode aging (E.L)

Accelerated aging tests in E.L mode have been performed – as explained by Higuchi H et al (1983) – at various currents and ambient temperatures. Fig.7 shows 3 typical behaviours of the variation of the 20°C

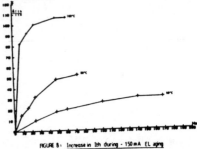

FIGURE 6 : INCREASE IN DRIVING CURRENT DURING 60°C APC AGING

threshold current as a function of time during EL test at 150mA operating conditions. These typical variations of Ith during EL mode are similar to the variations of driving current during APC mode.

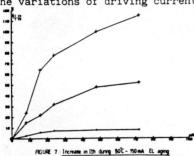

FIGURE 7 : Increase in Ith during 80°C - 150mA EL aging   FIGURE 8 : Increase in Ith during - 150 mA EL aging

After following the standard EL test the lasers are aged in APC mode. The lasers which have a quick variation in EL mode evolve with a similar variation in APC mode. The lasers which have a saturable variation or gradual variation in EL mode evolve with a gradual variation in APC mode, which confirms that EL and APC aging mode induce the same wearing mechanism in the lasers. The total drift of Ith and the stabilization time are a function of the temperature. Figure 8 shows an example of the type 2 (saturable variation) behaviour as a function of temperature, for lasers originating from the same wafer.
The EL aging test is a good way of accelerated aging and screening (getting rid of lasers aging in type 1 mode). This test is used in our fabrication process, as stabilization and screening test.

4.3 Electrical stress

– Resistance to reverse bias test is good since no degradation is observe after application of 10volts DC bias during 1mn.
– Normal production BH lasers display an excellent level of resistance to burst applied in forward bias. The first degradation being observed after 100 pulses per burst with an amplitude of 1 Amp and a duration of 1ms.
– Storage tests have shown that lasers are sustaining storage temperatures as high as 125°C without any damage.

4.4 Conclusions of the tests

The tests (A.P.C, E.L and electrical) have been performed on a large quantity of lasers ; 200 for the APC test, 50 for the E.L test and 100 for the

electrical tests. Thanks to this large number of devices tested, standard
screening tests have been defined.
These tests are being performed continuously on lasers of the production
for the APC and EL mode to select the component which could be immersed.

5. Conclusion

It has been shown that the fabrication process for index-guided buried heterostructure lasers (B.H) is now mastered. The particular critical fabrication steps such as inverse mesa etching or re-epitaxy have been brought
under control. The preliminary results of step-stress aging tests show a
good ruggedness of the laser structure.
The excellent thermal behaviour of the BH laser allows a reliability level
much better than the one of wide-ribbon high threshold current structures.
Then all our monomode pigtailed lasers are now being equipped with BH lasers.

Références

Higuchi H, Oomura E, Hirano R, Sakakibara Y, Namizaki N, Susaki W,
Fujikawa K 1983 Electronics letters Vol 19 n° 23 p. 976 - 977

Landreaux R, Boulard O, Jicquel J.P, Lebled H, Legouezigou L, Louis Y,
Magnabal J, Sigogne D, Benoît J 1982 Colloque C5 sup n° 12 T 43
Journal de physique Dcc 82.

Mizuishi KI, Hirao M, Tsuji S, Sato H, Nakahura M Tokyo V 1981 13th
conférence on solid state devices.

Mizuishi KI, Matsumoto K, Todoroki S, Sawai M, Hirao M, Takahashi T,
Nukamura M, Tokyo june 1983 4th international conférence on integrated
optics and optical communications.

Do substrate dislocations affect GaAs FET performance?

H.V. Winston, A.T. Hunter, H. Kimura, H.M. Olsen,
R.P. Bryan, R.E. Lee, and O.J. Marsh

Hughes Research Laboratories, Malibu, CA 90265, USA

Abstract. We have measured V_{th} for FETs in arrays fabricated by Si^+
implantation in LEC GaAs and In-alloyed GaAs, and found no correlation of
V_{th} with the FETs' proximity to dislocations. The lowest standard deviation (SD) found for 48 FETs in 1 mm^2 was 8.5 mV, for a substrate with an
almost constant dislocation density (DD) of about 10^5 cm^{-2}. The highest
SD values were for substrates with a normal W-shaped distribution of DD
ranging from 10^5 to less than 10^4 cm^{-2}. Low-dislocation In-alloyed GaAs
had an intermediate SD.

1.   Introduction

Dislocations are very obvious crystallographic nonuniformities in liquid-
encapsulated Czochralski (LEC) GaAs, readily revealed by etching or x-ray
topography. They are generally suspected of being responsible for
observed electrical and optical nonuniformities of devices fabricated on
SI substrates. There are many reports of correlations between the distributions of dislocation etch pit density (EPD) and of various electrical
and optical properties for Si LEC wafers (Matsumoto 1982, Matsumura 1983,
Mita 1983, Miyazawa 1982). Furthermore, it has been reported that the
sheet carrier concentration after Si implantation of such wafers correlates closely with EPD (Honda 1983), and that threshold voltage (V_{th}) and
saturated current (I_{dss}) for an array of Schottky barrier FETs fabricated
on a Si-implanted wafer have distributions correlating with EPD (Nanishi
1982). These studies seem to show that dislocations in the underlying
substrate are responsible for the nonuniform activation of implanted
dopants and the resulting nonuniformities of V_{th} in FET arrays. Miyazawa
and co-workers (1983) have presented evidence for Si-implanted GaAs substrates indicating that FETs closer than 20 or 30 μm to a dislocation have
more negative values of V_{th} than FETs further from a dislocation. They
proposed that the mechanism of this effect is the formation of a zone
around a dislocation denuded of compensating acceptor impurities.

However, we have conducted experiments on the influence of an FET's proximity to a dislocation on its V_{th} that do not support the picture just
outlined (Winston 1984). We found that when all the FETs in a local
region are studied, rather than FETs selected at random from a large
region sampling a range of different local dislocation densities, there is
no perceptible trend of V_{th} with dislocation proximity. In one case, we
found that FETs in a high-EPD region exhibited a more negative average V_{th}

than FETs in a lower-EPD region. We suggested that a scatter plot of V_{th} against distance to the nearest dislocation for a random sample of FETs spanning a whole wafer with a range of EPD values would show the most negative values of V_{th} at low distances because most of the low-distance points would come from high-EPD regions. Such a scatter plot could be interpreted as showing a proximity effect, although in fact it would only be a manifestation of a dislocation density effect.

2.   FET Experiments

We have now extended our measurements to additional LEC undoped substrates, including both GaAs and low-dislocation In-alloyed GaAs grown in our laboratory (Kimura 1984a) and GaAs obtained from commercial sources. The FETs formed on these substrates are arrayed on a 100-μm x 300-μm grid, aligned along <110> directions in the substrate surface, that allows space for three 75-μm square probe pads. The FET geometry is shown in Fig. 1. The 1-μm x 5-μm Schottky barrier gate electrodes and the source and drain electrodes, spaced 5 μm apart, along with the pads and interconnect lines, are formed by a four-mask-level planar process following inplantation of 3 x 10^{12} cm^{-2} Si^{28} ions at 60 keV and annealing at 830°C for 20 minutes under an SiO$_2$ cap. Inter-device isolation is provided by a shallow oxygen damage implant. These FETs contain no dielectrics and have undergone no channel etching.

We measured the FETs' V_{th} values using an Accutest Model 3000 automatic dc parameter test system. The FETs tested were either on a grid 1.5 mm x 1.5 mm spanning whole two-inch wafers and oriented along <110> directions in the (001) substrate, or in a stripe of four parallel columns along <110> spaced every 300 μm, with a gate-to-gate spacing of 100 μm along the columns. The groups of 48 FETs contained in successive 1.2 mm x 1.2 mm squares along the stripe have their gates included in an area that is almost exactly 1 mm^2. We calculated the average V_{th} ($\bar{V}_{th}$) and the standard deviation, (σV_{th}), for each successive group of 48 FETs along the stripe, and plotted them against position as in Figs. 2, 3 and 4. After the electrical measurements we exposed the wafers to molten KOH to form dislocation etch pits. The locations of gates and pads were still clearly visible, as in Fig. 1, and gate/nearest-dislocation distances and dislocation densities were easily measured.

3.   FET Results

A typical relief map of whole-wafer V_{th} results for the 1.5 mm grid appears in Fig. 5. The map is relatively smooth and flat except for a few points near the edges. The sample was from a Hughes-grown GaAs crystal, and exhibited a high, almost constant EPD, as shown in Fig. 2, which also shows $\bar{V}_{th}$ and σV_{th}. It is notable that neither $\bar{V}_{th}$ nor σV_{th} correlate with EPD, and also that the σ values are almost all lower than 20 mV, with some below 10 mV. In Fig. 3 we see the results for another Hughes-grown GaAs crystal with lower and more variable EPD. Again, there appears to be no correlation of $\bar{V}_{th}$ and σV_{th} with EPD, but the standard deviations are all greater than 20 mV, ranging up to 50 mV. The σ values for both substrates are considerably lower than the 60-120 mV recently reported (Ishii 1984) for LEC GaAs substrates.

In Figure 4 we exhibit the results for a Hughes-grown In-alloyed GaAs crystal that contained large areas completely free of dislocations. The

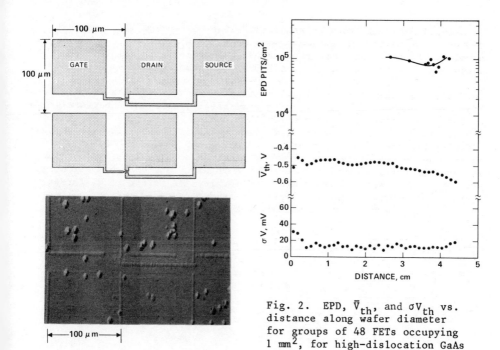

Fig. 1. Layout of FET and portion
of FET array after KOH etch.

Fig. 2. EPD, $\bar{V}_{th}$, and σV_{th} vs.
distance along wafer diameter
for groups of 48 FETs occupying
1 mm^2, for high-dislocation GaAs
substrate.

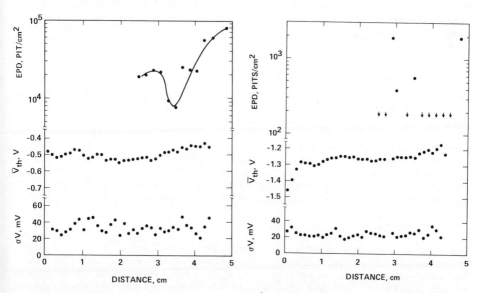

Fig. 3. EPD, $\bar{V}_{th}$, and σV_{th} vs.
distance along wafer diameter
for medium-dislocation GaAs
substrate.

Fig. 4. EPD, $\bar{V}_{th}$, and σV_{th} vs.
distance along wafer diameter
for low-dislocation In-alloyed
GaAs substrate.

In-alloyed material leads to standard deviations intermediate between the
two previous cases, but exhibits more negative V_{th} values than unalloyed
samples. We find that more negative V_{th}, implying more complete implant
activation, is typical of all the In-alloyed samples we have studied.

In Table 1 we summarize the local uniformity results for the samples of
Figs. 2 through 4, as well as for a commercially obtained substrate grown
by the low-pressure LEC method. In this table, the values of $\bar{V}_{th}$ and σV_{th}
are for that particular group of 48 contiguous FETs along the stripe that
exhibits the lowest value of σV_{th}. Such a group does not necessarily
correspond to one of the groups of 48 plotted in Figs. 2, 3, and 4, but
may overlap parts of two of them. The quantity U in the table was origin-
ally introduced as a way to compare uniformity for wafers of different
overall average V_{th} (McLevige 1982); it is defined as $\sigma V_{th}/(-\bar{V}_{th}+V_{BI})$,
where V_{BI} is the built-in voltage of the Schottky barrier gate.

Table 1. Comparison of Local Uniformity

	Hughes In-alloy	Hughes High EPD	Hughes Medium EPD	Commercial LP LEC
$\bar{V}_{th}$, V	−1.25	−0.485	−0.449	−0.879
σV_{th}, mV	17.3	8.5	18.7	18.9
V, %	0.84	0.66	1.50	1.13

A typical scatter plot of V_{th} vs. dislocation proximity for a group of 48
FETs in a 1 mm^2 region of the high-dislocation sample is shown in Fig. 6.
No perceptible trend appears; at least one FET is directly on a dislocation (gate-pit distance of 0) without any obvious effect, and the total
range of V_{th} values is less than 50 mV, with a standard deviation of less
than 12 mV.

4. Cathodoluminescence of In-Alloyed GaAs

In contrast to the band-edge cathodoluminescence (CL) of unalloyed Si LEC
GaAs, which shows bright areas within 20-30 µm of dislocations, we
observe for low-dislocation In-alloyed GaAs bright CL spots of up to
300 µm radius, with central dark spots up to 100 µm and sometimes a 10-µm
bright center, as shown in Fig. 7. Apparently, dislocations do have a
profound effect on luminescence in an extended neighborhood, which might
be related to the gettering of recombination centers. However, the
absence of a dislocation proximity effect on V_{th} suggests that significant
numbers of acceptors are not gettered by dislocations in our experiment.

5. Discussion

Our results indicate no effect of proximity to a dislocation on the V_{th} of
a FET and often no correlation of the distributions of V_{th} and EPD. On
the other hand, there is evidence for more subtle effects of dislocations
on FET behavior. We found that a GaAs substrate of high and uniform dislocation density led to the lowest observed standard deviations of V_{th} for
48 FETs in a 1 mm^2 area. The poorest uniformity was for a substrate of
GaAs with the usual W-shaped dislocation distribution, with EPD ranging
from 10^5 to below 10^4 cm^{-2}. Intermediate in uniformity was an array on

low-dislocation In-alloyed GaAs. The In-alloyed material, however, exhibited the most negative values of V_{th}, and thus the most complete implant activation, which may be connected with the absence of dislocations and dislocation cell networks in the nearly dislocation-free substrate.

The reasons for the disagreement between our results and those of other workers require further study. They may originate in slight differences in the material preparation; for example, much of the GaAs used in the other work contained small concentrations of Cr. Our implant process uses SiO_2 capping for the anneal, rather than Si_3N_4 as in the other studies. There may be other important differences as well in the FET fabrication process.

If dislocations do not affect V_{th} by a proximity effect or by way of the local dislocation density, what is the cause of the observed nonuniformities in V_{th}? The effects of dislocation networks, precipitates, and inclusions may be important. Growth striations are known to be present in LEC GaAs (Miyazawa 1982, Kimura 1984) and could cause variations in stoichiometry and lattice defect concentrations affecting implant activation efficiency and V_{th}.

Residual damage after polishing could be responsible for nonuniformity of activation, especially when it is considered that the active implantated layer is confined to within 0.1 or 0.2 µm of the surface. Some part of the V_{th} nonuniformity must arise from fluctuations in the microlithography; we frequently see large deviations and trends of V_{th} near wafer edges where microlithography is known to be unreliable, and there may be smaller lithography-generated deviations in the main part of the wafer. The reproducibility of the V_{th} measurements is probably no better than 2 mV.

The implantation fluence is carefully controlled, but it may vary by as much as 1.5% across the wafer. Such a variation could be responsible in part for long-range differences in V_{th}. Even an ideally random implantation fluence could lead to local variations in V_{th} because of the statistical fluctuations in the number of implanted ions under the small gate area. With only about 10^5 total implanted donors under a gate, this effect would introduce a relative standard deviation in the number of donors under a gate of about 0.3%. The quantity U, introduced in Table 1, might be expected to have the same relative standard deviation as the number of donors, corresponding to 4 or 5 mV for σV_{th}. Some of the σV_{th} values in Table 1 are not much larger than this fundamental uniformity limit.

6. Conclusions

FET arrays fabricated in our laboratory on LEC undoped GaAs and In-alloyed GaAs substrates do not exhibit a dependence of V_{th} on the distance of an individual FET gate from the nearest dislocation. In many cases they do not reveal a dependence of local average V_{th} on local average dislocation density. The values of σV_{th}, the standard deviation of V_{th} for the FETs in 1 mm^2, are much lower than those reported from other laboratories. Our best value of σV_{th}, 8.5 mV, is about twice the standard deviation expected on the basis of a Poisson distribution of the number of ions implanted under the area of a gate. This lowest σV_{th} value was observed for a substrate of more-or-less uniformly high dislocation density. The σV_{th} values on this wafer were almost all lower than 20 mV. A GaAs substrate with the usual W-shaped distribution of dislocation density across the wafer had σV_{th} values ranging from 20 to 50 mV, while a low-dislocation In-alloyed GaAs substrate had σV_{th} values between 15 and 30 mV. These results demonstrate that FET devices on LEC GaAs and In-alloyed GaAs substrates can have adequate local uniformity for such applications as A/D converters and memories.

References

Honda T, Ishii Y, Miyazawa S, Yamazaki H and Nanishi Y 1983
    Jap. J. Appl. Phys. <u>22</u> L270
Ishii Y, Miyazawa S and Ishida S 1984 IEEE Trans. El. Devices <u>ED-21</u> 1051
Kimura H, Afable C B, Olsen H M, Hunter A T and Winston H V 1984a Paper in
    Session VIA Sixth American Conf. on Cryst. Growth, to be published in J.
    Cryst. Growth
Kimura H, Afable C B, Olsen H M, Hunter A T, Miller K T and Winston H V
    1984b Paper B-1-1 in Extended Abstracts of 16th Int. Conf. on Sol. St.
    Devices and Mat. 59
Matsumoto Y and Watanabe H 1982 Jap. J. Appl. Phys. <u>21</u> L515
Matsumura T, Emori H, Terashima K and Fukuda T 1983 Jap. J.
    Appl. Phys. <u>22</u> L154
McLevige W V, Chang C T M and Duncan W M 1982 IEEE GaAs IC
    Symp. 127
Mita Y, Sugato S and Tsukuda N 1983 Appl. Phys. Lett. <u>43</u> 841
Miyazawa S, Ishii Y, Ishida S and Nanishi Y 1983 Appl. Phys. Lett. <u>43</u> 853
Miyazawa S, Mizutani T and Yamazaki H 1982 Jap. J. Appl. Phys. <u>21</u> L542
Miyazawa S 1982b J. Cryst. Growth <u>57</u> 459
Nanishi Y, Ishida S, Honda T, Yamazaki H and Miyazawa S 1982 Jap. J. Appl.
    Phys. <u>21</u> L335
Winston H V, Hunter A T, Olsen H M, Bryan R P and Lee R E 1984
    Appl. Phys. Lett. <u>45</u> 447

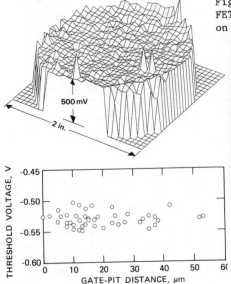

Fig. 5. Relief map of $-\nabla_{th}$, for FETs on a 1.5 mm square grid on high-dislocation GaAs substrate.

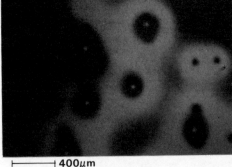

├────────┤ 400µm

Fig. 7. Cathodoluminescence image.

Fig. 6. Threshold voltage vs. dislocation proximity for 48 FETs on 1 mm^2 of a high-dislocation GaAs substrate.

Inst. Phys. Conf. Ser. No. 74: Chapter 7
Paper presented at Int. Symp. GaAs and Related Compounds, Biarritz, 1984

503

Substrate effects in submicronic gate low noise GaAs MESFETs

G. Salmer, M. Lefebvre, F. Héliodore
Centre Hyperfréquences et Semiconducteurs - L.A. C.N.R.S. n° 287 - Bât. P4
Université des Sciences et Techniques de Lille I - 59655 VILLENEUVE D'ASCQ
CEDEX - FRANCE.

O.L. El Sayed, K. Ismail, S. El Ghazali
Department of Electronics and Telecommunications - Faculty of Engineering,
Cairo University Giza, Cairo, Egypt.

1. Introduction

In submicronic gate GaAs MESFET's operating at minimum noise figure
substrate properties acquire a large importance. In fact at such gate
lengths, the gate fringing field results in increased carrier injection
into the substrate. This is further enhanced by the relatively large gate
bias required for low noise operation and the thin active layers used. In
such situations, the substrate (or buffer) might carry as much as 45 % of
the total current thereby affecting the whole device performance.

Although the importance of the role of the substrate was recognized very
early, the lack of an adequate physical characterization of the complex
trapping phenomena occuring in the different substrate types hindered an
adequate simulation of GaAs MESFET's. The purpose of this work is to assess
the effect of substrate properties on device performance in an attempt to
define an ideal substrate.

2. Model description and typical device features

The numerical model used has been described in detail elsewhere [Ibrahim
(1983) and El Ghazaly et al (1983)]. However for the sake of completeness
we will present its fundamental features.

It is basically a two-dimensional hydrodynamic model which incorporates the
effects of non-stationary electron dynamics specific in submicron structures. It is based on the particle, momentum and energy conservation
equations derived from Boltzmann transport equation, averaged over the
different valleys and subject to some simplifying assumptions. The main one
is the neglect of the inertial term in the momentum conservation
equation. This can be justified by the fact that the momentum relaxation
time is almost one order of magnitude smaller than the energy relaxation
time. This results in the following set of coupled equations where the main
transport parameters (mobility μ, electronic temperature T and energy
relaxation time τ_w) are energy dependent.

Work supported under contract by DAII-CNET

$$\nabla . J + q \frac{\partial n}{\partial t} = 0$$

$$J = \mu(w) \{ qnE - \nabla[kT(w)n]$$

$$\frac{\partial(nw)}{\partial t} = \frac{1}{q}(J.E) - \frac{1}{q}\nabla[kT(w)J] - \frac{1}{q}\nabla.(Jw) - n\frac{(w-w_0)}{\tau(w)}$$

In addition to Poisson's equation

$$\nabla^2 V = q\frac{(n - Nd)}{t}$$

This model follows the same basic ideas of Curtice and Yun (1981) and Cook and Frey (1983) with however a more accurate formulation of energy relaxation effects and diffusion current.

Fig. 1 gives a plot of the constant charge density and constant energy contours obtained using this model. These plots demonstrate the main features of submicronic gate GaAs MESFET operation namely

(i) The absence of a completely depleted layer under the gate and the gradual transition to the channel.

(ii) The maximum charge concentration in the channel under the gate never reaches the doping concentration on account of diffusion and injection in the substrate.

(iii) The presence of a stationary domain at the drain end of the gate that widens the effective channel and depresses the interface barrier thus enhancing injection in the substrate. The boundaries of this domain are defined by the energy contours for $w > 0.35$ eV.

(iv) Carriers are cooled off as they climb up the potential barriers whether that of the Schottky gate or that at substrate interface.

3. Effect of substrate properties

Two types of device structures were modeled. Table 1 gives the structural parameters of these devices. In both devices the low field mobility in the active layer was taken $4,000$ $cm^2/V.s.$ and the background doping in n type buffers or substrates 10^{14} cm^{-3}.

	Lg (μ)	a active layer	L_{gs}	L_{gd}	t_{buffer}	N_d active layer
Device 1	0.3	0.1	0.4	1	0.4	2×10^{17}
Device 2	0.3	0.1	0.24	0.72	0.15	2×10^{17}

TABLE 1

In device 1, four types of buffers were modeled : (i) high mobility (5 000 cm^2/V.s) thick buffer, (ii) very low mobility (100 cm^2/V.s) thick buffer, (iii) high mobility mini-buffer on Cr. doped substrate and (iv) a substrate with infinite interfacial barrier. Type (ii) is a limiting case which serves to study the effect of the mobility and could represent the case of an intentionally damaged buffer. Type iv represents the limiting case of heterojunction substrate with a sufficiently large barrier so as to suppress the injection in the substrate.

It is clearly seen from fig. 2 and 3, that the reduction in buffer mobility affects negatively both device current and transconductance. The reduction (compared to case i) becomes important as carrier injection in the buffer increases i.e. as V_{gs}, increases. It should be noted that low mobility results also in low diffusivity thereby limiting somewhat carrier injection. This explains the fact that the reductions in current and transconductance are not that drastic.

In case (iii), the interfacial barrier, simulated by a surface charge density of 10^{11} cm^{-2}, does not help confining the carriers. It only results in depleting the device resulting also in reduced current and transconductance. Case iv is characterized by a large reduction in current (50 % at V_{gs} = 0), an increase in transconductance at low V_{gs} and a sharp pinch off. This can be explained by the fact that the confinement of the carriers to the channel improves gate control. However at large V_{gs} the increased depletion rapidly offsets this action.

In device 2, a reference case consisting of a device with a thick high mobility buffer is simulated together with a device having a p type confinement layer above a thick buffer (or equivalently an unbuffered device with a Cr doped substrate). In this last type two values of surface charge densities are used, 10^{11} and 2×10^{11} cm^{-2}.

Again it is seen that the effect of the p type buffer is two fold : first it participates in depleting the channel and second it helps confining the carriers to the channel thereby reducing injection. These two effects join to reduce the current as compared to the reference case. At V_{gs} = 0, V_{ds} = 2 V the current is reduced by 13 and 26 % respectively for the two values of surface charge density. As far as the transconductance is concerned these effects are competitive. At low V_{gs}, carrier confinement predominates resulting in a larger g_m whereas for V_{gs} > 0.3V the depleting action becomes more important (fig. 4). As expected the effect of the p type layer on the gate-to-source capacitance is to reduce it on account of the enhanced depletion. However at sufficiently large V_{gs} it levels indicating that maximum depletion is reached (fig. 5). The resulting current cut-off frequency characteristic (fig. 5) shows that peak values of 88 GHz are reached as a result of better confinement. At a typical low noise operating point (I_{ds} = 0.2 I_{dss}) the p type layer with the smaller depth achieves the highest cut-off frequency (62.5 GHz). This indicates that the thickness of the p layer represents an optimization parameter.

The g_d-V_{gs} characteristics shown in fig. 7 corroborate the above physical interpretations.

Conclusion

Devices with different types of buffers and substrates have been simulated. It was demonstrated that buffers with low electron mobility and interfacial barriers at substrate-mini-buffer interface can have an adverse effect on GaAs MESFET performance. On the other hand a thin p-type confinement layer can result in a improved performance.

Acknowledgements

The authors wish to thank E. CONSTANT, from the Centre Hyperfréquences et Semiconducteurs, S. MOTTET, from the C.N.E.T., C. MAKRAM   Ebid   and A. ROCKI from the LEP for helpful discussions.

REFERENCES

Cook R.K. and Frey J., 1983, IEEE Trans. Electron Devices ED-29, 6, 970.

Curtice W.R. and Yun Y.H., 1981, IEEE Trans. Electron Devices ED-28, 8, 954

El Ghazaly et al, 1983, Proc. 13th ESSDERC Canterburry, p. 127.

Ibrahim M.M., 1983, M. Sc. Thesis Faculty of Engineering Cairo University.

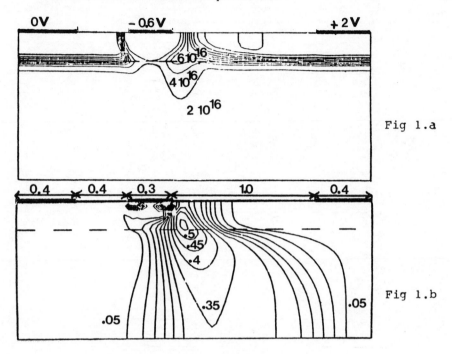

Fig 1.a

Fig 1.b

Fig 1.a Constant charge contours
Fig 2.b Constant energy contours

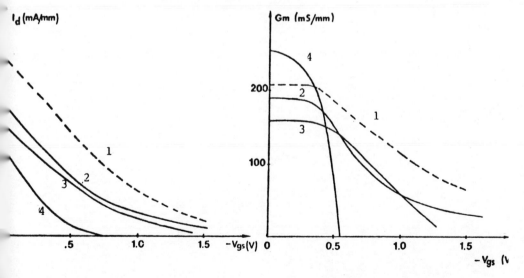

Fig.2 Output current as function
     of gate voltage

Fig.3 Transconductance as
      function of gate voltage

1. High mobility buffer
2. Mini buffer (0.4µ) on Cr doped SI substrate
3. Low mobility SI substrate
4. Ideal SI substrate

Inst. Phys. Conf. Ser. No. 74: Chapter 7
Paper presented at Int. Symp. GaAs and Related Compounds, Biarritz, 1984

509

Low resistive CVD W film as refractory gate metal for GaAs MESFET

S. Takahashi, M. Itoh, M. Akiyama and K. Kaminishi

Reseach Laboratory, OKI Electric Industry Co., Ltd
550-5 Higashiasakawa, Hachioji, Tokyo 193 Japan

Abstract. CVD W films on GaAs have been found to have the low resistivity and stability of Schottky contacts against high temperature treatment. The deposited W films showed a strong adhesion to GaAs and the resistivity of this film with a thickness of 0.7 μm decreased to 6 μΩcm after annealing at 800°C. Using this W film as a gate metal, self-alignment type GaAs MESFETs were fabricated on semi-insulating undoped LEC wafers. These MESFETs (Lg=0.9 μm and Wg=300 μm) showed the high transconductance of 150mS/mm, low gate resistance of 2.0 Ω and low source resistance of 1.6 Ω.

1. Introduction

The application of ion implantation for the formation of the channel layers of GaAs MESFETs has the advantages of good uniformity, low cost, reproducibility and the possibility of planar type devices. Therefore, this technology has been accepted as the most suitable fabrication process for GaAs ICs. In the fabrication processes used in implantation, the self-alignment process using the gate metals as a ion implantation mask is very useful. The gate metals for this process have to possess the refractoriness in high temperature treatment. Some kinds of metals have been studied for this purpose, for example, sputtered or evaporated W (Matsumoto et al 1982), Ti/W, W/Si (Yokoyama et al 1982, 1983), W/Al (Nakamura et al 1983) etc. These metals have disadvantages of high resistivity and/or poor adhesive properties to GaAs. Therefore, it is difficult to apply these materials to gate metals of microwave FETs for which the low gate resistance is required. However, since the self-alignment process has the advantages of simplicity, the low source and contact resistances of MESFETs, this process is very suitable for the application to microwave FETs, if the low resistive refractory metals are obtained. It were reported that CVD W films on Si substrates have good adhesive and low resistive properties (Melliar-Smith et al 1974) and that the W films on GaAs show good Schottky contacts (Batev et al 1978).

For obtaining more suitable gate metal to self-alignment process, the W films on GaAs deposited by CVD method were studied. It was found that CVD W films have the low resistivity, the strong adhesive properties to GaAs and the stable Schottky contacts in high temperature treatment up to 900°C. The self-alignment type GaAs MESFETs were fabricated using these

CVD W films as a gate metal on semi-insulating undoped LEC wafers.   These
MESFETs showed the high transconductance, low source and gate resistance.
In this paper, the deposition of the W films, the refractoriness of
CVD W/GaAs Schottky contacts, the resistivity and the characteristics of
self-alignment type GaAs MESFETs using the W films as a gate metal are
reported.

2.   Experimental

2.1 Deposition

The W films were deposited on GaAs through the reaction of WF_6 and H_2 in
an atmospheric pressure.   Schematic diagram of the deposition system is
shown in Fig. 1.   The system is a lamp heated horizontal type using graphite susceptor.   The temperature is measured by a thermocouple embedded in
the susceptor.   Since the purity of the gases are an important factor
affecting the properties of the films, the minimum purities of 99.99 % (N_2
and H_2) and 99.98 % (WF_6) were used.   The deposition temperature was
varied from 300°C to 500°C.   The deposition time dependence of the W film
was varied between 30 nm/min and 100 nm/min changing the substrate temperature and the flow rate of WF_6.   Most of the deposition experiments were
carried out by the use of hydrogen and nitrogen flow rate of 3000 CC/min
and 1000 CC/min, respectively, and WF_6 flow rate of 1-2 CC/min.   Under this
condition, the deposition rate of the W films was about 50 nm/min at 450°C.
The W films, 0.1-1.0 µm thick, have been deposited uniformly on GaAs substrates.   The deposited W films showed the strong adhesion and no scrolling
up appeared even if the layer was as thick as 1.0 µm.

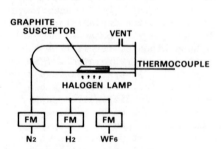

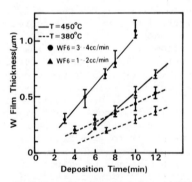

Fig. 1 Apparatus used for chemical
vapor deposition of tungsten

Fig. 2 Deposition time dependence of W film
thickness as the parameter of substrate
temperature and WF6 flow rate.

2.2 Etching

CVD W films can be easily etched in Reactive Ion Etching (RIE) by the use
of SF_6. As the thickness of W films were thick, up to 1.0 μm, etching was
done under the low pressure condition (2-5 Pa) to restrain the side etching.
The etching time dependence of etching depth of the W films is shown in
Fig. 3. The etching rate and etching uniformity were varied by changing
the rf power, the flow rate of SF_6 and the pressure. For the device fabrication, the etching was carried out at the rf power of 100 W and the SF_6
flow rate of 40 CC/min. Under this condition, etching rate of the W films
was 100 nm/min at the pressure of 2 Pa, and the ratio of the side etching
to depth was around 0.15.

2.3 Schottky contact

CVD W/GaAs Schottky contact properties were studied after annealing at
high temperature. The W film of 0.7 μm thickness was deposited on N-type
GaAs ($2 \times 10^{17}/cm^3$) substrate and Schottky contacts were patterned to 100 μmϕ
dot in RIE. The samples were annealed at various temperatures for 20 min
in an AsH_3 atmosphere. Then the ohmic contacts were formed by evaporating
AuGe/Ni/Au metals and sintering. The ideal factors and barrier heights of
the Schottky contact were obtained from forward I-V characteristics. With
these samples, annealing temperature dependence of Schottky contacts is
shown in Fig. 4. The ideal factors are almost constant even after the
annealing up to 900°C. Moreover, the barrier heights remain constant at a
value of 0.7 eV against the annealing up to 800°C.

2.4 Resistivity

Annealing temperature dependence of the W film resistivity is shown in
Fig. 5. The resistivity decreases as the annealing temperatures increase
and it saturated to constant value after annealing at higher than 800°C.
This decrease was, probably, owing to the increase in grain size of the
W film. After annealing at higher than 800°C for 20 min, the resistivity of
the W film with a thickness of 0.7 μm, decreased to as low as 6 μΩcm from
18 μΩcm which was the value of the as deposited film. This value is
nearly equal to that of the bulk value (5.4 μΩcm).

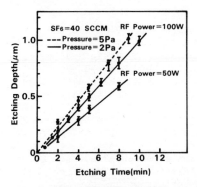

Fig. 3 Etching time dependene of etching
depth as the parameter of applied rf power
and a reactor pressure.

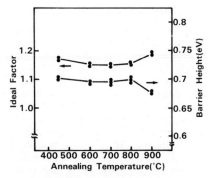

Fig. 4 Annealing temperature dependence of
the Schottky properties.

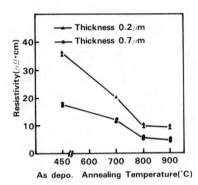

Fig. 5 Annealing temperature dependence of W
film resistivity as the parameter of film thickness.

So it is concluded that CVD W films can be sufficiently used for the low
resistive thermally stable gate metal of GaAs MESFETs.

3.   Device fabrication and characteristics

3.1 Device fabrication

Using this low resistive refractory CVD W film as a gate metal, self-
alignment type GaAs MESFETs were fabricated on (100)-oriented semi-
insulating undoped LEC wafers. The active channel layer was formed by
direct ion implantation with a Si ion dose of $2.5 \times 10^{12} \, cm^{-2}$ at 60 KeV.
After CVD W films with the thickness of 0.7 µm were deposited on GaAs
wafers, the gate patterns of Ti/Ni were formed by lift-off technique.
The W films were etched using Ti/Ni layer as an etching mask in RIE.
The source and drain layers of MESFETs were formed by selective ion
implantation with a Si ion dose of $1 \times 10^{13} \, cm^{-2}$ at 100 KeV using Ti/Ni
layer as an implantation mask. After removing Ti/Ni layers, annealing
was done at 800°C for 20 min in an AsH_3 atmosphere. The ohmic contacts
were formed by evaporating and sintering of AuGe/Ni/Au metals.

3.2 Characteristics

The fabricated GaAs MESFETs have the gate length of 0.9 µm and the gate
width of 300 µm, whose structure is shown in Fig. 6. This device consists
of a single gate with two gate pads and the source to drain spacing is
5.0 µm. The current-voltage characteristics of the obtained typical
MESFETs are shown in Fig. 7. The transconductance of these devices is
45 mS, which corresponds to 150 mS/mm. The drain to source saturation
current, Idss, is 60 mA and the pinch-off voltage Vp is -2 V. The nominal
channel resistance between source and drain is measured to be around 10.3 Ω
at zero gate voltage, and resistance between two gate pads which are 150 µm
apart is around 18 Ω. The source resistance and gate resistance are 1.6 Ω
and 2.0 Ω, respectively. The source resistance was obtained from measuring the nominal channel resistances of MESFETs formed by different gate
length and separating these to the channel resistance and source
resistance.

Fig. 6 Top view of 0.9×300μm FET device

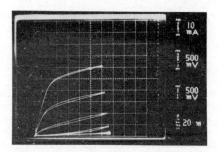

Fig. 7 I -V characteristics of a typical
0.9×300μm GaAs MESFET.

The noise figure and associated gain of these GaAs MESFETs were measured
at 10 GHz. The noise figure and associated gain as function of the drain-
to-source current at 3 V bias are shown in Fig. 8. The best device performance includes a noise figure of 3.5 dB with 9 dB associated gain at
10 GHz. The S-parameters of these MESFETs were measured by the use of an
automatic network analyzer in the 2-12 GHz range. From these paramters,
the current gain and the gain-band-product, f_T, were obtained. The
current gain as a function of frequency is plotted in Fig. 9. At the
minimum noise figure of device condition (Ids=10 mA) gain-band-product
was 16 GHz.

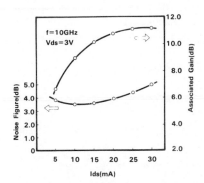

Fig.8 Noise figure and associated gain at 10GHz
as a function of drain to source current.

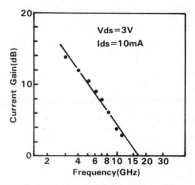

Fig. 9 Current gain as a function of frequency
for minimum noise-figure operation condition.

4. Conclusion

The CVD W films on GaAs were found to have the low resistivity, a strong
adhesive property and the stability of Schottky contacts against   high
temperature treatment. The resistivity of this W film with a thickness
of more than 0.7 µm was as low as 6 µΩcm. This resistivity was the lowest
value in the refractory Schottky metals reported before. Furthermore,
according to the refractoriness of CVD W/GaAs Schottky contact the W films
can be used as the gate metal of self-alignment type GaAs MESFETs and
expected to obtain highly reliable devices. The self-alignment type
GaAs MESFETs using CVD W films as a gate metal were fabricated on semi-
insulating undoped LEC wafers. These MESFETs (Lg=0.9 µm and Wg=300 µm)
showed the high transconductance of 150 mS/mm, low gate resistance of 2.0 Ω
and low source resistance of 1.6 Ω. RF performance of this MESFET has
obtained a noise figure of 3.5 dB with 9.0 dB associated gain at 10 GHz
and f_T of 16 GHz. These values are not so good. However, they will be
improved by optimizing the gate structure of the device, the fabrication
process and by using high purity epitaxial layers. These results made it
clear that CVD W film is a useful gate metal for the self-alignment type
GaAs MESFETs.

5. Acknowledgement

The authors would like to thank H. Nakamura and Y. Sano for their helpful
suggestions and discussions in the fabrication of MESFETs. They also
thank J. Shibata for microwave measurement.

6. Referrences

Batev P M, Ivanovitch M D, Kafediiska E and Simeonov S S   1978
   Phys, Stat, Sol, (a) 45   671

Matsumoto K, Hashizume N, Tanoue H and Kanayama T   1982
   Japanese Journal of Applied Physics Vol. 21, No. 6, L393

Melliar-Smith C M, Adams A C, Kaiser R H and Kushner R A   1974
   J. Electrochem. Soc. Solid-State Science and Technology Vol. 121,
   No. 2, 298.

Nakamura N. Sano Y, Nonaka T, Ishida T and Kaminishi K   1983
   GaAs IC Symp. Digest, 134.

Yokoyama N, Ohnishi T, Odani K, Onodera H and Abe M   1982
   IEEE Trans. Electron Devices ED-29 1541.

Yokoyama N, Ohnishi T, Onodera H, Shinoki and Shibatomi A   1983
   IEEE J. Solid State Circuits, 18 520.

Inst. Phys. Conf. Ser. No. 74: Chapter 7
Paper presented at Int. Symp. GaAs and Related Compounds, Biarritz, 1984

515

Effects of p-type barrier layer on characteristics of sub-micron gate self-aligned GaAs FET

K. MATSUMOTO, N. HASHIZUME, N. ATODA, and Y. AWANO

Electrotechnical Laboratory
1-1-4 Umezono, Sakura-mura, Niihari-gun, Ibaraki, Japan

Abstract

The substrate current under the channel layer in a sub-micron gate self-aligned GaAs FET is shown by Monte Carlo simulation to be the major cause of the short channel effects. By introducing a p-type layer between the channel layer and the substrate, the short channel effects of the FETs having sub- micron gate length are suppressed significantly.

1. Introduction

One of the best ways to enhance the performance of a GaAs FET is to adopt the self-alignment structure (MATSUMOTO 1982a) and simultaneously reduce the gate length. By reducing the gate length into the submicron range, however, short channel effects, that is, an increase of the drain conductance, and a large shift of the threshold voltage become significant (YAMASAKI 1982b). One of the reasons for this short channel effects is the space charge limited current which flows through the semi-insulating GaAs substrate under the channel layer due to a high electric field between the two adjacent source and drain n^+-regions. (AWANO 1983)

In the present paper, we will describe the effects of the substrate current on the FET characteristics based on the results obtained by Monte Carlo simulation. Then, we will describe an FET in which a p-type barrier layer is introduced between the channel layer and the substrate. The barrier layer is for preventing the substrate current.

Fig. 1, Model of submicron-gate self-aligned GaAs FET for Monte Carlo simulation. Gate is 0.25 μm in length and 20 μm in width. Channel is 0.1 μm in thickness and is doped to n=2x10^{17}/cm^3.

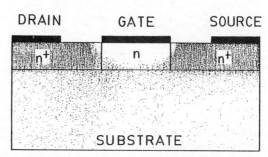

2. FET characteristics by Monte Carlo simulation

Characteristics of an self-aligned GaAs MES FET were calculated by Monte Carlo simulation using a model shown in Fig. 1. The channel is l_g= 0.25 μm in length, d= 0.1 μm in thickness and is doped to n= $2\times10^{17}/cm^3$. The source and drain n^+-regions are doped to $n^+=8\times10^{17}/cm^3$ and extend right up to the edges of the gate. The gate electrode is 0.25 μm in length, 20 μm in width. The thickness of an undoped substrate is 0.3 μm. The details of the calculation are described elsewhere (AWANO 1983).

Figure 2 (a) is the drain current-voltage characteristics of the FET. The FET shows normally-on characteristics. In order to check the effect of the substrate current, characteristics of an FET without the substrate was also calculated and shown in Fig. 2 (b). Comparing these two characteristics, following differences become clear. By adding the substrate to the FET, the drain conductance increases from 35 mS/mm to 85 mS/mm, the drain current increases from 6.2 mA to 8 mA, both at the drain voltage of V_{ds}=1.5 V, the gate bias of V_g= 0 V. The channel of the FET without the substrate can be almost pinched off at the applied gate bias of V_g=-0.8 V, V_{ds}=1.5 V, whereas the channel of the FET with the substrate can not. Moreover, the gate bias of the FET with the substrate becomes less effective at deep gate biases.

In Fig. 1, spatial distribution of electrons in the FET with the substrate is shown at V_g= -0.4 V, V_{ds}= 1.5 V. A substantial number of electrons flow through the substrate. The above results show. the importance of suppressing the substrate current for the suppression of the short channel effects.

3. Structure and fabrication process of FET
with p-type barrier layer

Based on the theoretical prediction in section 2, we placed a p-type barrier layer between the channel layer and the substrate as

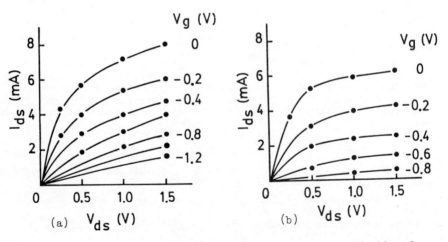

Fig. 2, Drain current - voltage characteristics of self-aligned GaAs FET with (a) and without (b) substrate by Monte Carlo simulation.

shown in Fig. 3. The p-type barrier layer is to suppress the substrate current by the p-n junction barrier. We fabricated two kinds of FETs: one is an FET with the p-type barrier layer and the other is an FET without the barrier layer. We compared the characteristics of these FETs. The fabrication process is as follows: An undoped semi-insulating LEC (100) wafer was used as a substrate. A channel layer for an FET with the p-type barrier layer was provided by selective ion implantation with $4 \times 10^{12}/cm^2$ Se at 160keV and $2 \times 10^{12}/cm^2$ Mg at 250keV, and a channel layer for a FET without the p-type barrier layer was provided by ion implantation with $2 \times 10^{12}/cm^2$ Se at 160keV. A $0.2\mu m$ thick tungsten gate, ranging from $0.19 \mu m$ to $1.55 \mu m$ in length and 20 μm in width, was delineated on the channel region by lift-off process. The tungsten gate served as a mask for the second ion implantation of $1.2 \times 10^{13}/cm^2$ Se ions at 160keV for the source and drain n^+ layers. Implanted layers were annealed in an infra-red lamp furnace at 950°C for 25 seconds without a cap, under arsenic pressure. Under that annealing condition, the lateral diffusion length of the n^+ impurities into the channel region can be held to less than $0.1 \mu m$ and the sheet resistance of the n^+ layer is 350 Ω /□. The ohmic electrodes, 4 μm apart, were formed by depositing a Au-Ge/Ni/Au film, delineating by lift-off, and alloying in H_2 ambient gas at 440°C for 30 seconds. An SEM photograph of FET thus fabricated is shown in Fig. 4. The gate length is $0.2 \mu m$ for this particular sample.

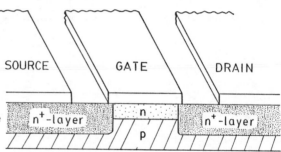

Fig. 3, Schematic structure of self-aligned GaAs FET with p-type barrier layer between channel layer and substrate.

Fig. 4, SEM photograph of self-aligned GaAs FET. Gate length is 0.2 μm.

4. Characteristics of FET

The carrier profiles of the channel region for the FET without and with the p-type barrier layer are shown in   Fig. 5 (a) and (b), respectively. The dashed line and dot-dash line are LSS theory curves for Se ions and Mg ions, respectively. The solid lines are actually obtained carrier profiles by capacitance-voltage measurements.

Figure 6 .(a) and (b) are the source current- drain source voltage characteristics of the FETs without and with the p-type barrier layer, respectively.   The   gate lengths are Lg=0.33 μm and Lg=0.23 μm, respectively. Comparing these two characteristics, similar tendency to that in Fig. 2 is observed. That is, though the FET of Fig. 6 (b) has smaller gate length than that of Fig. 6 (a), the drain conductance is almost half as much and the pinch off voltage is much smaller. The gate bias of the FET of Fig. 6 (b) is still effective at the deep gate biases. These effects are discussed in detail in the following.

Figure 7 (a), (b) show the source current in log scale versus gate bias characteristics of the FETs without (a) and with (b) the p-type barrier layer. We define the threshold voltage as the gate voltage at which the source current reaches 5×10^{-5} A. The drain bias is $V_{ds}=2$ V and the gate length is a parameter. With a reduction of the gate length

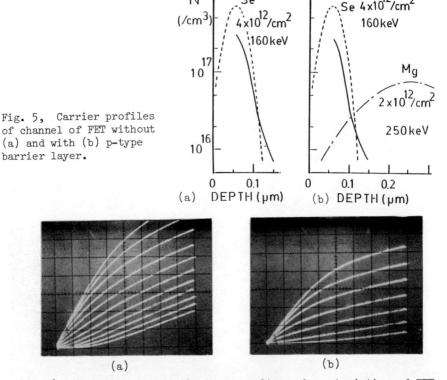

Fig. 5, Carrier profiles of channel of FET without (a) and with (b) p-type barrier layer.

(a)   (b)

Fig. 6, Source current-drain source voltage characteristics of FET without (a) and with (b) p-type barrier layer. (a) Gate length is Lg=0.33 μm, gate bias Vg=+0.4 ∿ -1.6. (b) Gate length is Lg=0.23 μm gate bias Vg=0 ∿ -1.2.   1 mA/div. 0.2 V/div.

from 1.55 μm to 0.44 μm, the threshold voltage shifts by about 1 V in the case of the FET without the barrier layer. In the case of the FET with the barrier layer, however, the shift of the threshold voltage is as small as 0.3 V with the reduction of the gate length from 1.3 μm to 0.4 μm.

Figure 8 and 9 show the dependences of the drain conductance and the transconductance on the gate length, respectively. The drain conductance is measured at V_{ds}=2 V and V_g=0.3 V and the transconductance at V_{ds}=2 V and V_g=-0.1V. White circle is for the FET with the p-type barrier layer, black circle for the FET without the barrier layer. By introducing the p-type barrier layer, the drain conductance reaches about one-half of that without the barrier for the gate length range of lg=0.8-1.5 μm, and increases less steeply in the sub-half micron gate length range. The steep increase of the drain conductance in the sub-half micron gate length range is partly due to the increase in the ratio between the channel thickness and length, which makes the gate bias less effective.

In Fig. 9, for the gate length range of lg=1-1.5 μm, the transconductance shows almost the same value for both cases. The transconductance of the FET without the barrier layer increases moderately when the gate length becomes short. However, the transconductance of the FET with the barrier layer shows a steep increase with the reduction of the gate length. The experimental points in this curve are the average values of eight FETs. The highest transconductance obtained for a single FET was 316 mS/mm at the gate length of 0.4 μm. When the gate length is reduced further, the transconductance decreases. The cause of this decrease is not clear so far. However, two possible reasons are considered. One of the reasons is the lateral diffusion of the n^+ impurities into the channel layer and the other reason is the increase in the ratio between the channel thickness and length.

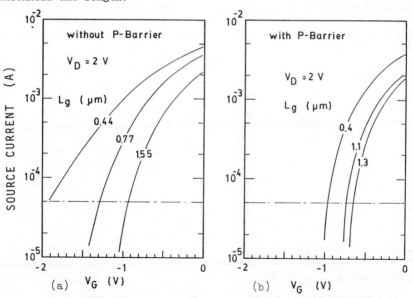

Fig. 7, Dependence of source current on gate bias of FET without (a) and with (b) p-type barrier layer. Drain voltage is V_D=2V and gate length is a parameter.

5. Conclusion

Using Monte Carlo simulation, we have shown that one of the causes for the short channel effects of a sub-micron gate self-aligned GaAs FET is the space charge limited current which flows through the substrate under the channel layer. In order to suppress the substrate current, we introduced a p-type layer between the channel layer and the substrate of the FET. As a result, the drain conductance reduced to about one half of that without the p-type barrier layer, the transconductance increased steeply even in the sub-half-micron gate length range, and the shift of the threshold voltage with the reduction of the gate length decreased. These results show the effectiveness of the p-type barrier layer for suppressing the short channel effects.

Acknowledgment

The authors would like to thank S. Kataoka for continuous encouragement. They are also indebted to K. Tomizawa, M. Iida, and T. Kurosu for helpful discussions, Y. Yamada, T. Endo, H. Yamazaki, H. Inage, and M. Yamada for technical assistance.

References

MATSUMOTO K, HASHIZUME N, ATODA N, TOMIZAWA K, KUROS T, IIDA M
    1982a Proceeding of Int. Symp. GaAs and Related Compounds.
    pp317-324
YAMASAKI K, KATO N, MATSUOKA Y, OhWADA K, 1982b Tech.Digest
    1982 Int. Electron Devices Meeting (IEEE New York 1982)
    pp166-169.
AWANO Y, TOMIZAWA K., HASHIZUME N., 1983 Tech. Digest 1983
    Int. Electron Devices Meeting (IEEE New York 1983)
    pp617-620.

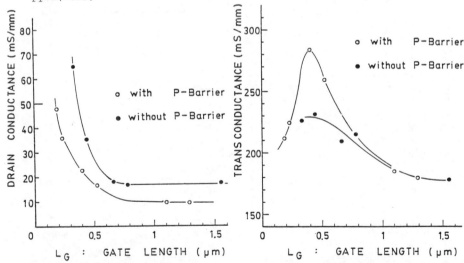

Fig. 8, Dependence of drain conductance on gate length of FET with and without p-type barrier layer.

Fig. 9, Dependence of transconductance on gate length of FET with and without p-type barrier layer.

Inst. Phys. Conf. Ser. No. 74: Chapter 7
Paper presented at Int. Symp. GaAs and Related Compounds, Biarritz, 1984

521

A new side-gating model for GaAs MESFETs based on surface avalanche breakdown

Hideki Hasegawa, Toshihiko Kitagawa, Takayuki Sawada and Hideo Ohno

Department of Electrical Engineering, Faculty of Engineering,
Hokkaido University, Sapporo, 060 Japan

Abstract. Side-gating of GaAs MESFETs on semi-insulating substrates is studied. Strong side-gating took place for negative side-gate potential at side-gate breakdown voltage V_T. Side-gating involved white light emission from source edge, long-range nature, light sensitivity, sensitivity to surface passivation conditions and linear dependence of V_T on side-gate separation. These observations cannot be explained by the previous SCLC model. A new model is proposed which includes avalanche injection due to field concentration by surface state filling.

1. Introduction

Packing density in GaAs LSI/VLSIs is ultimately limited by device isolation. Poor isolation is caused not only by surface breakdown between adjacent ohmic electrodes (Hasegawa et al 1983), but by the so-called side-gating effect (Goronkin 1984 and the references cited therein)where currents through MESFETs or HEMTs are modulated by voltages on nearby ohmic electrodes. Lee et al (1982) found a correlation between surface breakdown and side-gating and proposed a model, in which the breakdown is caused by the filling of the bulk traps and then the resultant trap-filled space charge limited current (SCLC) biases the active layer-substrate interface of the MESFET from backside.

The purpose of this paper is to study side-gating of GaAs MESFETs. Al-gate MESFETs were formed on OMVPE active layers grown on HB and LEC semi-insulating substrates. The observed relationship between the side-gate voltage and the drain current consisted of weak and strong side-gating regimes. In the former regime, the drain current gradually changed with the side-gate potential. In the latter regime, drain current decreased steeply when the negative side-gate potential exceeded a threshold value which was equal to the breakdown voltage between side gate and source in agreement with Lee et al (1982). However, new features such as light emission at the onset of strong side-gating, long-range nature and marked sensitivity to passivation conditions were found, that cannot be explained by the previous SCLC model.

A new model of side-gating is proposed in which the substrate current underneath the FET region changes the occupation function of dominant traps and therefore the space charge density. Weak side-gating is due to pre-breakdown substrate current. Strong side-gating is due to strong electron injection caused by surface avalanche breakdown. Filling of surface states produces a localized intense electric field at outer edge of source electrode owing to planar geometry and triggers avalanche.

2. Experimental

The structure employed in this study is shown in Fig.1.   Al-gate
MESFETs were formed on MOVPE active layers grown on Cr-O doped HB and
undoped LEC undoped substrate. Typical thickness, electron concentration
and mobility of the active layer were 2000 Å, 1 x 10^{17} cm^{-3} , 4,200 cm^2 V^{-1}
sec^{-1}., respectively. Size and spacing of AuGe/Ni source, drain and
side-gate electrodes were varied over wide ranges. Device isolation was
made by chemical etching of the active layer.

In order to study the effect of
surface passivation films on side-
gating, silicon dioxide and silicon
nitride films were deposited by
plasma CVD processes in SiH_4 + O_2 +
N_2 and SiH_4 + N_2 gas mixtures,
respectively, at the total pressure
of 0.4 -1.0 Torr. Wafers were kept
at 300 °C during deposition. The
refractive index of the films were in
the range of 1.4 - 1.5 for silicon
dioxide and 1.8 - 2.0 for silicon
nitride, respectively, depending on
deposition conditions. Resistivities
and breakdown field strength were
10^{13} - 10^{15} ohm-cm   and 5 - 8 x 10^6
V/cm, respectively, in both kinds of

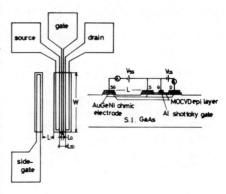

Fig.1 Sample Structure

films.   These properties are in good agreement with those of previously
reported plasma CVD films (Vossen and Kern 1978).

3. Side-gating and Surface Breakdown Characteristics

The observed side-gating behavior is summarized in Fig.2. Effect of light
illumination on side-gating is also shown in Fig.2. As seen in Fig.2,
the side-gating curves in the dark has the weak side-gating regime as
indicated by A and the strong side-gating regime as indicated by B. It
is also noted that the side-gating effect is of a surprisingly long range
nature,   since a side gate which is more than 200 μm away from the gate

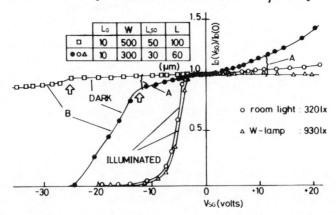

Fig.2 The observed side-gating behavior under dark and illuminated
      conditions.

can still modulate the FET characteristics. The corresponding I-V characteristics between the side-gate and
the FET source electrode of the
devices in Fig.2 are shown in Fig.3.
As indicated by arrows in Figs.2 and
3, the threshold voltage for onset
of strong side-gating in the dark is
equal to the surface breakdown voltage between the side-gate and the
source electrode of the FET in agreement with the obervation by Lee et al
(1982). In the case of Lee et al, the
side-gating threshold and the surface
breakdown voltages were shifted by
the magnitude of the drain bias. In
the present case, such a shift was
not observed. We believe that the
difference was due to the difference
in the geometry, because the side-
gate in Lee et al was perpendicular
to the gate of the MESFET.

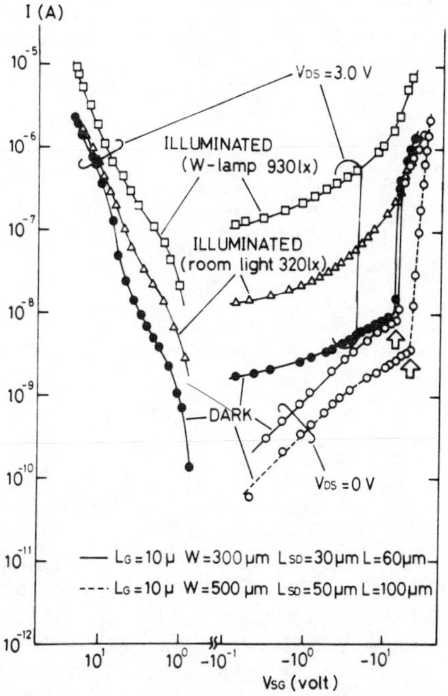

At the onset of strong side-gating,
visible white light emission took
place along the outer edge of source
electrode. The emission pattern was
either fairly uniform as shown in
Fig.4 (a) or spotty as in Fig.4(b),
depending on the various processing

Fig.3 I-V Characteristics

conditions. This light emission is
associated with the surface breakdown (Hasegawa et al 1984). Fig.5 shows
the emission intensity vs. side-gate leakage current. When the emission
was uniform, its intensity increased roughly in proportion to the square
of the leakage current. When the emission was spotty, the location of the
spot changed with current, resulting in the shifts of the characteristics
as seen in Fig.5. The spectrum of the emission is shown in Fig.6. A
special sample was used to measure the spectrum of the weak emission.

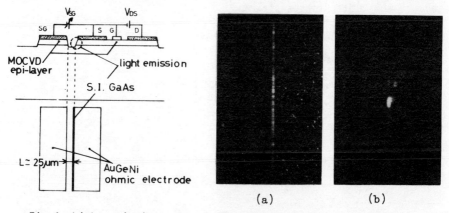

Fig.4   Light emission pattern from the edge of source electrode
        (a) uniform pattern   and (b) spotty pattern

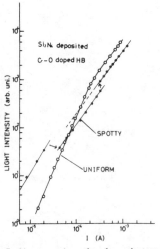

Fig.5 Observed emission intensity

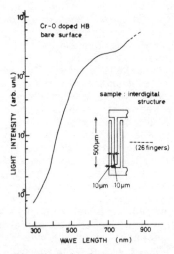

Fig.6   Emission spectrum

It was found that the side-gating threshold or the surface breakdown voltage was extremely sensitive to processing conditions for passivation. Under the same passivation conditions, however, V_T was proportional to the distance L between side gate and outer edge of source electrode as shown in Fig.7. The measured temperature dependence of the surface breakdown voltage is shown in Fig.8. It was also found that the surface leakage current in the ohmic region and its activation energy were extremely sensitive to the insulator formation and post-deposition processing conditions (Hasegawa et al 1983).   The measured resistance could be 1-2 orders of magnitude smaller than the theoretical resistance calculated from the bulk resistivity by conformal mapping.

On the other hand, no large quantitative difference in the side-gating and surface breakdown behavior was experienced beween Cr-O doped HB and undoped LEC substrates, although the prebreakdown leakage current and the magnitude of side-gating in the weak side-gating regime apparently depended on the substrate resistivity.

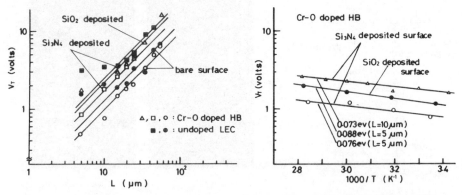

Fig.7 Dependence of side-gating   Fig.8 Dependence of side-gating
      threshold on spacing L   threshold on temperature

4. Side-Gating Mechanism

The previously proposed mechanism for side-gating (Lee et al 1982, Goronkin 1983) was that the side-gating threshold is given by the so-called trap-filled voltage and that the resultant trap-filled space charge limited current (SCLC) directly reverse-biases the substrate-active layer interface. This model cannot explain the observed light emission, the long range nature of the side-gating and the observed sensitivity to passivation conditions. In addition, the linear dependence of the side-gating threshold on the side-gate spacing L does not agree with the standard trap-filled SCLC theory which gives L^2 dependence. For the last point, however, one has to take into account   the planar geometry. In fact, our preliminary calculation of the trap-filled voltage pertinent to the present planar geometry, resulted in a linear dependence, but it also showed that the observed breakdown voltage is at least two or three orders of magnitude smaller than the theoretical value, if one assumes reasonable values of trap density of 10^{15}- 10^{17} cm^{-3}. On the other hand, the present light emission, its spectrum and the temperature dependence of V_T strongly indicate involvement of avalanche in the strong side-gating region. The weak dependence of the threshold voltage on temperature with a positive temperature coefficient shown in Fig.8 is also consistent with the avalanche mechanism.

Our model of the side-gating is the following. The side-gating is basically caused by the substrate current underneath the FET region, which changes the occupation function of the dominant traps, i.e., Cr acceptors in the Cr-O doped substrates and EL 2 donors in the case of undoped LEC substrates. It is known that the excess space charge near the substrate-active layer interface changes the drain current (Itoh and Yanai 1980, Kocot and Stolte 1982).   In the weak side-gating regime, side-gating is caused by the pre-breakdown substrate current. Even when the side-gate potential is zero,   there exists a finite substrate current towards drain for a positive drain bias, as is evident in Fig.3. Application of positive side-gate potential decreases the electron flow underneath the FET and reduces negative space charge on the substrate side. This makes the extension of the depletion layer into active layer smaller, resulting in increase of drain current. Obviously, application of negative potential has a reverse effect.

The strong side-gating is due to increase of electron flow into substrate by surface breakdown.   If there is a large number of trapping states near the surface, which is the case for GaAs, the filling of these states by electrons gives rise to intense local electric field at the anode edge due to the planar geometry. The situation is schematically shown in Fig.9(a). This intense electric field will trigger avalanche breakdown, which will lead to surface breakdown accompanied with white light emission. Previously, concentration of electric field lines has not been taken into consideration.   Owing to this avalanche breakdown, strong injection of high energy electrons from the avalanche plasma takes place, leading to subsequent electron trapping by increased electron flow, as schematically shown in Fig.9(b). Observed long range nature of side-gating is consistent with the high energy nature of avalanche plasma.

The trapping states could either be surface states or bulk states near the surface.   In either case, the observed linear dependecne of V_T on spacing can be explained equally well by a simple calculation. Recently, Chang et al (1984) discussed surface conduction in terms of SCLC mechanism

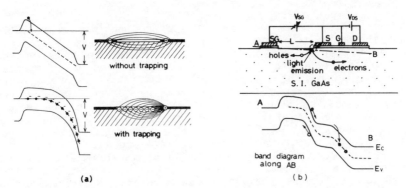

Fig.9 (a) Model for field concentration and (b) avalanche injection
        of carriers and subsequent trapping by bulk states.

due to filling of bulk traps. However, the observed sensitivity of backgating characteristics to surface processing conditions rather than to
substrate properties together with the correlation between ohmic current
and surface states properties (Hasegawa et al 1983), strongly suggests
that surface states control the surface breakdown. Effects of proton and
oxygen implantation in enhancing V_T (Lee et al 1982, Paulson et al 1982)
may be due to the fact that increase of bulk states near the surface
weakens the relative importance of surfaces states and relaxes field concentration by more diversified distribution of ionized trapping centers.

The observed light sensitivity is similar to that reported in literature
(Goronkin et al 1982, Lee et al 1982) and can be understood in terms of
electron-hole pair generation and photomultiplication by which the
effective side-gating threshold is reduced as shown in Fig.2.

Acknowledgement

The present work is supported in part by a Grant-in-Aid on "Nanometer
Structure Electronics" and by a Grant-in-Aid for Developmental Research
both from Ministry of Education, Science and Culture, Japan.

References

Chang M F, Lee C P, Hou L D, Varenkamp R P and Kirkpatrick C G 1984 Appl.
    Phys. Lett 44 869.
Goronkin H, Birrittella M S, Seelbach W C and Vaitkus R L 1982 IEEE
    Trans. Electron Devices ED-29 845.
Goronkin H Proc.1983 IEEE Cornell Conf. pp26-37
Hasegawa H, Sawada T and Kitagawa T 1983 GaAs IC Symp. Tech.Dig. pp145-148
Hasegawa H, Kitagawa T, Sawada T and H Ohno 1984 Electron. Lett.20 561
Itoh T and Yanai H 1980 IEEE Trans. Electron Devices ED-27 1037.
Kocot C and Stolte C A 1982 IEEE Trans. Electron Devices ED-29
    1059.
Lee C P, Vahrenkamp R, Lee S J, Shen Y D and Welch B M 1982 GaAs IC
    Symp. Tech. Dig. pp169-172.
Paulson W M, Birrittella M S , Miers T H and McLaughlin K L 1982 GaAs IC
    Symp. Tech. Dig. pp.166-168.
Vossen J L and Kern W 1978 Thin Film Processes (Academic Press, Inc. New
York) Part IV-1.

Inst. Phys. Conf. Ser. No. 74: Chapter 7
Paper presented at Int. Symp. GaAs and Related Compounds, Biarritz, 1984
527

A comparison of GaAs buffer layers and substrates for ion implanted MESFETs

D C Bartle, R Tayrani[*], C P Stewart[+] and J D Grange

GEC Research Laboratories, Hirst Research Centre, Wembley, UK
Present address:
[*] BP Research Centre, Chertsey Road, Sunbury-on-Thames, Middx, UK
[+] National Westminster Bank, Business Development Div., London, UK

Abstract.   Ultra low noise 1 μm gate length GaAs MESFETs have been
fabricated using ion implantation into $AsCl_3$ grown epilayers. This
paper compares material characteristics (electrical properties) and
device performance of Si implanted MESFET structures in nominally
undoped LEC semi-insulating (100) GaAs substrates and epitaxial buffer
layers.

Introduction

At present a substantial proportion of GaAs microwave and digital circuit
fabrication employs direct ion implantation into LEC grown semi-insulating
(SI) substrates.   The economics, reproducibility and flexibility of ion
implantation makes this an attractive route for LSI and VLSI GaAs
technology. This technique however is dependent on material properties and
although the 'quality' of commercially available SI substrate continues to
improve, it is still necessary to select crystals suitable for ion
implantation and annealing.   Currently many epitaxial growth techniques,
which may use the same SI substrates, rely on growing a high purity, high
resistivity undoped buffer layer between the substrate and active layer.
This is intended to improve the material quality at the active
layer/insulating layer interface by separating the metallurgical interface
from the electrical interface.   This has been shown to achieve improved
device results (Bonnet et al 1981, Nakanisi et al 1981). High performance
(low noise) small signal MESFETs have been realised by reducing gate
lengths, using self aligned gate structures and by using buffered epitaxy
(Chye and Huang 1982, Feng et al 1984, Wholey and Omori 1978).   This paper
reports on our investigation of ion implantation into high resistivity
buffer layers as a route to achieving low noise MESFETs.   We compare
material and device results with direct ion implantation into state-of-the-
art commercially available SI substrates.

Experimental

The undoped buffer layers used in these experiments were vapour phase $AsCl_3$
grown on (100) SI Cr-O doped Bridgman substrates.   The layers were 3 μm
thick, had a free electron concentration $<10^{14}$ cm^{-3} and sheet resistivity
of 3 x 10^5 Ω/square.   The SI (100) substrates used for direct ion
implantation were nominally undoped LEC grown with a free electron
concentration of $\sim 10^8$ cm^{-3}.   The substrate samples maintained a sheet
resistivity of $>10^7$ Ω/square after an 850°C 20 minute anneal, either in an

arsine atmosphere (Grange and Wickenden 1983) or encapsulated with silicon nitride (Bartle et al 1984). The SI substrates had also been qualified for implant activation.

All buffer layer and substrate samples were implanted and annealed under identical conditions. The implant schedule was designed to produce a doping profile suitable for a 1 μm gate length MESFET operating at 10 GHz. The free electron concentration profile has a shallow n^+ surface region to aid ohmic contact formation and an n channel of ~0.2 μm thickness. The implant schedule was $^{29}Si^+$: 4×10^{12} cm^{-2}; 200 keV, (n channel) and 8×10^{12} cm^{-2}; 25 keV (n^+ contact). Implantation was at nominally room temperature and in a non channelling direction. Post implantation annealing was capless in an arsine atmosphere at 850°C for 20 minutes (Grange and Wickenden 1983).

Capacitance-voltage profiling and differential Hall and strip measurements (Stewart et al 1984) were used to determine free electron concentration and Hall mobility profiles. Secondary ion mass spectrometry (SIMS) analysis has also been carried out to determine atomic profiles.

The active layers were processed into small signal depletion mode MESFETs. The 1 μm long gate was of Pi configuration and 300 μm wide. The gate was 1 μm from the source contact and 2 μm from the drain contact. Mesa etching was used to electrically isolate devices. Ohmic contacts were then formed by alloying Au-Ge eutectic and Ni evaporated layers. Saturation current was controlled by etching either a shallow channel recess or a gate recess prior to gate metallisation. The Ti/Au gate was defined by contact UV photolithography and lift off.

DC and RF measurements were made to assess the finished devices. The microwave measurements (at 10 GHz) were made using a computer corrected automatic network analyser. Minimum noise figure measurements were also carried out.

Results

Typical free electron concentration and Hall mobility profiles after implantation and annealing are illustrated in Figures 1 and 2. It can be seen that a higher free electron concentration is achieved in the buffer

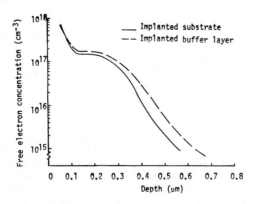

Fig 1: Free electron concentration profiles after implantation and annealing derived from CV measurements

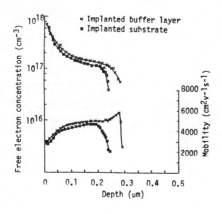

Fig 2: Free electron concentration and Hall mobility profiles obtained from Hall measurements

layer. The profiles in the buffer layers were also broader with a significant increase in free electron concentration in the tail region of the implant. This was observed from both CV and differential Hall measurements. The mobility in the implanted buffer layers is generally higher than in the implanted substrates. These higher values are maintained and more often than not observed to increase in the tail region of the implant (i.e. on approaching the active layer/buffer layer interface). Local Hall mobilities approaching 6000 $cm^2v^{-1}s^{-1}$ have been measured in this region of the implanted buffer layer as opposed to 4500 $cm^2V^{-1}s^{-1}$ in implanted substrates.

Figure 3 compares the SIMS derived atomic ^{29}Si concentration profiles for the implanted buffer layer and implanted substrate with the predicted LSS profile for reference. No significant difference is observed between the two profiles either before or after annealing. Figure 4 illustrates the C-V derived free electron and atomic ^{29}Si concentration profiles for an implanted and annealed buffer layer. These profiles are in good agreement with each other.

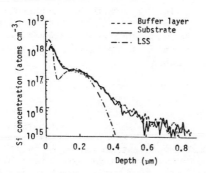

Fig 3: ^{29}Si atomic concentration profiles after implantation and annealing

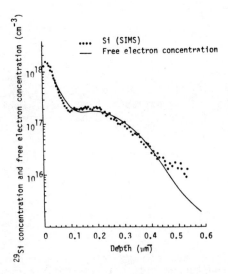

Fig 4: Free electron and ^{29}Si atomic concentration profiles for the implanted and annealed buffer layer

Figure 5 illustrates SIMS profiles of atomic Cr concentration in the buffer layer and into the substrate for the as-grown case and after annealing. The as-grown sample has an interface width of ~200 nm (from 10^{16} to 10^{15}) for the Cr concentration profile which is greater than the expected profile broadening due to atomic mixing under SIMS condition by a factor of at least 10. Therefore some out-diffusion of Cr occurred during the epilayer growth. This out-diffused Cr will reduce the free electron concentration in the buffer layer. Further out-diffusion of Cr is observed after the post implant anneal. This out-diffusion could be expected to play an important role in maintaining the low free electron concentration in the buffer layer during the anneal cycle.

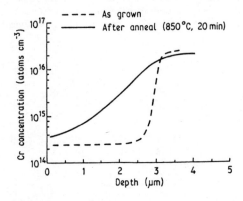

Fig 5: Atomic Cr concentration profile in the buffer layer/substrate interface, as grown and after 850°C, 20 minute anneal

MESFETs had either a gate recess or channel recess to attain similar I_{DSS} (~60 mA). DC measurements, at V_g = -1V, indicated that equivalent MESFETs fabricated from implanted buffer layers had a higher transconductance (~25% higher for channel recess, 50% for gate recess). The deep gate recessed buffer layer device gives gm = 41 ms for the device, corresponding to gm = 137 ms per mm. Microwave performance was also improved for the buffer layer devices. To obtain accurate information about the effect of ion implantation into a buffer layer on the intrinsic FET parameters, the universal equivalent circuit for a FET was used. The values for the elements of this circuit in both cases (ion implanted buffer layer and substrate) were obtained by fitting the measured S-parameters to this circuit. These values confirm the following differences in device performance observed: (a) The $|S_{21}|$ and MUG were generally higher for the buffer layer devices (measured at $\frac{1}{2}$ I_{DSS}); (b) The minimum noise figure and associated gain for buffer layer devices were also improved (V_g biased for minimum noise condition). These results are shown in Table (1).

| Material | Device | r.f. Performance at 10 GHz | | | | | | | Intrinsic E.q. circuit Parameters | | | | | |
| | | ($\frac{1}{2}$ I_{DSS}) | | | | | (min noise Condition) | | | | | | | |
		$|S_{11}|$ dB	$|S_{21}|$ dB	$|S_{12}|$ dB	$|S_{22}|$ dB	MUG dB	N_f dB	AG dB	R_{in} Ω	C_{in} pf	g_m mS	R_{dS} Ω	C_f pf	C_{dS} pf
Substrate) Channel		-1.5	2.2	-20	-1.7	13	2.7	4.5	9	.22	22	900	.024	.075
Buffer layer) Recess		-0.9	3.5	-24	-1.2	17	1.9	5.2	6	.2	28	1200	.016	.053
Substrate) Gate		-1.6	1.7	-17	-2.3	10	2.4	5	7	.26	23	600	.047	.07
Buffer layer) Recess		-0.8	4.2	-19	-3.6	13	1.9	6.5	2.5	.22	30	450	.03	.07
Buffer) deep Gate) Recess		-1.5	4	-16	-3.9	11	1.5	4.6	3.7	.24	32	400	.045	.07

Each data is average value from 10 devices.

Table 1 : Typical microwave device performance

Discussion

The more efficient electrical activation implies that the buffer layer is of higher purity than the substrate having fewer acceptor impurities or defects. The higher mobility in the implanted buffer layers also indicates that the buffer layer is a purer crystal lattice with less scattering centres. From Walukiewicz et al (1978) and using the mobility measured for the flat region of the channel layer ($n \approx 1.5 \times 10^{17}$ cm^{-3}) an estimate of the compensation ratio was made. Compensation ratios $N_a/N_d+N_a \sim 0.2$ and ~ 0.1 were estimated for the implanted substrate and buffer respectively. As observed experimentally in these and all other samples implanted and epitaxially grown, the mobility drops at the active layer/insulating layer interface. However the mobility profiles indicate that the implanted buffer layers maintain the mobility further into the interface. This drop in mobility near 'pinch off' has been explained by Wallis and Jay (1982) and Ford and Barrera (1982) as due to increased scattering by ionised donor impurities which are less efficiently screened. This occurs as the depleted region under the gate approaches the substrate/active layer interface; the density of ionised donors remains approximately constant but the density of free electrons decreases. In the case of the implanted buffer layer the thinner depletion layer at the buffer layer active layer interface and the less abrupt interface may well help to reduce this effect and account for the mobility profiles observed.

Although transconductance measurements made at high field cannot be directly related to low-field mobility measurements, our results support the commonly held view that larger g_m values are obtained from material with a higher mobility. The improved g_m for the buffer layer devices is attributed mainly to the high mobility in the tail of the implant. Devices with superior DC characteristics may be expected to yield improved RF performance. This has generally been seen to be the case. The lower noise figure for buffer layer devices is attributed to 'materials' properties as the devices are otherwise nominally identical. Contribution to noise due to carrier trapping may well be reduced if the buffer layer is a 'purer' matrix. Secondly the less abrupt free electron concentration profile may also reduce noise due to possibly reduced high field effects. Further work is required to determine if the shape of the free electron concentration profile is an important factor. This could be achieved by epitaxial growth to produce a 'graded' interface.

Conclusion

A comparison of the properties of LEC grown SI GaAs substrates and $AsCl_3$ grown buffer layers after ion implantation and annealing has been made. Implanted buffer layers have produced superior electrical properties which have resulted in improved microwave device performance. MESFETs have been fabricated using ion implantation into buffer layers with the lowest ever noise figure reported for a 1 µm gate length structure (1.5 dB with associated gain 4.6 dB at 10 GHz).

Acknowledgements

The authors are indebted to M G Dowsett (City of London Polytechnic) for SIMS and R H Wallis for technical discussions. They also thank Dr C E C Wood for encouragement and hepful comments during the execution of this programme.

This work has been carried out with the support of the Procurement Executive, Ministry of Defence, sponsored by DCVD.

References

Bartle D C, Andrews D C, Grange J D, Harris P G, Trigg A D, Wickenden D K, 1984, Vacuum 34, 315
Bonnet M, Visentin N, Bessonneau G and Duchemin J 1981, J. Crystal Growth. 55, 246
Chye P and Huang C 1982, IEEE EDL-3, No 12
Feng M, Eu V, Zielinski T, Kanber H and Henderson W 1984, IEEE EDL-5, No 1
Ford W, Barrera J 1982, Semi-Insulating III-V Materials Conf., Evian
Grange J D, and Wickenden D K 1983, Solid State Electron. 26, 313
Nakanisi T, Udagawa T, Tanaka A and Kamei K 1981, J. Crystal Growth, 55, 255
Stewart C P, Medland J D and Wickenden D K 1984, Semi-Insulating III-V Materials Conf., Oregon
Tayrani R 1982, Inst.Phys. Conf. Ser. No 65, Chapter 5
Wallis R and Jay P 1982, Semi-Insulating III-V Materials Conf, Evian
Walukiewicz W, Lagowski L, Jastrzebski L and Lichtensteiger M and Gatos H, 1979, J. Appl. Phys. 50, 2
Wholey J, and Omori M 1978, Inst.Phys. Conf. Ser. No. 45, Chapter 4

Inst. Phys. Conf. Ser. No. 74: Chapter 7
Paper presented at Int. Symp. GaAs and Related Compounds, Biarritz, 1984   533

Comparison between TEGFETs and FETs RF performances in the millimeter wave range

A. Cappy, A. Vanoverschelde, M. Shortgen, C. Versnaeyeen and G. Salmer
Centre Hyperfréquences et Semiconducteurs, U.E.R. d'I.E.E.A., Bât. P4
Université des Sciences et Techniques de Lille I - 59655 VILLENEUVE D'ASCQ
CEDEX - FRANCE.

Abstract. A theoretical modeling of the noise properties of TEGFET's that gives results in good agreement with experiment is presented. This modeling is used to study the influence of the main technological parameters on the noise figure. A comparison between MESFET's and TEGFET's is carried out. The limitations of the Fukui formula are shown. A new method for calculating the noise figure is then proposed.

1. Introduction

During the past three years the work on Two dimensional Electron Gas Field Effect transistor has demonstrated the very attractive possibilities of this structure in the field of low noise amplifiers.

With .35 gate length devices, noise figures as low as 1.5 dB and 2.7 dB were recently obtained at 18 GHz and 34 GHz with 10.5 and 5.9 dB associated gain (1). These results are comparable with the best quarter micron gate MESFET's. So, TEGFET's seem to have better performances than conventional MESFET's and this superiority is usually related to a lower k_F factor (1.6 for TEGFET's and 2.5 for MESFET's) in the Fukui formula of minimum noise figure.

Nevertheless, there is some problems concerning this explanation

     - k_F factors as low as 2.085 have been already obtained (2) in the case of MESFET's. In addition the good results previously given were achieved with k_F equal to 2.5.

     - the Fukui formula is based upon drastic assumptions. The gate-to-drain capacitance and the transit time are neglected as well as the higher terms in the frequency series expansion.

     - this formula is no more valid for low values of the source and gate resistances since the intrinsic noise is not taken into account.

For these different reasons, this simple formula can be unreliable, especially in the millimeter wave range and the purpose of this paper is to describe a theoretical modeling of TEGFET's in order to compare the RF performances of the two different devices.

The noise modeling of MESFET's has been previously described (3) and is based on the same main assumptions as the TEGFET's modeling.

2. The TEGFET noise modeling

In order to achieve the DC and small signal performances of the device, this one dimensional model is based on the main following assumptions (4) (Fig. 1).

(i) the longitudinal component of the electric field does not depend on the y coordinate in the GaAs layer.

(ii) the perpendicular component at the interface GaAs/AlGaAs is derived from the Delagebeaudeuf   model of the heterojunction (5).

(iii) the transport properties of electrons in the accumulated layer are deduced independently from a Monte Carlo calculation developed in our laboratory.

This simulation has shown that the evolution of average velocity, average energy and diffusion coefficient parallel to the interface are not very different as compared to the intrinsic bulk ones. The electron dynamics is then described using the relaxation time approximation.

Then, for given drain current and gate voltage the following set of equations are solved at each step from the source to the drain.

1. $I_{ds} = qZN_j v_j$

2. $(E_{x,j+1} - E_{x,j}).Y_s - E_{y,j}.\Delta x = qN_j \Delta x/\varepsilon$

3. $\varepsilon_{j+1} - \varepsilon_j = q.E_{x,j}.\Delta x - (\varepsilon_j - \varepsilon_o).\Delta x/\tau\varepsilon.v_j$

4. $m^*.v_j(v_{j+1} - v_j)/\Delta x = qE_{x,j} - m^*.v_j/\tau_m$

In these equations Z is the width of the structure $N_j.\Delta x.Z$ the number of carriers in the slice, Ex and Ey the components of the electric field, ε_j the average energy, v_j the average velocity, m^* the effective mass τ_m, and $\tau\varepsilon$ the momentum and energy relaxation times. This calculation is repeated with small I_{ds} and V_{gs} increments in order to give the small signal parameters g_m, g_d, C_{gs}, C_{gd}, R_i.

It is interesting to note that in this model, carriers can be injected in the GaAs layer (see the first term of the left hand side of equation 2). Thus, despite a one dimensional treatment, values of the output conductance in good agreement with experiment can be obtained by adjusting the value of the GaAs equivalent thickness y_s. This value is always close to $L_g/2 + A$. Moreover,   additional conduction in the AlGaAs layer is taken into consideration in both the source-to-gate and gate-to-drain space using a simple v(E) law.

When the small signal parameters are known for a bias point, the noise figure is computed as follows.

Uncorrelated noise sources are distributed along the source to drain axis. The mean square value and the spectral density of the current source in the jth section are given by :

$$\langle i^2 \rangle_j = q^2 N_j Z \langle v^2 \rangle/\Delta x$$

$$S_j = 4 \, q^2 \, D_{/\!/}(\varepsilon_j) \, N_j \, Z/\Delta x$$

where $\langle v^2 \rangle$ is the average quadratic fluctuation of the drift velocity and $D_{/\!/}$ the longitudinal diffusion coefficient. In order to take into account the non stationary electron dynamics on noise performances, this diffusion coefficient is assumed to be a function of the average carrier energy.

From the knowledge of these noise sources in each section, we have to determine the equivalent sources in the gate and drain circuits. For this purpose, we use the preceding model to numerically calculate the drain voltage and stored charge fluctuations ΔV_{ds} and ΔQ resulting from a small current fluctuation I_{ds} in a section of the channel.

The spectral densities of noise sources in the drain and gate circuits S_{vd} and S_{ig}, as well as the correlation coefficient can easily be computed by a quadratic summation (3).

From these three quantities and the whole set of small signal parameters, the noise figure is computed using the rigorous analysis of Rothe and Dahlke (6).

In order to check the validity of this work, we have compared some of the best noise figures reported for submicrometer gate TEGFET's with the typical evolution of the noise figure versus frequency (Fig. 2). The same comparison was made for MESFET's (Fig. 3) and we can conclude that the modeling of both TEGFET's and MESFET's give results in good agreement with experiments.

3. Influence of the main technological parameters

Since this modeling requires only short computation times, we have systematically studied the effect of the main technological parameters on the evolution of the noise figure and associated gain as a function of the frequency.

The influence of the gate length L_g is shown in figure 2. The strong decrease of the noise figure when the gate length diminishes is mainly due to the improvement of the cut-off frequency $g_m/2 \pi C_{gs}$. It is important to note that the use of quarter micron gate is necessary in the millimeter wave range.

The effects of the carrier mobility μ_0 and epilayer thickness A are presented in figure (4). This figure shows that a high mobility improves the noise figure and associated gain.

An increase of the AlGaAs layer thickness has three main effects on the noise figure :

  - an increase of the noise figure at low frequency

  - a decrease of the associated gain

  - a less important slope of the noise figure versus frequency.

These effects can be explained by an increase of the intrinsic noise and an increase of the output conductance since the ratio L_g/A is lower.

A comparison between FET's and TEGFET's is presented in figure (5). For each device two limiting values of the carrier mobility are considered. It must be pointed out that for the same value of carrier mobility the values of the noise figure are not very different. Thus, the better performance of TEGFET's is mainly due to the higher mobility and the resulting more important overshoot effect.

This modeling was also used to obtain the evolution of the noise figure as a function of the frequency. This evolution is shown in figure (6) where the noise figure and the associated gain are plotted with linear scales. According to Fukui's equation, the noise figure should increase linearly with the frequency. This is verified below 20 GHz in this case, but not at higher frequencies. Thus this formula seems to be no longer valid to predict **the noise** performances of FET's or TEGFET's in the millimeter wave range.

In order to overcome this difficulty, a new simple method was derived from the numerical model previously described.

4. New calculation of the noise figure

For this purpose, analytical formulas for the spectral densities of the drain and gate noise sources were deduced from the numerical model.

$$S_{vd} = \frac{4kTG_m L_g}{Gd^2 C_{gs}} (\alpha B + \beta I_{ds}) \qquad S_{id} = (Gd^2 + \omega^2 C_{gd}^2) S_{vd}$$

$$S_{ig} = 2kTC_{gs}^2/G_m \omega^2$$

In these expressions the two parameters ($\alpha = 2\ 10^{-3}$ and $\beta = 1.25\ 10^{-5}$ SI units) are the same for both TEGFET's and MESFET's. In addition the correlation coefficient between S_{vd} and S_{ig} is especially dependent on the ratio L_g/A (C # j.0.7 for L_g/A # 3 and C # j 0.8 for $L_g/A \geqslant 5$).

From the knowledge of the whole set of small signal parameters, the noise sources are calculated using these expressions. The noise figure as well as the other noise parameters can then be computed using a pocket computer. Using data given in the literature a comparison between the measured and calculated noise figure was made and the results are given in table 1.

These expressions are very useful to calculate the noise figure at any frequency but also to understand the influence of the different small signal parameters.

As a result, the spectral density S_{id} of the short circuit drain noise current increases strongly with frequency if C_{gd} is high. In addition, for a given value of C_{gd}, the spectral density S_{id} is lower for high values of the output conductance. This can explain why the better performances are always achieved for devices having high values of the output conductance (1, 2, 7, 8).

5. Conclusion

A noise modeling of TEGFET's that gives results in good agreement with ex-périment   has been described. This modeling allows us to study the influence of the main technological parameters on the values of noise figure and associated gain. The limits of validity of the Fukui   formula have been shown and in order to optimize both FET's and TEGFET's a new calculation of the noise figure based on analytical formulas for the noise sources has been proposed.

References

(1) J.J. Berenz et al. IEEE MTT-S digest, p. 83-86 (1984).

(2) M. Feng et al. Appl. Phys. Lett. 44 (2) January 1984.

(3) B. Carnez et al. IEEE Trans. Elec. Devices, vol. 28, n° 27, p. 784-789 (1981).

(4) A. Cappy. Proc. of IC Symposium, pp. 79-82 (1982).

(5) D. Delagebeaudeuf and N.T. Linh. IEEE Trans. Elec. Dev. 29, p. 950-955 (1982).

(6) H. Rothe and W. Dahlke. Proc. IRE, vol. 44, p. 811 (1956).

(7) P.W. Chye. Elect. Dev. Lett., vol. EDL-3, n° 12, Déc. 1982.

(8) M. Niori et al. ISSCC Digest of Technical papers, p. 198-199, 1983.

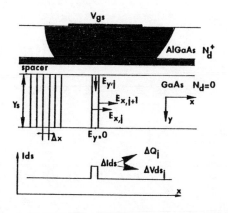

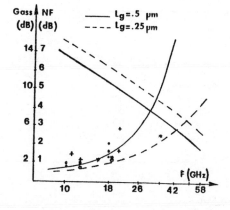

Fig. 1 : Model used to calculate the I_{ds} (Vgs, Vds) characteristics, the small signal equivalent circuit and the noise properties of TEGFET's.

Fig. 2 : Comparison between TEGFET's theoretical (Rs+Rg = 5 , Z = 300 μm) and experimental results. + Fujitsu ● Thomson LCR □ NEC ○ Thoshiba

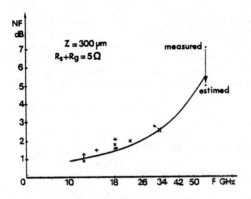

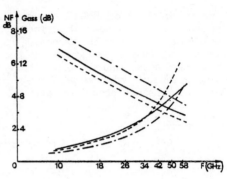

Fig. 3 : Comparison between MESFET's theoretical (Rs+Rg=5Ω, Z=300μm, Lg = 0,25μm) and experimental results + Plessey . Hughes x Avantek

Fig. 4 : Evolution of the noise figu-re and associated gain as a function of the low field mobility and AlGaAs layer thickness A (Z=300μm, Rs+RG=6Ω, Lg=0,25 m).
- ——— A = 900 A μ_0 = 8000 cm²/v.s
- —.— A = 500 A μ_0 = 8000 cm²/v.s
- — — — A = 500 A μ_0 = 4000 cm²/v.s

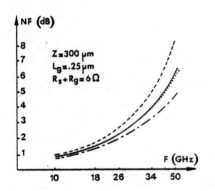

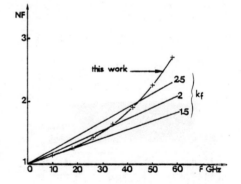

Fig. 5 : Comparison between TEGFET's and conventional MESFET's noise performances
- — — — MESFET μ_0 = 2000 cm²/v.s
- ..... MESFET μ_0 = 4000 cm²/v.s
- — — — TEGFET μ_0 = 4000 cm²/v.s
- —.— TEGFET μ_0 = 8000 cm²/v.s

Fig. 6 : Comparison with Fukui's formula.

TABLE 1

	F(GHz)	NF mes	NF calc.		F(GHz)	NF mes	NF calc.
Avantek	18	1.55	1.54	TRW	18	1.5	1.61
Lg = 0,25	21.7	1.98	1.81	Lg = 0,35			
	32	2.6	2.44	Fujitsu	8	1.3	1.23
Hughes	18	1.69	1.69	Lg = 0,5	11.3	1.7	1.77
Lg = 0,3					20	3.1	2.8
				Thomson CSF	17.5	1.5	1.64
		MESFET		Lg = 0,5		TEGFET	

Inst. Phys. Conf. Ser. No. 74: Chapter 7
Paper presented at Int. Symp. GaAs and Related Compounds, Biarritz, 1984   539

Ultra low noise and high frequency microwave operation of FETs made by MBE

M. Laviron, D. Delagebeaudeuf, J.F. Rochette, P.R. Jay, P. Delescluse, J. Chevrier and Nuyen T. Linh

THOMSON-CSF, DAG/DHM, Corbeville, BP 10, 91401 ORSAY (FRANCE)

Abstract. Both conventional FET structures and n-AlGaAs/GaAs TEGFET (Two-dimensional Electron Gas FET) heterostructures have been grown by molecular beam epitaxy on two inch diameter wafers.
TEGFET and MESFET structures with gatelengths of 0.5 microns have been made.
For 300 microns gatewidth TEGFETs, a 10 GHz noise figure of 1.10 dB with 10.6 dB of associated gain was obtained.
An optimisation of device geometry with gatewidths of 140 microns improves the room temperature performance to 1 dB noise figure at 10 GHz and 1.4 dB at 17.5 GHz.
Classical FET MBE structures using the optimised 140 microns geometry have given 2.1 noise figure with 10.3 dB of associated gain at 17.5 GHz.
The TEGFET results for 0.5 µm gatelength devices are comparable with those published for 0.25 µm gatelength MESFETs.

1. Introduction

Field effect transistors made using selectively doped AlGaAs/GaAs hetero-structures have demonstrated excellent high frequency performance at room temperature and show themselves to be potentially capable of surpassing the classical GaAs MESFET. The nature of the two-dimensional electron gas (TEG) channel offers advantages arising from the confinement of the conducting electrons, and an increased mobility which is particularly beneficial for low temperature operation.

This paper describes the design, evaluation and performance of TEGFET and MESFET devices made using epitaxial layers grown by molecular beam epitaxy on two inch diameter substrates. Two different device geometries are described in order to demonstrate how the device may be optimized for performance at higher frequencies.

2. Analysis of results

In addition to the epitaxial structure, the gatelength of the device and the parasitic access resistances play a major role in determining the noise figure of the transistor. A well-known expression by Fukui (1979) enables the minimum noise figure to be evaluated in terms of the various transistor parameters for a given operating frequency f :

$$NF = 1 + 2\pi f \, C_{GS} K_F (\, (R_S + R_G)/g_{mo} \,)^{1/2}$$

where K_F is a fitting factor related to the material quality and R_s, R_g, C_{gs} and g_{mo} are respectively the source resistance, gate resistance, gate-source capacitance and small-signal transconductance of the device. These values are obtained from s-parameter measurements.

A reduction of gatewidth Z is beneficial for F_{min} since the value of R_s is thereby minimized, although the total gatewidth chosen must take into account the intended frequency of operation from the point of view of impedance matching considerations. We have used gate widths of 300 µm and 140 µm for operation at 10 GHz and 17.5 GHz respectively, in each case a four-entry gate structure being used to assure low values of Rg. For most of the transistors described, the gate length was about 0.5 µm. For each geometry a comparison between classical GaAs and GaAlAs/GaAs hetero-structure epilayers was made. We will consider first the results obtained on 300 µm gatewidth devices.

3. 300 µm gatewidth

The submicron gatelength TEGFETs were made using direct-write electron-beam lithography on AlGaAs/GaAs heterostructures grown by MBE in a RIBER 2300 system using 2" diameter GaAs substrates. The mask design and the fabrication details have been described elsewhere (LAVIRON   1984).

Table 1 lists different characteristics of 4 TEGFET epilayers used, and shows that a reduction of the sheet resistance $R_\square$ (deduced from   TLM measurements on an ohmic contact test geometry) results in lower value of R_s. The increased transconductance is related both to the reduction of $R_\square$ and the decreased thickness of spacer layer.

Wafer n°	$R_\square$ (Ω)	R_s (Ω)	e (A)	g_{mo} (mS/mm)
A	990	6	75	150
B	820	7	80	150
C	260	3.6	80	200
D	240	2.5	20	230

Table 1

Microwave measurements on the devices made using the four layers (denoted by A,B,C,D) are given in table 2 for the measuring frequency of 10 GHz. It

Wafer n°	g_{mo} (mS/mm)	NF (dB)	G_{ass} (dB)	K_f
A	150	1.8	9.6	1.6
B	180	1.5	10.7	1.5
C	200	1.3	10.4	1.7
D	230	1.1	10.6	1.8

Table 2

may be seen that the higher values of transconductance result in lower noise figures on the devices measured, although the improvements in parasitic resistance (table 1) are also partly responsible. The application of the Fukui expression to these data gives values of the fitting factor K_F that are constant to within +/- 10% for these structures.

Using the same mask geometry and a classical GaAs $n^+/n/SI$ epilayer structure prepared by MBE in the same apparatus, a comparison of transistors using the two types of layer structure has been made. The results of this comparison are summarized in table 3. Since the parasitic resistance, and gatelengths involved are comparable, it is clear that the improved noise and gain behaviour in the TEGFET case arise directly from the higher

Wafer $n°$	R_s (Ω)	R_G (Ω)	g_{mo} (mS/mm)	NF (dB)	G_{ass} (dB)	K_F
D	2.5	2.3	230	1.1	10.6	1.8
E	3	2.2	150	1.8	8.5	2.6

Table 3 - (10 GHz)

Wafer $n°$	R_s (Ω)	R_G (Ω)	g_{mo} (mS/mm)	N_F (dB)	G_{ass} (dB)
D[1]	2.5	2.3	230	1.85	7.6
F[2]	3.8	0.92	240	1.5	1.4
G[3]	7.1	0.83	330	1.9	11.1

1) Z = 300 μm. e = 80 Angströms
2) Z = 140 μm. e = 80   "
3) Z = 140 μm. e = 20   "

Table 4 - (17.5 GHz)

transconductance and lower fitting factor K_f for the heterostructure layers.

While the above comparison was made for the same geometry, the potential of the TEGFET is of particular interest for higher frequency applications. For this propose we have chosen a shorter gatewidth geometry of 140 μm.

4. 140 μm gate width devices

A four entry gate width was retained to assure low values of gate resistance, and to clarify the comparison, the epilayer structure was basically unmodified. Table 4 compares data for wafers D and F which are the same structure realised in two different geometries and wafer G which was a layer in which the spacer layer thickness had been reduced to zero.

It is apparent that the reduced gate resistance of devices on layer F (in spite of a slight increase of source resistance) improves the noise figure at 17.5 GHz, and is also reflected in the higher gain.

As has already been noted in table 1, a reduction of the spacer thickness tends to be associated with an improvement in transconductance, and it may be seen that the trend is here continued with the wafer G. Unfortunately, the effect of an increase in source resistance for this wafer is reflected in the higher noise figure obtained for this device although an improvement in associated gain is nevertheless measured at 17,5 GHz.

In view of the interest in application of TEGFETs at higher frequencies, the experiment comparing the TEGFET and classical GaAs MBE structures was repeated for the 140 µm geometry, and the results obtained are given in table 5. In this case, the TEGFET noise figure and gain are only slightly better than for the GaAs case as a result of the high source resistance mentioned previously. Nevertheless, the comparison demonstrates that once again the heterostructure is beneficial in terms of the transconductance obtained and as in view of the systematically low values of K_f obtained on TEGFET devices, promises impressive performance when the various parasitic effects are minimized.

Wafer n°	$R_s + R_G$ (Ω)	g_{mo} (mS/mm)	N_F (dB)	G_{ass} (dB)	K_F
G^*	7.9	330	1.9	11	2.8
H	3.5	200	2.1	10.3	1.4

G^* = TEGFET

Table 5

5. Conclusions

By simply comparing the measured data from a number of transistors made on heterostructure and classical GaAs MBE epilayers, we have shown that the use of TEGFET wafers tends to result in better performance at microwave frequencies. The differences may be summarised as being due to the higher values of transconductance and systematically lower fitting coefficient K_f in the Fukui equation. Such differences can be enhanced by optimizing the epiwafer structure (reducing the spacer thickness) and using smaller gate-width geometries to minimise the parasitic gate resistance and permit better matching at higher frequencies.

In order to evaluate the future potential of these devices, we have shown that a performance around 1 dB N.F. and 14 dB N.F. associated gain at 17.5 GHz could be obtained with gatelength of 0.3 um in a 140 um wide TEGFET structure.

It is nevertheless interesting to note the performance achieved from classical GaAs FET devices realised on MBE epitaxial wafers, using 0.5 µm gatelength structures. Figure 3 shows the results of this work on 0.5 μm devices in comparison with the best published data on (0.25 μm) GaAs transistors. It may be concluded that the TEGFET structure has considerable potential for high frequency operation.

Acknowledgments

This work has been partly sponsored by DRET. The authors would like to thank A. Rannou, M. Chapuis, J. Frentzel and J.P. Levy for their technical assistance. The participation of J. Chaplart and that of the team of C. Rumelhard was appreciated.

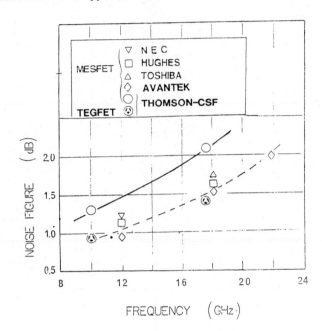

References :

Fukui H., 1979 IEEE Trans. Elect. Devices ED-26

Laviron M., Delagebeaudeuf D., Rochette J.F., Jay P.R., Delescluse P., Chevrier J. and Nuyen T. Linh, 1984 ESSDERC , Lille , France

Inst. Phys. Conf. Ser. No. 74: Chapter 7
Paper presented at Int. Symp. GaAs and Related Compounds, Biarritz, 1984

545

Low noise high electron mobility transistor

K. Kamei, S. Hori, H. Kawasaki, K. Shibata, H. Mashita* and Y. Ashizawa*

Komukai Works, *Toshiba Research and Development Center, Toshiba Corporation
1, Komukai Toshiba-cho, Saiwai-ku, Kawasaki 210, Japan

Abstract. Low noise high electron mobility transistors (HEMTs) have been developed on two types of epitaxial wafers, with and without an undoped $Al_{0.3}Ga_{0.7}As$ spacer layer, grown by MBE on undoped LEC substrates. HEMTs without an undoped $Al_{0.3}Ga_{0.7}As$ spacer layer have shown a noise figure of 0.95 dB with an associated gain of 11.8 dB at 12 GHz, and a noise figure of 1.4 dB with an associated gain of 9.4 dB at 18 GHz at room temperature. The noise figure increases at a rate of 0.01 dB/°C for a temperature change from -50 °C to 75 °C. The rate is close to that of GaAs FETs.

1. Introduction

For the past few years, remarkable progress has been achieved in low noise performance of GaAs FETs (Kamei et al. 1980, Watkins et al. 1983). Low noise microwave amplifiers, built using 0.25 μm to 1 μm gate GaAs FETs, are now becoming indispensable for various kinds of systems such as communication systems, radar systems and electronic warfare systems. Demands for transistors with still lower noise figures are strong from applications such as for satellite communication systems. Drastic improvement in GaAs FET performance, however, can not be expected, considering the fact that gate length and various device parasitics have already been reduced near to the technological limit.

Recently, extensive work has been done in many laboratories on high electron mobility transistors (HEMTs, otherwise called TEGFETs or MODFETs). HEMTs are expected to give much better performance than GaAs FETs due to their substantially higher electron mobility at $Al_xGa_{1-x}As$/GaAs interface than that of GaAs (Dingle et al. 1978). They have already shown microwave performance comparable to that of 0.25 μm gate GaAs FETs for last few months (Brenz et al. 1984, Ohata et al. 1984).

The purpose of this paper is to report on state-of-the-art microwave performance of HEMTs fabricated on epitaxial wafers grown by MBE. Another purpose of this work is to show various kinds of static characteristic and microwave performance important for practical applications.

2. Material and Device Fabrication

Two types of epitaxial wafers were grown by MBE to fabricate the HEMTs. The starting wafers were undoped LEC semi-insulating GaAs substrates with two-inch diameter and (100) orientation. One type of epitaxial wafer consists of a 1 μm thick undoped GaAs layer, a 300 Å thick n-$Al_{0.3}Ga_{0.7}As$ layer and a 500 Å thick n-GaAs layer. The n-type layers were Si doped with 2×10^{18} cm^{-3}. The growth temperature and the growth rate were 700 °C and ~1 μm/h, respectively. The other type of epitaxial wafer has a very similar structure to the one mentioned above, except that an undoped Al$_{0.3}$Ga$_{0.7}$As spacer layer is inserted between an undoped GaAs layer and an n-Al$_{0.3}$Ga$_{0.7}$As layer for the purpose of maintaining the high mobility of two dimensional electron gas.

The Hall mobility (μ) and the sheet electron concentration (n$_S$) were measured for the two types of epitaxial wafers after removing the n-GaAs layer and part of the n-Al$_{0.3}$Ga$_{0.7}$As layer. The wafer without undoped Al$_{0.3}$Ga$_{0.7}$As spacer layer showed μ=4900 cm^2/V.s with n$_S$=1.4 x 10^{12} cm^{-2} at room temperature, while μ at liquid nitrogen temperature was 25300 cm^2/V.s. The mobility of the wafer with undoped Al$_{0.3}$Ga$_{0.7}$As spacer layer was 5800 cm^2/V.s with n$_S$= 7.6 x 10^{11} cm^{-2} at room temperature, and 64400 cm^2/V.s at liquid nitrogen temperature. The room-temperature mobility of the wafer with undoped Al$_{0.3}$Ga$_{0.7}$As spacer layer was improved only by 20 % at the sacrifice of 50 % n$_S$ reduction.

The cross section of designed HEMTs without an undoped Al$_{0.3}$Ga$_{0.7}$As spacer layer is schematically shown in Fig.1. The gate electrode with a length of 0.4 μm and a width of 200 μm is formed by Al/Ti to a thickness of 0.6 μm. Two bonding pads are attached to the gate electrode to reduce the gate resistance. Hence, the effective gate width is 50 μm. A recess structure is formed at the midpoint between the source and drain electrodes, which are separated by 3 μm, to

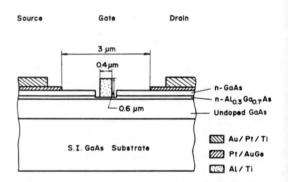

Fig. 1. Schematic cross section of HEMT without an Al$_{0.3}$Ga$_{0.7}$As spacer layer.

control the drain current by etching the n-GaAs layer and part of the n-Al$_{0.3}$Ga$_{0.7}$As layer before the gate electrode formation.

Two types of epitaxial wafers were processed under the same conditions in the same run. After 0.3 μm-high mesa formation by wet etchant, source and drain ohmic contacts were formed, by alloying evaporated Pt/AuGe at 450 °C for 2 minutes. Gate patterns were delineated by direct electron beam lithography by exposing a 6000 Å-thick AZ2415 resist film at a beam density of 70 μC/cm^2. After recess etch by monitoring the current between the

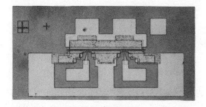
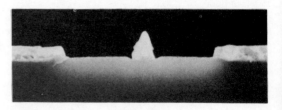

(a)   (b)

Fig. 2. Fabricated HEMT : a microphotograph of the chip (a) and a SEM photograph of the chip cross section near channel area (b).

source and the drain, Al/Ti was deposited by an e-beam evaporator. Then, the Al/Ti other than the gate electrode was removed away together with the resist film.

Figure 2 (a) shows a microphotograph of the fabricated chip. The chip size is 250 μm x 500 μm. Figure 2 (b) shows a SEM photograph of the chip cross section near the channel area. The shallow recess and penetration of ohmic contact metals into epitaxial layers can be seen in the SEM photograph. The gate electrode cross section is rather triangular against our expectation of being rectangular.

3. Static Characteristics

Figures 3 (a) and 3 (b) are the drain current/voltage characteristics of the HEMTs at room temperature, without (type A) and with (type B) an undoped $Al_{0.3}Ga_{0.7}As$ spacer layer, respectively. The left curves and the right curves were measured under dark and illuminated conditions, respectively. No differences are found between the dc characteristics under these conditions for both types of HEMTs. The maximum transconductance (g_m) of the type A HEMT is typically 250 mS/mm and that of the type B HEMT is 150 mS/mm at room temperature. The lower g_m of the type B HEMT is supposed to be due to the lower n_s at the $Al_{0.3}Ga_{0.7}As$/GaAs interface. Both types of HEMTs show large reduction of g_m near pinch-off, and this is conspicuous in the type A HEMT. Such reduction of g_m near pinch-off can be explained by the mechanism of channel current injection into undoped GaAs layer under high electric field suggested by Chandra and Eastman 1979 and shown by Drummond et al. 1982. We also examined the contribution of current injection into undoped GaAs layer, by fabricating HEMTs with a similar structure as the type A HEMT except for an undoped GaAs layer, where a multi-layer structure composed of an undoped GaAs layer (200 Å)/an undoped $Al_{0.3}Ga_{0.7}As$ layer (5000 Å)/an undoped GaAs layer (1 μm) was adopted as a substitute for a 1 μm thick undoped GaAs layer. The drain current/voltage characteristics of this type of HEMT are shown in Fig. 4. It can be clearly seen that the HEMT with such multi-layer structure has a lesser g_m reduction near pinch-off owing to the undoped $Al_{0.3}Ga_{0.7}As$ barrier, by comparing Fig.4

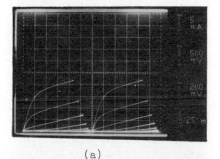

(a)

(b)

Fig. 3. Drain current/voltage characteristics of type A HEMT (a) and type B HEMT (b) at room temperature.

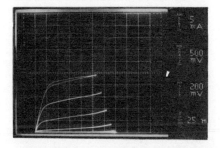

Fig. 4. Drain current/voltage characteristics of HEMT with multi-layer structure composed of undoped GaAs/undoped $Al_{0.3}Ga_{0.7}As$/undoped GaAs.

with Fig. 3 (a). The difference of
g_m reduction between type A and type
B HEMT is supposed to be due to the
difference between quality and/or
carrier concentration of the undoped
GaAs layers. Further study is needed
to make clear the g_m reduction phenomenon more quantitatively.

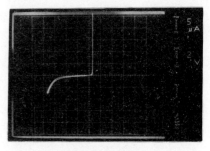

Fig. 5 shows the forward and the
reverse current/voltage characteristics
between the gate and the drain of the
type A HEMT. The breakdown voltage,
defined at a leakage current of 10 µA,
is -7 V, which is a comparable value
to that of GaAs FETs. The breakdown
voltage of the type B HEMT is -5 V.

Fig. 5. Foward and reverse
current/voltage characteristics
between the gate and the drain
of the type A HEMT.

In the early stage of GaAs FET development, the long-term drift of drain
current was one of the most serious problems in practical applications.
From this point of view, we have performed long-term drift measurements on
the drain current (initially set to 15 mA) for the developed HEMT at the
temperatures of 75 ℃, 25 ℃, 0 ℃ and -50 ℃. Both types of HEMTs have
not exhibited any measurable long-term drain current drift over the temperature range except that the drain current of the type B HEMT decreased by
0.5～1 mA at -50 ℃ during ten minutes after being switched on. To further
examine the current drift, several other lots of HEMTs with the structure
similar to that of type A HEMT have been fabricated and tested. It has been
found that, although most of HEMTs show no current drift, a few of them
exhibit a drift even at room temperature and the current drift after ten
minutes is at most 0.3 mA at the measured temperatures (-50 ℃～75 ℃).
It is our present speculation that the current drift originated from the
surface condition of HEMTs, even though the channel is isolated from the
surface by an n-GaAs layer or n-$Al_{0.3}Ga_{0.7}As$ layer. Further experiments,
however, should be continued to make this phenomenon clearer.

4. Microwave Performance

Microwave noise figure and associated gain have been measured at 12 GHz and
18 GHz on the HEMT chips mounted in microwave test fixtures. In Fig. 6,
the measured optimized noise figure and associated gain of the type A HEMT
are plotted as a function of drain current at 12 GHz and a drain voltage of
3 V at room temperature. The minimum noise figure is 0.95 dB with an associated gain of 11.8 dB at a drain current of 12 mA. Figure 7 shows the
drain current dependence of the optimized noise figure and associated gain
of type A HEMT at room temperature measured at 18 GHz, together with those
of a 0.25 µm gate GaAs FET developed in our laboratory. The minimum noise
figure of 1.4 dB and an associated gain of 9.4 dB are obtained from the type
A HEMT at a drain current of 16 mA. It is obvious from Fig. 7 that the 0.4
µm gate HEMT is much superior to the 0.25 µm gate GaAs FET. The improvement
in noise figure and associated gain of the HEMT are 0.3～0.4 dB and more
than 1 dB, respectively, in spite of its longer gate length. The similar
room temperature measurements have been performed on the type B HEMT.
The measured noise figures are, however, higher by 0.15～0.2 dB and the
associated gains are lower by 1～2 dB at both 12 GHz and 18 GHz than those
of the type A HEMT. These results are primarily due to smaller g_m of the
type B HEMT.

For practical application of HEMT chips
in microwave low noise amplifiers, it is
very important for the chips to be insensitive to light illumination. Hence, the
variation of noise figure and associated
gain have been compared at room temperature for both types of HEMTs under a dark
environment and a microscope-light illumination. It has been found that for both
cases noise figure and associated gain
vary very little by, for example, 0.02 dB
and 0.2 dB, respectively, for both types
of HEMTs at 12 and 18 GHz. This corresponds to the negligible drain current
variations for dark and illuminated conditions as previously shown in Fig. 3 (a)
and Fig. 3 (b). It is felt that the
light sensitivity of the fabricated
HEMTs is small enough for practical
applications.

The temperature dependences of noise
figure and associated gain have been
measured in a wide temperature range
from -50 ℃ to 75 ℃ at 18 GHz for both
types of HEMTs. It has been found that
both noise figure and associated gain
deteriorate at a rate of around 0.01 dB/℃
with the increase of ambient temperature.
The rate is very close to that of GaAs FETs.
Figure 8 shows an example of the measured
noise figure and associated gain of the
type A HEMT over -50 ℃ to 75 ℃.

Equivalent circuit element values have
been determined from S-parameter measurements on the type A HEMT over a 2 to 18
GHz frequency range using an automatic
network analyzer. The equivalent circuit
and its typical element values are shown
in Fig. 9. The values of g_m are in the range
of 45 ~ 55 mS. The corresponding value of
a GaAs FET with similar gate length and

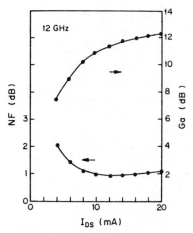

Fig. 6. Optimized noise
figure, associated gain
versus drain current of
type A HEMT at 12 GHz.

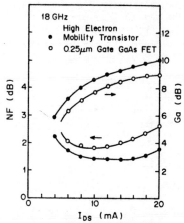

Fig. 7. Type A HEMT performance compared with 0.25 μm
gate GaAs FET at 18 GHz.

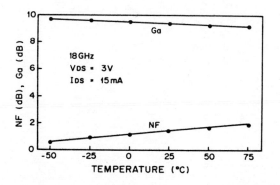

Fig. 8. Temperature dependence of noise figure and
associated gain of type A
HEMT at 18 GHz.

width is 25 mS. The gate
capacitance Cgs is deter-
mined to be 0.18 ~ 0.23 pF,
being close to that of a
0.4 μm gate GaAs FET. It
has been found, however,
that the parasitic resistances are much larger than
those of good GaAs FETs.
Reduction of source to
gate distance and formation
of a rectangular-shaped
cross-section gate instead
of the triangular shape
(see Fig. 2(b)) will enable
us to get better microwave
performance of HEMTs.

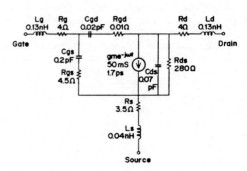

Fig. 9. Equivalent Circuit of type A
HEMT.

Moreover, reduction of gate length from 0.4 μm to 0.25 μm will highly
improve the microwave performance.

5. Conclusion

HEMTs with 0.4 μm-gate length have been successfully developed, by utilizing
epitaxial wafers grown by MBE technique. The measured noise figures at 12
GHz and 18 GHz were shown to be superior to hitherto reported values of
GaAs FETs and HEMTs. This work also showed that HEMTs are ready to be
introduced to low-noise amplifiers. Further improvement will be expected
by reductions of the gate length and the various device parasitics.

6. Acknowledgment

The authors would like to thank Dr. M. Ohtomo, S. Okano and Dr. T. Nakanisi
for their encouragement and helpful discussions. They are also grateful to
Dr. H. Tokuda for evaluating the epitaxial wafers.

References

Kamei K, Hori S, Kawasaki H, Chigira T and Kawabuchi K, 1980 IEDM Dig.
    Tech. Papers, 102
Watkins E T, Schellenberg J M, Hackett L H, Yamasaki H and Feng M, 1983
    IEEE MTT-s Digest, 145
Dingle R, Stormer H L, Gossard H L and Wiegmann W, 1978 Appl. Phys. Lett.
    33, 665
Berenz J J, Nakano K and Weller K P, 1984 IEEE MTT-s Digest, 98
Ohata K, Hida H, Miyamoto H, Ogawa M, Baba T and Mizutani T, 1984 IEEE
    MTT-s Digest, 434
Chandra A and Eastman L F, 1979 Electron. Lett. 15, 90
Drummond T J, Kopp W, Thorne R E, Fischer R and Morkoc H, 1982 Appl. Phys.
    Lett. 40, 879

Inst. Phys. Conf. Ser. No. 74: Chapter 7
Paper presented at Int. Symp. GaAs and Related Compounds, Biarritz, 1984

551

Planar AlGaAs/GaAs selectively doped structure with high performances and high stabilities

H. Hida, H. Miyamoto, K. Ohata, T. Itoh, T. Baba* and M. Ogawa*

Microelectronics Research Laboratories
*Fundamental Research Laboratories
NEC Corporation, 4-1-1, Miyazaki, Miyamae-ku, Kawasaki 213 Japan

Abstract. High performance planar-gate FETs with high stability have been newly developed utilizing an n-AlGaAs/GaAs selectively doped structure. The threshold voltage shift resulting from lowering the ambient temperature was reduced to less than half of that in the conventional FETs by optimizing the n-Al_xGa_{1-x}As layer thickness. Interfacial current transient spectroscopy was used to evaluate FET stabilities. 0.5μm-gate FETs exhibited markedly low noise performance of 1.2 dB noise figure at 20 GHz for -50°C operation.

1. Introduction

Field effect transistors based on an n-AlGaAs/GaAs selectively doped structure have been applied to high speed ICs (Linh et al 1982, Lee et al 1983, Nishiuch et al 1983 and Kiehl et al 1983) and low noise microwave devices (Ohata et al 1984, Berenz et al 1984 and Joshin et al 1984). Basically, those FETs are constructed with a Schottky barrier gate on the n-AlGaAs layer. However, there are high density of deep traps, so called DX centers in n-Al_xGa_{1-x} As ($x \gtrsim$ 0.2). This is possibly the cause of undesirable instabilities in FET characteristics, such as a large shift in the threshold voltage (V_T), a long term photoresponse, and a drain current variation resulting from applying a high electric field, when the FET is cooled down to 77 K. Furthermore, large AlAs content in n-AlGaAs layer degrades the quality of the alloyed ohmic contacts.

This paper reports a highly practical n-AlGaAs/GaAs selectively doped structure FET minimizing such instabilities as those described above. In the developed FET, the n-$Al_{0.3}Ga_{0.7}$As layer thickness is reduced to the minimum value, which is still sufficient for supplying the maximum two dimensional electron gas (2DEG) at the hetero-interface. The threshold voltage shift with decreasing the temperature from 300 K to 77 K was reduced to less than 0.2 V. In addition, 0.5μm gate FETs exhibited markedly high DC and RF performances. Interfacial current transient spectroscopy was also investigated in order to characterize deep traps in the n-AlGaAs/GaAs selectively doped structure FETs.

2. Device Structure

The structure of a conventional (non-optimized) FET is shown in Fig. 1. A noticeable feature of the FET, whether a planar type or a recessed one, is that only one doped layer of n-$Al_{0.3}Ga_{0.7}$As is employed as a semi-insulated semiconductor layer under the gate electrode. As described above, the n-

Al$_{0.3}$Ga$_{0.7}$As layer contains high density of DX centers. Therefore various undesirable instabilities in characteristics were observed in the FET as shown in Fig. 1. Figure 2 shows the energy band diagram in thermal equilibrium under the gate. Assuming a normally-on FET, two dimensional electron gas is accumulated in the undoped GaAs side near the hetero-interface. In the figure, t_{sb} is the thickness of the depletion layer due to a Schottky barrier, t_{int} is the layer thickness depleted by the band discontinuity at the hetero-interface to supply the maximum density of 2DEG and t_o is the sum of t_{sb} and t_{int}. The sheet carrier density (N_s) and individual depletion layer thickness as a function of donor density in an n-AlGaAs layer were calculated under the triangular potential approximation (Delagebeaudeuf et al 1982) without a spacer layer. The results are plotted in Fig. 3. It is desirable for the total thickness of the doped layer to just equal t_o in order to avoid both g_m compression due to existence of a neutral region in the doped layer and source resistance increase due to decrease in the sheet carrier density. It should be noted that only t_{int} is sufficient for supplying the maximum sheet carrier density at the donor density and consequently for realizing small source resistance.

Based on these points described above, the (optimized) FET shown in Fig.4 has been newly developed to eliminate the various instabilities in FET characteristics and also to achieve performance improvements. The main features of the developed FET structure are as follow. The n-Al$_{0.3}$Ga$_{0.7}$As layer thickness, needed to supply the maximum 2DEG density, is designed to be as small a value as t_{int}. The n-GaAs, which is expected to have much better quality

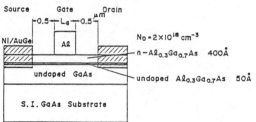

Fig. 1 Structure of the conventional FET.

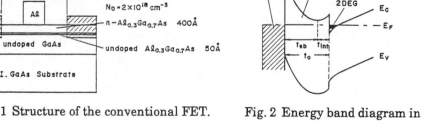

Fig. 2 Energy band diagram in thermal equilibrium under the gate.

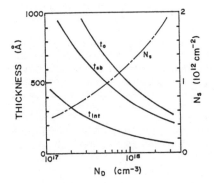

Fig. 3 Donor density vs. depletion layer thickness and sheet carrier density (Calculated).

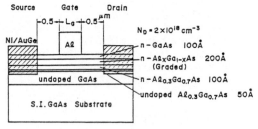

Fig.4 Structure of the developed FET

than n-AlGaAs, is employed for the surface depletion layer of the gate Schottky barrier. The donor doped layers consisted of a 100 Å n-Al$_{0.3}$Ga$_{0.7}$As, a 200 Å graded n-Al$_x$Ga$_{1-x}$As and a 100 Å n-GaAs. Each layer was intentionally doped to the level of $2x10^{18}$ cm^{-3}. The thickness of the n-Al$_x$Ga$_{1-x}$As ($x \gtrsim 0.2$) layer including high density of DX centers is reduced to less than one third of that in the conventional FETs, as shown in Fig. 1. In order to obtain high transconductance, super low noise and so forth, parasitic resistance reduction is particularly important. The small source resistance is realized by optimizing the n-Al$_{0.3}$Ga$_{0.7}$As layer thickness until it is just enough for supplying the maximum 2DEG and by employing the closely spaced electrode structure (Furutsuka et al 1982) for the developed FET.

3. DC Characteristics

The developed AlGaAs/GaAs selectively doped structure FET exhibited high DC performance. Fabrication steps for the FET are similar to those in the reference (Ohata et al 1984). Figure 5 shows typical drain current-voltage characteristics of a FET with 1μm gate length and 200μm width at room temperature (a) and at 77 K (b). Low saturation voltage and markedly high transconductance due to small source resistance can be clearly seen. The maximum transconductances were 280 mS/mm at 300 K and 450 mS/mm at 77 K for 0.5μm gate FETs. The collapse of drain I-V characteristics when cooled to 77 K without exposure to light, as reported by Fischer et al (1984), was not observed. In the non-optimized FETs shown in Fig. 1, ohmic contacts sometimes showed poor surface morphology and slightly poor I-V linearity at low temperature. These problems were noticeably eliminated in the newly developed FETs. This indicates that these problems are likely to be closely related to the AlGaAs layer with a large thickness and/or with large AlAs content. The temperature dependences of the threshold voltage V$_T$ for both types of FET are shown in Fig. 6. The gate length of the measured FET was 10μm. The threshold voltage was determined by extrapolating square root current versus gate voltage. For both types of device, V$_T$ was inclined to decrease as temperature increases from 77 K to 300 K with an abrupt change near 180 K, possibly due to electron detrapping of DX centers. A similar dependence of V$_T$ on temperature was reported by Valois et al (1983a). The difference between the threshold voltage at 300 K and that at 77 K was about 0.4 V for the non-optimized FETs and less than 0.2 V for the optimized FETs with the V$_T$ of about -0.8 V. It is clear that thinning the n-Al$_{0.3}$Ga$_{0.7}$As has a marked effect on reducing the V$_T$ shift. This indicates that the change in the occupation of deep traps in the n-Al$_{0.3}$Ga$_{0.7}$As layer may be dominant causing the observed large V$_T$ shift.

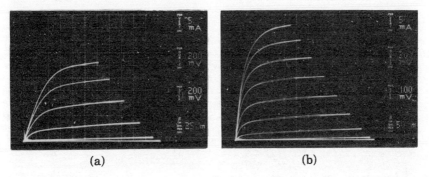

(a)   (b)

Fig. 5 Drain I-V characteristics of a 1μm gate FET. (a) 300 K and (b) 77 K.

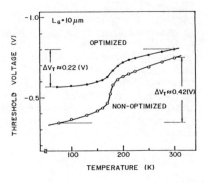

Fig. 6 Temperature dependence of threshold voltage for both types of FET.

4. Interfacial Current Transient Spectroscopy (ICTS)

It is very important to examine properties and influences of deep traps, not in diodes but in FETs, from the practical aspect. The ICTS technique can allow accurate quantitative characterization of deep traps in AlGaAs/GaAs selectively doped structure FETs directly, even when trap density is comparable to donor density (Valois et al 1983b, Takikawa and Ozeki 1984). The following reports the results of studies on instabilities in the FET characteristics using this effective technique. The threshold voltage depends on the occupancy of deep traps. Therefore under simple analysis a time response of V_T is as follows, as given by Valois et al (1983b)

$$V_T = V_T^\infty + \frac{qN_{To}d^2}{2\varepsilon} e^{-t/\tau} . \tag{1}$$

Here V_T^∞ is the value of V_T at time $t \to \infty$, d and ε are the thickness and the permittivity of the n-AlGaAs, N_{TO} is the trapped electron density at $t = 0$, and τ is the time constant for thermal emission. If a long gate FET measured is biased at a low drain voltage before current saturation, drain current I_{DS} is given by

$$I_{DS} = \frac{W_G\mu\varepsilon}{L_G d} V_{DS}(V_G - V_T) \quad for \quad V_{DS} << V_G - V_T \tag{2}$$

where W_G is the gate width, L_G is the gate length, μ is the electron mobility in the channel and V_{DS} is the drain voltage. From eqs. (1) and (2), the linear

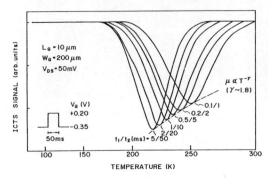

Fig. 7 Representative ICTS spectra for a non-optimized FET (normally-on).

relationship between $N_{TO}\exp(-t/\tau)$ and I_{DS} can be obtained. Therefore τ can be easily extracted using the ICTS technique without requiring $N_T << N_D$. Furthermore, assuming slow variation of N_{TO} and μ with temperature, the activation energy E_D and the capture cross section σ_n can be obtained with the usual analysis in DLTS (Lang et al 1974). N_T can be also estimated using the equations described above.

ICTS measurements were carried out for both types of FET with the same values of $L_G=10\mu m$, $W_G=200\mu m$ and $d=400$ Å. Typical spectra for a non-optimized FET are shown in Fig. 7. The FET was biased at $V_{DS}=50$ mV in the dark, the width of the trap filling pulse was 50 ms and the repetition rate was 4 Hz. The pulsed gate voltage was changed from at $+0.20$ V to at -0.35 V and the sampling time from $t_1/t_2=5/50$ to $0.1/1$ (ms) at a constant rate of 10. With an increase in temperature, the amplitude at the current peak temperature dereases in inverse proportion to $\sim T^{1.8}$, which may be caused by strong temperature dependence of the 2D electron mobility. By analyzing the spectra shown in Fig. 7, $E_D=0.52eV$, $\sigma_n=4 \times 10^{-12}$ cm^2 and $N_{TO}=5 \times 10^{17}$ cm^{-3} were estimated. These values are in reasonable agreement with previously reported values (Valois et al 1983a). The slight differences are possibly due to growth conditions. Figure 8 shows the spectra from both types of FET, an optimized FET (dashed line) and a non-optimized FET (solid line) with $t_1/t_2=2/20$ (ms). The spectra imply that there are at least three deep levels (ET1, ET2, ET3) in the layers, although they couldn't be fully resolved. It should be noted that the current amplitude in the optimized FET is fairly smaller than that in the non-optimized FET with regard to ET2. Considering that the electron mobility in the former FET was higher than that in the latter FET by 20 percent at the peak temperature, the current amplitude in the former is estimated to be less than half of that in the latter. This result directly demonstrates that the total number of the trap in the developed FET considerably decreased. In addition, this is in reasonable agreement with the result of the reduced V_T shift, shown in Fig. 6.

5. RF Performance

RF performances of planar n-AlGaAs/GaAs FET , which is suitable for practical use, were evaluated. Room temperature performance values for $0.5\mu m$ gate FETs with a $200\mu m$ gate width were 1.14 dB noise figure with 12 dB associated gain and 1.85 dB noise figure with 10.5 dB associated gain at 12 GHz and at 20 GHz, respectively. Moreover, marked noise performance of 1.2 dB noise figure at 20 GHz was obtained for -50°C operation. The RF characteristics of the developed FETs were very stable during operation which is suitable for practical applications.

Fig. 8 ICTS spectra for both types of FET.

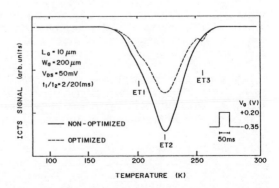

6. Conclusion

A planar AlGaAs/GaAs selectively doped structure FET with high performances and high stabilities, which is suitable for practical use, has newly developed. In the developed FET, n-$Al_{0.3}Ga_{0.7}As$ layer thickness is designed to be a minimum value while still being sufficient for supplying the maximum 2DEG. This makes the thickness of the n-AlGaAs layer which contains high density of DX centers ($x \gtrsim 0.2$) to be reduced to less than 1/3 of that in the conventional FET. This enabled various instabilities in FET characteristics caused by deep traps in the n-$Al_{0.3}Ga_{0.7}As$ to be greatly reduced. The difference between the threshold voltage at 300 K and that at 77 K was less than 0.2 V, which is less than half of the V_T shift in the conventional FETs with thick n-AlGaAs layer. Alloyed ohmic contacts were also noticeably improved, with respect to the surface morphology and the I-V linearity at low temperature. Moreover, interfacial current transient spectroscopy measurements clearly revealed a great decrease in the total number of deep traps. The new FET also exhibited high DC and RF performances. The maximum transconductances were as high as 280 mS/mm at 300 K and 450 mS/mm at 77 K for 0.5μm gate FETs. At -50°C, a 1.2 dB noise figure at 20 GHz was achieved.

Acknowledgment

The authors wish to thank Drs. T. Mizutani, H. Terao, H. Watanabe and S. Fukuda for their valuable discussions. They also thank Drs. N. Kawamura and Y. Takayama for their helpful suggestions and encouragement.

References

Berenz JJ, Nakano K and Weller KP 1984 IEEE INT. MTT-S Microwave Symp. (San Fran.: California) Digest Tech.Papers pp 98-101
Delagebeaudeuf D and Linh NT 1982 IEEE Trans. Electron Devices ED-29 955
Fischer R, Drummond T J, Klem J, Kopp W, Henderson TS, Prrachinone D and Morkoc H 1984 IEEE Trans. Electron Devices ED-31 1028
Furutsuka T, Tsuji T, Katano F, Kanamori M, Higashisaka A and Takayama Y 1982 Proc. 14th Int. Conf. on Solid State Devices (Tokyo) pp335-9
Joshin K, Yamashita Y, Niori M, Saito J, Mimura T and Abe M 1984 16th Int. Conf. Solid State Devices and Materials (Kobe: Japan) pp347-350
Kiehl RA, Feuer MD, Hendel RH, Hwang JCM, Keramidas V_G, Allyn CL and Dingle R 1983 IEEE Electron Device Lett. EDL-4 377
Lang DV 1974 J. Appl. Phys. 45 3023
Lee CP, Miller DL, Hou D and Anderson RJ 1983 41st Annual Device Res. Conf. (Vermont: Burlington) IIA-7
Linh NT, Tung PN, Delagebeaudeuf D, Delecluse P and Laviron M 1982 IEDM (San Fran.: California) Tech.Digest pp 582-5
Nishiuchi K, Mimura T, Kuroda S, Hiyamizu S, Nishi H and Abe M 1983 41stAnnual Device Res. Conf. (Vermont: Burlington) IIA-8
Ohata K, Hida H, Miyamoto H, Ogawa M, Baba T and Mizutani T 1984 IEEE Int. MTT-S Microwave Symp. (San Fran.: California) Digest Tech. papers pp434-6
Takikawa M and Ozeki M 1982 43th Autumn Meeting of Japan Society of Appl. Phys. (Inst. of Kyusyu Tech.) Digest (in Japanese) 406
Valois AJ, Robinson GY, Lee K and Shur MS 1983a J. Vac. Sci. Technol. B1 190
Valois AJ and Robinson GY 193b IEEE Trans. Electron Device Letters EDL-4 360

Inst. Phys. Conf. Ser. No. 74: Chapter 7
Paper presented at Int. Symp. GaAs and Related Compounds, Biarritz, 1984

n+-GaAs gate self-aligned GaAs MIS-like FET

K. MATSUMOTO, M. OGURA, T. WADA, T. YAO, N. HASHIZUME, Y. HAYASHI,
T. ENDO, and H. YAMAZAKI

Electrotechnical Laboratory
1-1-4 Umezono, Sakura-mura, Niihari-gun, Ibaraki, Japan

Abstract

Fundamental properties of a SIS FET having an n+-GaAs/ undoped GaAlAs/undoped GaAs structure is reported. The SIS FET is shown to withstand a high-temperature post-implantation annealing, which makes a self-aligned structure possible. The threshold voltage of the FET has very small dependence on the thickness of GaAlAs layer and on the ambient temperature.

1. Introduction

We report in this paper some fundamental properties of a SISFET (for Semiconductor Insulator Semiconductor FET). The SISFET was invented by the present authors (MATSUMOTO 1984) and employs n+-GaAs as a gate material and undoped AlGaAs as an insulator as shown in Fig.1 in a GaAs MIS-like heterojunction FET. It behaves much like a Si-MOSFET having a heavily doped polysilicon gate. It will be shown that the device has very promising features for LSI applications.

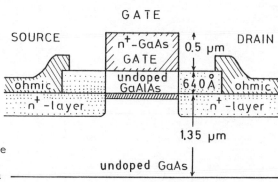

Fig. 1, Schematic structure of GaAs MIS-like FET (SIS FET). n+-GaAs gate is 2 µm in length, 0.5 µm in thickness.

2. Fabrication process.

The fabrication process started with the crystal growth of the following structure by MBE on an S.I. LEC (100) GaAs substrate:undoped GaAs (1.35μm)/undoped $Ga_{0.3}Al_{0.4}As$ (0.064μm)/n^+-GaAs (Si-doped,$1x10^{18}$/cm^3, 0.5μm). The top n^+-GaAs layer was selectively etched off using a chemical etchant ,leaving a strip of 20μm in width which worked as the gate. Using the n^+-GaAs gate as a mask,$2x10^{13}$/cm^2 Si ions were implanted at 100keV through the GaAlAs layer to form the self-aligned source and drain n^+-regions. The acceleration energy is so selected that the implanted impurity has the peak concentration near the GaAlAs/undoped GaAs interface. This is to ensure a good contact between the two-dimensional electron gas induced at the GaAs/GaAlAs interface and the source and drain n^+-regions. The n^+-GaAs gate is so thick that it can sufficiently protect the invasion of the implanted Si ions into the GaAlAs layer. (The projection range R_p and the standard deviationσ_p of Si in GaAs at E=100keV are R_p=840Å and σ_p=446Å , respectively.) The sample was then annealed in an infra-red lamp furnace at 800°C for 30 seconds without a cap, under an arsenic pressure. The sheet resistance of the n^+-regions thus obtained is 500 Ω /□. After selectively etching the GaAlAs layer, source and drain ohmic contacts,8μm apart,were formed with Au-Ge/Ni/Au by alloying at 440°C for 1 minute.

3. Fundamental properties of the SIS structure.

If an FET of a self-aligned structure is to be fabricated by ion implantation technique,then the FET has to experience a high annealing temperature of 800°C or above. In the case of a modulation doped hetero-structure (HEMT structure), high concentration impurities in the GaAlAs layer diffuse into the hetero-interface during the annealing and destroy the high electron mobility there. On the other hand, in the

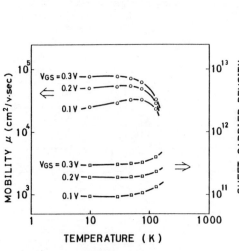

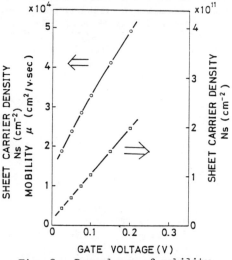

Fig. 2, Dependence of mobility and carrier concentration of SIS FET on temperature. Gate bias is a parameter.

Fig. 3, Dependence of mobility and carrier concentration of SIS FET on gate bias. Temperature is 77K.

case of a SIS structure, due to the lack of impurities in the GaAlAs layer, the electron mobility at the hetero-interface retains a high value even after the annealing.

The electron mobility at the hetero-interface was measured using a fat FET of 50x400 μm^2 which was fabricated using the same process as for the SIS FET described earlier. Fig.2 shows the electron mobility and the sheet carrier density versus temperature characteristics with the gate bias as a parameter. The higher the gate bias, the larger the electron mobility becomes. At the gate bias of V_g=+0.3 V, the electron mobility is as high as μ=80,000 cm^2/V.s at 77K. It increases still more with the decrease of the temperature, and shows saturation at around 20 K. At the small gate bias of V_g=+0.1V, however,the electron mobility shows a peak value at around 77 K and decreases with the decrease of the temperature. This may be due to the effect of the ion scattering by the unintentionally doped impurities in the GaAs layer. This is consistent with the results obtained by our C-V measurements which will be described later. The sheet carrier density at the hetero-interface increases with the increase of the gate bias and reaches the mid 10^{11}/cm^2 range, which is a sufficient value for the FET operation. The results obtained above show that the hetero-interface retains the good characteristics even after the high temperature annealing process.

Figure 3 shows the dependences of the sheet carrier density N_S and the electron mobility μ on the gate bias voltage V_g at 77 K. The sheet carrier density is linearly proportional to the applied gate bias. The capacitance calculated from the slope of the N_S vs.V_{gs} curve coincides with that of the GaAlAs layer, C_i obtained by the C-V measurement in Fig.5. Therefore,the sheet carrier density is represented approximately by N_S= C_i (V_g-V_{th})/q ,where V_{th} is the threshold voltage. The electron mobility increases linearly with V_g. This characteristic is quite different from that of the electron with constant mobility of a Si MOS FET.

Figure 4 shows the Shubnikov-de Haas (SdH) oscillations of the SIS structure. The applied gate bias is V_g =+0.5V, the sample temperature is T=1.25K. When the magnetic field is perpendicular to the hetero-interface, θ=90°,clear and large oscillations are observed. When the magnetic field is parallel to the hetero-interface, θ=0° ,however,

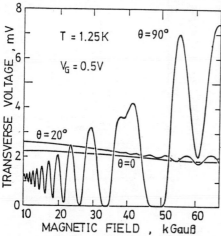

Fig. 4, Shubnikov-de Haas (SdH) oscillation of SIS structure.

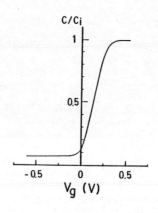

Fig. 5, Capacitance-voltage characteristics of SIS FET. Temperature is 77K.

there is no oscillation observed. This fact means that the electrons at the hetero-interface form the two dimensional electron gas.

Figure 5 shows the capacitance-voltage characteristics between the gate and the source electrodes of the SIS FET at 77 K in dark with the drain terminal open. The measured characteristic shows a clear MIS C-V curve, which consists of the accumulation region, the depletion region, and inversion region. The flat-band voltage is small, $V_{FB}=+0.16V$, as is expected by the use of the n^+-GaAs gate. From the C-V characteristic, the undoped GaAs layer is calculated to be unintentionally doped to $2.4\times10^{15}/cm^3$ at 77K.

4. FET characteristics

The source current I_s vs. drain-source voltage V_{ds} characteristics of a typical SIS FET measured at 77K in dark is shown in Fig. 6. The current is almost zero at the gate-source voltage V_g of zero for $V_{ds}=1V$. The increase of the current with a V_g step is almost constant up to $V_g=+0.8V$. The gate leak current is negligible up to this gate voltage. The threshold voltage, as determined by the extrapolation of I_s vs. V_g curve at $V_{ds}=1V$ is +0.04V. The transconductance at $V_g=+0.8V$, $V_{ds}=1V$ is 170mS/mm. The drain conductance at the same bias condition is as small as 8mS/mm. Thus the device operates as a normally-off type accumulation-mode FET of high performance.

The threshold voltages of 41 samples extended over a 10x5mm^2 wafer were measured. The distribution of the threshold voltage is shown in Fig.7. The average threshold voltage is 0.035V and the dispersion is 0.013V. Thus one of the most significant features of the present FET, the very low sensitivity of the threshold voltage to the material and process was confirmed.

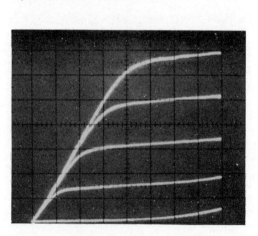

Fig. 6, Source current vs. drain-source voltage characteristics of SIS FET at 77K in dark. Gate bias is applied from 0V to +0.8V with 0.2V step. 0.4mA/div.; 0.2V/div.

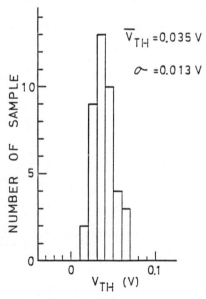

Fig. 7, Distribution of threshold voltage of 41 SIS FETs.

Figure 8 shows the source current I_s vs. drain-source voltage V_{ds} characteristics of a SIS FET at 279K (a) and at 10K (b). The thickness of the GaAlAs layer is 520 Å in this particular case. The gate bias of up to $V_g=+0.3V$ is applied with 0.1V steps. Comparing these characteristics,it is seen that the transconductance increases from 111mS/mm to 161.5mS/mm ,and the threshold voltage shifts from $V_{th}=-0.106V$ to $V_{th}=-0.016V$, when the temperature is decreased from 279K to 10K. However,neither the saturation drain current nor the drain conductance at $V_g=+0.3V$ show a large change. The transconductance and the saturation drain current of the FET do not increase very much by reducing the temperature. They are reduced by the high source resistance of 200-250 Ω , which is about three times as high as that of a typical self-aligined GaAs FET. The high source resistance is attributed to the not-optimized ion-implantation and annealing conditions.

Figure 9 shows the shift of the threshold voltage of the FET with the temperature and with the thickness of the GaAlAs layer. The thickness of the GaAlAs layer is 520Å and 250 Å. The shifts of the threshold voltage are as small as 0.05V and 0.1V for the FETs whose GaAlAs layer thicknesses are 250Å and 520Å,respectively, when the temperature is decreased from 279K to 77K. This is due to the lack of impurities, and the lack of DX centers in the GaAlAs layer. Though the difference of the thickness of GaAlAs layer for these two FETs is as large as 270 Å, the difference of the threshold voltage is as small as 50 mV at room temperature and 10 mV at 150K. Thus, it has been proved that the threshold voltage of the SIS FET has very small dependence on temperature and on the thickness of the GaAlAs layer.

5. Conclusion

In conclusion, we have demonstrated many attractive features of

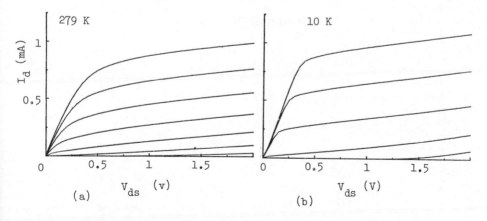

Fig. 8, Source current vs. drain-source voltage characteristics of SIS FET at 279K (a) and 10K (b).

the SISFET. It has been shown that the SISFET can withstand high temperature post-ion implantation annealing without degrading the channel electron mobility, and that its threshold voltage has very small dependence on the thickness of the GaAlAs layer and on the ambient temperature. Thus, the SISFET looks very attractive for high speed LSI applications.

Acknowledgment

The authors would like to thank S. Kataoka for continuous encouragement. They are also indebted to K. Tomizawa, M. Iida, and T. Kurosu for helpful discussions, and Y. Yamada, H. Inage and M. Yamada for technical assistance.

Reference

MATSUMOTO K., OGURA M., WADA T., HASHIZUME N., HAYASHI Y., 1984 Electronics Lett. 20 pp462-463.

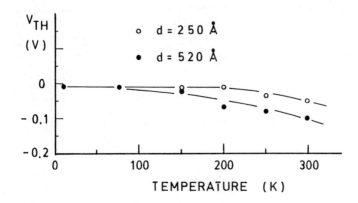

Fig. 9, Dependence of threshold voltage of SIS FET on temperature and thickness of GaAlAs layer.

Inst. Phys. Conf. Ser. No. 74: Chapter 7
Paper presented at Int. Symp. GaAs and Related Compounds, Biarritz, 1984

Novel deep depletion GaAs insulated gate FET

J.M. Dortu, E. Kohn

Thomson-CSF/DHM, Domaine de Corbeville, B.P 10, 91403 Orsay, France.

1.   Introduction

In this paper we present a novel FET structure suitable for a number
of III-IV compound semiconductor materials. A thin slightly conductive
dielectric spacer layer has been inserted between the gate metal and the
semiconductor. Thus the device can combine the advantages of MESFET and
MISFET characteristics to a wide extent.

MESFETs encounter often gate leakage problems, due to low metal
semiconductor barrier heights. And even on GaAs, gate leakage becomes a
limiting factor with the proximity of n^+ contact regions in self aligned
devices or when using high doping levels in order to avoid two
dimensional side effects in submicron gate FETs.

MISFETs on the other hand suffer from interface state and charge
storage instabilities especially at low frequency operation. Deep
depletion characteristics can only be achieved in the dynamic mode.

The transistor presented here, called Deep Depletion MISFET, exhibits
a MESFET like dc transfer characteristic and uses a dielectric spacer
layer between the gate and the channel for gate breakdown protection.

The feasibility of such a system has already been demonstrated with
the semi-insulated gate FET discussed by Pruniaux et al. (1972), the Lossy
Gate Dielectric FET presented by Andrade and Braslau (1983) and in a
vertical FET structure realized by Kohn et al. (1983).

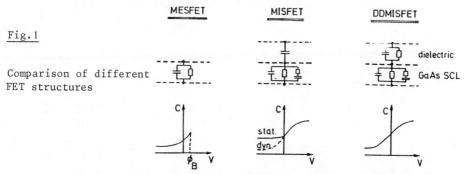

Fig.1

Comparison of different
FET structures

The electrical small signal equivalent circuit is presented in fig.1.
The DC behaviour is dominated by the resistances, whereas the RF behaviour
is determined by the capacitances. As a result, the spacer layer has to
be very thin and must have a permittivity as high as possible.

In this paper we will report on results of DDMISFET structures
fabricated on highly doped GaAs active layers with sub-µm channel lengths.

2. The GaAs DDMIS-SYSTEM

The aim is to realize a MIS system with a defined conductivity in the
dielectric. This conductivity has to support the minority carrier current
generated in the GaAs and at the interface while causing only a small
potential drop in the dielectric. Additionally, the dielectric should
have a high permittivity to be used on highly doped materials.

Si-rich silicon nitride has been chosen for application on GaAs. As
has been shown by Sasaki et al. (1982) conductivity, optical refractive
index and permittivity increase with Si-concentration. Two current
mechanisms have been identified as Poole-Frenkel and hopping conductivity.
A second reason for the use of silicon nitride is its application in
GaAs IC-technology, where this Si-rich SiN layer is used for device
passivation as reported by Kohn et al. (1984).
SiN films with a refractive index between 2.4 and 3.2 have been
deposited by a plasma assisted CVD process.
The analysis of MIS and MIM capacitors on the same wafer allow the
following general conclusions :
1) The conductivity in MIM diodes with Ti-electrodes is bulk-dominated
and of the Poole-Frenkel type.
2) The interface potential is 0.45V <$\emptyset B$ < 0.55V. For device analysis
a value of $\emptyset B$ = 0.50V. has been used.

Figures 2 and 3 demonstrate the CV behaviour of MIS diodes on n-type
GaAs. In reverse direction a deep depletion characteristic is seen ; in
forward (pos.) direction we note a frequency dispersion as typical for
GaAs MIS diodes. When comparing the positive branch with the corresponding
branch in the FET transfer characteristics (fig.4), one can see that this
capacitance represents surface state response. No variation in the space
charge layer width is observed here. At forward bias the leakage current
is not sufficient to short-circuit the high interface state density with
the metal plate. Fig.3 shows switching from a positive potential into
deep depletion condition. It is confirmed that (with 500 ms resolution)
a stable deep depletion position is obtained. No storage of injected
charge is observed up to +6V. This stable deep depletion characteristic
has been obtained in the full temperature range used for testing (-50°C
to + 150°C), see also FET characteristics in fig.5.

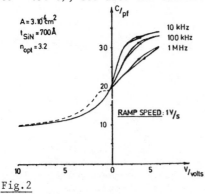

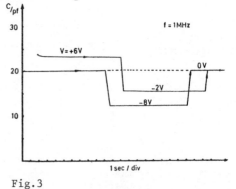

Fig.2   Fig.3

MIS CV-curve on N-GaAs   MIS diode switching behaviour.
For dashed line see text.

The IV characteristics of the SiN - film are essentially exponential, thus at low fields the current level is low. If in this case the leakage is not sufficient to support the minority carrier current from the GaAs, an accumulation of positive charge at the interface is observed. This manifests itself in a knee in the CV curve as indicated by the dashed line in fig.2. Within the plateau the deep depletion curve is only observed for fast ramp speeds. This effect has been avoided by minimizing the trap generation current at the interface and in the GaAs depletion layer by a special surface cleaning procedure prior to the deposition and by avoiding radiation damage in the GaAs caused by the deposition process.

3. FET Technology

Planar FETs have been fabricated using Si^+ implanted layers with peak doping levels between 10^{17} cm^{-3} and 10^{18} cm^{-3}. For the highly doped devices, closely spaced n^+ contact regions were implemented.

In the following we will discuss the fabrication process of the highly doped channel structure containing n^+ contact regions in detail :

After alignment key fabrication a 500 Å thick Si_3N_4 capping layer was deposited by RF sputtering. Next the active layer implant was performed at 50keV, 1.10^{13} cm^{-2} followed by the n^+ contact implant of 80 keV, 2×10^{13} cm^{-2} directly into the undoped semi-insulating substrate, activated by a 850°C anneal. The gap between the two contact implant regions was estimated to 0.6 μm. Then the Si_3N_4 layer was stripped off and ohmic contacts were deposited. At this point the entire surface was etched back to adjust the final FET saturation current. This was then immediately followed by the deposition of the 200 Å thick SiN spacer layer at 300°C in a planar plasma CVD system. Onto this structure the 1.0μm long TiPtAu gate metal strip was deposited using a classical lift-off process.

Thus, the structure contains gate overlap to the source and drain n^+ regions. It was basically designed to show the absence of gate breakdown, to demonstrate high static transconductance and to investigate sub-μm source drain contact spacing. A future μW-structure has to use a technology wherethe gate is self-aligned in respect to the n^+ contact areas. Work on the temperature stability of the SiN/GaAs system to support the n^+ contact implantation anneal is in progress.

Fig. 4

Cross section of DDMISFET structure with n^+ contact regions and sub-μm channel length.

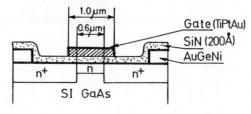

4. Measurement results.

4.1. Static measurements

Fig.5 shows the curve tracer output characteristics obtained on the highly doped device described above. The maximum static transconductance obtained is 200 mS/mm. Note that the maximum transconductance appears for Vg<0 and not Vg>0 like in MESFETs. As expected, the saturation voltage is

very low which is in agreement with sub-μm S-D spacing and the presence
of n⁺ contact regions.

   In general a saturation in the transfer characteristic is observed at
a knee voltage of Vg = -1.0V. By applying a positive gate bias no further
increase in the saturation current can be obtained, meaning that no
enhancement mode of operation is possible with the device. This fact is
in agreement with the CV measurements demonstrated in chap.2, where the
capacitance variation at positive gate bias has been associated with
surface state response and where a negative knee voltage has been related
to an insufficient leakage current in respect to the positive generation
current in the GaAs at low gate bias.

Fig.5

DDMISFET output characteristics
of highly doped channel device
with sub-μm SD contact layer
spacing.

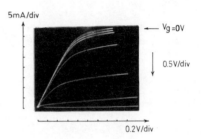

   The static deep depletion output characteristic has been observed
within the full temperature range of measurement as demonstrated with
Fig.6,where two characteristics measured at -50°C and +150°C are presented.
The small hysteresis observed at -50°C in the gate voltage saturation
region can again be correlated to the accumulation of positive charge due
to insufficient gate dielectric leakage at low gate bias and low
temperature.

Fig.6

MISFET output
characteristics at
- 50°C and + 150°C.

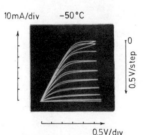

   The gate leakage current is smaller than 10^{-7}A for all bias points.
No gate to drain breakdown was observed by increasing Vsd, instead the
sub-μm channel devices exhibit a punch through behaviour as shown on
Fig.7. The punch through voltage is rather low due to the semi-insulating
substrate and can be improved essentially by a p-buffer implant underneath
the channel. This would also increase the output resistance largely.

Fig.7

Complete field of output characteristics
with limitation imposed by saturation
in output current for positive gate
bias and by punch through for high
drain bias.

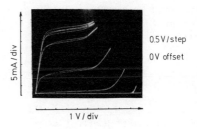

0.5V/step

0V offset

5mA/div

1 V/div

4.2. RF Measurements

Fig.8 shows the maximum transconductance as a function of frequency.
The curve shows two flat branches with a transition region at approx.
10 kHz. This dispersion is expected and represents the transition between
the low frequency MESFET like transconductance to high frequency MISFET
transconductance. The transition frequency is in agreement with values of
the dielectric capacitance and leakage current extracted from devices of
various gate lengths.

Fig.8

Dispersion of small signal transconductance showing the
transition between static and
RF behaviour.

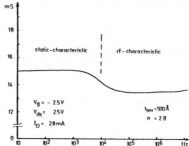

mS

18

16

static–characteristic   rf–characteristic

14

$V_g = -2.5V$
$V_{ds} = 2.5V$
$I_D = 28mA$

$I_{SiN} = 500 Å$
n = 2.8

12

0

10   10^2 10^3 10^4 10^5 10^6 Hz

From S-parameter measurements up to 18 GHz a maximum frequency of
oscillation of 30 GHz can be extrapolated. Extracting the parasitic gate
overlap capacitances, the performance of a self-aligned gate device can
be simulated. In this case a maximum frequency of oscillation of 50 GHz is
obtained.

5. Conclusion

The GaAs dielectric interface represents a very unfavorable case in
respect to high interface state densities and charge injection phenomena.
Nevertheless, widely stable conditions could be achieved with this new
leaky MIS-system, and the system has prooved its applicability in a high
transconductance FET on GaAs avoiding Schottky gate breakdown encountered
with highly doped channel MESFETs.

The gate dielectric deposition process however has still to be
optimised to withstand an n^+ implant anneal to make a self-aligned μm
FET technology possible.

A simple model has been used to evaluate the transconductance
expected on highlydoped channel DDMISFET's with 0.5μm channel length.
Taking into account the mobility-doping relationship as presented by
Hilsum (1974), static and rf-transconductances were computed, assuming a
spacer layer thickness of 100 Å with a dielectric permittivity of 10. The
pinch-off voltage is kept constant and equal to 1 V. A maximum DC
transconductance of 400 mS dropping to 300 mS/mm at RF in the 10^{18} cm^{-3}
doping level range is predicted (see Fig.9).

MIS-systems on other III-V materials do not suffer from pinning of
the surface potential at such high values as in n-GaAs. In InP and
InGaAs even electron accumulation at the interface is possible on n-type
channels permitting enhancement mode of operation. High output power has
been demonstrated by M. Armand et al. (1984) for InP MISFETs and high
cut-off frequencies are expected for InGaAs MISFETs as indicated by first
results of Cheng et al. (1984). However, the FET characteristics show in
general   strong hysteresis and large drift phenomena. It is expected that
the gate contact system presented here can also contribute to solving the
instability problems in these devices.

Fig.9

Maximum DDMISFET transconductance
vs. channel doping. Model
parameters are indicated.

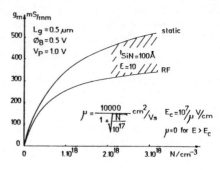

ACKNOWLEDGEMENTS.

We are delighted to acknowledge the technical assistance of A.
Maubert and D. Collignon in device fabrication. Special thanks are to J.
Magarshack for his steady support of this subject and to Prof. E.
Constant and his group at the CHS at University of Lille for his interest
in this work and the microwave measurements.

REFERENCES.

Andrade T L and Braslau N,1983, 39 th DRC Santa Barbara
    paper V A 5
Armand M, Bui D V, Chevrier J and Linh N T, 1984, Rev. Techn.
    Thomson-CSF, 16, 47-98
Cheng C L, Liao A S H, Chang T Y, Leny R F, Coldren L A, and
    Lalevie B, 1984, Electron Dev. Lett. EDL-5, 163-171
Kohn E, Lanteri J P, and Dortu J M, 1984, Europ. GaAs IC workshop,
    21-23 Feb. 1984, Nice, France
Kohn E, Magarshack J, Mishra U and Eastman L F, 1983, Electron.
    Lett. 19, 1021-1023
Sasaki G, Kondo S, Fujita S and Sasaki A, 1982, Jap. Appl. Phys.
    21, 1394-1399

Inst. Phys. Conf. Ser. No. 74: Chapter 7
Paper presented at Int. Symp. GaAs and Related Compounds, Biarritz, 1984

569

Mechanism and suppression of drain current drift in InP MISFETs

Hideki Hasegawa, Takayuki Sawada and Hideo Ohno

Department of Electrical Engineering, Faculty of Engineering,
Hokkaido University, Sapporo 060, Japan

Abstract The temperature dependence of the drain current drift in InP MISFETs is shown to be a measurement artefact. A model involving interface states below the conduction band is presented which explains the current drift quantitatively. It is shown that incorporation of thin native oxide interfacial layer is effective in suppressing the current drift.

1. Introduction

It has already been established that InP-insulator interfaces formed by a variety of dielectrics (Lile et al 1978, Sawada et al 1984 and references cited herein) allow formation of surface n-channel with a high mobility. Availability of semi-insulating substrates, high electron saturation velocity, high thermal conductivity, and compatibility with the long wavelength optoelectronic devices in InP makes the InP MISFET ICs very attractive as an approach for high-speed LSI/VLSI circuits. It has also been shown that InP MISFETs show superb microwave power capability (Itoh and Ohata 1983, Armand et al 1983). The most important unsolved issue for the practical exploitaion of InP MISFETs is the so-called drain current drift phenomenon.

This paper shows for the first time that the previously reported temperature dependence of the drain current drift is a measurement artefact, and disappears when the initial condition is kept the same. A model which involves tunneling assisted electron capture and emission by interface states lying <u>below</u> the conduction band edge E_c explains all the aspects of drift quantitatively. Also it is shown that the intentional incorporation of a thin (5-10 nm) high-quality native oxide layer at the interface reduces N_{ss} down into 10^{10} $cm^{-2}eV^{-1}$ range, and suppress the current drift to a small amount of below 15% together with increasing the channel mobility. The role of native oxide is explained by the "surface disorder model" for the origin of interface states (Hasegawa and Sawada 1983).

2. Fabrication and Performance of MISFETs

A cross-sectional view of the device is shown in Fig.1. The details of the fabrication process have been described in Sawada et al (1984). As the gate insulator, single-layer dielectric films of alumina and silicon nitride, and double-layer dielectric films of alumina-native oxide and

silicon nitride-native oxide were
prepared. Alumina and alumina-
native oxide layers were formed by
either electrolytic or plasma
anodization of Al films deposited
on InP in vacuum. By this
process, double layers can be
formed without interrupting the
anodization, and the thickness of
each layer can be precisely
controlled by monitoring the
voltage across the insulator
during anodization (Sawada and
Hasegawa 1983). For the formation
of the silicon nitride layer, a

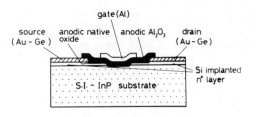

Fig.1 Cross-sectional view of the InP
MISFET with Al_2O_3-native oxide
insulator.

newly developed low-frequency (50 Hz), room temperature plasma CVD
process was employed (Shimozuma et al 1984). To form a silicon nitride-
native oxide double layer, a thin native oxide layer was formed by the
electrolytic anodization prior to the CVD process. In all the double-
layer systems, the thickness of the native oxide layer was maintained in
the range of 50-100 A. Linear relation between the square root of the
drain current vs. gate voltage in the saturation region was obtained.
Effective channel mobility and threshold voltage were determined from
these plots. Annealing was done on the processed devices in hydrogen at
350 C for 30 min.   It should be noted the annealing improves the
effective channel mobility very much in the devices with double layer
insulators.

3. Drain Current Drift Behavior and Drift Mechanism

3.1 Temperature Dependence of Drift Behavior

It has been commonly observed in
InP MISFETs that the drain current
reduces with time after
application of a gate voltage,
although the speed of decrease and
the degree of reduction depend
very much on the gate insulator
materials and on how they are
prepared. It was pointed out in a
number of papers that this drift
behavior was strongly temperature
dependent and the drift behavior
disappeared at low temperatures.
In order to explain this
temperature dependence, Okamura
and Kobayashi (1980), Lile and
Taylor (1983), and Yamaguchi et al
(1984) suggested thermally
activated tunneling of electrons
into defect states in the native
oxide energetically located above
E_C as shown in Fig.2. Goodnick et
al (1984) proposed that these
states were correlated with bulk-
like In_2O_3 incorporated into the
natural native oxide, whose

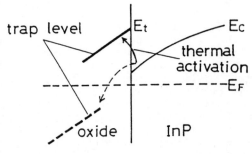

Fig.2 Previous model assuming
presence of defect states in
the native oxide above E_C.

conduction band was also
located slightly above E_C.

We have reexamined the
drift data to clarify and
to understand the drift
mechanism, since the drift
data are only meaningful
under the condition in
which the initial condition
of the trap occupancy
remains the same for each
temperature. This
condition was not met if
the measurements were done
by successively lowering
the temperature, where the
trapped electrons would not
be reemitted before the
next experiment. This means
that the next measurement
was done under
nonequilibrium condition
with excess electrons
occupying the interface
states. The number of such
occupied states increases
with lowering temperature
due to strong temperature
dependence of emission time

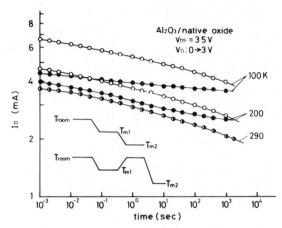

Fig.3 Temperature dependence of current
      drift behavior measured by
      successively reducing the
      temperature (closed circles) and by
      keeping the initial condition the
      same (open circles). The
      temperature sequences are also shown
      in the inset.

constant. If this is the case, the drift results arose only from the
measurement method, and do not reflect the true physical processes.

In order to check this point, the drift data was taken on the same sample,
using two different measurement sequences.The result is shown in Fig.3
together with the measurement sequence in the inset. For open circles the
measurements were done after
holding the device at room
temperature for several tens of
minutes prior to cooling the
device for each drift measurement
in order to achieve the
equilibrium initial state. It is
clear that the drift behavior is
independent of temperature. The
parallel shift of the current
level may be due to the
temperature variation of either
mobility or threshold voltage or
both.

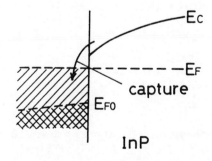

3.2 Drift Mechanism
The newly observed drift behavior
independent of temperature is
consistent with our electrical
modeling of GaAs and InP MIS
interfaces in terms of tunnel-

Fig.4 The proposed model for the
      current drift including
      tunneling assisted capture of
      carriers by interface states.

assisted capture and emission of carriers by energetically and spatially distributed interface states near the interface below E_c. This modeling was first proposed by the present group in 1979 for GaAs anodic MOS systems (Sawada and Hasegawa 1979). More recent refined discussion for GaAs and InP MIS systems is given in references Hasegawa and Sawada (1982) and Hasegawa and Sawada (1983). Recently, a similar model is discussed by Staa et al (1983) based on CCDLTS measurements on InP MIS capacitors. According to our model, the reducing type current drift can be explained by tunnel assisted trapping of carriers by interface states as shown in Fig.4. The reason why the drift behavior is independent of temperature is because both the capture by lower energy states and the tunneling process are almost temperature independent. If this model is correct, then, the variation of drain current caused by step decrease of gate voltage should correspond to emission, and should therefore be temperature dependent. The result of such a measurement displays a highly temperature-dependent and non-exponential behavior, which is consistent with the time constant dispersion due to energy and spatial distribution of interface states.

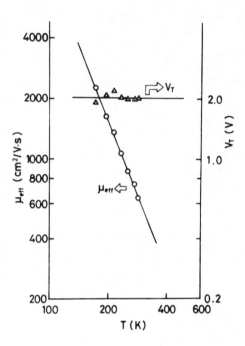

Fig.5 Temperature dependences of effective mobility and threshold voltage.

To clarify the mechanism for the shift of the current level with temperature, dependence of the current-voltage characteristics of the MISFETs were measured. The measured effective channel mobility and the threshold voltage are plotted vs. temperature in Fig.5. It is evident from Fig.5 that the shift of the current level is simply caused by the temperature variation of mobility. Note that the observed temperature dependence is approximately proportional to T^{-2}, which is typical of that of InP bulk, suggesting that the mobility is limited by polar optical phonon scattering in this temperature range.

3.3 Mathematical Modeling of Drain Current Drift

According to Heiman and Warfield (1965) tunnel-assisted filling of spatially distributed interface states can be described by the characteristic distance $x(t)$ from the interface which gives the boundary between filled states and empty states at time t. On the basis of our previous admittance analysis of MIS capacitors (Sawada and Hasegawa 1979, Hasegawa and Sawada 1980), it is assumed that the interace states $N_{sst}(x)$ to be filled by application of the gate bias step, is exponentially distributed with respect to distance for most of the cases.

Figure 6 compares the theoretical drift behavior thus derived with the experimentally observed drift behavior. As seen in Fig.6, good agreement between theory and experiment are seen, indicating the essential validity of the present model. Except for the sample indicated by triangles, in which a uniform distribution of the interface states is assumed, the exponentially decreasing distribution is assumed.

4. Suppression of Current Drift by Incorporation of Native Oxide Layer

Figure 6 also compares the drift behavior of the double-layer MISFETs and the single-layer MISFETs, which are normalized at $t = 10^{-4}$ sec. It is seen that decreasing-type current drift is always very much reduced by intentionally adding the thin intermediary native oxide layer. The drain current does almost reach its steady state after 10^3 - 10^4 sec. with keeping high current level. In the best sample, the reduction of the drain current was as small as 11 % after annealing. As shown in the previous section, current drift is caused by tunneling-assisted trapping of carriers by energetically and spatially distributed interface states. Therefore, the reduction of drift by incorporation of a native oxide layer should be related to reduction of interface state density. This was directly confirmed by MIS interface study. Figure 7 shows the energy distribution of interface state for annealed samples by Terman's method using 1 MHz C-V curves, indicating the presence of a native oxide layer reduces the state density about an order of magnitude. Since Terman's method measures all the states that are filled by the application of d.c. bias, the measured density includes states responsible for current drift.

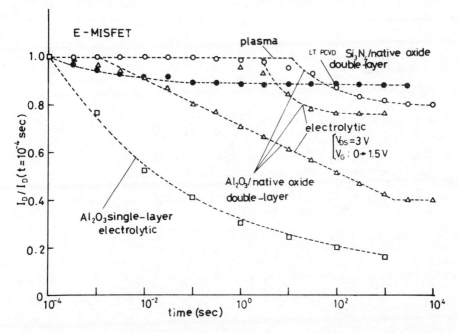

Fig.6 Comparison of experimantal drift behavior and the fitted theoretical behavior (broken lines). Note that the decreasing drift is very much reduced by the insertion of intermediary native oxide layer.

Recently, present authors have proposed a model for the origin of the interface states (Hasegawa and Sawada 1983), in which structural disorder of the semiconductor surface is responsible for the surface states.   If this model is valid, recovery of the crystalline order in the semiconductor surface region should lead to reduction of the interface state density.   For successful recovery of crystalline order in the semiconductor surface region during annealing,

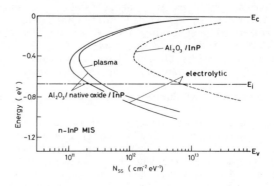

Fig.7 Interface state density distribution as determined by the Terman method.

(1) the disorder introduced during the processing step of insulator formation must be small enough and (2)the insulator must be well bonded to the semiconductor, chemically stable during annealing and flexible enough to allow recovery. The results obtained in this study appear to be consistent with above considerations.   E/D inverters and ring oscillators fabricated using the double layer technique showed excellent d.c. stability as reported in Sawada et al (1984).

Acknowledgements

This work was supported in part by a Grant-in-Aid for Scientific Research from the Ministry of Education, Science and Culture and partly by Hoso Bunka Foundation.

References

Armand M, Bui D V, Chevrier J and Linh N T 1983 Electron. Lett. 19 433.
Goodnick S M, Hwang T and Wilmsen C W 1984 Appl. Phys. Lett. 44 453.
Hasegawa H and Sawada T 1980 IEEE Trans. Electron Devices ED-27 1055.
Hasegawa H and Sawada T 1982 J. Vac. Sci. Technol. 21 457.
Hasegawa H and Sawada T 1983 Thin Solid Films 103 119.
Heiman F P and Warfield G 1965 IEEE Trans. Electron Devices ED-12 167.
Itoh T and Ohata K 1983 IEEE Trans. Electron Devices ED-30 811.
Lile D L, Collins D A, Meiners L G and Messick L 1978 Electron. Lett. 14 657.
Lile D L and Taylor M J 1983 J. Appl. Phys. 54 260.
Okamura M and Kobayashi T 1980 Jpn. J. Appl. Phys. 19 2143.
Sawada T and Hasegawa H 1979 Phys. Stat. Sol. (a) 54 689.
Sawada T and Hasegawa H 1982 Int. Symp. GaAs and Related Comp. Albuquerque U.S.A. (Inst. Phys. Conf. Ser. No. 65 1983) pp. 415-422.
Sawada T, Itagaki S, Hasegawa H and Ohno H 1984 IEEE Trans. Electron Devices ED-31 1038.
Shimozuma M, Sawada T, Ohno H, Hasegawa H and Tagashira H presented at 1984 Electronic Materials Conference (Santa Barbara U.S.A.).
Staa P, Rombach H and Kassing R 1983 J. Appl. Phys. 54 4014.
Yamaguchi E, Minakata M and Furukawa Y 1984 Jpn. J. Appl. Phys. 23 L49.

Inst. Phys. Conf. Ser. No. 74: Chapter 7
Paper presented at Int. Symp. GaAs and Related Compounds, Biarritz, 1984

575

Evidence for interfacial defects in metal–insulator–InP structures induced by the insulator deposition

B. SAUTREUIL, P. VIKTOROVITCH, R. BLANCHET

Laboratoire d'Electronique, Automatique et Mesures Electriques - Ecole Centrale de Lyon - BP 163 - 69131 ECULLY Cedex - France

Abstract : Photoluminescence (P.L.) intensity on n-type InP at room temperature is found to provide a convenient probe of the density of surface states in the upper part of the gap of InP. P.L. measurements are used to monitor the interaction phenomena occuring between an insulator (Al_2O_3, SiO_x) and the InP substrate during the first stages of its deposition. Despite the use of a soft deposition technique, the insulator induces systematically interfacial defects in metal–insulator-InP structures. The nature of the insulator and, to a larger extent, the InP surface preparation and the substrate temperature (although kept below 150°C) play a major role.

Since the first report by Casey and Buehler (1977) on a remarkably strong photoluminescence (P.L.) emitted by n-type InP surface and attributed to the low surface recombination velocity, several works have been published on the effect of surface-treatment on P.L. intensity of n-type InP (Nagai 1978, Tohru Suzuki 1979, Nagai 1979). More recently we have shown (Krawczyk 1984) that the P.L. intensity and the surface density of states in the upper part of the gap of n-type InP are in the inverse ratio which makes the measurement of the P.L. be a simple and efficient method for monitoring each individual technological step of fabrication of metal–insulator-semiconductor (MIS) devices on InP.

This remarkable property of the P.L. will be widely illustrated and used in this paper which addresses the problem of optimization of MIS devices on InP with a special emphasis on the interaction phenomena occuring between the insulator and the substrate during the first stages of its deposition. It is shown that the electronic properties of the surface of InP can be dramatically affected (drop of the P.L. intensity as a result of an increase of the density of surface states in the upper part of the gap) by the deposited insulator although the technique used for the deposition is rather soft (electron gun evaporation). It is further demonstrated that, even though the nature of the insulator is of some importance, its influence cannot be regarded independently of the other fabrication parameters ; among them, the chemical preparation of the substrate and the substrate temperature, although kept at a very low level (below 150°C in this study), play a major role.

Experiments are carried out on n-type, (100) -oriented InP substrates with a carrier density of about $2 \times 10^{16} cm^{-3}$. The InP substrates were prepared by first cleaning in a conventional solvent followed by a chemomechanical polishing in a bromine-methanol solution. Two types of chemical treatments, described in detail elsewhere (Sautreuil 1983), were then used : one resulting in the etching of the native oxide (HF aqueous solution), the other promoting the formation of a stable and indium-rich surface oxide (NH$_4$OH aqueous solution)(Guivarc'h 1984). The insulator (aluminium oxide Al$_2$O$_3$ and silicon oxide SiO$_x$) was deposited by electron-gun evaporation, the substrate being held at room temperature or at 150°C during the deposition. Its thickness (spanning the range 15 Å – 1000 Å) was monitored in situ and controlled ex situ by ellipsometry. The P.L. of the samples at 300 K was measured at different stages of their preparation. The conditions of measurement have been described in detail elsewhere (Krawczyk 1984). The density of interface states was measured on MIS devices with 1000 Å thick insulator using a mercury probe.

The P.L. intensity is systematically greater by a factor of about 6 for InP substrates treated in a HF solution as compared to those treated in a NH$_4$OH solution. This point is debated thoroughly in another publication (Krawczyk 1984). Figure 1 gives the normalized photoluminescence intensity of InP samples at different stages of the technological process : after heating at 150°C for about 2 hours, after deposition of Al$_2$O$_3$ or SiO$_x$ at room temperature or at 150°C and for different thicknesses, after post-deposition annealing at 300°C. For each sample the reference is the P.L. intensity of InP substrates measured immediately after the chemical treatment (HF or NH$_4$OH). The antireflecting effect of 1000 Å thick insulating layers tends to enhance the P.L. intensity by no more than 34 % which is the reflection coefficient of InP at $\lambda = 0.514$ μm (Aspnes 1983). In any event, this effect does not invalidate the conclusions of this paper, since the general trend observed is a systematic and often dramatic drop of the P.L. after deposition of a 1000 Å thick insulator. A simple heating of the InP substrates at 150°C for 2 hours results in an enhancement of the P.L. signal by a factor of 3 for the sample treated in a NH$_4$OH solution. On the other hand a dramatic drop of the P.L. by more than 2 orders of magnitude is observed for the samples treated in a HF solution. This indicates an extreme "fragility" of InP surface after chemical etching in a HF solution, whereas the stable oxide left by the NH$_4$OH treatment proves to be rather efficient in preventing the degradation of the

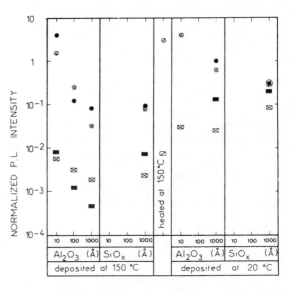

Fig.1 Normalized P.L. intensity for different insulator thicknesses and deposition conditions. For each sample the reference is the P.L. intensity measured just after the chemical treatment (HF or NH$_4$OH). Note that P.L. reference intensities are not the same for both treatments (see figure 2 for absolute values of the P.L. intensity). HF (⌀) and NH$_4$OH(∅) treated samples heated at 150°C ; HF (⊠) and NH$_4$OH (●) treated samples after insulator deposition ; HF(■) and NH$_4$OH(●) treated samples with insulator and after annealing at 300°C, 30 min under N$_2$ flow.

electronic properties of the InP surface at this temperature. This point is developed in more detail in another publication (Krawczyk 1984). The effect

of heating accounts for the major part of the degradation of the P.L. observed in HF samples after deposition of the insulator at 150°C (samples were
held at this temperature for about 2 hours in the process of deposition).
The extra degradation does not exceed a factor of 5 and gradually increases
for increasing thickness of the insulator.

A striking illustration of the specific influence of the insulator is provided by the behavior of NH_4OH sample processes at 150°C. It is clear that
the deposition of the insulator is solely responsible for the observed degradation of the P.L. Moreover, the degradation appears fully beyond a certain amount of deposited insulator. 10 Å of Al_2O_3 slightly decrease the P.L.
signal as compared to the value obtained after a simple heating of the sample. 100 Å of Al_2O_3 result in a drop of the P.L. by one order of magnitude
while 1000 Å further decrease the signal by an additional factor of about
10. 1000 Å of SiO_x deposited at 150°C result roughly in the same degradation.
Room temperature deposition is far less harmful for the electronic properties
of InP surface. Yet the detrimental influence of the insulator is clear with
NH_4OH samples if we compare the P.L. intensity measured after the deposition
of 10 Å and 1000 Å thick Al_2O_3. The observed improvement of the P.L. after
10 Å thick Al_2O_3 deposition is probably the result of the heating of the
sample by the radiating evaporation source. This effect of heating is stronger for Al_2O_3 deposition than for SiO_x deposition. On the other hand, this
heating effect is likely to be detrimental for HF samples and it is not
clear whether the drop of the P.L. after the deposition of the insulating
layers is due to the insulator. Thus, the deleterious effect of the insulator is fully observable provided that the substrate is heated during the deposition. Annealing the samples 30 minutes at 300°C under N_2 flow after the
deposition of the insulator generally improves the P.L. intensity. At the
present time, we cannot propose a definite interpretation of the effect of
annealing.

Figure 2 shows the P.L. intensity in arbitrary units of
several MIS structures (after
annealing). For comparison the
P.L. of InP substrate treated
in HF and in NH_4OH is also given. Figure 3 shows capacitance-voltage (C-V) and conductance-voltage (G-V) characteristics at different frequencies for three of the MIS
structures (Al_2O_3 deposited at
150°C on InP submitted to HF
or NH_4OH treatment, SiO_x deposited at room temperature on
InP treated in HF). A good correlation is observed between the
P.L. intensity and the density
of interface states N_{ss} (around
0.5 eV below the conduction band
as determined either from the
frequency dispersion of C-V
curves (when N_{ss} exceeds a few
$10^{11}cm^{-2}eV^{-1}$) or using the conductance technique (for $N_{ss} <$
$10^{11}cm^{-2}eV^{-1}$).

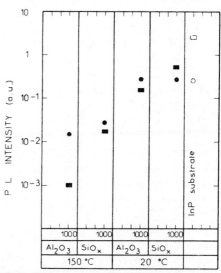

Fig.2 P.L. intensity in arbitrary units,
measured on MIS structure with 1000 Å
thick insulator, and just after the HF(□)
or NH_4OH(0) treatment ; (●, ■ as in figure
1).

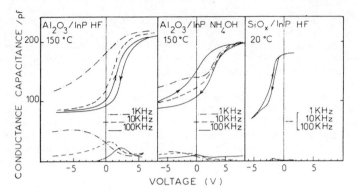

Fig. 3 Capacitance-voltage (top) and conductance-voltage (bottom) of three MIS structures of figure 2. Al_2O_3/InP HF : $N_{ss} = 10^{13}$ $eV^{-1}cm^{-2}$; Al_2O_3/InP NH_4OH : $N_{ss} = 4 \times 10^{11}$ $eV^{-1}cm^{-2}$; SiO_x/InP HF : $N_{ss} < 10^{11}$ $eV^{-1}cm^{-2}$

As shown in figure 4 the P.L. magnitude is a monotonic increasing function of the inverse of the density of states. We mention that independent measurements of the P.L. carried out on several insulator-InP structures showing up various surface band bending have led us to the conclusion that N_{ss}, rather than the surface electric field, is the main controlling factor of the P.L. intensity (Ando 1980, 1981). Aspnes (1983) interpreted P.L. data published in the literature (see Nagai 1979 for example) on the basis of the surface recombination velocity (S) depending solely on the surface Fermi level position. Our results suggest instead that the surface density of states (in the upper part of the gap) is the main controlling factor of S on n-type InP.

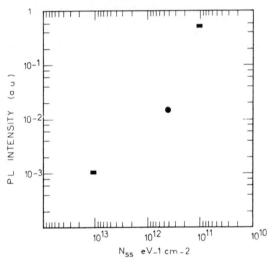

Fig.4 P.L. intensity in arbitrary units versus the inverse of the density of states estimated at 0.5 eV from the conduction band, for samples of figure 3.

The principal results of this study can be summerized as follows :

   i)- Chemical etching of InP in a HF solution results in a surface showing up remarkable electronic properties right after the treatment ; however this surface appears to be very "fragile". This statement can be extended to a wide variety of acid solutions (Krawczyk 1984).

   ii)- In any event the deposition of the insulator is deleterious. A strong interaction occurs between the InP surface and the insulator during the first stages of the deposition which systematically results in the degradation of the electronic properties of the surface.

   iii)- The observed degradation critically depends on the substrate temperature <u>during</u> the deposition. Thus, with the NH_4OH samples, the reduction of the P.L. is 10 times larger at a deposition temperature as low as 150°C by comparison with a room temperature deposition, despite the presence of the oxide (stable if simply heated at 150°C) left by the chemical treatment.

iv)- Although the nature of the insulators used in this study (Al_2O_3 and SiO_x) does not play a major role with NH_4OH samples, it appears that depositing SiO_x is significantly less harmful with samples chemically etched in HF.

v)- Our best results concern M-SiO_x-InP structures fabricated at <u>room temperature</u> after chemical etching of InP, which exhibit remarkably low density of states (below $10^{11}cm^{-2}eV^{-1}$).

Physical mechanisms responsible for the degradation observed after the insulator deposition are probably complex. Our understanding of these phenomena should be greatly improved in the light of extensive experimental and theoretical investigations published recently on Schottky barrier formation between metals and semiconductors and in particular InP. For example several authors have shown that the deposition of a metal on cleaved or chemically etched InP results in the breaking of In-P bonds owing to the heat of condensation (Williams 1979, Brillson 1981, 1982, Hökelek 1983). If the metal is not reactive (Au for example), an outdiffusion of In and P takes place resulting in an extension of the interface over tens of Angströms. On the other hand a reactive metal, such as Al for example, prevents the outdiffusion of P for forming AlP. Similar phenomena probably occur and both types of behavior may co-exist when an insulator such as Al_2O_3 is deposited. The evaporation of Al_2O_3 brings about elemental Al (Maissel 1970) which may react with InP to form AlP, whereas Al_2O_3 molecules are probably less reactive (AlO and Al_2O are also present). As it has been shown in this study, the presence of a surface oxide on the InP substrate has a deep influence on this phenomena. As a matter of fact, it has been reported by Wilmsen (1983) that Al may also react with the surface oxide. We believe that ultrahigh vacuum techniques of surface analysis, such as photoelectron spectroscopies, which have been extensively used recently for the investigation of metal-semiconductor interfaces, should provide an accurate picture of the physicochemical properties of the insulator-InP interface, very useful in the prospect of processing operative MIS devices.

Acknowledgments : This work was supported by the GRECO "Physique et Technologie des Dispositifs à Semiconducteurs Composés III-V" CNRS, France and by a contract DAII "Réalisation et Etude de Structures MIS sur InP". We are indebted to A. Regreny and R. Coquille from CNET Lannion, Lab. I.C.M., France, for providing InP substrates.

References

Ando K, Yamamoto A, Yamaguchi M 1980 J. Appl. Phys. vol. 5 n° 12 p.6432
Ando K, Yamaguchi M 1981 Jpn. J. Appl. Phys. vol. 20 n° 7 p.1335
Aspnes D E, Studna A A 1983 Physical Review B vol. 27 n° 2 p.985
Aspnes D E 1983 Surface Science 132 p.406-421
Brillson L J, Brucker C F, Katnani A D, Stoffel N G and Margaritondo G 1981 Appl. Phys. Lett. 38 n° 10 p.784
Brillson L J 1982 Thin Solid Films 89 p.461-469
Casey H C and Buehler E 1977 Appl. Phys. Lett. 30, 247
Guivarc'h A, L'haridon H, Pelous H, Hollinger G and Pertosa P 1984 J. Appl. Phys. vol. 55 n° 4 p.1139
Hökelek E and Robinson G Y 1983 J. Appl. Phys. vol. 54 n° 9 p.5199
Krawczyk S, Bailly B, Sautreuil B, Blanchet R and Viktorovitch P 1984 Electron. Lett. vol. 20 n° 6 p.255
Krawczyk S, Hollinger G 1984 to be published in Appl. Phys. Lett.

Maissel L I, Glang R 1970 Handbook of thin film technology Chap.1 p.66 Mc
  Graw Hill, New York
Nagai H and Noguchi Y 1978   Appl. Phys. Lett. 33 312
Nagai H and Noguchi Y 1979 J. Appl. Phys. vol. 50 n° 3 p.1545
Sautreuil B, Bailly B, Blanchet R, Garrigues M and Viktorovitch P 1983 Revue
  de Physique Appliquée 18 p.769-773
Tohru Suzuki and Masaki Ogawa 1979 Appl. Phys. Lett. 34 447
Williams R H, Varma R R and Montgomery V 1979 J. Vac. Sci. Technol. vol.16
  n° 5 p.1418
Wilmsen C W, Wager J F, Geib K M, Hwang T and Fathipour M 1983 Thin Solid
  Films 103 p.47-52

Inst. Phys. Conf. Ser. No. 74: Chapter 7
Paper presented at Int. Symp. GaAs and Related Compounds, Biarritz, 1984

581

Very low resistance ohmic contact fabrication for $Al_{0.25}Ga_{0.75}As$/GaAs and $Al_{0.48}In_{0.52}As$/$Ga_{0.47}In_{0.53}$ as lateral devices: a comparison

P.Zwicknagl, S D Mukherjee, W L Jones, H Lee, P M Capani, T Griem, J D Berry, L Rathbun and L F Eastman

School of Electronic   Engineering and National Research and Resource Facility for Submicron Structures, Cornell University Ithaca NY 14853

Abstract: For short gate MODFET's to operate at high frequencies, the resistance from the ohmic contact metallization to the 2-dimensional electron gas needs to be reduced to its minimum. Alloyed ohmic contacts based on AuGeNiAg metallizations with transfer resistances $\rho_t \leq 0.1\,\Omega$mm have recently been fabricated in $Al_{0.25}Ga_{0.75}As$/GaAs and $Al_{0.48}In_{0.52}As$/$Ga_{0.47}In_{0.53}As$ MODFET structures. We shall discuss the extremely low ρ_t results based on transmission line measurements, metallization penetration depth studies based on sputter Auger analysis and the resulting smooth surface morphology, grain structure and composition with the help of SEM with energy dispersive x-ray analysis.

1. Introduction

For short gate modulation doped FET's (MODFET's, HEMT's) to operate at high frequencies one needs to reduce the resistance from the ohmic contact metallization to the 2-dimensional e-lectron gas (2DEG) to its minimum. To achieve this the metalli-zation should (1.) penetrate through the high band gap materi-als ($Al_{0.25}Ga_{0.75}As$ or $Al_{0.48}In_{0.52}As$) to reach the low band-gap materials (GaAs or $Ga_{0.47}In_{0.53}As$) underneath, and then (2.) form extremely low resistance ohmic contacts with the low bandgap material, and (3.) maintain a surface morphology smooth enough to facilitate easy alignments with no sideways spreading which could short out small geometry devices.

Alloyed ohmic contacts based on AuGeNiAg have recently been fabricated in both systems by controlling the alloy cycle and the Ge/Au and Ag/Au ratios. In this paper we shall compare Au-GeNiAg based ohmic contacts made to AlGaAs/GaAs and AlInAs/Ga-InAs systems with the help of transmission line measurements (TML), sputter Auger analysis,scanning electron microscopy with energy dispersive x-ray analysis investigations· We shall also attempt to describe why the two alloy temperature ranges differ from each other by about 100-150 K using simple metall-urgical considerations.

2. Experimental

Typical MBE grown MODFET heterostructures for the two systems are shown in the inserts of Fig. 1a and b. Hall mobilities and sheet carrier concentrations for AlGaAs/GaAs were: μ_H=6000 -

$8000cm^2/Vs$; n_s=5-6x10^{11}cm^{-2} at 300K, μ_H = 50000-90000cm^2/Vs; n_s= 5-6x10^{11}cm^{-2} at 77K, and for the AlInAs/GaInAs structures were μ_H = 50000-90000cm^2/Vs; n_s= 1.6x10^{12}cm^{-2} at 300K, μ_H= 40000-50000cm^2/Vs, n_s=1.5x10^{12}cm^{-2} at 77K. In one other experiment the AlGaAs/GaAs wafers were implanted with 80KeV Si$^+$ and annealed at 915°C for 15 min. Activation of ≈40%, ≈30% and ≈15% were obtained for doses 2,5 and 10x10^{13} ions/cm^2 with the corresponding sheet resistivities being 400 $\Omega/\square$, 220 $\Omega/\square$ and 400 $\Omega/\square$.

These two experiments were aimed at establishing metallurgies and alloy conditions for contacting the n$^+$ low bandgap material beneath the large bandgap material so that the optimized process   could be applied to MODFET structures.

Based on ohmic metallization used by Braslau et.al (1967) for GaAs,a modified system of Ni(100Å)/Au-Ge(20at%)(800Å)/Ag(1000 Å)/Au(1000Å)-hitherto referred to as M1- was used for the AlGaAs/GaAs system.The Ag was used ostensibly as a diffusion barrier between the AuGe layer close to the semiconductor and the Au layer on top.It turned out that according to Mukherjee et. al(1984) Ag alloyed extensively with all the Au in the layer while inhibiting Ga incorporation in and outdiffusion through Au.

In an earlier experiment the combination M1/AlInAs/GaInAs was investigated by Capani et.al(1984). The well-known Ag-In affinity was found to cause extensive In outdiffusion and Ag indiffusion causing substantial change in stoichiometry of the semiconductor beneath the metallization causing very poor contacts. Consequently a new metallization, M2, was developed for contacting the AlInAs/GaInAs system: Ni(100Å)/Ge(350-450Å)/Au(800Å)/ Ag(220Å)/Au(800Å). An attempt to use M2 for the AlGaAs/GaAs case yielded very poor results: due to the high Ge concentration (20-25at% with respect to Au) Ge-Au liquid eutectic formation at 450°-500°C caused severe surface degradation accompanied by poor ohmic contact formation.

Alloying was performed in a transient alloy furnace with temperature T_f between 600 and 800°C. The maximum sample temperature T_{max} depended upon the time the sample spent inside the furnace. After alloying, TLM measurements were done with 40 μm wide mesas and 50μm long ohmic contact pads using a four point technique.

After single alloy cycles depth profiles were obtained by AES/ sputtering using a Perkin-Elmer PHI scanning Auger microscope having a 500eV Ar sputter gun, with an sputter etch rate of 30Å/min. Scanning electron micrographs and EDX spectra were taken before and after AES sputtering to inspect surface morphologies after alloying and subsurface grain structures revealed by different sputter rates of various alloys/compounds at the bottom and the sides of the sputtered crater.

3. Results and Discussion

3.1. Alloy temperatures

Following Reeves and Harrison (1982) we contend that the specific transfer resistance ρ_t(Ωmm) is more useful a parameter for FET evaluation than the reported specific contact resistance ρ_c(Ωcm^2): ρ_t=R_txW. R_t is half the y-intercept in a TLM

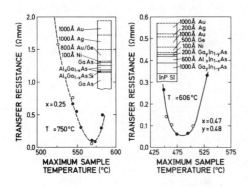

Fig.1a)Specific transfer resistance ρ_t vs. max. sample temperature T_{max} at furnace temperature $T_F = 750°C$ for as grown M1/AlGaAs/GaAs
b)ρ_t vs T_{max} for M2/AlInAs/GaInAs at $T_F = 605°C$ (O= single alloy cycle, ●= sequential alloying of a single sample)
Insert: Schematic diagram of MBE grown heterostructure with deposited metals

plot, W is the mesa width. In order to avoid the confusion and also because our experiments did not yield values of the transfer length L_T or the sheet resistance $R_{\Box C}$ beneath the metallization pad, -information necessary to evaluate ρ_c in a correct way- we express all our data as $\rho_t(\Omega mm)$ plotted against T_{max} (°C) in Fig.1.

There are two metallurgical significant observations. Firstly, the optimum alloy temperature is about 100K higher for the M1/AlGaAs/GaAs system compared with the M2/AlInAs/GaInAs system. Secondly, on interchanging the metallization very poor contacts were formed. In attempting to explain both, we suggest that the critical temperature ranges are indicative of the temperatures at which the ohmic metals are capable of penetrating the high bandgap material. A good indicator could be the melting points of the various binary alloys involved in the alloying process, as at these temperatures the various metals can easily penetrate into each other. For M1/AlGaAs/GaAs some possible alloys with their melting points obtained from Hultgren et.al (1973) are: Al-3%Ni(640°C), Au-20%Al(527°C), Ge-25%Al(424°C) etc. The average temperature e.g. ≃540°C, coincides rather well with the temperature of the onset for good ohmic contact formation ≃550°C($\rho_t < 0.3 \Omega mm$)(Fig.1a). For M2/AlInAs/GaInAs the possible alloys are: In-30-40%Au(450°C), Ge-30%Al(424°C, In-5% Ni(410°C)etc., their biased average being around 430°C close to the ohmic contact onset temperature ≃440°C.

3.2. Depth profiles and grain structures

For the M1/AlGaAs/GaAs system, AES/sputter depth profiles indicate complete Au/Ag alloying, Ni/Ge peak formation at the metal-semiconductor interface and almost no Ga outdiffusion for $T_{max} \leq 550°C$ ($\rho_t \leq 0.2\Omega mm$) with excellent surface morphology. For $T_{max} = 600°C$ with about $\rho_t \geq 0.06\Omega mm$, 7-10% atomic Ga is incorporated in the AuAg layer and the interface becomes rather diffused. For the latter sample the SEM/EDX of the surface and at ≃1500Å, ≃3000Å and≃3800Å depths are shown in Fig.2. EDX of the surface region indicate about 12% more Ag and 50% more Ga in the dark areas than in the brighter ones, which have more Au. At ≃1500Å depth the grain sizes are 2-5µm laterally and <1000Å in the vertical direction. EDX of dark areas indicate presence of Ni and Ge in Ga and As. The brighter areas at both ≃1500Å and≃3000Å depths have mostly Au and Ag.

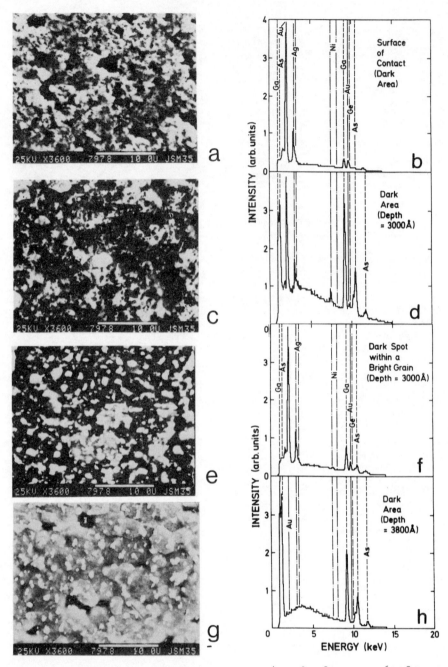

Fig.2 M1/AlGaAs/GaAs alloyed contact (T_F=800°C,T_{max}=600°C, ρ_t<
0.1 Ωmm). SEM micrographs a) surface of contact, c) grain stru
cture ≃2000Å, e)≃3000Å and g) ≃3800Å deep in the contact pad.
Selected EDX profiles of b) a dark spot at the surface, d) a
dark area and f) a dark spot within a bright grain at ≃3000Å
depth and h) a dark area at ≃3800Å depth

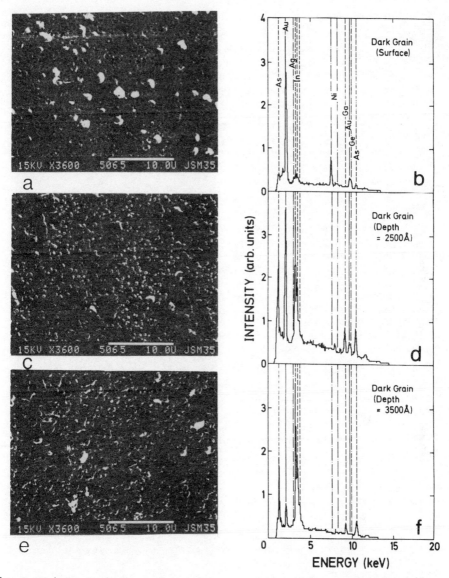

Fig.3 M2/AlInAs/GaInAs alloyed contacts at $T_F=600°C$; $T_{max}=500°C$ $\rho_t = 0.06 \ \Omega$mm. SEM micrographs a) surface c) lateral grain structure at $\simeq 2500$Å and f) at $\simeq 3500$Å depth. Selected EDX profiles of b) a dark spot at the surface, d) dark grain at $\simeq 2500$ Å and f) at $\simeq 3500$Å depth

At 4000Å the grey area is GaAs with some Au/Ag bright spots.

The grain structure depth profile corresponds well with that found in the TEM studies of Kuan et.al(1983) of AuGeNi contacts on GaAs. A good ohmic contact formation was found to be accompanied by Ni_2GeAs complex formation near the interface,in agreement with our EDX results at 3000Å depths.

AES/sputtering depth profiles for the alloyed M2/AlInAs/GaInAs
systems indicate complex interdiffusion patterns with deep Au
penetration, with Ge and Ni diffusion into larger depths. For
T_{max}=525°C, the ρ_t value increases and this is accompanied by
the dissociation of the thin (1000Å)n$^+$-GaInAs while the metals
start interacting with the InP underneath.
SEM/EDX studies of the sputter crater for the sample with T_{max}
=500°C(ρ_t≈0.06 mm) in Fig.3 show 2μm lateral diameter grains.
The bright areas are rich in Au,and yield higher Ni signal
than As and Ge. At 2500Å and 3500Å depth the dark areas consist of Au,In,As with small quantities of Al,Ga and Ge. The reaction front is much more complex and since no TEM cross section results exist, it isn't easy to draw conclusions.

Conclusion

The two systems differ from each other in a number of respects:
i) About 100K higher T_{max} is needed to contact AlGaAs/GaAs
   than AlInAs/GaInAs and/or GaAs. This is explained by a
   simplistic model based on the fact that the alloyed products have melting points differing roughly by 100K.
ii)The presence of In in the AlInAs/GaInAs system warrants considerable reduction in Ag in the ohmic metallization, the
   presence of Ag for the AlGaAs/GaAs case being required for
   the sake of reducing excessive Ga outdiffusion at higher
   temperatures.
iii) Since GaInAs is thermally less stable than GaAs,the metallization M2 was chosen for lower temperature alloying. This
   was achieved by increasing Ge:Au ratio from ≈10at% for AlGa
   As/GaAs to ≈24at% for AlInAs/GaInAs.
iv)The grain sizes are much larger than 2-5μm for the AlGaAs/
   GaAs contacts than in the AlInAs/GaInAs case (<2μm).
Further work is in progress to fabricate contacts that are substantially more temperature stable and that allow no surface
diffusion at FET operating conditions. A detailed,extended paper including various depth profiles and TML data for MODFET
structures of different kinds will be published elsewhere.

Acknowledgement

Thanks are due to G Wicks and A Chandra for stimulating discussions. This work was supported by the Joint Services Electronics Program under contract no FA 9620-81-C-0082,the Army Research Office through contract no DAAG 29-82-K-011, by an IBM
contract,two IBM Fellowship and a Bell Laboratories grant.

References

Braslau N, Gunn J B, Staples J L 1967 Solid State Electron 10
381
Capani P M, Mukherjee S D, Zwicknagl P, Berry J D, Griem H T,
Rathbun L, Eastman L F 1984 Electr.Lett. 20 446
Hultgren R, Desai P D, Hawkins D T, Gleiser M, Kelley K K 1973
Selected Values of Thermodynamic Properties of Binary Com-
pounds (American Society for Metals, Metals Park, Ohio)
Kuan T S, Batson P E, Jackson T N, Rupprecht H, Wilkie E L
1983 J.Appl.Phys. 54 6952
Mukherjee S D, Zwicknagl P, Lee H, Lepore A, Eastman L F 1984
paper presented at WOCSEMMAD'84 San Francisco
Reeves G K and Harrison H B 1982 IEEE Electr.Dev.Lett. 3 111

Inst. Phys. Conf. Ser. No. 74: Chapter 7
Paper presented at Int. Symp. GaAs and Related Compounds, Biarritz, 1984

587

Study of a real space hot-electron transfer in AlGaAs/GaAs heterostructure

A Kastalsky [1], S Luryi [2], A C Gossard [2] and R Hendel [2]

[1]Bell Communications Research Inc., Murray Hill, New Jersey 07974, USA
[2]AT&T Bell Laboratories, Murray Hill, New Jersey 07974, USA

We have studied the new effect based on real-space electron transfer
between two conducting layers in semiconducting heterostructures.
The charge injection occurs due to heating of electrons in one
of the layers separated from the other layer by a potential
barrier and contacted separately. The hot-electron layer was
realized as a high mobility two-dimensional channel at a modulation
doped GaAs/$Al_{0.3}Ga_{0.7}As$ interface. The second conducting layer
(heavily doped n^+ GaAs substrate) was isolated from the channel by
an undoped graded $Al_xGa_{1-x}As$ barrier of thickness $\sim$1500Å.

We report the results of our investigation of the real-space hot-electron
transfer between two conducting layers in an AlGaAs/GaAs heterostructure
separated by a potential barrier and contacted individually. The
physical mechanism is based on charge injection over the barrier
controlled by the electron temperature in one of the layers. It can be
illustrated (Fig. 1) by comparison with a vacuum diode whose cathode
temperature is varied by an input electrode. In such a diode the anode
current as a function of the anode voltage saturates at a level
determined by the cathode temperature and work function. Our structure,
schematically shown on the left side of Fig. 1 represents a solid-state
analog of such a hypothetical diode: the first conducting layer
("channel") plays the role of a
cathode whose temperature is
modulated by the source (S) to
drain (D) field. The second
conducting layer is the anode.
Application of a voltage V_{SD}
heats the channel electrons and
leads to an exponential
enhancement of charge injection
over the barrier (whose height
corresponds to the cathode work
function). In contrast to the
vacuum diode, charge injection in
our structure is governed by the
electron temperature T_e which
can be modulated very rapidly.

The investigated structure along
with its energy band diagram and

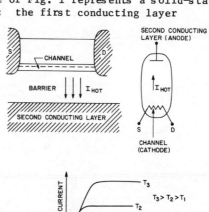

Fig. 1

contact layout is presented in Fig. 2. Details of its manufacture (MBE growth and processing) can be found in ref. 1. The channel represents a two-dimensional electron gas formed at the GaAs/AlGaAs interface (the modulation-doped heterostructure transistor). The second conducting layer is implemented as an n^+ GaAs substrate and it is separated from the channel by a graded AlGaAs barrier. It is important that the S and D contacts to the channel are insulated from the substrate.

Typical results for charge injection at 77K are shown in Fig. 3 for different substrate voltages. Consider the curve corresponding to $V_{SUB} = 1.0V$. We see that the substrate current I_{SUB} exhibits a sharp minimum when $V_{SD} \rightarrow 0$.

For $V_{SD} \leqslant 0.15V$ we have $I_{SUB} < 10^{-11}A$, which proves a good contact separation from the substrate. As the heating voltage increases, I_{SUB} rises by more than 8 orders of magnitude. The fact that the current polarity is the same for both polarities of V_{SD} and corresponds to electrons injected into the substrate is direct evidence of its hot-electron nature. Qualitatively similar characteristics were obtained at T = 4.2K.

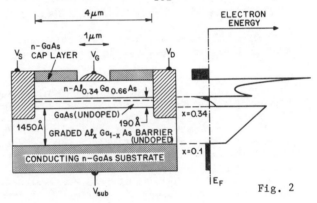

Fig. 2

Detailed analysis[1] of the characteristics in Fig. 3 allows us to determine the electron temperature T_e as a function of the heating voltage V_{SD}. For low V_{SUB} we found (Fig. 4) that $T_e \propto (V_{SD})^2$ and at a fixed V_{SD}, the T_e increases with V_{SUB}. These simple dependences no longer hold for $V_{SUB} \geqslant 1V$.

The effect of hot-electron injection in double-layered heterostructures underlies several new semiconductor device concepts.[2] One of these devices — the charge injection transistor or CHINT — is the direct solid-state analog of the hypothetical vacuum diode with controlled cathode temperature (as discussed above in connection with Fig. 1). It is a three-terminal device whose output (substrate) current is a function of the input voltage V_{SD}. Fig. 5 displays the output characteristics in CHINT with the heating voltage V_{SD} as a parameter. We have experimentally demonstrated[1] the existence of a power gain in this device. By the physical principle involved, CHINT is different from all previous three-terminal

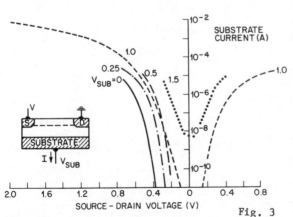

Fig. 3

semiconductor devices – all
of which can be classified
in either of the two
groups: potential modulation
and charge modulation
transistors. In the first
group (which includes the
bipolar transistor, all
analog transistors, and
also the vacuum triode) the
transistor action results
from modulating the height
of a potential barrier by a
controlling electrode. The
second group, containing a
great variety of FETs is
based on the modulation of
charge in a resistive
channel due to the
screening of an applied
field.

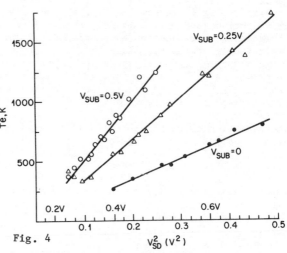

Fig. 4

In CHINT the control of output current is effected by a
modulation of T_e resulting in the charge injection over a barrier of
fixed height.

The hot-electron injection in our structure is accompanied by a strong
negative differential resistance (NDR) in the channel circuit. As
discussed below, this effect arises due to electron accumulation outside
the high-mobility channel. The idea of an NDR effect due to real-space
transfer was first proposed be Hess et al[3], as a generalization of
the Gunn effect: hot electrons transferred into a low mobility layer
give rise to a depletion of the high-mobility layer and the overall
lowering of the current. In our case the electron transfer occurs
between two conducting GaAs layers contacted separately.

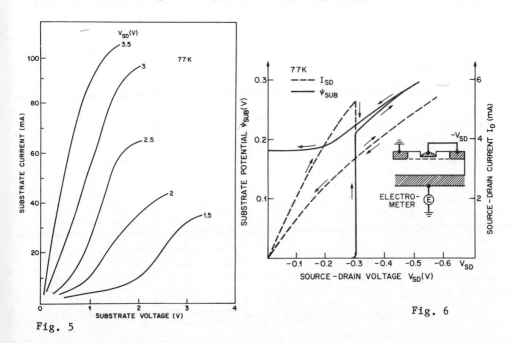

Fig. 5

Fig. 6

Consider first the situation in which the second conducting layer
(substrate) is floating, see Fig. 6. Hot-electron injection charges up
the substrate and raises its potential ψ_{SUB}. At the same time the
current I_{SD} drops due to the channel depletion. The observed curves
are strongly hysteretic indicating a memory effect. At T = 77K the
retention time of the injected charge (relaxation of ψ_{SUB} is
controlled by the leakage through the electrometer E of resistance
$10^{14}\Omega$. In this configuration the second conducting layer acts as a
"giant trap" for electrons.[4, 5]

The situation is quite different if the second conducting layer is kept
at a fixed positive voltage. In this case there is no charge
accumulation in this layer. Nevertheless, we observe[6] a strong NDR
both at low temperatures (4.2-77K) and at T = 300K.

Typical IV characteristics in the channel circuit are shown in
Fig. 7 for different substrate biases. Higher V_{SUB} enhances the
electron concentration in the channel (backgate action). The NDR appears
for V_{SUB} 2V and it is strongly affected by V_{SUB}. We believe that
the physical mechanisms of this behavior consists in the dynamical
screening effect predicted earlier.[2] Injected electrons move in the
conduction band of the AlGaAs barrier constituting a space-charge limited
current. The associated space-charge potential screens the positive
backgate voltage V_{SUB} and thus depletes the channel.

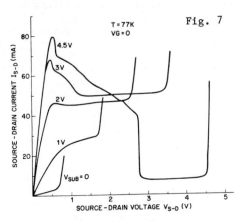

Fig. 7

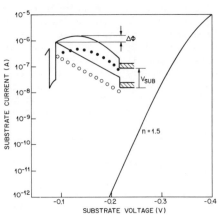

Fig. 8

The dynamical screening mechanism of NDR is extremely fast –
intrinsically limited only by the electron transit time over high-field
regions of the sample (distances of order 10^{-5} cm). The same speed
limitation applies to the operation of CHINT, as discussed in ref. 1.

A parasitic speed limitation may also result from electron trapping in
the barrier region. Indeed, we have independent evidence for the
existence of donor-like traps in the nominally undoped AlGaAs barrier at
the level of
$1-2 \times 10^{16}$ cm^{-3}. This evidence comes from the analysis of the
forward-bias IV characteristic of the substrate-to-channel diode,
Fig. 8. For an ideal triangular barrier of the form shown in Fig. 1 one
would expect an ideality factor n ~ 1 (neglecting the channel thickness).
However, the curve in Fig. 8 corresponds to n = 1.5 at low currents.
This curve was measured in the dark and after prebiasing the diode at

0.5V in the forward direction. Under these conditions the curve is
stable and non-hysteretic. We believe that after the initial forward
biasing the barrier traps are fully charged changing the shape of the
barrier as shown in the insert to Fig. 8. The experimental curve can be
fitted by assuming a uniform density of electrons trapped in the barrier
equal to $\sim 1.5 \times 10^{16}$ cm^{-3}. Besides, the curve in Fig. 8 allows us
to estimate the barrier height $\Phi \cong 0.32$ eV which is somewhat higher than
one would expect for a GaAs/Al$_{0.34}$Ga$_{0.66}$As heterojunction (assuming
the recently proposed[7] conduction-band discontinuity rule $E_c \approx$
$0.6 \, \Delta E_{gap}$ the expected $\Phi \cong 0.25$ eV). The difference $\Delta \Phi \approx 0.07$ eV can
thus be attributed to the space-charge potential of trapped electrons.
Since such a situation persists for a long time (hours) it means that the
release of electrons from traps cannot contribute to a hot-electron
transfer at T = 77K and therefore cannot impede the speed of NERFET. On
the other hand, at T = 300K the reemission of electrons from traps is
very efficient and the total amount of charge trapped on these centers is
therefore negligible.

Finally, we performed first experiments on the microwave generation in
the NERFET.[8] In an untuned microwave circuit at 77K, the NERFET was
found to generate wideband noise to frequencies up to 2.3 GHz. In a
tunable resonant circuit, stable microwave oscillations were observed at
frequencies as high as 1.45 GHz with efficiency of $\sim 5\%$. We believe that
the frequency cutoff in these experiments was due to our high-frequency
setup rather than to an intrinsic device limitation.

REFERENCES

1. S Luryi, A Kastalsky, A C Gossard, R Hendel, IEEE Trans. on Electron
   Devices, ED-31, p.832 (1984).

2. A Kastalsky and S Luryi, IEEE Electron Device Letters, EDL-4, p.334
   (1984).

3. K Hess, H Morkoc, H Shichijo, B G Streetman, Appl. Phys. Lett. 35,
   469 (1979).

4. P J Price, IEEE Trans. on Electron Devices, ED-28, p.911 (1981).

5. M Keever, K Hess, M Ludowise, IEEE Electron Device Letters, EDL-3,
   p.297 (1984).

6. A Kastalsky, S Luryi, A C Gossard, R Hendel, IEEE Electron Device
   Letters, EDL-5, p.57 (1984).

7. R C Miller, D A Kleinman and A C Gossard, Phys. Rev. 29, 7085, (1984).

8. A Kastalsky, R A Kiehl, S Luryi, A C Gossard and R Hendel,
   IEEE-EDL-5, 321 (1984).

Inst. Phys. Conf. Ser. No. 74: Chapter 7
Paper presented at Int. Symp. GaAs and Related Compounds, Biarritz, 1984

593

Luminescence of hot carriers in the base of an AlGaAs/GaAs HBT

Tadao ISHIBASHI, Hiroshi ITO, and Takayuki SUGETA

Atsugi Electrical Communication Laboratory, N.T.T.
1839 Ono, Atsugi-shi, Kanagawa 243-01, Japan

<u>Abstract</u> Luminescence from the bases of AlGaAs/GaAs HBTs has been measured for devices with both abrupt and graded emitter structures. The carrier temperatures, evaluated at room temperature, from the high energy exponential tails of the luminescence spectra have dependencies on the background hole concentration and the conduction band discontinuity, ΔEc, at the emitter base interface. A carrier temperature of ~ 500 K has been estimated for the abrupt emitter HBT with 5×10^{18} /cm^3 doped base. Effective energy relaxation time is as short as 100 fsec for a 1000 Å base layer thickness.

1. Introduction

In AlGaAs/GaAs heterojunction bipolar transistors (HBTs) with abrupt emitter base structures, high velocity electron injection into the base through a potential drop at the interface is expected to reduce the base transit time. The average electron velocity in such an injection mechanism has been calculated based on "near ballistic" motion of electrons which takes account of optical phonon scattering as a main interaction process (Ankri et al. 1982). It has been predicted that injected ballistic electrons over a potential spike will reach a velocity of $\sim 10^8$ cm/sec. This is far higher than the $10^6 \sim 10^7$ cm/sec by diffusion across the base. High frequency measurements have demonstrated electron velocity enhancement in the bases of AlGaAs /GaAs HBTs, although the effect of the hot carriers was not as significant as that expected for "ballistic" motion (Ankri et al. 1983, Ito et al. 1984a). On the other hand, in highly photoexcited bulk AlGaAs, extremely fast energy relaxation time of excited states due to carrier carrier and optical phonon scattering has been observed by Tang et al. (1983). The measured life time for the states 160 meV above the conduction band edge in AlGaAs was below 50 fsec for excited carrier density near 10^{19} /cm^3, at room temperature. Background carrier density dependence of carrier temperature in photoexcited InGaAs has also been reported (Sah et al. 1982). These results suggest the importance of carrier-carrier scattering in heavily doped GaAs base layers of HBTs.

In this work, the luminescence spectrum of recombination radiation in the base, which provides the electron energy distribution , was measured to investigate electron transport in the base.

2. Experimental

Epitaxial layers of HBTs were grown by MBE on semi-insulating GaAs substrates with (100) orientation at a substrate temperature of 650 °C. The dopants used were Si and Be. In order to prevent the Be diffusion into the AlGaAs emitter layer, we employed an undoped GaAs spacer layer at the emitter base interface. The epitaxial layer parameters of the HBTs are listed in table I. Three types of HBTs, in which Type I and II have abrupt emitter structures and Type III has a graded one, were fabricated. In the graded emitter HBTs, Type III, the parabolic grading of AlAs fraction (Hayes et al. 1983) from x= 0.1 to x= 0.3 was utilized. To make contacts for the emitter, base and collector, the HBT wafers were recessed by chemical etching, then AuGe /Ni/Ti/Au for the emitter and collector and Cr/Au for the base were evaporated, followed by metal sintering at 370 °C for 30 sec (Ito et al.1984b). Isolation between the electrodes was obtained by ion implantation of protons or by deposition of SiN film. The emitter base junction areas were 4.5x40 μm^2 or 16x90 μm^2. Maximum current gains obtained for the HBTs were 30~120 depending on base doping, device type and device size. Emission from the HBT was measured in normal direction to the device plane through the emitter layer without ohmic metal. To detect luminescence, a GaAs photocathode photomultiplier was used with a lock-in amplifier.

Table I Epitaxial layer parameters of fabricated HBTs.

Type	I	II	III
n^+ GaAs cap	0.2 (μm) 3 E18 (/cm^3)	0.2 3 E18	0.2 3 E18
N AlGaAs emitter	x = 0.3 0.1 (μm) 5 E17 (/cm^3)	0.3 0.1 5 E17	0.3 0.07 5 E17
N AlGaAs grading	-------	------	x = 0.3 to 0.1 0.03 (μm)
undoped GaAs	100 (Å)	50	100
p^+ GaAs base	900 (Å) 2 E19 (/cm^3)	950 5 E18	900 1 E19
n GaAs collector	0.3 (μm) 1 E17 (/cm^3)	0.3 1 E17	0.3 1 E17
n^+ GaAs buffer	1.0 (μm) 3 E18 (/cm^3)	0.7 3 E18	1.0 3 E17

3. Results and Discussion

The luminescence spectra observed at room temperature for Type I and Type II HBTs with abrupt emitters, for constant base current in the common emitter bias condition is shown in Fig.1. The spectra are from near band

to band recombination emission of GaAs. Luminescence associated with
AlGaAs has not been observed, which implies that hole injection from the
base into the AlGaAs emitter is negligibly small. As the electron density
injected into the base is low ($\sim 10^{15}$/cm^3) in the measured region compared
with that of background holes, the luminescence intensity reflects the
energy distribution of the injected electrons averaged over the neutral
base region. As seen in the Fig.1, the spectra show a Maxwellian carrier
distribution with high energy exponential tails, which are characterized by a
carrier temperature Tc (Shah 1978). It must be remembered, however, that
the luminescence is contributed from the whole base area and that the
carrier density distribution is not uniform. In Fig.1 (a), the spectra for
a Type I HBT with a p = 2 x 10^{19}/cm^3 doped base is shown, where the base
current was kept constant at I_B = 60 μA and the collector bias voltage,
Vce, was varied. When Vce is 0.33 V, which is just above the collector
turn on voltage of 0.30 V, Tc is 296 K. Upon increasing Vce, Tc rapidly
increases then saturates at around 410 K for Vce above 0.5 V. For a Type
II HBT with base doping of p = 5 x 10^{18}/cm^3, a similar Tc increase was
observed as shown in Fig.1 (b), where the saturated Tc value of 503 K is
much higher than that in the Type I HBT.

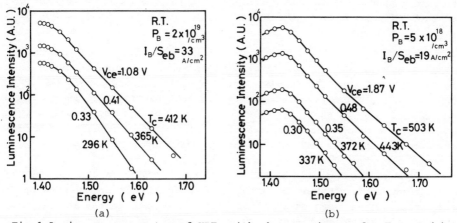

Fig.1 Luminescence spectra of HBTs with abrupt emitters for Type I (a)
and Type II (b). Seb is emitter base junction area.

These saturation tendencies of Tc during Vce variation is more clearly seen
in Fig.2. Here, the collector currents of the HBTs are also shown versus
Vce. The effect of lattice temperature increase is negligibly small because
the Tc in the saturation region shows little change with Vce variation
shown in Fig.2 and also with the collector current. Therefore,the observed
broadening of the spectra is attributed to the carrier heating in the base
due to the effect of high energy electron injection from the AlGaAs emitter
into the base through the potential spike at the interface. It is found
that the saturated Tc of Type II HBT with base doping of p_3 = 5 x 10^{18}/cm^2
is much higher than that of Type I HBT with 2 x 10^{19}/cm^3 doping. This
doping dependence of Tc indicates the difference in energy loss rate of

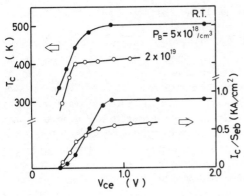
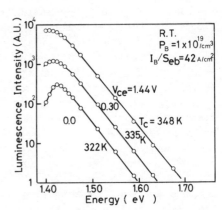

Fig.2 Tc and Ic dependencies on Vce   Fig.3 Luminescence spectra of HBTs
for Type I (o) and Type II (●).   with graded emitters, Type III.

injected electrons by electron-hole interaction. The transition range in
Tc curves correspond to the saturation region of transistors. At these
regions, observed spectra include the emission around the base electrode in
a forward biased collector base junction. Therefore, measured luminescence
from an HBT biased only in the active region gives pure information about
internal base layers.

In graded emitter HBTs with a lower ΔEc at the emitter base interface
(Type III), no drastic Tc increase for Vce variation has been observed as
shown in Fig.3. For Vce variation from 0 V to 1.44 V, the Tc increase was
from 322 K to 348 K. The lower value of Tc for Type III HBTs compared with
that for the abrupt emitter HBTs can be explained by a smaller conduction
band discontinuity of Δ Ec = 0.11 eV at the interface.

We also measured the spectra of HBTs at 77 K. The spectra of Type I HBTs,
where Vce was kept constant at 0.8 V in the active region can be seen in
Fig.4. At 77 K, the spectra do not show the Maxwellian carrier
distribution. Although the exponential fall off for the main peak (1.48
eV) retains its spectral shape for the base current increase, the weak
emission peaks in high energy shift toward higher energy. The reason for
this is not clear at present. When the base current is 32 µA, this peak is
around 1.65 eV. This higher energy band emission is not due to the near
band to band emission in the AlGaAs emitter as its band gap energy at 77 K
is 1.88 eV. Such a weak emission peak is probably due to the high energy
electron injection.

As described above, the carrier temperature in the base, Tc, depends on the
background hole density. We will estimate the effective energy relaxation
time, $\mathcal{T}$ eff.,for the base charge by a simple analysis. At steady state,
the kinetic equation for the total electron energy E in the base is given
as follows

$$\left(\frac{dE}{dt}\right)_{injection} + \left(\frac{dE}{dt}\right)_{scatt.} + \left(\frac{dE}{dt}\right)_{diff.} = 0 \qquad (1)$$

For simplicity, we approximate the hot carrier distribution as a Maxwellian with carrier temperature Tc, throughout the base region. Energy flow from the emitter through emitter current Je, and energy loss by collector current Jc are given as

$$\left(\frac{dE}{dt}\right)_{injection} = \frac{Je \ (\Delta Ec - qV')}{q} \qquad (2)$$

and

$$\left(\frac{dE}{dt}\right)_{diff.} = - \frac{Jc \ (1.5 \ k \ Tc)}{q} \qquad (3)$$

where, q is electron charge and V', the band bending in the GaAs base at the emitter base interface. The energy loss rate of the hot carriers in the base can be approximated as

$$\left(\frac{dE}{dt}\right)_{scatt.} = - \frac{1.5 \ k \ (Tc - Tlatt.)}{\mathcal{T} \ eff.} \ \frac{Qb}{q} \qquad (4),$$

where Tlatt. is the lattice temperature. Base charge Qb is given as

$$Qb = Je \ W^2 / 2 \ De$$

where W is neutral base layer thickness and De, the diffusion constant for hot carriers,

$$De = \mu \ k \ Tc \ / \ q$$

where μ is electron mobility in the base and k, Boltzmann's constant. Substituting eqs. (2) to (4) into eq. (1) and approximating Je = Jc under the condition of low electron hole recombination rate, obtains

$$\frac{1}{\mathcal{T} \ eff.} = \frac{(\Delta Ec - 1.5 \ k \ Tc)}{1.5 \ k \ (Tc - T_{latt.})} \ \frac{2 \ \mu \ k \ Tc}{W^2 \ q}.$$

In fabricated HBTs, there is small uncertainty in qV', because we did not measure the precise Be doping profile in the base. Here, we use a collector turn on voltage of 0.3 V in evaluating the effective potential

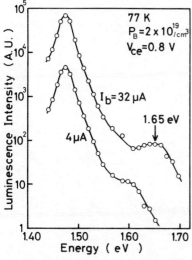

Fig.4 Luminescence spectra of HBTs, Type I measured at 77 K.

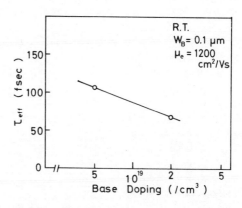

Fig.5 Effective energy relaxation time $\mathcal{T}$ eff. in the base of abrupt emitter HBT versus hole concentration.

height. Electron energy relaxation time τ eff. at room temperature in the base is deduced to be 68 fsec for Type I HBT and 107 fsec for Type II HBT assuming μ = 1200 cm^2/V sec as shown in Fig.5. This τ eff. dependence on doping is reasonable, because both a lower background hole density and a higher carrier temperature give a smaller energy loss rate for the hot carriers according to carrier carrier interaction (Conwell 1967). The less sensitive hole concentration dependence of τ eff compared with the four fold concentration change of holes in the HBTs is presumably due to the presence of both electron – hole and electron – LO phonon interaction processes.

4. Summary

In this investigation, the luminescence spectra from the bases of AlGaAs /GaAs HBTs were measured. At room temperature, the spectra showed high energy exponential tails associated with hot electrons. For HBTs with abrupt emitter structures, carrier heating up to as high as 500 K was observed in contrast with the case for the graded emitter HBTs. This result indicates hot carrier injection through the potential spike at the emitter base interface. Carrier temperature dependence on hole concentration in the base was also demonstrated showing the existence of electron-hole interaction. At 77 K, the spectra did not show Maxwellian carrier distribution, but did show two peak structures, where the higher energy peak was at around 1.65 eV. Effective energy relaxation time in the base τ eff. was evaluated from the observed carrier temperature by simple analysis. τ eff. was 68 fsec and 107 fsec for HBTs with 2 x 10^{19} and 5 x 10^{18}/cm^3 doped bases, respectively.
In conclusion, it has been shown for HBTs with abrupt emitter structures that shorter base transit times due to hot electrons is expected although "ballistic" motion of electrons can hardly be realized for heavily doped bases even with a 1000 Å base layer thickness.

Acknowledgement   The authors wish to thank T. Honda and Y. Yamauchi for HBT fabrication. They would also like to thank M. Fujimoto, T. Ikegami and T. Izawa for useful discussions and suggestions.

References
Ankri D and Eastman L F 1982 Electron. Lett. 18 750
Ankri D, Schaff W J, Smith P and Eastman L F 1983 Electron. Lett. 19 147
Ito H, Ishibashi T and Sugeta T 1984a IEEE EDL5 214
Tang C L and Elskine D L 1983 Phys. Rev. Lett. 51 840
Shah J, Nahory R E, Leheny R F, Degani J and DiGivanni A E 1982 Appl.
    Phys. Lett. 40 505
Ito H, Ishibashi T and Sugeta T 1984b Jpn. J. Appl. Phys. 23 L635
Shah J 1978 Sol. St. Electron. 21 43
Conwell E M 1967 Sol. St. Phys. Suppl. 9 (Academic Press) 12

Inst. Phys. Conf. Ser. No. 74: Chapter 7
Paper presented at Int. Symp. GaAs and Related Compounds, Biarritz, 1984

599

Optimization of the injection efficiency (γ) of the HBT by studying electroluminescence (EL) of MBE heterojunction diodes

P. Enquist, L.M. Lunardi, D.F. Welch, G.W. Wicks,
J.R. Shealy, L.F. Eastman and A R Calawa*
School of Electrical Engineering, Phillips Hall,
Cornell University, Ithaca, NY 14853 USA
* Cornell lecturer, permanent address: MIT Lincoln Laboratory, Lexington
MA 02173, USA

The injection efficiency of $nAl_{0.22}Ga_{0.78}As/p^+GaAs$ heterojunction
diodes is experimentally investigated with electroluminescence (EL).
It is shown that simple $nAl_{0.22}Ga_{0.78}As/p^+GaAs$ heterojunctions do
not perform as well as expected theoretically due to surface
diffusion of Be into the $nAl_{0.22}Ga_{78}As$ (AlGaAs) during growth. The
introduction of 200 Å of undoped GaAs at the heterojunction provides
a buffer against surface diffusion of Be, increasing the injection
efficiency (γ) to that expected for a simple $nAlGaAs/p^+GaAs$
heterojunction.

Introduction

The $nAlGaAs/p^+GaAs$ emitter/base heterojunction is the cornerstone of
the heterojunction bipolar transistor because it avoids the tradeoff
between low base resistance and high γ. Low base resistance is
necessary to obtain power gain at microwave frequencies while high γ is
needed for current gain. High γ and low base resistance can be
obtained because the barrier to hole injection is increased by the
valence band discontinuity. This barrier exponentially decreases the
hole current relative to the electron current by (Kromer, 1982)

$$\frac{I_n}{I_p} = \frac{N_{AlGaAs}}{N_{GaAs}} \frac{V_{GaAs}}{V_{AlGaAs}} e^{\Delta E_v/kT} \qquad (1)$$

where

$I_{n(p)}$ = injected electron (hole) current
$N_{AlGaAs(GaAs)}$ = doping of AlGaAs (GaAs)
$V_{AlGaAs(GaAs)}$ = velocity of holes (electrons) in AlGaAs (GaAs)
$\Delta E_{v(c)}$ = valence (conduction band) discontinuity

This ratio can be increased by $\sim e^{\Delta E_c/kT}$ if the AlGaAs at the
heterojunction is properly graded since this lowers the barrier for
injected electrons. Neglecting recombination at the junction, the
ratio of electron to hole current determines γ for an abrupt
heterojunction.

$$\gamma = (1 + \frac{I_p}{I_n})^{-1} \qquad (2)$$

Although the theoretical expectation of a high γ is well understood,
experimental verification of γ is difficult due to the problem of
distinguishing between hole and electron current. Previous

determination of γ has depended upon measurement of the current gain
with appropriate assumptions about the base transport factor. This
problem is overcome by using EL as a measure of γ. This technique
consists of identifying AlGaAs EL with undesired current and GaAs EL
with desired electron current injected into the GaAs.

EL spectra of simple nAlGaAs/p+GaAs heterojunctions has shown more
AlGaAs EL than is expected theoretically. A high γ emitter with hole
current two orders of magnitude less than the electron current should
exhibit at least four orders of magnitude less AlGaAs EL than the GaAs
EL because of the lower internal quantum efficiency of the AlGaAs
compared to the GaAs. The reason for the large amount of AlGaAs EL
observed in this study is shown to be due to a thin layer of p+AlGaAs
at the heterojunction. This undesired p+AlGaAs is formed by surface
diffusion of Be into the nAlGaAs, destroying γ by converting an
efficient heterojunction into an inefficient homojunction. Due to the
greater doping in p-type vs n-type material, less than 50 Å of the p-
type material is depleted. The undesired formation of an AlGaAs
homojunction thus occurs with less than 50 Å of conversion from n-type
to p-type AlAs.

Experimental
All material was grown in a Varian Gen II MBE machine. Si and Be were
used as dopants. Zn doped $1 \times 10^{18}/cm^3$ and Si doped $1 \times 10^{18}/cm^3$
horizontal Bridgeman substrates were used. Substrate preparation was
standard and substrate temperature during growth was $620^oC \pm 10^oC$. CW
EL and photoluminescence spectra were taken at room temperature from a
cleaved facet.

Experimental Results
The EL spectra of an abrupt nAlGaAs/p+ GaAs heterojunction diode is
shown in Fig. 1. The source of the
AlGaAs EL was investigated by altering
the abrupt heterojunction structure to
include either an undoped GaAs layer or
a thin $4 \times 10^{18}/cm^3$ p+AlGaAs layer.
The ratio of AlGaAs to GaAs EL (R_C) vs.
current density for these structures is
shown in Fig. 2. The ideality (n) of
the diodes is also included. A thinner
undoped GaAs layer results in an increased R_c as does a thicker p+AlGaAs
layer. The increase of R_C at high current densities is most likely due to a
heating effect. The dependence of R_c
on the doping of the GaAs is shown in Fig.
3. The experimental data do not have the
linear dependence on base doping expected
from the theoretical relation (Eq. 1),

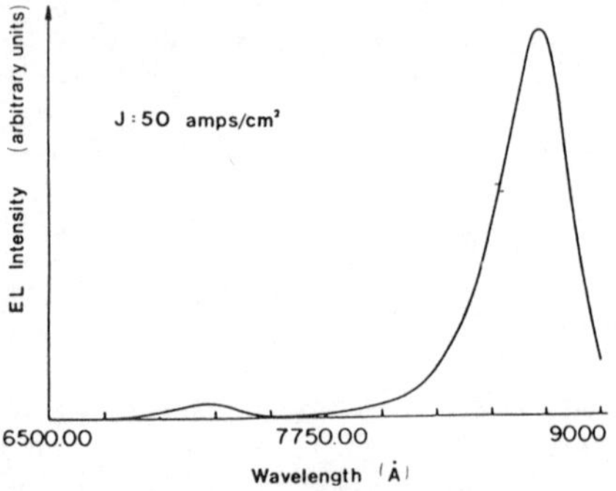

J:50 amps/cm^3

6500.00   7750.00   9000
Wavelength (Å)

Fig. 1. EL spectra of
abruptly doped diode.

indicating that the AlGaAs EL is not explained by the ideal
heterojunction. Fig. 4 shows the EL spectra from abruptly doped
heterojunction diodes with and without 100 Å of AlGaAs grading at the
junction. The p+GaAs was doped at $1 \times 10^{19}/cm^3$. The same AlGaAs
grading on diodes with GaAs doping of $5 \times 10^{18}/cm3$ succeeded in
eliminating AlGaAs EL. Fig. 5 compares the EL seen from an abrupt

nAlGaAs on p+GaAs hetero-
junction to the inverted
structure, growing p+
GaAs on nAlGaAs. The
elimination of AlGaAs EL
by inverting the struc-
ture suggests an asym-
metry in the growth of
the heterojunction. An
indication of the cause
of this asymmetry is
seen in SIMS profiles of
Be doping spikes in GaAs
(Fig. 6) (Enquist et al
1984).

The relative internal
quantum efficiencies of
the GaAs and AlGaAs were
estimated by a PL
measurement. A He-Ne
laser was used for
excitation resulting in
AlGaAs and GaAs peaks
similar to those in Fig.
1. After correcting for the
amount of light absorbed in
the AlGaAs and GaAs, relative
internal quantum efficiencies
of about 250 to 1 (GaAs to
AlGaAs) were obtained.

Fig. 2. Ratio of AlGaAs to GaAs El
vs. current density as a function
of the heterointerface.

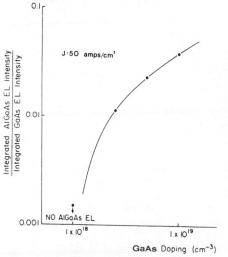
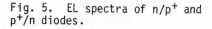

Fig. 3. Ratio of AlGaAs to GaAs EL
vs. GaAs doping.

Fig. 5. EL spectra of n/p+ and
p+/n diodes.

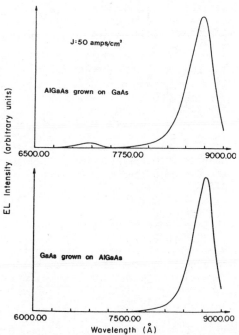

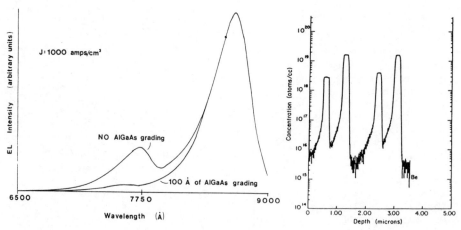

Fig. 4. EL spectra of diodes with and without 100 A° of AlGaAs grading.

Fig. 6. SIMS profiles of Be spikes in GaAs.

Discussion

As indicated earlier, the amount of AlGaAs EL observed is greater than that expected for an ideal heterostructure. The most likely source of the AlGaAs EL is p^+AlGaAs resulting from diffusion of Be from the p^+ GaAs into the nAlGaAs. The band diagram for this diode is shown in Fig. 7, assuming 200 Å of 4×10^{18}/cm^3 doped p^+AlGaAs at the hetero-junction. Since the p^+AlGaAs is doped much heavier than the nAlGaAs the depletion depth into p^+AlGaAs is less than 50 Å at zero bias. The heterojunction is

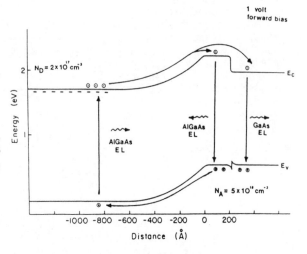

Fig. 7. Band diagram of diode with 200 Å of p^+AlGaAs at heterointerface.

thus degraded to an n/p$^+$ AlGaAs homojunction in series with a p^+AlGaAs/p^+GaAs heterojunction. The γ of this structure is severely decreased in two ways. Because the junction has now become a homojunction, the barrier for hole injection is decreased by the removal of the valence band discontinuity. There is also the added possibility of recombination of electrons injected from the nAlGaAs into the p^+AlGaAs.

The trends in Figs. 2 and 3 support the above explanation indicating

that the observed AlGaAs is due to p+AlGaAs. Fig. 2 indicates
increased AlGaAs EL for decreased undoped GaAs thickness. A 200 Å
undoped region of GaAs at the heterojunction is sufficient to eliminate
the AlGaAs EL. This indicates that under these growth conditions less
than 200 Å of nAlGaAs is converted to p+AlGaAs by movement of Be. It
also supports the assertion made earlier that AlGaAs recombination in
the ideal structure is not detectable. As the undoped GaAs layer is
thinned, Be diffusion into the nAlGaAs increases resulting in a
thicker p+AlGaAs region which increases AlGaAs EL. Intentionally
growing p+AlGaAs at the heterojunction simulates increased Be diffusion
and continues this trend. The increasing ideality (n) factor for these
diodes with decreasing spacer thickness also suggests increasing
recombination at the heterojunction. Fig. 3 can also be explained by
diffusion of Be into the nAlGaAs if one realizes that diffusion of Be
is concentration dependent (Enquist et al 1984). Increasing the
concentration of Be at the heterojunction results in a thicker p+AlGaAs
region which increases AlGaAs EL.

The dependence of the EL spectra on AlGaAs grading at the
heterojunction also suggests that a large portion of AlGaAs EL is
occurring in p+AlGaAs. Any p+AlGaAs EL should be decreased because
electrons will be accelerated over the graded region, decreasing their
chance for recombination. In addition, EL should be spread in energy
due to the graded energy gap. Since the AlGaAs EL has these features,
p+AlGaAs EL is proably responsible.

The mechanism of Be diffusion responsible for p+AlGaAs EL is shown in
Figs. 5 and 6. It has been shown that an abruptly grown
heterojunction, nAlGaAs on p+GaAs, results in large amounts of AlGaAs
EL. However if the structure is inverted, p+GaAs on nAlGaAs, no AlGaAs
EL is seen (Fig. 5). This suggests that the diffusion of Be into the
AlGaAs does not occur by solid state diffusion but is by surface
diffusion. This theory is in agreement with SIMS profiles of doping
spikes in GaAs (Fig. 6). Abruptly grown Be spikes are abrupt on the
substrate side but exhibit tails toward the surface after the Be
shutter is closed. These tails are greater than the doping in the
nAlGaAs used in this study. This causes conversion of nAlGaAs to
p+AlGaAs which lowers the injection efficiency by degrading a
nAlGaAs/p+GaAs heterojunction to an n/p+AlGaAs homojunction. These
tails thus appear to be responsible for the AlGaAs EL seen in abruptly
grown nAlGaAs on p+GaAs heterojunction diodes by MBE.

The γ of the diodes is approximated by converting the EL ratio to a
current ratio. A first order correction includes the relative system
response to optical signals at GaAs and AlGaAs wavelengths (~ 70%) and
the relative internal quantum efficiencies of GaAs and AlGaAs (~ 250),
leading to a correction factor of ~ 175. Due to the evidence of
undesired recombination near the heterojunction (p+AlGaAs EL), it is
necessary to redefine the γ as

$$\gamma = (1 + \frac{\text{electrons injected into GaAs}}{\text{Total current}})^{-1} = (1+175 \ R_c)^{-1}$$

The results are shown in Fig. 8 where γ is plotted against the thickness of p⁺AlGaAs. It has been assumed that 200 Å of p⁺AlGaAs has been formed by Be surface diffusion. Note that the γ drops sharply if there is between 50 Å and 100 Å of p⁺AlGaAs. This is in reasonable agreement with the earlier estimate of the ability of ~ 50 Å of p⁺AlGaAs to destroy the heterojunction. However, any AlGaAs EL is undesirable. The addition of 200 Å of undoped GaAs at the heterojunction is therefore necessary to obtain a high γ.

Conclusion

We have observed a large amount of AlGaAs EL in nAlGaAs/p⁺GaAs heterojunction diodes grown by MBE at 620°C. The amount of AlGaAs EL is larger than theoretically expected for such a structure. The excess AlGaAs EL arises from the decreased γ resulting from the deterioration of a nAlGaAs/p⁺GaAs heterojunction to a n/p⁺AlGaAs homojunction. The AlGaAs El has been eliminated by the introduction of 200 Å of undoped GaAs. The success of the undoped spacer is readily explained as a buffer against surface diffusion of Be.

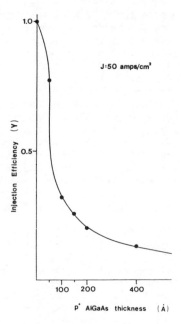

Fig. 8. Estimation of γ vs. p⁺AlGaAs thickness.

Acknowledgements

The authors thank Cornell MBE personnel for MBE support, J. Berry for technical support, W. Schaff for assistance with software and E. Weaver for preparation of the manuscript. This work was funded by the Office of Naval Research under contract no. N00014-75-C-0739.

References

Enquist P, Lunardi L M, Wicks G W, Eastman L F, Hitzman C 1984 Int. MBE Conf. to be published.
Kromer H 1982 Proceedings of the IEEE 70 1.

Inst. Phys. Conf. Ser. No. 74: Chapter 7
Paper presented at Int. Symp. GaAs and Related Compounds, Biarritz, 1984

A novel npn InGaAs bipolar transistor with a wide gap cadmium oxide (CdO) emitter

L.M. Su[*], N. Grote, H.G. Bach, W. Döldissen and M. Rosenzweig

Heinrich-Hertz-Institut für Nachrichtentechnik Berlin GmbH
Einsteinufer 37, D-1000 Berlin 10 (West)

[*] on leave from Peking Electron Tube Factory, Peking, China.

Abstract

A CdO film was deposited on an InGaAs pn junction to form an npn bipolar transistor, which can be used also as a phototransistor. The high conductivity of CdO ($\sigma = 5 \times 10^3 / \Omega cm$) reduces the emitter contact resistance and with a refractive index of $n \simeq 2.2$ at $1.15 \mu m$ wavelength the transparent film can serve as an antireflective window.
The processing of the transistor is described. A current gain of $h_{fe} = 10$ ($V_{CE} = 3V$, $I_C = 1mA$) and an emitter-collector breakdown voltage of $V_{CEO} = 6V$ were obtained.
Other results including $h_{fe}-I_C$ characteristic, dark current, I-V characteristic of the CdO-InGaAs emitter and spectral responses are presented and discussed.

1. Introduction

For new infrared ($\lambda = 1.0 \ldots 1.6 \mu m$) optical communication systems heterojunction phototransistors based on InGaAs(P)/InP materials have been considered as alternatives to avalanche photodiodes (APD's) because of low bias voltage, high gain and elimination of excess noise which results from the avalanche process in APD's.

However, all the previously reported InGaAs/InP phototransistors were fabricated as back-illuminated devices (Campbell et al. 1980, Campbell et al. 1981) mainly because of the epitaxial melt back problem, preventing the use of front-illuminated structures. Another advantage of such devices is the effective reduction of the emitter contact resistance due to the large contact area on the substrate side. However, their design flexibility for optoelectronic circuits is severely limited. Furthermore, the device process is also somewhat complicated due to the requirement of processing both sides of the wafer.

In this paper a front-illuminated InGaAs phototransistor is presented. The main feature of the transistor is the use of a non-epitaxial transparent conductor film (CdO) as the emitter (TCE) which can be considered as the further development of two previously reported devices (Su et al. 1984).

2. Device structure and fabrication

The structure of the npn CdO-InGaAs transistor is shown schematically in fig. 1.

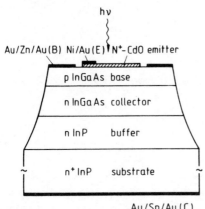

Fig. 1:

Schematic drawing of the npn CdO-InGaAs heterojunction bipolar transistor (HBT).

A 2μm thick n-InP buffer layer ($n=2\times10^{16}$ cm^{-3}), a 3μm thick n-InGaAs collector layer ($n=2\times10^{16}$ cm^{-3}) and a 0.4μm thick p-InGaAs base layer ($p=7\times10^{17}$ cm^{-3}, Zn doped) were grown sequentially on a (100) oriented n$^+$-InP substrate ($n=2\times10^{18}$cm^{-3}, Sn doped) by LPE.

A high doping level of the base reduces the base resistance r_B and consequently will improve the high frequency behaviour of the device.

Referring to the conventional wide gap emitter transistor the emitter-base capacitance C_{EB} and breakdown voltage V_{EBO} are mainly determined by the emitter doping level ($N_E<<N_B$). However, in the CdO-InGaAs wide gap emitter transistor due to the high electron concentration in the CdO emitter ($n\sim3\times10^{20}$cm^{-3}) the upper limit of the base doping level is controlled by the minimum emitter-base breakdown voltage V_{EBO}. $V_{EBO}>2-3V$ necessitates a base doping level of $N_B<7-8\times10^{17}$cm^{-3} because of the narrow band gap of InGaAs material.

Ohmic contact formation to base and collector (substrate side) was achieved by Au/Zn/Au and Au/Sn/Au evaporation, respectively, and subsequent alloying in N$_2$ at 400°C for 1 min.

The emitter was produced by sputtering of a 180 nm thick CdO film followed by an Ni/Au metallization pad. The emitter area (0.64×10^{-4} cm^2) was defined using a lift-off technique.

With a refractive index of $n\cong2.2$ (Krauser, priv. comm.) a CdO layer of d=180 nm serves as a $\lambda/4$-antireflective coating reducing the reflectivity to R=2.2% at normal incidence. The absorption of the CdO film is dependent on carrier concentration and is estimated as $\alpha\cong5\times10^3$ cm^{-1} at a wavelength of $\lambda=1.5$μm (Kocka and Konák 1971).

The specific contact resistances R_C between CdO and different metallization layers after optimum annealing conditions (in N$_2$, T=300°C, t=3min) are shown in fig. 2. For evaluation the well-known transmission line model was used.

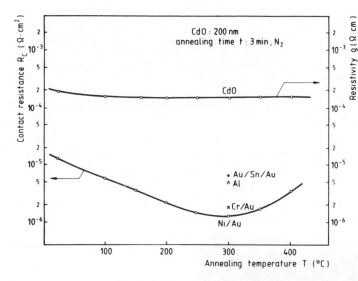

Fig. 2:

Resistivity of CdO (upper curve) and specific contact resistance R_C (lower curve) versus annealing temperature T.

Among these Ni(20nm)/Au(200nm) provides the lowest specific contact resistance of $R_C=1.5\times10^{-6}$ Ωcm^2. The resistivity of sputtered CdO, which was magnetron sputtered from a CdO-target using Ar, versus annealing temperature is also given in the upper part of fig. 2. These data combined with the carrier concentration of $n=3\times10^{20}$ cm^{-3} obtained from Van der Pauw Hall measurements yield a mobility $\mu_{nCdO}=100cm^2/Vs$.

Prior to sputtering the p-InGaAs surface in the emitter window was prepared by wet chemical etching with $HF:H_2O$ (1:100) followed by ion beam etching with $Ar:O_2$ (1:1) gas mixture at an accelerating voltage of 500 V (current density = 0.35mA/cm^2). The etching rate was about 100 Å/min. Finally wet etching using $H_2SO_4:H_2O:H_2O_2$ (5:1:1) was used to define the collector mesa ($A_c=4\times10^{-4}cm^2$).

3. Experimental results and discussion

The I-V characteristics of the CdO-InGaAs transistor are shown in fig. 3.

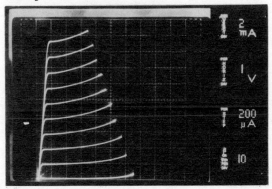

Fig. 3:

I-V characteristics of the npn CdO-InGaAs HBT.

A current gain $h_{fe}=10$ ($V_{CE}=3V$, $I_C=1mA$) was obtained. The emitter-collector breakdown voltage V_{CEO} is about 6V. The current gain ranging from unity to ten depends strongly on the surface preparation prior to CdO sputtering. Wet etching only using $HF:H_2O$ (1:100) was usually found to result in smaller current gains ($h_{fe}\sim 1...3$) which, however, can be improved by a factor of 3 by annealing in N_2 at 200°C for 5 min.
Ion beam dry etching resulted in higher current gain values ($h_{fe}\sim 5...10$) which, however, proved to be only weakly affected by annealing procedures.
Similar to other types of wide gap emitter transistors (Beneking and Su, 1982) a turn on voltage of about 80 mV (fig. 3) is observed indicating that the built-in voltage of the CdO-InGaAs emitter-base hetero-junction is larger than that of the InGaAs collector homojunction. The knee voltage is 0.3 V ($I=100\mu A$) for the collector junction and 0.4V ($I=100\mu A$) for the emitter junction as shown in fig. 4.

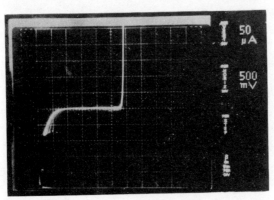

Fig. 4:

I-V characteristic of a CdO-InGaAs emitter junction

The emitter-base breakdown voltage is about 2V ($I_E=10\mu A$). The ideality factor n of the emitter heterojunction as derived from forward I-V characteristic is 1.5.
Fig. 5 shows the h_{fe}-I_C characteristic of the transistor exhibiting a constant behaviour over more than three decades of collector current ($I_C=5\times 10^{-4}A...10^{-1}A$). At low injection levels h_{fe} decreases probably due to the recombination at the interfacial states in the amorphous-crystalline interface. Nevertheless a current gain of $h_{fe}=6$ ($V_{CE}=2V$) is still obtained at very low injection ($I_C=1\mu A$) implying a surprisingly low recombination in the emitter heterojunction interface.

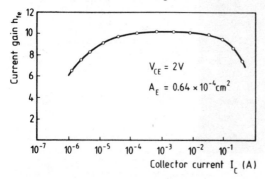

Fig. 5:

Current gain h_{fe} versus collector current I_C of the npn CdO-InGaAs HBT.

The maximum current of the transistor, on the other hand, is about $I_C=200mA$ corresponding to a current density $j_c=3100A/cm^2$. The optical gain G, defined by $G = \eta.h_{fe}=h\nu I_C/qPi$, was 3.5 and 5 at an incident optical power of $P_i=1\mu W$ and $P_i=1mW$, respectively, yielding a quantum efficiency of η =50-60% for the base-collector InGaAs diode (at λ =1.3μm and V_{CE}=2V).

The dark current I_d versus emitter-collector voltage V_{CE} is depicted in fig. 6.

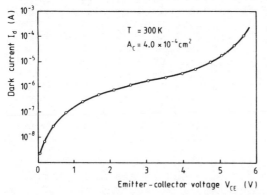

Fig. 6:

Plot of dark current I_d versus emitter-collector voltage of the npn CdO-InGaAs HBT.

Its relatively high value is due to the large base-collector junction area ($A_c=4x10^{-4}cm^2$) and should be improved when proper passivation of the sidewall of the mesa is applied. The spectral response of the transistor is presented in fig. 7 by curve A.

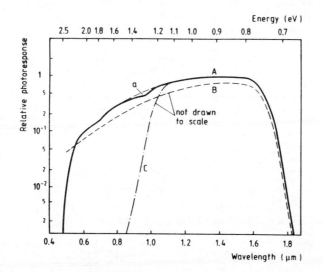

Fig. 7:

Spectral responses;
Curve A: npn CdO-InGaAs HBT
Curve B: for the same device of curve A after CdO emitter removed
Curve C: InP-InGaAs phototransistor (only for comparison).

The long wavelength cut-off is determined by the absorption edge of the narrow bandgap base and collector depletion regions. For InGaAs ternary material, this corresponds to a wavelength of approximately 1.65 μm. At short wavelengths the photoresponse is limited by absorption within the CdO emitter film caused by direct band transitions which corresponds to a wavelength of $0.54\mu m$ indicating a direct "optical" band gap $E_{gd} = 2.30eV$ in good agreement with the CdO data of Koffyberg (1976). Curve C shows the photoresponse of the ternary base transistor in the case of using n-InP as the emitter instead of CdO, thus the short wavelength cut-off is at $\lambda = 0.92 \mu m$.

A noticeable feature of curve A is the existence of a soft dip indicated by "a" located at a wavelength of $\lambda = 0.95\mu m$ ($\hat{=} E = 1.3eV$). It is supposed that this notch might be caused either by weak absorption of the indirect band gap of CdO or by standing wave interferences in the transparent conductor emitter. The value $E_{gi}=1.3eV$ would be in good agreement with the CdO band gap data $E_{gi}(\Sigma_3-\Gamma)=1.05-1.45eV$ (Koffyberg, 1976) and fits also well with the data of Su et al., 1984, where an indirect band gap of no less than 1.25eV was concluded from gain measurements on TCE-InP diffusion transistors. If the CdO is removed, the remaining photoresponse of the base-collector diode is now monotonous in this region (curve B) which indicates that the notch is associated with optical effects of the CdO emitter film.

To summarize, we have demonstrated a new InGaAs HBT using a transparent conductor (CdO) serving as a wide gap emitter. Transistors of this type can be used as discrete phototransistors at wavelengths ranging from $0.54-1.65\mu m$ and they can also be expected to exhibit good properties in microwave applications because of the high electron mobility of InGaAs. The advantages of simple processing and front-illumination render the TCE-InGaAs transistor a suitable candidate in the development of optoelectronic integrated circuits.

4. Acknowledgements

The skilled assistance of Mrs I. Tiedke in device fabrication is gratefully acknowledged.
This work was supported by the Federal Ministry for Research and Technology (BMFT) and the Senate of Berlin (West).

References

Beneking H., and Su L.M., 1982 Electron. Lett. **18** 25

Campbell J.C., Dentai A.G., Burrus C.A. and Ferguson J.F., 1980 Electron. Lett. **16** 713

Campbell J.C., Dentai A.G., Burrus C.A. and Ferguson J.F., 1981 IEEE Trans. **17** 264

Kocka J., Konák C., 1971 Phys. Stat. Sol. (b) **43** 731

Koffyberg F.P., 1976 Phys. Rev. B **13**, No. 10, 4470

Su L.M., Grote N. and Schmitt F., 1984 Electron. Lett. **20** 716

Su L.M., Grote N., Kaumanns R., Katzschner W. and Bach H.G., submitted for publication, 1984.

Inst. Phys. Conf. Ser. No. 74: Chapter 7
Paper presented at Int. Symp. GaAs and Related Compounds, Biarritz, 1984

611

Low frequency noise in GaAs MESFETs

Christos TSIRONIS*, Jacques GRAFFEUIL**, Frank HENZE***, Zahia HADJOUB**

  * Laboratoires d'Electronique et de Physique Appliquée
    3, avenue Descartes, 94450 LIMEIL-BREVANNES, France
  ** LAAS-CNRS et Université Paul Sabatier
    7, avenue du Colonel Roche, 31400 TOULOUSE, France
 *** University of Braunschweig (RFA)

INTRODUCTION

GaAs MESFETs are well known to exhibit large low frequency noise (LF noise) in excess of the thermal noise of the channel. Below a transition frequency which is between 20 MHz and 300 MHz, depending on the device characteristics, the LF noise is several orders of magnitude larger than the intrinsic thermal noise. The applications of GaAs MESFETs as ultra-wideband or DC amplifier, low phase noise oscillators and sensitive low IF frequency mixers are therefore severely limited. As an example (figure 1), a device with a good microwave minimum noise figure (0.8 dB at 4 GHz) exhibits a minimum noise figure in excess of 15 dB at 10 kHz. Moreover, if this LF noise is referred to the input terminal (gate) of the device, it can be experimentally proven (Graffeuil 1982) that a noise voltage of amplitude $\sqrt{S_{en}(f)}$, where $S_{en}(f)$ is the noise spectral density at a frequency f, produces the same frequency shift on a FET oscillator at a frequency f off the carrier as a low frequency sinusoïdal signal of equal amplitude at the same baseband frequency. Therefore LF noise is responsible for the large FM noise observed in usual FET oscillators.

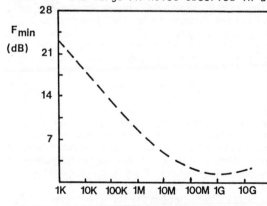

FREQUENCY / Hz

Figure 1 : Variation of the minimum noise figure versus frequency for a low microwave noise MESFET (L = 1 μm, Z = 200 μm).

In the last two years, some papers were concerned with the origins of LF noise but their conclusions are somewhat different. Let us consider the cross section of a typical MESFET shown in figure 2. There are four main regions where the noise can be expected to be generated, i.e. :

1. The free surface (with or without protection between contacts),
2. the metal semiconductor interface of the gate,
3. the bulk epilayer beyond the gate,
4. the interface between the highly doped material and the undoped region beyond.

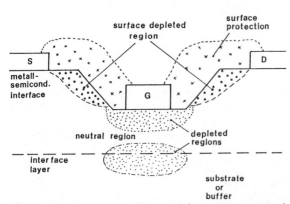

surface depleted region

surface protection

S

metall-semicond. interface

G

D

neutral region

depleted regions

interface layer

substrate or buffer

Figure 2 : Cross section showing the expected sources of LF noise.

Among the well established results, it can be mentioned that :
1. The noise is independent of gate technology (Chung-Yi-Su 1983),
2. The noise depends on the protection technique used for the free surface between contacts (Pucel 1983)
3. The noise is enhanced when no buffer is used between the highly doped material and the substrate or undoped region below (Chung-Yi-Su 1983, Ahmed 1982).

However, no clear distinction has been made up to now between the contribution of each region to the overall noise. It is the objective of this paper to establish some properties of GaAs MESFET's LF noise sources in connection with their localization and with the bias voltages.

EXPERIMENTAL SET-UP

An important goal of this work was to develop a fast and reliable characterization method of LF noise on wafer. This has been done using a point contact test fixture and a transistor noise analyzer HP 4470 A. The chip carrier is temperature controlled in the range - 10°C to 80°C. The measuring process is remotely controlled and evaluated continuously by a CBM microcomputer. The measured output noise is automatically referred back to the input terminals of the device and separated into current noise and voltage noise sources since it is essential to distinguish between them.

Indeed these two noise sources have different origins at low frequency : present paper deals with noise voltage which is the main contributor to the overall noise as far as the audio impedance connected on the gate is lower than approximately 1 MΩ or less depending on the device.

For measurements at very low drain voltages (0.1 V) a separate set-up has been used based on an ultra low noise amplifier followed by a wave analyzer. All measurements have been performed between 100 Hz and 300 kHz in a screened room.

EXPERIMENTAL RESULTS

A typical noise voltage spectrum is given in figure 3. It can be seen that it behaves as $1/f$ noise above 1 kHz and $1/f^2$ noise below 1 kHz. The $1/f^2$ noise probably denotes a Lorentzian spectrum associated with trapping and detrapping on a discrete level. Nevertheless on the present paper we will focus mostly on the noise above 1 kHz which is the more detrimental phenomenon for FET applications.

This noise amplitude, for usual bias conditions, is rather bias independent and compares well with its theoretical value given by Graffeuil (1982).

$$S_{en}(f) = \frac{V_P^2}{2 \cdot N_D \cdot Z \cdot a \cdot L} \cdot \frac{\alpha_c}{f} = \frac{q \cdot N_D \cdot a^3}{8 \cdot \varepsilon^2 \cdot Z \cdot L} \cdot \frac{\alpha_c}{f} \qquad (1)$$

where α_c is an experimental noise coefficient, Z and L are the gate width and length, a is the effective thickness of the highly doped layer, N_D the doping density, ε the GaAs permittivity, and V_p the pinch-off voltage ($V_p = q \, N_D \, a^2/2$).

It is found that the value of the noise coefficient α_c scatters between 5.10^{-4} and 5.10^{-5} (Graffeuil, 1984). To check the validity of the relationship (1) noise measurements have been performed on chips elaborated on the same wafer with different gate lengths or different gate widths. As indicated on tables I and II, the noise voltage e_n (f) in a bandwidth of 1 Hz (e_n (f) = $\sqrt{S_{en}(f)}$) scales fairly with $1/\sqrt{L}$ and $1/\sqrt{Z}$ as predicted by relationship (1).

These results support the assumption used to establish relationship (1) that the noise is generated by carrier density fluctuations in the bulk. These fluctuations are probably caused by traps and defects mostly located in the Debye region adjacent to the neutral channel (Kandiah 1983). However, since the noise parameter α_c is always in the same range 5.10^{-4} - 5.10^{-5} whatever are the technological processes involved, the question arises if some other noise mechanism independent of traps but for example in connection with any collective motion of carriers (Hooge 1969) does not occur.

Moreover, in most cases for unusual bias conditions such as near zero drain voltage or near threshold gate voltage the observed noise does not obey the present model apparently due to additional noise sources located at the active layer boundaries.

Noise ratio	Theoret. value	Measured values at		
		300 Hz	1 kHz	3 kHz
$\dfrac{e_n \ (L = 1 \ \mu m)}{e_n \ (L = 9 \ \mu m)}$	3.0	2.98	3.58	2.88
$\dfrac{e_n \ (9 \ \mu m)}{e_n \ (17 \ \mu m)}$	1.37	1.27	1.31	1.37
$\dfrac{e_n \ (1 \ \mu m)}{e_n \ (17 \ \mu m)}$	4.12	3.78	4.69	3.95

Table I : Influence of gate length on low frequency noise of implanted GaAs FET at $I_{DS} = 0.74 \ I_{DSS}$, $V_{DS} = 3$ V

Noise ratio	Theoret. value	Experim. value at	
		1 KHz	3 kHz
$\dfrac{e_n \ (Z=300 \ \mu m)}{e_n \ (Z=600 \ \mu m)}$	1.41	1.36	1.40
$\dfrac{e_n \ (150 \ \mu m)}{e_n \ (300 \ \mu m)}$	1.41	1.29	1.40
$\dfrac{e_n \ (150 \ \mu m)}{e_n \ (600 \ \mu m)}$	2.0	1.74	1.9

Table II : Influence of gate width on low frequency noise of implanted GaAs FETs at $V_{GS} = - 2$ V, $V_{DS} = 3$ V.

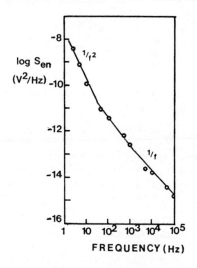

Figure 3 : Typical noise spectrum of a MESFET between 1 Hz and 100 kHz ($V_{DS} = 0.1$ V, $V_{GS} = 0$ V)

ADDITIONAL NOISE SOURCES

The bulk noise generated for a given V_p in the conductive region below the gate roughly increases with $1/V$ where V is the volume of this region. In lateral regions between contacts, the free surface depleted region modulates the current and is therefore responsible for an additional noise proportional to $1/V_s$ where V_s is the volume of the conductive region below the free surface. Therefore at normal bias conditions where V is many times smaller than V_s, the contribution of the additional noise cannot usually been observed. However, at low drain voltage and near zero gate voltage, where V and V_s are comparable, it can be seen on figure 4 that a sharp increase of noise appears. The increase is smoother at higher V_{DS}. Moreover variations of noise after a short chemical etching of the chip are observed mostly at V_{GS} and V_{DS} near zero. Finally, a slight decrease of LF noise has been obtained by increasing the contact spacings.

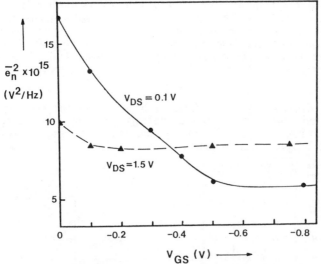

Figure 4 : LF noise voltage variations versus V_{GS}. The contribution of lateral region's noise is essential at low V_{DS} and low V_{GS}.

These observations support the assumption that this noise is generated in lateral regions in connection with the free surface. As a consequence, recessing the gate should minimize this noise contribution but a given protection would minimize this noise only if depleted region adjacent to the free surface is properly controlled. On the other hand, a sharp increase in noise is also observed for near threshold gate bias (figure 5) and present model cannot explain this statement.

Therefore, it is likely that the second space charge layer present at the interface between the highly doped layer and the high resistivity material below is responsible for this noise : if many traps are present at the edge of this space charge in the conductive region an additional noise is generated. Figure 5 shows that this noise is two orders of magnitude larger for an active layer directly implanted in a Cr doped substrate than for an epitaxial device isolated by a 2 μm thick buffer from the substrate. Since more traps are expected in a Cr doped substrate, the observed behaviour is in agreement with the previous discussion.

Further investigations of noise variations versus temperature show the existence of discrete trapping levels in the unbuffered device whereas no similar observations have been clearly made in a buffered device. It has also been found that the LF noise is influenced by the backgating voltage.

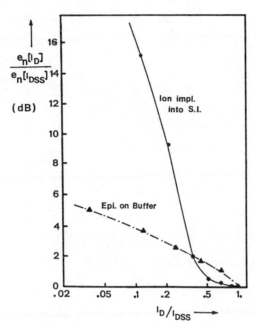

Figure 5 : LF noise voltage variations at constant V_{DS} = 3 V versus I_{DS}/I_{DSS} (I_{DS} drain current, I_{DSS} drain current at zero gate voltage) for an epitaxial buffered device and an ion implanted device into Cr-doped substrate. The contribution of interface noise at near threshold V_{GS} is essential in the ion implanted device.

CONCLUSION
From present investigations it can be deduced that in order to minimize LF noise in GaAs MESFETs great care must be taken to ensure a conductive layer as free of traps and defects as possible. The disturbed region at the interface to the substrate is also very detrimental on noise for near threshold bias conditions. Additional noise of lateral regions can also contribute mostly at low drain and near zero gate voltage but its influence can be minimized either by recessing the gate or increasing contact spacing.

Moreover a minimization is also possible by increasing gate length and gate width as far as operation frequency limitations do not occur.

Finally the origins of the major noise sources for different bias conditions are displayed on figure 6.

It is therefore evident why LF noise particularly affects the spectral purity of FET oscillators since the operation point successively goes through each region with specific noise sources.

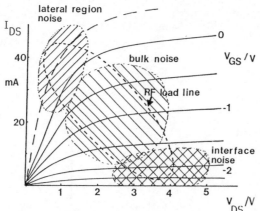

<u>Figure 6</u> : Output voltage current characteristics showing the principal
different noise origins for each position of the quiescent point.

REFERENCES

M.K. Ahmed
"Zwischenschichteinfluss auf das Rauschverhalten epitaxialer GaAs
MESFET's"
Phd Thesis, Techn. Univ. of Aachen, 1982.

Chung-Yi-Su, H. Rohdin and C. Stolte
"1/f noise in GaAs MESFETs"
Digest I.E.D.M. 1983.

J. Graffeuil, D. Sauvage, A. Amana, M. Gueguen
"Le bruit basse fréquence dans le TEC GaAs : une limitation sévère
pour certaines applications"
Digest "Journées Nationales Microondes", Lannion, France, June 1984.

J. Graffeuil, K. Tantrarongroj and J.F. Sautereau
"Low frequency noise physical analysis for the improvement of the
spectral purity of GaAs FETs oscillators"
Solid State Electronics, <u>25</u>, 5, pp. 367-374, 1982.

R.A. Pucel, and J. Curtis
"Near carrier noise in FET oscillators",
IEEE MTT-S Digest 1983.

F.N. Hooge
"1/f noise is no surface effect"
Phys.Lett. <u>29A</u>, pp. 139-140, 1969.

K. Kandiah
"Energy levels of bulk defects responsible for LF noise in Si JFETs"
Proceedings of the 7th International Conference on noise in Physical
Systems, Montpellier, France, 1983.

Inst. Phys. Conf. Ser. No. 74: Chapter 7
Paper presented at Int. Symp. GaAs and Related Compounds, Biarritz, 1984

Charge trapping in GaAs/AlGaAs modulation doped FETs

P. M. Mooney, P. M. Solomon, and T. N. Theis

IBM Thomas J. Watson Research Center, P.O. Box 218, Yorktown Heights, NY 10598 USA

Abstract

Charge trapping at low temperature which has been linked to the D-X center in n-type AlGaAs has been identified as a major problem in GaAs/n-AlGaAs selectively doped FETs. We report measurements of the pinch-off voltage shift due to charge trapping and find that our data can be interpreted by assuming a distribution of capture cross sections for the D-X center. The distribution is narrower for a lightly doped sample and the center of the distribution is a lower value. Our values of the capture cross section in samples doped to 1×10^{18} cm^{-3} are lower than those previously reported. Our results in lightly doped samples are consistent with Lang's model for the D-X center.

1. Introduction

Charge trapping, especially at low temperature, in GaAs/n-AlGaAs selectively doped FETs grown by MBE has been identified as a major problem for this device which could be a limitation for its use in large scale integration (Drummond et. al. 1983). It has been linked to the D-X center in AlGaAs (Valois et. al. 1983), which is believed to be a complex defect involving the n-type dopant species (Si, Te, etc.) and an unknown defect (X), which might be an arsenic vacancy (Lang et. al. 1979). The D-X center is found in concentrations nearly equal to the dopant concentration in n-type AlGaAs grown by MBE (Kunzel et. al. 1983), MOCVD (Wagner et. al. 1980), and LPE (Lang et. al. 1979) and its concentration varies only slightly with the crystal growth conditions. Thus this trapping center, which effectively increases the ionization energy of the donors for AlAs mole fraction > 25%, appears to be intrinsic to the doped AlGaAs system.

The D-X center is believed to be responsible for the persistent photoconductivity observed in AlGaAs. The capture cross section has an exponential dependence on the temperature so that at temperatures of about 77 K or lower the electrons cannot be recaptured once the trap has been photo-ionized. Lang et. al. (1979) have described both the thermal and optical properties of D-X centers using a configuration coordinate diagram. From the principle of detailed balance, the activation energy for thermal emission of an electron from a trap is the sum of the Hall activation energy (the binding energy of the carrier in the level) plus the activation energy of the capture cross section. The emission energy is thus much larger than the binding energy of the electron. When the trap captures an electron, the lattice distorts around the defect moving the level deeper into the gap. Thus the photo-ionization energy is larger than the thermal emission energy.

C-V measurements were used to measure the shift in pinch-off voltage due to charge trapping in long channel FETs. We show that the voltage shift depends on the bias condition while the FET is cooled to low temperature. We have also used a constant capacitance technique to monitor changes in pinch-off voltage due to charge trapping and present detailed measurements of the capture kinetics in these devices. The data can be fit by assuming a distribution of capture cross sections for the defect.

2. Sample Preparation

The modulation doped FETs used for these measurements were fabricated from MBE grown material consisting of a semi-insulating substrate with a 1 μm thick undoped GaAs layer followed by 100 Å of undoped $Al_{0.35}Ga_{0.65}As$ followed by 600-1000 Å of $Al_{0.35}Ga_{0.65}As$ doped with 1×10^{18} cm^{-3} silicon. The FETs have large area gates (175 μm by 125 μm) which give a zero bias capacitance of about 50 pF. A more lightly doped sample was prepared in the same MBE system by growing a 0.3 μm thick $Al_{0.35}Ga_{0.65}As$ layer doped with 1×10^{17} cm^{-3} Si on top of a layer of GaAs with the same doping on a conducting substrate. Schottky diodes were fabricated on this wafer. A sample of $Al_{0.35}Ga_{0.65}As$ doped with 7×10^{16} cm^{-3} Si which was prepared at the University of Illinois was also used for these experiments. DLTS spectra for all samples showed a dominant peak with thermal activation energy of ~ 0.4 eV, which has been identified as the D-X center in Si-doped $Al_xGa_{1-x}As$ (Wanatabe 1984, Zhou 1982).

3. C-V Measurements

Figure 1 shows C-V curves for a long channel FET. At 300K there is no hysteresis and the pinch-off voltage for this device is -2.7 V. The band diagram for this device at 300 K and 0 V bias is shown in Fig. 2(a). The lack of hysteresis is expected because the D-X center is thermally ionized and the carriers in the band follow the 1 MHz signal. DLTS measurements show that any other deep traps are low in concentration compared to the D-X center and can be ignored. When the FET is cooled with no applied bias, the pinch-off voltage shifts toward zero bias since electrons are trapped in the AlGaAs. Fig. 2(b) shows the band diagram for a FET under these conditions. The D-X centers are no longer thermally ionized so there are only a few carriers in the conduction band in the AlGaAs. The density of free carriers will depend on the number of shallow silicon donors and the number of shallow acceptors in the AlGaAs layer. Cooling the sample under these conditions reduces the ionized charge in the AlGaAs layer by the number of D-X centers which have trapped electrons and hence the pinch-off voltage is much smaller. The C-V curve is unchanged unless the FET is forward biased allowing additional D-X centers closer to the gate to capture electrons causing a further shift in the pinch-off voltage. The shift in pinch-off voltage can be reversed to the 300 K value by shining GaAs band gap light on the sample or by heating to above 150 K.

Charge trapping and hence a shift in pinch-off can be prevented by biasing the FET beyond pinch-off during cooling to 77 K. As shown in Fig. 2(c), when the FET is biased beyond pinch-off, carriers have been swept out of the conduction band by the electric field and are thus not available to be trapped during cool-down. The C-V curve is traced by increasing the bias to zero or positive bias (Fig. 1). When the voltage is returned to a large negative value a shift in the pinch-off voltage is observed. Reducing the reverse bias allows electrons to be trapped by D-X centers, the number trapped (shift in pinch-off) being determined by the value of the bias voltage, the time that capture is permitted to occur, and also the sample temperature, since the capture cross section of the D-X center is temperature dependent (Lang et. al. 1979).

When the sample is cooled at large forward bias, the total charge trapped can be calculated from from the shift in pinch-off from when it is cooled fully depleted. We find the amount is approximately equal to the number of silicon atoms in the AlGaAs layer. Thus, the majority of the Si atoms are incorporated as D-X centers. Depletion mode devices show more trapping than enhancement mode devices because, in the enhancement mode device, the AlGaAs is always depleted under normal bias conditions.

4. Capture Kinetics of the D-X Center

The above measurements suggest a method for studying the kinetics of the charge trapping process in a FET. Since the pinch-off voltage shift is a measure of the total amount of charge in the AlGaAs, we measure the change of the pinch-off voltage as a function of the time that the FET is biased at zero volts. The sample is cooled to the measurement temperature at a reverse

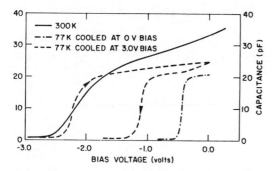

Fig. 1. C-V curves for a modulation doped FET.

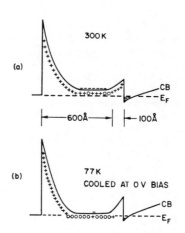

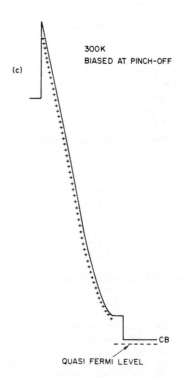

Fig. 2. Band diagrams for a modulation doped FET.

bias just smaller than that required to pinch off the channel to give a nonzero capacitance. A feedback circuit is used to return the FET to a constant capacitance after the bias voltage has been pulsed to zero and the shift in the voltage required to achieve that capacitance is measured after each pulse using a gated boxcar integrator. In the case of the Schottky diode, the sample was cooled at large constant reverse bias and then switched into the constant capacitance mode. The most lightly doped sample was photo-ionized after cooling down to the measurement temperature. Otherwise the measurements were the same. Measurements were done at sample temperatures low enough so that no charges were emitted from the D-X centers once they had been captured. The zero bias pulse duration was varied from 1 μsec to 2 msec allowing measurement times ranging from 10^{-6} to 10^2 seconds. Capture in the Debye tail of the free carrier distribution will be ignored in the discussion of our results. We know that it is negligible except perhaps for very long capture times since the value of the zero bias capacitance (and hence the zero bias depletion depth) changes only very slightly when the sample is held at zero bias for several minutes (Fig. 2). Similarly when the capture pulse is turned off, the bias voltage required to give a constant capacitance remains constant until the pulse is turned on again.

Data taken for a FET at several different sample temperatures are shown in Fig. 3. After an initial short time, the decrease in voltage required to return to the desired capacitance value, i. e. the decrease in the concentration of ionized traps, is proportional to the logarithm of the time at zero bias, i. e. the time during which capture occurs. For a trap which is present in large concentration, the fraction which is ionized determines the electron concentration in the conduction band. If this trap were characterized by a single capture cross section should have a decrease in the concentration of ionized traps should decrease in proportion to $1/t$. Thus it appears that our results are inconsis-

tent with a single value for the capture cross section. For the D-X center, the capture cross section has the form
$\sigma = \sigma_\infty \exp(-E/kT)$, where E is the activation energy. Assuming a uniform distribution of activation energies for the ensemble of traps, between energies E_1 and E_2, we find a decrease in the concentration of ionized traps which is proportional to the logarithm of the capture time. Details of this model will be published elsewhere. Due to the factor of kT in the exponent, the distribution narrows and shifts to larger values as the sample temperature is increased.

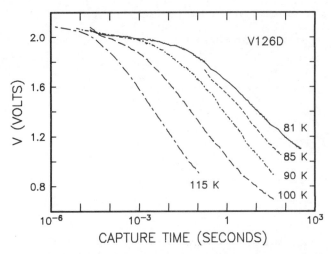

Fig. 3. Pinch-off voltage vs. trapping time for a FET at several different temperatures. The doping concentration in the AlGaAs is 1×10^{18} cm^{-3}.

Examining our data we see that the range of capture times for the linear portion of the data decreases with increasing temperature as does the value of the shortest capture time. The value of the longest capture time on the linear part of the curve can be estimated only for the data taken at higher temperatures since the filling process takes much longer at low temperatures and was not completed over the time measured. The capture time for a given voltage decrease, i. e. the time to fill the same number of traps at each temperature, decreases exponentially with temperature. This is expected if D-X centers are present, since the capture cross section for the D-X center increases exponentially with temperature (Lang et. al. 1979). The thermal activation energy of the time for a given voltage decrease increases with larger values of the voltage shift. For this sample the range is from 0.16 to 0.21 eV. Taking 1×10^{-15} cm^2 for σ_∞, this gives 8×10^{-25} to 6×10^{-28} cm^2 for the range of capture cross sections at 80 K.

Several experiments were performed to try to understand what factors influence the range of capture cross sections we observed. One possibility we investigated was the compositional variation in the AlGaAs layer due to the rotation of the sample during MBE growth. A sinusoidal variation in composition has been observed by TEM, with the range of compositions increasing with the distance from the center of rotation of the wafer (Kuan 1984). For $Al_{0.35}Ga_{0.65}As$ <100> oriented substrates, the composition ranges from about x=0.32 to x=0.38. To see if this effects the distribution of capture cross sections, FETs from the center of a wafer were measured and the data were compared with those of other FETs. Fig. 4 shows this comparison for two different measurement temperatures. The uncertainty in the temperature is about 1 K. Because 300 K pinch-off voltages were different for the two FETs, the data were normalized to the same voltage at 10^{-3} sec. As can be seen in Fig. 4, the ranges of the capture times measured are identical. Thus the variation in the alloy composition does not seem to effect the distribution of capture cross sections.

The effect of randomness in alloy composition on the DLTS emission energy has been studied in $GaAs_{1-x}P_x$ (Omling 1983). It was shown that the DLTS peak is broadened and that the single value of the activation energy for thermal emission must be replaced by a Gaussian distribution of activation energies in this case. For the case of a thermally activated capture cross section the DLTS emission energy should be the sum of the electron binding energy and the activation energy of the capture cross section. Thus this distribution of emission energies would show up as a distribution of capture cross sections or binding energies or both. An investigation of both the distribution of capture energies and DLTS emission energies as a function of the alloy composition is needed to determine the effect of alloy broadening on the D-X center. However, in light of the data shown in Fig. 4, we suspect the effect of alloy broadening to be small.

High doping concentration may also effect the capture cross section for a trap. Many defects in GaAs exhibit an enhancement of the thermal emission rate in heavily doped material due to the electric field (Pons 1979). Not only are the emission transients not exponential for highly doped material but the apparent thermal emission energy decreases with increased doping concentration. Evidence for this effect on the D-X center is that both the thermal emission energy of the D-X center and the hall activation energy have been found to decrease in samples with very high silicon concentrations (Künzel 1983). However at lower doping concentrations the trend for the thermal activation energy was reversed (Künzel 1984). To study the effect of doping concentration on the distribution of capture cross sections a sample with a room temperature carrier concentration of 1×10^{17} cm^{-3} grown in the same MBE system and a sample

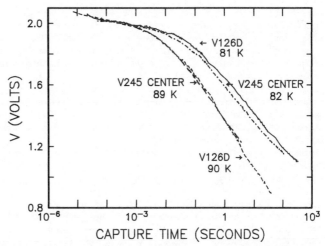

Fig. 4. Pinch-off voltage vs. trapping time for two FETs at two different temperatures. V245 was selected form the center of the wafer where the compositional variation is a minimum. Curves are normalized to the same pinch-off voltage at 10^{-3} sec.

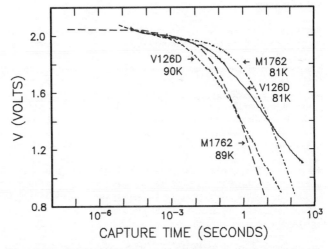

Fig. 5. Pinch-off voltage vs. trapping time for a FET and for a lightly doped (7×10^{16} cm^{-3}) sample (M1762). The data were normalized at 10^{-3} sec.

with 7×10^{16} cm^{-3} from the university of Illinois were also measured. Capture data for a FET and for a lightly doped sample at two different temperatures are shown in Fig. 5. As can be seen there is a significant difference between the lightly doped sample and the FET which had a Si concentration of about 1×10^{18} cm^{-3}. The distribution of activation energies for the capture cross section observed in the lightly doped sample ranges from 0.23 to 0.25 eV corresponding to 3×10^{-29} to 2×10^{-30} cm^2 for the range of the capture cross section at 80 K. It is much narrower than for the more heavily doped FET and the mean value is slightly higher.

According to the Lang (1979) model, the thermal emission energy should equal the sum of the Hall activation energy and the capture energy. The thermal emission energies are 0.38 eV for the lightly doped sample and 0.43 eV for the heavily doped FETs. This is consistent with the trend reported by Künzel et. al. (1984). Hall activation energies are not available for these samples, however published values of the Hall activation energy for other samples grown by the university of Illinois group (Chand 1984) give 0.11 eV for a $Al_{0.35}Ga_{0.65}As$ sample doped with 2.5×10^{17} cm^{-3}. Since the Hall activation energy is found to increase with decreasing Si concentration (Künzel 1984, Künzel 1983), this number is a lower limit to the actual value in the lightly doped sample. The agreement for this sample is good. In the heavily doped FETs the sum of the Hall energy and the capture energy would be smaller than the thermal emission energy by about 0.1 eV. The reason for this discrepancy is not known. The emission transient is highly non-exponential in the heavily doped samples due to the high trap concentration and because the depletion depth is changing due to carrier freeze-out during the measurement. Thus great care must be taken with DLTS measurements in this case.

Values of 0.33 eV and 0.29 eV have been reported for the thermal activation energy of the capture cross section for D-X centers in samples doped to about 1×10^{18} cm^{-3} (Wanatabe 1984, Zhou 1982). A single value for the capture cross section was assumed in those experiments. In both those cases the sum of the Hall activation energy and the activation energy for the capture cross section were in good agreement with the thermal emission energy. Our value for the center of the distribution of the capture cross sections for the FET at 115 K is about 100 times larger than that reported by Zhou, et. al. (1982). The raw data, however, show a difference in the capture time of   about three orders of magnitude indicating that at least some of this discrepancy is in the interpretation of the data.

5. Conclusions

The data presented here show charge capture in modulation doped FETs which is in general consistent with Lang's (1979) model for the D-X center. We have shown that for heavily doped $Al_xGa_{1-x}As$ the capture process must be described by a distribution of capture cross sections rather than a single capture cross section as had been assumed previously. The distribution of activation energies for the capture cross section centers around a value which is lower than what was reported previously. The width of the distribution of capture cross sections is narrower in lightly doped samples and is shifted to smaller values. The effect of alloy broadening on the width of the distribution is smaller than the effect of high dopant concentration.

Acknowledgments

We thank S. Wright who grew most of the MBE material used for these experiments, D. LaTulipe who fabricated the FETs, and D. Decain who fabricated the Schottky diodes. We also thank Prof. H. Morkoc and his students at the University of Illinois who supplied one of the samples.

References

Drummond T J, et. al. 1983 IEEE Transactions on Electron Devices ED-30 1806
Chand N, et. al. 1984 to appear in Phys Rev. B
Kuan T S 1984 private communication
Künzel H, et. al. 1983 Applied Phys. A 32 69
Künzel H, et. al. 1984 J. Electron. Mat. 13 281
Lang D V et. al. 1979 Phys. Rev. B. 19 1015
Omling P, et. al. 1983 J. Appl. Phys. 54 5117
Pons D and Makram-Ebeid S 1979 J. de Physique 40 1161
Valois A J and Robinson G Y 1983 IEEE Electron Device Letters EDL-4 360
Wagner E E, et. al. 1980 J. Appl. Phys. 51 5434
Wanatabe M O, et. al. 1984 J. J. Appl. Phys. 23 L103
Zhou B L et. al. 1982 Applied Phys. A 28 223

Inst. Phys. Conf. Ser. No. 74: Chapter 7
Paper presented at Int. Symp. GaAs and Related Compounds, Biarritz, 1984

623

Electrical performances of GaAs permeable base ballistic electron transistors

Y. AWANO*, K. TOMIZAWA*, and N. HASHIZUME

Electrotechnical Laboratory
1-1-4 Umezono, Sakura-mura, Niihari-gun, Ibaraki, Japan

Abstract. Electrical performances of GaAs Permeable Base Transistors (PBTs) having a quarter micron emitter-collector distance is studied by two-dimensional Monte Carlo particle simulation. Dependence on the base thickness, base spacing, and doping density of the electrical performances is discussed.

1. Introduction

This paper reports Monte Carlo particle simulation of GaAs Permeable Base Transistors (PBTs) having a quarter micron emitter-collector distance. The electron transport and the electrical performances of the device will be described.

A detailed theoretical study on submicron GaAs PBTs requires full consideration of the ballistic nature of electron transport. Therefore, the mobility/diffusion approximation, which has been applied to the case of GaAs PBTs having an emitter-collector distance of $1\mu m$ or more (Bozler et al. 1980, Marty et al.1983) is not appropriate for this case. The most accurate method for the simulation of such a submicron GaAs device at present stage would be to use Monte Carlo particle simulation of electron transport in combination with a calculation of self-consistent electric field distribution. We have already applied the Monte Carlo method to a two-dimensional simulation of submicron -channel (and gate) GaAs MESFETs, and demonstrated that the electron transport in such a device is almost ballistic and that the FET exhibits extremely high electrical performances (AWANO et al. 1983a,b).

2. Model

The two-dimensional PBT model employed is shown in Fig. 1. The base grating of submicron periodicity is entirely embedded in a continuum of semiconductor. The base thickness L_B is 400A. Both the base width and the base spacing are 2d=2000A. The active n-layer with a doping density

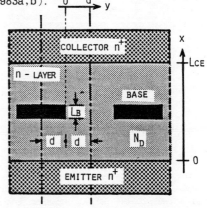

Fig.1 PBT model used in the simulation

of $N_d = 7 \times 10^{16} cm^{-3}$ is sandwiched between two n^+-layers each with a doping density of $N_d{}^+ = 5 \times 10^{17} cm^{-3}$. The emitter-collector distance L_{CE} is 0.25μm. A unit structure sandwiched between two dot-dash lines in Fig.1 is used for the calculation.

The two-dimensional Poisson's equation was solved using the FACR method and the capacity matrix method (Hockney 1970). The mesh spacing and the time increment used are 0.01μm × 0.0032μm and 5 femto sec, respectively. The Dirichlet boundary conditions were applied to all metal-semiconductor contacts, and the Neumann boundary conditions (zero normal derivative of the potential) were applied at the dot-dash lines.

The Monte Carlo particle method employed is the same as that employed by Hockney et al.(Hockney et al. 1974). The total number of particles employed is about 25,000. The parameters used in the present simulation are the same as those used by Fawcett et al.(Fawcett et al.1970). Phonon scatterings of both intervalley and intravalley types, and ionized impurity scattering (Ruch et al. 1970) are taken into account. The nonparabolicity of the (000)-valley of the conduction band is included (Fawcett et al. 1970). The lattice temperature is 77K.

3. Results and Discussion

A. Electrical Performances and Electron Motion in the PBT

Fig.2 shows the collector current I_{CE} vs. collector-to-emitter voltage V_{CE} characteristics obtained from the present simulation. The base-to-emitter voltage V_{BE} includes the built-in potential of ≒ -0.8V (for the actual base-to-emitter voltage add ≒ 0.8V to the value indicated). The collector current I_{CE} and the transconductance g_m at $V_{BE} = -0.2V$ and $V_{CE} = 0.8V$ are about 6.4mA/20μm and 500mS/mm, respectively. The small-signal short-circuit unity-current-gain frequency $f_T (= g_m/2\pi C_{BE})$ was calculated to be about 150GH$_z$, where C_{BE} is the base-to-emitter capacitance. These figures are extremely high when compared with the values obtained experimentally with a GaAs PBT having L_{CE} of 2.3μm ($f_T = 37$GH$_z$) (Bozler et el. 1980). This suggests a possibility of future improvement of this device.

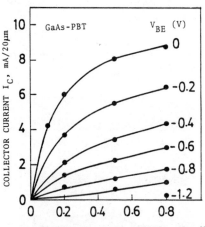

Fig.3(a) and (b) show the distributions of the electron kinetic energies and the lateral component of the electron velocities u_x, respectively, of the electrons found in the range between y=0.09μm and 0.1μm at $V_{BE} = -0.2V$ and $V_{CE} = 0.8V$. In Fig.3(a) the profile of the potential times the elementary charge q is also shown. It is seen from Fig.3(a) that the electron energies are distributed close to the potential line, showing that the electrons lose only a small fraction of their kinetic energies in the

Fig.2 Collector current versus emitter-to-collector voltage characteristics. V_{BE} includes the built-in potential of about -0.8V.

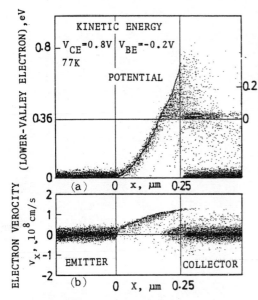

Fig.3 Distributions in the PBT (between y=0.09 and 0.1μm). (a) Distribution of electron kinetic energies. (b) Distribution of lateral components of electron velocity.

scattering processes. It is seen from Fig.3(b) that the electron velocities are densely populated near the upper envelope of the distribution across the entire length of the n-layer. Therefore, it is concluded that the near ballistic nature of the electron transport is maintained throughout the n-layer. These distributions are very much like those of a GaAs quarter-micron MESFET (Fig.3, Awano et al. 1984).

Fig.2 also shows a relatively large collector conductance g_C $(=\partial I_{CE}/\partial V_{CE})$ of about 120mS/mm at $V_{BE}=-0.2V$, $V_{CE}=0.8V$ and a large deviation of the threshold voltage V_{th} $(\doteqdot -1.4V$ at $V_{CE}=0.8V$) from the pinch-off voltage V_P $(=-qN_dd^2/2\varepsilon \doteqdot -0.5V$, ε:the dielectic constant of GaAs). This result means that the controllability of the base is rather poor.

In order to understand the physical basis for the base controllability, it is useful to investigate the potential distribution and the spatial electron distribution in the device. Figs.4(a),(b) show the equi-potential lines (a) and the spatial electron distribution (b) at a constant emitter-to-collector voltage $(V_{CE}=0.8V$) for three different emitter-to-base voltages $(V_{BE}=-0.2,-0.4,$ and $-0.8V)$. It should be noted that the depletion region extends more easily towards the n^+-layers rather than towards the channel. This is due to the two-dimensional nature of the depletion region. This nature becomes more pronounced as the aspect ratio of the conducting channel $(\doteqdot L_B/d)$ becomes smaller, as will be described later.

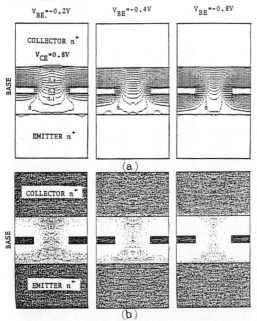

Fig.4 (a) Equipotential lines in the PBT for different V_{BE}'s. (b) Spatial distributions of electrons in the PBT for different V_{BE}'s.

B. Base Thickness Dependence

Fig.5(a) shows the equi-potential lines in the PBT at a fixed bias of V_{BE} =-0.8V and V_{CE} =0.8V for three different base thicknesses L_B (=200A, 400A, and 1000A). Fig.5(b) shows the spatial distributions of electrons corresponding to Fig.5(a). From these figures, it can be seen that as the base thickness increases, the potential barrier becomes larger, and thereby the crosssectional area of the channel becomes smaller.

In Fig.6, the threshold voltage V_{th} at V_{CE} =0.8V, the transconductance g_m, the base-to-emitter capacitance C_{BE}, and the unity-current-gain frequency f_T at V_{CE} =0.8V, V_{BE} =-0.2V are plotted as a function of L_{BE} in Fig.6. V_{th} decreases sharply as the base thickness decreases towards 200A. This is primarily due to the two-dimensional nature of the depletion region. V_{th} for L_B =200A is about -1.9V. f_T $(=g_m/2\pi C_{BE})$ is nearly constant for $L_B \leq 1000A$. This is due to the fact that the increase of g_m with the increase of L_B is compensated for by the increase of C_{BE}, as shown in Fig.6. For $L_B \geq 1000A$, f_T decreases with the increase of L_B due to the rapid increase of C_{BE}.

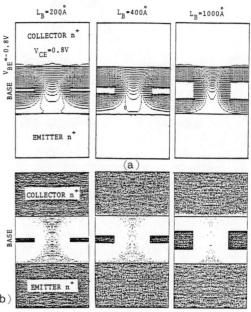

(a)

(b)

Fig.5 (a) Equipotential lines in the PBT for different L_{BE}'s. (b) Spatial distributions of electrons in the PBT for different L_{BE}'s.

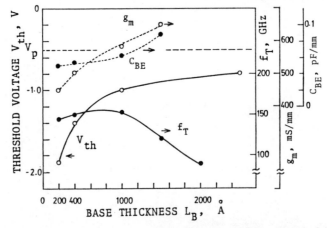

Fig.6 Dependences of the unity-current-gain frequency, threshold voltage, transconductance, and base-to-emitter capacitance of the PBT on the base thickness. (V_{BE} =-0.2V, V_{CE} =0.8V).

C. Donor Density Dependence

Fig.7 shows the dependences of the collector current I_{CE}, the transconductance g_m, the unity-current-gain frequency f_T and the threshold voltage V_{th} on the donor density. In the simulation, the base grating period, 4d, was also changed in such a way that the pinch-off voltage V_p ($=-qN_dd^2/2\varepsilon$) was kept constant at $-0.5V$. The PBT shows the increases in I_{CE}, g_m, and f_T ($=g_m/2\pi C_{BE}$) from 9.0mA/20μm to 15.5mA/20μm, from 500mS/mm to 1100mS/mm, and from 150GHz to 300GHz, respectively, with the increase in N_d from 7×10^{16} cm^{-3} to 2.8×10^{17} cm^{-3}. However, the threshold voltage V_{th} exhibits only a slight change with the increase of N_d.

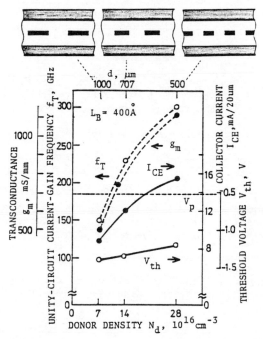

Fig.7 Dependences of the unity-current-gain frequency, collector current, transconductace, and threshold voltage of the PBT on the donor density. L_B =400A, V_{BE} =-0.2V (g_m and f_T), V_{BE} = 0 V (I_{CE}), V_{CE} =0.8V.

4. Conclusion

Electrical performances of GaAs PBTs having a quarter micron emitter-collector distance have been studied by Monte Carlo particle simulation. A GaAs PBT with N_d=7$\times 10^{16}$ cm^{-3}, base width and spacing, 2d, of 2000A, and base thickness L_B of 400A shows extremely high performances of g_m=500mS/mm, f_T=150GHz, and I_{CE}=6.4mA/20μm at 77K. The penalties that must be paid are a relatively large g_C (=120mS/mm) and a large deviation of V_{th} (=-1.4V) from V_p (=-$qN_dd^2/2\varepsilon$ =-0.5V). These are primarily due to the two-dimensional nature of the depletion layer. However, it has been shown that these figures can be improved by increasing L_B (L_B=1000A; V_{th}=-1.0V). When the doping density N_d and the base spacing 2d are changed without changing the pinch-off voltage V_p, f_T increases rapidly as N_d increases (N_d=2.8× $10^{17} cm^{-3}$,d=500A; f_T=300GH$_z$). The near ballistic electron transport in these devices have also been confirmed.

Acknowledgments

The authors would like to thank S.Kataoka for continuous encouragement and M.Kawashima, K.Ohta, N.J.Kawai, T.Nakagawa, and K.Matsumoto for helpful discussions. Y.Awano would also like to express his thanks to Prof. M.Mukaidono of Meiji University for further continuous encouragment.

REFERENCES

Awano Y, Tomizawa K, Hashizume N, Kawashima M, and Kanayama T 1983a
    Tech. Digest 1983 Int. Electron Device Meeting (IEEE New York 1983) 617
Awano Y, Tomizawa K, Hashizume N, Kawashima M 1983b *Electron. Lett.* **19**
    20
Awano Y, Tomizawa K, and Hashizume N 1984 *IEEE Trans. Electron Devices*
    31 448
Bozler C O, and Alley G D 1980 *IEEE Trans. Electron Devices* **27** 1128
Fawcett W, Boardman A D, and Swain S 1970 *J. Phys. & Chem. Solids* **31**
    1963
Hockney R W 1970 *IBM Res. Report* RC-2870
Hockney R W, Warriner R A, and Reiser M 1974 *Electron. Lett.* **10** 484
Marty A,Clarac J, Bailbe J P, and Reg G 1983 *IEE Proc. Part I* **130** 24
Ruch J and Fawcett W 1970 *J. Appl. Phys.* **41** 3843

* Permanent Address : School of Engineering, Meiji University, 1-1-1 Higashimita, Tama-ku, Kawasaki-shi, Kanagawa, Japan 314

Inst. Phys. Conf. Ser. No. 74: Chapter 7
Paper presented at Int. Symp. GaAs and Related Compounds, Biarritz, 1984

629

GaAs surface oriented microwave PIN diodes

R Tayrani, D C Bartle* and N J Barrett*

BP Research Centre, Chertsey Road, Sunbury-on-Thames, Surrey, UK
* GEC Research Laboratories, Hirst Research Centre, Wembley, UK

Abstract. This paper describes the design and fabrication techniques for successful realisation of two types of GaAs PIN diodes which lend themselves to integration into GaAs ICs. PIN diodes with a new shallow trench structure exhibit superior microwave performance to the interdigitated structure.

1 Introduction

Until now, the need for microwave control devices have been mostly met by silicon PIN diodes. The construction of these devices has reached the state of the art during the last 20 years and their properties have been thoroughly investigated and well documented (J F White). Although other GaAs devices like Impatts, MESFETs, etc, have improved to their present state, little effort has been reported towards GaAs PIN diodes. As phased array radars and communications systems become increasingly complex and GaAs monolithic microwave integrated circuits (MMIC) are becoming more of a reality, there is a growing need for a compatible microwave control device, preferably with performance at least comparable to the silicon PIN diode. Previously, FETs have been used for control purposes in GaAs MMICs with inferior performance due to the FETs inherent high 'on' resistance, low power handling capability and complex matching network for wide band applications.

The first successful realisation of GaAs microwave PIN diodes in both 'vertical' and 'planar' structures, with superior performance to FETs, were reported by the authors (Tayrani, 1983, 1984). This paper will present the results of our continuation work on two types of diode designs (i.e. interdigitated and shallow trench gap structures) which can be readily integrated into a monolithic GaAs IC.

2 Device designs

For diodes operating in the series mode, there are two suitable diode designs which are suitable for integration into a monolithic GaAs IC. These are the planar multi-electrode interdigitated structure shown in Figure 1 and the shallow trench surface orientated gap structure shown in Figure 2 (both with 50 Ω transmission lines providing the input-output terminals). The advantages of the latter device are that it is simpler to fabricate such a structure (i.e. 1–2 μm deep trenches), and it has superior microwave performance.

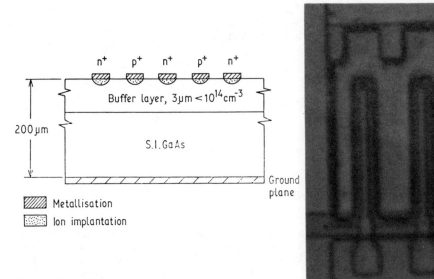

Fig 1: Interdigitated structure

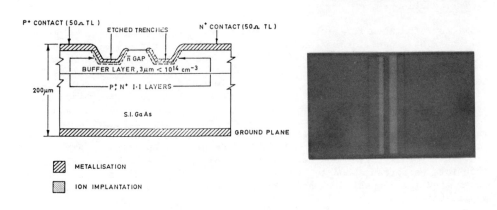

Fig 2: Shallow trench gap structure

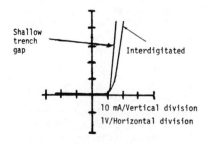

Fig 3: IV characteristics

2.1   Interdigitated structure

The junction capacitance of the interdigitated structure shown in Figure 1 results almost entirely from the fringing fields existing between device electrode regions with some contribution from the fringing fields between the transmisison lines. This is based on the assumption that the inter-electrode spacing is fully depleted of carriers which should be the case for electrodes of $W=S=2$ μm on $n^- <10^{-14}$ cm^{-3} buffer layer of 2–3 μm thickness.

The fringing capacitance between a pair of coplanar electrodes on a dielectric substrate is rigorously analysed elsewhere (Lim 1968). However, it can be shown (Wilson) that the capacitance of a multi-electrode diode can be calculated from

$$\frac{C}{A} = \frac{0.58 \left(\frac{W}{S}\right)^{0.33}}{(W+S)} \qquad \text{where } A = \text{total diode area} = N\,(W+S)\ell. \tag{1}$$

The above empirical relation agrees with the exact calculation (5) to better than 3% over $0.1 < \frac{W}{S} < 10$ which is the range of interest. For N = 4, W = S = 2 μm and ℓ = 50 μm. The junction capacitance would be $C_j \simeq 20$ fF (assuming full depletion). This would produce an 'isolation' value of around 19 dB at 10 GHz, monotonically decreasing at a rate of 6 dB/octave at zero d.c. bias voltage.

The forward resistance of the diode (R_f) which controls the 'insertion loss' of the device is the sum of the current independent resistance (R) and current dependent resistance (r_f). The former is the sum of resistances due to the P^+ and N^+ regions together with contributions from the metal contacts. The current dependent resistance is due to the injection of the holes and electrons into the n^- region from forward bias P^+ and N^+ electrodes. Assuming an ideal case where the plasma generation recombination is limited to the n^- region of the diode, this resistance is given by

$$r_f = \frac{S^2}{2\bar{\mu}\,\tau\,I_{DC}} \qquad \text{where } \begin{array}{ll} S & \text{interelectrode spacing} \\ \bar{\mu} & \text{average electron and hole mobility} \\ \tau & \text{effective minority carrier lifetime} \end{array} \tag{2}$$

It is reasonable to expect that the above equation predicts an optimistic value for (r_f), since the recombinations in P^+, N^+ and the device surface regions have been ignored.

2.2 Shallow trench surface orientated gap structure

Silicon diodes of similar structure but with deep 'pockets' (i.e. >10 μm) have been reported earlier (Emery 1968); however, due to the difficulties in processing, this technique has not been widely used. The present structure shown in Figure 2 does not suffer from the above, since shallow trenches of 1-2 μm depth have been utilized. Ignoring the minimal shunt capacitance associated with the series gap, the capacitance of the structure consists of two parts: the gap capacitance in the transmission line, which is described in (Hammerstad, Bekkadel, 1975); and a parallel plate capacitance due to the trench depths. In our case, for a gap of 2 μm and width of 148 μm (i.e. $\frac{W}{h} = \frac{148}{200}$ for 50 Ω T.L.), the total junction capacitance was calculated to be $C_j = 35$ fF. This can be reduced further, either by using a higher impedance transmission line system (i.e. reducing T.L. width will reduce the gap capacitance contribution) and/or increasing the gap width, which will reduce the parallel plate capacitance. However, it is worth mentioning that although the theoretical calculation of C_j favours the interdigitated structure, it will be shown later that the microwave measurements indicate otherwise, and both structures have similar junction capacitance (i.e. $C_j \simeq 35$ fF). This is mainly due to unaccounted parasitic capacitances in the interdigitated structure.

The r.f. series resistance of the shallow trench gap device is complex to calculate, as it depends on the shape and geometry of the lateral current injecting structures and on the actual P^+ and N^+ ion implanted profiles. The latter is not easy to predict or determine for the sloped wall structures shown in Figure 2. However, first order calculation predicts that such a structure should have an 'on' resistance as low as half of the interdigitated structure. This reduction can be attributed to the higher lateral current injection into the n^- region, thereby modulating the conductivity of this region more efficiently when compared to the interdigitated structure.

3 Material growth and device fabrication

The devices were fabricated on 3 μm thick undoped buffer layers grown by AsCl$_3$ vapour phase epitaxy on Cr-0 doped semi-insulating Bridgman substrates.

The buffer layer which formed the n^- region of the device, had a free electron concentration $<10^{14}$ cm^{-3}. The n^+ (10^{18} cm^{-3}) and p^+ (10^{19} cm^{-3}) regions were selectively ion implanted using silicon nitride and photoresist masking layers. The implant schedules were $^{29}Si^+$; 4 x 10^{13} ions cm^{-3}; 150 keV for N^+ and $^9Be^+$; 10^{15} ions cm^{-2}; 40 keV for P^+ regions. The implants were simultaneously activated using a transient annealing technique which reduces the diffusion of Be and increases the activation efficiency of high dose implants. The samples were rapidly heated to 1000°C in a N$_2$ ambient with a silicon wafer as an encapsulant using an in house built optical furnace (Barrett et al 1984). Selective area zinc diffusion has also been used for P^+ formation in the interdigitated device producing a similar surface hole concentration. The diffusion was carried out at 710°C for 30 minutes using silicon nitride as a diffusion mask.

The fabrication of both device structures begins with the deposition of 1000 Å plasma enhanced Si$_3$N$_4$ at 250°C and in the case of the gap structure, this was followed by the trench etching. Next, N^+ selective area ion implantation was performed, for both structures. These stages were repeated for the P^+ implantation. After rapid thermal annealing of both implants, P^+ ohmic contacts were formed by evaporating Au-Zn followed by Au and then N^+ ohmic contacts by evaporating Au-Ge of eutectic composition followed by Ni. In both cases the excess metals were removed by lift off and the contacts were then alloyed in forming gas. To reduce the contact resistance and also to provide input-output 50 Ω transmission lines for both devices, Au was evaporated to Ti thickness of 1 μm. The ground plane loss was minimized by polishing/etching the GaAs substrates prior to the ground plane metallisation and plating.

4 Device results

The D.C. current voltage characteristics for the interdigitated and the shallow trench gap structures are shown in Figure 3. The forward current characteristic of these diodes were found to follow the parallel plane diode equation given by

$$I = I_o \left[\exp\left(\frac{qV}{nkT}\right) - 1 \right] \tag{3}$$

over three orders of magnitude. The values of I_o (saturation current) and n (diode quality factor) were found to be I_o = 150 pA, n = 3 for the interdigitated structure and I_o = 200 nA, n = 5 for the trench gap device. The high values of n observed are due to the surface recombination effects which are inherent in the surface orientated devices, the higher value obtained in the gap structure is mainly due to the unpassivated surface of the device.

It is clear from figures 3 and 4 that as predicted, the shallow gap structure has yielded a device with a factor of two lower forward series resistance. Figure 4 is a plot of the total forward resistance of the diode (R_f) against $1/I_{DC}$ measured at 400 MHz, showing an unexpected non-linear behaviour. This effect has not been observed in the vertical GaAs PIN devices which have a linear relationship over a wide current range (2).

The small signal s-parameters of both PIN diodes, which have already been assessed at D.C., were measured using a computer corrected automatic network analyser (ANA) over a frequency range of 5-15 GHz. The measured input-output return loss, the magnitude of the insertion loss and the isolation for both devices are shown in Figure 5. The insertion loss of the shallow trench gap structure is <1 dB compared to the ≈2 dB for the other device. The isolation of both devices are the same and have a 6 dB/octave fall corresponding to a junction capacitance of 35 fF. This value of capacitance is in very good agreement with the calculated values for the shallow trench gap design described earlier. However, for the interdigitated device, the isolation measurement indicates a factor of two increase from the calculated value i.e. (C_j = 20 fF), which is mainly attributed to the unaccounted parasitic capacitance arising from the fringing fields between the transmission lines. This can be partly overcome by chamfering the transmission lines. To obtain accurate equivalent circuits of both device structures, their measured S-parameters were fitted to the circuits shown in Figures 6 and 7. The element of the circuits were found using (compact) computer program optimisation routines. These are in good agreement with the measured values obtained from the device analysis described earlier. The switching speeds of the diodes from the low loss i.e. (forward bias 30 mA) to the high loss state and vice versa were measured to be less than one nanosecond. The very fast switching observed was obtained by ensuring that the width of n^- region is less than twice the ambipolar diffusion length L_D, which is strongly dependent on the quality of the n^- region and subsequent processing.

5 Conclusions

Two types of GaAs surface orientated microwave PIN diodes which are compatible with integration into monolithic ICs have been successfully designed and fabricated. It is shown that, the new 'shallow trench gap structure' PIN device, has a superior microwave performance to the interdigitated structure. Such a device, mounted as a series SPST switch in 50 Ω system, typically gives 18 dB isolation at 5 GHz i.e. (C_j = 35 fF) and less than 1 dB insertion loss up to 15 GHz. The switching speeds of these devices are measured to be less than 1 nanosecond.

The factor of two improvement in the series resistance of the new diode, coupled with its low junction capacitance, renders it especially suitable for monolithic phase shifter applications. The significance of these results is that GaAs surface orientated PIN diodes can now be considered for use in MMICs in place of FETs for microwave switching and control purposes.

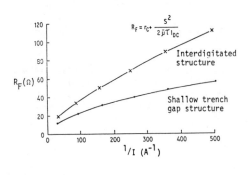

Fig 4: Resistance vs $1/I_{DC}$ at 400 MHz

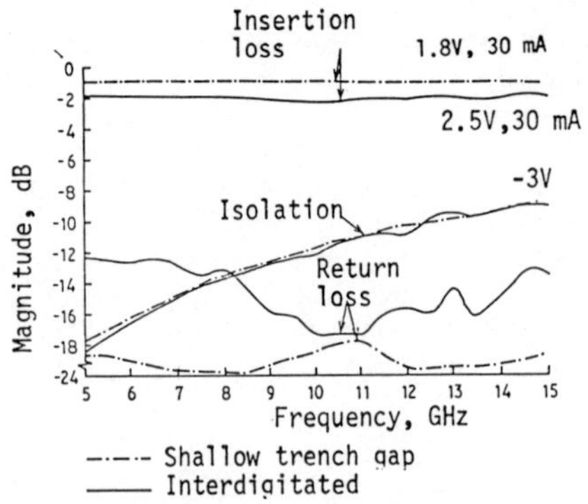

Fig 5: R.F. performance

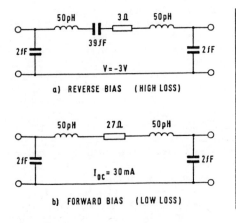

Fig 6: Equivalent circuit of integrated PIN diode

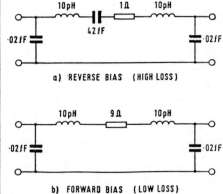

Fig 7: Equivalent circuit of shallowtrench gap structure

References

Barret, N J, Bartle D C and Grange J D. Optical furnace annealing of Be implanted GaAs, 11th International Symposium on GaAs and Related Compounds, Biarrtiz, France Inst.Phys.Conf.Ser. 74 77

Emery, F E. Monolithic microwave integrated circuitry, NERM Conference, Digest 1968

Hammerstad E O and Bekkadal F. A microstrip handbook, ELAB Report, University of Trondheim Norway 1975

Lim, Y C and More R A. Properties of alternately charge coplanar parallel strips by conformal mapping, IEEE Trans, ED-15, No 3, Mach 1968

Tayrani, R, and Glew, R W. GaAs PIN diodes, WOCSEMMAD 83, San Antonio, Texas

Tayrani, R, and Glew, R W. Ultra fast GaAs microwave PIN diode, Electronics Letters, Vol 19, No 13, 23 June 1983

Tayrani, R. GaAs surface orientated microwave PIN diode, WOCSEMMAD 84, San Francisco, USA

White J F, Semiconductor Control. Artech House

Wilson K. Interdigital capacitors for MMIC, Private Communication

Inst. Phys. Conf. Ser. No. 74: Chapter 7
Paper presented at Int. Symp. GaAs and Related Compounds, Biarritz, 1984

635

Quasiballistic transport in GaAs at 77 K: a comparison between Monte-Carlo simulations and photoconductivity measurements in submicron structures

P. HESTO, C. BRU, S. LAVAL, R. CASTAGNE

Institut d'Electronique Fondamentale, CNRS LA 22, Université Paris Sud, Bât. 220, (F) 91405 ORSAY CEDEX (FRANCE).

Abstract. Current-voltage characteristics obtained from Monte-Carlo simulations and from photocurrent measurements in submicron GaAs structures at 77 K are compared. Both are linear at low bias voltage and present a break for an applied voltage value increasing with the active region length. This break is related to the drift velocity decrease associated with the electron transfer in the L valleys. The difference between the potential energy corresponding to the break voltage and the Γ-L transfer energy corresponds to the emission of a limited number of polar optical phonons. Then for voltages slightly lower than the break value, electrons have a quasiballistic behavior through the device.

1. Introduction

The analysis of results obtained from measurements on submicron electronic devices is often complicated by the unavoidable mixing of various physical parameters which interfere. Monte-Carlo simulation is an efficient way to get physical insight into transport properties, in particular when non-stationary conditions are involved, and it is essential for the interpretation of experimental results. Calculated current-versus-voltage characteristics of GaAs submicron structures at 77 K are compared to experimentally recorded ones from photoconduction measurements. Simulation results point out the influence of quasiballistic transport on the characteristic shape.

2. Monte-Carlo simulation of N^+NN^+ submicron structures

Conduction in GaAs N^+NN^+ structures at 77 K has been simulated by the Monte-Carlo method. The doping levels were chosen to be $5 \cdot 10^{16}$ cm^{-3} in the N^+ regions and 10^{14} cm^{-3} in the central N region whose length is adjusted from 0.1 to 0.5 µm. The Monte-Carlo algorithm was described in ref 1 and the values of the parameters introduced in the calculation are those given by Littlejohn [2].

The current density variation as a function of the bias voltage across the structure is drawn in figure 1 for a 0.5 µm long N region. The current increases linearly for low bias values, and a bending appears in the curve for an applied voltage about .65 V. The shape of the I(V) characteristic is similar for any N region length, and the voltage beyond which the slope changes increases with the thickness of the N zone.

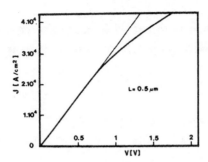

Fig. 1. I(V)characteristic from Monte-Carlo simulation for a 0.5 μm long N^+NN^+ structure at 77 K.

In order to characterize electron behavior at low voltage, the mean kinetic energy $< \varepsilon_{cin}>$, and the drift velocity v_d of the electrons, versus distance are given in figure 2 for a low bias voltage (V = 0.25 V).

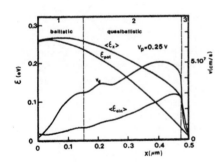

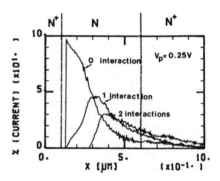

Fig. 2. Variations versus distance of the mean kinetic energy $< \varepsilon_{cin} >$ drift velocity V_d, potential energy ε_{pot} and of the sum of potential energy and kinetic energy fraction corresponding to electron motion in the field direction $< \varepsilon_x >$.

Fig. 3. Contribution to the current of electrons which encounter 0,1 or 2 interactions along the structure.

Variations of the potential energy ε_{pot} are also plotted in the same figure, together with the sum of ε_{pot} and of the fraction of the kinetic, energy corresponding to the electron velocity parallel to the field direction, $< \varepsilon_x >$, averaged over the whole electron population.

Three regions can be distinguished. In the first one, $< \varepsilon_x >$ remains roughly constant while V_d and $< \varepsilon_{cin} >$ increase. This corresponds to ballistic flights for the electrons: the mean kinetic energy is still lower than 35 meV which is the optical phonon energy in GaAs, so phonon emission cannot occur and as the other scattering probabilities are very low, most of the electrons move under free flight conditions. In the second region, part of the electrons have an energy higher than 35 meV and polar optical phonon emission probability becomes important. This induces electron energy losses and $< \varepsilon_x >$ decreases. However, as in this scattering mechanism the velocity direction is only slightly deviated, the drift velocity goes on increasing. This corresponds to quasiballistic transport. In the third region electrons reach the accumulation zone due to electron diffusion from the N^+ anode, then $< \varepsilon_x >$, $< \varepsilon_{cin} >$ and v_d decrease drastically.

The evolution versus distance of the current fractions corresponding respectively to electrons which suffered 0,1 or 2 interactions along the N layer are plotted in fig. 3. In the first region, 75% of the current is due to ballistic electrons. At the beginning of the N^+ anode layer, still 35% of the current arises from electrons which emitted at the most 2 polar optical phonons.

When the bias voltage becomes higher than about 0.37 V, that is the energy between the Γ and L valleys plus the intervalley phonon energy, electrons can gain enough energy to allow the ΓL intervalley transfer. Such a transfer induces a drastic increase of the electron effective mass and a randomization of the velocity direction.

The respective populations in the three valleys, along the device, are reported in figure 4 for a 1V bias voltage and the electron drift velocity in each valley is plotted in figure 5.

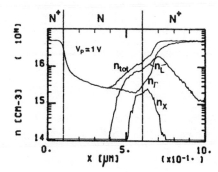
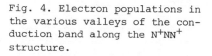
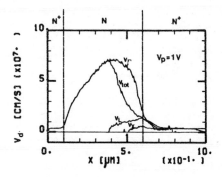

Fig. 4. Electron populations in the various valleys of the conduction band along the N^+NN^+ structure.

Fig. 5. Drift velocity for electrons in the Γ, L or X valleys and total drift velocity as a function of distance.

At the end of the N layer, the number of electrons in the L valley becomes larger than in the Γ one and the total drift velocity decreases (fig. 5). This results in bending of the I(V) characteristics when the bias voltage increases.

3. Photoconduction experiments

Photoconduction experiments have been performed on planar devices consisting of two ohmic contacts separated by a submicron gap (fig. 6).

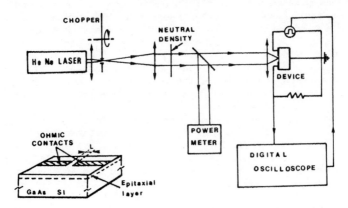

Fig. 6. Experimental set up

They are realized on a non-intentionally doped GaAs epitaxial layer
($N_D \simeq 10^{14}$ cm^{-3}) deposited on a semi-insulating substrate. Several structures which only differ by their gap lengths L have been tested
(L = 0.08 μm, 0.2 μm, 0.4 μm).

The photocurrent induced by a laser illumination in the device has been
measured as a function of the bias voltage (fig. 6). To prevent surface
breakdown , the bias voltage is only applied during the laser illumination.
To minimize diffraction effects at the edges of the metal strips the HeNe
laser beam is linearly polarized perpendicularly to the electrode edges.
The incident light power is accurately controlled. Measurements are performed using a digital storage programmable oscilloscope monitored by a
minicomputer. The advantages of such a photoconduction method have been
discussed previously [3].

The photocurrent versus voltage characteristics $I_{ph}(V)$ recorded at 77 K
for devices whose gap lengths are 0.08, 0.2 and 0.4 μm are plotted in
fig. 7. The illumination conditions were chosen in such a way that the
geometrical effects related to the surface space charge, which depends on
the gap length, are minimized in order to allow a comparison between the
various sample characteristics.

As in simulation results, the measured photocurrent increases linearly
with voltage at low bias level, and bending of the characteristic is observed beyond a given voltage. The voltage value V_s corresponding to the
slope change increases with the interelectrode distance. For the smaller
device, i.e. L = 0.08 μm, this occurs for a bias voltage about 370 mV,
which corresponds to the minimum energy required for intervalley transfer.
According to simulation results this can be interpreted as an experimental
proof for ballistic transport for bias voltages slightly lower than 370 mV.
For longer interelectrode gaps, electrons can undergo a few interactions
with optical phonons and a larger potential is required to reach transfer
energy. For example, for L = 0.2 μm, V_s is about 430 mV and this roughly
corresponds to mean energy loss through 2 interactions with polar optical

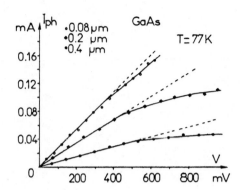

Fig. 7. Experimental photocurrent versus voltage characteristics at 77 K for various submicron device lengths.

phonons; for L = 0.4 μm, V_s ≃ 520 mV and this would correspond to 5 interactions on average. This is quasiballistic transport.

Conclusion

Experimental and simulation results show that even in presence of ballistic or quasi-ballistic transport the I(V) characteristics are linear at
low voltages, and bend when a significant fraction of the electron population has transferred to the upper valleys of the conduction band. The increase of the bending voltage with the interelectrode distance yields
indications about the mean number of interactions occurring in the device
according to its length.

Acknowledgements

The authors are indebted to C. Arnodo from LCR-Thomson-CSF for the submicron device realization and to D. Pascal, J-L. Pelouard and J-F. Pône
for helpful discussions.

References.

1. P.Hesto,Surface Science, <u>132</u>, (1983) 623.
2. M.A. Littlejohn, J.R. Hauser and T.H. Glisson,
    J. Appl. Phys. 48 (1977) 4587.
3. C. Bru, Thèse 3ème cycle, Orsay, Feb. 1982.

GaAs MIS Schottky barrier and its application in GaAs FETs with high-doped channel

Zhou Mian and Wang Weiyuan

Shanghai Institute of Metallurgy, Academia Sinica
865 Chang Ning Road, Shanghai 200050, China

Abstract. A measurement of I-V and I-1/T relationship for Al-I(80Å)-nGaAs(10^{18} cm^3) SB is made and is in good agreement with theory. GaAs MIS SB FET with high-doped channel is presented and discussed.

1. Introduction

The ideal schottky barrier (SB) theory about M-S contact and carrier transport has been well established. Because of a thin layer of natural oxide films exists between M and S, the practical GaAs MS SB is non-ideal. The first part of this paper deals with the effects on carrier transport mechanism due to thin films which are not well understood.

The GaAs MESFET is widely used in microwave and IC. Hasegawa (1982) showed that improved device characteristics can be obtained in high-doped channel. But it is not easy to fabricate GaAs FETs with high-doped channel due to difficulty in obtaining good SB. The second part of this paper is on the fabrication, measurement and simulation of FET using MIS SB gate, and in fact, an application of the first part.

2. Experiments

For MIS SB research, Si doped GaAs substrates with $N_D \sim 10^{18}$ cm^{-3} and <100> orientation were used. Two kinds of oxide layer were fabricated by anodization with thickness of 80Å: GaAs native oxide and Al_2O_3. The anolyte was a solution of anhydrous tartaric acid and ethylene glycol (NSG). The anodization of GaAs or electron beam evaporated Al thin films on GaAs was proceeded under room temperature, magnetic stirring and constant current density of 50-500 μAcm^{-2}. After anodization, the wafers were annealed in H_2 at 270°C for 30 min. The Al barrier metal was formed by evaporation. The back ohmic contact was Ag-Sn-Cu alloy. The fabricated M-I(80Å)-S SB diode was mounted in ceramic package for electrical measurements at a temperature of 77-400K without illumination. The whole system was well connected to the ground.

For MIS SB FET with high-doped channel, Cr-doped SI GaAs substrates with $R \sim 10^8 \Omega/\square$ were used. After implantation with $^{28}Si^+$ at energy of 50-80 KeV and dose of 8-6×10^{12} cm^{-2} in the orientation of <100> about 7° and capless annealing at 700-800°C for 30 min to form active and contact layer, the wafer was anodized in NSG solution to grow GaAs thin oxide films. Au-Ge-Ni was used as ohmic contact and Al as gate contact. The dual gate device with gate length L=2 μm, width Z=400 μm and D-S distance L_{DS}=9 μm were used.

3. Carrier Transport in Al-I(80Å)-nGaAs SB

The experimental forward and reverse logI-V relationship of two kinds of oxide films at room and low temperature are shown in Fig. 1(a) and (b), respectively. Fig. 2 shows current changes with temperature at a given voltage. The solid curves in Fig. 1 and 2 are theoretical computations which will be stated below.

Padovani (1971) had deduced J-V expressions for various carrier transport mechanisms and applicable ranges in SB. The solid line 1 and 2 shown in Fig. 1(a) are calculated from Padovani's expressions and exhibit several orders higher than experimental results. Similarly, the computated $logI_R-V_R$ curves not shown in Fig. 1(b) are also much higher than experiments. The reason for this serious deviation is that there is no interface layer between M and S considered by Padovani. The modifications are necessary in order to count the interface layer of $\sim 10^2$Å in MIS SB. Here, only the main points are mentioned and the detailed expressions will be published elsewhere (Zhou and Wang 1984).

First, the electrons tunneled through the barrier of interface layer can be described in transmission coefficient P. For GaAs-oxide film system P= $exp(-0.26\chi^{\frac{1}{2}}\delta)$ (Card and Rhoderick 1971, Ashok et al 1979), in which χ and δ is interface layer barrier height (eV) and layer thickness (Å), respectively. The value of χ or P can only be obtained by means of coincidence with experimental curve (Card and Rhoderick 1971) and is considered in J-V and J-T calculations shown in Fig. 1 and 2. Second, due to the dependence of voltage drop across interface layer on external bias, the effective

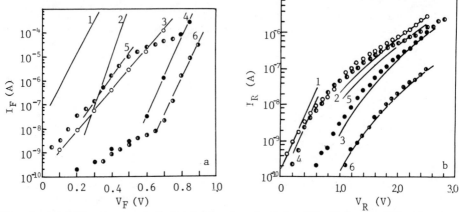

Fig. 1 LogI-V curves of Al-I-highly doped nGaAs SBD (I layer 80Å, Al area 3.2×10^{-5} cm^2)
Experimental results

O 300K \
● 85K } GaAs native oxide   ◐ 300K \
◑ 85K } Al$_2$O$_3$

Theoretical computation (solid line)

(a)					
1	MS	300K	T-F*		
2	MS	85K	F		
3	MIS	300K	T-F	χ 0.11	n 1.59
4	MIS	85K	F	χ 0.20	n 1.15
5	MIS	300K	T-F	χ 0.14	n 1.24
6	MIS	85K	F	χ 0.30	n 1.23

(b)				
1	MIS	300K	T-F	χ 0.11
2	MIS	300K	F	χ 0.23
3	MIS	85K	F	χ 0.20
4	MIS	300K	T-F	χ 0.14
5	MIS	300K	F	χ 0.26
6	MIS	85K	F	χ 0.30

* T-F and F are thermionic field and field emission mechanism respectively.

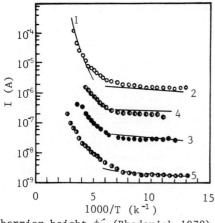

Fig. 2 LogI-1/T curves of
Al-I-highly doped nGaAs SBD
(I layer 80Å, Al area 3.2×10^{-5} cm^2)
Experimental results
  ○ V_F 0.7V $\Big\}$ GaAs native oxide
  ● V_R -1.5V
  ◐ V_F 0.75V $\Big\}$ Al$_2$O$_3$
  ◑ V_R -1.5V
MIS theoretical computation
1 V_F 0.7V T-F χ 0.11 n 1.59
2 V_F 0.7V F χ 0.20 n 1.15
3 V_R -1.5V F χ 0.20
4 V_F 0.75V F χ 0.30 n 1.23
5 V_R -1.5V F χ 0.30

barrier height ϕ_b' (Rhoderick 1978) is used to substitute ϕ_b at reverse bias
in Fig. 1(b). At last, an ideal factor n classically used in thermionic
emission (T) under forward bias (Rhoderick 1978) is supposed to be applicable in thermionic field (T-F) and field emission (F) shown in Fig. 1(a).

In order to calculate the theoretical curves shown in Fig. 1-2, we substitute the constants, experimental data and reasonably chosen figures
into the J-V expressions. Here, we chose the Richardson constant $A^* =$
$120T^2$ Acm^{-2} for GaAs and $A^{**} = 40T^2$ for Al (Herring and Nichols 1949), the
permittivity $\varepsilon_s = 1 \times 10^{-12}$ fcm^{-1} for GaAs and $\varepsilon_i = 0.35 \times 10^{-12}$ for oxide layer,
the barrier height $\phi_b = 0.8$ eV (85-300K) for MS SB and $\phi_{bo} = 0.74$ (300K), 0.78
(85K) for MIS SB at zero bias (Rhoderick 1978), and the interface state
$D_s = 4 \times 10^{13}$ cm^{-2}eV^{-1} (Yamasaki and Sugano 1979, Streever et al 1980, Morante
et al 1983). The χ and n are chosen by coinciding with experimental data.

The theoretical calculations show that the forward and reverse logI-V
curves at room or low temperature are in good agreement with experimental
results shown in Fig. 1, in which the reverse logI$_R$-V$_R$ curves at room
temperature are described by two curves due to different transport mechanisms. The low temperature theoretical logI-1/T curves at $V_F = 0.7$ and
0.75V or $V_R = -1.5V$ shown in Fig. 2 are also consistent with experimental
results. At room temperature, except curve 1, we have not drawn curves on
the upper part of curve 4, 3 and 5 because of the I$_F$ saturate at $V_F = 0.75V$
(refer to curve 5 in Fig. 1(a)) and lack of quantitatively calculated
method to consider both T-F and F mechanism (refer to curve 1, 2 and 4, 5
in Fig. 1(b)), respectively. The conclusions are that we have to modify the
Padovani's expressions in order to agree with experimental results; for
Al-I(10 Å)-highly doped GaAs SBD at 77K both forward and reverse conduction
mechanism is F emission; at 300K the forward and lower reverse conduction
is T-F emission, while at higher reverse bias the mechanism is F emission.

The decrease of current with increasing   barrier height due to interface
layer was pointed out in literature . For T-F or F mechanism, if we suppose P as the increased barrier height $\Delta\phi_b$, then we have

$$\Delta\phi_{b,T-F} = 0.26\chi^{\frac{1}{2}}\delta E_o \quad \text{and} \quad \Delta\phi_{b,F} = 0.26\chi^{\frac{1}{2}}\delta E_{oo} \qquad (1)$$

Combining the average χ of 0.12 and 0.25 eV for T-F and F mechanism, respectively, with $E_o = 0.0314$ eV (300K) and $E_{oo} = 0.021$ eV (300-85K) (Padovani
1971), for 60Å oxide films on GaAs we have $\Delta\phi_{b,T-F} = 0.17$ eV and

$\Delta\phi_{b,F} = 0.16$ eV which are close to experimental results $\Delta\phi_b = 0.11$ eV (300–77K) of Ashok et al (1979). Having known InP $m^* = 0.7 \times 10^{31}$ kg (Chamberlain et al 1971) and corresponding $P = \exp(-0.28\chi^{\frac{1}{2}}\delta)$, similar to Eq (1), we have $\Delta\phi_{b,T}$ of T mechanism

$$\Delta\phi_{b,T} = 0.28\chi^{\frac{1}{2}}\delta KT \tag{2}$$

If the χ of GaAs is adopted for InP native oxide films, we obtain $\Delta\phi_{b,T} = 0.025$ or 0.036 eV/10Å (at 300K) which is also close to $\Delta\phi_b = 0.032$ eV/10Å for $\delta = 40$–80Å experimentally reported by Wada et al (1982).

As regards ideal factor n, Laflère et al (1982) gave $n = 1.12$–1.24 for $\delta = 20$Å, Ashok et al (1979) reported $n = 1.1$–2 for $\delta = 80$Å, we have $n = 1.15$–1.59 for $\delta = 80$Å. Because D_{sa} and D_{sb} are large value (Rhoderick 1978), we have $qD_{sb} > \varepsilon_s/W$ and $qD_{sa} > \varepsilon_i/\delta$ and $n \cong 1+(D_{sb}/D_{sa})$. Taking $D_{sb}/D_{sa} = 0.1$–0.6, we have $n = 1.1$–1.6, which is close to all published data.

4. GaAs MIS SB FET with High-doped Channel

The carrier profiles with peak concentration of 10^{18} cm^{-3} in active layer after 700–800°C annealing have been measured. Due to the presence of thin oxide layer in MIS SB, at a given current, the voltage drop V_{MIS} in MIS must be larger than drop V_{MS} in MS, and the difference between them is the voltage drop across the oxide layer V_i. The relationship between V_i and V_{MIS} obtained from experimental I-V curves of MIS and MS is shown in Fig. 3 and can be expressed as

$$V_i = aV_{MIS} + b \tag{3}$$

a and b are constant depending on oxide layer thickness but not on voltage.

Fig. 4 shows the MIS FET DC and SBD characteristics. The diode exhibits reverse breakdown voltage of 4–5V, forward threshold voltage of 0.8V and is typically SB junction. The FET DC characteristics show saturation current of 90 mA, pinch-off voltage V_p of 4.3V and maximum transconductance $g_{m,o}$ of 25 mS. The gate-source capacitance at zero bias $C^*_{gs,o}$ is 0.6 pf. The normally fabricated MESFETs (same mask pattern, peak concentration 1–2×10^{17} cm^{-3}, no thin oxide films) with $g_{m,o}$ of 15 mS were obtained in the same period. It is obvious that high-doped channel has higher transconductance.

The FET DC and small signal performances were calculated on computer simulation using two-region model (Pucel et al 1975) and the results of Fig. 3. The optimum noise figure F_o was calculated using semi-empirical equation (Fukui 1979), and the maximum available gain MAG was obtained by equivalent network equations (Pucel et al 1975). When an external bias of V_{gs} is applying on gate, the effective gate bias V^*_{gs} can be represented from Fig. 3 as follows

$$V^*_{gs} = (1-a)V_{gs} - b \tag{4}$$

Considering the oxide film capacitance C_i in series with depletion layer capacitance C_s, in addition, effective gate-source capacitance C^*_{gs} is

$$C^*_{gs} = (\frac{1}{C_s}+\frac{1}{C_i})^{-1} \tag{5}$$

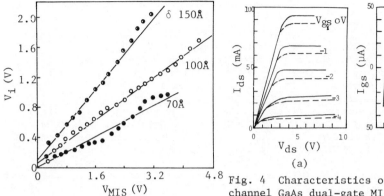

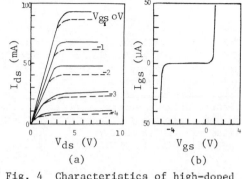

Fig. 3 Change of voltage drop across oxide films with external voltage

Fig. 4 Characteristics of high-doped channel GaAs dual-gate MIS SB FET (a) and diode (b) (I layer 70Å; implanted E 80 KeV, ϕ 8×10^{12} cm^{-2})
— experimental; --- theoretical

where $C_i = ZL\varepsilon_i/\delta$. By multiplying a constant to flat distribution (Shur and Eastman 1980), a modified flat distribution proved by exact numerical computation by Cheng and Wang (1981) was used instead of the Gaussian distribution. In computing the Fig. 4(a), the implanted activation of 60%, E_s = 2900 Vcm^{-1} and V_s = 1.3×10^7 cms^{-1} (Pucel et al 1975), R_s = 8Ω and R_d = 22Ω were taken.

Theoretically computed results and experimental data shown in Fig. 4(a) are in good agreement. The calculated V_p of device without oxide films (MESFET) is 3V, while that of device with 70Å oxide films increases to

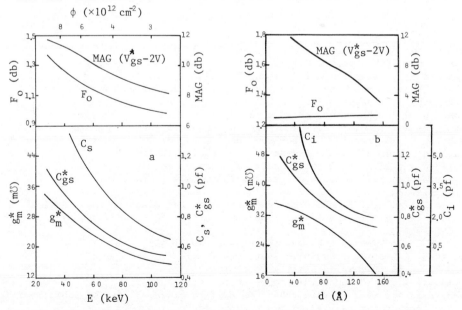

Fig. 5 Calculated parameters of high-doped channel GaAs MIS SB FET (V_p -3.8V, L 1.5 μm, Z 500 μm, L_{DS} 10 μm, other parameters see text) (a) I layer 100Å   (b) Implanted E 50 KeV, ϕ 6.1×10^{12} cm^{-2}

4.26V which well agrees with measured value of 4.3V. Under equal pinch-off voltage, the calculated effective transconductance at zero gate bias $g_{m,o}^*$, $C_{gs,o}^*$, F_o and MAG increase with increasing ϕ as shown in Fig. 5. Still, $C_{gs,o}^*$ lowers to 2/3 of C_s, because of series effect of C_i. Therefore, there must be a comprehensive consideration for selection of implanted dose or energy in order to obtain larger gain and cut-off frequency and lower noise. Under a given implanted condition, with decreasing oxide film thickness, the $g_{m,o}^*$, $C_{gs,o}^*$ and MAG increase but the F_o remains constant (Fig. 5). Without oxide films, the high-doped channel FET would fail to work, hence, the devices with δ about 50Å will have maximum $g_{m,o}$ and MAG and minimum F_o. Of course, we must pay attention to reverse breakdown behavior between G and S. For MIS SB FET shown in Fig. 5, the calculated $C_{gs,o}^*$ is in good agreement with measured results.

Some comparisons between GaAs high-doped channel MIS SB FET with normally fabricated MESFET have been given. Both MIS SB FET and MESFET can be simulated by two-region model, but effective gate bias and gate-source capacitance must be adopted in the former in order to take account of the effects of thin oxide films. For microwave application, the former has higher microwave gain and cut-off frequency, while the noise remains constant. On device technology, the normal MESFET must be implanted twice to form active channel and drain-source contact region, but for MIS SB FET, only single energy implantation is needed to form two regions simultaneously. However, anodization must be used for the MIS SB FET. Using high-doped active layer, the device parameters will be uniform due to less affect by SI substrate which may have important significance for GaAs IC.

References

Ashok S et al 1979 Solid-State Electronics 22 621
Card H C and Rhoderick E H 1971 J. Phys. D Appl. Phys. 4 1589
Chamberlain J M et al 1971 J. Phys. C 4 L38
Cheng Zhaonian and Wang Weiyuan 1981 National Conference on GaAs and
    Related Compounds p 302
Fukui H 1979 IEEE Trans. ED-26 1032
Hasegawa H 1982 GaAs FET Principles and Technology (Artech House) p 177
Herring C and Nichols M H 1949 Rev. Mod. Phys. 21 185
Laflère W H et al 1982 Solid-State Electronics 25 389
Morante J R et al 1983 Solid-State Electronics 26 537
Padovani F A 1971 Semiconductors and Semimetals 7A (Academic Press)
    Chapter 2
Pucel R A et al 1975 Advances in Electronics and Electron Phys. 38 195
Rhoderick E H 1978 Metal-Semiconductor Contacts (Clarendon Press)
    Chapters 2-3
Shur M S and Eastman L F 1980 IEEE Trans. ED-27 455
Streever R L et al 1980 Solid-State Electronics 23 863
Wada O et al 1982 Solid-State Electronics 25 381
Yamasaki K and Sugano T 1979 Appl. Phys. Lett. 35 932
Zhou Mian and Wang Weiyuan 1984 Acta Physica Sinica 33 (to be published)

Inst. Phys. Conf. Ser. No. 74: Chapter 8
Paper presented at Int. Symp. GaAs and Related Compounds, Biarritz, 1984

647

Gallium arsenide integrated circuits

Masamichi Ohmori

Atsugi Electrical Communication Laboratory, NTT, Kanagawa, Japan

Abstract   In this paper GaAs LSI performance and process technology are
surveyed.   Recently, FET threshold voltage standard deviation was
reduced to 20 mV over a 2-inch wafer by applying a dislocation-free LEC
crystal and fine lithography with a wafer stepper.   Using this
technology, a GaAs 16 Kb static RAM with 4.1 ns access time and a fully
operational 4 Kb SRAM were fabricated.   GaAs LSI technology has the
potential for transition from laboratory research to industrial
application.

1. Introduction

GaAs IC complexity development since 1974 is shown in Fig. 1.   Fabrication
of the first GaAs MESFET IC was reported by Hewlett Packard in 1974.
Logic gate integration was first advanced at HP and later at Rockwell
International Limited.   In 1980, the first GaAs LSI having 8 x 8 parallel
multipliers was fabricated.   After that, a 3,000-gate 16 x 16 parallel
multiplier using low power dissipation logic, DCFL, was built by Fujitsu
Limited.   A static RAM, utilizing 6 transistors per memory cell, was
developed for the memory circuit.   The first 16-bit static RAM was
reported by NTT in 1981 (Asai et al. 1981).   Static RAM integration has
been progressing rapidly.   Recently, a 16 Kb static RAM integrating
102,000 devices was fabricated in our laboratory.

GaAs IC development can be divided into four generations as shown in Fig.
1.   The first generation (1974-1977) gave us SSIs and MSIs.   The circuits
were fabricated using epitaxial layers with mesa device isolation.   The
next generation (1978-1980) also gave us MSIs.   However, the fabrication
process was greatly improved using selective ion-implantation in a semi-
insulating substrate.   A planar and uniform process was developed.   The
third generation (1980-1984) is a LSI generation.   Low power dissipation
logic, DCFL, has been applied.   Several kinds of self-aligned FET
structures have also been investigated.

The next generation will start around 1985.   GaAs LSI factory production
will be attained.   In the research field, VLSI and submicron gate
technologies are developing.

2. GaAs IC Performance

2.1 Logic gates

Propagation delay time versus year for commonly used GaAs logic gates is
shown in Fig. 2.   In the early development stage, normally-on logics, such

as Buffered FET Logic (BFL) and Schottky Diode FET Logic (SDFL), were
investigated primarily. A normally-off type DCFL has been successively
developed. The propagation delay time was reduced to 11 ps (with a power
dissipation of 20 mW) (Yamasaki 1984) from the early stage of 300 ps
(Ishikawa et al. 1977) in about 7 years. HEMT performance at room
temperature operation and heterojunction bipolar transistor ECL and NTL
performances are also plotted. GaAs DCFL logic is currently the fastest
semiconductor device. In GaAs MESFET logic, power dissipation increases
in sequence from DCFL, SDFL, BFL, and SCFL.

2.2 Logic integrated circuits

GaAs logic IC propagation delay time versus IC complexity is shown in Fig.
3. Silicon bipolar and MOS gate performance are also shown for
comparison. In the small scale integration region, speed is comparable
for GaAs and bipolar gates. However, a propagation delay time of 150-160
ps per gate was obtained in the LSI region for SDFL 8 x 8 (Lee et al.
1981) and DCFL 16 x 16 multipliers (Nakayama et al. 1983). This value is
faster than bipolar logic gate circuits.

2.3 Memory integrated circuits

The performance progress of GaAs static memory is shown in Fig. 4 with
current silicon bipolar and MOS static RAMs performance. Recently, a GaAs
4 Kb static RAM with 2 ns access time and a 16 Kb static RAM with 4.1 ns
access time have been fabricated. These values are much faster than
silicon bipolar and MOS memories. GaAs memories should aim at performance
outside silicon technology. GaAs performance is now approaching this
goal.

3. LSI-Directed MESFETs

Among frequently used GaAs logic gates, DCFL is the most suitable for the
LSI approach owing to its circuit simplicity and low power dissipation.
However, as DCFL logic voltage swing is limited to about 0.5 V due to the
MESFET Schottky barrier potential of about 0.7 V, it is imperative that
the device have uniform characteristics. GaAs MESFETs conditions for DCFL
LSIs are 1) a high speed of less than 100 ps per gate for the loaded
condition, 2) high current transconductance drivability of more than 200
mS/mm, 3) high packaging density with design rules of 2 μm and gate length
of 1 μm, and most importantly 4) uniformity of threshold voltage. To meet
these requirements, several kinds of LSI-directed MESFETs have been
developed. These MESFETs have closely spaced ohmic electrode structures
(Furutsuka et al. 1981), Pt buried gate structures (Hojo et al. 1981),
self-alignment structures to form the n$^+$ region by ion-implantations
(Yokoyama et al. 1981), as well as other structures. These devices are
developed to avoid the surface depletion layer effect caused by surface
trapped electrons and to reduce series resistance. Recently, a variety of
refractory metals, such as W-silicide, W-nickel, and W-aluminum, have
been used at several laboratories, in a self-alignment process using a
refractory metal gate. However, a disadvantage of this process is
parasitic gate capacitance increase due to lateral spreading of n$^+$ regions
under the gate metal. Another self-alignment process, named SAINT
(Self-Aligned Implantation for N$^+$-layer Technology), was developed at our
laboratory (Yamasaki et al. 1982). In this process, plasma-CVD silicon
nitride film is first deposited on an n-layer. Next, a multilevel resist
consisting of AZ-photoresist, sputter-deposited silicon-dioxide, and

AZ-photoresist is deposited. Patterning and reactive ion etching the
multiresist results in a T-shaped resist. Then, n^+-layer implantation is
carried out. After this, silicon dioxide is sputtered, and the T-shaped
resist is lifted off. Finally, thermal annealing is performed and ohmic
contacts and the Schottky gate are formed.

There are two significant advantages to this technology. The first is
that since the annealing is carried out without a metal gate, any desired
metal can be selected. The second is that the distance between the gate
and the n^+-region can be precisely controlled by selecting the multilayer
resist under-cutting depth. Thus, FET series resistance can be reduced
without increasing the parasitic gate capacitance. Using the SAINT FETs,
transconductances of 420 mS/mm for a 0.2 μm gate length and 280 mS/mm for
a 1 μm gate length were obtained.

4. Recent Results of GaAs Static RAMs

More recent results of GaAs static RAMs developed in our laboratory are
described in this section. GaAs 4 and 16 Kb static RAMs and monitoring
enhancement- and depletion-mode FETs for evaluating threshold voltage
scattering were fabricated, using the SAINT process and dislocation-free
2-inch crystals. For the purposes of comparison, conventional LEC
crystals with EPD of above 10,000 cm^{-2} were also used. A 10 to 1 wafer
stepper lithography was used.

4.1 Threshold voltage standard deviation

There are approximately 500 E- and D-FET pairs with 1 μm gate length SAINT
FETs for monitoring. The distribution of threshold voltage of an E- and
D-FET pair located with a 8.5 μm gate-to-gate distance is plotted in Fig.
5. In the figure, (a) and (b) correspond to cases of a dislocation-free
and a conventional wafer with a 10 to 1 wafer stepper, and (c) is the
results obtained in a former 4 Kb SRAM fabricated using a conventional LEC
wafer and conventional contact lithography. The threshold voltage
scattering reduced in order from (c) to (b) and quite drastically to (a).
Threshold voltage standard deviation for E- and D-FETs are 20 and 24 mV in
(a), 74 and 104 mV (b), and 90 and 117 mV in (c), respectively. The
hypothesis in a previous paper (Ohmori 1983) that the threshold voltage
scattering was mainly caused by gate length scattering and crystal
dislocation effect is verified quite well in this experiment.

4.2 GaAs 4 and 16 Kb SRAMs

The 4 and 16 Kb memories were designed with organizations of 1 K x 4 b and
4 K x 4 b, respectively. Process parameters are the same for both
memories and are listed in Table 1. A fully operational 4 Kb SRAM was
obtained by a marching pattern tested for the first time for a
dislocation-free LEC crystal.   Minimum and maximum access times were 4
and 14 ns, respectively, with the most frequent time being 5 ns. The
applied voltages were 1.5, 1.0 and 1.1 V for the cell, periphery logic,
and pull-up circuit, respectively, resulting in total power dissipation of
504 mW. A minimum access time of 2 ns was obtained with 0.9 W power
dissipation in another chip. A microphotograph of a 16 Kb static RAM is
shown in Fig. 6. The chip size is 7.18 mm x 6.24 mm. The chip integrates
102 K devices, which is the largest scale for GaAs ICs. The size of the
memory cells is 41 μm x 32.5 μm. The FET gate is arranged in one
direction to avoid orientation dependence of device characteristics. For

the 16 Kb SRAM, minimum address access time of 4.1 ns with power
dissipation of 1.46 W was obtained. 4.1 ns access time is about 4 times
faster than that of silicon bipolar and MOS equivalent circuits. To
obtain high speed, source-followers are used at the word driver circuits
to increase current drivability through word lines. Normally-on type FETs
are used for memory cell transfer-gate FETs, reducing series resistance.

5. Yield

For a long time, the weakest point of GaAs integrated circuits was low
fabrication yield due to large threshold voltage scattering. However, the
situation is changing. Recent results for chip yield per wafer versus
element number per chip are shown in Fig. 7. Here, chip yield per wafer
means the number of fully functional operating chips divided by the number
of total chips per wafer. In the small scale integrated region, yields up
to 83 and 58 percent were obtained for the divide-by-two and divide-by-
eight frequency counter (Takada et al. 1984). Although the substrates
used were conventional LEC crystals with EPD of above 10,000 cm^{-2}, their
circuitry was designed using source coupled FET logic, in which the logic
voltage swing was increased to 2.8 V. In the LSI region, DCFL 8 x 8
parallel multipliers, fabricated using MOCVD-grown high resistive layers
for an ion-implantation, were operated above 10 percent yields. The
threshold voltage standard deviation was less than 20 mV over a 2-inch
wafer (Kaminishi 1984). A DCFL 4 Kb static RAM fabricated using a
dislocation-free crystal was operated, with a corresponding yield of 6
percent.

6. Conclusion

GaAs LSI development was surveyed. GaAs LSIs of 4 to 64 Kb static RAMs
and 1 to 10 K gate logics appear to offer advantages over silicon LSIs.
Threshold voltage standard deviation was greatly reduced to 20 mV over
2-inch wafers using a dislocation-free crystal. A fully operational 4 Kb
SRAM and a 16 Kb SRAM integrating 102,000 devices were demonstrated. The
conclusion is that GaAs LSIs are promising for performance and fabrication
technology. GaAs LSI cost-performance relative to silicon LSI's will be
investigated and industry applications will be developed.

References

Asai K, Ino M, Kurumada K, Kawasaki Y and Ohmori M 1981 GaAs and Related
    Compounds pp 533-8
Furutsuka T, Tsuji T, Katano F, Higashisaka A and Kurumada K 1981
    Electron. Lett. 17 pp 944-5
Hojo A, Toyoda N, Mochizuki M, Mizoguchi T and Nii R 1981 GaAs IC Symp. 11
Ishikawa H, Kusakawa H, Suyama K and Fukuta M 1977 ISSCC pp 200-1
Kaminishi K 1984 private communication
Lee F S, Eden R C, Long S I, Welch B M and Zucca R 1980 ICCC pp 697-700
Nakayama Y, Suyama K, Shimizu H, Yokoyama N, Shibatomi A and Ishikawa H
    1983 ISSCC pp48-9
Ohmori M 1983 Conf. Solid State Devices and Materials, Tokyo pp61-4
Takada K, Togashi M, Kato N and Ida M 1984 ibid. Kobe
Yamasaki K, Asai K, Mizutani T and Kurumada K 1982 Electron. Lett. 18
    pp 119-121
Yamasaki K 1984 to be published
Yokoyama N, Mimura T, Fukuta M and Ishikawa M 1981 ISSCC pp 218-9

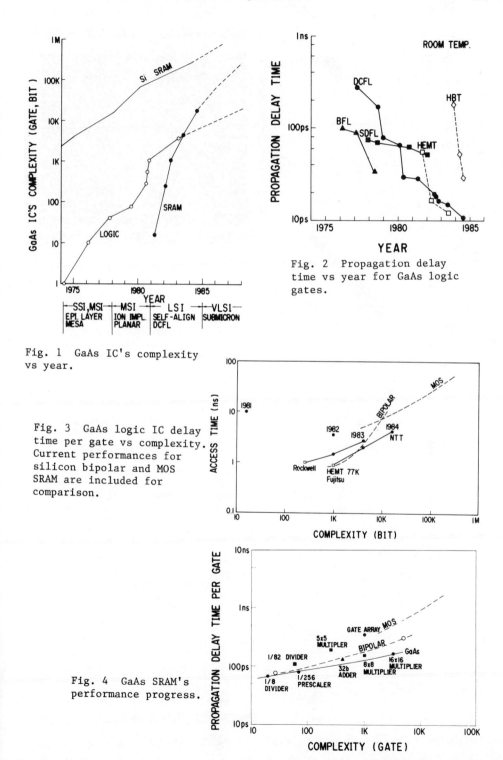

Fig. 1 GaAs IC's complexity
vs year.

Fig. 2 Propagation delay
time vs year for GaAs logic
gates.

Fig. 3 GaAs logic IC delay
time per gate vs complexity.
Current performances for
silicon bipolar and MOS
SRAM are included for
comparison.

Fig. 4 GaAs SRAM's
performance progress.

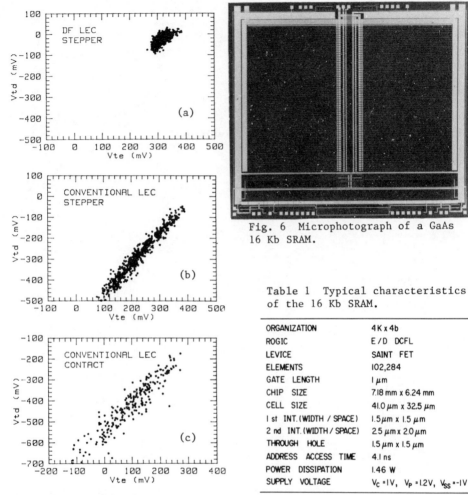

Fig. 5 Threshold voltage distribution tor E-D pair FETs for a dislocation-free (a) and conventional (b, c) 2 inch LEC waters.

Fig. 6 Microphotograph of a GaAs 16 Kb SRAM.

Table 1 Typical characteristics of the 16 Kb SRAM.

ORGANIZATION	4 K x 4b
ROGIC	E / D DCFL
LEVICE	SAINT FET
ELEMENTS	102,284
GATE LENGTH	1 μm
CHIP SIZE	7.18 mm x 6.24 mm
CELL SIZE	41.0 μm x 32.5 μm
1 st INT.(WIDTH / SPACE)	1.5 μm x 1.5 μm
2 nd INT.(WIDTH / SPACE)	2.5 μm x 2.0 μm
THROUGH HOLE	1.5 μm x 1.5 μm
ADDRESS ACCESS TIME	4.1 ns
POWER DISSIPATION	1.46 W
SUPPLY VOLTAGE	V_C =1V, V_P =1.2V, V_{SS} =-1V

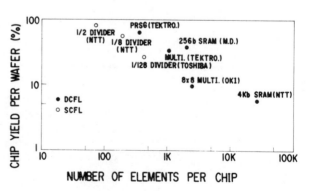

Fig. 7 GaAs IC chip yields.

Inst. Phys. Conf. Ser. No. 74: Chapter 8
Paper presented at Int. Symp. GaAs and Related Compounds, Biarritz, 1984

Planar p⁺-gate E/D technology for n-AlGaAs/GaAs selectivity doped high speed ICs

K. Ohata, M. Ogawa*, H. Hida and H. Miyamoto

Microelectronics Research Laboratories
*Fundamental Research Laboratories
NEC Corporation, 4-1-1 Miyazaki, Miyamae-ku Kawasaki, 213 Japan

Abstract.   A new planar enhancement/depletion FET technology without gate recessing for n-AlGaAs/GaAs selectively doped structure ICs has been developed. This E/D structure consists of an enhancement FET with a p⁺-GaAs gate and a depletion FET with a Schottky gate on a common n-AlGaAs/GaAs selectively doped structure. The p⁺-gate has a higher barrier potential than that of the Schottky gate, so that an enhancement-mode can be achieved. 1μm gate FETs exhibited high transconductance of 280 mS/mm and 220 mS/mm for enhancement and depletion, respectively.

1. Introduction

An n-AlGaAs/GaAs selectively doped structure is very attractive for high speed applications because of its high electron mobility and velocity. Many attempts to fabricate ICs up to LSI level utilizing FETs based on the selectively doped structure have been conducted and have proved its high speed capabilities (Nishiuchi et al 1984, Linh et al 1982, Feuer et al 1893, Lee and Wang 1983). In those ICs reported previously, DCFL circuits are employed, where driver enhancement FETs usually have a recessed gate structure. In this structure, a n-type GaAs surface layer is used outside the gate on the n-type AlGaAs layer, in order to keep the   two dimensional electron gas (2DEG) outside the gate when there are   no electrons under the gate.   However, this recessed gate structure has such disadvantages as requiring precise etching and, as a result, degrading threshold voltage uniformity, decreasing gate breakdown voltage, possibly increasing gate fringing capacitance and degrading the surface flatness.

This paper reports a new planar enhancement/depletion FET technology for realizing high speed ICs based on n-AlGaAs/GaAs selectively doped structure without gate recessing. The new E/D structure consists of an enhancement FET with a p⁺-gate and a depletion FET with a metal Schottky gate on a common n-AlGaAs/GaAs selectively doped structure, which solves disadvantages for the recessed gate structure mentioned above. This paper describes device design, device structure and self-alignment fabrication process. Marked FET performances, such as 430 mS/mm transconductance for enhancement FETs at 77K are also shown.

2. Design of Planar E/D-FETs

Figure 1 shows the basic structure of the new planar E/D-FETs based on the n-AlGaAs/GaAs selectively doped structure.   The   left   hand   part   shows   an enhancement FET (E-FET). The right hand part shows a depletion FET (D-FET). The E-FET has a p+-gate and the D-FET has a Schottky gate on a common n-AlGaAs/GaAs selectively doped structure. The p+-layer is made with AlGaAs or GaAs. Energy band diagrams for both FETs in thermal equilibriu ı are shown in Fig.2. Energy band diagrams between the source and the gate for both FETs are mostly the same as shown in Fig. 2(b), due to surface Fermi level pinning. If acceptor density N_A for the p+-layer is much higher than donor density N_D for the n-AlGaAs layer, the depletion layer in the n-AlGaAs under the p+-gate is wider than that under the Schottky gate, due to a large diffusion potential of the p+-n junction. Therefore, when there are no electrons under the gate for the E-FET, there is still much 2DEG in the source and drain regions for the E-FET and in the D-FET. Therefore, high performance enhancement-mode operation can be expected. Although a planar E-FET with a Schottky gate was reported (Delagebeaudeuf et al 1982), it is necessary to lower the surface potential outside the gate and the FET would be apt to have a large source resistance.

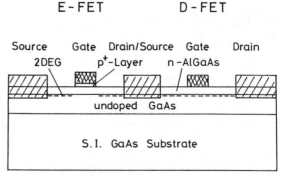

Fig. 1   Basic structure of planar p+-gate E/D-FETs.

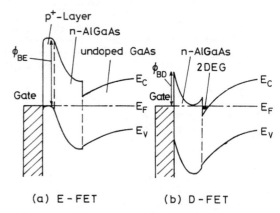

Fig. 2   Energy band diagram under the gate in thermal equilibrium.

When the gate barrier heights for the E-FET and the D-FET are designated as ϕ_{BE} and ϕ_{BD}, the threshold voltage difference between the E-FET and the D-FET is approximately equal to $(\phi_{BE}-\phi_{BD})$, as deduced from the theoretical expression given by Delagebeaudeuf and Linh (1982). Thus, the threshold voltages for both E-FET and D-FET are determined only by the donor density and as grown thickness of the common n-AlGaAs layer. Therefore, threshold voltage uniformity is guaranteed by good uniformity of a MBE grown layer and is not degraded by processing. The p+-layer can be so thin with very high doping that the planar gate structure can be realized. Moreover, the gate potential barrier is so high that this E/D structure has further advantages of large logic swing, resulting in large threshold votlage tolerance, small parasitic capacitance due to large depletion layer width and planar gate structure, and high reliability for the gate electrode.

3. Device Structure

Figure 3 shows the E-FET structure of fabricated E/D-FETs. The D-FET structure is similar to that of the E-FET except that a p+-layer for the E-FET gate is eliminated. The p+-gate for the E-FET is composed with a p+-GaAs thin layer and an overspreading Al electrode. The n-$Al_{0.3}Ga_{0.7}As$ layer is thinned to 100 A which is sufficient for supplying 2DEG. The surface n-GaAs layer is used for the gate depletion layer in order to minimize unfavorable effects of using thick n-AlGaAs layer (Hida et al 1984). The spacing between the source and the gate was also reduced in order to achieve small source resistance (Ohata et al 1984). The p+-GaAs layer is so highly doped and thin that the increase in the gate resistance compared to Schottky gate can be avoided. The gate p+-n junction was made with GaAs. Therefore, the ϕ_{BE} and ϕ_{BD} are about 1.4 V and 0.7 V, respectively, which leads the threshold voltage difference between the E-FET and the D-FET to be about 0.7 V. The doping concentration and the total thickness of the common n-layers are 2×10^{18} cm^{-3} and 350 A. These values gave threshold voltages of +0.1 V and -0.6 V for the E-FET and the D-FET, respectively.

4. Fabrication Process

The new E/D-FET structure can be fabricated in a self-alignment process as shown in Fig. 4. The fabrication process is as follows. (1) p+-GaAs (200 A)/n-GaAs (50 A)/n-$Al_xGa_{1-x}As$ (graded composition) (200 A)/n-$Al_{0.3}Ga_{0.7}As$ (100 A) /undoped $Al_{0.3}Ga_{0.7}As$ (spacer layer) (50 A)/undoped GaAs (1μm) layers were successively grown on a semi-insulating substrate by MBE. (2) The p+-GaAs layer in the D-FET region was etched off. (3) An Al film for the gate metal was evaporated, then gate electrodes for both E-FETs and D-FETs were formed by side etching using photoresist masks for source and drain contacts. (4) Ni/Au-Ge ohmic metals were evaporated through the photoresist masks used in step (3). After a lift-off process, ohmic contacts were formed by alloying. (5) The p+ -layer outside the E-FET gate was etched, protecting the D-FET with photoresist. (6) The E/D structure was completed after removing the photoresist.

Therefore, source and drain contacts for both E-FETs and D-FETs can be formed simultaneously in self-alignment with the gates. The dopants for the p+-layer and n-layers are Be and Si. The growth temperature was 650°C, and the electron mobilities were 6500 and 89000 cm^2/V·s at room temperature and at 77 K, respectively.

5. E/D-FETs Performance

Figure 5 shows I-V characteristics of 1μm gate enhancement and depletion FETs at room temperature. The gate width is 200μm. The threshold voltages for the E-FET and the D-FET are +0.1 V and -0.6 V, respectively. High performances of high g_m and low saturation voltage were achieved, especially for the E-FET. Maximum transconductances of the E-FET and the D-FET are 280 mS/mm and 220 mS/mm, respectively. In an E/D inverter circuit, load current to drive current ratio is about 1/2 for 1.0 V logic swing and the same gate width, which is adequate for the high speed operation.

Figure 6 shows the 1μm gate E-FET characteristics at 77 K. The g_m increased and the maximum value of 430 mS/mm was obtained. Table I shows a summary of FET characteristics. Threshold voltages, g_m and gate breakdown voltages are listed. The gate breakdown voltage was high of 1.2 V in the forward direction and 9 V in the reverse direction for the E-FET.

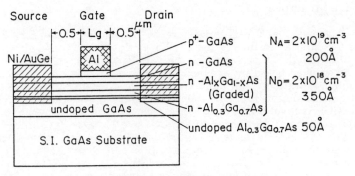

Fig. 3   Developed E-FET structure.

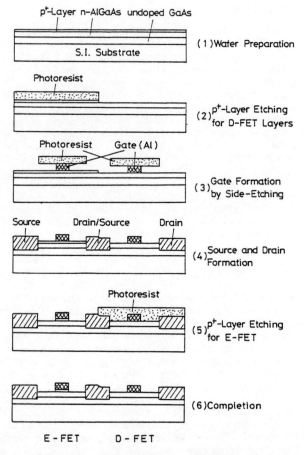

Fig. 4   Fabrication Process for new E/D-FET structure.

Lg= 1 μm , Z =200μm

E – FET

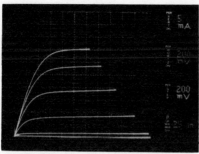

D – FET

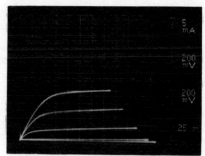

V: 5mA/div , H: 0.2V/div
Gate: +0.2V step
V_{gmax} = +1.0 V

V: 5mA/div , H: 0.2V/div
Gate: –0.2V step
V_{gmax} = 0V

Fig. 5   I-V characteristics for 1μm gate E/D-FETs at room temperature.

6. Conclusion

A new planar enhancement/
depletion FET technology for n-
AlGaAs/GaAs selectively doped
structure high speed ICs without
gate recessing has been developed.
This new E/D structure consists of
an enhancement FET with a p^+-
gate and a depletion FET with a
metal Schottky gate on a common n-
AlGaAs/GaAs selectively doped
structure. The p^+-gate has a higher
barrier potential than that of the
Schottky gate, so that there is much
2DEG in the source and drain
regions of the E-FET and in the D-
FET when there are no electrons
under the E-FET gate. This enables
high performance E-FET and D-FET
operation. The threshold voltages
for both E-FET and D-FET are
determined only by the donor
density and as grown thickness of
the common n-layer, so that
degradation in the threshold voltage
uniformity by recessing can be
prevented. The E/D-FETs were
fabricated simultaneously in a self-
alignment process. 1μm gate FETs
exhibited 280 mS/mm and 220
mS/mm transconductances at room

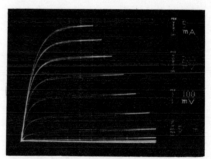

V: 5mA/div , H: 0.2V/div
Gate: +0.1V step
V_{gmax} = + 1.0 V

Fig. 6   1μm gate E-FET characteristics
at 77 K.

Table I   Summary of E/D-FETs
characteristics.

Type	V_T(V)	g_{mmax}(mS/mm)		BV$_G$ (V)	
		R.T.	77K	Forward	Reverse
E–FET	+0.1	280	430	+1.2	– 9
D–FET	–0.6	220	400	+0.7	– 8

temperature and 430 mS/mm and 400 mS/mm at 77 K for enhancement and depletion, respectively. These results indicated that this p$^+$-gate E/D technology is very promising for realizing high performance ICs.

Acknowledgement

The authors wish to thank Dr. T. Itoh, T.Baba, Dr. T. Mizutani, Dr. H. Terao and Dr. H. Watanabe for their valuable discussions. They also thank Dr. Y. Takayama and Dr. N. Kawamura for their helpful suggestions and encouragement.

References

Delagebeaudeuf D et al 1982 Electron Lett. 18 103
Delagebeaudeuf D and Linh N T 1982 IEEE Trans. Electron Devices ED-29 955
Feuer M D et al 1983 Electron Device Lett. EDL-4 306
Hida et al 1984 GaAs and Related Compounds 1984 Inst.Phys.Conf.Ser. 74 545
Lee C P and Wang W I 1983 Electron Lett. 19 155
Linh N T et al 1982 IEDM Digest Tech. Papers pp.582-585
Nishiuchi K et al 1984 ISSCC Digest Tech. Papers pp.48-49
Ohata K et al 1984 Int'l MTT-S Microwave Symp. Digest pp.434-436

Inst. Phys. Conf. Ser. No. 74: Chapter 8
Paper presented at Int. Symp. GaAs and Related Compounds, Biarritz, 1984

659

N-on versus N-off GaAs MESFETs for an ultra high speed comparator

Thierry DUCOURANT, Didier MEIGNANT, Michel BINET

* Laboratoires d'Electronique et de Physique Appliquée
  3, avenue Descartes, 94450 LIMEIL-BREVANNES, France

INTRODUCTION

Telecommunications and future signal processing systems will rapidly require analogue to digital converter (ADC) which should be able to work at gigabit sampling rates with a medium resolution (4 bits) and with a reasonable power consumption ($P \leqslant 500$ mW).

In order to design such a converter, we first put the stress on the comparator which is its most critical component. D. Meignant and M. Binet (1983) reported a first approach utilizing N-ON GaAs MESFETS with a standard V_T of - 2 V and resulting in good operation up to 1.8 GHz. Meanwhile, this comparator appeared to be highly sensitive to parasitic effects and exhibited a very strong LF hysteresis.

We now present a second version of the same comparator using recessed gate N-OFF GaAs MESFETs. The aim of this paper is to give a complete and comparative characterisation of both devices. Experimental results are detailed and conclusions are given which often go further than the simple comparator's case.

CIRCUIT DESCRIPTION

Figure 1 depicts the two comparator circuits : both are subdivided into a differential amplifier controlled by a clock (H) or a DC bias and a

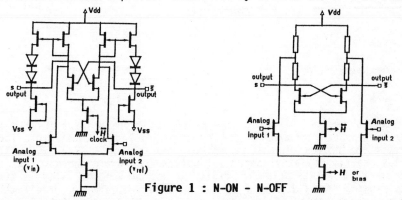

Figure 1 : N-ON - N-OFF

*This work is carried out with financial support from the D.A.I.I. (Direction aux Affaires Industrielles et Internationales).

complementary clocked flip-flop acting as a regenerative amplifier. Active loads (N-ON) have just been replaced by resistors (N-OFF). Direct level compatibility (Direct Coupled FET Logic) greatly simplifies the design of the N-OFF circuit.

In response to any rising edge of the clock $\overline{H}$, both comparators will rapidly amplify the input voltage difference and deliver useful logic signals. Logic swing is 2.5 V (N-ON) and 0.6 V (N-OFF).

PROCESSES COMPARISON

The following table gives the main technology steps for both versions :

	ACTIVE LAYER	ACTIVE REGION ISOLATION	GATE	V_T	ISOLATION BETWEEN LEVELS	IDSS mA/mm	GM mS/mm
N –ON	Se implantation	Boron implantation	Al. 0.6μm (minimum under etching)	- 2 v	1 μm' Si O$_2$	230	130 (Vgs=ov)
N·OFF	Si "	"	(recessed) Ti/Pt/Au 1μm (aligned)	+0,05v	"	35	110 (Vgs=0.4v)

Table I : Process   comparison

LF HYSTERESIS MEASUREMENTS

1. Normally-on :

The existence of non negligible low frequency input hysteresis (LF hysteresis) turned out to be the most important problem concerning the use of N-ON GaAs comparators. To characterize this parasitic phenomenon, a complete series of measurements has been carried out :

Comparing a very slowly variable input voltage (V_{in}) crossing up and down (figure 2) DC voltage reference (V_{ref}), the hysteresis amplitude can be plotted versus V_{ref} for various clock frequencies (FH). Figure 3 shows the effect of light on the device's operation : Hysteresis changes from a mean value of 100 mV (illuminated) up to 600 mV (in darkness).

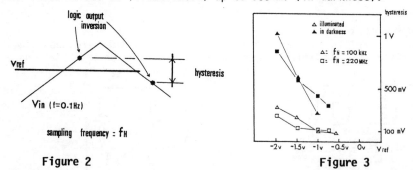

Figure 2   **Figure 3**

This phenomenon can be related to the existence of deep traps located at the interface between the channel and the semi-insulating substrate : looping effects in IDS(VDS) curves at medium frequencies (10 kHz) are well-known and S. Makram (1983) reported a low frequency drain current lag, in response to a large variation of drain to source voltage (VDS) : time constants ranging from 0.1 s to 1 µs can be observed, and are drastically affected by a negative biasing of the semi-insulating substrate with respect to the MESFET source potential (kinetic backgating). Great improvement is obtained when the amplitude of the VDS variation is reduced.

This can be roughly modelled (figure 4, Rocchi, 1984) by a D-MOSFET (in parallel with the MESFET) with its gate controlled by an R-C network. Under

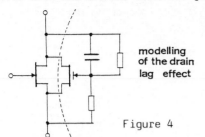

modelling
of the drain
lag effect

Figure 4

these conditions, the drain current of a GaAs FET pulsed on its drain (with constant gate to source polarisation) presents a "dynamic" value which may differ from its steady state one, depending on :
- the pulse amplitude,
- the pulse frequency,
- the presence or not of near-by negatively biased ohmic contacts.

In the particular case of the comparator, this effect results in a dynamic input offset voltage and consequently, in an input hysteresis : as a matter of fact, the internal flip-flop generates a large voltage transient on the drain of one of the input transistors (in order to force a logic "zero").

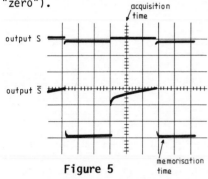

Figure 5

This large VDS variation gives a time dependent extra current **dI** on the corresponding branch of the differential amplifier, and results in a "dynamic" input offset voltage **dV**. Figure 5 shows the complementary output waveforms (S) and ($\overline{S}$) for a low sampling frequency (330 Hz) and a zero differential input. A slow current lag during acquisition time can be noticed on ($\overline{S}$). Output (S), presenting a smaller pulse is not affected and reaches instantaneously its final value.

In the case of high frequency sampling the offset is generated by the mean electrical state of the traps and is proportional to the difference of the VDS pulse amplitude on the two input transistors.

When logic inversion is achieved (for V_{in} = V_{ref} + dV) and provided that enough time is left before logic re-inversion, the roles of (S) and ($\overline{S}$) are progressively exchanged ; then, the re-inversion is obtained for V_{in} = V_{ref} - dV, giving a total hysteresis of 2 dV. Meanwhile, the hysteresis can be cancelled provided that the time between two logic inversions is smaller than the smallest traps'time constant (1 µs). In order to verify that, we designed a voltage reference alternatively reset at a high and a low level as shown in figure 6. A reduction by a factor of 30 of the hysteresis and 80 % improvement of the useful V_{ref} range have been obtained (figure 7).

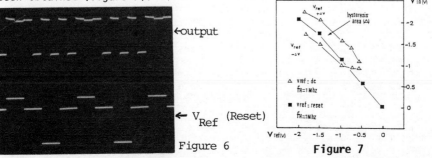

←output

← V_{Ref} (Reset)

Figure 6

Figure 7

2. Normally-off :

On the other hand, the hysteresis of normally-off comparators has been measured and the results, plotted on figure 8 gave a mean hysteresis value of m (Hyst) = 15 mV, with a standard deviation of (Hyst) = 8 mV.
Let us add that these results are completely light-independent.

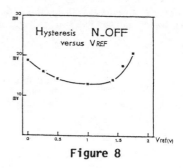

Figure 8

In fact, this can be explained as follows :
- the amplitudes of VDS swings (and then of electric fields) are nearly 10 times greater in the N-ON version than in the N-OFF one and thus more efficiently activate filling and emptying of the traps. Under these conditions, the traps remain "quasi frozen" or un-activated, in N-OFF devices.
- there is no negative power supply (N-OFF),
- even if the amplitudes of voltage variations in N-OFF devices remain large enough to involve a parasitic extra-current, the actual effect on input offset voltage and corresponding hysteresis is not the same in a N-OFF FET than in a N-ON one (this will be developed in the offset analysis).

INPUT OFFSET VOLTAGE

This parameter is particularly important for a comparator using MESFET differential amplifier and presently limits to 4 bits the resolution of a theoretical GaAs A to D converter. It can be attributed either to non-uniformity of the active layer doping profile or to slight deviation in gate length. It has been estimated for the N-ON version on a separated differential amplifier consisting of two 50 μm width FETs and external resistors. Results are given in the following table :

VGS (V)	VDS (V)	VT (V)	M(OFFSET) (mV)	σ(OFFSET) (mV)	SOURCES
-0.5	1.5	-1.2	37	50	Czochralski : wafer 1
-0.1	5.0	-2	9	40	" " 2
-0.6	1.5	-2.5	10	55	" " 3
-0.1	4	-2	7	90	Czochralski
-1	4	-2	- 2	82	Dislocation free
-0.1	4	-2	-15	45	Bridgeman : wafer 1

Table II

So, mean value = 10 mV with standard deviation = 60 mV.
It is clear that the parameter which must be taken into account is the offset standard deviation. Let us notice that dislocation free material do not bring a great improvement, and that gate length scattering is responsible to a large extent, for the voltage standard deviation.
For recessed gate N-OFF comparators, we have chosen a more exacting test and measured directly the total input offset voltage of the device by comparing two DC signals (V_{in}) and (V_{ref}) sampled at a 1 MHz clock frequency. The results obtained on a complete wafer of 2" are the following :

VT (V)	M (OFFSET) (mV)	σ(OFFSET) (mV)	SOURCE
0.03	20	30	Czochralski

These good results can be explained : for a same geometrical variation (for example gate length deviation) between two N-ON and between two N-OFF

transistors, different drain current's offsets will be observed which are
dependent on the channel depth and then on the saturation current Idss. Effective input offset voltage is easily deduced by dividing these current
deviations dI, by the transconductance gm.
dI (N-ON)   = K.IDSS (N-ON)   and dI (N-OFF) = K.IDSS (N-OFF)
dV_{in} (N-ON) = dI (N-ON)/gm (N-ON) and dV_{in} (N-OFF) = dI (N-OFF)/gm (N-OFF)

     Since typical values are :
IDSS (N-ON) = 23 mA/mm (VT = - 2) and IDSS (N-OFF) = 3 mA/mm (VT = 0.05)
gm (N-ON)   = 130 mS/mm   and gm (N-OFF)   = 110 mS/mm.
we can assume that the N-OFF offset voltage should be around five times
better than the N-ON one.
     We obtained only a ratio of 2 on the standard deviations but as we
mentioned above, the two measurements methods were not the same and certainly to the advantage of N-ON devices.

PERFORMANCE EVALUATION
     To get a good idea of the comparator potentiality, both in speed and
precision, the sensitivity parameter (S), has been evaluated on the two different versions. This sensitivity can be defined as the smallest variation
voltage of the input signal (V_{in}) around DC reference voltage (V_{ref}) causing the logic inversion of the outputs.

1. N-ON devices
To get rid of any parasitic effect the amplitude of the input signal
has to be larger than the mean hysteresis voltage (in the other case, good
but very unstable operation has been obtained). (S) was deduced from the
minimum slope of V_{in} and from the sampling frequency (FH) : (S) = P/FH. Under these conditions, a 30 mV sensitivity has been observed at FH = 1 GHz.
Complete measurements are reported on figure 10. Proper operation has been
obtained up to 1.8 GHz.

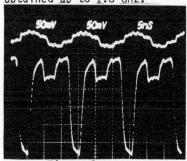

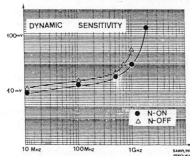

       Figure 9   Figure 10

2. N-OFF devices :
The quasi-absence of hysteresis enables us to use the classical measurement test called (F-F/2) where V_{in} is a small signal at half the clock
frequency crossing up and down the voltage reference, and whose amplitude
can be lowered to the value of (S) with an error-free decision (figure 9).
     (S) was better than 15 mV at 100 MHz and was still about 50 mV at a
1 GHz sampling frequency.

POWER CONSUMPTION
     This power consumption falls down from 60 mW (N-ON) to 1 mW (N-OFF).

CONCLUSION

All the experimental results concerning both comparators are shown in the following table in order to make a simple comparison :

	V_t	DVR	LSB$=\frac{DVR}{15}$	DC HYST.	σ (OFFSET)	PW	SENSITIVITY (1GHZ)
N.ON	-2.4 v	1.7 v	113 mV	100 mV best case	60 mV	60 mW	30 mV
N.OFF	+0.05v	1.5 v	100mv	15 mv	30 mv	1 mW	50 mV

Table III

We can easily deduce the expected performances of a complete 4 bit-ADC

	LSB	$\frac{\sigma \text{ (OFFSET)}}{LSB}$	$\frac{HYST.}{LSB}$	$\frac{SENS. (1GHZ)}{LSB}$	PW
N-ON	113 mV	0.6	1	0.3	1.5 w
N-OFF	100 mV	0.32	0.16	0.53	<50 mW

Table IV

Though N-ON comparator seem well adapted for ultra high speed operation, its high LF hysteresis and its heavy power consumption are limiting it to special applications like ultra-high-speed decision circuits.

On the contrary, first results on N-OFF comparator indicate that it is a real break through for a high speed ADC : ultra low power consumption, reduced offset voltage, and good 1 GHz-sensitivity enable us to reasonably foresee the design and fabrication of a 1 GHz-4 bit flash A to D converter.

ACKNOWLEDGMENTS

The authors are very grateful to J.C. Baelde, C. Rocher and M. Fichelson for processing the wafers.

REFERENCES

Akiri Tuka Wa, "A highly sensitive strobed comparator", IEEE Solid State, April 1981.

Makram-Ebeid S., P. Minondo, "Effets parasites dans les transistors à effet de champ sur AsGa", Acta Electronica, 25, 3, 1983, pp. 241-260).

Meignant D., Binet M., "A high performance 1.8 GHz strobed comparator for A/D converter", GaAs Symposium, october 1983.

Rocchi M., "Status of the surface and bulk parasitic effects limiting the performances of GaAs IC's", ESSDERC 84, Lille, septembre 1984.

Lin H.C., "Comparison of input offset voltage of differential amplifiers using bipolar and Field-Effect Transistors", IEEE Journal of S.S.C., June 1970.

Inst. Phys. Conf. Ser. No. 74: Chapter 8
Paper presented at Int. Symp. GaAs and Related Compounds, Biarritz, 1984

665

High-speed E/D DCFL frequency divider using high transconductance GaAs enhancement-mode MESFET

T. Onuma, K. Nagano, A. Tamura, T. Uenoyama,
H. Tsujii, K. Nishii, and T. Sakashita
Central Research Laboratory
Matsushita Electric Industrial Co., Ltd.
Moriguchi, Osaka, 570 Japan

Abstract. Very-high-speed divide-by-252/256 prescaler IC's have been
fabricated by using newly developed high transconductance GaAs enhancement-mode MESFET's (E-FET). The E-FET was improved by reducing
parasitic series resistance and interface states of the Schottky gate
and by using a semi-insulating GaAs substrates with high electron
mobility. The transconductance of 300 ms/mm has been obtained for
the 1 μm gate length GaAs E-FET.
The measured performance of the GaAs frequency divider is as follows:
1) the maximum counting frequency is 3.7 GHz with power dissipation
of 180 mW, 2) the total power dissipation as low as 42 mW and 34 mW
are achieved for 3.0 GHz and 2.5 GHz operation, respectively.

1. Introduction

High-speed and low-power dissipation variable-modulus dividers find
numerous applications to such as prescalers for frequency counters and
frequency synthesizers   (Walton, 1982). Especially, phase locked loop
(PLL) frequency synthesizers which are able to operate at frequencies
above 1 GHz are strongly demanded for automobile telephones and CATV to
increase their channels. It is also required to reduce the power dissipation of the synthesizers for battery drive. Among the GaAs logic
circuits, a Direct Coupled FET logic (DCFL) is considered to be one of the
candidates for high-speed and low-power IC's   (Mizutani, 1982, Mitsui,
1982). In order to realize an excellent performance on DCFL GaAs IC's,
enhancement-mode FET with high-transconductance is required. To obtain
the high-transconductance enhancement-mode FET, it is necessary to obtain
the channel layer with high mobility, to reduce source-to-gate parasitic
resistance and to reduce the interface states of the metal Schottky
barrier gate.
To reduce source-to-gate parasitic resistance, several attempts have
been made including the use of recess structure (Suyama, 1980), self-
aligned ion-implantation technology (Yokoyama, 1981, Yamasaki, 1982), Pt
buried gate technology (Toyoda, 1981), and the method of reducing the
source-gate and drain-gate spacing   (Furutsuka, 1981).
As for interface states, on the contrary, there have been few reports on
the reduction of interface states at the GaAs Schottky barrier, even
though, interface states for the GaAs metal-oxide semiconductor (MOS)
structure (Kamieniecki, 1980) and for the Si Schottky barrier (Muret,
1982) have been measured by transit capacitance spectroscopy.
To obtain the channel layer with high electron mobility, it is necessary
to employ a high quality   semi-insulating substrate because the
channel layers formed by direct ion implantation into semi-insulating
GaAs are strongly dependent upon the quality of the substrates

(Eisen, 1981, Onuma, 1981).

In this paper, we describe: 1) the performance of the GaAs enhancement-mode FET fabricated by using high dose ion implantation and Pt buried gate technology, 2) circuit design, fabrication process and measurement aspects of a 252/256 variable-modulus divider based on DCFL GaAs IC's using high transconductance enhancement-mode FET.

2. Fabrication technology of high-transconductance enhancement-mode GaAs MESFET

Substrates used in this work were undoped semi-insulating GaAs (100) wafers, grown by LEC method. An n-type channel region, n$^+$ type source and drain regions were formed by selective Si ion implantation. Then it was annealed at 850°C for 20 min. in AsH$_3$, H$_2$ and Ar atmosphere without encapsulant. AuGe/Ni ohmic contacts and Pt/Ti/Au (1000Å/500Å/2500Å) Schottky gates were formed by using conventional mask-alignment and lift-off methods. The threshold-voltages were adjusted by controlling the sintering time at 400°C in Ar atmosphere. Gate length, gate width and source-drain spacing were 1, 20 and 3μm respectively.

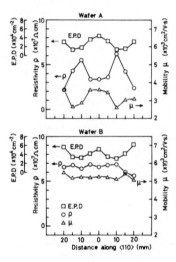

Fig.1. Mobility, resistivity and etch-pit density along <110> direction.

Semi-insulating substrate

Figure 1 shows the typical Hall mobility, resistivity and etch pit density distributions along <110> direction of the wafers supplied by different suppliers. A correlation between the etch pit densities and the mobility was observed for wafer A, but no distinct correlation was observed for wafer B. It was also observed that Hall mobility and doping efficiency after ion implantation were also influenced by the quality of substrate and a higher mobility substrate results in a higher mobility and higher doping efficiency in an implanted layer. The threshold voltage distribution of the GaAs MESFET's made on an entire 2 inch wafer were also compared. Standard deviation (σ Vth) was around 100 – 150 mV for wafer A and 60 – 80 mV for wafer B after adjusting the average threshold voltage ∿0.1V. Therefore, we used semi-insulating GaAs substrate named wafer B in Fig.1 in the following experiments.

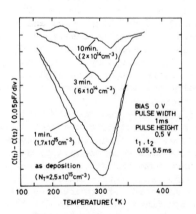

Fig.2. DLTS spectra of Pt-GaAs Schottky barrier diode as a function of sintering time. Sintering is performed at 400°C in Ar atmosphere.

Reduction of interface state density in Pt-GaAs buried Schottky barrier

Figure 2 shows deep level transient spectroscopy (DLTS) spectra for Schottky diodes with a diameter of 200 μm after the deposition and sintering. The deep level near the Schottky barrier was measured at bias voltage of 0V and a pulse height of 0.5V. The location of this level has been found less than 0.1 μm from the interface of the Schottky barrier by using C - V technique. The activation energy is found to be 0.46 eV and the corresponding capture cross section is about 4.4×10^{-15} cm^{-2} for both as deposited and sintered samples. With increasing the sintering time for the Pt-GaAs Schottky barrier, trap density decreases as shown in Fig.2. The reduction of interface states of the Pt-GaAs Schottky barrier by sintering would be caused by the formation PtAs$_2$ front, that is, the formation of the buried Pt-Schottky barrier, with a result that nearly ideal interfaces of PtAs$_2$-GaAs Schottky barrier would be formed.

Properties of high-transconductance E-FET

Figure 3 shows the saturated drain currents at drain voltage of V_{ds}=1.2V with gate biases of V_{gs}=0V and V_{gs}=0.7V as a function of threshold voltage for a large number of processed wafers. These curves show a strong correlation between saturation current and threshold voltage with the point falling on nearly parabolic curves. According to the ordinary square-law fitting of I_{ds}=K(V_{gs}-Vth)^2, K values are 2.2 ms/V and 3.2 ms/V at V_{gs}=0.7V for as deposited and sintered Schottky gate, respectively. In other words, K values were improved by a factor of 1.45 at V_{gs}=0.7V for sintered Schottky gate, but K values remain constant at V_{gs}=0V. This would be due to the fact that the electrical characteristics of Schottky gate interface has been improved by Pt-GaAs reaction.

In order to reduce the source and drain parasitic resistances, it is necessary to decrease ohmic contact resistance as well as surface depletion between the gate and ohmic contact edges. Table 1 shows the typical values of an enhancement-mode FET implanted various doses into ohmic regions. The drain conductance in the linear region, the saturated drain current, and the transconductance were improved by increasing the doses of implantation into the ohmic regions.

3. High-speed and low-power dissipation 252/256 DCFL divider

Circuit design

The unit cell of the modulus divider was composed of master/slave flip-flop circuit comprising 8 NOR gates consisting of an enhancement-mode driver and a depletion-mode load GaAs MESFET. Figure 4 shows a schematic diagram of the 252/256

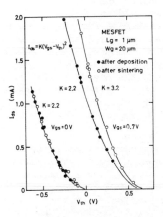

Fig.3. Saturated drain current (V_{ds}=1.2V) versus threshold voltage.

Table 1. Comparison of typical values of enhancement-mode FET implanted various doses into ohmic regions.

		Wg=1mm	
OHMIC DOSE (cm^{-2})	Rs (Ω)	gm (mS)	Ron (Ω)
5×10^{12}	2.00	168	7.0
1×10^{13}	1.82	185	6.4
5×10^{13}	1.02	240	5.2
1×10^{14}	0.90	265	5.1
5×10^{14}	0.74	270	5.0
1×10^{15}	0.60	300	4.6
Vth =0.1 V	Vgs = 0.7 V		

modulus divider circuit
consisting of eight
stages of unit cells.
The gate width of the
driver FET was designed
to be 40 μm for the
first stage and 20 μm
for the second stage in
order to achieve high-
speed performance in
these stages.   10 μm
wide FET's were employed
in the following stages

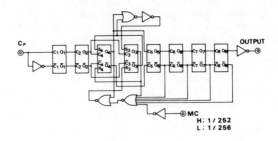

Fig.4. Block diagram of 252/256
variable-modulus frequency divider.

for reduction of the
total power dissipation.
The gate widths of the
depletion-mode FET's are one half of the width of the corresponding
drivers.   The selection of divide-by-252 or divide-by-256 is controlled
by a mode control input.   When the mode control line is kept to be a low
level, the divide-by-256 mode is selected and when the mode control line
is high, the divide-by-252 is selected.   By computer-aided simulation it
was determined that the threshold voltage of the E-MESFET and D-MESFET
should be $V_{thE}=0 \sim +0.4V$ and $V_{thD}=-0.5 \sim -1.2V$, respectively, for operation
above 3 GHz.

Fabrication
The schematic cross section
is shown in Fig.5. The FET's
fabrication process was the
same as that described above.
Si ion implantation for the
formation of E-MESFET and D-
MESFET were performed at an
energy of 100 keV with doses
of $\sim 4 \times 10^{12}$ cm^{-2} and at 150
keV with $\sim 3 \times 10^{12}$ cm^{-2},
respectively.   The first and
second interconnection lines
were formed with about 2 μm-
thick Ti-Au. A 2 μm-thick
polyimide layer was formed
as the crossover insulator
in order to reduce the
interconnection capacitance.
The transconductance gm
($V_{gs}=+0.7V$, $V_{ds}=1.0V$) of the
driver FET was about 250
ms/mm. Microphotograph of
the chip is shown in Fig.6.
The chip size is 1.65x1.2
mm^2, and it has 86 gates
including buffer FET's.

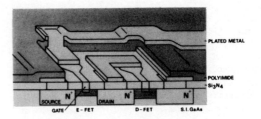

Fig.5. A schematic cross
section of GaAs IC.

Circuit performance
Figure 7 shows output
waveforms of the divide-by-4
and divide-by-256 mode for

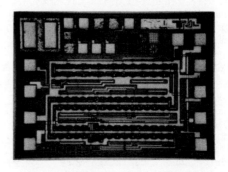

Fig.6. Microphotograph of chip.

an input frequency of 3.7 GHz.
Figure 8 shows the output
waveform of the variable-
modulus divider for an input
frequency of 2.5 GHz. This
device was operated at
frequencies as high as 3.3
GHz. Figure 9 shows the
dependence of the total power
dissipation on the input
frequency under minimum operation supply voltage. A
total power dissipation as
low as 42 mW was obtained at
input frequencies below 3 GHz.
However, it increased rapidly
above 3 GHz due to the increase of supply voltage required to maintain the
circuit in operation.
Typical performance of the
GaAs modulus frequency divider
are summarized in Table 2.

4. Conclusion
Very-high-speed and
low-power dissipation
GaAs modulus frequency
dividers have been
achieved by using high
transconductance GaAs
enhancement-mode MESFET.
Summary of this study
is as follows.
(1) Very high-trans-
    conductance
    enhancement-mode
    FET's have been
    fabricated by using
    a semi-insulating
    substrate with high
    mobility, by reducing
    the deep trap density
    near the interface
    of the Schottky gate
    and by decreasing the
    gate to n^+ spacing
    by high doses ion
    implantation into the
    ohmic region.
(2) High-speed character-
    istics were obtained
    and maximum clock
    frequency of 3.7 GHz
    were achieved, by
    using 1 μm gate length

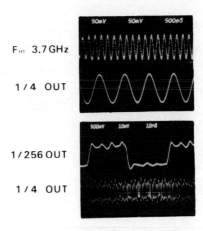

Fig.7. Output waveform of
divide-by-4 and divide-by-256.

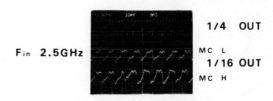

Fig.8. Output waveform of
variable-modulus divider.

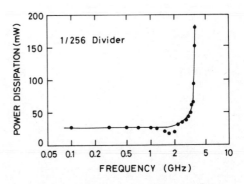

Fig.9. Total power dissipation
against input clock frequency.

driver FET's. The
total power dissi-
pation were reduced
to   as low as 42
mW at 3 GHz. The
device will have an
impact on the per-
formance of systems
in the field of
communication.

Table 2. Typical performance of
252/256 variable-modulus divider.

Supply Voltage	0.9 ~ 5.0 V
Power Dissipation	⟨ 30 mW (⟨ 2.0 GHz)
	34 mW (2.5 GHz)
	42 mW (3.0 GHz)
Maximum Clock Freq.	3.7 GHz (180 mW)
Input Sensitivity	-40dBm (2.0 GHz)
	-13 dBm (2.5 GHz)

Acknowledgments
The authors wish to
thank S. Horiuchi
and T. Kajiwara for continuous encouragement through this work. They also
wish to acknowledge Y. Hori for helpful discussions.

References
Eison, F.H., 1981 GaAs and Related Compounds.
Furutsuka, T., Tsuji, T., Katano, F., Higashisaka, A., and Kurumada, K.,
1981 Electron. Lett. 17, 281.
Kamieniecki, K., Kazior, T.E., Lagowski, J., Gato, H.C., 1980 J. Vac.
Sci. Technol., 17, 1041.
Mitsui, Y., Nishitani, K., Ishihara, O., Nakatani, H., 1982 GaAs IC
Symposium, p.p.115.
Mizutani, T., Kato, N., Osafune, K., and Ohmori, M., 1982 IEEE Trans.
ED-29, 1116.
Muret, P., 1982 J. Appl. Phys. 53 6300.
Onuma, T., Hirao, T., and Sugawa, T., 1981 J. Appl. Phys. 52 6128.
Suyama, K., Kusakawa, H., Fukuta, M., 1980 IEEE Trans. ED-27, 1092.
Toyoda, N., Mochizuki, M., Mizoguchi, T., Nii, R., and Hojo, 1981 GaAs
and Related Compounds.
Walton, E.R., Shen, E.K., Lee, F.S., Zucca, R., Shen, Y.D., Welch, B.M.,
Dikshit, R., 1982 IEEE Trans. ED-29, 1116.
Yamasaki, K., Asai, A., Mizutani, T., Kurumada, K., 1982 Electron. Lett.,
18 119.
Yokoyama, N., Mimura, T., Fukuta, M., Ishikawa, H., 1981 Solid-State
Circuit Conf. Dig. Tech. Papers p.p.281.

Inst. Phys. Conf. Ser. No. 74: Chapter 8
Paper presented at Int. Symp. GaAs and Related Compounds, Biarritz, 1984

671

Bipolar structures on GaAs for integrated circuits

G. REY, J.P. BAILBE, A. MARTY, C. ANTARASANA, A. CAZARRE, J. TASSELI

L.A.A.S. du C.N.R.S.
7, avenue du Colonel Roche - 31077 TOULOUSE CEDEX - France

Introduction

The bipolar integrated circuits realization on GaAs requires a careful study of the discrete circuit elements which are essentially : GaAlAs/GaAs single heterojunction transistors, GaAlAs/GaAs/GaAlAs double heterojunction transistors, phototransistors and homojunction transistors. We are going to study the potentialities and the results obtained for each one of these Mesa type structures produced by LPE.

1. Single heterojunction transistor

A basic advantage of the heterojunction transistors is the very high value of the injection efficiency of the emitter-base junction; the intrinsic current gain is practically equal to the base transport factor, that means, for an NPN transistor,

(1) $$h_{FE}^{\bullet} \simeq \tau_n / \tau_B$$

where τ_B is the base transit time and τ_n the electron lifetime. Under these conditions the current gain does not depend upon the emitter doping level and is sensitive to the base doping level only through the variations of τ_n. It will be possible therefore to shape doping profiles such that N_{DE} is much smaller than N_{AB} so as to reduce both the base transversal resistance $r_{BB'}$ and the emitter transition capacitance C_{TE}. By so doing, one can upgrade the device high-frequency performances, such as cut-off frequency f_T and power gain G whose expressions are given by :

(2) $$\frac{1}{2\pi f_T} = \frac{U_T}{I_C}(C_{TE} + C_{TC}) + r_{CC'}C_{TC} + \tau_C + \tau_B$$

(3) $$G = \frac{f_T}{8\pi f^2 r_{BB'} C_{TC}}$$

with

$$(4) \begin{cases} C_{TE} = A_E \sqrt{\dfrac{q\epsilon N_{DE}}{2(V_D - V_{BE})}} & C_{TC} = A_C \sqrt{\dfrac{q\epsilon N_{DC}}{2(V_D + V_{CB})}} \\[4mm] r_{CC'} = \dfrac{\rho_C W_C}{A_E} = \dfrac{W_C}{A_E q \mu_n N_{DC}} & r_{BB'} = \dfrac{1}{8\pi \mu_p Q_B} \\[4mm] \tau_B = \dfrac{W_B^2}{2 D_{nB}} & \tau_C = \dfrac{W_{TC}}{2 v_l} \end{cases}$$

These expressions also show how the high electron mobility in GaAs will greatly contribute to the transistor performance improvement : an increase of μ_n causes a decrease of both $\mathcal{C}_B$ and $r_{CC'}$, which results in an increase of f_T and consequently of G.

The high electron mobility in GaAs in conjunction with a high base doping level will have a supplementary beneficial impact on the device performances. Parasitic mechanisms, such as emitter current crowding and base widening in the collector region, whose occurrence causes a significant fall-off of the transition frequency value, are rejected away into the very large collector current area. Indeed, the critical values J_{ECC} and J_{BWC} of the collector current density given by the relations :

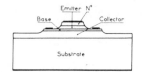

Figure 1

$$(5) \quad \begin{cases} J_{ECC} = \dfrac{8}{l^2} D_{PB} Q_B h_{FE}^* \\[2ex] J_{BWC} = q\mu_{nC} N_{DC} \dfrac{V_{CB}}{W_C} \end{cases}$$

and above which emitter current crowding and base widening effects take place, become higher as μ_n and N_{AB} get larger. Consequently, it is advantageous to overdope the base region and to use a material with large mobility values.

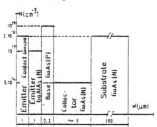

Figure 2

The doping profiles and a cross-sectional view of the transistor are represented on figures 1 and 2 .The best results obtained are f_T=4 GHz, $f_{max} \simeq 3$ GHz, $h_{FE} \simeq 500$, $V_{BCE_O} \simeq 130$ V (Figure 3 and 4).

2. Double heterojunction transistors

Within the framework of the elaboration of integrated circuits, double heterojunction transistors (DHBT's) offer advantages in relation to single heterojunction transistors (SHBT's). Such advantages concern the suppression of hole injection from base into collector in digital switching transistors under conditions of saturation, the emitter/collector interchangeability in IC's, the reduction of the emitter/collector offset voltage.

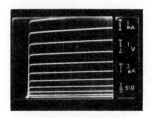

Figure 3

The theoretical analysis of the double heterojunction bipolar transistor has been considered as a generalisation of the single heterojunction transistor study. The corresponding analytical model has been already presented in the ESSDERC 84. This one permits to analyse the influence of the major electrical and technological characteristics which condition the behavior of the DHBT.

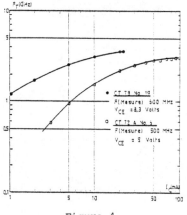

Figure 4

For example we report here the aluminium concentration influence in the collector GaAlAs.

The results of the calculations plotted in Figure 5 a lead us to state that the collector current is limited by the collector spike junction when the aluminium rate is high (x > 0.2). Conversely, when x < 0.2, the collector-base heterojunction does not influence the transfer characteristic. Let us note as well that, as predicted by the model, the ideality coefficient remains equal to the unit whatever the value of the aluminium rate.

With respect to the intrinsic current gain h_{FE}^* , (Figure 5.b), its variations which are those of the transport factor (γ remaining very high, in relation to δ) can be interpreted directly ; for x < 0.2 it is identical to the transport factor of a homojunction transistor (HBT) or single heterojunction emitter-base transistor (S H B T) ; for x > 0.2 h_{FE}^* is drastically altered.

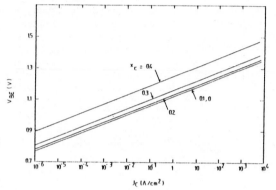

Figure 5a

The qualitative explanation concerning the results obtained can be simply derived from Figure 5.c where the barrier height of the potential B defined by the spike, decreases when the aluminium rate in the collector diminishes.

This study is illustrated by the Figure 6a and 6b showing the Common emitter-current voltage characteristics of two series of DHBT's. It can be observed that the type A transistor, characterized by a weak collector doping and a high

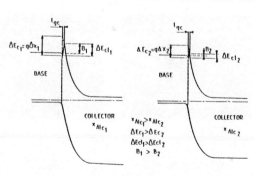

Figure 5c

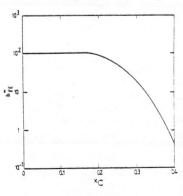

Figure 5b

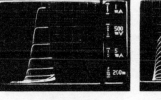

Figure 6a

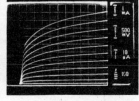

Figure 6b

aluminium content has a current gain in the order of 0.4 preventing thus its use. The type B transistor, elaborated with a lower aluminium content behaved normally ($h_{FE} \simeq 80$).

3. Phototransistors

GaAlAs/GaAs heterojunction phototransistors benefit, on one hand, from the GaAlAs "window effect" which makes it possible to light the active zone through the emitter without the need for overdimensioning the lateral zone as this would penalize the frequential behavior (Figure 7), and, on the other, from the high current gain allowed by the heterojunction which increases the sensitivity of the device and leads to a reduction of the surface which is lighted. In this device, the gain is given by the transport factor which variations versus the base region width and doping are reported in the Figure 8. The good cut-off frequency of GaAs bipolar transistors naturally enhances the velocity of phototransistors. Finally, together with a laser, they could be used to elaborate an integrated light amplifier. Current gains of 50 000 have been obtained (Figure 9) which induce a sensitivity of 15 000 A/W ; rise times in the other of 5 ns have been observed with a optical sensitivity about 30 A/W (Figure 10).

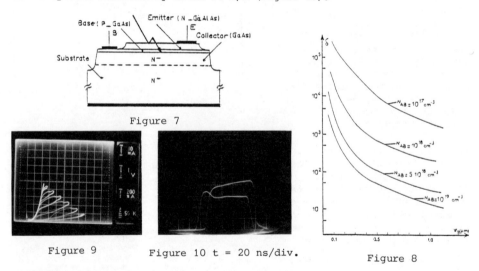

Figure 7

Figure 9   Figure 10 t = 20 ns/div.   Figure 8

4. Double homojunction transistor

The heterojunction transistors fabricated on gallium arsenide profit by the possibility of overdoping the base in relation to the emitter. If a high current gain is not required, this characteristic can still be taken advantage of for the design of homojunction bipolar transistors. In effect, in the case of an inverted profile where the base is more doped than the emitter, the degeneracy mechanisms contribute to maintaining a sufficient injection efficiency for many applications i.e. H.F. amplification, integration potential into fast logic circuits.

The injection efficiency of the homotransistor with a low level of current is given by the analytical expression :

(6) $\gamma = \mu_n \, n_{iB}^2 \, Q_E \Big/ \mu_p \, n_{iE}^2 \, Q_B$

where Q_E and Q_B refer respectively to the doping integrals in the emitter and base regions, μ_{pE} and μ_{nB} to the minority carriers mobilities in these regions, n_{iE} and n_{iB} to the intrinsic concentrations.

At low doping, the n_{iB}^2/n_{iE}^2 ratio is equal to   unity. However, when the impurity concentration injected into the semiconductor increases, the degeneracy mechanisms occur in the highly doped areas.

n_i increases with the level of doping and when the base is more doped than the emitter (i.e. inverted profile), the n_{iB}^2/n_{iE}^2 ratio is higher than unity. The degeneracy mechanisms which normally degrade the performances of the devices are beneficial in this particular case.

Figure 11 illustrates the variations of the n_{iB}^2/n_{iE}^2 ratio as a function of the base doping for different emitter dopings.

Moreover, the ratio μ_n/μ_p induces a gain increase.

In order to highlight the degeneracy mechanisms we have carried out two series of investigation whose doping profiles are given in Table 1.

TABLE 1	Emitter-doping (cm^{-3})	Base-doping (cm^{-3})	Collector-doping (cm^{-3})	W_E μm	W_B μm	W_C μm
A. Series	$2.3 \quad 10^{18}$	$1.4 \quad 10^{19}$	$3.9 \quad 10^{15}$	1	0.7	2
B. Series	$1.2 \quad 10^{18}$	$0.6 \quad 10^{18}$	$1.5 \quad 10^{15}$	0.65	0.65	6

Figure 12 illustrates the output characteristics $I_C(V_{CE})$ of each of these series. A comparison of the experimental results with those deduced from the analytical calculation is given in Table 2.

TABLE 2	μ_n/μ_p	Q_E/Q_B	n_{iB}^2/n_{iE}^2	$\frac{Q_E}{Q_B} \cdot \frac{\mu_n}{\mu_p} = \gamma$ (without degeneracy)	γ_{total} relat(1)	γ_{max} (experimental)
A. Series	12.6	0.23	4	2.9	11.5	12
B. Series	22.1	2	0.4	44	18.1	25

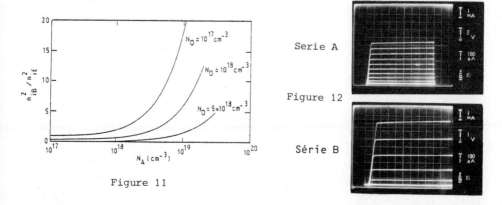

Serie A

Figure 12

Série B

Figure 11

It has been found that for the A series for which the base is more doped than the emitter the addition of the degeneracy mechanisms constitutes a significant part of the gain obtained ; conversely, in the case of the B series, for which the values of base and emitter dopings are almost similar, the degeneracy becomes unfavorable and the gain is given by the mobility ratio.

To conclude, it appears that the influence of degeneracy mechanisms must be taken advantage of in order to elaborate homojunction bipolar structures with an inverted doping profile capable of maintaining the potentialities of heterojunction transistors in the frequential field.

Conclusion

The performances obtained for each one of the studied devices permit us to define some applications. In the case of single heterojunction transistors, it seems to be normal to turn towards the ECL circuit realization, whereas the double heterojunction transistors are expected to provide some advantages for the I^2L Logic circuits. The double homojunction transistors and FETs fabrication technologies are compatible which permit us to consider their integration. At last the phototransistor integration with a Laser diode could be used to elaborate a light amplifier.

LP-MOCVD growth of high purity indium phosphide and high power high efficiency 94 GHz Gunn diodes

M.A. di FORTE-POISSON, C. BRYLINSKI, G. COLOMER, D. OSSELIN, J.P. DUCHEMIN*

F. AZAN, D. LE CHEVALIER, J. LACOMBE **

THOMSON-CSF -
DOMAINE DE CORBEVILLE - BP N° 10
91401 ORSAY Cedex (France)
* L.C.R.
** D.H.M.

The highest purity Indium phosphide was grown by low pressure metalorganic chemical vapor deposition. Non intentionally doped layers exhibited net residual donors as low as 2.10^{14} cm^{-3} with hall mobility as high as 5300 cm^2 V^{-1} S^{-1} at room temperature and 147000 cm^2 V^{-1} S^{-1} at liquid nitrogen temperature. The observed mobility values were higher than the theoretical mobility calculations of Walukiewicz for $N_A/N_D = 0$.

The use of such calculation is not precise enough to obtain exact compensation ratios and acceptor levels of these layers.

But we are allowed to suppose that very little compensation is present in these layers.

This result was found to be in very good agreement with photoluminescence measurements at 1.7°K. The observed PL. spectrum shown in Fig. 1 is typical for high purity n type material. Transitions such as free exciton (X), exciton bound to shallow impurities (A°,X), (D°,X) and donor level to valence band (D°,h) are exhibited. So Indium phosphide appears as a valuable material to fabricate high frequency microwave devices. 94 GHz Indium phosphide gunn diodes were fabricated. These diodes were based on a three layer N$^+$ N N$^+$ structure as shown in Fig. 2 where Si (from silane) was used as the dopant. This dopant allows a precise doping control of doping profiles.

Fig.3 illustrates variation of hall mobility with doping level for Non intentionally doped and Si doped InP layers. An integral heat sink technology was used to fabricate InP Gunn diodes, and diode chips were mounted in standard W5 packages.

Power levels in excess of 50 mW were obtained at 94 GHz with an efficiency of 3% while an optimized structure operating at 75 mW has been obtained with an efficiency of 2.4%.

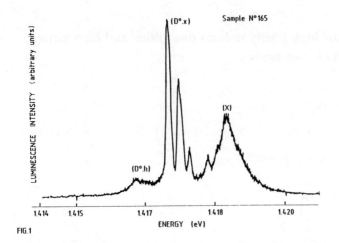

FIG.1

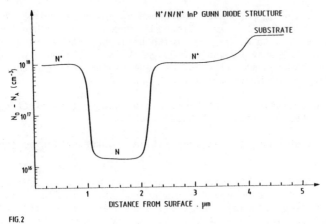

FIG.2

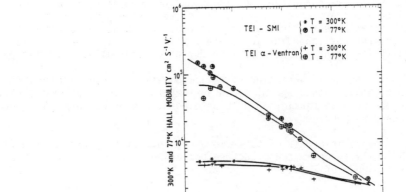

FIG.3

Inst. Phys. Conf. Ser. No. 74: Chapter 9
Paper presented at Int. Symp. GaAs and Related Compounds, Biarritz, 1984

679

$Ga_xIn_{1-x}As_yP_{1-y}$ – InP, DH lasers emitting at 1.3 μm grown on GaAs substrates by LP-MOCVD growth technique

M. Razeghi, R. Blondeau , J.P. Duchemin
THOMSON-CSF - Laboratoire Central de Recherches
Domaine de Corbeville - BP N° 10
91401 ORSAY CEDEX (FRANCE)

Long distance optical links use lasers emitting at 1-3 µm fabricated by material grown on InP substrate.

Unfortunately the technology of IC's(integrated circuits) for signal treatment is a lot more difficult on this material than on GaAs substrate.

A solution will be to combine the advantages of these two materials which is against nature, due to the large lattice mismatch.

However we have shown that it is possible with metal organic chemical vapor deposition at reduced pressure LP-MOCVD growth technique, to do it.

LP-MOCVD has been used for the growth of good quality GaAs layers on InP substrates and InP layers on GaAs substrates.

The morphology of these layers were excellent and x-ray diffraction measurements show that the InP or GaAs layers are monocrystalline (see fig.1).

The near gap luminescence spectrum ofan InP layer grown on a GaAs substrate by LP-MOCVD is shown in Fig.(2), which is exciton related. GaInAsP-InP Buried ridge structure (BRS) lasers (see RAZEGHI et al, this conference) emitting at 1.3 µm have been fabricated on GaAs substrates, using the LP-MOCVD growth technique (see Fig.3). Pulse threshold current of 190 mA at room temperature has been measured with an output power up to 10 mW. Fig.4 presents the light current characteristic of this device.

REFERENCE

M. RAZEGHI, R. BLONDEAU, J.C. BOULAY, B. de CREMOUX and J.P. DUCHEMIN LP-MOCVD growth and CW operation of high quality SLM and DFB semiconductor $Ga_xIn_{1-x}As_yP_{1-y}$-InP lasers.
11th International Conference on GaAs and related compounds
25-28 Septembre 1984 - BIARRITZ (FRANCE). Inst.Phys.Conf.Ser. 74 445

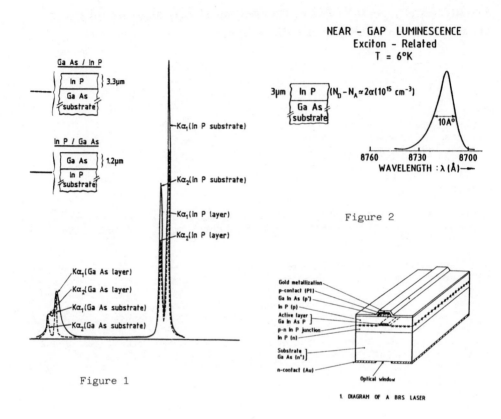

Figure 1

NEAR - GAP LUMINESCENCE
Exciton - Related
T = 6°K

Figure 2

Figure 3

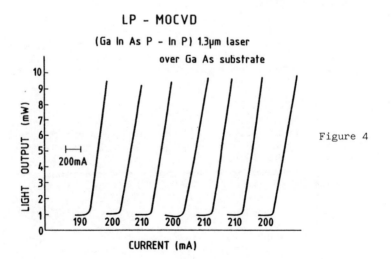

LP - MOCVD

(Ga In As P - In P) 1.3µm laser

over Ga As substrate

Figure 4

Inst. Phys. Conf. Ser. No. 74: Chapter 9
Paper presented at Int. Symp. GaAs and Related Compounds, Biarritz, 1984
681

Growth of (Ga,In)As/InP superlattice by adduct MOVPE

R. H. Moss, P. C. Spurdens and E. A. D. White
British Telecom Research Laboratories, Martlesham Heath,
Ipswich, IP5 7RE, U. K.

Introduction

The preparation of quantum well structures for compositions in the system Ga-In-P-As presents severe problems compared with those in the Al-Ga-As system because of the need to achieve close lattice matching while retaining accurate control of the energy gap, E_g. In MOVPE growth, the greater complexity of the In precursor pyrolysis imposes an additional problem (Moss et al. 1984). To date the only successful MOVPE growth of (Ga,In)As/InP superlattices has been achieved by Duchemin and co-workers, who adopted a low pressure route for their preparation (Razeghi and Duchemin, 1983). This approach evidently alleviates some of the complexities associated with gas composition switching and eddy formation in the reaction chamber. However, the need to achieve fast and accurate switching has been amply demonstrated in recent publications (eg. Thrush et al. 1984), and is reflected in the designs of vent/run fast switching gas manifolds that are emerging. The latter have been shown to be effective for GaAs/(Al,Ga)As superlattices (Roberts et al. 1984), but this approach is limited and will depend on various factors, including dead space volume, surface areas in the gas handling system, and the nature of the surfaces present. A more positive solution to the problem is described here.

Growth Technique

Movement of the substrate between two well-established gas flows of different compositions provides an effective means of altering the epilayer composition in heterostructure growth by VPE (Yoshida et al. 1978). We have applied this principle to MOVPE. The approach has the clear advantage that it makes growth independent of the presence of eddies in the reactor system, provided that they do not interfere with the decomposition process. Similarly sorption/desorption effects experienced on changing partial pressures become unimportant, and do not detract from the sharpness of the interfaces achieved. In fact the only limiting effect will be a slight perturbation of the gas flows at the moment of transfer, which will disturb growth to an extent appreciably less than 1 atomic layer in thickness.

Using this approach we have grown simple heterostructures and superlattice specimens of (Ga,In)As lattice matched to InP. Preformed adduct materials – $Me_3Ga.PEt_3$ and $Me_3In.PEt_3$ – were used as the group III precursors (Moss, 1984); these have the particular advantages from the point of view of superlattice growth that it is simple to vary the temperature of the sources to achieve slow growth rates, while maintaining the advantages of liquid sources. Both the Ga and In sources can also be held in the same temperature controlled bath.

Characterisation

Sharp compositional changes in GaInPAs heterostructures can be demonstrated by X-ray diffraction and TEM photomicrographs of sections through the interface. An interesting comparison of X-ray double-crystal diffraction linewidths (which provide an indication of lattice perfection in the epilayers), for specimens of (Ga,In)As on InP grown with and without vent/run fast gas switching, is given in the accompanying table.   A dramatic decrease in the diffraction half-peak widths can be seen, starting from growth with manual operation of gas valves, through vent/run with the use of an InP buffer layer, to the use of the new 'substrate transfer' technique.

Typical X-ray Linewidths for (Ga,In)As Epilayers on InP Substrates.

Manual operation without fast switching		c. 1000 arc s
Manually operated vent/run switching		400–600 arc s
Manually operated vent/run)		130–170 arc s
Automatic fast switching)with InP buffer layer		50–150 arc s
'Substrate transfer')		20–60 arc s

An X-ray double crystal diffraction scan of the (002) reflection for a (Ga,In)As/InP superlattice structure grown by substrate   transfer between streams of differing gas compositions appropriate for the growth of each of the two constituents of the superlattice, clearly showed satellite superlattice diffraction lines, indicative of good periodicity in the structure.   The spacing of these lines corresponded to a structural periodicity of c. 170Å.

Photoluminescence from the superlattice under stimulation from a Nd:YAG laser was encouragingly intense.   The emitted radiation showed a spread in wavelength on scanning the specimen, indicative of some grading in the thickness of the superlattice.   The wavelength range observed is, from current theoretical considerations, consistent with a quantum well size of c. 60 Å.

Finally, a 10 layer (Ga,In)As/InP MQW laser structure has been grown and emission at 1·55 μm under pulsed injection at room temperature has been demonstrated.

Conclusions

The results presented here are important for three reasons:
1   They confirm that the preformed adduct route for III-V epitaxy is viable for superlattice structure preparation.
2   They show that growth at atmospheric pressure does not detract from the formation of sharp interfaces.
3   They introduce a novel compositional switching method for heterostructure formation which decouples problems associated with precursor adsorption on gas pipes, and gas flow patterns in reactors, both of which are troublesome in conventional MOVPE.

Acknowledgements

The Authors wish to thank Mrs M.A.G. Halliwell and Dr M.H. Lyons for the X-ray results, and Mrs S. O'Brien and Dr M. Hockly for the P/L and TEM results respectively.   They also acknowledge permission from the Director of Research of British Telecommunications plc to publish this work.

References

1   R.H. Moss and E.A.D. White 1984, British Telecom Tech. J. 2 74.
2   M. Razeghi and J.P. Duchemin 1983, J. Vac. Sci. Tech. B1 262.
3   E.J. Thrush, J.E.A. Whiteaway, G. Wale-Evans, D.R. Wight and A.G. Cull 1984, J. Crystal Growth 68 in press.
4   J.S. Roberts, N.J. Mason and M. Robinson 1984, J. Crystal Growth 68 in press.
5   M. Yoshida, R. Mizutani, H. Watanabe and Y. Seki 1978, Extended Abstract 207, 152nd Electrochemical Soc. Meeting, Seattle, WA Electrochem. Soc., Princeton, New Jersey.
6   R.H. Moss 1984, J. Crystal Growth 68 in press.

Inst. Phys. Conf. Ser. No. 74: Chapter 9
Paper presented at Int. Symp. GaAs and Related Compounds, Biarritz, 1984

683

Photoluminescence and electrical characterisation of very high purity GaAs by low-pressure MOVPE

J. Black, P. Norris, E. Koteles, and S. Zemon
GTE Laboratories, Incorporated
Waltham, MA 02254

We have studied undoped GaAs epitaxial layers (5.5 µm to 17 µm) grown by low-pressure metalorganic vapor phase epitaxy (LP-MOVPE) by 4.2 K photolumines-cence (PL), Hall effect (300 K and 77 K), and carrier concentration profiling (N vs x at 300 K). Growth temperatures were varied from 575° to 700°C and the ratio of group V to group III flows from 17.3 to 54.4 in order to generate a matrix of samples in growth parameter space. The LP-MOVPE system and other growth conditions have been previously described.[1] In one region of growth parameter space, undoped epilayers were grown with 77 K electron mobilities of up to 175,000 cm^2/V-s (measured with weak tungsten illumination).

Epilayer Characterization

4.2 K PL (100 mW/cm^2 excitation intensity) and concentration profiling meas-urements were performed on all epilayers. Hall measurements were performed on mesa-etched cloverleaf samples using the van der Pauw method. In all cases "f" values were 0.99 or greater, and magnetic fields of 1 kG or less were used. When carrier concentration profiles indicated totally depleted epilay-.ers, no Hall measurements were attempted.

Results

The salient aspects of the matrix investigation are as follows.

   1. In a region of growth parameter space defined by $T_G \geq 650°C$ and V/III $\geq$ 35, donor luminescence dominates and the epilayers are n-type with free car-rier concentrations in the mid-10^{14} cm^{-3} range and with electron mobilities of about 80,000 cm^2/V-s.

   2. For $T_G \leq 600°C$, donor luminescence is weak and the epilayers are to-tally depleted. The PL spectra is dominated by free exciton emission rather than by impurity-associated transitions. This is the first observation of free exciton dominated PL spectra in MOVPE-grown GaAs. Such spectra have pre-viously been reported only for high quality MBE-grown GaAs.[2,3]

   3. Between these two regions (T_G = 625°C, V/III > 27), the PL spectra were dominated by sharp (X,D°) transitions. The free electron concentrations ranged between 1 - 2 x 10^{14} cm^{-3} (corrected for depletion effects), and mobil-ities lay between 130 - 140,000 cm^2/V-s. For lower V/III ratios (V/III = 27), even thick epilayers (17 µm) were totally depleted and no Hall measurements were possible in the dark. However, with weak tungsten illumination, 77 K electron mobilities of up to 175,000 cm^2/V-s were measured, the highest values reported to date for MOVPE-grown GaAs. Weak illumination of lightly compen-sated, high purity n-type (n = 2 x 10^{14}, $N_D/N_A \simeq 6$) samples resulted in only a slight (< 10%) increase in mobility, indicating that screening effects are probably not significant at these illumination levels.

The excitonic region of the PL spectrum of a high-purity epilayer, 17 μm thick, which was totally depleted at 300 K (via N-x profiling) is shown in Figure 1. The prominence of sharp excitonic features [(X,D°) FWHM = 0.0875 nm)] and excited states (X,D°) is indicative of high purity. The intensities of the acceptor-related peak (which is carbon), (e,A°) and (D°,A°) are less than 10% of the intensity of (X,D°). For increasing V/III ratios, the (X,A°) intensity decreases rapidly as reported previously[4] with similar behavior observed when T_G is increased.

A further indication of the quality of some of the samples is given by the effect of epilayer thickness on exciton-polariton scattering. A step-etched sample which was grown in region 2 ($T_G \leq 600°C$) was prepared with three regions of decreasing thickness; 5.5 μm, 3.5 μm, and 2.0 μm. 4.9 K PL spectra of this sample are shown in Figure 2. The reduction of the magnitude of the "dip" seen in the emission line shape of the exciton-polariton $X_{n=1}$ as the epilayer thickness is decreased occurs because the layer thickness is comparable to the scattering length of the exciton-polariton mode in the bulk. Such long scattering lengths occur in samples with low residual donor concentrations and have previously been observed only in high quality MBE-grown GaAs.[5]

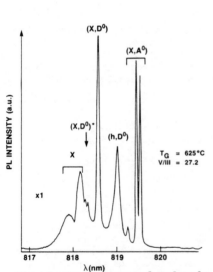

Fig. 1--PL Spectrum of Undoped MOVPE-grown GaAs Epilayer (17 μm)

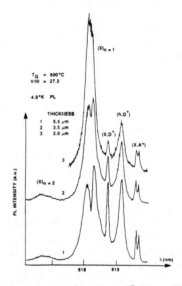

Fig. 2--PL Spectra of Step-Etched GaAs Epilayer (5.5, 3.5, and 2.0 μm)

References

1. P. Norris, J. Black, S. Zemon, and G. Lambert, Proceedings of ICMOVPE II, Sheffield, UK (1984), J. Cryst. Growth (in press).

2. M. Heiblum, E. E. Mendez, and L. Osterling, J. Appl. Phys. 54, 6982 (1983).

3. E. S. Koteles, J. P. Salerno, W. Bloss, and E. M. Brody, paper presented at International Physics of Semiconductors Conference, 1984.

4. S. Takagishi and H. Mori, Jap. J. Appl. Phys. 23, L100 (Feb. 1984).

5. W. L. Bloss, E. S. Koteles, E. M. Brody, B. S. Sowell, J. P. Salerno, and J. V. Gormley, to be published.

Inst. Phys. Conf. Ser. No. 74: Chapter 9
Paper presented at Int. Symp. GaAs and Related Compounds, Biarritz, 1984

685

Enhanced electroabsorption in GaAs/GaAlAs multiple quantum wells and its application to opto-electronic devices

T. H. Wood and C. A. Burrus
AT&T Bell Laboratories, Crawford Hill, NJ 07733 USA

D. A. B. Miller, D. S. Chemla, and T. C. Damen
AT&T Bell Laboratories, Holmdel, NJ 07733 USA

A. C. Gossard and W. Wiegmann
AT&T Bell Laboratories, Murray Hill, NJ 07974 USA

ABSTRACT

An enhanced electroabsorption effect has been observed in multiple quantum wells(MQW) of GaAs/GaAlAs. Recent calculations indicate that the effect arises from a large change in the confinement energies of the electrons and holes in the wells with the application of an electric field perpendicular to the layers. Excellent fits to the measured energy shifts can be achieved with no adjustable parameters. This effect is used to make an optical modulator that has low insertion loss and high speed. Optical pulses 131 ps long have been generated from 121 ps electrical pulses.

The optical properties of excitons at room temperature in MQW have attracted a great deal of interest, because of their novel physical properties and their potential usefulness in optoelectronic devices.[1,2] Recently, we showed that by the application of an electric field perpendicular to the quantum well layers, a strong shift could be introduced into the absorption edge, and that this shift could be used to make an optical modulator. In this paper, we present the theory of the magnitude of the shift, and demonstrate that the optical modulator can be operated at high speed.

In our device, the MQW are fabricated in the undoped region of a *pin* diode, which is operated back-biased. This configuration is important, because it allows us to apply a well-known electric field quickly and with a simple electronic drive. As described elsewhere[2], the absorption spectrum of the MQWs are recorded as a function of the voltage applied to the device. The observed energy of the heavy and light-hole transitions shift to lower energy as the electric field is increased.

In order to calculate the shift in the absorption energy with electric field, we first calculate the change in the "particle-in-a-box" confinement energies by a variational calculation[3], a numerical tunneling resonance calculation, or a modified infinite well calculation. The three approaches give very similar results. Then, the change in binding energy of the exciton itself is calculated by a 1-parameter variational procedure in which the size of the exciton parallel to the layers is the free parameter. The experimentally observed peak shifts and the results of these calculations are shown in Fig. 1. The data points show the experimentally observed peak energies, and the lines indicate the result of the calculation with and without the change in the binding energy of the exciton. As there are no adjustable parameters in the calculation, the agreement between theory and experiment is impressive. A similar calculation has been carried out to explain electroreflectance shifts[4], but, as the electric field was not well know in that sample, it was necessary to arbitrarily adjust the field strength by 2/3 to obtain agreement between theory and experiment.

In order to operate this device as a high speed modulator, it is necessary to reduce the size of the device so it will have low capacitance. We have fabricated a 95 μ diameter mesa device, with an uncontacted capacitance of about 0.5 pF. Fig. 2 shows the result of operation of this device at high speed. The upper trace is the electrical drive to the device, which has a deconvoluted FWHM of 121 ps. The lower trace is the observed optical signal; it has a deconvoluted FWHM of 131 ps. This indicates the high speed of the device. Note that these times are already much faster than the 20 ns carrier lifetime.[1]

We have demonstrated that the observed shift in the exciton peaks can be predicted from first principles. Furthermore, we have demonstrated high speed operation of a device based on these principles. In the future, higher speed devices should be possible.

REFERENCES

1. D. A. B. Miller, D. S. Chemla, D. J. Eilenberger, P. W. Smith, A. C. Gossard, and W. T. Tsang, Appl. Phys. Lett. *41,* 679(1982).

2. T. H. Wood, C. A. Burrus, D. A. B. Miller, D. S. Chemla, T. C. Damen, A. C. Gossard, and W. Wiegmann, Appl. Phys. Lett. *44,* 16(1984).

3. G. Bastard, E. E. Mendez, L. L Chang, and L. Esaki, Phys. Rev. *B28,* 3241(1983).

4. M. Erman, P. Frijlink, C. Alibert, and S. Gaillard, presented at the 11th Int. Symp. on GaAs and Related Compounds, Biarritz, France, 1984, Inst.Phys.Conf.Ser. 74 327

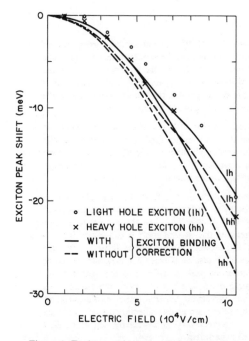

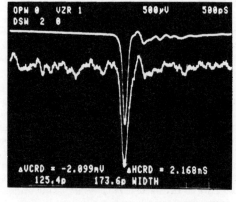

Figure 1: Exciton peak energy shifts vs. applied electric field. The data points are experimental observations and the lines are theoretical predictions.

Figure 2: Sampling oscilloscope traces for high speed modulator. Upper trace is the electrical drive; lower trace is the detected optical output.

Inst. Phys. Conf. Ser. No. 74: Chapter 9
Paper presented at Int. Symp. GaAs and Related Compounds, Biarritz, 1984

687

Long wavelength, room temperature observation of excitons and 2-dimensional electron–hole states in multiple quantum wells (MQWs)

T. H. Wood and C. A. Burrus
AT&T Bell Laboratories, Crawford Hill, NJ 07733 USA

J. S. Weiner, D. S. Chemla, D. A. B. Miller, and T. C. Damen
AT&T Bell Laboratories, Holmdel, NJ 07733 USA

D. L. Sivco and A. Y. Cho
AT&T Bell Laboratories, Murray Hill, NJ 07974 USA

ABSTRACT

For the first time, room temperature exciton resonances have been observed at long wavelength. MQWs of $Ga_{.47}In_{.53}As/Al_{.48}In_{.52}As$ were prepared in a *pin* doped configuration by molecular beam epitaxy. Absorption spectroscopy from room to Helium temperature reveals 5 features, identified as transitions between subbands and as confined exciton resonances. We deduce the conduction and valence band discontinuities to be 0.44 and 0.29 eV, respectively. These results suggest that quantum well devices at long wavelength should be possible.

Multiple quantum wells(MQWs) of GaAs/GaAlAs have been shown recently to have novel optical and electronic properties. For example, room temperature observation of exciton resonances has allowed the demonstration of optoelectronic devices, such as a high-speed modulator.[1,2] However, much of the interest in these devices is in devices which would operate at long wavelength(i. e. 1.55 μ), where optical fibers are most transparent. In this work, we report on the first room temperature observation of exciton resonances at long wavelength.

The material system we use is MQW of $Ga_{.47}In_{.53}As/Al_{.48}In_{.52}As$ grown lattice-matched to InP by molecular beam epitaxy. Most of our samples are fabricated with the quantum wells in the intrinsic region of a *pin* diode, although some undoped samples have also been studied. This *pin* configuration allows us to apply a very well known electric field to the MQWs, for measurement of the electroabsorption effect or collection of photocurrent.

The measured absorption spectrum at room temperature for a sample of 50 periods of approximately 110 Å wells and 110 Å barriers is shown in Fig. 1. A clear absorption resonance peak is seen on the fundamental absorption edge at 0.78 eV, along with a second peak at 0.92 eV, and 3 absorption plateaus. We have assigned these 2 peaks to heavy-hole excitonic resonances and the plateaus to the characteristic densities-of-states for 2-dimensional sub-bands. This assignment is made by comparison with the results of a numerical calculation of the optical transition energies from a simple 1-dimensional potential model of the quantum wells. This model, which uses the literature values for effective masses and the measured value for the well width, takes only the relative conduction band/valence band discontinuity and the InGaAs band gap as adjustable parameters. The calculated transition energies are shown in Fig. 1 by the arrows. This fit is produced by assuming that 60% of the band discontinuity is taken up in the conduction band.

The very strong nature of the excitonic peaks should be pointed out. In bulk InGaAs, only very weak exciton resonances are seen in absorption, and only at Helium temperature.[3] The exciton resonances are greatly enhanced by the presence of the quantum wells, as in the GaAs MQWs[1]. This effect should make practical, long wavelength MQW devices possible.

Although these resonances are visible at room temperature, the spectra are much sharper at low temperature. Fig. 2 shows the absorption spectrum of this sample at 30 K, where the light-hole exciton resonances can be seen, in addition to the heavy-hole excitons and the absorption plateaus. The excellent quality of these samples can be seen by the narrow linewidths. At 77 K, we measure a half-width of the fundamental exciton line of 4.2 meV in absorption; similar measurements of the luminescence linewidths in other samples have given results approximately a factor of 2 wider.[4]

In summary, we have observed sharp exciton resonances in long-wavelength material for the first time. This observation suggests that the extension of our short-wavelength devices to long-wavelength should be possible.

REFERENCES

1. D. A. B. Miller, D. S. Chemla, D. J. Eilenberger, P. W. Smith, A. C. Gossard, and W. T. Tsang, Appl. Phys. Lett. *41*, 679(1982).

2. T. H. Wood, C. A. Burrus, D. A. B. Miller, D. S. Chemla, T. C. Damen, A. C. Gossard, and W. Wiegmann, Appl. Phys. Lett. *44*, 16(1984).

3. Y. S. Chen, O. K. Kim, J. Appl. Phys. *52*, 7392(1981).

4. L. Goldstein, A. M. Jean-Louis, J. Y. Marzin, M. Allovon, C. Allibert, and S. Gaillard, presented at the 11th Int. Symp. on GaAs and Related Compounds, Biarritz, France, 1984 Inst.Phys. Conf.Ser. 74 133

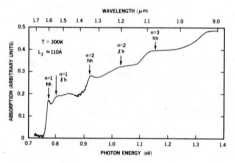

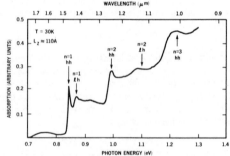

Figure 1: Absorption spectrum of 110 Å MQW at room temperature.

Figure 2: Absorption spectrum of 110 Å MQW at 30 K.

Inst. Phys. Conf. Ser. No. 74: Chapter 9
Paper presented at Int. Symp. GaAs and Related Compounds, Biarritz, 1984

689

The 'charged insulated gate field effect transistor' (CIGFET), a new normally-off logic with depletion mode FETs

H. Thim[+], T. Neugebauer, R. Chabicovsky, L. Palmetshofer[+] and
K. Lübke

Technische Universität Wien, Gusshausstrasse 27, A-1040 Wien
[+]Universität Linz, Altenbergerstrasse 69, A-4040 Linz, Austria

Abstract. A new normally-off field effect transistor having a relatively large channel doping times channel thickness product typical for normally-on depletion mode FETs
is proposed. The normally-off behaviour is obtained by u-
sing an insulating gate with a permanently charged floating
gate embedded in the insulating layer underneath the control gate. The amount of negative charge injected into the
floating gate is chosen sufficiently large to fully deplete
the channel at zero control gate voltage. Switching into
the on-state occurs at positive control gate voltage thus
making this device suitable for direct coupled fet logic.
As a consequence of the uncritical thickness of the active
n-layer good threshold voltage uniformity is expected over
large wafer area not easily obtainable with E-MESFETs. However, additional processing steps are needed for fabricating the charged floating gate structure which at present
is a well established process only in silicon technology.

1. Theoretical aspects

It is well known that D-MESFET based GaAs ICs have given excellent performance but they require two power supplies and some
kind of level shifting circuitry to generate the negative gate
voltages needed for switching D-MESFETs off. The additional
circuitry leads to higher power dissipation and uses up more
chip area making D-MESFET technology less suitable for VLSI applications. The E-MESFET avoids the requirement for dual power
supply and level shifting circuitry by having a positive pinch-
off voltage ("normally-off") due to its relatively thin channel
region. As a consequence of this, E-MESFETs can be coupled directly to each other (DCFL) making this technology much better
suited for VLSI. However, logic voltage swings are limited to
about 0.5 volts above which excess gate current begins to flow.
Because of the small logic swing the pinch-off voltage must be
uniform to within, say, 25 mV which is extremely difficult to
control. Another serious problem is the high series resistance
of the channel between the gate and the source or drain. Self-
aligned gate fabrication is one way to reduce these series resistances, the HEMT - technology (TEGFET) is an attractive alternative approach for keeping series resistances small.

The purpose of this paper is to describe a new normally-off logic with field-effect transistors having high channel doping
and channel thickness as used in normally-on D-MESFETs. The
normally-off behaviour is obtained by using an insulated control gate with a permanently charged floating gate underneath
as shown in Fig.1. The amount of negative charge is chosen to
fully deplete the channel region at zero control gate voltage.
Consequently, it can be used in direct coupled fet logic and
thus combines the simple processing requirement of D-MESFETs
with the advantage of simple circuitry of E-MESFETs. Admitted-
ly, the gate structure needs additional processing as it is

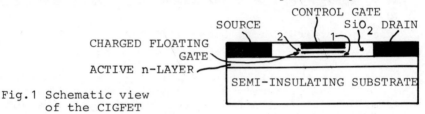

Fig.1 Schematic view
     of the CIGFET

more complicated than the Schottky gate of MESFETs. The additional processing steps are:
  - deposition of insulating layer 1
  - deposition of the floating gate
  - deposition of insulating layer 2
  - charging of the floating gate through the gate insulator
An additional requirement pertinent to a MIS device is to keep
the density of energy states at the insulator/semiconductor
interface sufficiently low in order to prevent Fermi level pinning. Fortunately, InP, GaInAs and, of course, Si have been
shown to exhibit low interface state densities when SiO_2 or
Al_2O_3 are used as insulating materials.

The new device described in this paper has a number of advantages which are summarized below:

  - it can be used in direct coupled fet logic circuits
  - only a single power supply is needed
  - easy control of V_T due to the relatively thick n-layer
  - channel resistance between gate and source/drain is small
  - gate current is blocked by the insulating layer
  - high electron mobility and velocity since electrons move
    within the bulk rather than at the semiconductor/insulator
                                                    interface

2. Experimental results

Successful fabrication of this device depends primarily on the
quality of the gate insulator. There is no doubt that thermally
grown SiO_2/Si CIGFETs will work satisfactorily (EPROM). On III-
V-materials CVD grown oxids must be used eventually. We are
presently using SiO_2 deposited on Si and InP by "spin-on" technique. A 2 μm long and 10 nm thick titanium film was embedded
in the oxide layer and charged by an electron beam using a SEM.
The electron beam was switch off when the drain current vanish-
ed. No charge decay was observed in the vacuum chamber of the
SEM. However, discharging occurred in open air within a few
hours which was probably caused by humidity penetrating into
the insulating silicon dioxide layer.

Author Index